D0164409

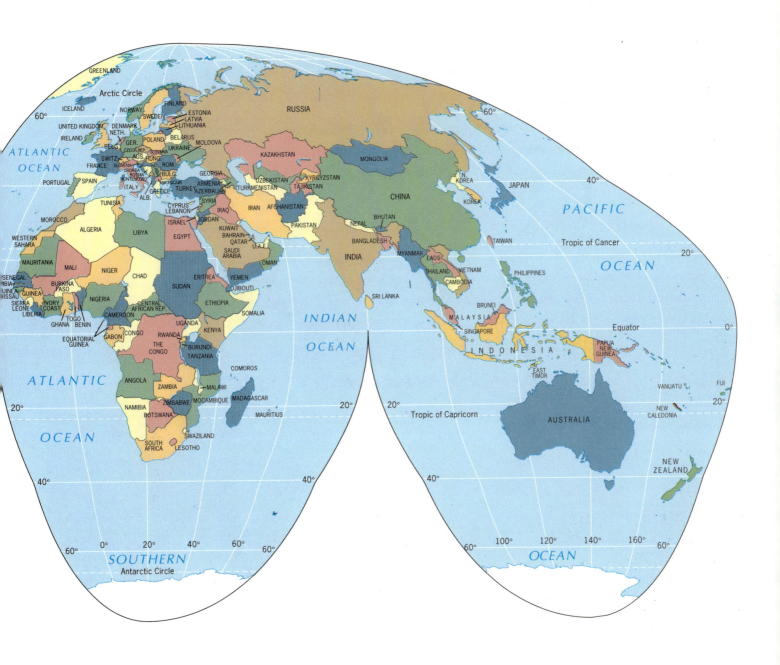

The Wiley Faculty Network

Where Faculty Connect

The Wiley Faculty Network is a faculty-to-faculty network promoting the effective use of technology to enrich the teaching experience. The Wiley Faculty Network facilitates the exchange of best practices, connects teachers with technology, and helps to enhance instructional efficiency and effectiveness. The network provides technology training and tutorials, including *Wiley PLUS* training, online seminars, peer-to-peer exchanges of experiences and ideas, personalized consulting, and sharing of resources.

Connect with a Colleague

Wiley Faculty Network mentors are faculty like you, from educational institutions around the country, who are passionate about enhancing instructional efficiency and effectiveness through best practices. You can engage a faculty mentor in an online conversation at **www.WhereFacultyConnect.com**

Participate in a Faculty-Led Online Seminar

The Wiley Faculty Network provides you with virtual seminars led by faculty using the latest teaching technologies. In these seminars, faculty share their knowledge and experiences on discipline-specific teaching and learning issues. All you need to participate in a virtual seminar is high-speed internet access and a phone line. To register for a seminar, go to **www.WhereFacultyConnect.com**

Connect with the Wiley Faculty Network

Web: **www.WhereFacultyConnect.com**

Phone: 1-866-4FACULTY

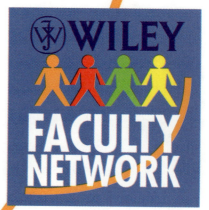

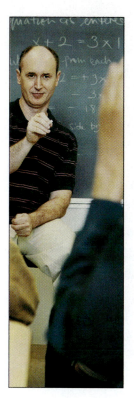

The first person to invent a car that runs on water...

... may be sitting right in your classroom! Every one of your students has the potential to make a difference. And realizing that potential starts right here, in your course.

When students succeed in your course—when they stay on-task and make the breakthrough that turns confusion into confidence—they are empowered to realize the possibilities for greatness that lie within each of them. We know your goal is to create an environment where students reach their full potential and experience the exhilaration of academic success that will last them a lifetime. *WileyPLUS* can help you reach that goal.

WileyPLUS is an online suite of resources—including the complete text—that will help your students:

- come to class better prepared for your lectures
- get immediate feedback and context-sensitive help on assignments and quizzes
- track their progress throughout the course

"I just wanted to say how much this program helped me in studying... I was able to actually see my mistakes and correct them. ... I really think that other students should have the chance to use *WileyPLUS*."

Ashlee Krisko, *Oakland University*

www.wiley.com/college/wileyplus

80% of students surveyed said it improved their understanding of the material.*

FOR INSTRUCTORS

WileyPLUS is built around the activities you perform in your class each day. With WileyPLUS you can:

Prepare & Present

Create outstanding class presentations using a wealth of resources such as PowerPoint™ slides, image galleries, interactive simulations, and more. You can even add materials you have created yourself.

Create Assignments

Automate the assigning and grading of homework or quizzes by using the provided question banks, or by writing your own.

Track Student Progress

Keep track of your students' progress and analyze individual and overall class results.

Now Available with WebCT and Blackboard!

"It has been a great help, and I believe it has helped me to achieve a better grade."

Michael Morris,
Columbia Basin College

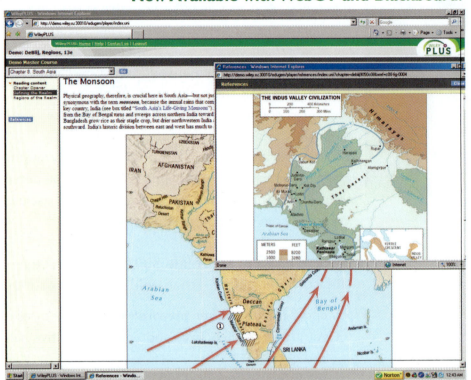

FOR STUDENTS

You have the potential to make a difference!

WileyPLUS is a powerful online system packed with features to help you make the most of your potential and get the best grade you can!

With WileyPLUS you get:

- A complete online version of your text and other study resources.

- Problem-solving help, instant grading, and feedback on your homework and quizzes.

- The ability to track your progress and grades throughout the term.

For more information on what *WileyPLUS* can do to help you and your students reach their potential, please visit www.wiley.com/college/*wileyplus*.

76% of students surveyed said it made them better prepared for tests. *

*Based on a survey of 972 student users of *WileyPLUS*

GEOGRAPHY

Realms, Regions, and Concepts

1807

WILEY

2007

THE WILEY BICENTENNIAL–KNOWLEDGE FOR GENERATIONS

Each generation has its unique needs and aspirations. When Charles Wiley first opened his small printing shop in lower Manhattan in 1807, it was a generation of boundless potential searching for an identity. And we were there, helping to define a new American literary tradition. Over half a century later, in the midst of the Second Industrial Revolution, it was a generation focused on building the future. Once again, we were there, supplying the critical scientific, technical, and engineering knowledge that helped frame the world. Throughout the 20th Century, and into the new millennium, nations began to reach out beyond their own borders and a new international community was born. Wiley was there, expanding its operations around the world to enable a global exchange of ideas, opinions, and know-how.

For 200 years, Wiley has been an integral part of each generation's journey, enabling the flow of information and understanding necessary to meet their needs and fulfill their aspirations. Today, bold new technologies are changing the way we live and learn. Wiley will be there, providing you the must-have knowledge you need to imagine new worlds, new possibilities, and new opportunities.

Generations come and go, but you can always count on Wiley to provide you the knowledge you need, when and where you need it!

WILLIAM J. PESCE
PRESIDENT AND CHIEF EXECUTIVE OFFICER

PETER BOOTH WILEY
CHAIRMAN OF THE BOARD

THIRTEENTH EDITION

GEOGRAPHY

Realms, Regions, and Concepts

H. J. de Blij

John A. Hannah Professor of Geography
Michigan State University

Peter O. Muller

Professor, Department of Geography and Regional Studies
University of Miami

John Wiley & Sons, Inc.

PUBLISHER	Jay O'Callaghan
EXECUTIVE EDITOR	Ryan Flahive
ASSISTANT EDITOR	Courtney Nelson
EDITORIAL ASSISTANT	Erin Grattan
MARKETING MANAGER	Jeffrey Rucker
MARKETING ASSISTANT	Rachel Cirone
PRODUCTION EDITOR	Barbara Russiello
SENIOR ILLUSTRATION EDITOR	Sigmund Malinowski
ELECTRONIC ILLUSTRATIONS	Mapping Specialists, Ltd.
SENIOR PHOTO EDITOR	Jennifer MacMillan
SENIOR MEDIA EDITOR	Lynn Pearlman
DIRECTOR CREATIVE SERVICES	Harry Nolan
SENIOR DESIGNER	Kevin Murphy
INTERIOR DESIGN	Nancy Field
COVER DESIGN	David Levy
FRONT COVER PHOTO	Maggie Steber
BACK COVER PHOTOS	H. J. de Blij
WILEY BICENTENNIAL LOGO DESIGNER	Richard Pacifico

This book was set in 10.5/12.6 Times Roman by GGS Book Services and
printed and bound by RR Donnelley. The cover was printed by RR Donnelley.

This book is printed on acid free paper. ∞

Copyright © 2008 by H. J. de Blij and Peter O. Muller. All rights reserved. No part of this
publication may be reproduced, stored in a retrieval system or transmitted in any form
or by any means, electronic, mechanical, photocopying, recording, scanning or
otherwise, except as permitted under Sections 107 or 108 of the 1976 United States Copyright
Act, without either the prior written permission of the Publisher, or authorization through
payment of the appropriate per-copy fee to the Copyright Clearance Center, Inc. 222 Rosewood
Drive, Danvers, MA 01923, website www.copyright.com. Requests to the Publisher for
permission should be addressed to the Permissions Department, John Wiley & Sons, Inc.,
111 River Street, Hoboken, NJ 07030-5774, (201)748-6011, fax (201)748-6008, website
http://www.wiley.com/go/permissions.

To order books or for customer service, please call 1-800-CALL WILEY (225-5945).

Library of Congress Cataloging in Publication Data:

de Blij, Harm J.
    Geography : realms, regions, and concepts / H. J. de Blij, Peter O. Muller.—Thirteenth ed.
      p. cm.
    ISBN 978-0-470-12905-0 (cloth)
      1. Geography—Textbooks.   I. Muller, Peter O.   II. Title.

G128.D42 2008
910—dc22

                                                                   2007035566

Printed in the United States of America

10 9 8 7 6 5 4 3 2 1

To Gil Grosvenor,
whose tireless, nationwide advocacy of geography
inspires us all

PREFACE

FOR NEARLY FOUR decades, *Geography: Realms, Regions, and Concepts* has reported (and sometimes anticipated) trends in the discipline of Geography and developments in the world at large. In twelve preceding editions, *Regions*, as the book is generally called, has explained the modern world's great geographic realms (i.e., the largest regional units the world can be divided into) and their physical and human contents, and has introduced geography itself, the discipline that links the study of human societies and natural environments through a fascinating, spatial approach. From old ideas to new, from environmental determinism to expansion diffusion, from decolonization to devolution, *Regions* has provided geographic perspective on our transforming world.

The book before you, therefore, is an information highway to geographic literacy. The first edition appeared in 1971, at a time when school geography in the United States (though not in Canada) was a subject in decline. It was a precursor of a dangerous isolationism in America, and geographers foresaw the looming cost of geographic illiteracy. Sure enough, the media during the 1980s began to report that polls, public surveys, tests, and other instruments were recording a lack of geographic knowledge at a time when our world was changing ever faster and becoming more competitive by the day. Various institutions, including the National Geographic Society, banks, airline companies, and a consortium of scholarly organizations, mobilized to confront an educational dilemma that had resulted substantially from a neglect of the very topics this book is about.

Before we can usefully discuss such commonplace topics as our "shrinking world," our "global village," and our "distant linkages," we should know what the parts are, the components that do the shrinking and linking. This is not just an academic exercise. You will find that much of what you encounter in this book is of immediate, practical value to you—as a citizen, a consumer, a traveler, a voter, a jobseeker. North America is a geographic realm with intensifying global interests and involvements. Those interests and involvements require countless, often instantaneous decisions. Such decisions must be based on the best possible knowledge of the world beyond our continent. That knowledge can be gained by studying the layout of our world, its environments, societies, resources, policies, traditions, and other properties—in short, its regional geography.

REALMS AND REGIONS . . .

This book is organized into thirteen chapters. The Introduction discusses the world as a whole, outlining the physical stage on which the human drama is being played out, providing environmental frameworks, demographic data, political background, and economic-geographical context. Each of the remaining twelve chapters focuses on one of the world's major geographic realms.

The first part of each chapter addresses the dominant geographic aspects of the realm under discussion, including considerable emphasis on its environmental geography. The second part focuses on the realm's component regions and countries, highlighting the often-fascinating diversity that marks the physical and human geography of every area of the planet. Here we get a chance to report on the hopes and the problems of peoples facing forces of change over which all too many have little or no control.

. . . AND CONCEPTS

To achieve this, we introduce as many as 150 geographic concepts placed in their regional settings. Most of these concepts are indeed primarily geographical, but others are ideas about which, we believe, students of geography should have some knowledge. Although such concepts are listed on the opening page of every chapter, we have not, of course, enumerated every geographic notion used in that chapter. Many colleagues, we suspect, will want to make their own realm-concept associations, and as readers will readily perceive, the book's organization is quite flexible. It is possible, for example, to focus almost exclusively on substantive regional material or, alternatively, to concentrate mainly on conceptual issues.

WHAT'S NEW ABOUT THE THIRTEENTH EDITION?

This Thirteenth Edition of *Regions* sustains a tradition of comprehensive revision that is a hallmark of this book's context and currency. As those who have long been familiar with it are aware, new editions of *Regions* are not merely updates of previous information but reflect major recent developments in the world as well as in the discipline and respond to the comments we receive from colleagues and students. We are finding that *Regions* is increasingly a global book, and since the publication of the Twelfth Edition we have received numerous messages from colleagues and students in many parts of the world on broad issues as well as specific topics. Several of these recent comments came from India, contributing to the major revision of the chapter on the South Asian realm; others drew our attention to devolution in Europe, environmental problems in Australia, the absence of a world-languages map in the Introduction, agricultural developments in North America, the need to "go metric" in the field of regional geography, and a host of additional items. We respond personally to every one of these concerns and, as this Thirteenth Edition shows, we consider all of them in our revision.

We have indeed reversed British and metric units, so that kilometers (the global standard of long-distance measurement) now precede miles, and elevations are shown in meters first, feet second. Get used to square kilometers and hectares rather than square miles and acres! These changes also affected Table G-1, the much-used data source at the end of the Introduction (pp. 38–44), and maps throughout the book. A conversion table appears in Appendix A.

New Openers, New Layout

The opening pages of each chapter look very different from those of earlier editions, again in response to readers' comments, one of which said that there was "nothing modern" about the previous versions. Our superb cartographers, Mapping Specialists of Madison, Wisconsin, have prepared a completely new set of opener maps that reflect more accurately the global regional framework on which *Regions* is based. The new design combines structural and conceptual content with hints (In This Chapter) of what readers will encounter, augmented by some representative photography.

As will be clear to anyone scanning the book, another change involves the page itself. The number of heads and subheads has been significantly increased to guide you through the narrative. This makes studying easier and self-testing more effective.

We have retained the Regional Issue feature that has proven remarkably popular with students, who continue to "vote" their opinion and, we gather, debate these contentious topics in class and among themselves. In this edition, we have added a new feature that, we hope, will elicit similar reaction under the title Points to Ponder—brief statements about things of geographic import happening in a realm or region that may give one pause.

About the Maps

The world continues to change rapidly. Those of us living in the United States might find it unimaginable that a State would split up or a corner of the country would announce its decision to vote on secession, but all this is happening elsewhere, even in the birthplace of the "nation-state." Since the Twelfth Edition appeared, we have had to revise all of our maps of Europe because a new country gained independence: Montenegro. And even as we complete this Thirteenth Edition, still another European territory appears to be headed for sovereignty: Kosovo (although Islamic-majority Kosovo may have a tougher road ahead than Montenegro did). Meanwhile, India may not be finished awarding Statehood to some of its claimants, Sudan completely redrew its administrative map, Iraq's provincial map may not survive its violent transition, and other changes of importance are in the works. Our maps reflect the state of affairs in mid-2007, and in our text we report what these ongoing changes mean to the people involved and affected.

Our readers were right: the language issue (will English become the Latin of the latter day?) is of growing importance worldwide, and we need a world map and text to introduce this important topic in the book's Introduction (see pp. 22–23). As long-time readers are aware, the realmwide sections of many of the chapters already contain larger-scale language (or ethnolinguistic) maps, which can now be put in global context. North America's agricultural transformation is displayed in a new map in Chapter 3 (Fig. 3-15), and Mexico's dangerous political-geographic division, emphasized by its 2006 presidential election, is shown in Chapter 4 (Fig. 4-10). Two important new maps appear in Chapter 5: South America's modern ethnocultural pattern (Fig. 5-5) and the realm's evolving agricultural systems (Fig. 5-4). Also in South America, we have introduced a new map of Bolivia (Fig. 5-11) showing this important country's internal economic and cultural divisions, deepening in this time of indigenous assertiveness.

In Chapter 7, a population cartogram of Iraq (Fig. 7-14) reveals dramatically why Baghdad is the key to the country's elusive stability; we also introduce in this edition new maps of two countries of growing importance and concern,

Turkey (Fig. 7-19) and Iran (Fig. 7-20). In Chapter 9, a new map of China's regional disparities (Fig. 9-17), as reflected by the income of its provinces, underscores one of this giant country's most serious and growing problems. And in Chapter 10 still another new map shows the complex regional and cultural geography of one of the world's most severely misgoverned countries, Myanmar (Fig. 10-11), the story of whose new capital is one of the ironies of the era.

In addition, hundreds of changes were made in existing maps, including Sudan's provincial restructuring (Fig. 7-10 —in anticipation of still another vote on independence, this time by its southern sector) and Israel's expanding "Security Barrier" (Fig. 7-16) that now carves up its neighborhood. The growing European Union (Fig. 1-11) and the evolving crisis in Sri Lanka (Fig. 8-18) are just some of the developments requiring updated cartographic representation. Overall, the book before you records the state of the world in maps and forecasts the future.

The Narrative

It would be impractical to enumerate the revisions contained in the text; suffice it to say that the narrative manuscript (not counting updates) measured about 50,000 words. In the Introduction, new text accompanies the new map of world languages, and new material covers population and cultural issues as well as climate change. Table G-1 is revised and updated.

For Europe (Chapter 1), new text addresses the "ins" and "outs" of the European Union (EU), coverage of the Benelux countries is expanded, and new text addresses Austria, Spain, and Greece. Major revisions address the evolving political and economic geography of Eastern Europe, as well as the implications of the EU accession of Romania and Bulgaria in context of the prospect of Turkey's admission.

Russia (2) is given a new introduction, and a stronger justification accompanies the Defining the Realm rubric in view of the changing character of the Russian state and its peripheries. Russia's physical geography receives new attention in the context of global warming (also on the maps), and significant new text deals with the country's demographic dilemmas. This chapter is also reorganized structurally to facilitate the flow of ideas.

North America (3) has been strengthened by an entirely new coverage of U.S. agricultural activities, which highlights the transforming regional system that is mapped in Figure 3-15. This analysis continues in the discussion of the Continental Interior region, which includes new material on biofuels and the rapid expansion of ethanol production. Canada's political geography has been updated, particularly the latest developments involving Quebec's devolutionary pressures. Revisions in the physical geography sec-

tion include the impact of Hurricane Katrina, augmented by before-and-after aerial photographs. Urban geography has been updated throughout, as well as the contents of each of the realm's nine regions.

Middle America (4) has a revised and strengthened introduction that more effectively justifies this realm's discrete identity. The regions are more clearly defined, and coverage of physiography and cultural geography has been augmented. Major revision of the text on Mexico deals with the indigenous question, the evolving regional split, energy and industrial prospects, and the geography of inequality (one of Mexico's key concerns). The new discussion on Mexico's future should be of interest to all North American students. In Central America, social problems ranging from persistent poverty to gang violence are seen against a background of political polarization; new text discusses the Panama Canal enlargement and the explosive growth of Panama City. The latest developments in the Caribbean Basin are also highlighted.

South America (5) has three new maps and major revisions dealing with cultural fragmentation, agricultural transformation, cultural identity (notably the rising strength of indigenous movements), and economic integration. Also new are updates on urbanization. The Three Guianas, formerly discussed in a separate box, are now integrated into the running text for reasons of growing integration with their neighbors (not all of it desirable). All of the major states, in northern as well as western South America, receive new and refocused attention, including Ecuador's oil economy and Bolivia's threatened devolution. The text on Paraguay is completely recast, and the three countries of the Southern Cone also have expanded coverage. Brazil, which was extensively revised for the Twelfth Edition, is given lighter treatment in the form of updates.

Subsaharan Africa's (6) environmental problems are newly addressed in the first part of this chapter, and developments in individual regions and countries are more extensively dealt with in the second part. Included here is major new text on Angola and Zambia, Tanzania and Uganda; the oil boom in Chad, Equatorial Guinea, and Gabon also receives attention. New text on Nigeria and its election (2007) plus oil and religious issues is followed by an entirely new section on a country not previously given its own coverage: Guinea. The ongoing crisis in failed-state Somalia, eastern anchor of the Islamic Front stretching across Africa, is brought up to date.

Almost the entire introduction to North Africa/Southwest Asia (7) is revised and rewritten, and the map of the realm's oil reserves and production is reconfigured. Additional narrative on Sudan (and the Darfur tragedy), neighboring Chad, and the changing economic geography of this area centers on the involvement of outsiders. The prospect of Sudan fragmenting along ethnoreligious lines is raised in the text as well as on the map. The Israel-Lebanon conflict of 2006 and its aftermath are given geographic perspective. The smaller Arabian Peninsula states are highlighted for the first time. With the addition of two new maps, coverage of Turkey and Iran is strengthened significantly.

The chapter on South Asia (8) is revised so extensively that we invite comparisons with the Twelfth Edition, especially in the context of India. No brief summary would do justice to this extensive revision, which introduces a new perspective on India while retaining the strengths of the narrative on the giant state's periphery. India's meteoric rise is transforming its regional geography, but in competition with China it faces major regional and cultural obstacles.

China (9) was revised extensively for the Twelfth Edition, but in this edition new text more effectively introduces this realm. Also new is text on Tibet (Xizang) and the Qaidam Basin, and additional narrative revises the Pacific Rim material in areas of rapid change.

Southeast Asia (10) is significantly revised in this Thirteenth Edition. While some new material appears in the first part of this chapter, the bulk of the new narrative is in the second part, ranging from a clearer delineation of the regions to detailed discussions of Vietnam's economic geography, Thailand's cultural-regional problems, the misrule (and consequences) of Myanmar accompanied by a new map, an important modification of the coverage of Singapore (accompanied by a revised map), new text on Indonesia, and a complete rewrite of the text on East Timor. Standard updates affect all countries in this realm, as is the case throughout the book.

Australia (11) is strengthened especially in the discussion of Aboriginal rights and issues as well as the country's deepening drought.

The Pacific Realm (12) was revised in the Twelfth Edition, so that updating here exceeds other changes.

This Thirteenth Edition of *Regions* also includes new photography, some of it again accompanied by Field Notes from the senior author's fieldwork, new captions, statistical tables and data updated to the second half of this decade (including the latest census data and projections for the subnational units of India and China), and updated bibliographies, pronunciation guide, and other supporting features. We hope that you will find this Thirteenth Edition as challenging and productive as you have our earlier ones.

Data Sources

For all matters geographical, of course, we consult *The Annals of the Association of American Geographers, The Professional Geographer, The Geographical Review, The Journal of Geography*, and many other academic journals published regularly in North America—plus an array of similar periodicals published in English-speaking countries from Scotland to New Zealand.

As with every new edition of this book, all quantitative information was updated to the year of publication and checked rigorously. In addition to the major revisions described above, hundreds of other modifications were made, many in response to readers' and reviewers' comments. Some readers found our habit of reporting urban population data within the text disruptive, so we continue to tabulate these at the beginning of each chapter's Regions of the Realm section. The stream of new spellings of geographic names continues, and we pride ourselves in being a reliable source for current and correct usage.

The statistical data that constitute Table G-1 (pp. 38–44) are derived from numerous sources, and each data category is explained in the box on page 45. As users of such data are aware, considerable inconsistency marks the reportage by various agencies, and it is often necessary to make informed decisions on contradictory information. For example, some sources still do not reflect the rapidly declining rates of population increase or life expectancies in AIDS-stricken African countries. Others list demographic averages without accounting for differences between males and females in this regard. In formulating Table G-1, we have used among our sources the United Nations, the Population Reference Bureau, the World Bank, the Encyclopaedia Britannica *Books of the Year*, the *Economist* Intelligence Unit, the *Statesman's Year-Book*, and *The New York Times Almanac*.

The urban population figures—which also entail major problems of reliability and comparability—are mainly drawn from the most recent database published by the United Nations' Population Division. For cities of less than 750,000, we developed our own estimates from a variety of other sources. At any rate, the urban population figures used here are estimates for 2008, and they represent metropolitan-area totals unless otherwise specified.

Photography

The map undoubtedly is geography's closest ally, but there are times when photography is not far behind. Whether from space, from an aircraft, from the tallest building in town, or on the ground, a photograph can, indeed, be worth a thousand words. When geographers perform field research in some area of the world, they are likely to maintain a written record that correlates with the photographic one.

This Thirteenth Edition revision is not confined to the text and maps. As readers of *Regions* will note, the illustrations program was substantially overhauled with numerous new photographs, many by the authors, new accompanying "Field Notes" where appropriate, and new, detailed captions.

Pedagogy

We continue to devise ways to help students learn important geographic concepts and ideas, and to make sense of our complex and rapidly changing world. Our newest special feature, *Points to Ponder*, is discussed on p. xii. A list of the continuing special features now follows.

Opener Maps. As in previous editions, a comprehensive map of the realm opens each chapter. These maps, as mentioned above under "New Openers, New Layout," were specially created according to our specifications, and we proudly unveil them in this Thirteenth Edition. As in the most recent editions, these maps are assigned the first figure number in each chapter.

Concepts, Ideas, and Terms. Each chapter begins with a boxed sequential listing of the key geographic concepts, ideas, and terms that appear in the pages that follow. These are noted by numbers in the margins that correspond to the introduction of each item in the text.

Two-Part Chapter Organization. To help the reader to logically organize the material within chapters, we have divided the regional chapters into two distinct parts. First, "Defining the Realm" includes the general physiographic, historical, and human-geographic background common to the realm. The second section, "Regions of the Realm," presents each of the distinctive regions within the realm (denoted by the symbol ●).

List of Regions. Also on the chapter-opening page, a list of the regions within the particular realm provides a preview and helps to organize the chapter. For ease of identification, the ● symbol that denotes the regions list here also appears beside each region heading in the chapter.

Major Geographic Qualities. Near the beginning of each realm chapter, we list, in boxed format, the major geographic qualities that best summarize that portion of the Earth's surface.

Sidebar Boxes. Special topical issues are highlighted in boxed sections. These boxes allow us to include interesting and current topics without interrupting the flow of material within chapters.

Regional Issues (Boxes). In each regional chapter, a controversial issue has been singled out for debate by two opposing voices. As indicated above, these commentaries are drawn from the senior author's field notes and correspondence and are intended to stimulate classroom discussion. Individual readers may "vote" their opinions by e-mail, as indicated at the bottom of each box.

Among the Realm's Great Cities (Boxes). This feature reflects the growing process and influence of urbanization worldwide. More than thirty profiles of the world's leading cities are presented, each accompanied by a specially drawn map.

Major Cities of the Realm Population Tables. Near the beginning of every Regions of the Realm section of each chapter, we have included a table reporting the most up-to-date urban population data (based on 2008 estimates drawn from the sources listed earlier). Readers should find this format less disruptive than citation of the population when the city is mentioned in the text.

What You Can Do (Boxes). At the end of each chapter, we have provided a box that focuses on practical suggestions, ideas to get involved with, and ways to contribute. The first element is a Recommendation to help you get involved in and contribute to projects and/or organizations related to world regional geography. The second element, entitled Geographic Connections, offers a set of challenges of a practical nature that asks you to visualize how you would use what you have learned in a business, government, or educational setting.

From the Field Notes. In the Eighth Edition we introduced a new feature that has proven effective in some of our other textbooks. Many of the photographs in this book were taken by the senior author while doing fieldwork. The more extensive captions and From the Field Notes provide valuable insights into how a geographer observes and interprets information in the field.

Appendices, Glossary, References, and Indexes. In previous editions, we included a number of features at the end of the book that enriched and/or supplemented the main text. These have all been preserved and updated, and we have even added a new feature to this Thirteenth Edition. The increasing pressures of space, however, allow only three sections to remain in the book itself: (1) Appendix A, a conversions table linking metric and British units of measurement (the new section mentioned in the previous sentence); (2) an extensive Glossary of terms; and (3) a comprehensive general index. Four additional sections, which should not be overlooked, now appear exclusively on the book's website **www.wiley.com/college/deblij**: (1) Appendix B, a guide to Using the Maps; (2) Appendix C, an overview of Opportunities in Geography; (3) Appendix D, a Pronunciation Guide; and (4) a detailed bibliography—under the title References and Further Readings, and organized by chapter—that introduces the wide-ranging literature of the discipline and World Regional Geography.

Ancillaries

A broad spectrum of print and electronic ancillaries are available to accompany the Thirteenth Edition of *Regions*. (Additional information, including prices and ISBNs for ordering, can be obtained by contacting John Wiley & Sons.) These ancilliaries are described next.

THE TEACHING AND LEARNING PACKAGE

This Thirteenth Edition of *Geography: Realms, Regions, and Concepts* is supported by a comprehensive supplements package that includes an extensive selection of print, visual, and electronic materials.

Wiley/National Geographic *College Atlas of the World*
Edited by H. J. de Blij and Roger Downs

Wiley is proud to offer the *College Atlas of the World* to your students through our exclusive partnership with the National Geographic Society. State-of-the-art cartographic technology plus award-winning design and content make this affordable, compact, yet comprehensive atlas the ultimate resource for every geography student. The past decade has seen an explosion of innovative and sophisticated digital mapping technologies—and the Society has been at the forefront of developing these powerful new tools to create the finest, most functional, and informative atlases available anywhere. This powerful resource can be purchased individually or at a significant discount when packaged with any Wiley textbook.

Resources That Help Teachers Teach

NEW PowerPoint Presentations. Antoinette Winkler-Prins of Michigan State University has created a fresh collection of PowerPoint presentations that will help you bring the striking photographs, precise maps, and vivid illustrations from *Regions* into your classroom. We understand that each professor has unique needs for their classroom presentations, so we offer three different types of presentation for each chapter:

- PowerPoints with just the text art.
- PowerPoints with text art and presentation notes.
- Media-Integrated PowerPoints with links to videos and animations.

Videos and Podcasts. We have created a series of streaming video resources and podcasts to support *Geography: Realms, Regions, and Concepts 13e*. Our streaming videos provide short lecture-launching video clips that can be used to introduce new topics, enhance your presentations, and stimulate classroom discussion. To help you integrate these resources into your syllabus, we have developed a comprehensive library of teaching resources, study questions, and assignments.

Image Gallery. We provide online electronic files for the line illustrations and maps in the text, which the instructor can customize for presenting in class (for example, in handouts, overhead transparencies, or PowerPoints).

Wiley Faculty Network. This peer-to-peer network of faculty is ready to support your use of online course management tools and discipline-specific software/learning systems in the classroom. The WFN will help you apply innovative classroom techniques, implement software packages, tailor the technology experience to the needs of each individual class, and provide you with virtual training sessions led by faculty for faculty.

The Regions 13e Instructors' Site. This comprehensive website includes numerous resources to help you enhance your current presentations, create new presentations, and employ our pre-made PowerPoint presentations. These resources include:

- A complete collection of PowerPoint presentations (prepared by Antoinette WinklerPrins). Three for each chapter.
- All of the line art from the text.
- Access to photographs.
- A comprehensive collection of animations.
- A comprehensive test bank.

Instructor's Manual and Test Bank. This manual includes a test bank, chapter overviews and outlines, and lecture suggestions. Prepared and updated by co-author Peter Muller, the *Test Bank* for *Regions* contains over 3,000 test items including multiple-choice, fill-in, matching, and essay questions. It is distributed via the secure Instructor's website as electronic files, which can be saved into all major word processing programs. The *Instructor's Manual* is also available on the password protected Instructor's website.

Overhead Transparency Set. All of the book's maps and diagrams are available as print-on-demand transparencies in beautifully rendered, 4-color format, and have been resized and edited for maximum effectiveness in large lecture halls.

Course Management. Online course management assets are available to accompany *Regions*.

Resources That Help Students Learn

GeoDiscoveries Website www.wiley.com/college/deblij
This easy to use and student-focused website helps reinforce and illustrate key concepts from the text. It also provides interactive media content that helps students prepare for tests and improve their grades. This website provides additional resources that complement the textbook and enhance your students' understanding of geography and improve their grade by using the following resources:

- **Videos** provide a first-hand look at life in other parts of the world.
- **Interactive Animations and Exercises** allow you to explore key concepts from the text.
- **Virtual Field Trips** invite you to discover different locations through photos, and help you better understand how geographers view the world.
- **Podcasts, Web Cams, and Live Radio** let you see and hear what is happening all over the world in real time.
- **Flashcards** offer an excellent way to drill and practice key concepts, ideas, and terms from the text.
- The **Learning Styles Survey** will help you identify your unique way of understanding and processing information.
- With the **Study Skills Chart**, you can create a customized learning plan that best suits your personal learning style.
- **Map Quizzes** help students master the place names that are crucial to their success in this course. Three game-formatted place name activities are provided for each chapter.
- **GeoDiscoveries Modules** allow students to explore key concepts in greater depth using videos, animations, and interactive exercises.
- **Chapter Review Quizzes** provide immediate feedback to true/false, multiple choice, and short answer questions.
- **Concepts, Ideas, and Terms Interactive Flashcards** help students review and quiz themselves on the concepts, ideas, and terms discussed in each chapter.
- **Interactive Drag-and-Drop Exercises** challenge students to correctly label important illustrations from the textbook.
- **Audio Pronunciation** is provided for over 2000 key words and place names from the text.
- **Annotated Web Links** put useful electronic resources into context.
- **Area and Demographic Data** are provided for every country and world realm.

Student Study Guide. Text co-author Peter Muller and his geographer daughter, Elizabeth Muller Hames, have written a popular Study Guide to accompany the book that is packed with useful study and review tools. For each chapter in the textbook, the Study Guide gives students and faculty access to chapter objectives, content questions-and-answers, outline maps of each realm, sample tests, and more.

Social Geographies of the Modern World. This free 60-page supplement includes chapters that discuss:

- Global Disparities in Nutrition and Health;
- Geographies of Inequality: Race and Ethnicity;
- Gender Inequities in Geographic Perspective.

To arrange for your students to receive this free supplement with their copy of *Regions*, please contact your local Wiley sales representative.

ACKNOWLEDGMENTS

Over the 37 years since the publication of the First Edition of *Geography: Realms, Regions, and Concepts*, we have been fortunate to receive advice and assistance from literally hundreds of people. One of the rewards associated with the publication of a book of this kind is the steady stream of correspondence and other feedback it generates. Geographers, economists, political scientists, education specialists, and others have written us, often with fascinating enclosures. We make it a point to respond personally to every such letter, and our editors have communicated with many of our correspondents as well. Moreover, we have considered every suggestion made and many who wrote or transmitted their reactions through other channels will see their recommendations in print in this edition.

STUDENT RESPONSE

A major part of the correspondence we receive comes from student readers. We would like to take this opportunity to extend our deep appreciation to the several million students around the world who have studied from the first 15 editions of our World Regional Geography texts. In particular, we thank the students from more than 130 different colleges across the United States who took the time to send us their opinions.

Students told us they found the maps and graphics attractive and functional. We have not only enhanced the map program with exhaustive updating but have added a number of new and updated maps to this Thirteenth Edition as well as making significant changes in many others. Generally, students have told us that they found the pedagogical devices quite useful. We have kept the study aids the students cited as effective: a boxed list of each chapter's key concepts, ideas, and terms (numbered for quick reference in both the box and text margins); a box summarizing each realm's major geographic qualities; and an extensive Glossary.

FACULTY FEEDBACK

In assembling the Thirteenth Edition, we are indebted to the following people for advising us on a number of matters:

THOMAS L. BELL, University of Tennessee
RICHARD A. BERRYHILL, East Central Community College (Mississippi)
KATHLEEN BRADEN, Seattle Pacific University
DEBORAH CORCORAN, Southwest Missouri State University
EDMAR BERNADES DASILVA, Broward Community College (Florida)
WILLIAM V. DAVIDSON, Louisiana State University
GARY A. FULLER, University of Hawai'i
RICHARD J. GRANT, University of Miami
MARGARET M. GRIPSHOVER, University of Tennessee
SHIRLENA HUANG, National University of Singapore
RICHARD LISICHENKO, Fort Hays State University (Kansas)
DALTON E. MILLER, Mississippi State University
JAN NIJMAN, University of Miami
VALIANT C. NORMAN, Lexington Community College (Kentucky)
PAI YUNG-FENG, New York City
IWONA PETRUCZYNIK, Mercyhurst College (Pennsylvania)
RINKU ROY CHOWDHURY, University of Miami
SHOURASENI SEN ROY, University of Miami
IRA SHESKIN, University of Miami
RICHARD SLEASE, Oakland, North Carolina
CATHY WEIDMAN, Austin, Texas
ANTOINETTE WINKLERPRINS, Michigan State University

In addition, several faculty colleagues from around the world assisted us with earlier editions, and their contributions continue to grace the pages of this book. Among them are:

JAMES P. ALLEN, California State University, Northridge
STEPHEN S. BIRDSALL, University of North Carolina
J. DOUGLAS EYRE, University of North Carolina
FANG YONG-MING, Shanghai, China
EDWARD J. FERNALD, Florida State University
RAY HENKEL, Arizona State University
RICHARD C. JONES, University of Texas at San Antonio
GIL LATZ, Portland State University (Oregon)
IAN MACLACHLAN, University of Lethbridge (Alberta)
MELINDA S. MEADE, University of North Carolina
HENRY N. MICHAEL, late of Temple University (Pennsylvania)
CLIFTON W. PANNELL, University of Georgia
J. R. VICTOR PRESCOTT, University of Melbourne (Australia)
JOHN D. STEPHENS, University of Washington
CANUTE VANDER MEER, University of Vermont

Faculty members from a large number of North American colleges and universities continue to supply us with vital feedback and much-appreciated advice. Our publishers arranged several feedback sessions, and we are most grateful to the following professors for showing us where the text could be strengthened and made more precise:

MARTIN ARFORD, Saginaw Valley State University (Michigan)
DONNA ARKOWSKI, Pikes Peak Community College (Colorado)
GREG ATKINSON, Tarleton State University (Texas)
DENIS BEKAERT, Middle Tennessee State University
DONALD J. BERG, South Dakota State University
DAVID COCHRAN, University of Southern Mississippi
MARCELO CRUZ, University of Wisconsin at Green Bay
LARRY SCOTT DEANER, Kansas State University
JASON DITTMER, Georgia Southern University
JAMES DOERNER, University of Northern Colorado
STEVEN DRIEVER, University of Missouri-Kansas City
ELIZABETH DUDLEY-MURPHY, University of Utah
DENNIS EHRHARDT, University of Louisiana-Lafayette
WILLIAM FORBES, Stephen F. Austin State University (Texas)
BILL FOREMAN, Oklahoma City Community College
ERIC FOURNIER, Samford University (Alabama)
RANDY GABRYS ALEXSON, University of Wisconsin-Superior
WILLIAM GARBARINO, Community College of Allegheny County (Pennsylvania)
JON GOSS, University of Hawai'i
SARA HARRIS, Neosho County Community College (Kansas)
INGRID JOHNSON, Towson State University (Maryland)
KRIS JONES, Saddleback College (California)
THOMAS KARWOSKI, Anne Arundel Community College (Maryland)
ROBERT KERR, University of Central Oklahoma
ERIC KEYS, University of Florida
JACK KINWORTHY, Concordia University-Nebraska
CHRISTOPHER LAINGEN, Kansas State University
UNNA LASSITER, Stephen F. Austin State University (Texas)
CATHERINE LOCKWOOD, Chadron State College (Nebraska)
GEORGE LONBERGER, Georgia Perimeter College
DALTON E. MILLER, Mississippi State University
FERNANDO F. MINGHINE, Wayne State University (Michigan)
VERONICA MORMINO, Harper College (Illinois)
TOM MUELLER, California University (Pennsylvania)
IRENE NAESSE, Orange Coast College (California)
RICHARD OLMO, University of Guam
J. L. PASZTOR, Delta College (Michigan)
PAUL E. PHILLIPS, Fort Hays State University (Kansas)
ROSANN POLTRONE, Arapahoe Community College (Colorado)
JEFF POPKE, East Carolina University
DAVID PRIVETTE, Central Piedmont Community College (North Carolina)
RHONDA REAGAN, Blinn College (Texas)
A. L. RYDANT, Keene State College (New Hampshire)
NANCY SHIRLEY, Southern Connecticut State University
DEAN SINCLAIR, Northwestern State University (Louisiana)
SUSAN SLOWEY, Blinn College (Texas)
DEAN B. STONE, Scott Community College (Iowa)
JAMIE STRICKLAND, University of North Carolina at Charlotte
RUTHINE TIDWELL, Florida Community College at Jacksonville

IRINA VAKULENKO, Collin County Community College (Texas)
KIRK WHITE, York College of Pennsylvania
THOMAS WHITMORE, University of North Carolina at Chapel Hill
KEITH YEARMAN, College of DuPage (Illinois)
LAURA ZEEMAN, Red Rocks Community College (Colorado)

We also received input from a much wider circle of academic geographers. The list that follows is merely representative of a group of colleagues across North America to whom we are grateful for taking the time to share their thoughts and opinions with us:

MEL AAMODT, California State University-Stanislaus
WILLIAM V. ACKERMAN, Ohio State University
W. FRANK AINSLEY, University of North Carolina, Wilmington
SIAW AKWAWUA, University of Northern Colorado
DONALD P. ALBERT, Sam Houston State University
TONI ALEXANDER, Louisiana State University
R. GABRYS ALEXSON, University of Wisconsin-Superior
KHALED MD. ALI, Geologist, Dhaka, Bangladesh
NIGEL ALLAN, University of California-Davis
JOHN L. ALLEN, University of Wyoming
KRISTIN J. ALVAREZ, Keene State College (New Hampshire)
DAVID L. ANDERSON, Louisiana State University, Shreveport
KEAN ANDERSON, Freed-Hardeman University
JEFF ARNOLD, Southwestern Illinois College
JERRY R. ASCHERMANN, Missouri Western State College
JOSEPH M. ASHLEY, Montana State University
PATRICK ASHWOOD, Hawkeye Community College
THEODORE P. AUFDEMBERGE, Concordia College (Michigan)
JAIME M. AVILA, Sacramento City College
EDWARD BABIN, University of South Carolina-Spartanburg
ROBERT BAERENT, Randolph-Macon College (Virginia)
MARVIN W. BAKER, University of Oklahoma
GOURI BANERJEE, Boston University (Massachusetts)
MICHELE BARNABY, Pittsburg State University (Kansas)
J. HENRY BARTON, Thiel College (Pennsylvania)
CATHY BARTSCH, Temple University (Pennsylvania)
STEVEN BASS, Paradise Valley Community College (Arizona)
THOMAS F. BAUCOM, Jacksonville State University (Alabama)
KLAUS J. BAYR, Keene State College (New Hampshire)
DENIS A. BEKAERT, Middle Tennessee State University
JAMES BELL, Linn Benton Community College (Oregon)
KEITH M. BELL, Volunteer State Community College (Tennessee)
WILLIAM H. BERENTSEN, University of Connecticut
DONALD J. BERG, South Dakota State University
ROYAL BERGLEE, Morehead State University (Kentucky)
RIVA BERLEANT-SCHILLER, University of Connecticut
LEE LUCAS BERMAN, Southern Connecticut State University
RANDY BERTOLAS, Wayne State College (Nebraska)
THOMAS BITNER, University of Wisconsin-Marshfield/ Wood County
WARREN BLAND, California State University, Northridge
DAVIS BLEVINS, Huntington College (Alabama)
HUBERTUS BLOEMER, Ohio University
S. BO JUNG, Bellevue College (Nebraska)
R. DENISE BLANCHARD-BOEHM, Texas State University
MARTHA BONTE, Clinton Community College (Idaho)
GEORGE R. BOTJER, University of Tampa (Florida)

R. LYNN BRADLEY, Belleville Area College (Illinois)
KEN BREHOB, Elmhurst College (Illinois)
JAMES A. BREY, University of Wisconsin-Fox Valley
ROBERT BRINSON, Santa Fe Community College (Florida)
REUBEN H. BROOKS, Tennessee State University
PHILLIP BROUGHTON, Arizona Western College
LARRY BROWN, Ohio State University
LAWRENCE A. BROWN, Troy State-Dothan (Alabama)
ROBERT N. BROWN, Delta State University (Mississippi)
STANLEY D. BRUNN, University of Kentucky
RANDALL L. BUCHMAN, Defiance College (Ohio)
MICHAELE ANN BUELL, Northwest Arkansas Community College
DANIEL BUYNE, South Plains College
MICHAEL BUSBY, Murray State University (Kentucky)
MARY CAMERON, Slippery Rock University (Pennsylvania)
MICHAEL CAMILLE, University of Louisiana at Monroe
MARY CARAVELIS, Barry University (Florida)
DIANA CASEY, Muskegon Community College (Michigan)
DiANN CASTEEL, Tusculum College (Tennessee)
BILL CHAPPELL, Keystone College
MICHAEL SEAN CHENOWETH, University of Wisconsin-Milwaukee
STANLEY CLARK, California State University, Bakersfield
JOHN E. COFFMAN, University of Houston (Texas)
DWAYNE COLE, Cornerstone University (Michigan)
JERRY COLEMAN, Mississippi Gulf Coast Community College
JONATHAN C. COMER, Oklahoma State University
BARBARA CONNELLY, Westchester Community College (New York)
FRED CONNINGTON, Southern Wesleyan University
WILLIS M. CONOVER, University of Scranton (Pennsylvania)
OMAR CONRAD, Maple Woods Community College (Missouri)
ALLAN D. COOPER, Otterbein College
WILLIAM COUCH, University of Alabama, Huntsville
BARBARA CRAGG, Aquinas College (Michigan)
GEORGES G. CRAVINS, University of Wisconsin-La Crosse
ELLEN K. CROMLEY, University of Connecticut
JOHN A. CROSS, University of Wisconsin-Oshkosh
SHANNON CRUM, University of Texas at San Antonio
WILLIAM CURRAN, South Suburban College (Illinois)
KEVIN M. CURTIN, University of Texas at San Antonio
JOSE A. DA CRUZ, Ozarks Technical Community College (Arkansas)
ARMANDO DA SILVA, Towson State University (Maryland)
MARK DAMICO, Green Mountain College (Vermont)
DAVID D. DANIELS, Central Missouri State University
RUDOLPH L. DANIELS, Morningside College (Iowa)
SATISH K. DAVGUN, Bemidji State University (Minnesota)
WILLIAM V. DAVIDSON, Louisiana State University
CHARLES DAVIS, Mississippi Gulf Coast Community College
JAMES DAVIS, Illinois College
JAMES L. DAVIS, Western Kentucky University
PEGGY E. DAVIS, Pikeville College
ANN DEAKIN, State University of New York, College at Fredonia
KEITH DEBBAGE, University of North Carolina at Greensboro
MOLLY DEBYSINGH, California State University, Long Beach
DENNIS K. DEDRICK, Georgetown College (Kentucky)
JOEL DEICHMANN, Bentley College
STANFORD DeMARS, Rhode Island College
TOM DESULIS, Spoon River College
THOMAS DIMICELLI, William Paterson College (New Jersey)
MARY DOBBS, Highland Community College, Wamego

Scott Dobler, Western Kentucky University
Daniel F. Doeppers, University of Wisconsin-Madison
James Doerner, University of Northern Colorado
Ann Doolen, Lincoln College (Illinois)
Steven Driever, University of Missouri-Kansas City
William Druen, Western Kentucky University
Alasdair Drysdale, University of New Hampshire
Keith A. Ducote, Cabrillo Community College (California)
Elizabeth Dudley-Murphy, University of Utah
Walter N. Duffet, University of Arizona
Mike Dunning, University of Alaska Southeast
Christina Dunphy, Champlain College (Vermont)
Anthony Dzik, Shawnee State University (Kansas)
Dennis Edgell, Bowling Green State University, Firelands (Ohio)
Ruth M. Ediger, Seattle Pacific University
James H. Edmonson, Union University (Tennessee)
M. H. Edney, State University of New York-Binghamton
Harold M. Elliott, Weber State University (Utah)
James Elsnes, Western State College
Robert W. Evans, Fresno City College (California)
Everson Dicken, California State University, San Bernardino
Dino Fiabane, Community College of Philadelphia
G. A. Finchum, Milligan College (Tennessee)
Ira Fogel, Foothill College (California)
Richard Foley, Cumberland College
Robert G. Foote, Wayne State College (Nebraska)
Ronald Foresta, University of Tennessee
Ellen J. Foster, Texas State University
G. S. Freedom, McNeese State University (Louisiana)
Edward T. Freels, Carson-Newman College
Philip Friend, Inver Hills Community College (Minnesota)
James Fryman, University of Northern Iowa
Owen Furuseth, University of North Carolina at Charlotte
Richard Fusch, Ohio Wesleyan University
Gary Gaile, University of Colorado-Boulder
Evelyn Gallegos, Eastern Michigan University & Schoolcraft College
Gail Garbrandt, Mount Union College, University of Akron
Richard Garrett, Marymount Manhattan College (New York)
Jerry Gerlach, Winona State University (Minnesota)
Mark Gismondi, Northwest Nazarene University
Lorne E. Glaim, Pacific Union College (California)
Sharleen Gonzalez, Baker College (Michigan)
Daniel B. Good, Georgia Southern University
Gary C. Goodwin, Suffolk Community College (New York)
S. Gopal, Boston University (Massachusetts)
Marvin Gordon, Lake Forest Graduate School of Business (Illinois)
Robert Gould, Morehead State University (Kentucky)
Mary Graham, York College of Pennsylvania
Gordon Grant, Texas A&M University
Paul Gray, Arkansas Tech University
Donald Green, Baylor University (Texas)
Gary M. Green, University of North Alabama
Stanley C. Green, Laredo State University (Texas)
Raymond Greene, Western Illinois University
Mark Greer, Laramie County Community College (Wyoming)
Walt Guttinger, Flagler College (Florida)
John E. Gygax, Wilkes Community College (Pennsylvania)
Lee Ann Hagan, College of Southern Idaho

Ron Hagelman, University of New Orleans
W. Gregory Hager, Northwestern Connecticut Community College
Ruth F. Hale, University of Wisconsin-River Falls
John W. Hall, Louisiana State University-Shreveport
Peter L. Halvorson, University of Connecticut
David Hansen, Pennsylvania State University, Harrisburg & Schuylkill
Mervin Hanson, Willmar Community College (Minnesota)
Robert C. Harding, Lynchburg College (Virginia)
Michelle D. Harris, Bob Jones University (South Carolina)
Scott Harris, CGCS Drury University
Robert J. Hartig, Fort Valley State College (Georgia)
Suzanna Hartley, Shelton State Community College
Truman A. Hartshorn, Georgia State University
Carol Hazard, Meredith College
Harlow Z. Head, Barton College
Doug Heffington, Middle Tennessee State University
James G. Heidt, University of Wisconsin Center-Sheboygan
Catherine Helgeland, University of Wisconsin-Manitowoc
Norma Hendrix, East Arkansas Community College
James E. Herrell, Otero Junior College (Colorado)
James Hertzler, Goshen College (Indiana)
John Hickey, Inver Hills Community College (Minnesota)
Thomas Higgins, San Jacinto College (Texas)
Eugene Hill, Westminster College (Missouri)
Miriam Helen Hill, Indiana University Southeast
Suzy Hill, University of South Carolina-Spartanburg
Robert Hilt, Pittsburg State University (Kansas)
Sophia Hinshalwood, Montclair State University (New Jersey)
Priscilla Holland, University of North Alabama
Mark R. Hooper, Freed-Hardeman University
R. Hostetler, Fresno City College (California)
Erick Howenstine, Northeastern Illinois University
Lloyd E. Hudman, Brigham Young University (Utah)
Janis W. Humble, University of Kentucky
Juana Ibanez, University of New Orleans
Tony Ijomah, Harrisburg Area Community College (Pennsylvania)
William Imperatore, Appalachian State University (North Carolina)
Richard Jackson, Brigham Young University (Utah)
Mary Jacob, Mount Holyoke College (Massachusetts)
Gregory Jeane, Samford University (Alabama)
Scott Jeffrey, Catonsville Community College (Maryland)
Jerzy Jemiolo, Ball State University (Indiana)
Nils I. Johansen, University of Southern Indiana
David Johnson, University of Southwestern Louisiana
Ingrid Johnson, Towson State University (Maryland)
Richard Johnson, Oklahoma City University
Sharon Johnson, Marymount College (New York)
Jeffrey Jones, University of Kentucky
Kris Jones, Saddleback College (California)
Marcus E. Jones, Claflin College (South Carolina)
Mohammad S. Kamiar, Florida Community College, Jacksonville
Matti E. Kaups, University of Minnesota-Duluth
Jo Anne W. Kay, Brigham Young University, Idaho
Philip L. Keating, Indiana University
David Keeling, Western Kentucky University

COLLEEN KEEN, Gustavus Adolphus College (Minnesota)
ARTIMUS KEIFFER, Wittenberg University (Ohio)
GORDON F. KELLS, Mott Community College (Michigan)
KAREN J. KELLY, Palm Beach Community College, Boca Raton (Florida)
RYAN KELLY, Lexington Community College (Kentucky)
VIRGINIA KERKHEIDE, Cuyahoga Community College and Cleveland State University (Ohio)
TOM KESSENGER, Xavier University (Ohio)
MASOUD KHEIRABADI, Maryhurst University
SUSANNE KIBLER-HACKER, Unity College (Maine)
CHANGJOO KIM, Minnesota State University
JAMES W. KING, University of Utah
JOHN C. KINWORTHY, Concordia College (Nebraska)
ALBERT KITCHEN, Paine College (Georgia)
TED KLIMASEWSKI, Jacksonville State University (Alabama)
ROBERT D. KLINGENSMITH, Ohio State University-Newark
LAWRENCE M. KNOPP, JR., University of Minnesota-Duluth
LYNN KOEHNEMANN, Gulf Coast Community College
TERRILL J. KRAMER, University of Nevada
JOE KRAUSE, Community College of Indiana at Lafayette
BRENDER KREKELER, Northern Kentucky University; Miami University (Ohio)
ARTHUR J. KRIM, Cambridge, Massachusetts
MICHAEL A. KUKRAL, Rose-Hulman Institute of Technology
ELROY LANG, El Camino Community College (California)
RICHARD L. LANGILL, Saint Martin's University (Washington)
CHRISTOPHER LANT, Southern Illinois University
A. J. LARSON, University of Illinois-Chicago
PAUL R. LARSON, Southern Utah University
LARRY LEAGUE, Dickinson State University (North Dakota)
DAVID R. LEE, Florida Atlantic University
WOOK LEE, Texas State University
JOE LEEPER, Humboldt State University (California)
YECHIEL M. LEHAVY, Atlantic Community College (New Jersey)
SCOTT LEITH, Central Connecticut State University
JAMES LEONARD, Marshall University (West Virginia)
ELIZABETH J. LEPPMAN, St. Cloud State University (Minnesota)
JOHN C. LEWIS, Northeast Louisiana University
DAN LEWMAN JR., Southwest Mississippi Community College
CAEDMON S. LIBURD, University of Alaska-Anchorage
T. LIGIBEL, Eastern Michigan University
Z. L. LIPCHINSKY, Berea College (Kentucky)
ALLAN L. LIPPERT, Manatee Community College (Florida)
RICHARD LISICHENKO, Fort Hays State University (Kansas)
JOHN L. LITCHER, Wake Forest University (North Carolina)
LEE LIU, Central Montana State University
LI LIU, Stephen F. Austin State University (Texas)
WILLIAM R. LIVINGSTON, Baker College (Michigan)
CATHERINE M. LOCKWOOD, Chadron State College (Nebraska)
GEORGE E. LONGENECKER, Vermont Technical College
CYNTHIA LONGSTREET, Ohio State University
JACK LOONEY, University of Massachusetts, Boston
TOM LOVE, Linfield College (Oregon)
K. J. LOWREY, Miami University (Ohio)
JAMES LOWRY, Stephen F. Austin State University (Texas)
MAX LU, Kansas State University
ROBIN R. LYONS, University of Hawai'i-Leeward Community College
SUSAN M. MACEY, Texas State University

MICHAEL MADSEN, Brigham Young University, Idaho
RONALD MAGDEN, Tacoma Community College (Washington)
CHRISTIANE MAINZER, Oxnard College (California)
LAURA MAKEY, California State University, San Bernardino
HARLEY I. MANNER, University of Guam
ANTHONY PAUL MANNION, Kansas State University
GARY MANSON, Michigan State University
CHARLES MANYARA, Radford University (Virginia)
JAMES T. MARKLEY, Lord Fairfax Community College (Virginia)
SISTER MAY LENORE MARTIN, Saint Mary College (Kansas)
KENT MATHEWSON, Louisiana State University
PATRICK MAY, Plymouth State College (New Hampshire)
DICK MAYER, Maui Community College (Hawai'i)
SARA MAYFIELD, San Jacinto College, Central (California)
DEAN R. MAYHEW, Marine Maritime Academy
J. P. MCFADDEN, Orange Coast College (California)
BERNARD MCGONIGLE, Community College of Philadelphia
PAUL D. MEARTZ, Mayville State University (North Dakota)
DIANNE MEREDITH, California State University-Sacramento
GARY C. MEYER, University of Wisconsin-Stevens Point
JUDITH L. MEYER, Southwest Missouri State University
MARK MICOZZI, East Central University (Oklahoma)
JOHN MILBAUER, Northeastern State University
DALTON E. MILLER, Mississippi State University
RAOUL MILLER, University of Minnesota-Duluth
ROGER MILLER, Black Hills State University (South Dakota)
JAMES MILLS, State University of New York, College at Oneonta
INES MIYARES, Hunter College, CUNY (New York)
BOB MONAHAN, Western Carolina University
KEITH MONTGOMERY, University of Wisconsin-Madison
DAN MORGAN, University of South Carolina, Beaufort
DEBBIE MORIMOTO, Merced College (California)
JOHN MORTRON, Benedict College (South Carolina)
ANNE MOSHER, Syracuse University
BARRY MOWELL, Broward Community College (Florida)
TOM MUELLER, California University (Pennsylvania)
DONALD MYERS, Central Connecticut State University
ROBERT R. MYERS, West Georgia College
GARY NACHTIGALL, Fresno Pacific University (California)
YASER M. NAJJAR, Framingham State College (Massachusetts)
KATHERINE NASHLEANAS, University of Nebraska-Lincoln
JEFFREY W. NEFF, Western Carolina University
DAVID NEMETH, University of Toledo (Ohio)
ROBERT NEWCOMER, East Central University (Oklahoma)
WILLIAM NIETER, St. John's University (New York)
WILLIAM N. NOLL, Highland Community College
VALIANT C. NORMAN, Lexington Community College (Kentucky)
RAYMOND O'BRIEN, Bucks County Community College (Pennsylvania)
PATRICK O'SULLIVAN, Florida State University
DIANE O'CONNELL, Schoolcraft College
JOHN ODLAND, Indiana University
DOUG OETTER, Georgia College and State University
ANNE O'HARA, Marygrove College (Michigan)
PATRICK OLSEN, University of Idaho
JOSEPH R. OPPONG, University of North Texas
LYNN ORLANDO, Holy Family University (Pennsylvania)
MARK A. OUMETTE, Hardin-Simmons University (Texas)
RICHARD OUTWATER, California State University, Long Beach
CISSIE OWEN, Lamar University (Texas)

MARY ANN OWOC, Mercyhurst College (Pennsylvania)
STEVE PALLADINO, Ventura College (California)
BIMAL K. PAUL, Kansas State University
SELINA PEARSON, Northwest-Shoals Community College (Alabama)
MAURI PELTO, Nichols College (Massachusetts)
JAMES PENN, Southeastern Louisiana University
LINDA PETT-CONKLIN, University of St. Thomas (Minnesota)
DIANE PHILEN, Lower Brule Community College (South Dakota)
PAUL PHILLIPS, Fort Hays State University (Kansas)
MICHAEL PHOENIX, ESRI, Redlands, California
JERRY PITZL, Macalester College (Minnesota)
BRIAN PLASTER, Texas State University
ARMAND POLICICCHIO, Slippery Rock University (Pennsylvania)
ROSANN POLTRONE, Arapahoe Community College (Colorado)
BILLIE E. POOL, Holmes Community College (Mississippi)
GREGORY POPE, Montclair State University (New Jersey)
JEFF POPKE, East Carolina University
VINTON M. PRINCE, Wilmington College (North Carolina)
GEORGE PUHRMANN, Drury University (Missouri)
DONALD N. RALLIS, Mary Washington College (Virginia)
RHONDA REAGAN, Blinn College (Texas)
DANNY I. REAMS, Southeast Community College (Nebraska)
JIM RECK, Golden West College (California)
ROGER REEDE, Southwest State University (Minnesota)
JOHN RESSLER, Central Washington University
JOHN B. RICHARDS, Southern Oregon State College
DAVID C. RICHARDSON, Evangel University (Missouri)
GRAY RINGLEY, Virginia Highlands Community College
SUSAN ROBERTS, University of Kentucky
CURT ROBINSON, California State University, Sacramento
WOLF RODER, University of Cincinnati
JAMES ROGERS, University of Central Oklahoma
PAUL A. ROLLINSON, AICP, Southwest Missouri State University
JAMES C. ROSE, Tompkins/Cortland Community College (New York)
THOMAS E. ROSS, Pembroke State University (North Carolina)
THOMAS A. RUMNEY, State University of New York, College at Plattsburgh
GEORGE H. RUSSELL, University of Connecticut
BILL RUTHERFORD, Martin Methodist College
RAJAGOPAL RYALI, Auburn University at Montgomery (Alabama)
PERRY RYAN, Mott Community College (Michigan)
JAMES SAKU, Frostburg State University (Maryland)
DAVID R. SALLEE, University of North Texas
RICHARD A. SAMBROOK, Eastern Kentucky University
EDUARDO SANCHEZ, Grand Valley State University (Michigan)
JOHN SANTOSUOSSO, Florida Southern College
GINGER SCHMID, Texas State University
BRENDA THOMPSON SCHOOLFIELD, Bob Jones University (South Carolina)
ADENA SCHUTZBERG, Middlesex Community College (Massachusetts)
ROGER M. SELYA, University of Cincinnati
RENEE SHAFFER, University of South Carolina
WENDY SHAW, Southern Illinois University, Edwardsville
SIDNEY R. SHERTER, Long Island University (New York)
HAROLD SHILK, Wharton County Jr. College (Texas)
NANDA SHRESTHA, Florida A&M University
WILLIAM R. SIDDALL, Kansas State University

DAVID SILVA, Bee County College (Texas)
JOSE ANTONIO SIMENTAL, Marshall University (West Virginia)
MORRIS SIMON, Stillman College (Alabama)
ROBERT MARK SIMPSON, University of Tennessee at Martin
KENN E. SINCLAIR, Holyoke Community College (Massachusetts)
ROBERT SINCLAIR, Wayne State University (Michigan)
JIM SKINNER, Southwest Missouri State University
BRUCE SMITH, Bowling Green State University (Ohio)
EVERETT G. SMITH, JR., University of Oregon
PEGGY SMITH, California State University, Fullerton
RICHARD V. SMITH, Miami University (Ohio)
JAMES SNADEN, Charter Oak State College (Connecticut)
DAVID SORENSON, Augustana College (Illinois)
SISTER CONSUELO SPARKS, Immaculata University (Pennsylvania)
CAROLYN D. SPATTA, California State University, Hayward
M. R. SPONBERG, Laredo Junior College (Texas)
DONALD L. STAHL, Towson State University (Maryland)
DAVID STEA, Texas State University
ELAINE STEINBERG, Central Florida Community College
D. J. STEPHENSON, Ohio University Eastern
HERSCHEL STERN, Mira Costa College (California)
REED F. STEWART, Bridgewater State College (Massachusetts)
NOEL L. STIRRAT, College of Lake County (Illinois)
JOSEPH P. STOLTMAN, Western Michigan University
DEAN B. STONE, Scott Community College
WILLIAM M. STONE, Saint Xavier University, Chicago
GEORGE STOOPS, Minnesota State University
DEBRA STRAUSSFOGEL, University of New Hampshire
JAMIE STRICKLAND, University of North Carolina at Charlotte
WAYNE STRICKLAND, Roanoke College (Virginia)
PHILIP STURM, Ohio Valley College, Vienna
PHILIP SUCKLING, Texas State University
ALLEN SULLIVAN, Central Washington University
SELIMA SULTANA, University of North Carolina at Greensboro
RAY SUMNER, Long Beach City College (California)
CHRISTOPHER SUTTON, Western Illinois University
T. L. TARLOS, Orange Coast College (California)
WESLEY TERAOKA, Leeward Community College
MICHAEL THEDE, North Iowa Area Community College
DERRICK J. THOM, Utah State University
CURTIS THOMSON, University of Idaho
BEN TILLMAN, Texas Christian University
CLIFF TODD, University of Nebraska, Omaha
STANLEY TOOPS, Miami University (Ohio)
RICHARD J. TORZ, St. Joseph's College (New York)
HARRY TRENDELL, Kennesaw State University (Georgia)
ROGER T. TRINDELL, Mansfield University (Pennsylvania)
DAN TURBEVILLE, East Oregon State College
NORMAN TYLER, Eastern Michigan University
GEORGE VAN OTTEN, Northern Arizona University
GREGORY VEECK, Western Michigan University
C. S. VERMA, Weber State College (Utah)
KELLY ANN VICTOR, Eastern Michigan University
SEAN WAGNER, Tri-State University (Indiana)
GRAHAM T. WALKER, Metropolitan State College of Denver
MONTGOMERY WALKER, Yakima Valley Community College (Washington)
DEBORAH WALLIN, Skagit Valley College (Washington)
MIKE WALTERS, Henderson Community College (Kentucky)
LINDA WANG, University of South Carolina, Aiken

J. L. Watkins, Midwestern State University (Texas)
David Welk, Reedley College (California)
Kit W. Wesler, Murray State University (Kentucky)
Peter W. Whaley, Murray State University (Kentucky)
Macel Wheeler, Northern Kentucky University
P. Gary White, Western Carolina University (North Carolina)
W. R. White, Western Oregon University
Gary Whitton, Fairbanks, Alaska
Rebecca Wiechel, Wilmington College (North Carolina)
Mark Wiljanen, Eastern Kentucky University
Gene C. Wilken, Colorado State University
Forrest Wilkerson, Texas State University
P. Williams, Baldwin-Wallace College (Ohio)
Stephen A. Williams, Methodist College (North Carolina)
Deborah Wilson, Sandhills Community College (Nebraska)
Morton D. Winsberg, Florida State University
Roger Winsor, Appalachian State University (North Carolina)
William A. Withington, University of Kentucky
A. Wolf, Appalachian State University (North Carolina)
Joseph Wood, University of Southern Maine
Richard Wood, Seminole Junior College (Florida)
George I. Woodall, Winthrop College (North Carolina)
Stephen E. Wright, James Madison University (Virginia)
Leon Yacher, Southern Connecticut State University
Kyong Yup Chu, Bergen Community College (New Jersey)
Firooz E. Zadeh, Colorado Mountain College
Donald J. Zeigler, Old Dominion University (Virginia)
Robert C. Ziegenfus, Kutztown University (Pennsylvania)

PERSONAL APPRECIATION

For assistance with the map of North American indigenous people (Fig. 3-3), we are greatly indebted to Jack Weatherford, Professor of Anthropology at Macalester College (Minnesota); Henry T. Wright, Professor and Curator of Anthropology at the University of Michigan; and George E. Stuart, President of the Center for Maya Research (North Carolina). The map of Russia's federal regions (Fig. 2-9) could not have been compiled without the invaluable help of David B. Miller, Senior Edit Cartographer at the National Geographic Society, and Leo Dillon of the Russia Desk of the U.S. Department of State. The map of Russian physiography (Fig. 2-2) was updated thanks to the suggestions of Mika Roinila of the State University of New York, College at New Paltz. And special thanks also go to Charles Pirtle, Professor of Geography at Georgetown University's School of Foreign Service for his advice on Chapter 4; to Antoinette WinklerPrins of the Department of Geography at Michigan State University for her recommendations on Chapter 5; and to Charles Fahrer of Georgia College and State University for his suggestions on Chapter 6.

We also record our appreciation to those who ensured the quality of this book's ancillary products: Antoinette WinklerPrins (Michigan State University) prepared the PowerPoint presentation new to this edition, and Elizabeth Muller Hames (M.A. in Geography, University of Miami) co-authored the Study Guide. At the University of Miami's Department of Geography and Regional Studies, Peter Muller is most grateful for the advice and support of faculty colleagues Doug Fuller, Richard Grant, Larry Kalkstein, Mazen Labban, Jan Nijman, Rinku Roy Chowdhury, Shouraseni Sen Roy, and Ira Sheskin, and particularly GIS Lab Manager Chris Hanson.

We are privileged to work with a team of professionals at John Wiley & Sons that is unsurpassed in the college textbook publishing industry. As authors we are acutely aware of these talents on a daily basis during the crucial production stage, especially the outstanding coordination and leadership skills of Production Editor Barbara Russiello, Senior Illustration Editor Sigmund Malinowski, and Senior Photo Editor Jennifer MacMillan. Others who played a leading role in this process were Senior Designer Kevin Murphy and Interior Designer Nancy Field, copyeditor Betty Pessagno, Joyce Franzen of GGS Book Services in Atlantic Highlands, New Jersey, photo researcher Lisa Passmore, and Don Larson and Mary Swab of Mapping Specialists, Ltd., in Madison, Wisconsin. We much appreciated the leadership of Executive Geosciences Editor Ryan Flahive, who was the prime mover in launching and guiding this latest edition, and was superbly assisted throughout the revision process by Courtney Nelson. Our College Marketing Manager, Jeffrey Rucker, gave us considerable attention throughout the book's preparation, and his advice and enthusiasm were most welcome. We also thank Lynn Pearlman, Nicole Ferrato, and Kelly Tavares for their help and support. Beyond this immediate circle, we acknowledge the support and encouragement we have received over the years from many others at Wiley including Vice-President for Production Ann Berlin and Publisher Jay O'Callaghan.

Finally, and most of all, we express our gratitude to our wives, Bonnie and Nancy, for seeing us yet again through the challenging schedule of our sixteenth collaboration in the past 25 years.

H.J. de Blij
Boca Grande,
Florida

Peter O. Muller
Coral Gables,
Florida

August 7, 2007

BRIEF TABLE OF CONTENTS

CONTENTS

CHAPTER 6

SUBSAHARAN AFRICA   288

CHAPTER 7

NORTH AFRICA/
SOUTHWEST ASIA   342

Introduction

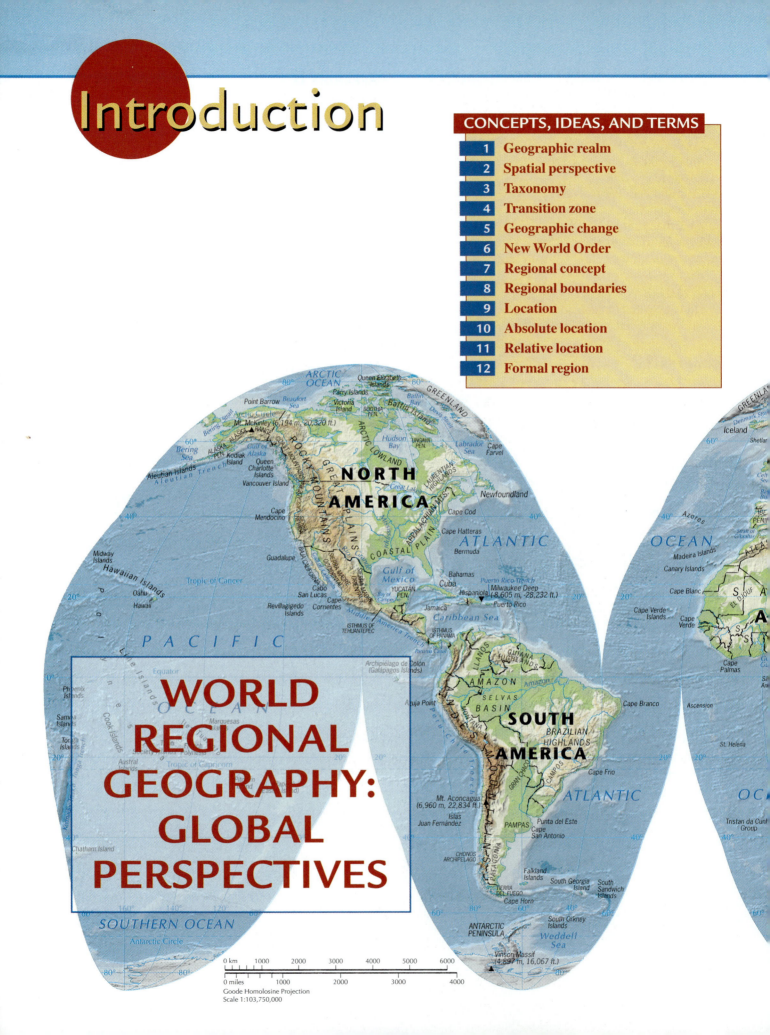

WORLD REGIONAL GEOGRAPHY: GLOBAL PERSPECTIVES

Goode Homolosine Projection
Scale 1:103,750,000

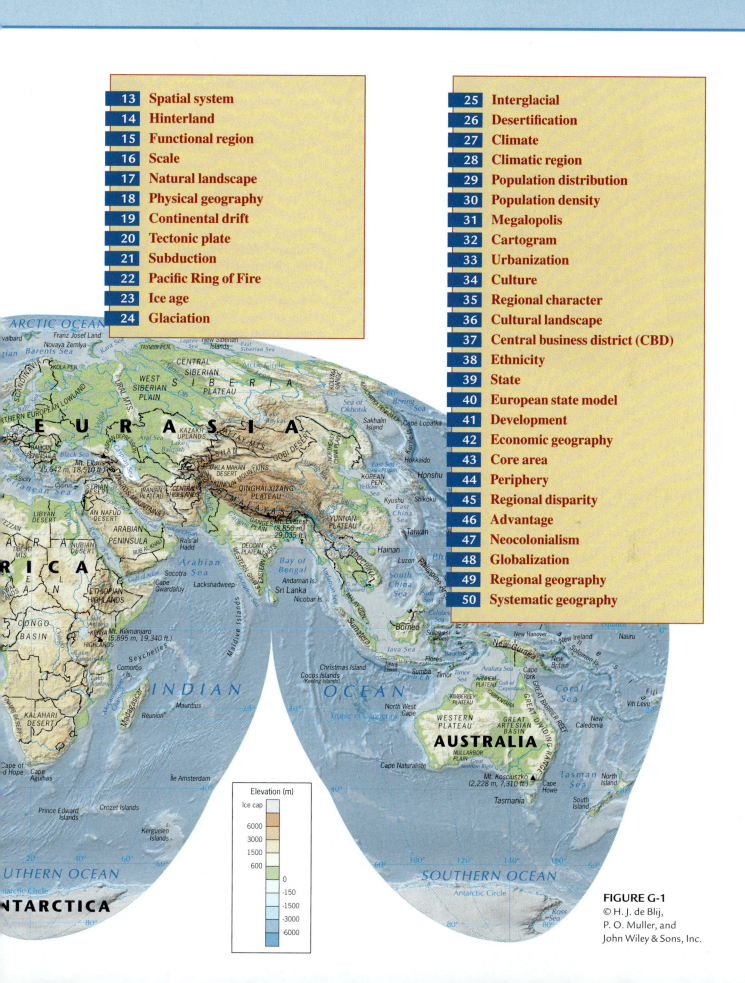

13	Spatial system
14	Hinterland
15	Functional region
16	Scale
17	Natural landscape
18	Physical geography
19	Continental drift
20	Tectonic plate
21	Subduction
22	Pacific Ring of Fire
23	Ice age
24	Glaciation

25	Interglacial
26	Desertification
27	Climate
28	Climatic region
29	Population distribution
30	Population density
31	Megalopolis
32	Cartogram
33	Urbanization
34	Culture
35	Regional character
36	Cultural landscape
37	Central business district (CBD)
38	Ethnicity
39	State
40	European state model
41	Development
42	Economic geography
43	Core area
44	Periphery
45	Regional disparity
46	Advantage
47	Neocolonialism
48	Globalization
49	Regional geography
50	Systematic geography

FIGURE G-1

© H. J. de Blij,
P. O. Muller, and
John Wiley & Sons, Inc.

Where were these photos taken? Find out at **www.wiley.com/college/deblij**

In This Chapter

- Maps in our minds
- Why *tsunamis* happen and where they get started
- The world's most dangerous places to live
- Global warming and the sudden drying of the Sahara
- How countries become mired in debt
- Is globalization good or bad?

Photos: © H. J. de Blij.

You could not have chosen a better time to take a course in world regional geography than the present. Those of us in North America live in, or in the shadow of, the country that has become the world's sole superpower. Those who live elsewhere face forces unleashed by the United States—military, economic, cultural—and felt around the globe. As the 2004 presidential election approached, foreign newspapers and other national media commented that the result would directly affect not just the United States but their countries as well. Some countries even conducted surveys and polls to determine which candidate was perceived as preferable. "When the United States votes for president," editorialized one European newspaper, "every citizen in the world has a stake in the outcome. You can't say that about any other nation."

The United States has emerged from the twentieth century as the planet's most powerful country, capable of influencing nations and peoples, lives and livelihoods from pole to pole. That power confers on Americans the responsibility to learn as much as they can about those nations and livelihoods, so that their decisions will be well-informed and carefully considered. But in this respect, the United States is no superpower. Geographic literacy is a measure of international comprehension and awareness, and Americans' geographic literacy ranks low among countries of consequence. For the world, that is not a good thing, because such geographic fogginess tends to afflict not only voters but also the representatives they elect, from the school board to the White House.

So you are about to strike a blow against geographic illiteracy, and at the end of this term you will find yourself among a still-small minority of Americans who have spent even one semester studying what you will study in the weeks ahead. When you get into discussions about world affairs, you will be able to bring to the table a dimension not usually weighed during such conversations.

GEOGRAPHIC PERSPECTIVES

In this book, we take a penetrating look at the geographic framework of the contemporary world, the grand design that is the product of thousands of years of human achievement and failure, movement and stagnation, revolution and stability, interaction and isolation. Ours is an interconnected world of travel and trade, tourism and television, a global village—but the village still has neighborhoods. Their names are Europe, South America, Southeast Asia, and others familiar to all of us. We call such global neighborhoods **geographic realms**, and when we subject these realms to geographic scrutiny, we find that each has its own identity and distinctiveness.

Geographers study the location and distribution of features on the Earth's surface. These features may be the landmarks of human occupation, the properties of the natural environment, or both; one of the most interesting themes in geography is the relationship between natural environments and human societies, which is why the first map in this introductory chapter summarizes the prominent natural (physical) features of the continents we inhabit (Fig. G-1). Geographers investigate the reasons for these distributions. Their approach is guided by a **spatial perspective**. Just as historians focus on chronology, geographers concentrate on space and place. The spatial structure of cities, the layout of farms and fields, the networks of transportation, the system of rivers, the pattern of climate—all these and more go into the study of a geographic realm. As you will discover, geography is full of spatial terms: area, distance, direction, clustering, proximity, accessibility, isolation, and many others that we will encounter in the pages that follow.

In this book we use the geographic perspective and geography's spatial terminology to investigate the world's great geographic realms. We will find that each of these realms possesses a special combination of cultural, organizational, and environmental properties. These characteristic qualities are imprinted on the landscape, giving each realm its own traditional attributes and social settings. As we come to understand the human and natural makeup of those geographic realms, we learn not only *where* they are located (always a crucial question in geography, and often the answer is not as simple as it may appear), but also *why they are located where they are*, how they are constituted, and what their future is likely to be in our changing world.

It is not enough, however, to study the world from a geographic viewpoint without learning something about the discipline of geography itself. Beginning in this introductory chapter, we introduce the fundamental ideas and concepts that make modern geography what it is. Not only will these geographic ideas enhance your awareness of the many dimensions of our complex, multicultural, interconnected world, but also many of them will remain

FROM THE FIELD NOTES

"From the observation platform atop the Seoul Tower one would be able to see into North Korea except for the range of hills in the background: the capital lies in the shadow of the DMZ (demilitarized zone), relic of one of the hot conflicts of the Cold War. The vulnerable Seoul-Inchon metropolitan area ranks among the world's largest, its population approaching 20 million. I asked my colleague, a professor of geography at Seoul National University, why the central business district, the cluster of buildings in the center of this photo, does not seem to reflect the economic power of this city (note the sector of traditional buildings in the foreground, some with blue roofs, the culture's favorite color). 'Turn around and look across the [Han] River,' he said. 'That's the new Seoul, and there the skyline matches that of Singapore, Beijing, or Tokyo.' Height restrictions, disputes over land ownership, and congestion are among the factors that slowed growth in this part of Seoul and caused many companies to build in the Samsung area. Looking in that direction, you cannot miss the large U.S. military facility, right in the heart of the urban area, on some of the most valuable real estate and right next to an upscale shopping district. In the local press, the debate over the presence of American forces dominated the letters page day after day."
© H. J. de Blij

www.conceptcaching.com

Concept Caching

useful to you long after you have closed this book. Welcome to geography . . . realms, regions, *and* concepts.

REALMS AND REGIONS

Geographers, like other scholars, seek to establish order from the countless data that confront them. Biologists have established a system of classification, or **taxonomy**, to categorize the many millions of plants and animals into a hierarchical system of seven ranks. In descending order, we humans belong to the animal ***kingdom***, the ***phylum*** (division) named chordata, the ***class*** of mammals, the ***order*** of primates, the ***family*** of hominids, the ***genus*** designated

Homo, and the *species* known as *Homo sapiens*. Geologists classify the Earth's rocks into three major (and many subsidiary) categories and then fit these categories into a complicated geologic time scale that spans hundreds of millions of years. Historians define eras, ages, and periods to conceptualize the sequence of the events they study.

Geography, too, employs systems of classification. When geographers deal with urban problems, for instance, they use a classification scheme based on the sizes and functions of the places involved. Some of the terms in this classification are part of our everyday language: megalopolis, metropolis, city, town, village, hamlet.

In regional geography, which is the focus of this book, our challenge is different. We, too, need a hierarchical framework for the areas of the world we study, from the largest to the smallest. But our classification scheme is horizontal, not vertical. It is *spatial*. Our equivalent of the biologists' overarching kingdoms (of plants and animals) is the Earth's natural partitioning into landmasses and oceans. The next level is the division of the inhabited landmasses into geographic realms based on human as well as natural properties. The great global realms, in turn, subdivide into regions.

Criteria for Geographic Realms

In any classification system, criteria are the key. Not all animals are mammals; the criteria for inclusion in that biological class are more specific and restrictive. A dolphin may look and act like a fish, but both anatomically and functionally dolphins belong to the class of mammals.

Geographic realms are based on sets of spatial criteria. First, they are the largest units into which the inhabited world can be divided. The criteria on which such a broad regionalization is based include both physical (that is, natural) and human (or social) yardsticks. South America is a geographic realm, for example, because physically it is a continent and culturally it is dominated by a set of social norms. The realm called South Asia, on the other hand, lies on a Eurasian landmass shared by several other geographic realms; high mountains, wide deserts, and dense forests combine with a distinctive social fabric to create this well-defined realm centered on India.

Second, geographic realms are the result of the interaction of human societies and natural environments, a *functional* interaction revealed by farms, mines, fishing ports, transport routes, dams, bridges, villages, and countless other features that mark the landscape. According to this criterion, Antarctica is a continent but not a geographic realm.

Third, geographic realms must represent the most comprehensive and encompassing definition of the great clusters of humankind in the world today. China lies at the heart of such a cluster, as does India. Africa constitutes a geographic realm from the southern margin of the Sahara (an Arabic word for "desert") to the Cape of Good Hope and from its Atlantic to its Indian Ocean shores.

Figure G-2 displays the 12 world geographic realms based on these criteria. As we will show in more detail later, waters, deserts, and mountains as well as cultural and political shifts mark the borders of these realms. We will discuss the position of these boundaries as we examine each realm.

Geographic Realms: Margins and Mergers

• Where geographic realms meet, **transition zones**, not sharp boundaries, mark their contacts. **4**

We need only remind ourselves of the border zone between the geographic realm in which most of us live, North America, and the adjacent realm of Middle America. The line on Figure G-2 coincides with the boundary between Mexico and the United States, crosses the Gulf of Mexico, and then separates Florida from Cuba and the Bahamas. But Hispanic influences are strong in North America north of this boundary, and the U.S. economic influence is strong south of it. The line, therefore, represents an ever-changing zone of regional interaction. Again, there are many ties between South Florida and the Bahamas, but the Bahamas resemble a Caribbean more than a North American society.

In Africa, the transition zone from Subsaharan to North Africa is so wide and well defined that we have put it on the world map; elsewhere, transition zones tend to be narrower and less easily represented. In these early years of the twenty-first century, such countries as Belarus (between Europe and Russia) and Kazakhstan (between Russia and Muslim Southwest Asia) lie in inter-realm transition zones. Remember, over much (though not all) of their length, borders between realms are zones of regional change.

Geographic Realms: Changing Times

• **Geographic** realms **change** over time. **5**

Had we drawn Figure G-2 before Columbus made his voyages (1492–1504), the map would have looked different: Amerindian territories and peoples would have determined the boundaries in the Americas; Australia and New Guinea would have constituted one realm, and New Zealand would have been part of the Pacific Realm. The colonization, Europeanization, and Westernization of the world changed that map dramatically. During the four decades after World War II relatively little change took place, but since 1985 far-reaching realignments have again been occurring.

As we try to envisage what a **New World Order** will **6** look like on the map, note that the 12 geographic realms can be divided into two groups: (1) those dominated by one major political entity, in terms of territory or population or both (North America/United States, Middle America/Mexico, South America/Brazil, South Asia/India, East Asia/China, Southeast Asia/Indonesia, as well as Russia and Australia), and (2) those that contain many countries but no dominant state (Europe, North Africa/Southwest Asia, Subsaharan Africa, and the Pacific Realm). For several decades two major powers, the United States and the former Soviet Union, dominated the world and competed for global influence. Today, the United States is dominant in what has become a unipolar world. What lies ahead? Will China and/or some other power challenge U.S. supremacy? Is our map a prelude to a multipolar world? We will address such questions in many chapters.

Criteria for Regions

The spatial division of the world into geographic realms establishes a broad global framework, but for our purposes a more refined level of spatial classification is needed. This brings us to an important organizing concept **7** in geography: the **regional concept**. To continue the analogy with biological taxonomy, we now go from phylum to order. To establish regions within geographic realms, we need more specific criteria.

Let us use the North American realm to demonstrate the regional idea. When we refer to a part of the United States or Canada (e.g., the South, the Midwest, or the Prairie Provinces), we employ a regional concept—not scientifically but as part of everyday communication. We reveal our *perception* of local or distant space as well as our mental image of the region we are describing.

But what exactly is the Midwest? How would you draw this region on the North American map? Regions are easy to imagine and describe, but they can be difficult to outline on a map. One way to define the Midwest is to use the borders of States: certain States are part of this region, others are not. You could also use agriculture as the chief criterion: the Midwest is where corn and/or soybeans occupy a certain percentage of the farmland. Each method results in a different delimitation; a Midwest based on States is different from a Midwest based on farm production. Therein lies an important principle: regions are scientific devices that allow us to make spatial generalizations, and they are based on artificial criteria we establish to help us construct them. If you were studying the geography behind politics, then a Midwest region defined by State boundaries would make sense. If you were studying agricultural distributions, you would need a different definition.

Area

Given these different dimensions of the same region, we can identify properties that all regions have in common.

To begin with, all regions have *area*. This observation would seem obvious, but there is more to this idea than meets the eye. Regions may be intellectual constructs, but they are not abstractions: they exist in the real world, and they occupy space on the Earth's surface.

Boundaries

It follows that regions have *boundaries*. Occasionally, nature itself draws sharp dividing lines, for example, along the crest of a mountain range or the margin of a forest. More often, **regional boundaries** are not self-evident, and **8** we must determine them using criteria that we establish for that purpose. For example, to define a citrus-growing agricultural region, we may decide that only areas where more than 50 percent of all farmland stands under citrus trees qualify to be part of that region.

Location

All regions also have **location**. Often the name of a region **9** contains a locational clue, as in Amazon Basin or Indochina (a region of Southeast Asia lying between India and China). Geographers refer to the **absolute location** of a place or region **10** by providing the latitudinal and longitudinal extent of the region with respect to the Earth's grid coordinates. A far more practical measure is a region's **relative location**, that **11** is, its location with reference to other regions. Again, the names of some regions reveal aspects of their relative locations, as in **Eastern** Europe and **Equatorial** Africa.

Homogeneity

Many regions are marked by a certain *homogeneity* or sameness. Homogeneity may lie in a region's human (cultural) properties, its physical (natural) characteristics, or both. Siberia, a vast region of northeastern Russia, is marked by a sparse human population that resides in widely scattered, small settlements of similar form, frigid climates, extensive areas of permafrost (permanently frozen subsoil), and cold-adapted vegetation. This dominant uniformity makes it one of Russia's natural and cultural regions, extending from the Ural Mountains in the west to the Pacific Ocean in the east. When regions display a measurable and often visible internal homogeneity, they are called **formal regions**. But not all for- **12** mal regions are visibly uniform. For example, a region may be delimited by the area in which, say, 90 percent of the people or more speak a particular language. This cannot be seen in the landscape, but the region is a reality, and we can use this criterion to draw its boundaries accurately. It, too, is a formal region.

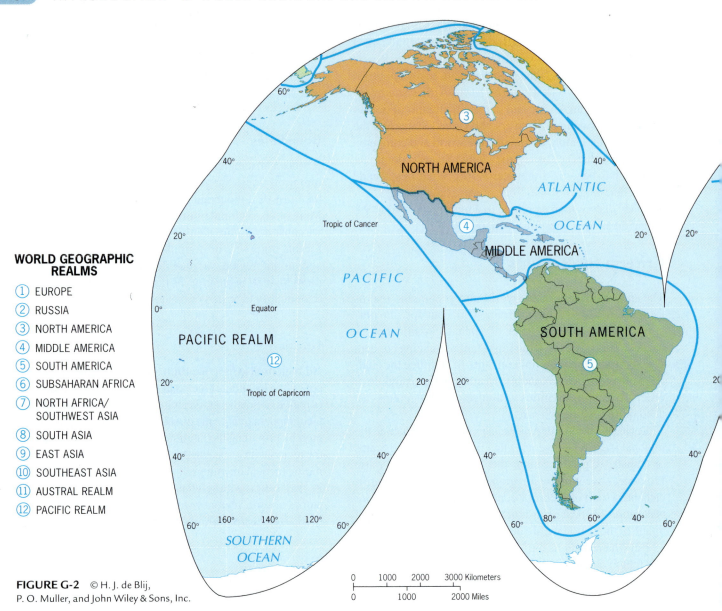

WORLD GEOGRAPHIC REALMS

1. EUROPE
2. RUSSIA
3. NORTH AMERICA
4. MIDDLE AMERICA
5. SOUTH AMERICA
6. SUBSAHARAN AFRICA
7. NORTH AFRICA/ SOUTHWEST ASIA
8. SOUTH ASIA
9. EAST ASIA
10. SOUTHEAST ASIA
11. AUSTRAL REALM
12. PACIFIC REALM

FIGURE G-2   © H. J. de Blij, P. O. Muller, and John Wiley & Sons, Inc.

Regions as Systems

Other regions are marked *not* by their internal sameness but by their functional integration—that is, the way they work. These regions are defined as **spatial systems** and are formed by the areal extent of the activities that define them. Take the case of a large city with its surrounding zone of suburbs, urban-fringe countryside, satellite towns, and farms. The city supplies goods and services to this encircling zone, and it buys farm products and other commodities from it. The city is the heart, the *core* of this region, and we call the surrounding zone of interaction the city's **hinterland**. But the city's influence wanes on the outer periphery of that hinterland, and there lies the boundary of the functional region of which the city is the focus. A **functional region**, therefore, is forged by a structured, urban-centered system of interaction. It has a core and a pe-

riphery. As we shall see, core-periphery contrasts in some parts of the world are becoming strong enough to endanger the stability of countries.

Interconnections

All human-geographic regions are **interconnected**, being linked to other regions. We know that the borders of geographic realms sometimes take on the character of transition zones, and so do neighboring regions. Trade, migration, education, television, computer linkages, and other interactions blur regional boundaries. These are just some of the links in the fast-growing interdependence among the world's peoples, and they reduce the differences that still divide us. Understanding these differences will lessen them further.

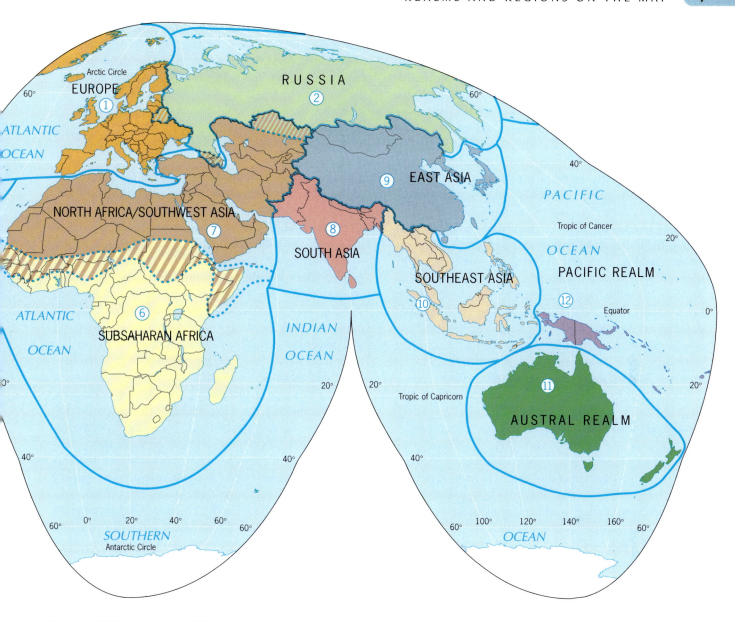

REALMS AND REGIONS ON THE MAP

Just a casual glance at the pages that follow reveals a difference between this and other textbooks: there are almost as many maps as there are pages. Geography is more closely identified with maps than any other discipline, and we urge you to give as much (or more!) attention to the maps in this book as you do to the text. It is often said that a picture is worth a thousand words, but a map can be worth a million. When we write "see Figure XX," we really mean it. . . . and we hope that you will get into the habit. We humans are territorial creatures, and the boundaries that fence off our 200 or so countries reflect our divisive ways. But other, less visible borders—between religions, languages, rich and poor—partition our planet as well. When political and cultural boundaries are at odds, there is nothing like a map to summarize the circumstances. Just look at the one on page 374. How clear were the implications of this map to those who made the decision to send American troops into war?

The Map Revolution

The maps in this book show larger and smaller parts of the world in various contexts, some depicting political frameworks, others displaying ethnic, cultural, economic, environmental, and other features unevenly distributed across our world. But cartography (the making of maps) has undergone a dramatic technological revolution—a revolution that continues. Earth-orbiting satellites with special onboard scanners and television cameras transmit information to computers on the surface, recording the expansion of deserts, the shrinking of glaciers, the depletion of forests, the growth of cities, and myriad other geographic phenomena. The Earthbound computers possess ever-expanding capabilities not only to sort this information but also to display it graphically. This allows geographers to develop geographic information systems (GIS), presenting on-screen information within seconds that would have taken months to assemble just a few decades ago.

Nevertheless, satellites—even spy satellites—cannot record everything that occurs on the Earth's surface. Sometimes the transition zones between ethnic groups or cultural sectors can be discerned by satellites, for example, in changing types of houses or religious shrines, but this kind of information tends to require on-the-ground verification through field research and reporting. No satellite view of Iraq could show you the distribution of Sunni and Shia Muslim adherents. Many of the boundaries you see on the maps in this book cannot be seen from space because long stretches are not even marked on the ground. So the maps you are about to "read" have their continued uses. They summarize complex situations and allow us to begin forming lasting *mental maps* of the areas they represent.

Scale and Detail

Maps are the language of geography. But some maps convey much more local detail than others, and this is a function of *scale*. Obviously an atlas map of the world, at a small scale, cannot show you anything about the street layout of your hometown: for that, you need a large-scale map covering just the municipal area where you live. In this book, we must use mostly small-scale maps because regions are large and our page size is small, but from time to time we use a larger-scale map to highlight a particular regional issue or problem. And we also include maps of many of the world's great cities, employing a larger scale still.

The map is the geographer's strongest ally. It does for geography what taxonomic (and other) classification systems do for the other sciences. Maps display enormous amounts of information. They can suggest spatial relationships, they pose questions for researchers, and they often suggest where the answers may lie. Maps represent the surface of the Earth at various levels of generalization. A map's **16** **scale** is the ratio of the distance between two locations on a map and the actual distance between those two locations on the Earth's surface. That ratio is expressed as a fraction.

In Figure G-3, note how that fraction grows larger as the level of detail increases—and the area displayed shrinks. Map A shows almost all of North America at the small scale of 1:103,000,000. The scale of map B is almost twice as large but shows only the northeast quadrant of the realm. At 1:24,000,000 (map C) only the Canadian province of Quebec can mostly be shown; at 1:1,000,000

(map D) you can see details of the urban area of Montreal at a scale about 100 times as large as map A. (Scale is a complicated concept, with more to it than areal size and coverage; some insights are provided in Appendix A.)

THE PHYSICAL SETTING

This book focuses on the geographic realms and regions produced by human activity over thousands of years. But we should never forget the natural environments in which all this activity took place because we can still recognize the role of these environments in how people make their living. Certain areas of the world, for example, presented opportunities for plant and animal domestication that other areas did not. The people who happened to live in those favored areas learned to grow wheat, rice, or root crops, and to domesticate oxen, goats, or llamas. We can still discern those early "patterns of opportunity" on the map in the twenty-first century. From such opportunities came adaptation and invention, and thus arose villages, towns, cities, and states. But people living in more difficult environments found it much harder to achieve this organization. Take tropical Africa. There, the human communities could not domesticate any of the many species of wildlife, from gazelles and zebras to giraffes and buffalo. Wild animals were a threat, not an opportunity. Eventually humans domesticated only one African animal, the guinea fowl. Early African peoples faced environmental disadvantages that persist today. The modern map carries many imprints of the past.

Natural (Physical) Landscapes

The landmasses of Planet Earth present a jumble of **natu-** **17** **ral landscapes** ranging from rugged mountain chains to smooth coastal plains (Fig. G-1). The map of natural landscapes reminds us that the names of the continents and their mountain backbones are closely linked: North America and its Rocky Mountains, South America and the Andes, Eurasia with its Alps (in the west) and Himalayas (toward the east), Australia and its Great Dividing Range. There is, however, a prominent exception: Africa, the Earth's second-largest landmass, has peripheral mountains in the north (the Atlas) and south (the Cape Ranges), but no linear chain from coast to coast. We will soon discover why.

In his book *Guns, Germs and Steel*, UCLA geographer Jared Diamond argues that the orientation of these continental "axes" has had "enormous, sometimes tragic, consequences [that] affected the rate of spread of crops and livestock, and possibly also of writing, wheels, and other inventions. That basic feature of geography thereby con-

EFFECT OF SCALE

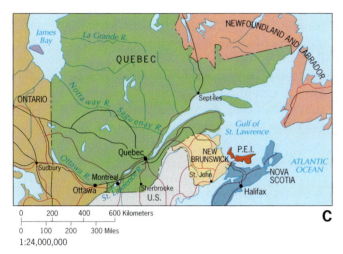

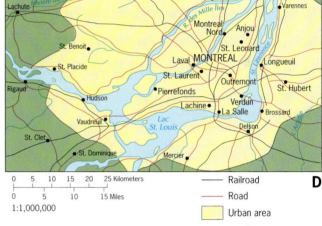

FIGURE G-3 © H. J. de Blij, P. O. Muller, and John Wiley & Sons, Inc.

tributed heavily to the very different experiences of Native Americans, Africans, and Eurasians in the last 500 years." Inevitably it also affected the evolution of the regional map we are set to investigate. And it affected not only its evolution, but also its future. Mountain ranges formed barriers to movement but, as Diamond stresses, also channeled the spread of agricultural and technical innovations. Rugged mountains obstructed contact but also protected peoples and cultures against their enemies (today al-Qaeda, Taliban, and Chechnyan fugitives are using remote and rugged mountain refuges for such purposes). As we study each of the world's geographic realms, we will find that physical landscapes continue to play significant roles in this modern world. River basins in Asia still contain several of the planet's largest population concentrations: the advantages of fertile soils and ample water that first enabled clustered human settlement now sustain hundreds of millions in crowded South and East Asia. But river basins in Africa and South America support no such numbers because the combination of natural and historical factors is different. That is one reason the study of world regional

geography is important: it puts the human map in environmental as well as regional perspective.

Continents and Plates

Studying natural landscapes is part of **physical geography**. [18] Nearly a century ago a physical geographer, Alfred Wegener, used spatial analysis to explain the details he saw on maps like Figure G-1. He studied the outlines of the continents and, marshalling vast evidence, theorized that the Earth's landmasses are pieces of a giant supercontinent that existed hundreds of millions of years ago. He named that postulated supercontinent *Pangaea*, and Africa lay at the heart of it. When North and South America split away from Eurasia and Africa, the Atlantic Ocean opened—but you can still see how the opposite coastlines fit together.

Wegener's hypothesis of **continental drift** seemed to [19] explain much of what Figure G-1 shows, including the global distribution of major mountain ranges. As South America "drifted" westward, away from Africa, the rocks of its leading edge crumpled like folds of an accordion, creating the Andes Mountains. On the opposite side of Africa, Australia moved eastward, so that its major mountain chain now lies near its eastern margin. Africa itself moved little, which is why no major linear mountain range formed there.

Wegener's hypothesis, based on spatial evidence, pointed the way for the research of others. Eventually geologists, building on the geographic evidence, formulated the notion of plate tectonics. The continental landmasses, they reasoned, rest upon great slabs of crust called **tectonic plates**. These plates are mobile, moving slowly a few centimeters (an inch or so) a year, propelled by heat-driven convection cells in the molten rock below.

Where these giant plates collide, the plate that consists of lighter (less heavy) rock rides up over the weightier plate, which is pushed downward in a process called **subduction**. Such gigantic collisions create the mountain chains that girdle the Earth, causing earthquakes and volcanic eruptions in the process. In Figure G-4, note that the Pacific Ocean is encircled by a nearly continuous plate-collision zone called the **Pacific Ring of Fire**. Disastrous earthquakes, volcanic eruptions, landslides, and other threats are facts of daily life here (Fig. G-5).

Natural Hazards

The danger associated with these plate-collision zones, in the Pacific as well as elsewhere, was underscored when, on December 26, 2004, a major earthquake beneath the Indian Ocean off western Sumatera (Sumatra) generated a sea wave called a ***tsunami*** that radiated outward to strike coastal Indonesia, Malaysia, Thailand, Myanmar, Sri Lanka, India, the Maldives, and even Somalia and Kenya in Africa. The site of this earthquake was the subduction zone where the Eurasian Plate collides with the Indian and Australian Plates (upper end of the pink plate fragment in Fig. G-4). It was the most powerful earthquake recorded on the planet in 40 years, and eyewitnesses reported that the resulting wall of water was as high as 15 meters (50 ft) when it smashed into the coast in Sumatera and Sri Lanka. The exact death toll will never be known, but estimates exceed 300,000; the victims ranged from fishermen and farmers to hotel employees and tourists representing more than 50 countries. A massive international relief effort could only partially alleviate the misery of millions of survivors who lost their families and livelihoods; the devastation will take decades to clear.

Sudden, catastrophic events of this kind that change the natural landscape occur fairly frequently in geologic terms, but other processes are more gradual. The rocks of the Earth's crust vary in their resistance to breakdown by ***weathering*** and removal by ***erosion***, in which rivers, glaciers, coastal waves, and desert winds participate. This results in the highly diverse and often spectacular scenery that uniquely marks our planet's surface.

Climate

A key factor in the geography of realms and regions is the prevailing climate (as we will discover, some regions are essentially *defined* by climate). But we should always re-

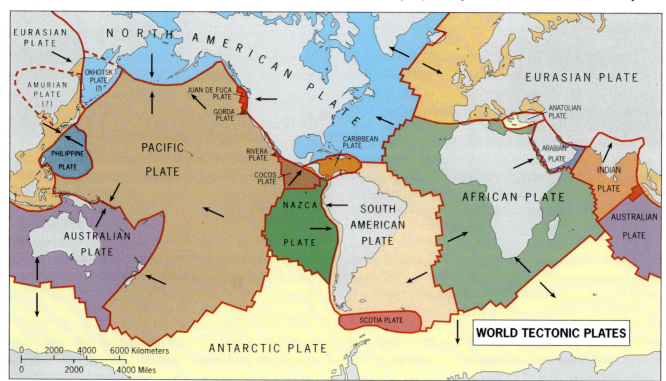

FIGURE G-4 © H. J. de Blij, P. O. Muller, and John Wiley & Sons, Inc.

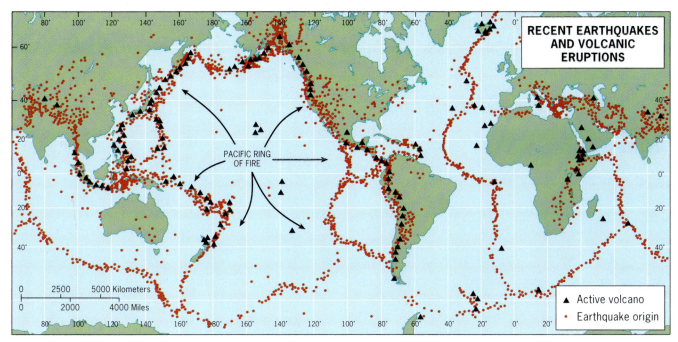

RECENT EARTHQUAKES
AND VOLCANIC
ERUPTIONS

PACIFIC RING
OF FIRE

▲ Active volcano
• Earthquake origin

FIGURE G-5   © H. J. de Blij, P. O. Muller, and John Wiley & Sons, Inc.

member that climates change and that the climate predominating in a certain region today may not be the climate prevailing there several thousand years ago. Any map of climate, including the ones in this chapter, is but a still-picture of our always-changing world.

Ice Ages and Climate Change

Climatic conditions have swung back and forth for as long **[23]** as the Earth has had an atmosphere. Periodically, an **ice age** lasting tens of millions of years chills the planet and causes massive ecological change. One such ice age occurred while Pangaea was still in one piece, between 250 and 300 million years ago. Another started about 35 million years ago, and we are still experiencing it. The current epoch of this ice age, on average the coldest yet, is called the *Pleistocene* and has been going on for nearly two million years.

In our time of global warming this may come as a surprise, but we should remember that an ice age is not a period of unbroken, bitter cold. Rather, an ice age consists of surges of cold, during which glaciers expand and living space shrinks, separated by warmer phases when the ice recedes and life spreads poleward again. The cold phases **[24]** are called **glaciations**, and they tend to last longest, although milder spells create some temporary relief. The truly warm phases, when the ice recedes poleward and **[25]** mountain glaciers melt away, are known as **interglacials**. We are living in one of these interglacials today. It even has a geologic name: the *Holocene*.

Imagine this: just 18,000 years ago, great icesheets had spread all the way to the Ohio River Valley, covering most of the Midwest, the zenith of a glaciation that had lasted about

100,000 years, the *Wisconsinan Glaciation* (Fig. G-6). The Antarctic Icesheet was bigger than ever, and even in the tropics, great mountain glaciers pushed down valleys and onto plateaus. But then Holocene warming began, the continental and mountain glaciers receded, and ecological zones that had been squeezed between the advancing glaciers now spread north and south. In Europe, particularly, where humans had arrived from Africa via Southwest Asia during one of the milder spells of the Wisconsinan Glaciation, living space expanded and human numbers grew.

Global Warming

Consider the transformation that our planet—and our ancestors—experienced as the ice receded. Huge slabs of ice thousands of miles across slid into the oceans from Canada and Antarctica. Rivers raged with silt-clogged meltwater and formed giant deltas. Sea level rose rapidly, submerging vast coastal plains. Animals and plants migrated into newly opened lands at higher elevations and latitudes. As climatic zones shifted, moist areas fell dry and arid areas turned wet. In Southwest Asia, where humans had begun to cultivate crops and towns signaled the rise of early states, such sudden changes often destroyed what had been achieved. In northern tropical Africa, where rivers flowed and grasses fed wildlife, a region extending from the Atlantic shores to the Red Sea fell dry in a very short time around 5000 years ago. We know it today as the Sahara. We call the process **desertification**. **[26]**

Temperatures reached their present-day levels about 7000 years ago, but the effects of this Holocene interglacial warmth took more time to take hold, as the Sahara's later

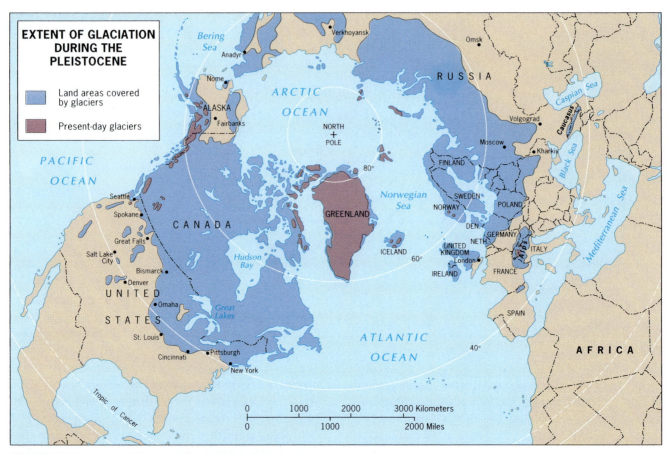

FIGURE G-6   © H. J. de Blij, P. O. Muller, and John Wiley & Sons, Inc.

desiccation shows. Soils formed on newly exposed rock, but generally faster in warm and moist areas than in dry and still-cooler zones. Pine forests that had migrated southward during the Wisconsinan Glaciation now moved north, and equatorial rainforests expanded in all directions. The world's first empires formed in areas that had been inhospitable during the Wisconsinan: the Roman Empire at the western end of Eurasia, the Han Empire at the eastern end. Europeans and Chinese traded with each other along the Silk Route, now open for business.

A great deal of this environmental history remains imprinted in Figure G-2 today. The great river basins of East Asia, where humans exploited agricultural opportunities thousands of years ago, still anchor populous China today. The great Ganges Basin, where one of early humankind's great population explosions may have taken place, remains the core of a huge modern state, India. The Roman Empire's legacies infuse much of European culture.

The past 1000 years have witnessed some troubling environmental developments. Around 1300, the first warning signals of a return to cooler conditions caused crop failures and social dislocation, first in Europe and later in China. This episode, which worsened during the 1600s and has come to be known as the *Little Ice Age*, had major impacts on the human geography of Eurasia and on peoples elsewhere in the world as well, accompanied as it was by long-term changes in climate. A return to pre-Little Ice Age conditions commenced in the early nineteenth century, but then the exploding human population began to have a stronger impact on the atmosphere, and humanity itself became a factor in global climate change. Today, global warming accelerates as a result of a combination of natural rebound and anthropogenic (human-source) causes, and a leading question is how long the relative stability of the past 7000 years will survive.

Climatic Regions

We have just learned how variable climate can be, but in a human lifetime we see little evidence of this variability. We talk about the *weather* (the immediate state of the atmosphere) in a certain place at a given time, but as a technical term **climate** defines the aggregate, total record of weather conditions at a place, or in a region, over the entire period during which records have been kept.

The key records are those of temperature and precipitation, their seasonal variations and extremes, but other data (humidity, wind direction, and the like) also come into play at the larger scale. Precipitation across the planet, for example, is quite variable (Fig. G-7), but this map cannot

27

FROM THE FIELD NOTES

"Walking the hilly countryside anywhere in Indonesia leaves you in no doubt about the properties of *Af* climate. The sweltering sun, the hot, humid air, the daily afternoon rains, and the lack of relief even when the sun goes down—the still atmosphere lies like a heavy blanket on the countryside to make this a challenging field trip. Deep, fertile soils form rapidly here on the volcanic rocks, and the entire landscape is green, most of it now draped by rice fields and terraced paddies. The farmer whose paddies I photographed here told me that his land produces three crops per year. Consider this: the island of Jawa, about the size of Louisiana, has a population of more than 130 million— its growth made possible by this combination of equatorial circumstances . . . Alaska, almost a dozen times as large as Jawa, has a population under three-quarters of a million. Here climates range from *Cfc* to *E*, soils are thin and take thousands of years to develop, and the air is arctic. We sailed slowly into Glacier Bay, in awe of the spectacular, unspoiled scenery, and turned into a bay filled with ice floes, some of them serving as rafts for sleeping seals. Calving (breaking up) into the bay was the Grand Pacific Glacier, with evidence of its recent recession all around. Less than 300 years ago, all of Glacier Bay was filled with ice; today you can sail miles to the Johns Hopkins (shown here) and Margerie tidewater glaciers' current outer edges. Global warming in action—and a reminder that any map of climate is a still picture of a changing world."   © H. J. de Blij

www.conceptcaching.com

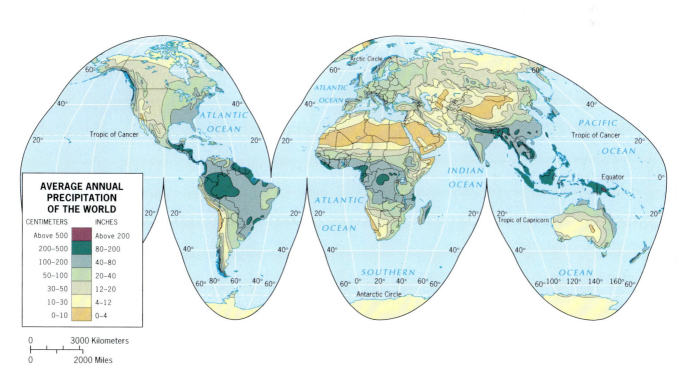

AVERAGE ANNUAL PRECIPITATION OF THE WORLD

CENTIMETERS	INCHES
Above 500	Above 200
200–500	80–200
100–200	40–80
50–100	20–40
30–50	12–20
10–30	4–12
0–10	0–4

0   3000 Kilometers

0   2000 Miles

FIGURE G-7   © H. J. de Blij, P. O. Muller, and John Wiley & Sons, Inc.

show when, seasonally, this precipitation arrives. Average temperatures, too, mean very little: what is needed at minimum is information on daily and monthly ranges.

For our discussion of global realms and regions, a comprehensive map is enormously important and useful. Efforts to create such a map have gone on for about a century, and Figure G-8, based on a system devised by Wladimir Köppen and modified by Rudolf Geiger, is one of the most useful because of its comparative simplicity.

28 The Köppen-Geiger map represents **climatic regions** through a system of letter symbols. The first (capital) letter is the critical one: the *A* climates are humid and tropical, the *B* climates are arid, the *C* climates are mild and humid, the *D* climates show increasing extremes of seasonal heat and cold, and the *E* climates reflect the frigid conditions at and near the poles.

Figure G-8 merits your attention because familiarity with it will help you understand much of what follows in this book. The map has practical utility, too. Although it depicts climatic regions, daily weather in each color-coded region is relatively standard. If, for example, you are familiar with the weather in the large area mapped as *Cfa* in the southeastern United States, you will feel at home in Uruguay (South America), Kwazulu-Natal (South Africa), New South Wales (Australia), and Fujian Province (China). Let us look at the world's climatic regions in some detail.

Humid Equatorial (A) Climates

The humid equatorial, or tropical, climates are characterized by high temperatures all year and by heavy precipita-

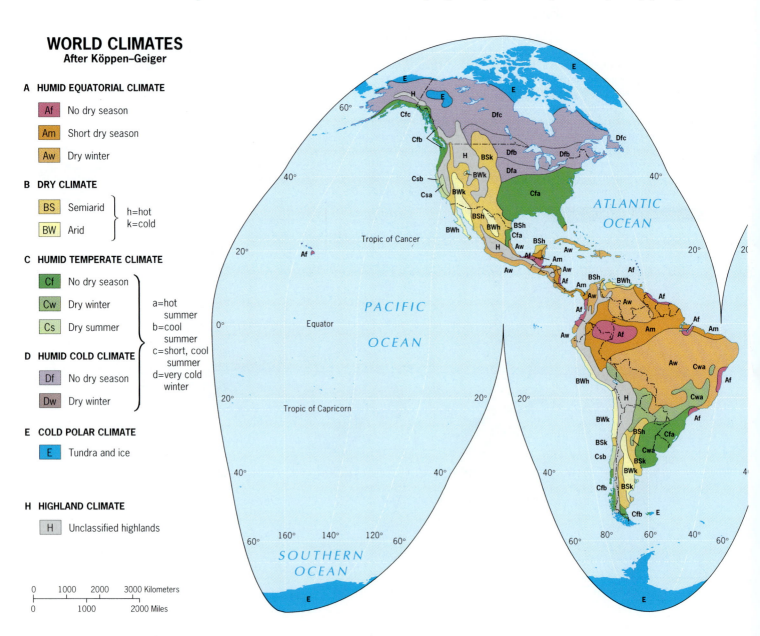

WORLD CLIMATES
After Köppen–Geiger

A HUMID EQUATORIAL CLIMATE
Af — No dry season
Am — Short dry season
Aw — Dry winter

B DRY CLIMATE
BS — Semiarid
BW — Arid
h=hot
k=cold

C HUMID TEMPERATE CLIMATE
Cf — No dry season
Cw — Dry winter
Cs — Dry summer
a=hot summer
b=cool summer
c=short, cool summer
d=very cold winter

D HUMID COLD CLIMATE
Df — No dry season
Dw — Dry winter

E COLD POLAR CLIMATE
E — Tundra and ice

H HIGHLAND CLIMATE
H — Unclassified highlands

0 1000 2000 3000 Kilometers
0 1000 2000 Miles

tion. In the *Af* subtype, the rainfall arrives in substantial amounts every month; but in the *Am* areas, the arrival of the annual wet *monsoon* (the Arabic word for "season" [see p. 406]) marks a sudden enormous increase in precipitation. The *Af* subtype is named after the vegetation that develops there—the *tropical rainforest*. The *Am* subtype, prevailing in part of peninsular India, in a coastal area of West Africa, and in sections of Southeast Asia, is appropriately referred to as the monsoon climate. A third tropical climate, the *savanna* (*Aw*), has a wider daily and annual temperature range and a more strongly seasonal distribution of rainfall. As Figure G-7 indicates, savanna rainfall totals tend to be lower than those in the rainforest zone, and savanna seasonality is often expressed in a "double maximum." Each year produces two periods of increased rainfall separated by pronounced dry spells. In many sa-

vanna zones, inhabitants refer to the "long rains" and the "short rains" to identify those seasons; a persistent problem is the unpredictability of the rain's arrival. Savanna soils are not among the most fertile, and when the rains fail hunger looms. Savanna regions are far more densely peopled than rainforest areas, and millions of residents of the savanna subsist on what they cultivate. Rainfall variability is their principal environmental problem.

Dry (*B*) Climates

Dry climates occur in both lower and higher latitudes. The difference between the *BW* (true *desert*) and the moister *BS* (semiarid *steppe*) varies but may be taken to lie at about 25 centimeters (10 in) of annual precipitation. Parts of the central Sahara in North Africa receive less than 10 centimeters

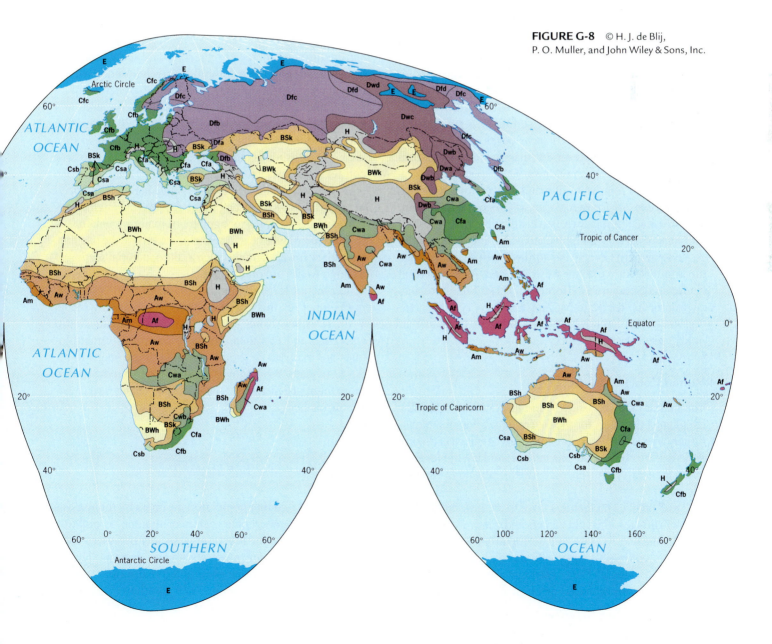

FIGURE G-8   © H. J. de Blij, P. O. Muller, and John Wiley & Sons, Inc.

(4 in) of rainfall. Most of the world's arid areas have an enormous daily temperature range, especially in subtropical deserts. In the Sahara, there are recorded instances of a maximum daytime shade temperature of more than 50°C (122°F) followed by a nighttime low of less than 10°C (50°F). Soils in these arid areas tend to be thin and poorly developed; soil scientists have an appropriate name for them—aridisols.

Humid Temperate (C) Climates

As the map shows, almost all these mid-latitude climate areas lie just beyond the Tropics of Cancer and Capricorn (23.5° North and South latitude, respectively). This is the prevailing climate in the southeastern United States from Kentucky to central Florida, on North America's west coast, in most of Europe and the Mediterranean, in southern Brazil and northern Argentina, in coastal South Africa, in eastern Australia, and in eastern China and southern Japan. None of these areas suffers climatic extremes or severity, but the winters can be cold, especially away from water bodies that moderate temperatures. These areas lie midway between the winterless equatorial climates and the summerless polar zones. Fertile and productive soils have developed under this regime, as we will note in our discussion of the North American and European realms.

The humid temperate climates range from moist, as along the densely forested coasts of Oregon, Washington, and British Columbia, to relatively dry, as in the so-called Mediterranean (dry-summer) areas that include not only coastal Southern Europe and northwestern Africa but also the southwestern tips of Australia and Africa, central Chile, and Southern California. In these Mediterranean environments, the scrubby, moisture-preserving vegetation creates a natural landscape different from that of richly green Western Europe.

Humid Cold (D) Climates

The humid cold (or "snow") climates may be called the continental climates, for they seem to develop in the interior of large landmasses, as in the heart of Eurasia or North America. No equivalent land areas at similar latitudes exist in the Southern Hemisphere; consequently, no **D** climates occur there.

Great annual temperature ranges mark these humid continental climates, and cold winters and relatively cool summers are the rule. In a **Dfa** climate, for instance, the warmest summer month (July) may average as high as 21°C (70°F), but the coldest month (January) might average only −11°C (12°F). Total precipitation, much of it snow, is not high,

ranging from about 75 centimeters (30 in) to a steppe-like 25 centimeters (10 in). Compensating for this paucity of precipitation are cool temperatures that inhibit the loss of moisture from evaporation and evapotranspiration (moisture loss to the atmosphere from soils and plants).

Some of the world's most productive soils lie in areas under humid cold climates, including the U.S. Midwest, parts of southern Russia and Ukraine, and Northeast China. The winter dormancy (when all water is frozen) and the accumulation of plant debris during the fall balance the soil-forming and enriching processes. The soil differentiates into well-defined, nutrient-rich layers, and substantial organic humus accumulates. Even where the annual precipitation is light, this environment sustains extensive coniferous forests.

Cold Polar (E) and Highland (H) Climates

Cold polar (**E**) climates are differentiated into true ice-cap conditions, where permanent ice and snow keep vegetation from gaining a foothold, and the tundra, which may have average temperatures above freezing up to four months of the year. Like rainforest, savanna, and steppe, the term *tundra* is vegetative as well as climatic, and the boundary between the **D** and **E** climates in Figure G-8 corresponds closely to that between the northern coniferous forests and the tundra.

Finally, the **H** climates—unclassified highlands mapped in gray (Fig. G-8)—resemble the **E** climates. High elevations and the complex topography of major mountain systems often produce near-Arctic climates above the tree line, even in the lowest latitudes such as the equatorial section of the high Andes of South America.

Let us not forget an important qualification concerning Figure G-8: this is a still-picture of a changing scene, a single frame from an ongoing film. Climate is still changing, and less than a century from now climatologists are likely to be modifying the climate maps to reflect new data. Who knows: we may have to redraw even those familiar coastlines. Environmental change is a never-ending challenge.

REALMS OF POPULATION

Earlier we noted that population numbers by themselves do not define geographic realms or regions. Population distributions, and the functioning society that gives them common ground, are more significant criteria. That is why we can identify one geographic realm (the Austral) with just 25 million people and another (East Asia) with more than 1.5 billion inhabitants. Neither population numbers nor territorial size alone can delimit a geographic realm. Nevertheless, the map of world population distribution (Fig. G-9) suggests the relative location of several of the world's geographic realms, based on the strong clustering of population in certain areas. Before we examine these clusters in some detail, remember that the Earth's human population now totals well over 6 billion—6 thousand million people confined to the landmasses that constitute less than 30 percent of our planet's surface, much of which is arid desert, rugged mountain terrain, or frigid tundra. (Remember that Figure G-9 is an early twenty-first-century still-picture of an ever-changing scene: the rapid growth of humankind continues.) After thousands of years of slow growth, world population during the nineteenth and twentieth centuries expanded at an increasing rate. That rate has recently been slowing down somewhat, but consider this: it took about 17 centuries after the birth of Christ for the world to add 250 million people to its numbers; now we are adding 250 million about every *four years*.

Demographic (population-related) issues will arise repeatedly in later chapters as we survey the world's most crowded realms. For the moment, we will confine ourselves to the overall global situation. For that purpose let us compare the map of world **population distribution** to the demographic data shown in Table G-1 (pp. 38–44). This table provides information about the total populations in each geographic realm as well as their individual countries. Also, if you compare Figure G-9 to the maps of terrain (Fig. G-1) and climate (Fig. G-8), you will note that some of the largest population concentrations lie in the fertile basins of major rivers (China's Huang He and Chang Jiang, India's Ganges). We live in a modern world, but old ways of life still prescribe where hundreds of millions on this Earth live.

Major Population Clusters

The world's greatest population cluster, **East Asia**, lies centered on China and includes the Pacific-facing Asian coastal zone from the Korean Peninsula to Vietnam. The map indicates that the number of people per unit area—the **population density**—tends to decline from the coastal zone toward the interior. Note, however, the ribbon-like extensions marked **A** and **B** (Fig. G-9). As the map of world natural landscapes (Fig. G-1) confirms, these are populations concentrated in the valleys of China's major rivers, the Huang He (Yellow) and the Chang Jiang (Yangzi), respectively. Here in East Asia, the majority of people are farmers, not city-dwellers. There are great cities in China

(such as Beijing and Shanghai), but their total population still is far outnumbered by the farmers—those who live and work on the land and whose crops of rice and wheat feed not only themselves but also the people in those cities.

The **South Asia** population cluster lies centered on India and includes the populous neighboring countries of Bangladesh and Pakistan. This huge agglomeration of humanity focuses on the broad plain of the lower Ganges River (**C** in Fig. G-9). This cluster is nearly as large as that of East Asia and at present growth rates will overtake East Asia in less than 20 years. As in East Asia, most people are farmers, but in South Asia pressure on the land is even greater, and farming is less efficient. As we note in Chapter 8, the population issue looms large in this realm.

The third-ranking population cluster, **Europe**, also lies on the world's biggest landmass but at the opposite end from China. The European cluster, including western Russia, counts over 700 million inhabitants, which puts it in a class with the two larger Eurasian concentrations—but there the similarity ends. In Europe, the key to the linear, east-west orientation of the axis of population (**D** in Fig. G-9) is not a fertile river basin but a zone of raw materials for industry. Europe is among the world's most highly urbanized and industrialized realms, its human agglomeration sustained by industries and offices rather than paddies and pastures. The three world population concentrations just discussed (East Asia, South Asia, and Europe) account for more than 3.6 billion of the world's 6.7 billion people. No other cluster comes close to these numbers. The next-ranking cluster, **Eastern North America**, is only about one-quarter the size of the smallest of the Eurasian concentrations. As in Europe, the population in this area is concentrated in major metropolitan complexes; the rural areas are now relatively sparsely settled. The heart of the North American cluster lies in the urban complex that lines the U.S. northeastern seaboard from Boston to Washington, D.C., and includes New York, Philadelphia, and Baltimore—the great multimetropolitan agglomeration that urban geographers refer to as **Megalopolis**.

The quantitative dominance of Eurasia's population clusters is revealed dramatically by a special map transformation called a **cartogram** (Fig. G-10). Here the map of the world is redrawn, so that the area of each country reflects *not* territory but population numbers. China and India stand out because, unlike the European population cluster, their populations are not fragmented among many national political entities.

Neither Figure G-9 nor Figure G-10 can tell us how the populations they exhibit are changing. As we will see, varying rates of population growth mark the world's realms and regions; some countries' populations are growing rapidly, while populations in other countries are actually shrinking.

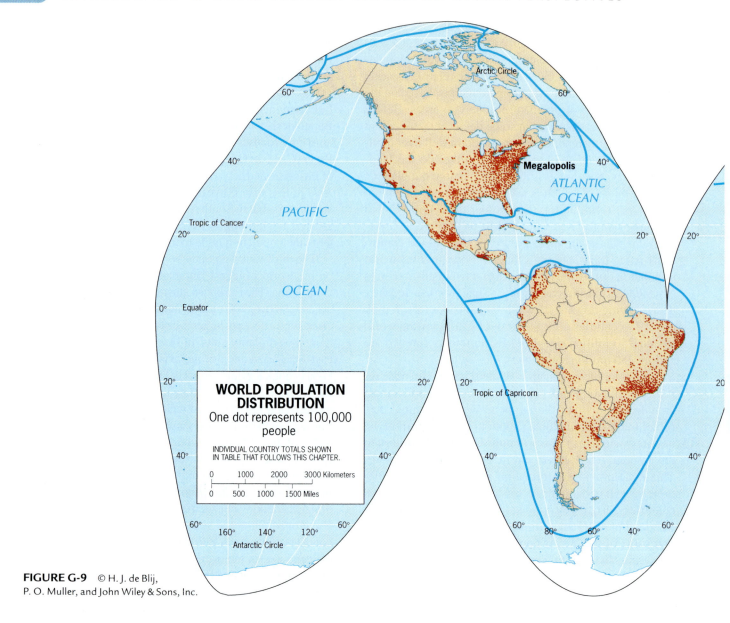

FIGURE G-9 © H. J. de Blij, P. O. Muller, and John Wiley & Sons, Inc.

33 Another key factor is **urbanization**, the percentage of the population living in cities and towns. Not only is there much geographic variation in the level of urbanization among realms and regions, but some realms are urbanizing much more rapidly than others. Urbanization is a defining property of geographic regions, and we will explain why they differ so markedly in this respect.

REGIONS AND CULTURES

Whenever we explore a geographic realm or region, we should assess the physical stage that forms its base—its total physical geography or *physiography*. Nonetheless, the realms and regions we discuss in the chapters that follow are defined primarily by human-geographic criteria, and one criterion, culture, is especially significant. Therefore, we should carefully consider the culture concept in a regional context.

When anthropologists define **culture**, they tend to **34** concentrate on abstractions: learning, knowledge and its transmission, and behavior. Geographers are most interested in how culture and the patterns of behavior associated with it are imprinted on the landscape. Thus we will examine how members of a society perceive and exploit their available resources, how they maximize the opportunities and adapt to the limitations of their natural environment, and how they organize their portion of the Earth. Some human works remain etched on the Earth's surface for a long time: the Egyptian pyramids, the Great Wall of China, and Roman roads and bridges are still in

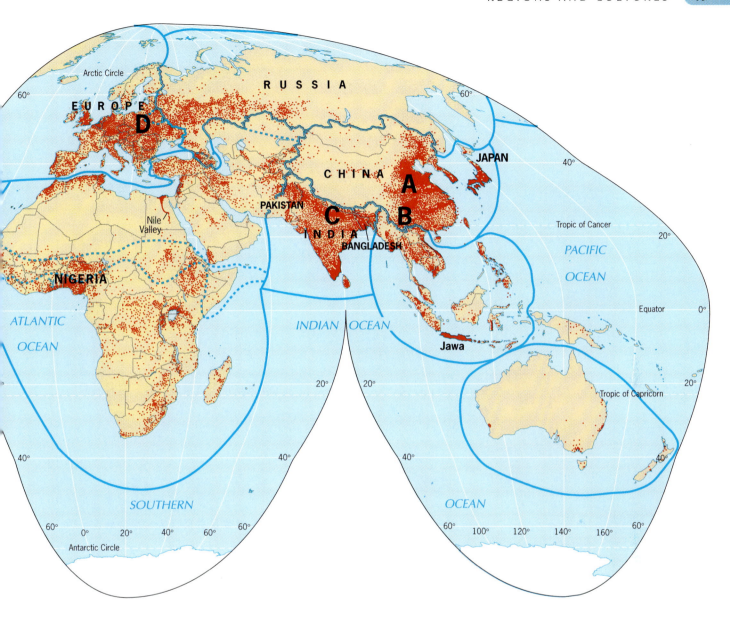

place millennia after they were built. Over time, regions take on dominant qualities that collectively constitute a **35** **regional character**, a personality, a distinct atmosphere. Regional character is a crucial criterion in ascertaining how we divide the human world into major geographic realms and regions.

Cultural Landscapes

As we noted, geographers are particularly concerned with the impress of culture on the Earth's physical surface. Culture gives visible character to a region in many ways. Often a single scene in a photograph or picture can reveal, in general terms, where the photo was taken (such as those on p. 21). The architecture, forms of transportation, goods

being carried, and clothing of the people (all these are part of culture) help us guess the region in the photo. We can do this because the people of any culture are active agents of change: they transform the land they occupy by building structures, creating lines of transport and communication, parceling out fields, and tilling the soil (among countless other activities).

The composite of human imprints on the Earth's surface is called the **cultural landscape**, a term that came **36** into general use in geography during the 1920s. Carl Ortwin Sauer (a professor of geography at the University of California, Berkeley) developed a school of cultural geography that focused on the concept of cultural landscape. In a paper titled "Recent Developments in Cultural Geography" (1927), Sauer proposed his most straightforward definition of the cultural landscape:

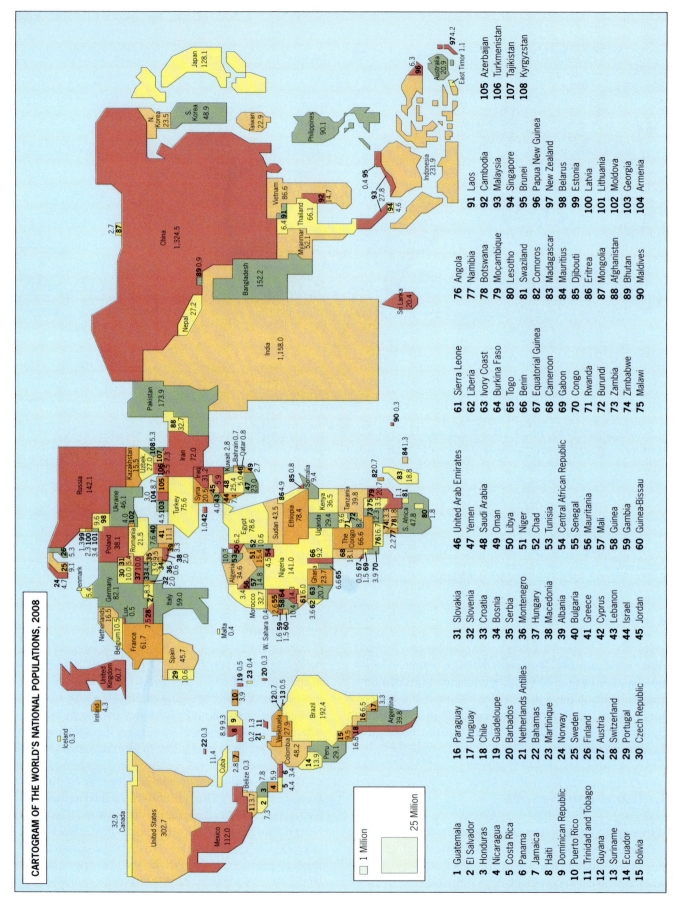

CARTOGRAM OF THE WORLD'S NATIONAL POPULATIONS, 2008

1 Guatemala	16 Paraguay	31 Slovakia	46 United Arab Emirates	61 Sierra Leone
2 El Salvador	17 Uruguay	32 Slovenia	47 Yemen	62 Liberia
3 Honduras	18 Chile	33 Croatia	48 Saudi Arabia	63 Ivory Coast
4 Nicaragua	19 Guadeloupe	34 Bosnia	49 Oman	64 Burkina Faso
5 Costa Rica	20 Barbados	35 Serbia	50 Libya	65 Togo
6 Panama	21 Netherlands Antilles	36 Montenegro	51 Niger	66 Benin
7 Jamaica	22 Bahamas	37 Hungary	52 Chad	67 Equatorial Guinea
8 Haiti	23 Martinique	38 Macedonia	53 Tunisia	68 Cameroon
9 Dominican Republic	24 Norway	39 Albania	54 Central African Republic	69 Gabon
10 Puerto Rico	25 Sweden	40 Bulgaria	55 Senegal	70 Congo
11 Trinidad and Tobago	26 Finland	41 Greece	56 Mauritania	71 Rwanda
12 Guyana	27 Austria	42 Cyprus	57 Mali	72 Burundi
13 Suriname	28 Switzerland	43 Lebanon	58 Guinea	73 Zambia
14 Ecuador	29 Portugal	44 Israel	59 Gambia	74 Zimbabwe
15 Bolivia	30 Czech Republic	45 Jordan	60 Guinea-Bissau	75 Malawi

76 Angola	91 Laos
77 Namibia	92 Cambodia
78 Botswana	93 Malaysia
79 Mocambique	94 Singapore
80 Lesotho	95 Brunei
81 Swaziland	96 Papua New Guinea
82 Comoros	97 New Zealand
83 Madagascar	98 Belarus
84 Mauritius	99 Estonia
85 Djibouti	100 Latvia
86 Eritrea	101 Lithuania
87 Mongolia	102 Moldova
88 Afghanistan	103 Georgia
89 Bhutan	104 Armenia
90 Maldives	105 Azerbaijan
	106 Turkmenistan
	107 Tajikistan
	108 Kyrgyzstan

FIGURE G-10 © H. J. de Blij, P. O. Muller, and John Wiley & Sons, Inc.

20

FROM THE FIELD NOTES

"The Atlantic-coast city of Bergen, Norway displayed the Norse cultural landscape more comprehensively, it seemed, than any other Norwegian city, even Oslo. The high-relief site of Bergen creates great vistas, but also long shadows; windows are large to let in maximum light. Red-tiled roofs are pitched steeply to enhance runoff and inhibit snow accumulation; streets are narrow and houses clustered, conserving warmth . . . The coastal village of Mengkabong on the Borneo coast of the South China Sea represents a cultural landscape seen all along the island's shores, a stilt village of the Bajau, a fishing people. Houses and canoes are built of wood as they have been for centuries. But we could see some evidence of modernization: windows filling wall openings, water piped in from a nearby well."   © H. J. de Blij

www.conceptcaching.com

the forms superimposed on the physical landscape by the activities of man. Such forms result from cultural processes—causal forces that shape cultural patterns—that unfold over a long time and involve the cumulative influences of successive occupants.

When it comes to mapping a cultural landscape, its material, tangible properties are the easiest to observe and record. Take, for instance, the urban "townscape"— a prominent element of the overall cultural landscape— and compare a major U.S. city with, say, a major Japanese city. Photographs of these two metropolitan scenes would reveal the differences quickly, of course, but so would maps. The American city with its square-grid **[37]** layout of the **central business district (CBD)** and its widely dispersed, now heavily urbanized suburbs contrasts sharply with the clustered, space-conserving Japanese metropolis. In a rural example, the spatially lavish subdivision and ownership patterns of American farmland look unmistakably different from the traditional African countryside, with its irregular, often tiny patches of land surrounding a village. Still, the totality of a cultural landscape can never be captured by a photograph or map because the personality of a region involves more than its prevailing spatial organization. One must also include the region's visual appearance, its noises and odors, the shared experiences of its inhabitants, and even their pace of life.

Culture and Ethnicity

Language, religion, and other cultural traditions often are durable and persistent. Culture is not necessarily based on **ethnicity**, however, and as we study the world's **[38]** human-geographic regions, we should be aware that peoples of different ethnic stocks can achieve a common cultural landscape, while people of the same ethnic background can be divided along cultural lines.

The recent events in the former Eastern European country of Yugoslavia are a case in point. As the old Yugoslavia broke up in the early 1990s, its component parts were first affected and then engulfed by what was often described as "ethnic" conflict. When the crisis reached Bosnia-Herzegovina in the heart of Yugoslavia, three groups fought a bitter civil war. These groups were identified as Bosnian Muslims, Serbs, and Croats—but in fact all were ethnic Slavs (Yugoslavia means "Land of the South Slavs"). What distanced them from one another, and what kept the conflict going, was cultural tradition, not ethnicity. The Bosnian Muslims and the Serbs had developed different communities in different parts of the former Yugoslavia. The Muslims were descendants of Slavs who, a century ago or more, had been converted by the Turks (who once ruled here) from Christianity to Islam. Each of these groups feared domination by the others, and so South Slav turned against South Slav in what was, in truth, a culture-based conflict.

Even though the post–Cold War world is still young, it has already witnessed numerous intraregional conflicts; in later chapters, we will explain some of these conflicts in geographic context. Not all of them have ethnicity at their roots. Culture is a great unifier; it can also be a powerful divider.

The Geography of Language

Language and religion are two of these great unifiers and simultaneously powerful dividers, as we will see in the chapters that follow. Here we take early note of the importance of language in the cultural fragmentation of the world, because language plays a crucial role at all levels of generalization. No small-scale map such as Figure G-11 can adequately reflect how complex the global pattern is, but the map does reveal how the more than 15 dominant *language families* (groups of languages with a shared but usually distant origin) are distributed across the world. The most widely distributed language family, the Indo-European, prevails in Europe, the Americas, parts of Asia and Australia—these were the languages of colonization

LANGUAGE FAMILIES OF THE WORLD
Majority Speakers

- INDO-EUROPEAN
- AFRO-ASIATIC
- NIGER-CONGO
- SAHARAN
- SUDANIC
- KHOISAN
- URALIC
- ALTAIC
- SINO-TIBETAN
- JAPANESE AND KOREAN
- DRAVIDIAN
- AUSTRO-ASIATIC
- AUSTRONESIAN
- TRANS-NEW GUINEA AND AUSTRALIAN
- AMERINDIAN
- OTHERS
- UNPOPULATED AREAS

0   1000   2000   3000 Kilometers
0   1000   2000 Miles

FIGURE G-11   © H. J. de Blij, P. O. Muller, and John Wiley & Sons, Inc.

(and now of globalization). But do not assume that the languages in a language family are mutually understandable: English, Russian, Hindi (India), and Farsi (Iran) are all Indo-European languages.

At a different scale we are reminded how much Figure G-11 cannot reveal. In West Africa, Nigeria seems to lie almost entirely within the linguistic Niger-Congo area, but Nigeria's 140-plus million people speak more than 300 languages, most of them mutually unintelligible. Here as in many other former colonies, the colonists' language—in this case English—has become the *lingua franca* (the language of national politics, commerce, and trade). In India, too, language is a contentious issue, a force behind regionalism as well as nationalism, but English is the bridge.

At a still-larger scale, you find a disturbing trend: many languages, some of them ancient and venerable, are in danger of disappearing, and others have already become extinct. In this era of globalization, small groups living in often-remote locales are on the losing end of a continuum that favors the powerful and well-connected. As we will discover, efforts to protect such languages and to sustain others that may still be stronger today but at risk in the future can create conflicts between national governments and local peoples. Many governments have tried to suppress the use of minority languages in misguided attempts to forge national unity, thereby provoking rebellion and strife. We will see troubling examples of this practice, reminding us that the geography of language is a challenge to governments around the world.

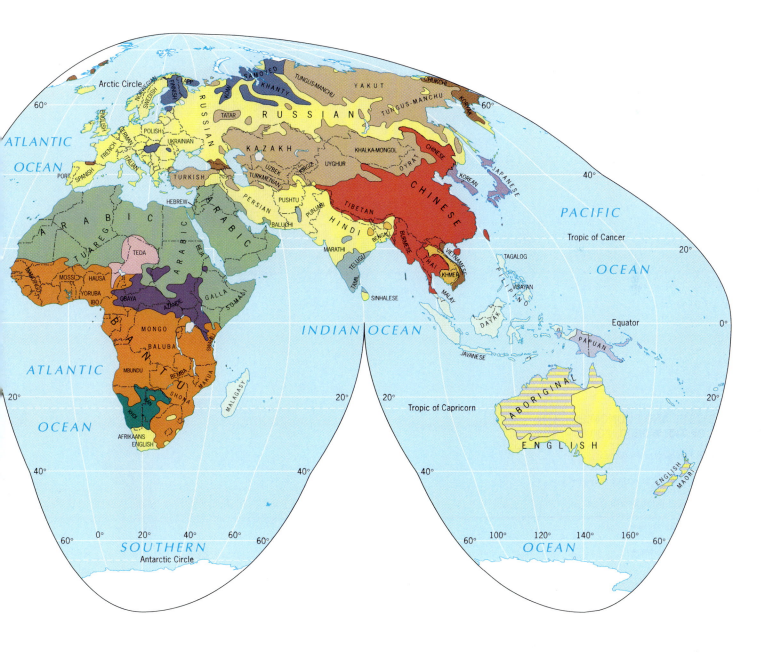

REALMS, REGIONS, AND STATES

Our analysis of the world's regional geography requires data, and it is crucial to know the origin of these data. Unfortunately, we do not have a uniformly sized grid to superimpose over the globe: we must depend on the world's 180-plus countries to report vital information. Irregular as the boundary framework shown on the world map (see Front Endpaper or Fig. G-12) may be, it is all we have. Fortunately, all large and populous countries are subdivided into provinces, States, or other internal entities, and their governments provide information on each of these subdivisions when they conduct their census.

Although we often refer to the political entities on the world map as *countries*, the appropriate geographic term for them is *states*. To avoid confusion with subdivisions

of the same name, we capitalize internal States (as in the preceding paragraph). Thus we refer to the State of California and the State of Victoria (Australia), but to the state of Japan and the state of Argentina.

The State

The **state** as a political, social, and economic entity has **39** been developing for thousands of years, ever since agricultural surpluses made possible the growth of large and powerful towns that could command hinterlands and control peoples far beyond their walls. But the modern state is a relatively recent phenomenon. The boundary framework we see on the world political map today substantially came about during the nineteenth century, and the independence of dozens of former colonies (which made them states as

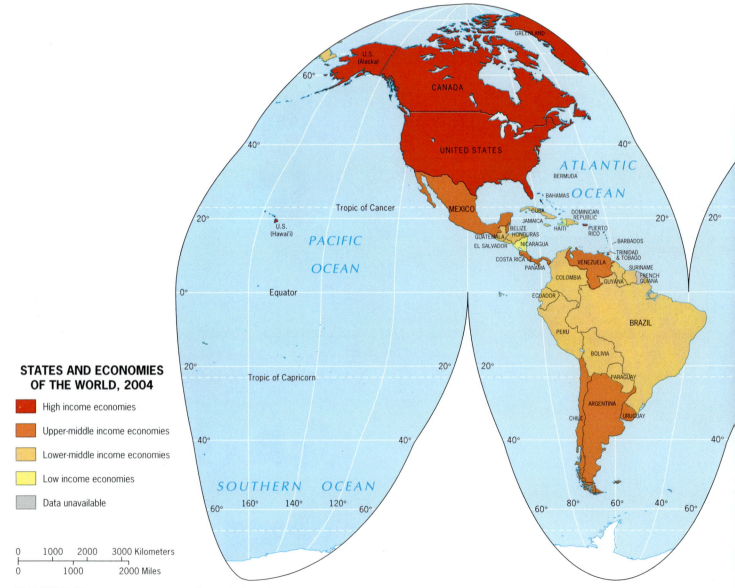

**STATES AND ECONOMIES
OF THE WORLD, 2004**

- High income economies
- Upper-middle income economies
- Lower-middle income economies
- Low income economies
- Data unavailable

0   1000   2000   3000 Kilometers
0   1000   2000 Miles

FIGURE G-12   © H. J. de Blij, P. O. Muller, and John Wiley & Sons, Inc.

40 well) occurred during the twentieth century. Today, the **European state model**—a clearly and legally defined territory inhabited by a population governed from a capital city by a representative government—prevails in the aftermath of the collapse of colonial and communist empires.

States and Realms

As Figures G-2 and G-12 suggest, geographic realms are mostly assemblages of states, and the borders between realms frequently coincide with the boundaries between countries—for example, between North America and Middle America along the U.S.-Mexico boundary. But a realm boundary can also cut *across* a state, as does the one between Subsaharan Africa and the Muslim-dominated realm of North Africa/Southwest Asia. Here the boundary

takes on the properties of a wide transition zone, but it still divides states such as Chad and Sudan. The transformation of the margins of the former Soviet Union is creating similar cross-country transitions. Newly independent states such as Belarus (between Europe and Russia) and Kazakhstan (between Russia and Muslim Southwest Asia) lie in zones of regional change.

Most often, however, geographic realms consist of groups of states whose boundaries also mark the limits of the realms. Look at Southeast Asia, for instance. Its northern border coincides with the political boundary that separates China (a realm practically unto itself) from Vietnam, Laos, and Myanmar (Burma). The boundary between Myanmar and Bangladesh (which is part of the South Asian realm) defines its western border. Here, the state boundary framework helps delimit geographic realms.

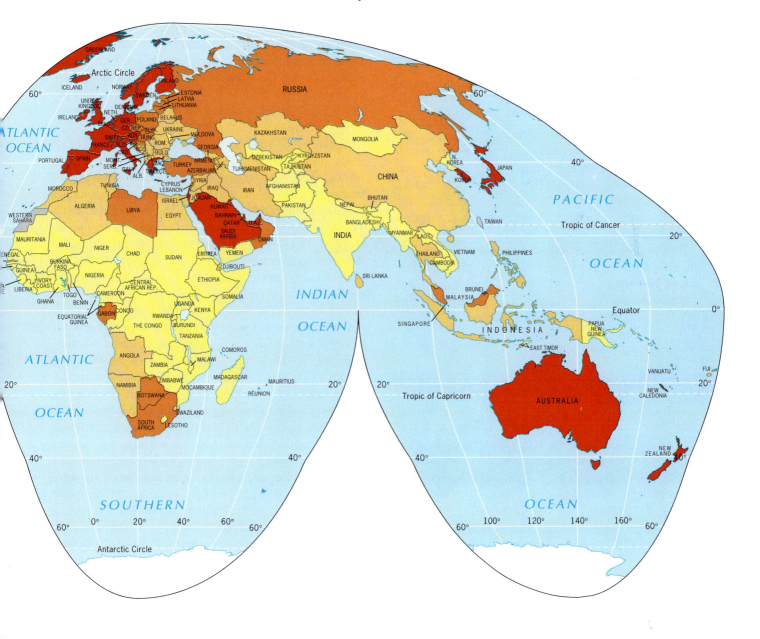

States and Regions

The global boundary framework is even more useful in delimiting regions *within* geographic realms. We shall discuss regional divisions every time we introduce a geographic realm, but an example is appropriate here. In the Middle American realm, we recognize four regions. Two of these lie on the mainland: Mexico, the giant of the realm, and Central America, which consists of the seven comparatively small countries located between Mexico and the Panama-Colombia border (which is the boundary with the South American realm). Central America often is misdefined in news reports; the correct regional definition is based on the politico-geographical framework.

Political Geography

To our earlier criteria of physiography, population distribution, and cultural geography, therefore, we now add political geography as a determinant of world-scale geographic regions. In doing so, we should be aware that the global boundary framework continues to change and that boundaries are created (as in 1993 between the Czech Republic and Slovakia) as well as eliminated (e.g., between former West and East Germany in 1990). But the overall system, much of it resulting from colonial and imperial expansionism, has endured, despite the predictions of some geographers that the "boundaries of imperialism" would be replaced by newly negotiated ones in the postcolonial period.

Toward the end of this book, we will discuss a recent, ominous development in boundary-making: the extension of boundaries onto and into the oceans and seas. This process has been consuming the last of the Earth's open frontiers, with uncertain consequences.

PATTERNS OF DEVELOPMENT

Finally, we turn to economic geography for criteria that allow us to place countries, regions, and realms in regional groupings based on their level of **development**. **41** **Economic geography** focuses on spatial aspects of how people make a living; thus it deals with patterns of production, distribution, and consumption of goods and services. As with all else in this world, these patterns exhibit considerable variation. Individual states report their imports and exports, farm and factory production, and many other economic data to the United Nations and other international agencies. From such information we can determine the comparative economic well-being of the world's countries.

Classification Schemes

The World Bank, one of the agencies that monitor economic conditions, classifies countries into four groups: (1) high income, (2) upper-middle income, (3) lower-middle income, and (4) low income countries. These groupings display regional clustering when mapped (Fig. G-12). The higher-income economies are concentrated in Europe, North America, and along the western Pacific Rim, most notably in Japan and Australia. The lower-income countries dominate in Africa and parts of Asia.

As Figure G-12 shows, certain geographic-realm boundaries can be discerned on this economic map, including those between North America and Middle America, as well as Australia and Southeast Asia. Within the great geographic realms, you can also see some distinct regional boundaries: between Western and Eastern Europe, between the oil-rich Arabian Peninsula and the countries of the Middle East to its northwest, and between South America's Southern Cone (Chile-Argentina-Uruguay) and its less affluent northern neighbors.

Figure G-12 reveals the level of economic development among the world's countries—based, we should remind ourselves, on the political framework as the reporting grid. Until recently, economists and economic geographers used this basis to divide the world into developed countries (DCs) and underdeveloped countries (UDCs). But that distinction has lost much of its relevance, for reasons Figure G-12 cannot show. Today, the world has **core areas** where the richer countries are clustered. **43** But many countries, whatever their level of income, display cores of development (where they often resemble the richest societies) as well as **peripheries** of **44** poverty and underdevelopment (see box titled "Core-Periphery Relationships"). Even comparatively wealthy countries still have areas of severe underdevelopment. Other countries, mapped as low-income economies, have given rise to bustling, skyscrapered cities whose streets are jammed with luxury automobiles and shops selling opulent goods. As the rich cores get richer and the poor peripheries get poorer, the averages reported as "national" economic statistics lose much of their geographic meaning. Perhaps an index of **regional** **45** **disparity** within each economy would give us a truer picture.

It is therefore no longer appropriate to divide the world geographically into developed and underdeveloped realms. East Asia is dominated by lower-income economies, but it includes Japan, one of the world's richest states. Europe is one of the world's highest-income realms, but not in Albania, Moldova, or Ukraine.

Not long ago, countries also were categorized as being part of a First World (capitalist), Second World (socialist/communist), Third World (underdeveloped), Fourth World (severely underdeveloped), or even Fifth World (poorest). This classification, too, has lost most of its significance in the twenty-first century. The most meaningful distinction we can make is based on the no-

46 tion of **advantage**: some countries have it and many others do not. Such advantage takes many forms: geographic location (to be on the coast is generally better than to be landlocked!), raw materials, government, political stability, productive skills, and much more. As the gap between the advantaged and the disadvantaged states widens (and it *is* widening), the stability of the political world is at risk, and with it the advantages of the "haves" over the "have nots."

As we proceed with our investigation of the world's geographic realms, the concept of development will be examined in a regional context and from several viewpoints. Figure G-12 reflects events that began long before the Industrial Revolution in the late eighteenth and nineteenth centuries: Europe, even by the middle of the eighteenth century, had laid the foundations for its colonial expansion. The Industrial Revolution then magnified Europe's demands for raw materials, and its manufactures increased its imperial control. Western countries thereby gained an enormous head start, while colonial dependencies continued to supply raw materials and to consume Western industrial products. Thus was born a system of international exchange and capital flow that changed little when the colonial period ended. Developing countries, well aware of their predicament, accused the developed world of perpetuating its long-term advan-

47 tage through **neocolonialism**—the entrenchment of the old system under a new guise.

Symptoms of Underdevelopment

Although it is no longer appropriate to divide the world into developed and underdeveloped geographic realms, make no mistake: there still *are* regions within realms, and countries within regions, that suffer from underdevelopment. As we will discover during our global journey, underdevelopment takes many forms, displays many symptoms, and has many causes. To grasp some of the symptoms, see Table G-1 (pp. 38–44) and compare, say, Bangladesh or Moçambique with Japan or Canada. Population growth rates in the periphery are higher, life expectancies shorter, urbanization rates lower, incomes smaller. As we will find, the poorer countries suffer from

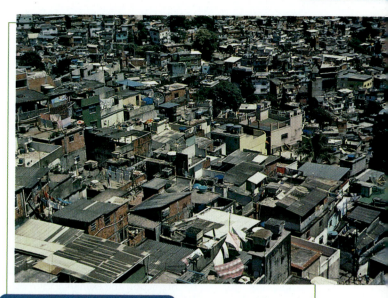

FROM THE FIELD NOTES

"February 1, 2003. A long-held hope came true today: thanks to a Brazilian intermediary I was allowed to enter and spend a day in two of Rio de Janeiro's hillslope *favelas*, an eight-hour walk through one into the other. Here live millions of the city's poor, in areas often ruled by drug lords and their gangs, with minimal or no public services, amid squalor and stench, in discomfort and danger. And yet life in the older *favelas* has become more comfortable as shacks are replaced by more permanent structures, electricity is sometimes available, water supply, however haphazard, is improved, and an informal economy brings goods and services to the residents. I stood in the doorway of a resident's single-room dwelling for this overview of an urban landscape in transition: satellite-television disks symbolize the change going on here. The often blue cisterns catch rainwater; walls are made of rough brick and roofs of corrugated iron or asbestos sheeting. No roads or automobile access, so people walk to the nearest road at the bottom of the hill. Locals told me of their hope that they will some day have legal rights to the space they occupy. During his campaign for president of Brazil, Lula da Silva suggested that long-term inhabitants should be awarded title, and in 2003 his government approved the notion. It will be complicated: as the photo shows, people live quite literally on top of one another, and mapping the chaos will not be simple (but will be made possible with geographic information systems). This would allow the government to tax residents, but it would also allow residents to obtain loans based on the value of their *favela* properties, and bring millions of Brazilians into the formal economy. The hardships I saw on this excursion were often dreadful, but you could sense the hope for and anticipation of a better future." © H. J. de Blij

www.conceptcaching.com

high infant and child mortality rates, poor overall health and sanitation, inefficient farming, insufficient diets, overcrowded cities, and many other ills. Many countries in the periphery are trapped in a global economic system in which the export of unfinished or partially finished raw materials is their only source of outside income.

Core-Periphery Relationships

OURS IS A WORLD of uneven development. The advantages lie with the developed countries (DCs), and economic indicators show that the gap between the DCs and many underdeveloped countries (UDCs) is growing. How has this situation become so extreme? And what perpetuates it?

Regional contrasts in development have been attributed to many factors. Climate has been blamed (as a conditioner of human capacity), as has cultural heritage (e.g., resistance to innovation), colonial exploitation, and, more recently, neocolonialism. Obviously, the distribution of accessible natural resources and the territorial location of countries play a major part. Certain areas have always had more opportunities for interaction and exchange than others. Isolation from the mainstreams of change has always been a disadvantage. These factors, to a greater or lesser degree depending on location, have been important.

One hallmark of national spatial organization is the evolution of *core areas*, foci of human activity that function as the leading regions of control and change. Today we tend to associate this regional concept with a country's heartland—its largest population cluster, most productive and influential region, the area possessing the greatest centrality and accessibility that usually contains a powerful capital city.

This is nothing new. Archeologists have unearthed evidence of the earliest states, proving that city-dominated cores formed their foci. These were small urban centers, but later they were recast as city-states, culture hearths, seats of empires, and sources of modern technological revolutions. But such cores would not have developed without contributions from their surrounding areas. One of the earliest developments in ancient cities was the creation of organized armed forces to help rulers secure taxes and tribute from the people living in the countryside. Thus core and periphery (margin) became functionally tied together: the core had requirements, and the periphery supplied them.

Have modern times ended this core-periphery relationship? The answer not only is no—the situation has worsened. You can see core (have) and periphery (have-not) relationships in every country in the world. In many African countries (Kenya, Senegal, and Zimbabwe, to name examples in three different regions), the former colonial capital now lies at the focus of the national core area. Its tall buildings and government centers symbolize the concentration of privilege, power, and control. Here are the trappings of development, the paved roads and streetlights, hospitals and schools, corporate headquarters and businesses, industries and markets. Farms producing high-priced goods, especially perishables, for the city-dwellers who can afford to buy them surround the capital. But farther out into the countryside, life changes. In the periphery, the core area is distant, even ominous—always a concern. How much will markets in the core pay for crops? How much will goods made there, goods that the countryside needs, cost? How high will taxes be? This core-periphery system keeps regional development contrasts high. The core may appear to be a thriving metropolis and look like a city in a developed country. But in the periphery underdevelopment reigns.

Looking at the world from this perspective, we see that core-periphery systems encompass not just individual countries but the entire globe. Today the whole of Western Europe, North America, and Japan function as core areas. Therefore, what we said about that powerful city-region in an otherwise underdeveloped country also applies, at the macroscale, to these global core areas. Power, money, influence, and decision-making organizations are concentrated in Europe, North America, and Japan.

Imagine that you, as a geographer, are asked to locate a factory (say, a textile plant) to sell products in core-area markets. Because salaries and wages in the core area are high, you look for an alternative site in the periphery, where wages are much lower. Raw materials also can be bought locally. So you advise the corporation to establish its factory in a certain country in the periphery, thereby initiating a relationship between core-based business and periphery-based producers. Now, however, the corporate management is concerned about political stability in that periphery country. Soon you are talking to government

The Specter of Debt

All of us are aware of the dangers of going too deeply into debt. Borrow too much money, and payments of interest and principal may leave too little for routine needs such as food and clothing, not to mention repairs and replacement of equipment. Individuals, families, villages, and cities must devise budgets and balance incomes against expenditures.

So it is with countries. National governments, like individuals and families, must sometimes borrow to make ends meet. If a drought curtails domestic food production, a government may borrow against future oil or ore exports to pay for urgently needed staples. Should domestic harvests fail in the following year as well, further borrowing will be necessary. This means that the government's future budgets will be burdened by ever-higher "debt retirement" payments, limiting its ability to spend on schools, roads,

representatives about guarantees, and the economic connection also becomes a political one.

Is all this necessarily disadvantageous to the countries in the periphery, the mainly underdeveloped countries, the countries of the disadvantaged world? After all, the core-area corporations and businesses make investments, stimulate economies, and put people to work in the countries of the periphery. However, a dependency relationship develops that soon diminishes any such advantage. The profits earned return mostly to the core area and benefit the periphery only minimally. When you see that ultramodern high-rise hotel towering above the sandy beaches of a Caribbean island and watch its staff at work, you observe both sides of it. An investment was made, and jobs were created. But the hotel's profits go back to the multinational corporation's bank account—in New York, Tokyo, or London.

Another problem concerns the sociopolitical effects. That textile factory we mentioned needs a manager and other administrative personnel. These people tend to be (or become) part of the elite, the "upper class" in that periphery country, a kind of cadre of the world core area in the disadvantaged realms. Because they represent core-area interests, these people have divided loyalties: what is best for core-area enterprises is not always best for their home country. Such examples of core-periphery interactions abound, and the entire world is now more interconnected than many of us realize. Superimposed on the traditional cultural landscapes of the disadvantaged countries is a growing global network of interaction that reaches into even the smallest village store in the remotest part of a UDC. The numerous geographic realms and regions we will come to recognize in this book are, above all, enmeshed in a worldwide economic-geographic system—a system tightly oriented to the world's core areas.

Thus the countries of the periphery find themselves locked within a global economic system over which they have no control. Even countries that possess commodities in high demand in the core (such as the oil of the OPEC states) have difficulty converting their temporary advantage into longer-term parity with core-area powers. Countries whose income depends heavily on the export of raw materials such as strategic minerals (or single crops like bananas or sugar) are at the mercy of those who do the buying. But there are other reasons for the depressing condition of many periphery countries. Like the contemporary city, the world core area constantly siphons off the skilled and professional people from the periphery. How many doctors, engineers, and teachers from India are working in England and the United States? And how badly India now needs such trained people! Every loss is magnified—but oppressive political circumstances in periphery countries contribute to this "brain drain."

The continuing underdevelopment of periphery countries also involves traditional cultural attitudes and values. The introduction of core-area tentacles into outlying regions where traditional culture is strong may cause a reaction, perhaps a resurgence of fundamentalism. This further restricts economic change. The invasions of modern ways can be unsettling. Do not forget that the long traditions of societies in the periphery have generated large bureaucracies that are threatened by the change development entails. Occasionally, the tangible presence of an external (core) installation such as an office building, airline agency, or retail establishment is violently attacked. Such incidents symbolize ideological opposition to the intrusion they represent.

The countries of the periphery, therefore, face severe problems. They are pawns in a global economic game whose rules they cannot change. Their internal problems are intensified by the aggressive involvement of core-area interests. The way they use resources (such as whether to produce food for local consumption or to produce export crops) is strongly affected by foreign interference. They suffer far more from environmental degradation, overpopulation, and mismanagement than core-area countries do. Their inherited disadvantages have grown, not lessened, over time. The widening gaps that result between enriching cores and persistently impoverished peripheries threaten the future of the world.

and clinics. Should the decision be made to raise taxes to help make the payments, the result may be a slowdown in the economy. The cycle of debt is hard to escape.

Underlying Causes

All too often it is not just a food emergency or some other unavoidable problem, but a government's mismanagement that leads to excessive debt. In the postcolonial period, many governments of former colonies sought shortcuts to boost their economies, building dams, factories, and ports that did not justify their huge investments. Of course, the governments and corporations of the former colonial powers were all too willing to do the building, making large profits and even lending the new governments money to build still more.

Soon, many of the civilian governments that took over from the colonial powers, weakened by such errors and losing the confidence of their citizens, fell prey to military takeovers. And the military dictatorships that replaced them needed weapons to maintain their hold on power—weapons readily sold for credit by the arms manufacturers in the wealthy core countries. During the Cold War, dictatorial regimes in the periphery became close allies of both superpowers, and proxy wars were fought in Asia, Africa, and the Americas. By the time the Cold War ended and representative government began to return to some (though not all) of these countries, they were so deeply in debt to the rich core countries that they had no prospect of recovery.

The Uneven Debt Burden

External debt is not confined to the countries in the periphery, and even the richer countries (including the United States) have national debts. What matters is a country's ability to service its debt and still have the capacity to pay for its other needs. South Korea, for example, currently has the highest per-capita debt in the world; but its per-capita *gross national income* (or *GNI*, the total income earned from all goods and services produced by the citizens of a country, within or outside of its boundaries, during a calendar year) is more than three times as high. Nicaragua, a Middle American Cold War victim, is much worse off: its per-capita debt is nearly *four times* its per-capita GNI. When a destructive hurricane struck Nicaragua in 1998, the country was left totally dependent on outside help. It had no reserves, nor, given its debts, did it have the ability to borrow.

Several Middle American countries suffer from heavy debt burdens, but the realm most severely afflicted in this respect is Subsaharan Africa. As we note in Chapter 6, postcolonial tropical Africa suffers from a combination of conditions and circumstances that mires most of its countries in debt-ridden poverty.

Globalization

If a geographic concept can arouse strong passions, **globalization** is it. To leading economists, politicians, and businesspeople, this is the best of all worlds: the march of international capitalism, open markets, and free trade. To millions of poor farmers and powerless citizens in debt-mired African and Asian countries, it represents a system that will keep them in poverty and subservience forever. Economic geographers can prove that global economic integration allows the overall economies of poorer countries to grow faster: compare their international trade to their national income, and you will find that the GNI of those that engage in more international trade (and thus are more "globalized") rises, while the GNI of those with

less international trade actually declines. The problem is that those poorest countries contain more than 2 billion people, and their prospects in this globalizing world are worsening, not rising. In countries like the Philippines, Kenya, and Nicaragua, income per person has shrunk, and there globalization is seen as a culprit, not a cure.

One-Way Street?

Globalization in the economic sphere is proceeding under the auspices of the World Trade Organization (WTO), of which the United States is the leading architect. To join, countries must agree to open their economies to foreign trade and investment and to adhere to a set of economic rules discussed annually at ministerial meetings. By 2004, the WTO had approximately 150 member-states, all expecting benefits from their participation. But the driving forces of globalization, including European countries and Japan, did not always do their part when it came to giving the poorer countries the chances the WTO seemed to promise. The case of the Philippines is often cited: joining the WTO, its government assumed, would open world markets to Filipino farm products, which can be brought to those markets cheaply because farm workers' wages in the Philippines are very low. Economic planners forecast a huge increase in farm jobs, a slow rise in wages, and a major expansion in produce exports. Instead they found themselves competing against American and European farmers who receive subsidies toward the production as well as the export of their products—and losing out. Meanwhile, low-priced, subsidized American corn appeared on Filipino markets. As a result, the Philippine economy lost several hundred thousand farm jobs, wages went down, and WTO membership had the effect of severely damaging its agricultural sector. Not surprisingly, the notion of globalization is not popular among rural Filipinos.

This story has parallels among banana and sugar growers in the Caribbean, cotton and sugar producers in Africa, and fruit and vegetable farmers in Mexico. Economic globalization has not been the two-way street most WTO members, especially the poorer countries, had hoped. Nor is rich-country protectionism confined to agriculture. When cheap foreign steel began to threaten what remains of the U.S. steel industry in 2001, the American government erected tariffs to guard domestic producers against unwanted competition.

An Open World

When meetings of economic ministers take place in major cities, they attract thousands of vocal and sometimes violent opponents who demonstrate and occasionally riot to

FROM THE FIELD NOTES

"It was an equatorial day here today, in hot and humid Singapore. Walked the three miles to the Sultan Mosque this morning and observed the activity arising from the Friday prayers, then sat in the shade of the large ficus tree on the corner of Arab Street and watched the busy pedestrian and vehicular traffic. An Indian man saw me taking notes and sat down beside me. For some time he said nothing, then asked whether I was writing a novel. 'If only I could!' I answered, and explained that I was noting locations and impressions for photos and text in a geography book. He said that he was Hindu but liked living in this area because some of what he called 'Old Singapore' still survived. 'No problems between Hindus and Muslims here,' he said. 'The police maintain strict security here around the mosques, especially after September 2001. The government tolerates all religions, but it doesn't tolerate religious conflict.' He offered to show me the Hindu shrine where he and his family joined others from Singapore's Hindu community. As we walked down Arab Street, he pointed skyward. 'That's the future,' he said. 'The symbol of globalization. Do you mention it in your book? It's in the papers every day, supposedly a good thing for Singapore. Well, maybe. But a lot of history has been lost to make space for what you see here.' I took the photo, and promised to mention his point."   © H. J. de Blij

express their opposition. This happens because globalization is seen as more than an economic strategy to perpetuate the advantages of rich countries: it is also viewed as a cultural threat. Globalization constitutes Americanization, these protesters say, eroding local traditions, endangering moral standards, and menacing the social fabric. Such views are not confined to the disadvantaged, poorer countries still losing out in our globalizing world: in France the leader of a so-called Peasant Federation became a national hero after his sympathizers destroyed a McDonald's restaurant in what he described as an act of cultural self-defense. McDonald's restaurants, he argued, promote the consumption of junk food over better, more traditional fare. Elsewhere the target is violent American movies or the lyrics of certain American popular music.

Nevertheless, the process of globalization—in cultural as well as economic spheres—continues to change our world, for better and for worse. (In the United States, opponents of globalization bemoan the loss of American jobs to foreign countries where labor is cheaper.) In theory, globalization breaks down barriers to international trade, stimulates commerce, brings jobs to remote places, and promotes social, cultural, political, and other kinds of exchanges. High-tech workers in India are employed by computer firms based in California. Japanese cars are assembled in Thailand. American shoes are made in China. Fast food restaurant chains spread standards of service and hygiene as well as familiar (and standard) menus from Tokyo to Tel Aviv. If wages and standards of employment are lower in newly involved countries than in the global core, the gap is shrinking, employment conditions are more open to scrutiny, and global openness is a byproduct of globalization.

A New Revolution

We should be aware that the current globalization process is a revolution—but also that it is not the first of its kind.

POINTS TO PONDER

- In less than four years, the global human population will reach and exceed 7 billion.

- In 2008, more than 900 million people continued to survive on less than one American dollar a day.

- The planet appears to be entering a period of increasing environmental fluctuation, posing growing threats to future economic and political stability.

- Globalization is glorified as well as denigrated, but its march is inexorable. Those who support it say that its "rising tide lifts all boats."

The first "globalization revolution" occurred during the nineteenth and early twentieth centuries, when Europe's colonial expansion spread ideas, inventions, products, and habits around the world. Colonialism transformed the world as the European powers built cities, transport networks, dams, irrigation systems, power plants, and other facilities, often with devastating impact on local traditions, cultures, and economies. From goods to games (soap to soccer) people in much of the world started doing similar things. The largest of all colonial empires, that of Britain, made English a worldwide language, a key element in the current, second globalization process.

The current globalization is even more revolutionary than the colonial phase because it is driven by more modern, higher-speed communications. When the British colonists planned the construction of their ornate Victorian government and public buildings in (then) Bombay, the architectural drawings had to be prepared in London and sent by boat to India. When the Chinese government in the 1990s decided to create a Manhattan-like commercial district on the riverfront in Shanghai, the plans were drawn in the United States, Japan, and Western Europe and transmitted to Shanghai via the Internet. Today, one container ship carrying products from China to the U.S. market hauls more cargo than a hundred colonial-era boats.

And, as the chapters that follow will show frequently, the world's national political boundaries are becoming increasingly porous. Economic alliances enable manufacturers to send raw materials and finished products across borders that once inhibited such exchanges. Groups of countries forge economic unions whose acronyms (NAFTA, Mercosur) stand for freer trade. The ultimate goal of the World

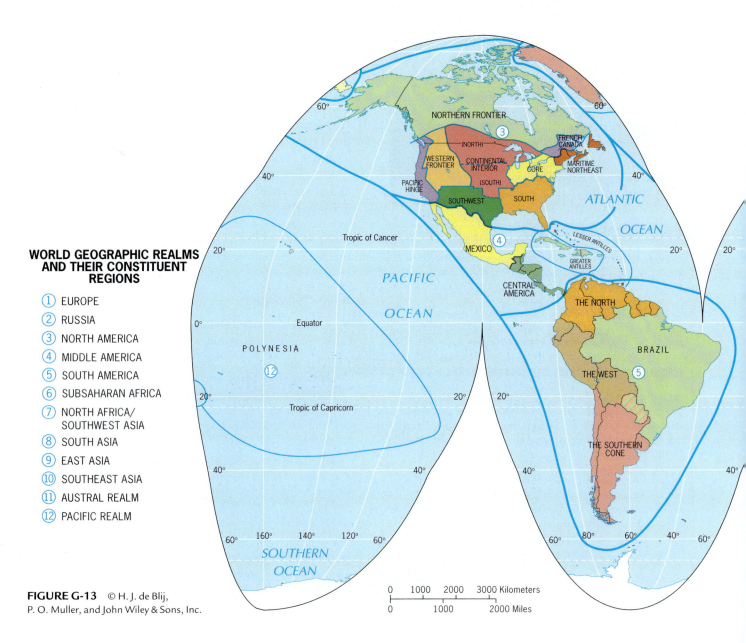

WORLD GEOGRAPHIC REALMS AND THEIR CONSTITUENT REGIONS

1. EUROPE
2. RUSSIA
3. NORTH AMERICA
4. MIDDLE AMERICA
5. SOUTH AMERICA
6. SUBSAHARAN AFRICA
7. NORTH AFRICA/ SOUTHWEST ASIA
8. SOUTH ASIA
9. EAST ASIA
10. SOUTHEAST ASIA
11. AUSTRAL REALM
12. PACIFIC REALM

FIGURE G-13 © H. J. de Blij, P. O. Muller, and John Wiley & Sons, Inc.

Trade Organization is to lower remaining barriers the world over, thereby boosting not just regional commerce but also global trade.

The Future

As with all revolutions, the overall consequences of the ongoing globalization process are uncertain. Critics underscore that one of its outcomes is a growing gap between rich and poor, a polarization of wealth that will destabilize the world. Core-periphery contrasts are intensified, not lessened, by globalization as the poor in peripheral societies are exploited by core-based corporations. Proponents argue that, as with the Industrial Revolution, it will take time for the benefits to spread—but that globalization's ultimate effects will be advantageous to all.

Indeed, the world is functionally shrinking, and we will find evidence for this throughout the book. But the "global village" still retains its distinctive neighborhoods, and two revolutionary globalizations have failed to erase their particular properties. In the chapters that follow we use the vehicle of geography to visit and investigate them.

THE REGIONAL FRAMEWORK

At the beginning of this Introduction, we outlined a map of the great geographic realms of the world (Fig. G-2). We then addressed the task of dividing these realms into regions, and we used criteria ranging from physical geography to economic geography. The result is Figure G-13.

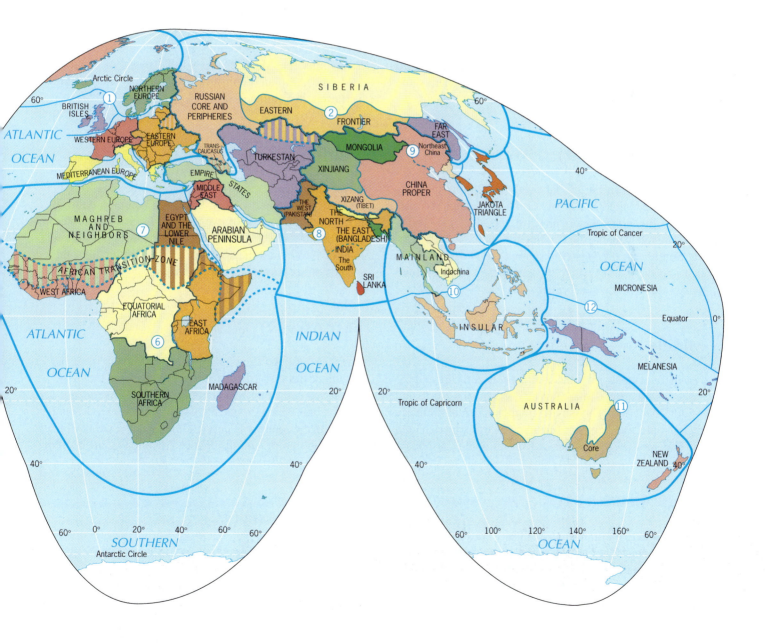

Before we begin our survey, here is a summary of the 12 geographic realms and their regional components.

Europe (1)

Territorially small and politically fragmented, Europe remains disproportionately influential in global affairs. A core geographic realm, Europe has five regions: Western Europe, the British Isles, Northern (Nordic) Europe, Mediterranean Europe, and Eastern Europe.

Russia (2)

Territorially enormous and politically unified, Russia was the dominant force in the former Soviet Union that disbanded in 1991. Undergoing a difficult transition from dictatorship to democracy and from communism to capitalism, Russia is geographically complex and changing. We define four regions: the Russian Core and Peripheries, the Eastern Frontier, Siberia, and the Far East.

North America (3)

Another realm in the global core, North America consists of the United States and Canada. We identify nine regions: the North American Core, the Maritime Northeast, French Canada, the Continental Interior, the South, the Southwest, the Western Frontier, the Northern Frontier, and the Pacific Hinge. Five of these regions extend across the U.S.-Canada border.

Middle America (4)

Nowhere in the world is the contrast between core and periphery as sharply demarcated as it is between North and Middle America. This small, fragmented realm clearly divides into four regions: Mexico, Central America, and the Greater Antilles and Lesser Antilles of the Caribbean.

South America (5)

The continent of South America also defines a geographic realm in which Iberian (Spanish and Portuguese) influences dominate the cultural geography but Amerindian imprints survive. We recognize four regions: the North, composed of Caribbean-facing states; the Andean West,

with its strong Amerindian influences; the Southern Cone; and Brazil, the realm's giant.

Subsaharan Africa (6)

Between the African Transition Zone in the north and the southernmost Cape of South Africa lies Subsaharan Africa. The realm consists of five regions: Southern Africa, East Africa, Equatorial Africa, West Africa, and the African Transition Zone.

North Africa/Southwest Asia (7)

This vast geographic realm has several names, extending as it does from North Africa into Southwest and, indeed, Central Asia. Some geographers call it *Naswasia* or *Afrasia*. There are six regions: Egypt and the Lower Nile Basin plus the Maghreb in North Africa; the Middle East, the Arabian Peninsula, the Empire States, and Turkestan in Southwest Asia.

South Asia (8)

Physiographically, South Asia is one of the most clearly defined geographic realms, but has a complex cultural geography. It consists of five regions: India at the center, Pakistan to the west, Bangladesh to the east, the mountainous North, and the southern region that includes the islands of Sri Lanka and the Maldives.

East Asia (9)

The vast East Asian geographic realm extends from the deserts of Central Asia to the tropical coasts of the South China Sea and from Japan to Xizang (Tibet). We identify five regions: China Proper, including North Korea; Xizang (Tibet) in the southwest; desert Xinjiang in the west; Mongolia in the north; and the Jakota Triangle (*Ja*pan, South *Ko*rea, and *Tai*wan) in the east.

Southeast Asia (10)

Southeast Asia is a varied mosaic of natural landscapes, cultures, and economies. Influenced by India, China, Europe, and the United States, it includes dozens of religions and hundreds of languages plus economies representing both core and periphery. Physically, Southeast Asia consists of a broad peninsular mainland and an arc consisting of thousands of islands. The two regions (Mainland and Insular) are based on this distinction.

Austral Realm (11)

Australia and its neighbor New Zealand form the Austral geographic realm by virtue of continental dimensions, insular separation, and dominantly Western cultural heritage. The regions of this realm are defined by physical as well as cultural geography: in Australia, a highly urbanized, two-part core and a vast, desert-dominated interior; and in New Zealand, two main islands that exhibit considerable geographic contrast.

Pacific Realm (12)

The vast Pacific Ocean, larger than all the landmasses combined, contains tens of thousands of islands large and small. Dominant cultural criteria warrant three regions: Melanesia, Micronesia, and Polynesia.

THE PERSPECTIVE OF GEOGRAPHY

As this introductory chapter demonstrates, our world regional survey is no mere description of places and areas. We have combined the study of realms and regions with a look at geography's ideas and concepts—the notions, generalizations, and basic theories that make the discipline what it is. We continue this method in the chapters ahead so that we will become better acquainted with the world and with geography. By now you are aware that geography is a wide-ranging, multifaceted discipline. It is often described as a social science, but that is only half the story: in fact, geography straddles the divide between the social and the physical (natural) sciences. Many of the ideas and concepts you will encounter have to do with the many interactions between human societies and natural environments.

Regional geography allows us to view the world in [49] an all-encompassing way. As we have seen, regional geography borrows information from many sources to create an overall image of our divided world. Those sources are not random but form topical or **systematic geography**. [50] Research in the systematic fields of geography makes our world-scale generalizations possible. As Figure G-14

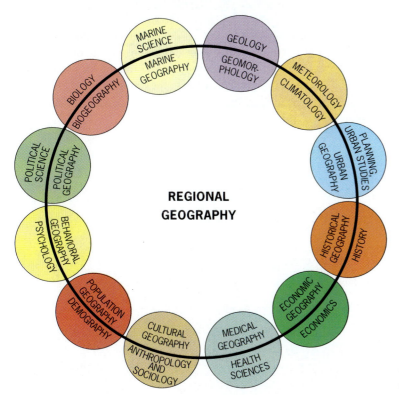

THE RELATIONSHIP BETWEEN REGIONAL AND SYSTEMIC GEOGRAPHY

REGIONAL GEOGRAPHY

MARINE SCIENCE / MARINE GEOGRAPHY

GEOLOGY / GEOMORPHOLOGY

METEOROLOGY / CLIMATOLOGY

PLANNING, URBAN STUDIES / URBAN GEOGRAPHY

HISTORY / HISTORICAL GEOGRAPHY

ECONOMICS / ECONOMIC GEOGRAPHY

HEALTH SCIENCES / MEDICAL GEOGRAPHY

ANTHROPOLOGY AND SOCIOLOGY / CULTURAL GEOGRAPHY

DEMOGRAPHY / POPULATION GEOGRAPHY

BEHAVIORAL PSYCHOLOGY / BEHAVIORAL GEOGRAPHY

POLITICAL SCIENCE / POLITICAL GEOGRAPHY

BIOLOGY / BIOGEOGRAPHY

FIGURE G-14   © H. J. de Blij, P. O. Muller, and John Wiley & Sons, Inc.

shows, these systematic fields relate closely to those of other disciplines. Cultural geography, for example, is allied with anthropology; it is the spatial perspective that distinguishes cultural geography. Economic geography focuses on the spatial dimensions of economic activity; political geography concentrates on the spatial imprints of political behavior. Other systematic fields include historical, medical, behavioral, environmental, agricultural, and coastal geography. We will also draw on information from biogeography, marine geography, population geography, and climatology (as we did earlier in this chapter).

These systematic fields of geography are so named because their approach is global, not regional. Take the geographic study of cities, urban geography. Urbanization is a worldwide process, and urban geographers can identify certain human activities that all cities in the world exhibit in one form or another. But cities also display regional properties. The typical Japanese city is quite distinct from, say, the African city. Regional geography, therefore, borrows from the systematic field of urban geography, but it injects this regional perspective.

In the following chapters we call upon these systematic fields to give us a better understanding of the world's realms and regions. As a result, you will gain insights into the discipline of geography as well as the regions we investigate. This will prove that geography is a relevant and practical discipline when it comes to comprehending, and coping with, our fast-changing world (see box titled "What Do Geographers Do?" and Appendix B for a discussion of career opportunities).

What Do Geographers Do?

A SYSTEMATIC SPATIAL perspective and an interest in regional study are the unifying themes and enthusiasms of geography. Geography's practitioners include physical geographers, whose principal interests are the study of geomorphology (land surfaces), research on climate and weather, vegetation and soils, and the management of water and other natural resources. There also are geographers whose research and teaching concentrate on the ecological interrelationships between the physical and human worlds. They study the impact of humankind on our globe's natural environments and the influences of the environment (including such artificial contents as air and water pollution) on human individuals and societies.

Other geographers are regional specialists, who often concentrate their work for governments, planning agencies, and multinational corporations on a particular region of the world. Still other geographers—who now constitute the largest group of practitioners—are devoted to topical or systematic subfields such as urban geography, economic geography, and cultural geography (see Fig. G-14). They perform numerous tasks associated with the identification and resolution (through policy-making and planning) of spatial problems in their specialized areas. And, as in the past, there remain many geographers who combine their fascination for spatial questions with technical know-how.

Computerized cartography, geographic information systems, remote sensing, and even environmental engineering are among the specializations listed by the 10,000-plus professional geographers of North America. In Appendix B, you will find considerable information on the discipline, how one trains to become a geographer, and the many exciting career options that are open to the young professional.

What You Can Do

RECOMMENDATION: If you are not already doing so, please allow us to urge you to begin keeping a diary. Several decades from now, the most interesting and perhaps even useful parts of your diary may not be time and place, but your feelings and opinions about events and experiences you are having and will have over the years to come. What is important enough to warrant recording your own views? The intervention in Iraq? The earthquake-generated tsunami disaster in Indonesia and elsewhere? The impact of globalization? The effect on you of an unusually stimulating course? The comments of a particularly opinionated visiting speaker? Many years from now you may be surprised about your concerns and perceptions in this first decade of our new century.

GEOGRAPHIC CONNECTIONS

1 If you were asked by someone without geographic background to define the geographic characteristics of your home region, how would you respond? Where do you perceive its boundaries to lie, on what criteria do you base these, and are they sharp or transitional? (You will be tempted to draw a sketch map to support this part of your answer.) Where is your home with respect to the urban framework in your region, and do you consider yourself a resident of a city, suburbia, exurbia, or a rural hinterland? What elements of the formal and the functional can you discern in your regional experience? How would you summarize your region's natural environment(s), and, in your lifetime, have you noticed changes that might be considered trends and if so, how have these affected you?

2 Imagine that you are working in your university's study-abroad office. Students will be planning to study for a semester in numerous countries, and they know that you have some geographic experience. How would you answer their questions about the climate and weather they are likely to experience in those countries, using Figure G-8? What should a student from Vancouver in western Canada's British Columbia expect in Auckland, New Zealand? Would a student from Charleston, South Carolina feel at home (weatherwise) in Shanghai, China? What can you tell a student from San Francisco about the weather in Cape Town, South Africa? How would you prepare a student from Phoenix, Arizona for the weather in Paris, France?

3 A well-known economist once wrote the following: "Virtually all of the rich countries of the world are outside the tropics, and virtually all the poor countries are in them . . . climate, then, accounts for quite a significant proportion of the cross-national and cross-regional disparities of world income." Comment on this assertion, using Figures G-8 and G-12 as well as the text and box on pp. 28–29. How might a geographer's perspective on such an issue differ from an economist's?

Geographic Literature: The key introductory works on the discipline of geography were authored by de Blij; de Blij, Muller, & Williams; de Blij, Murphy, & Fouberg; Gaile & Willmott; Livingstone; Martin; and Wheeler et al. These works, together with a comprehensive listing of noteworthy books and recent articles on global geography, can be found in the *References and Further Readings* section of this book's website at **www.wiley.com/college/deblij**.

TABLE G-1

Area and Demographic Data for the World's States (Categories explained on page 45)

| | Land Area | | Population | | Population Density | | Birth Rate | Death Rate | Natural Increase % | Infant Mortality per 1,000 | Child Mortality per 1,000 | Life Expectancy | | Percent Urban Pop. | Literacy | | Corruption Index | Big Mac Price ($US) | Per Capita GNI ($US) |
	Sq km	Sq mi	2008 (Millions)	2025 (Millions)	Arithmetic	Physiologic						Male (years)	Female (years)		Male %	Female %			
WORLD	134,134,451	51,789,601	6707.4	7940.0	50	432	21	9	1.2	52	76	65	69	50	83.4	69.9	—		$9,190
Europe	5,919,355	2,284,950	590.1	587.9	32	254	10	12	-0.1	9	11	71	79	75	99.3	98.0	—		$21,120
Albania	28,749	11,100	3.3	3.5	113	452	14	6	0.8	16	18	72	79	45	95.5	88.0	2.6		$5,420
Austria	83,859	32,378	8.3	8.7	99	550	9	9	0.0	4	5	76	82	66	100.0	100.0	8.6		$33,140
Belarus	207,598	80,154	9.6	9.4	46	149	9	15	-0.6	10	12	63	75	72	99.7	99.2	2.1		$7,890
Belgium	30,528	11,787	10.5	10.8	345	1,379	11	10	0.1	4	5	76	82	97	100.0	100.0	7.3		$32,640
Bosnia	51,129	19,741	3.9	3.7	76	587	9	9	0.1	13	15	71	77	43	96.5	76.6	2.9		$7,790
Bulgaria	110,908	42,822	7.6	6.6	69	167	9	15	-0.5	12	15	69	76	70	99.1	98.0	4.0		$8,630
Croatia	56,539	21,830	4.4	4.3	78	298	9	11	-0.2	6	7	71	78	60	99.4	97.3	3.4		$12,750
Cyprus	9,249	3,571	1.0	1.1	109	681	11	7	0.4	4	5	75	80	66	98.7	95.0	5.6		$22,230
Czech Republic	78,860	30,448	10.3	10.2	130	303	10	11	-0.1	3	4	73	79	74	100.0	100.0	4.8	$2.51	$20,140
Denmark	43,090	16,637	5.4	5.6	126	225	12	10	0.2	4	5	76	80	85	100.0	100.0	9.5	$5.08	$33,570
Estonia	45,099	17,413	1.3	1.2	29	106	11	13	-0.2	6	7	66	78	71	99.9	99.6	6.7		$15,420
Finland	338,149	130,560	5.3	5.4	16	225	11	9	0.2	3	4	75	82	62	100.0	100.0	9.6		$31,170
France	551,497	212,934	61.7	63.4	112	320	13	9	0.4	4	5	77	84	76	98.9	98.7	7.4		$30,540
Germany	356,978	137,830	82.1	82.0	230	657	8	10	-0.2	4	5	76	82	88	100.0	100.0	8.0		$29,210
Greece	131,960	50,950	11.1	11.4	84	280	10	10	0.0	4	5	77	81	61	98.6	96.0	4.4		$23,620
Hungary	93,030	35,919	10.0	9.6	108	196	10	13	-0.3	7	8	69	77	65	99.5	99.3	5.2	$3.33	$16,940
Iceland	102,999	39,768	0.3	0.3	3	296	14	6	0.8	2	3	79	83	93	100.0	100.0	9.6	$7.44	$34,760
Ireland	70,279	27,135	4.3	4.5	61	304	15	7	0.8	5	6	75	80	60	100.0	100.0	7.4		$34,720
Italy	301,267	116,320	59.0	58.7	196	529	10	10	0.0	4	4	78	83	90	98.9	98.1	4.9		$28,840
Latvia	64,599	24,942	2.3	2.2	35	122	9	14	-0.5	9	11	67	77	68	99.8	99.6	4.7	$2.52	$13,480
Liechtenstein	161	62	0.1	0.1	252	1,006	11	6	0.5	3	4	79	82	21	100.0	100.0	—		—
Lithuania	65,200	25,174	3.4	3.1	52	112	9	13	-0.4	7	9	66	78	67	99.7	99.4	4.8	$2.45	$14,220
Luxembourg	2,587	999	0.5	0.5	195	779	12	8	0.4	4	5	75	81	91	100.0	100.0	8.6		$65,340
Macedonia	25,711	9,927	2.0	2.1	78	300	11	9	0.2	15	17	71	76	59	94.2	83.8	2.7		$7,080
Malta	321	124	0.4	0.4	1,250	3,678	9	7	0.2	5	6	77	81	91	91.4	92.8	6.4		$18,960
Moldova	33,701	13,012	4.0	3.8	118	179	11	12	-0.2	14	16	65	72	45	99.6	98.3	3.2		$2,150
Montenegro	13,812	5,333	0.6	0.6	44	156	13	9	0.3	12	15	—	—	—	100.0	100.0	—		—

Country	Land Area Sq km	Land Area Sq mi	Population 2008 (Millions)	Population 2025 (Millions)	Pop. Density Arithmetic	Pop. Density Physiologic	Birth Rate	Death Rate	Natural Increase %	Infant Mortality per 1,000	Child Mortality per 1,000	Life Exp. Male (years)	Life Exp. Female (years)	Percent Urban Pop.	Literacy Male %	Literacy Female %	Corruption Index	Big Mac Price ($US)	Per Capita GNI ($US)
Netherlands	40,839	15,768	16.5	16.9	405	1,499	12	8	0.3	4	5	77	81	90	100.0	100.0	8.7	—	$32,480
Norway	323,878	125,050	4.7	5.2	15	487	12	9	0.3	3	4	78	83	78	100.0	100.0	8.8	$6.63	$40,420
Poland	323,249	124,807	38.1	36.7	118	251	10	10	0.0	6	7	71	79	62	99.8	99.8	3.7	$2.51	$13,490
Portugal	91,981	35,514	10.6	10.4	115	397	10	10	0.1	4	5	75	81	55	94.8	90.0	6.6	—	$19,730
Romania	238,388	92,042	21.5	18.1	90	205	10	12	-0.2	16	19	68	75	55	99.1	97.3	3.1	—	$8,940
Serbia	88,357	34,115	9.5	9.2	108	284	13	12	0.1	12	15	69	75	52	100.0	100.0	3.0	—	—
Slovakia	49,010	18,923	5.4	5.2	110	324	10	10	0.0	7	8	70	78	56	100.0	100.0	4.7	$2.13	$15,760
Slovenia	20,251	7,819	2.0	2.0	99	705	9	9	0.0	3	4	74	81	49	98.6	96.8	6.4	—	$22,160
Spain	505,988	195,363	45.7	46.2	90	231	11	9	0.2	4	5	77	84	78	100.0	100.0	6.8	—	$25,820
Sweden	449,959	173,730	9.1	9.9	20	289	11	10	0.1	3	4	78	83	84	100.0	100.0	9.2	$4.86	$31,420
Switzerland	41,290	15,942	7.5	7.4	182	1,520	10	8	0.2	4	5	79	84	68	99.5	97.4	9.1	$5.20	$37,080
Ukraine	603,698	233,089	46.1	41.7	76	129	9	17	-0.8	13	17	63	74	68	100.0	100.0	2.8	—	$6,720
United Kingdom	244,878	94,548	60.7	65.8	248	954	12	10	0.2	5	6	76	81	89	99.0	99.0	8.6	$4.01	$32,690
Russia	17,075,323	6,592,819	140.6	130.0	8	137	10	16	-0.6	14	18	59	72	73	99.8	99.2	2.5	$2.03	$10,640
Armenia	29,800	11,506	3.0	3.4	101	507	13	9	0.4	26	29	67	75	65	99.4	98.1	2.9	—	$5,060
Azerbaijan	86,599	33,436	8.7	9.7	100	456	17	6	1.1	74	89	70	75	52	98.9	95.9	2.4	—	$4,890
Georgia	69,699	26,911	4.4	3.9	63	422	12	11	0.1	41	45	69	75	52	99.7	99.4	2.8	—	$3,270
North America	19,599,647	7,567,466	335.6	387.0	17	138	14	8	0.6	6	7	75	81	79	95.7	95.3		—	$40,980
Canada	9,970,600	3,849,670	32.9	37.6	3	66	11	7	0.3	5	6	77	82	80	95.7	95.3	8.5	$3.68	$32,220
United States	9,629,047	3,717,796	302.7	349.4	31	157	14	8	0.6	6	7	75	80	79	95.7	95.3	7.3	$3.41	$41,950
Middle America	2,712,660	1,047,364	194.8	234.2	72	444	23	6	1.7	24	31	70	75	68	81.9	83.8		—	$7,475
Antigua and Barbuda	440	170	0.1	0.1	233	1,292	18	6	1.3	11	12	69	74	37	—	—	—		$11,700
Bahamas	13,880	5,359	0.3	0.3	22	2,205	19	9	1.0	13	15	67	73	89	95.4	96.8	—		$15,800
Barbados	430	166	0.3	0.3	706	1,811	14	8	0.6	11	12	70	74	50	98.0	96.8	6.7		—
Belize	22,960	8,865	0.3	0.4	14	341	27	5	2.3	15	17	67	74	50	76.7	77.1	3.5		$6,740
Costa Rica	51,100	19,730	4.4	5.6	86	959	17	4	1.3	11	12	77	81	61	95.5	95.7	4.1	$2.18	$9,680
Cuba	110,859	42,803	11.4	11.8	103	251	11	7	0.4	6	7	75	79	76	96.5	96.4	3.5		—
Dominica	751	290	0.1	0.1	135	676	15	7	0.8	13	15	71	77	71	—	—	4.5		$5,560
Dominican Republic	48,731	18,815	9.3	11.6	191	616	23	6	1.7	26	31	66	69	64	84.0	83.7	2.8		$7,150
El Salvador	21,041	8,124	7.3	9.1	346	887	26	6	2.0	23	27	67	73	59	81.6	76.1	4.0		$5,120

TABLE G-1 (Continued)

| | Land Area | | Population | | Population Density | | Birth Rate | Death Rate | Natural Increase % | Infant Mortality per 1,000 | Child Mortality per 1,000 | Life Expectancy | | Percent Urban Pop. | Literacy | | Corruption Index | Big Mac Price ($US) | Per Capita GNI ($US) |
	Sq km	Sq mi	2008 (Millions)	2025 (Millions)	Arithmetic	Physiologic						Male (years)	Female (years)		Male %	Female %			
Grenada	339	131	0.1	0.1	302	943	19	7	1.2	17	21	—	—	39	—	—	3.5	—	$7,260
Guadeloupe	1,709	660	0.5	0.5	298	1,989	16	6	1.0	—	—	75	82	100	89.7	90.5	—		—
Guatemala	108,888	42,042	13.7	20.0	126	701	34	6	2.8	32	43	63	71	44	76.2	61.1	2.6		$4,410
Haiti	27,749	10,714	8.9	13.0	321	971	36	13	2.3	84	120	51	54	36	51.0	46.5	1.8		$1,840
Honduras	112,090	43,278	7.8	10.7	69	385	31	6	2.5	31	40	67	74	47	72.5	72.0	2.5		$2,900
Jamaica	10,989	4,243	2.8	3.0	252	1,008	19	6	1.3	17	20	69	73	52	82.5	90.7	3.7		$4,110
Martinique	1,101	425	0.4	0.4	368	1,842	14	7	0.7	—	—	76	82	95	96.0	97.1	—		—
Mexico	1,958,192	756,062	112.0	129.4	57	409	22	5	1.7	22	27	73	78	75	93.1	89.1	3.3	$2.69	$10,030
Netherlands Antilles	800	309	0.2	0.2	252	2,524	13	8	0.5	—	—	72	79	69	96.6	96.6	—		—
Nicaragua	129,999	50,193	5.9	7.7	45	205	29	5	2.4	30	37	66	70	59	64.2	64.4	2.6		$3,650
Panama	75,519	29,158	3.4	4.2	45	502	22	5	1.7	19	24	73	78	62	92.6	91.3	3.1		$7,310
Puerto Rico	8,951	3,456	3.9	4.1	441	4,899	13	7	0.6	—	—	73	81	94	93.7	94.0	—		—
Saint Lucia	619	239	0.2	0.2	333	1,189	20	5	1.5	12	14	72	77	28	89.5	90.6	—		$5,980
St. Vincent & the Grenadines	391	151	0.1	0.1	261	933	18	7	1.1	17	20	68	74	45	—	—	—		$6,460
Trinidad and Tobago	5,131	1,981	1.3	1.3	256	1,068	14	8	0.6	17	19	67	73	74	99.0	97.5	3.2		$13,170
South America	**17,855,070**	**6,893,881**	**389.0**	**465.2**	**22**	**348**	**21**	**6**	**1.4**	**25**	**28**	**69**	**76**	**80**	**89.9**	**88.6**			**$8,210**
Argentina	2,780,388	1,073,514	39.8	46.4	14	143	18	8	1.1	15	18	71	78	90	96.9	96.9	2.9	$2.67	$13,920
Bolivia	1,098,575	424,162	9.5	12.1	9	433	31	8	2.2	52	65	62	66	63	92.1	79.4	2.7	$3.61	$2,740
Brazil	8,547,360	3,300,154	192.4	228.9	23	322	21	6	1.4	31	33	68	76	83	85.5	85.4	3.3		$8,230
Chile	756,626	292,135	16.8	19.1	22	738	16	5	1.0	8	10	75	81	87	95.9	95.5	7.3	$2.97	$11,470
Colombia	1,138,906	439,734	48.2	58.3	42	1,058	20	5	1.5	17	21	69	75	77	91.8	91.8	3.9	$3.06	$7,420
Ecuador	283,560	109,483	13.9	17.5	49	444	27	6	2.1	22	25	71	77	61	93.6	90.2	2.3		$4,070
French Guiana	89,999	34,749	0.2	0.3	2	234	31	4	2.7	—	—	72	79	75	83.6	82.3	—		—
Guyana	214,969	83,000	0.7	0.7	3	167	22	9	1.3	47	63	72	80	36	99.0	98.1	2.5		$4,230
Paraguay	406,747	157,046	6.5	8.6	16	267	22	5	1.7	20	23	69	73	57	94.4	92.2	2.6	$1.90	$4,970
Peru	1,285,214	496,224	29.1	34.1	23	756	19	6	1.3	23	27	67	72	74	94.7	85.4	3.3	$3.00	$5,830
Suriname	163,270	63,039	0.5	0.5	3	315	21	7	1.4	30	39	66	73	74	95.9	92.6	3.0		—
Uruguay	177,409	68,498	3.3	3.5	19	268	15	10	0.5	14	15	71	79	93	97.4	98.2	6.4	$2.17	$9,810
Venezuela	912,046	352,143	27.9	35.2	31	765	22	5	1.7	18	21	70	76	88	93.3	92.7	2.3	$3.45	$6,440

| | Land Area | | Population | | Population Density | | Birth Rate | Death Rate | Natural Increase % | Infant Mortality per 1,000 | Child Mortality per 1,000 | Life Expectancy | | Percent Urban Pop. | Literacy | | Corruption Index | Big Mac Price ($US) | Per Capita GNI ($US) |
	Sq km	Sq mi	2008 (Millions)	2025 (Millions)	Arithmetic	Physiologic						Male (years)	Female (years)		Male %	Female %			
Subsaharan Africa	**21,786,509**	**8,411,818**	**761.1**	**1,090.2**	**35**	**464**	**40**	**16**	**2.4**	**98**	**164**	**47**	**49**	**34**	**72.7**	**55.2**			**$1,970**
Angola	1,246,693	481,351	16.7	25.9	13	446	49	22	2.6	154	260	39	42	35	55.6	28.5	2.2		$2,210
Benin	112,620	43,483	9.2	14.3	82	511	41	12	2.9	89	150	53	55	40	47.8	23.6	2.5		$1,110
Botswana	581,727	224,606	1.8	1.7	3	309	26	27	-0.1	87	120	35	33	54	74.4	79.8	5.6		$10,250
Burkina Faso	274,000	105,792	14.3	23.2	52	435	44	19	2.5	96	191	48	49	18	31.2	13.1	3.2		$1,220
Burundi	27,829	10,745	8.2	14.0	296	689	46	18	2.7	114	190	44	45	10	56.3	40.5	2.4		$640
Cameroon	475,439	183,568	18.1	24.3	38	238	37	14	2.3	87	149	50	52	52	81.8	69.2	2.3		$2,150
Cape Verde Islands	4,030	1,556	0.5	0.7	130	1,185	30	5	2.5	26	35	68	74	55	84.3	65.3	—		$6,000
Central African Republic	622,978	240,533	4.5	5.5	7	238	37	19	1.7	115	193	43	44	43	59.6	34.5	2.4		$1,140
Chad	1,283,994	495,753	10.6	17.2	8	274	48	20	2.8	124	208	43	45	25	66.9	40.8	2.0		$1,470
Comoros	2,230	861	0.7	1.0	333	628	37	7	2.9	53	71	62	66	33	63.5	49.1	—		$2,000
Congo	341,998	132,046	3.9	5.9	11	1,139	40	14	2.6	81	108	50	52	51	87.5	74.4	2.2		$810
Congo, The	2,344,848	905,351	66.6	108.0	28	711	45	14	3.1	129	205	49	52	35	86.6	67.7	2.0		$720
Djibouti	23,201	8,958	0.8	1.1	36	3,580	31	12	1.9	88	133	52	54	82	65.0	38.4	—		$2,240
Equatorial Guinea	28,050	10,830	0.5	0.8	19	207	43	20	2.3	123	205	43	44	39	92.5	74.5	2.1		$7,580
Eritrea	117,598	45,405	4.9	7.4	41	1,033	39	11	2.8	50	78	53	57	19	43.9	33.4	2.9		$1,010
Ethiopia	1,104,296	426,371	78.4	107.8	71	646	39	15	2.4	109	164	48	50	15	83.7	70.0	2.4		$1,000
Gabon	267,668	103,347	1.5	1.8	5	272	33	13	2.0	60	91	53	55	81	79.8	62.2	3.0		$5,890
Gambia	11,300	4,363	1.6	2.4	140	665	38	12	2.7	97	137	52	55	50	43.8	29.6	2.5		$1,920
Ghana	238,538	92,100	23.7	32.7	99	431	33	10	2.3	68	112	57	58	45	79.5	61.2	3.3		$2,370
Guinea	245,860	94,927	10.4	15.2	42	702	41	13	2.8	98	150	54	54	30	55.1	27.0	1.9		$2,240
Guinea-Bissau	36,120	13,946	1.5	2.4	41	316	50	20	3.0	124	200	44	46	48	53.0	21.4	—		$700
Ivory Coast	322,459	124,502	20.7	27.1	64	279	39	14	2.5	118	195	49	53	48	54.6	38.5	2.1		$1,490
Kenya	580,367	224,081	36.5	49.4	63	785	40	15	2.5	79	120	49	47	40	89.0	76.0	2.2		$1,170
Lesotho	30,349	11,718	1.8	1.7	60	542	28	25	0.3	102	132	35	36	13	73.6	93.6	3.2		$3,410
Liberia	111,369	43,000	3.6	5.8	32	808	50	21	2.9	157	235	41	44	45	69.9	36.8	—		$130
Madagascar	587,036	226,656	18.8	28.2	32	641	40	12	2.7	74	119	53	57	30	87.7	72.9	3.1		$880
Malawi	118,479	45,745	13.5	23.8	114	542	44	18	2.6	79	125	44	47	17	74.5	46.7	2.7		$650
Mali	1,240,185	478,838	14.8	24.0	12	298	50	18	3.2	120	218	48	49	27	47.9	33.2	2.8		$1,000
Mauritania	1,025,516	395,954	3.4	5.0	3	330	42	14	2.8	78	125	53	55	51	50.6	29.5	3.1		$2,150
Mauritius	2,041	788	1.3	1.4	647	1,245	15	7	0.8	13	15	69	76	42	87.7	81.0	5.1		$12,450
Moçambique	801,586	309,494	20.7	27.6	26	647	41	20	2.0	100	145	41	42	34	59.9	28.4	2.8		$1,170

| | Land Area | | Population | | Population Density | | Birth Rate | Death Rate | Natural Increase % | Infant Mortality per 1,000 | Child Mortality per 1,000 | Life Expectancy | | Percent Urban Pop. | Literacy | | Corruption Index | Big Mac Price ($US) | Per Capita GNI ($US) |
	Sq km	Sq mi	2008 (Millions)	2025 (Millions)	Arithmetic	Physiologic						Male (years)	Female (years)		Male %	Female %			
Namibia	824,287	318,259	2.2	2.5	3	262	29	15	1.4	46	62	47	47	33	82.9	81.2	4.1		$7,910
Niger	1,266,994	489,189	15.4	26.4	12	304	55	21	3.4	150	256	44	44	21	23.5	8.3	2.3		$800
Nigeria	923,766	356,668	141.0	199.5	153	449	43	19	2.4	100	194	43	44	43	72.3	56.2	2.2		$1,040
Réunion	2,510	969	0.8	1.0	328	2,185	19	5	1.4	—	—	72	80	89	84.8	89.2	—		—
Rwanda	26,340	10,170	9.6	13.8	364	866	43	17	2.7	118	203	46	48	17	73.7	60.6	2.5		$1,320
São Tomé and Príncipe	961	371	0.2	0.2	219	509	34	9	2.5	75	118	62	64	38	70.2	39.1	—		
Senegal	196,720	75,954	12.6	17.3	64	534	39	10	2.9	77	136	55	58	50	47.2	27.6	3.3		$1,770
Seychelles	451	174	0.1	0.1	226	1,509	18	8	1.0	12	13	66	76	50	82.9	85.7	3.6		$15,940
Sierra Leone	71,740	27,699	6.0	8.7	83	1,039	46	23	2.3	165	282	39	42	36	50.7	22.6	2.2		$780
Somalia	637,658	246,201	9.4	14.9	15	739	46	17	2.9	133	225	46	50	34	85.8	84.5	—		—
South Africa	1,221,034	471,444	47.8	48.0	39	301	23	18	0.5	55	68	45	49	58	68.3	46.0	4.6	$2.22	$12,120
Swaziland	17,361	6,703	1.1	1.0	63	577	29	28	0.1	110	160	33	35	23	80.9	78.7	2.5		$5,190
Tanzania	945,087	364,900	39.8	53.6	42	843	42	17	2.5	76	122	44	45	32	84.1	66.6	2.9		$730
Togo	56,791	21,927	6.6	9.6	117	272	38	12	2.6	78	139	53	57	33	72.2	42.6	2.4		$1,550
Uganda	241,040	93,066	29.4	55.5	122	359	47	16	3.1	79	136	47	47	12	77.7	57.1	2.7		$1,500
Zambia	752,607	290,583	12.3	16.4	16	234	41	23	1.9	102	182	38	37	35	85.2	71.2	2.6		$950
Zimbabwe	390,759	150,873	13.3	14.4	34	378	30	23	0.7	81	132	37	34	34	95.5	89.9	2.4		$1,940
North Africa/Southwest Asia	**19,318,887**	**7,459,064**	**581.0**	**754.1**	**30**	**328**	**26**	**7**	**1.8**	**47**	**51**	**66**	**70**	**55**	**76.2**	**56.9**	**—**		**$5,674**
Afghanistan	652,086	251,772	32.7	50.3	50	418	48	22	2.6	165	257	41	42	23	51.0	20.8	—		
Algeria	2,381,730	919,591	34.6	43.1	15	485	21	4	1.7	34	39	74	76	81	75.1	51.3	3.1		$6,770
Bahrain	689	266	0.7	1.0	1,053	11,700	21	3	1.8	9	11	73	75	71	91.0	82.7	5.7		$21,290
Egypt	1,001,445	386,660	78.6	101.1	78	2,616	27	6	2.1	28	33	67	72	43	66.6	43.7	3.3	$1.68	$4,440
Iran	1,633,182	630,575	72.0	89.0	44	401	18	6	1.2	31	36	69	72	67	83.5	70.4	2.7		$8,050
Iraq	438,319	169,236	31.2	44.7	71	547	36	10	2.6	102	125	57	60	68	70.7	45.0	1.9		—
Israel	21,059	8,131	7.4	9.3	353	1,681	21	5	1.5	5	6	78	82	91	97.9	94.3	5.9		$25,280
Jordan	89,210	34,444	5.9	7.9	66	1,316	29	5	2.4	22	26	71	72	82	94.9	84.4	5.3		$5,280
Kazakhstan	2,717,289	1,049,151	15.5	16.0	6	48	18	10	0.8	63	73	61	72	57	99.1	96.1	2.6		$7,730
Kuwait	17,819	6,880	2.8	3.9	157	15,672	19	2	1.7	9	11	77	79	97	84.3	79.9	4.8		$24,010

Country	Land Area Sq km	Land Area Sq mi	Population 2008 (Millions)	Population 2025 (Millions)	Pop. Density Arithmetic	Pop. Density Physiologic	Birth Rate	Death Rate	Natural Increase %	Infant Mortality per 1,000	Child Mortality per 1,000	Life Expectancy Male (years)	Life Expectancy Female (years)	Percent Urban Pop.	Literacy Male %	Literacy Female %	Corruption Index	Big Mac Price ($US)	Per Capita GNI ($US)
Kyrgyzstan	198,499	76,641	5.3	6.6	27	385	21	7	1.4	58	67	64	72	35	98.6	95.5	2.2		$1,870
Lebanon	10,399	4,015	4.0	4.6	386	1,244	19	5	1.5	27	30	70	74	87	92.3	80.4	3.6		$5,740
Libya	1,759,532	679,359	6.2	8.3	4	351	27	4	2.4	18	19	74	78	86	90.9	67.6	2.7		$5,530
Morocco	446,548	172,413	32.7	38.8	73	332	21	6	1.6	36	40	68	72	55	61.9	36.0	3.2		$4,360
Oman	212,459	82,031	2.7	3.1	13	1,273	24	4	2.0	10	12	73	75	71	80.4	61.7	5.4		$14,680
Palestinian Territories	6,260	2,417	4.2	7.1	665	33,240	37	4	3.3	21	23	71	74	57	—	—	—		—
Qatar	11,000	4,247	0.8	1.2	75	7,508	18	2	1.6	18	21	71	76	100	80.5	83.2	6.0		—
Saudi Arabia	2,149,680	829,996	25.4	35.6	12	591	30	3	2.7	21	26	70	74	86	84.1	67.2	3.3	$2.40	$14,740
Sudan	2,505,798	967,494	43.5	61.3	17	248	36	9	2.6	62	90	57	59	37	36.0	14.0	2.0		$2,000
Syria	185,179	71,498	20.5	28.1	111	369	29	4	2.5	14	15	71	75	50	88.3	60.4	2.9		$3,740
Tajikistan	143,099	55,251	7.3	9.3	51	730	30	8	2.2	59	71	61	66	26	99.6	98.9	2.2		$1,260
Tunisia	163,610	63,170	10.3	11.6	63	197	17	6	1.1	20	24	71	75	65	81.4	60.1	4.6		$7,900
Turkey	774,816	299,158	75.6	86.0	98	257	19	6	1.3	26	29	69	74	62	93.6	76.7	3.8	$3.66	$8,420
Turkmenistan	488,099	188,456	5.5	6.6	11	281	25	8	1.6	81	104	58	67	47	98.8	96.6	2.2		—
United Arab Emirates	83,600	32,278	5.0	7.1	60	6,027	15	1	1.3	8	9	75	80	74	75.5	79.5	6.2	$2.72	$24,090
Uzbekistan	447,397	172,741	27.0	33.0	60	504	23	7	1.6	57	68	63	70	36	98.5	96.0	2.1		$2,020
Western Sahara	252,120	97,344	0.4	0.7	2	4	28	8	2.0	—	—	62	66	93	—	—	—		
Yemen	527,966	203,849	23.0	38.8	44	1,452	41	9	3.2	76	102	59	62	26	67.4	25.0	2.6		$920
South Asia	**4,487,760**	**1,732,734**	**1,532.9**	**1,841.9**	**342**	**682**	**25**	**8**	**1.7**	**58**	**76**	**62**	**64**	**30**	**65.9**	**39.5**			**$3,330**
Bangladesh	143,998	55,598	152.2	190.0	1,057	1,678	27	8	1.9	54	73	61	62	23	51.7	29.5	2.0		$2,090
Bhutan	47,001	18,147	0.9	1.3	20	655	20	7	1.3	65	75	62	64	31	61.1	33.6	6.0		—
India	3,287,576	1,269,340	1,158.0	1,363.0	352	618	24	8	1.7	56	74	62	63	29	68.6	42.1	3.3		$3,460
Maldives	300	116	0.3	0.4	1,029	10,287	18	3	1.5	33	42	70	70	27	96.3	96.4	—		$4,635
Nepal	147,179	56,826	27.2	36.2	185	879	31	9	2.2	56	74	62	63	14	59.1	21.8	2.5		$1,530
Pakistan	796,098	307,375	173.9	228.8	218	753	33	9	2.4	79	99	61	63	38	57.6	27.8	2.2	$2.31	$2,350
Sri Lanka	65,610	25,332	20.4	22.2	311	1,073	19	6	1.3	12	14	71	77	20	94.5	88.9	3.1	$1.75	$4,520
East Asia	**11,773,125**	**4,545,629**	**1,550.6**	**1,699.4**	**132**	**1,068**	**12**	**7**	**0.5**	**21**	**24**	**71**	**75**	**43**	**93.3**	**80.4**			**$9,050**
China	9,572,855	3,696,100	1,324.5	1,476.0	138	988	12	7	0.6	23	27	70	74	37	92.3	77.4	3.3	$1.45	$6,600
Japan	377,799	145,869	128.1	121.1	339	2,607	9	8	0.0	3	4	79	86	79	100.0	100.0	7.6	$2.29	$31,410
Korea, North	120,541	46,541	23.5	25.8	195	1,219	16	7	0.9	42	55	68	73	60	99.0	99.0	—		—

TABLE G-1 (Continued)

| | Land Area | | Population | | Population Density | | Birth Rate | Death Rate | Natural Increase % | Infant Mortality per 1,000 | Child Mortality per 1,000 | Life Expectancy | | Percent Urban Pop. | Literacy | | Corruption Index | Big Mac Price ($US) | Per Capita GNI ($US) |
	Sq km	Sq mi	2008 (Millions)	2025 (Millions)	Arithmetic	Physiologic						Male (years)	Female (years)		Male %	Female %			
Korea, South	99,259	38,324	48.9	49.8	493	2,592	9	5	0.4	5	5	74	81	82	99.2	96.4	5.1	$3.14	$21,850
Mongolia	1,566,492	604,826	2.7	3.1	2	170	18	6	1.2	39	49	64	68	57	99.2	99.3	2.8		$2,190
Taiwan	36,180	13,969	22.9	23.6	634	2,536	9	6	0.3	—	—	73	79	78	97.6	90.2	5.9	$2.28	$16,250
Southeast Asia	4,494,792	1,735,449	581.7	682.0	129	621	21	6	1.4	30	39	66	71	39	92.8	85.7	—		$4,530
Brunei	5,770	2,228	0.4	0.5	72	3,585	20	3	1.7	8	9	72	77	72	94.7	88.2	—		—
Cambodia	181,040	69,900	14.7	19.6	81	369	30	9	2.1	98	143	57	63	19	79.7	53.4	2.1		$2,490
East Timor	14,869	5,741	1.1	1.9	71	591	42	15	2.7	52	61	54	57	22	65.0	52.0	2.6		$600
Indonesia	1,904,561	735,355	231.9	263.7	122	716	20	6	1.4	28	36	67	72	42	91.9	82.1	2.4	$1.75	$3,720
Laos	236,800	91,429	6.4	8.7	27	899	36	13	2.3	62	79	53	56	21	73.6	50.5	2.6		$2,020
Malaysia	329,750	127,317	27.8	34.6	84	351	20	4	1.6	10	12	72	76	64	91.5	83.6	5.0	$1.60	$10,320
Myanmar/Burma	676,577	261,228	52.1	59.0	77	482	21	10	1.1	75	105	57	63	29	89.0	80.6	1.9		—
Philippines	299,998	115,830	90.1	115.7	300	910	27	5	2.1	25	33	67	72	59	95.5	95.2	2.5	$1.85	$5,300
Singapore	619	239	4.6	5.2	7,357	367,860	10	4	0.6	3	3	78	82	100	96.4	88.5	9.4	$2.59	$29,780
Thailand	513,118	198,116	66.1	70.2	129	322	14	7	0.7	18	21	68	75	34	97.2	94.0	3.6	$1.78	$8,440
Vietnam	331,689	128,066	86.6	102.9	261	1,186	19	5	1.3	16	19	70	73	27	95.7	91.0	2.6		$3,010
Austral Realm	8,011,714	3,093,340	25.0	29.2	3	44	13	6	0.7	5	6	78	83	91	100.0	100.0	—		$29,352
Australia	7,741,184	2,988,888	20.9	24.6	3	39	13	6	0.6	5	6	78	83	91	100.0	100.0	8.7	$2.95	$30,610
New Zealand	270,529	104,452	4.2	4.6	15	128	14	7	0.7	5	6	77	81	86	100.0	100.0	9.6	$5.89	$23,030
Pacific Realm	975,341	376,804	9.0	11.4	24.5	757	30	9	2.0	44	58	59	61	22	69.1	59.0	—		$2,540
Federated States of Micronesia	699	270	0.1	0.1	149	286	26	6	2.0	34	42	67	67	22	67.0	87.2	—		—
Fiji	18,270	7,054	0.8	0.9	45	282	21	6	1.4	16	18	66	71	46	95.0	90.9	—		$5,960
French Polynesia	3,999	1,544	0.3	0.3	77	962	18	5	1.3	—	—	72	77	53	94.9	95.0	—		—
Guam	549	212	0.2	0.2	377	1,712	21	4	1.6	—	—	75	81	93	99.0	99.0	—		—
Marshall Islands	179	69	0.1	0.1	597	3,512	38	5	3.3	51	58	—	—	68	92.4	90.0	—		—
New Caledonia	18,581	7,174	0.2	0.3	11	1,102	17	5	1.2	—	—	71	77	71	57.4	58.3	—		—
Papua New Guinea	462,839	178,703	6.3	8.2	14	1,351	32	11	2.1	55	74	55	56	13	63.4	50.9	2.4		$2,370
Samoa	2,841	1,097	0.2	0.2	74	171	29	6	2.4	24	29	72	74	22	100.0	100.0	—		$6,480
Solomon Islands	28,899	11,158	0.5	0.7	18	607	34	8	2.6	24	29	62	63	16	62.4	44.9	—		$1,880
Vanuatu	12,191	4,707	0.2	0.4	17	172	31	6	2.5	31	38	66	69	21	57.3	47.8	—		$3,170

TABLE G-1 on pages 38–44 is a valuable resource, and should be consulted throughout your reading. Like all else in this book, Table G-1 is subject to continuous revision and modification. In this 13th edition, we have deleted some indices, elaborated others, and introduced new ones. For example, in a world with ever-slower population growth, the so-called Doubling Time index—the number of years it will take for a population to double in size based on its current rate of natural increase—has lost most of its utility. On the other hand, when it comes to **Life Expectancy** and **Literacy**, general averages conceal significant differences between male and female rates, which in turn reflect conditions in individual countries, so we report these by gender. Also in this edition, we continue to use the **Corruption Index**, not available for all countries but an important reflection of a global problem. The **Big Mac Price** index, a measure introduced by the journal *The Economist*, tells you much more than what a hamburger with all the trimmings would cost in real dollars in various countries of the world—it also reflects whether those countries' currencies are overvalued or undervalued. And the final column, in which we formerly used to reported the per-capita GNP (Gross National Product), now reveals the GNI, that is, the Gross National Income per person and what this would buy in each country. In the language of economic geographers, this is called the GNI-PPP, the per-capita Gross National Income in terms of its Purchasing Power Parity.

Indexes that may not be immediately obvious to you include **Arithmetic Population Density**, the number of people per square kilometer in each country; **Physiologic Population Density**, the number of people per square kilometer of agriculturally productive land; **Birth** and **Death Rates** per thousand in the population, resulting in the national population's rate of **Natural Increase**; a population's **Infant Mortality**, the number of deaths per thousand in the first year of life, thus reflecting largely the number of deaths at birth; **Child Mortality**, the deaths per thousand of children in their first five years; the **Corruption Index**, based on Transparency International data in which 10.0 is perfect and 0.1 is the worst; the **Big Mac Price** index, which tells you why China in 2007 was the best place to buy a hamburger in U.S. dollars; and the **Per Capita GNI ($US)**, the GNI-PPP index referred to above, which tells you how spendable income varies around the globe. For additional details on sources and data, please consult the Data Sources section of the Preface.

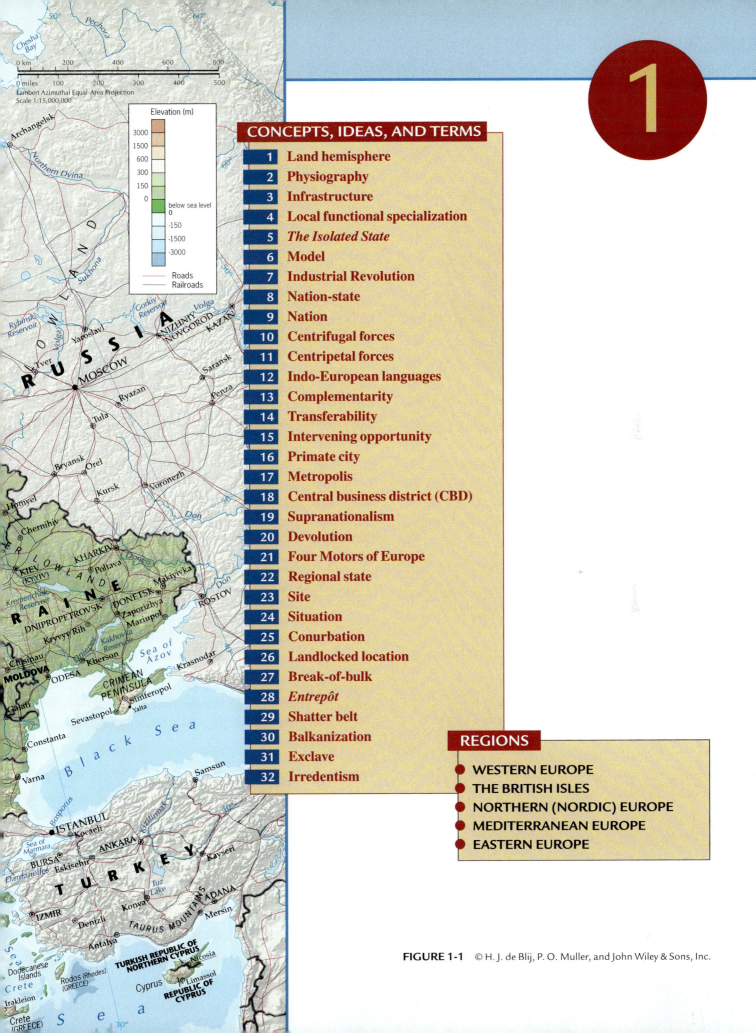

CONCEPTS, IDEAS, AND TERMS

1	Land hemisphere
2	Physiography
3	Infrastructure
4	Local functional specialization
5	*The Isolated State*
6	Model
7	Industrial Revolution
8	Nation-state
9	Nation
10	Centrifugal forces
11	Centripetal forces
12	Indo-European languages
13	Complementarity
14	Transferability
15	Intervening opportunity
16	Primate city
17	Metropolis
18	Central business district (CBD)
19	Supranationalism
20	Devolution
21	Four Motors of Europe
22	Regional state
23	Site
24	Situation
25	Conurbation
26	Landlocked location
27	Break-of-bulk
28	*Entrepôt*
29	Shatter belt
30	Balkanization
31	Exclave
32	Irredentism

REGIONS

- WESTERN EUROPE
- THE BRITISH ISLES
- NORTHERN (NORDIC) EUROPE
- MEDITERRANEAN EUROPE
- EASTERN EUROPE

FIGURE 1-1   © H. J. de Blij, P. O. Muller, and John Wiley & Sons, Inc.

Where were these photos taken? Find out at www.wiley.com/college/deblij

In This Chapter

Photos: © H. J. de Blij.

IT IS APPROPRIATE to begin our investigation of the world's geographic realms in Europe because over the past five centuries Europe and Europeans have influenced and changed the rest of the world more than any other realm or people has done. European empires spanned the globe and transformed societies far and near. European colonialism propelled the first wave of globalization. Millions of Europeans migrated from their homelands to the Old World as well as the New, changing (and sometimes nearly obliterating) traditional communities and creating new societies from Australia to North America. Colonial power and economic incentive combined to impel the movement of millions of imperial subjects from their ancestral homes to distant lands: Africans to the Americas, Indians to Africa, Chinese to Southeast Asia, Malays to South Africa's Cape, Native Americans from east to west. In agriculture, industry, politics, and other spheres, Europe generated revolutions—and then exported those revolutions across the world, thereby consolidating the European advantage.

But throughout much of that 500-year period of European hegemony, Europe also was a cauldron of conflict. Religious, territorial, and political disputes precipitated bitter wars that even spilled over into the colonies. And during the twentieth century, Europe twice plunged the world into war. The terrible, unprecedented toll of World War I (1914–1918) was not enough to stave off World War II (1939–1945), which ended with the first-ever use of nuclear weapons in Japan. In the aftermath of that war, Europe's weakened powers lost most of their colonial possessions and a new rivalry emerged: an ideological Cold War between the communist Soviet Union and the capitalist United States. This Cold War lowered an Iron Curtain across the heart of Europe, leaving most of the east under Soviet control and most of the west in the American camp.

MAJOR GEOGRAPHIC QUALITIES OF

Europe

1. The European realm lies on the western extremity of the Eurasian landmass, a locale of maximum efficiency for contact with the rest of the world.

2. Europe's lingering and resurgent world influence results largely from advantages accrued over centuries of global political and economic domination.

3. The European natural environment displays a wide range of topographic, climatic, and soil conditions and is endowed with many industrial resources.

4. Europe is marked by strong internal regional differentiation (cultural as well as physical), exhibits a high degree of functional specialization, and provides multiple exchange opportunities.

5. Manufacturing brought prosperity to European economies, but high-technology and service industries are changing the pattern. Protectionist policies help sustain agriculture, which is still strong in the west where economic development and change have been slower.

6. Europe's nation-states emerged from durable power cores that formed the headquarters of world colonial empires. A number of those states are now plagued by internal separatist movements.

7. Europe's rapidly aging population is generally well off, highly urbanized, well educated, and enjoys long life expectancies.

8. A growing number of European countries are experiencing population declines; in many of these countries, the natural decrease is partially offset by immigration.

9. Europe has made significant progress toward international economic integration and, to a lesser extent, political coordination.

Western Europe proved resilient, overcoming the destruction of war and the loss of colonial power to regain economic strength. Meanwhile, the Soviet communist experiment failed at home and abroad, and in 1990 the last vestiges of the Iron Curtain were lifted. Since then, a massive effort has been underway to reintegrate and reunify Europe from the Atlantic coast to the Russian border, the key geographic story of this chapter.

Defining the Realm

GEOGRAPHICAL FEATURES

As Figure 1-1 shows, Europe is a realm of peninsulas and islands on the western margin of the world's largest landmass, Eurasia. It is a realm of 590 million people and 40 countries, but it is territorially quite small. Yet despite its modest proportions it has had—and continues to have—a major impact on world affairs. For many centuries Europe has been a hearth of achievement, innovation, and invention.

Europe's Eastern Border

The European realm is bounded on the west, north, and south by Atlantic, Arctic, and Mediterranean waters, respectively. But where is Europe's eastern limit? Some scholars place it at the Ural Mountains, deep inside Russia, thereby recognizing a "European" Russia and, presumably, an "Asian" one as well. Our regional definition places Europe's eastern boundary between Russia and its numerous European neighbors to the west. This definition is based on several geographic factors including European-Russian contrasts in territorial dimensions, population size, cultural properties, and historic aspects, all discernible on the maps in Chapters 1 and 2.

Resources

Europe's peoples have benefited from a large and varied store of raw materials. Whenever the opportunity or need arose, the realm proved to contain what was required. Early on, these requirements included cultivable soils, rich fishing waters, and wild animals that could be domesticated; in addition, extensive forests provided wood for houses and boats. Later, coal and mineral ores propelled industrialization. More recently, Europe proved to contain substantial deposits of oil and natural gas.

Climates

From the balmy shores of the Mediterranean Sea to the icy peaks of the Alps, and from the moist woodlands and moors of the Atlantic fringe to the semiarid prairies north of the Black Sea, Europe presents an almost infinite range of natural environments (Fig. 1-2). Compare Western Europe and eastern North America in Figure G-8 (pp. 14–15) and you will see the moderating influence of the warm ocean current known as the North Atlantic Drift and its onshore windflow.

Human Diversity

The European realm is home to peoples of numerous cultural-linguistic stocks, including not only Latins, Germanics, and Slavs but also minorities such as Finns, Magyars (Hungarians), Basques, and Celts. This diversity of ancestries continues to be an asset as well as a liability. It has generated not only interaction and exchange, but also conflict and war.

Locational Advantages

Europe also has outstanding locational advantages. Its *relative location*, at the heart of the **land hemisphere**, creates maximum efficiency for contact with much of the rest of the world (Fig. 1-3). A "peninsula of peninsulas," Europe is nowhere far from the ocean and its avenues of seaborne trade and conquest. Hundreds of kilometers of navigable rivers, augmented by an unmatched system of canals, open the interior of Europe to its neighboring seas and to the shipping lanes of the world.

Also consider the scale of the maps of Europe in this chapter. Europe is a realm of moderate distances and close proximities. Short distances and large cultural differences

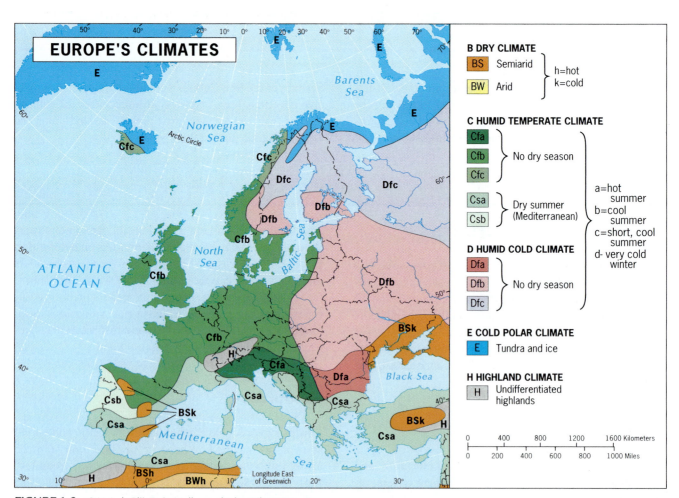

FIGURE 1-2 © H. J. de Blij, P. O. Muller, and John Wiley & Sons, Inc.

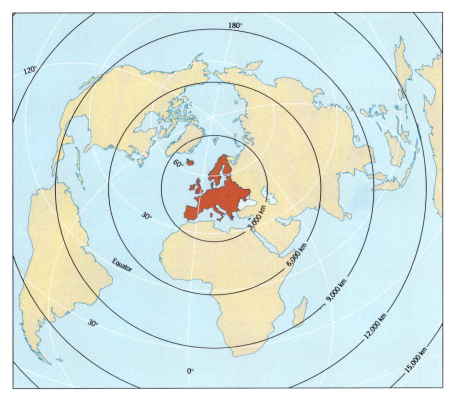

3000 kilometers=ca. 1900 miles

FIGURE 1-3 © H. J. de Blij, P. O. Muller, and John Wiley & Sons, Inc.

border to the Balkan Mountains near the Black Sea, and include Italy's Appennines and Eastern Europe's Carpathians.

The *Western Uplands*, geologically older, lower, and more stable than the Alpine Mountains, extend from Scandinavia through western Britain and Ireland to the heart of the Iberian Peninsula in Spain.

The *North European Lowland* extends in a lengthy arc from southeastern Britain and central France across Germany and Denmark into Poland and Ukraine, from where it continues well into Russia. Also known as the Great European Plain, this has been an avenue for human migration time after time, so that complex cultural and economic mosaics developed here together with a jigsaw-like political map. As Figure 1-4 shows, many of Europe's major rivers and connecting waterways serve this populous region, where many of Europe's leading cities (London, Paris, Amsterdam, Copenhagen, Berlin, Warsaw) are located.

make for intense interaction, the constant circulation of goods and ideas. That has been the hallmark of Europe's geography for over a millennium.

LANDSCAPES AND OPPORTUNITIES

In the Introduction we noted the importance of physical geography in the definition of geographic realms. The natural landscape with its array of landforms (such as mountains and plateaus) is a key element in the total physical geography—or **physiography**—of any part of the terrestrial world. Other physiographic components include climate and the physical features that mark the natural landscape, such as vegetation, soils, and water bodies.

Europe's area may be small, but its landscapes are varied and complex. Regionally, we identify four broad units: the Central Uplands, the southern Alpine Mountains, the Western Uplands, and the North European Lowland (Fig. 1-4).

The *Central Uplands* form the heart of Europe. It is a region of hills and low plateaus loaded with raw materials whose farm villages grew into towns and cities when the Industrial Revolution transformed this realm.

The *Alpine Mountains*, a highland region named after the Alps, extend from the Pyrenees on the French-Spanish

HISTORICAL GEOGRAPHY

Modern Europe was peopled in the wake of the Pleistocene's most recent glacial retreat and global warming—a gradual warming that caused tundra to give way to deciduous forest and ice-filled valleys to turn into grassy vales. On Mediterranean shores, Europe witnessed the rise of its first great civilizations, on the islands and peninsulas of Greece and later in what is today Italy.

Ancient Greece and Imperial Rome

Ancient Greece lay exposed to influences radiating from the advanced civilizations of Mesopotamia and the Nile Valley, and in their fragmented habitat the Greeks laid the foundations of European civilization. Their achievements in political science, philosophy, the arts, and other spheres have endured for 25 centuries. But the ancient Greeks never managed to unify their domain, and their persistent conflicts proved fatal when the Romans challenged them from the west. By 147 BC, the last of the

FIGURE 1-4 © H. J. de Blij, P. O. Muller, and John Wiley & Sons, Inc.

sovereign Greek intercity leagues (alliances) had fallen to the Roman conquerors.

The center of civilization and power now shifted to Rome in present-day Italy. Borrowing from Greek culture, the Romans created an empire that stretched from Britain to the Persian Gulf and from the Black Sea to Egypt; they made the Mediterranean Sea a Roman lake carrying armies to distant shores and goods to imperial Rome. With an urban population that probably exceeded 1 million, Rome was the first metropolitan-scale urban center in Europe.

The Romans founded numerous other cities throughout their empire and linked them to the capital through a vast system of highway and water routes, facilitating political control and enabling economic growth in their provinces. It was an unparalleled **infrastructure**, much of which long outlasted the empire itself.

Triumph and Collapse

Roman rule brought disparate, isolated peoples into the imperial political and economic sphere. By guiding (and often forcing) these groups to produce particular goods or materials, the Romans launched Europe down a road for which it would become famous: **local functional specialization**. The workers on Elba, a Mediterranean island, mined iron ore. Those near Cartagena in Spain produced silver and lead. Certain farmers were taught irrigation to produce specialty crops. Others raised livestock for meat or wool. The ***production of particular goods by particular people in particular places*** became and remained a hallmark of the realm.

The Romans also spread their language across the empire, setting the stage for the emergence of the ***Romance languages***; they disseminated Christianity; and they established durable systems of education, administration, and commerce. But when their empire collapsed in the fifth century, disorder ensued, and massive migrations soon brought Germanic and Slavic peoples to their present positions on the European stage. Capitalizing on Europe's weakness, the Arab-Berber Moors from North Africa, energized by Islam, conquered most of Iberia and penetrated France. Later the Ottoman Turks invaded Eastern Europe and reached the outskirts of Vienna.

Rebirth and Royalty

Europe's revival—its ***Renaissance***—did not begin until the fifteenth century. After a thousand years of feudal turmoil marking the "Dark" and "Middle" Ages, powerful monarchies began to lay the foundations of modern states. The discovery of continents and riches across the oceans opened a new era of ***mercantilism***, the competitive accumulation of wealth chiefly in the form of gold and silver. Best placed for this competition were the kingdoms of Western Europe. Europe was on its way to colonial expansion and world domination.

THE REVOLUTIONS OF MODERNIZING EUROPE

Even as Europe's rising powers reached for world domination overseas, they fought with each other in Europe itself. Powerful monarchies and land-owning ("landed") aristocracies had their status and privilege challenged by ever-wealthier merchants and businesspeople. Demands for political recognition grew; cities mushroomed with the development of industries; the markets for farm products burgeoned; and Europe's population, more or less stable at about 100 million since the sixteenth century, began to increase.

The Agrarian Revolution

We know Europe as the focus of the Industrial Revolution, but before this momentous development occurred another revolution was already in progress: the ***agrarian revolution***. Port cities and capital cities thrived and expanded, and their growing markets created economic opportunities for farmers. This led to revolutionary changes in land ownership and agricultural methods. Improved farm practices, better equipment, superior storage facilities, and more efficient transport to the urban markets marked a revolution in the countryside. The colonial merchants brought back new crops (the American potato soon became a European staple), and market prices rose, drawing more and more farmers into the economy.

Agricultural Market Model

The transformation of Europe's farmlands reshaped its economic geography, producing new patterns of land use and market links. The economic geographer Johann Heinrich von Thünen (1783–1850), himself an estate farmer who had studied these changes for several decades, published his observations in 1826 in a pioneering work entitled ***The Isolated State***, chronicling the geography of Europe's agricultural transformation.

Von Thünen used information from his own farmstead to build what today we call a **model** (an idealized representation of reality that demonstrates its most important properties) of the location of productive activities in Europe's farmlands. Since a model is an abstraction that must always involve assumptions, von Thünen postulated a self-contained area (hence the "isolation") with a single market center, surrounded by flat and uninterrupted land without impediments to cultivation or transportation. In such a situation, transport costs would be directly proportional to distance.

Location Theory

Von Thünen's model revealed four zones or rings of land use encircling the market center (Fig. 1-5). Innermost and directly adjacent to the market would lie a zone of intensive farming and dairying, yielding the most perishable and highest-priced products. Immediately beyond lay a zone of

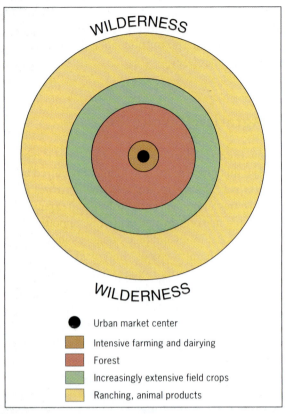

FIGURE 1-5 © H. J. de Blij, P. O. Muller, and John Wiley & Sons, Inc.

market for a realmwide, macroscale "Thünian" agricultural system (Fig. 1-6).

The Industrial Revolution

The term **Industrial Revolution** suggests that an agrarian Europe was suddenly swept up in wholesale industrialization that changed the realm in a few decades. In reality, seventeenth- and eighteenth-century Europe had been industrializing in many spheres, long before the chain of events known as the Industrial Revolution began. From the textiles of England and Flanders to the iron farm implements of Saxony (in present-day Germany), from Scandinavian furniture to French linens, Europe had already entered a new era of local functional specialization. It would therefore be more appropriate to call what happened next the period of Europe's *industrial intensification*.

British Primacy

In the 1780s, the Scotsman James Watt and others devised a steam-driven engine, which was soon adopted for numerous industrial uses. At about the same time, coal (converted into carbon-rich coke) was recognized as a vastly superior

forest used for timber and firewood (still a priority in von Thünen's time). Next there would be a ring of field crops, for example, grains or potatoes. A fourth zone would contain pastures and livestock. Beyond lay wilderness, from where the costs of transport to market would become prohibitive.

In many ways, von Thünen's model was the first analysis in a field that would eventually become known as *location theory*. Von Thünen knew, of course, that the real Europe did not present the conditions postulated in his model. But it did demonstrate the economic-geographic forces that shaped the new Europe, which is why it is still being discussed today. More than a century after the publication of *The Isolated State*, geographers Samuel van Valkenburg and Colbert Held produced a map of twentieth-century European agricultural intensity, revealing a striking, ring-like concentricity focused on the vast urbanized area lining the North Sea—now the dominant

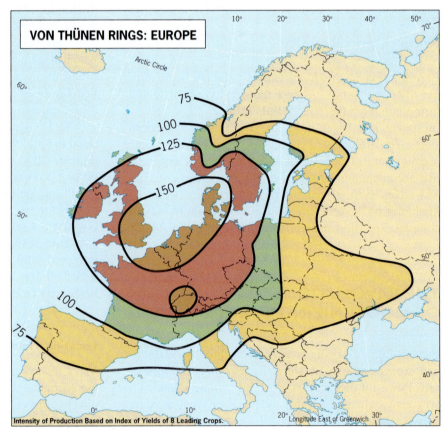

FIGURE 1-6 © H. J. de Blij, P. O. Muller, and John Wiley & Sons, Inc.

substitute for charcoal in smelting iron. These momentous innovations had a rapid effect. The power loom revolutionized the weaving industry. Iron smelters, long dependent on Europe's dwindling forests for fuel, could now be concentrated near coalfields. Engines could move locomotives as well as power looms. Ocean shipping entered a new age.

Britain had an enormous advantage, for the Industrial Revolution occurred when British influence reigned worldwide and the significant innovations were achieved in Britain itself. The British controlled the flow of raw materials, they held a monopoly over products that were in global demand, and they alone possessed the skills necessary to make the machines that manufactured the products. Soon the fruits of the Industrial Revolution were being exported, and the modern industrial spatial organization of Europe began to take shape. In Britain, manufacturing regions, densely populated and heavily urbanized, developed near coalfields in the English Midlands, at Newcastle to the northeast, in southern Wales, and along Scotland's Clyde River around Glasgow.

Resources and Industrial Development

In mainland Europe, a belt of major coalfields extends from west to east, roughly along the southern margins of the North European Lowland, due eastward from southern England across northern France and Belgium, Germany (the Ruhr), western Bohemia in the Czech Republic, Silesia in southern Poland, and the Donets Basin (Donbas) in eastern Ukraine. Iron ore is found in a broadly similar belt, and the industrial map of Europe reflects the resulting concentrations of economic activity (Fig. 1-7). Another set of manufacturing zones emerged in and near the growing urban centers of Europe, as the same map demonstrates. London—already Europe's leading urban focus and Britain's richest domestic market—was typical of these developments. Many local industries were established here, taking advantage of the large supply of labor, the ready availability of capital, and the proximity of so great a number of potential buyers. Although the Industrial Revolution thrust other places into prominence, London did not lose its primacy: industries in and around the British capital multiplied.

Consequences of Clustering

The industrial transformation of Europe, like the agrarian revolution, became the focus of geographic research. One of the leaders in this area was the economic geographer Alfred Weber (1868–1958), who published a spatial analysis of the process titled *Concerning the Location of Industries* (1909). Unlike von Thünen, Weber focused on activities that take place at particular points rather than over large areas. His model, therefore, represented the factors of industrial location, the clustering or dispersal of places of intense manufacturing activity.

One of Weber's most interesting conclusions has to do with what he called **agglomerative** (concentrating) and **deglomerative** (dispersive) forces. It is often advantageous for certain industries to cluster together, sharing equipment, transport facilities, labor skills, and other assets of urban areas. This is what made London (as well as Paris and other cities that were not situated on rich deposits of industrial raw materials) attractive to many manufacturing plants that could benefit from agglomeration and from the large markets that these cities anchored. As Weber found, however, excessive agglomeration may lead to disadvantages such as competition for increasingly expensive space, congestion, overburdening of infrastructure, and environmental pollution. Manufacturers may then move away, and deglomerative forces will increase.

The Industrial Revolution spread eastward from Britain onto the European mainland throughout the middle and late nineteenth century (see inset map, Fig. 1-7). Population skyrocketed, emigration mushroomed, and industrializing cities burst at the seams. European states already had acquired colonial empires before this revolution started; now colonialism gave Europe an unprecedented advantage in its dominance over the rest of the world.

Political Revolutions

Revolution in a third sphere—the political—had been going on in Europe even longer than the agrarian or industrial revolutions. Europe's **political revolution** took many different forms and affected diverse peoples and countries, but in general it headed toward parliamentary representation and democracy.

Historical geographers point to the Peace (Treaty) of Westphalia in 1648 as a key step in the evolution of Europe's state system, ending decades of war and recognizing territories, boundaries, and the sovereignty of countries. This treaty's stabilizing effect lasted until 1806, by which time revolutionary changes were again sweeping across Europe.

Most dramatic was the French Revolution (1789–1795), but political transformation had come much earlier to the Netherlands, Britain, and the Scandinavian countries. Other parts of Europe remained under the control of authoritarian (dictatorial) regimes headed by monarchs or despots. Europe's patchy political revolution lasted into the twentieth century, and by then **nationalism** (national

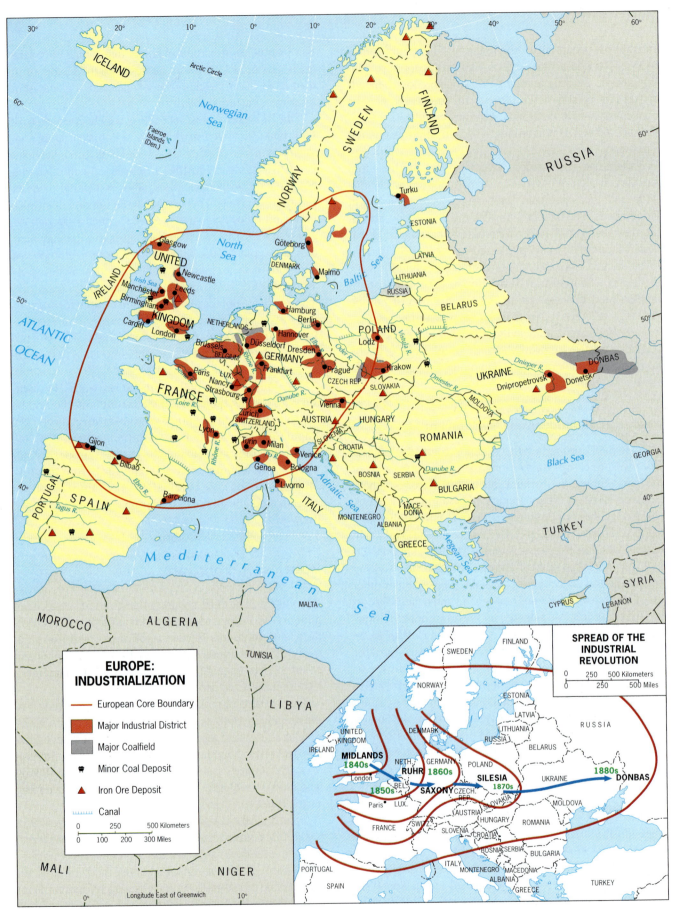

FIGURE 1-7 © H. J. de Blij, P. O. Muller, and John Wiley & Sons, Inc.

spirit, pride, patriotism) had become a powerful force in European politics.

A Fractured Map

When you look at the political map of Europe, the question that arises is how did so small a geographic realm come to be divided into so many political entities? Europe's map is a legacy of its feudal and royal periods, when powerful kings, barons, dukes, and other rulers, rich enough to fund armies and powerful enough to exact taxes and tribute from their domains, created bounded territories in which they reigned supreme. Royal marriages, alliances, and conquests actually simplified Europe's political map. In the early nineteenth century there still were 39 German states; Germany as we know it today did not emerge until the 1870s.

State and Nation

Europe's political revolution produced a form of political-territorial organization known as the **nation-state**, a state embodied by its culturally distinctive population. But what is a nation-state and what is not? The term **nation** has multiple meanings. In one sense it refers to a people with a single language, a common history, a similar ethnic background. In the sense of *nationality* it relates to legal membership in the state, that is, citizenship. Very few states today are so homogeneous culturally that the culture is conterminous with the state. Europe's prominent nation-states of a century ago—France, Spain, the United Kingdom, Italy—have become multicultural societies, their nations defined more by an intangible "national spirit" and emotional commitment than by cultural or ethnic homogeneity. Today, Poland, Hungary, and Sweden are among the few states that still satisfy the definition of the nation-state in Europe.

Division and Unity

Mercantilism and colonialism empowered the states of Western Europe; the United Kingdom (Britain) was the superpower of its day. But all countries, even Europe's nation-states in their heyday, are subject to divisive stresses. Political geographers use the term **centrifugal forces** to identify and measure the strength of such division, which may result from religious, racial, linguistic, political, or regional factors. In the United States, racial issues form a centrifugal force; during the Vietnam (Indochina) War (1964–1975) politics created strong and dangerous disunity.

Centrifugal forces are measured against **centripetal forces**, the binding, unifying glue of the state. General satisfaction with the system of government and administration, legal institutions, and other functions of the state (notably including its treatment of minorities) can ensure stability and continuity when centrifugal forces threaten. In the recent case of Yugoslavia, the centrifugal forces unleashed after the end of the Cold War exceeded the weak centripetal forces in that relatively young state, and it disintegrated.

Europe's political revolution continues. Today a growing group of European states is trying to create a realmwide union that might some day become a European superstate.

CONTEMPORARY EUROPE

If you were to arrive by air or ship in Europe for a visit, you would probably be impressed by the modernity of its infrastructure. Airports tend to be modern and spacious, and security operates smoothly. Public transport, from high-speed intercity trains to local buses, is efficient and seems to reach even the most remote places. Superhighways connect major cities, although Europe's relatively compact, old, and densely-built-up urban areas tend to delay incoming automobiles and cause traffic jams that will look familiar. Ultramodern high-rise architecture and historic preservation give Europe a cultural landscape welding the future to the past. From nuclear power plants (France) and offshore oil platforms (Norway) to flood-control technology (Netherlands) and mega-ship construction (Finland), as well as in countless other enterprises, Europe is in the vanguard of the world. In 2006, nine of Europe's national economies ranked among the top 20 in the world; collectively, Europe's economies outrank even the United States.

Collectively—but that's the problem. Europe does not have a long-term history of collective action. Europe has historically been a crucible of political revolution and evolution. Its nation-states, enriched and empowered by their colonial conquests, fought with each other at home and abroad, dragging distant peoples into their wars and plunging the world into global conflict twice during the twentieth century. After the end of World War II in the mid-1940s, Europeans saw their ravaged realm divided in the Cold War between Western and Soviet spheres. The countries on the western side of the "Iron Curtain" embarked on a massive effort not only to recover, but also to put old divisions permanently behind them. Assisted by the United States, they moved far along the road to unification—especially in the economic arena. But, as we will see later in this chapter, Europe's old fractiousness has not been overcome.

Language and Religion

It is worth remembering that Europe's territory is just over 60 percent the size of the United States, but that the population of Europe's 40 countries is about twice as large as America's. This population of more than 590 million speaks numerous languages, almost all of which belong to **12** the **Indo-European language family** (Figs. 1-8, G-11). But most of those languages are not mutually understandable; some, such as Finnish and Hungarian, are not even members of this Indo-European family. When the unification effort began, one major problem was to determine which languages to recognize as "official." That problem still prevails, although English has become the realm's unofficial *lingua franca*. During your visit to Europe, though, you would find that English is more commonly usable in Western Europe than farther east. Europe's multilingualism remains a major barrier to integration.

Nevertheless, the unification effort has gone so far that Europeans in recent years have been debating the terms of a proposed Constitution. And in this debate, another divisive element of the regional culture has come to the fore: religion. Europe's cultural heritage is steeped in Christian traditions, but Christianity has not unified Europeans. Sectarian strife between Catholics and Protestants that previously plunged Europe into bitter conflict still divides communities and, as in Northern Ireland, still can arouse violence. Religion and politics remain closely connected, and some national and regional political parties call themselves "Christian Democrats" (as in Germany). Eastern Europe has a history of strife between Christian Orthodox and Islamic faiths. Today, a new factor roils the religious landscape: the rise of Islam in Western Europe, resulting mainly from the recent immigration of millions of Muslims from North Africa and other parts of the Muslim world. When some of the framers of the European Constitution wanted to state that the realm's cultural heritage was "Judeo-Christian-Islamic," they were outvoted by those who preferred no reference to religion at all. Meanwhile, as mosques overflow with the faithful, churches stand near empty as secularism is on the rise. Some scholars have gone so far as to describe Europe as being in a "post-Christian" era.

For so small a realm, Europe's cultural geography is sharply varied. The popular image of Europe tends to be formed by British pageantry, French countrysides, German cities—but go beyond this core, and you will find isolated Slavic communities in the mountains facing the Adriatic, Muslim towns in poverty-mired Albania, Roma (Gypsy) villages in the interior of Romania, farmers using traditional methods unchanged for centuries in rural Poland. That map of 40 countries does not begin to reflect the diversity of European cultures.

FROM THE FIELD NOTES

"It seemed the quickest way to go from central London to central Paris: the Eurostar 'Chunnel' Express from Waterloo Station to the Gare du Nord in less than three hours. But it was also a study in contrasts. On the British side of the English Channel, the train chugged along, sometimes at less than 65 kph (40 mph). Once in the tunnel beneath the Channel, the track was smoother and the speed picked up. On the French side, we moved at 250 kph (155 mph), not fast enough, however, to make up for lost time. The conductor's announcement pointedly referred to delays on the Angleterre segment of the journey being the cause of the late arrival at the Paris terminal. This money-losing link between the United Kingdom and the continent is nevertheless symbolic of unifying Europe, but it reveals some other contrasts: the dreadful breakfast out of London and the delightful dinner out of Paris, the chaotic security and immigration scene at Waterloo and the smooth process at Gare du Nord (wasn't this supposed to be a domestic intra-EU excursion?), and both ways the complete lack of passenger information—no map, no brochure on the train and its manufacture, no history of the link. Who's in charge here?"
© H. J. de Blij

Concept Caching

www.conceptcaching.com

Spatial Interaction

If not culture, what does unify Europe? The answer lies in this realm's outstanding opportunities for productive interaction. The ancient Romans realized it, but they developed a system of production that focused on Rome, not primarily on regional exchange. Modern Europeans seized on the same opportunities to create a regionwide network of spatial interaction whose structure intensified as time went on, linking places, communities, and countries in countless ways. The geographer Edward Ullman envisaged this process as operating on three principles:

- **Complementarity** When one area produces a surplus **13** of a commodity required by another area, regional complementarity prevails. The mere existence of a particular resource or product is no guarantee of trade: it

LANGUAGES OF EUROPE

0   200   400   600 Kilometers
0   100   200   300 Miles

MAJOR INDO-EUROPEAN BRANCHES

GERMANIC GROUP

WESTERN GERMANIC
1 Dutch
2 German
3 Frysian
4 English

NORTHERN GERMANIC
5 Danish
6 Swedish
7 Norwegian
8 Icelandic
9 Faeroese

ROMANCE GROUP
10 Portuguese   16 Rhaeto-Romansch
11 Spanish   17 Romanian
12 Catalan   18 Corsican-Italian
13 Provençal   19 Sardinian-Italian
14 French   20 Walloon
15 Italian

SLAVIC GROUP

WEST SLAVONIC
21 Polish
22 Slovak
23 Czech
24 Lusatian

EAST SLAVONIC
25 Russian
26 Ukrainian
27 Belarussian

SOUTH SLAVONIC
28 Slovene
29 Serbo-Croatian
30 Macedonian
31 Bulgarian

OTHER INDO-EUROPEAN BRANCHES

CELTIC GROUP

BRITANNIC
32 Breton
33 Welsh

GAELISH
34 Irish Gaelic
35 Scots Gaelic

BALTIC GROUP
36 Latvian   37 Lithuanian

HELLENIC
38 Greek

THRACIAN/ILLYRIAN GROUP
39 Albanian

INDO-IRANIAN GROUP
40 Romani (dispersed)

URALIC LANGUAGE FAMILY

FINNO-UGRIC GROUP
41 Finnish   44 Estonian
42 Karelian   45 Hungarian
43 Saami   46 Komi

SAMOYEDIC GROUP
47 Samoyedic

ALTAIC LANGUAGE FAMILY

TURKIC GROUP
48 Turkish

OTHER LANGUAGES

BASQUE
49 Basque

Areas with significant concentrations of other languages (usually adjacent national languages)

Boundary between languages

FIGURE 1-8   © H. J. de Blij, P. O. Muller, and John Wiley & Sons, Inc.

must be needed elsewhere. When two areas each require the other's products, double complementarity occurs. Europe has countless examples of this, from local communities to entire countries. Industrial Italy needs coal from Western Europe; Western Europe needs Italy's farm products.

14 ● **Transferability** The ease with which a commodity can be transported by producer to consumer defines its transferability. Distance and physical obstacles can raise the cost of a product to the point of unprofitability. But Europe is small, distances are short, and Europeans have built the world's most efficient transport system of roads, railroads, and canals linking navigable rivers.

15 ● **Intervening Opportunity** The third of Ullman's spatial interaction principles holds that potential trade between two places, even if they are in a position of strong complementarity and high transferability, will develop only in the absence of a closer, intervening source of supply. To use our earlier example, if Switzerland proved to contain large coal reserves close to the Italian border, Italy would avail itself of that opportunity, reducing or eliminating its imports from more distant Western Europe.

Athens anchors and dominates Greece. Europe's primate cities have also grown disproportionately since the end of World War II, gaining rather than losing primacy in this still-urbanizing realm.

Cityscapes

Primate cities tend to be old, and in general the European cityscape looks quite different from the American. Seemingly haphazard inner-city street systems impede traffic; central cities may be picturesque, but they are also cramped. The urban layout of the London region (Fig. 1-9) reveals much about the internal spatial structure of the European **metropolis** (the central city and its suburban ring). The **17** metropolitan area remains focused on the large city at its center, especially the downtown **central business district** **18** **(CBD)**, which is the oldest part of the urban agglomeration and contains the region's largest concentration of business, government, and shopping facilities as well as its wealthiest and most prestigious residences. Wide residential sectors radiate outward from the CBD across the rest of the central city, each one home to a particular income group. Beyond the central city lies a sizeable suburban ring, but residential densities are much higher here than in the United States because the European tradition is one of set-

A Highly Urbanized Realm

Such opportunities for spatial interaction and exchange contribute to an ever-increasing level of urbanization in Europe. Overall, about three of every four Europeans live in towns and cities, an average that is far exceeded in the west (see Table G-1) but not yet attained in much of the east. Large cities are production centers as well as marketplaces, and they also form the crucibles of their nations' cultures. By several measures, European cities are more diverse than their American counterparts.

The geographer Mark Jefferson during the 1930s studied the pivotal role of great cities in the development of national cultures. He postulated a "law"
16 of the **primate city**, which stated that "a country's leading city is always disproportionately large and exceptionally expressive of national culture." This obviously debatable "law" still has relevance: Paris personifies France in countless ways; nothing in the United Kingdom rivals London;

FIGURE 1-9   © H. J. de Blij, P. O. Muller, and John Wiley & Sons, Inc.

ting aside recreational spaces (in "greenbelts") and living in apartments rather than in detached single-family houses. There is also a greater reliance on public transportation, which further concentrates the suburban development pattern. That has allowed many nonresidential activities to suburbanize as well, and today ultramodern outlying business centers increasingly compete with the CBD in many parts of urban Europe (see photo below).

A Changing Population

When a population urbanizes, average family size declines, and so does the overall rate of natural increase. There was a time when Europe's population was (in the terminology of population geographers) exploding, sending millions to the New World and the colonies and still growing at home. But today Europe's indigenous population, unlike most of the rest of the world's, is actually shrinking. To keep a population from declining, the (statistically) average woman must bear 2.1 children. For Europe as a whole, that figure was 1.4 in 2006. Several large countries recorded 1.3, including

Germany and Italy. And no fewer than 4 Eastern European countries recorded 1.2—the lowest ever seen in any human population.

Such **negative population growth** poses serious challenges for any nation. When the population pyramid becomes top-heavy, the number of workers whose taxes pay for the social services of the aged goes down, leading to reduced pensions and dwindling funds for health care. Governments that impose tax increases endanger the business climate; their options are limited. Europe, and especially Western Europe, is experiencing a **population implosion** that will be a formidable challenge in decades to come.

Immigrant Infusions

Meanwhile, **immigration** is partially offsetting the losses European countries face. Millions of Turkish Kurds (mainly to Germany), Algerians (France), Moroccans (Spain), West Africans (Britain), and Indonesians (Netherlands) are changing the social fabric of what once were unicultural nation-states. One key dimension of this change is the spread of Islam in Europe (Fig. 1-10).

Booming *suburban downtowns* (see Chapter 3) are transforming the U.S. metropolitan landscape, and such "edge cities" are now multiplying in Western Europe as well. The biggest and most important of these complexes is *La Défense*, located just west of Paris, which has grown so robustly that it is now continental Europe's single largest business district, whose annual transactions exceed the GNI of the Netherlands! Besides its economic prowess, the cubic Grande Arche that forms the centerpiece of *La Défense* is an architectural jewel that is the western anchor of Paris's "Historic Axis" which includes the *Arc de Triomphe* (top center of photo), *Champs Elysées, Place de la Concorde,* and the *Louvre.* © Yann Arthus-Bertrand/Photo Researchers.

EUROPE: MUSLIMS AS PERCENTAGE OF NATIONAL POPULATION, 2008

- Less than 1%
- 1–4%
- 4–7%
- 7–12%
- 12–25%
- More than 25%

0   250   500 Kilometers
0   100   200   300 Miles

FIGURE 1-10   © H. J. de Blij, P. O. Muller, and John Wiley & Sons, Inc.

62

Muslim populations in Eastern Europe (such as Albania's, Kosovo's, and Bosnia's) are local, Slavic communities converted during the period of Ottoman rule. The Muslim sectors of Western European countries, on the other hand, represent more recent immigrations.

The vast majority of these immigrants are intensely devout, politically aware, and culturally insular. They continue to arrive in a Europe where native populations are stagnant or declining, where religious institutions are weakening, where secularism is rapidly rising, where political positions often appear to be anti-Islamic, and where cultural norms are incompatible with Muslim traditions. Many young men, unskilled and uncompetitive in a European Union (EU) where unemployment is already high, get involved in petty crime or the drug trade, are harassed by the police, and turn to their faith for solace and reassurance. Unlike most other immigrant groups, Muslim communities also tend to resist assimilation, making Islam the essence of their identity. In Britain alone there are more than 1500 mosques; and wherever they appear, local cultural landscapes are transformed (see photo this page).

The Growing Multicultural Challenge

In truth, European governments have not done enough to foster the very integration they see Muslims rejecting. The French assumed that their North African immigrants would aspire to assimilation as Muslim children went to public schools from their urban public-housing projects; instead they got into disputes over the dress codes of Muslim girls. The Germans for decades would not award German citizenship to children of immigrant parents born on German soil. The British found their modest efforts at integration stymied by the Muslim practice of importing brides from Islamic countries, which had the effect of perpetuating segregation. In many European cities, the housing projects that form the living space of Muslim residents (by now many of them born and retired there) are dreadful, barren, impoverished environments unseen by Europeans whose images of Paris, Manchester, or Frankfurt are far different.

The social and political implications of Europe's demographic transformation are numerous and far-reaching. Long known for tolerance and openness, European societies are attempting to restrict immigration in various ways; political parties with anti-immigrant platforms are gaining ground. Multiculturalism poses a growing challenge in a changing Europe.

FROM THE FIELD NOTES

"After taking the train from Schiphol Airport to Rotterdam and checking into my downtown hotel, I took a train from the 'Central Station' (which was in the middle of one huge construction site) to what the Dutch call 'Hoek van Holland' ('Corner of Holland') on the North Sea coast, where huge ferries disembark thousands of travelers from England and elsewhere daily. A colleague had advised me to get off at Maassluis and walk the streets for a sense of the history as well as the modern changes going on here. Certainly the growing Islamic presence is transforming cultural landscapes in Europe, but not just in the cities. Even in small towns like this, the minarets of mosques rise between the spires of churches and, as is the case here, the local mosque is the centerpiece for a multipurpose Muslim cultural center that serves as a social (and political) center in ways that churches do not. A couple of generations ago, the people living in those apartments might have expected to overlook a church, but today their vista is quite different." © H. J. de Blij.

www.conceptcaching.com

EUROPE'S MODERN TRANSFORMATION

At the end of World War II, much of Europe lay shattered, its cities and towns devastated, its infrastructure wrecked, its economies ravaged. The Soviet Union had taken control over the bulk of Eastern Europe, and communist parties seemed poised to dominate the political life of major Western European countries.

In 1947, U.S. Secretary of State George C. Marshall proposed a European Recovery Program designed to help counter all this dislocation and to create stable political conditions in which democracy would survive. Over the next four years, the United States provided about $13 billion in assistance to Europe (almost $100 billion in today's money). Because the Soviet Union refused U.S. aid and forced its Eastern European satellites to do the same, the Marshall Plan applied solely to 16 European countries, including defeated (West) Germany and Turkey.

European Unification

The Marshall Plan did far more than stimulate European economies. It confirmed European leaders' conclusion that their countries needed a joint economic-administrative structure not only to coordinate the financial assistance, but also to ease the flow of resources and products across Europe's mosaic of boundaries, to lower restrictive trade tariffs, and to seek ways to improve political cooperation.

Benelux Precedent

For all these needs Europe's governments had some guidelines. While in exile in Britain, the leaders of three small countries—Belgium, the Netherlands, and Luxembourg—had been discussing an association of this kind even before the end of the war. There, in 1944, they formulated and signed the Benelux Agreement, intended to achieve total economic integration. When the Marshall Plan was launched, the Benelux precedent helped speed the creation of the Organization for European Economic Cooperation (OEEC), which was established to coordinate the investment of America's aid (see box below, "Supranationalism in Europe").

Supranationalism

Soon the economic steps led to greater political cooperation as well. In 1949, the participating governments created the Council of Europe, the beginnings of what was

Supranationalism in Europe

1944	Benelux Agreement signed.	1985	Greenland, acting independently of Denmark, withdraws from EC.
1947	Marshall Plan created (effective 1948–1952).		
1948	Organization for European Economic Cooperation (OEEC) established.	1986	Spain and Portugal admitted as members of EC, creating "The Twelve." Single European Act ratified, targeting a functioning European Union in the 1990s.
1949	Council of Europe created.		
1951	European Coal and Steel Community (ECSC) Agreement signed (effective 1952).	1987	Turkey and Morocco make first application to join EC. Morocco is rejected; Turkey is told that discussions will continue.
1957	Treaty of Rome signed, establishing European Economic Community (EEC) (effective 1958), also known as the Common Market and "The Six." European Atomic Energy Community (EURATOM) Treaty signed (effective 1958).	1990	Charter of Paris signed by 34 members of the Conference on Security and Cooperation in Europe (CSCE). Former East Germany, as part of newly reunified Germany, incorporated into EC.
1959	European Free Trade Association (EFTA) Treaty signed (effective 1960).	1991	Maastricht meeting charts European Union (EU) course for the 1990s.
1961	United Kingdom, Ireland, Denmark, and Norway apply for EEC membership.	1993	Single European Market goes into effect. Modified European Union Treaty ratified, transforming EC into EU.
1963	France vetoes United Kingdom EEC membership; Ireland, Denmark, and Norway withdraw applications.	1995	Austria, Finland, and Sweden admitted into EU, creating "The Fifteen."
1965	EEC–ECSC–EURATOM Merger Treaty signed (effective 1967).	1999	European Monetary Union (EMU) goes into effect.
1967	European Community (EC) inaugurated.	2001	Denmark's voters reject EMU participation by 53 to 47 percent.
1968	All customs duties removed for intra-EC trade; common external tariff established.	2002	The euro is introduced as historic national currencies disappear in 12 countries.
1973	United Kingdom, Denmark, and Ireland admitted as members of EC, creating "The Nine." Norway rejects membership in the EC by referendum.	2003	First draft of a European constitution is published to mixed reviews from member-states.
1979	First general elections for a European Parliament held; new 410-member legislature meets in Strasbourg. European Monetary System established.	2004	Historic expansion of EU from 15 to 25 countries with the admission of Cyprus, the Czech Republic, Estonia, Hungary, Latvia, Lithuania, Malta, Poland, Slovakia, and Slovenia.
		2005	Proposed EU Constitution is disapproved by voters in France and the Netherlands.
1981	Greece admitted as member of EC, creating "The Ten."	2007	Romania and Bulgaria are admitted into EU, raising its membership to 27 countries.

to become a European Parliament meeting in Strasbourg, France. Europe was embarked on still another political revolution, the formation of a multinational union involving a growing number of European states. Geographers **19** define **supranationalism** as the voluntary association in economic, political, or cultural spheres of three or more independent states willing to yield some measure of sovereignty for their mutual benefit. In later chapters we will encounter other supranational organizations, including the North American Free Trade Agreement (NAFTA), but none has reached the plateau achieved by the European Union (EU).

Policies and Priorities

In Europe, the key initiatives arose from the Marshall Plan and lay in the economic arena; political integration came more haltingly. Under the Treaty of Rome, six countries joined to become the European Economic Community (EEC) in 1957, also called the "Common Market." In 1973 the United Kingdom, Ireland, and Denmark joined, and the renamed European Community (EC) had nine members. As Figure 1-11 shows, membership reached 15 countries in 1995, after the organization had been renamed yet one more time to become the European Union (EU).

The European Union is not just a paper organization for bankers and manufacturers. It has a major impact on the daily lives of its member countries' citizens in countless ways. (You will see one of these ways when you arrive at an EU airport and find that EU-passport holders move through inspection at a fast pace, while non-EU citizens wait in long immigration lines.) Taxes tend to be high in Europe, and those collected in the richer member-states are used to subsidize growth and development in the less prosperous ones. This is one of the burdens of membership that is not universally popular in the EU, to say the least. But it has strengthened the economies of Portugal, Greece, and other national and regional economies to the betterment of the entire organization. Some countries also

object to the terms and rules of the Common Agricultural Policy (CAP), which, according to critics, supports farmers far too much and, according to others, far too little. (France in particular obstructs efforts to move the CAP closer to consensus, subsidizing its agricultural industry relentlessly while arguing that this protects its rural cultural heritage as well as its farmers.)

New Money

Ever since the first steps toward European unification were taken, EU planners dreamed of a time when the EU would have a single currency not only to symbolize its strengthening unity but also to establish a joint counterweight to the mighty American dollar. A European Monetary Union (EMU) became a powerful objective of EU planners, but many observers thought it unlikely that member-states would in the end be willing to give up their historic marks, francs, guilders, liras, and escudos. Yet it happened. In 2002, twelve of the (then) 15 EU countries withdrew their currencies and began using the euro, with only the United Kingdom (Britain), Denmark, and Sweden staying out (Fig. 1-11). Add nonmember Norway to this group, and not a single Scandinavian country has converted to the euro. On the other hand, Slovenia adopted the euro in 2007, and Cyprus and Malta are slated to follow in 2008. The European Monetary Union marches onward, with or without Scandinavia.

FROM THE FIELD NOTES

"No place in Europe displays the flag of the European Union as liberally as does the Dutch city of Maastricht, where the European Union Treaty of 1991 was signed. The flag, seen here at City Hall, shows twelve yellow stars (representing the signatories to the Treaty) against a blue background. Although three additional states joined the EU in 1995 [and 12 more by 2007], the flag will remain as is, and will not, as the U.S. flag does, change to reflect changing times." © H. J. de Blij

Concept Caching

www.conceptcaching.com

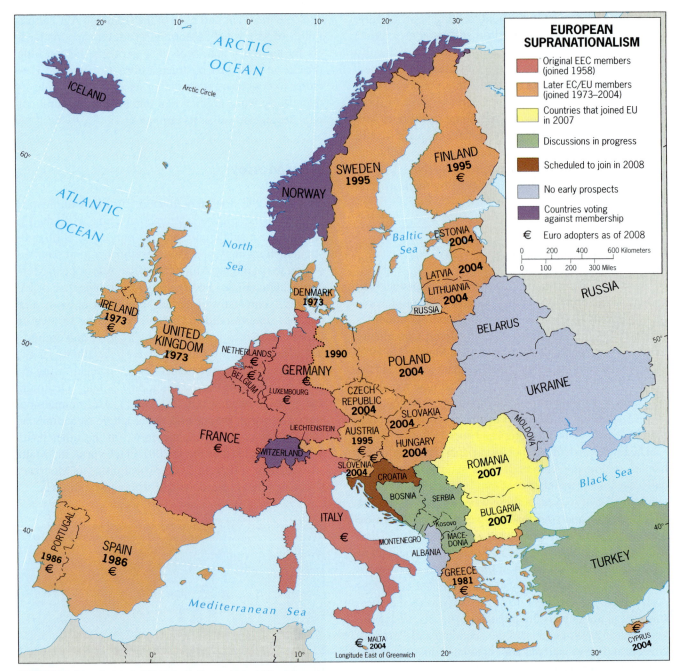

FIGURE 1-11 © H. J. de Blij, P. O. Muller, and John Wiley & Sons, Inc.

Momentous Expansion

Expansion has always been an EU objective, and the subject has always aroused passionate debate (see the Issue Box titled "Europe: How Desirable Is Economic and Political Union?"). Will the incorporation of weaker economies undermine the strength of the whole? Despite such misgivings, negotiations to expand the EU have long been in progress and still continue. In 2004 a momentous milestone was reached: ten new members were added, creating a greater European Union with 25 member-states. Geographically, these ten came in three groups: three

Baltic states (Estonia, Latvia, and Lithuania); five contiguous states in Eastern Europe extending from Poland and the Czech Republic through Slovakia, Hungary, and Slovenia; and two Mediterranean island-states, the ministate of Malta and the still-divided state of Cyprus. And in 2007, both Romania and Bulgaria were added, raising the number of members to 27 and extending the EU to the shores of the Black Sea (Fig. 1–11).

Numerous structural implications arise from this, affecting all EU countries. A common agricultural policy is now even more difficult to achieve, given the poor condition of farming in most of the new members. Also,

Europe: How Desirable Is Economic and Political Union?

IN SUPPORT OF UNION

"As a public servant in the French government I have been personally affected by the events of the past half-century. I am in favor of European unification because Europe is a realm of comparatively small countries, all of which will benefit from economic and political union. Europe's history is bedeviled by its fragmentation—its borders and barriers and favored cores and disadvantaged peripheries, its different economies and diverse currencies, its numerous languages and varied cultures. Now we have the opportunity and determination to overcome these divisions and to create an entity whose sum is greater than its parts, a supranational Europe that will benefit all of its members.

"In the process we will not only establish a common currency, eliminate trade tariffs and customs barriers among our members, and enable citizens of one member-state to live and work in any of the others. We will also make European laws that will supersede the old "national" laws and create regulations that will be adhered to by all members on matters ranging from subsidies to help small farms survive to rules regarding environmental protection. In order to assist the poorer members of this European Union, the richer members will send billions of euros to the needy ones, paid for from the taxes levied on their citizens. This has already lifted Spain and Ireland from poverty to prosperity. But to ensure that "national" economies do not get out of line, there will be rules to govern their fiscal policies, including a debt limit of 3 percent of GDP.

"The EU will require sacrifices from people and governments, but the rewards will make these worth it. Our European Parliament, with representatives from all member countries, is the harbinger of a truly united Europe. The European Commission deals with the practical problems as we move toward unification, including the momentous EU enlargement of 2004. A half-century after the first significant steps were taken, the EU has 27 member-states including former wartime enemies, reformed communist economies, and newly independent entities. Rather than struggling alone, these member-states will form part of an increasingly prosperous and economically powerful whole.

"I hope that administrative coordination and economic union will be followed by political unification. The ultimate goal of this great movement should be the creation of a federal United States of Europe, a worthy competitor and countervailing force in a world dominated by another federation, the United States of America."

IN OPPOSITION TO THE EUROPEAN UNION

"I am a taxi driver in Vienna, Austria, and here's something to consider: the bureaucrats, not the workers, are the great proponents of this European Union plan. Let me ask you this: how many member-countries' voters have had the chance to tell their governments whether they want to join the EU and submit to all those confounded regulations? It all happened before we realized the implications. At least the Norwegians had common sense: they stayed out of it, because they didn't want some international body to tell them what to do with their oil and gas and their fishing industry. But I can tell you this: if the workers of Europe had had the chance to vote on each step in this bureaucratic plot, we'd have ended it long ago.

"So what are we getting for it? Higher prices for everything, higher taxes to pay for those poor people in southern Italy and needy Portugal, costly environmental regulations, and—here's the worst of it—a flood of cheap labor and an uncontrolled influx of immigrants. Notice that this recent 2004 expansion wasn't voted on by the Germans (they're too small to matter much, but Germany is the largest EU member), and do you know why? Because the German government knew well and good that the people wouldn't support this ridiculous scheme. This whole European Union program is concocted by the elite, and we have to accept it whether we like it or not.

"Does make me wonder, though. Do these EU-planners have any notion of the real Europe they're dealing with? They talk about how those "young democracies" joining in 2004 will have their democratic systems and their market reforms locked into place by being part of the Union, but I have friends who try to do business in Poland and Slovakia, and they tell me that there may be democracy, but all they experience is corruption, from top to bottom. I'll bet that those subsidies we'll be paying for to help these "democracies" get on their feet will wind up in Swiss bank accounts. By the way, the Swiss, you'll notice, aren't EU members either. Wonder why?

"A United States of Europe? Not as far as I am concerned. I am Austrian, and I am and will always be a nationalist as far as my country is concerned. I speak German, but already English is becoming the major EU language, though the Brits were smart enough not to join the euro. I'm afraid that these European Commission bureaucrats have their heads in the clouds, which is why they can't see the real Europe they're putting in a straitjacket."

Regional ISSUE

Vote your opinion at www.wiley.com/college/deblij

some of the former EU's poorer states, which were on the receiving end of the subsidy program that aided their development, now will have to pay up to support the much poorer new eastern members. And disputes over representation at EU's Brussels headquarters arose quickly. Even before the newest members' accession, Poland was demanding that the representative system favor medium-sized members (such as Poland and Spain) over larger ones such as Germany and France.

The Remaining Outsiders

This momentous expansion is having major geographic consequences for all of Europe, and not just the EU. As Figure 1-11 shows, following the admission of Romania and Bulgaria in 2007, two groups of countries and peoples are still left outside the Union:

1. *States emerging from the former Yugoslavia plus Albania.* In this cluster of western Balkan states, most of them troubled politically and economically, only Slovenia has achieved full membership. The remainder rank among Europe's poorest and ethnically fractured states, but they contain nearly 25 million people whose circumstances could worsen outside "the club." Better, EU leaders reason, to move them toward membership by demanding political, social, and economic reforms, although the process will take many years. Discussions with Croatia began in 2005 and with newly independent Montenegro in 2006, but the aftermath of war is holding up Serbia's invitation. In 2007, Macedonia was arguing that if Bulgaria can be admitted, then Macedonia's case has merit as well. Ethnically and politically divided Bosnia seemed far from ready for negotiations, Albania was even less prepared, and the Kosovo issue (see p. 104) remained unresolved. Note that this area of least-prepared countries also is the most Islamic corner of Europe.

2. *Ukraine and its neighbors.* Four former Soviet republics in Europe's "far east" may some day join the EU, although there is no interest whatsoever today in one of them, Belarus. The government of Ukraine, however, has shown interest, although its electorate is strongly (and regionally) divided between a pro-EU west and a pro-Russian east. As we shall see later, Moldova views the EU as a potential supporter in a struggle against devolution along its eastern flank, and in 2006 a country not even on the map—Georgia, across the Black Sea from Ukraine—proclaimed its interest in EU membership for similar reasons: to find an ally against Russian meddling in its internal affairs.

An Islamic EU Member?

We should take note of still another potential candidate for EU membership: Turkey. EU leaders would like to include a mainly Muslim country in what Islamic states sometimes call the "Christian Club," but Turkey needs progress on social standards, human rights, and economic policies before its accession can be contemplated. Still, the fact that the EU has now reached deeply into Eastern Europe, encompasses 27 members, has a common currency and a parliament, is developing a constitution, and is even considering expansion beyond the realm's borders constitutes a tremendous and historic achievement in this fragmented, fractious part of the world. The EU now has a combined population of nearly 500 million constituting one of the world's richest markets; its member-states account for more than 40 percent of the world's exports.

It is remarkable that all this has been accomplished in little more than half a century. Some of the EU's leaders want more than economic union: they envisage a United States of Europe, a political as well as an economic competitor for the United States. To others, such a "federalist" notion is an abomination not even to be mentioned (the British in general are especially wary of such an idea). Whatever happens, Europe is going through still another of its revolutionary changes, and when you study its evolving map you are looking at history in the making.

Centrifugal Forces

For all its dramatic progress toward unification, Europe remains a realm of geographic contradictions. Europeans are well aware of their history of conflict, division, and repeated self-destruction. Will supranationalism finally overcome the centrifugal forces that have so long and so frequently afflicted this part of the world?

Devolutionary Pressures

Even as Europe's states have been working to join forces in the EU, many of those same states are confronting severe centrifugal stresses. The term **devolution** has come **20** into use to describe the powerful centrifugal forces whereby regions or peoples within a state, through negotiation or active rebellion, demand and gain political strength and sometimes autonomy at the expense of the center. Most states exhibit some level of internal regionalism, but the process of devolution is set into motion when a key centripetal binding force—the nationally accepted idea of what a country stands for—erodes to the point that a regional drive for autonomy, or for outright secession, is launched.

The Case of the United Kingdom

As Figure 1-12 shows, numerous European countries are affected by devolution. States large and small, young and old, EU members, and non-EU members, must deal with the problem. Even the long-stable United Kingdom is affected. England, the historic core area of the British Isles, dominates the UK in terms of population as well as political and economic power. The country's three other entities—Scotland, Wales, and Northern Ireland—were acquired over several centuries and attached to England (hence the "United" Kingdom). But neither time nor representative democratic government was enough to eliminate all latent regionalism in these three components of the UK. During the 1960s and 1970s, the British government confronted a virtual civil war in Northern Ireland and rising tides of nationalism in Scotland and Wales. In 1997, the government in London gave the Scots and Welsh the opportunity to vote for greater autonomy in new regional parliaments that would have limited but significant powers over local affairs. The Scots voted overwhelmingly in favor, and the Welsh by a slim majority—and thus a major devolutionary step was taken in one of Europe's oldest, most durable, and most unified states.

FIGURE 1-12   © H. J. de Blij, P. O. Muller, and John Wiley & Sons, Inc.

Devolution Elsewhere

Even while this devolutionary process was ongoing, the British government still joined the European Union on behalf not only of the English, but also of the Scots, Welsh, and Northern Irelanders seeking a new relationship with London. As the map shows, the UK is not alone in such contradiction. Spain faces severe devolutionary forces in its Basque Region and lesser ones in Catalonia and Galicia; France contends with a secessionist movement on its island of Corsica; Belgium is riven by Flemish-Walloon separatism; Italy confronts devolutionary pressures in South Tyrol and Lombardy. In recent decades Eastern Europe has been a cauldron of devolution as Yugoslavia and Czechoslovakia collapsed, Moldova fragmented, and Montenegro seceded from Serbia-Montenegro in 2006.

The EU's New Economic Geography

Earlier in this chapter, we used as our frame of reference the old designations of Western and Eastern, Northern and Southern Europe. Certainly those regional names still have some validity: Western Europe remains the core of the EU; the cultural landscapes of Northern and Southern Europe, though converging, still differ markedly; Eastern Europe still is, on average, the poorer, lagging part of the realm. But the birth and growth of the European Union have generated a new economic landscape that transcends the old.

Engines of Growth

By investing heavily in regional infrastructure and by smoothing the flows of money, labor, and products, European planners have dramatically reduced the divisive effects of their national boundaries. And by acknowledging demands for greater freedom of action by their provinces, States, departments, and other administrative units of their countries, European leaders unleashed a wave of economic energy that transformed some of these units into powerful engines of growth. In the next section of this chapter, when we look at regions and states, we will have to pay attention to these emerging powerhouses, which tend to be centered on major cities.

Four of these growth centers are especially noteworthy, **21** to the point that geographers refer to them as the **Four Motors of Europe**: (1) France's ***Rhône-Alpes Region***, centered on the country's second-largest city, Lyon; (2) ***Lombardy*** in northern Italy, focused on the industrial city of Milan; (3) ***Catalonia*** in northeastern Spain, anchored by the cultural and manufacturing center of Barcelona; and (4) ***Baden-Württemberg*** in Germany, headquartered at the high-tech city of Stuttgart. Often, the local governments in these subregions simply bypass the governments in their national capitals, dealing not only with each other but even with foreign governments as their business networks span the globe. In this they are imitated by other provinces, all seeking to foster their local economies and, in the process, strengthen their political position relative to the state.

Regional States

The lowering of political barriers to trade, and the resulting increase in regional transferability, is having a further effect on Europe's economic map: States or provinces on opposite sides of political boundaries can now cooperate to achieve common economic goals. Already, the hinterlands of the cities anchoring the Four Motors extend into neighboring countries. The Japanese economist Kenichi Ohmae termed these cross-border growth engines **regional states**, **22** new phenomena on the old map of Europe. Elsewhere, notably along the borders between the 15 pre-2004 EU member-states and the newly admitted 12, cross-border investment is creating start-up regional states that will further alter the economic landscape.

All these developments underscore how far-reaching and irreversible Europe's current transformation is. European leaders understood the need for it; the Marshall Plan jump-started it; good economic times sustained it; and the end of the Cold War galvanized it. But Europe's ultimate supranational goals have not yet been attained. The problems of the old Eastern Europe remain evident in Table G-1. The sense of have and have-not, core and periphery, still pervades EU negotiations. Many Europeans, perhaps a majority, feel that the EU project is a bureaucratic edifice constructed by leaders who are out of touch with grassroots Europe. Europe's quest for enduring unity continues.

POINTS TO PONDER

- New states are still forming in Europe, where the modern state originated; Kosovo may be next, but cultural passions may endanger the project.

- A resurgence of Scotland's independence movement may imperil the unity of the "United" Kingdom.

- Public support in Islamic Turkey in favor of joining the (mostly Christian) European Union has dropped from more than 70 to less than 30 percent.

- Members of an Iranian parliamentary delegation visiting Belgium in 2006 refused to shake the hand of the female president of the Senate, then declined to attend a welcome lunch because alcohol was served.

Regions of the Realm

Our objective in this and later regional discussions is to become familiar with the regional and national frameworks of the realms under investigation, and to discover how individual states fit into and function within the larger mosaic. Even (perhaps especially) in this era of globalization, and even in integrating Europe, the state still plays a key role in the process. Where these states are located, how they interact, and what their prospects may be in this changing world are among the key questions we will address.

Europe presents us with a particular challenge because of its numerous countries and territories (the latter including a territory in limbo, Kosovo, plus a set of *microstates* such as Monaco, San Marino, and Andorra that do not qualify as full-fledged countries). So it is not surprising that the old geographic division into five regions remains helpful. In Figure 1-13, the colors represent (1) Western Europe, (2) the British Isles, (3) Northern (Nordic) Europe, (4) Mediterranean Europe, and (5) Eastern Europe. This regionalization scheme employs the formal-region concept, but we can also approach Europe in other ways. Focusing on Europe's supranational and subnational development in a functional-region context, we recognize a European core (delimited by the red line in Fig. 1-13) and a highly varied periphery beyond it. We employ a combination of perspectives in the discussion that follows.

WESTERN EUROPE

Is it still appropriate to view "Western" Europe as a regional concept? Consider this: the three Benelux countries started the ball rolling toward postwar European integration; France and Germany have been the main proponents and remain the most influential forces in the European Union; and the ideals of the EU are nowhere closer to achievement than in Western Europe. For example, the Schengen Agreement, which makes border-crossings even easier than EU regulations stipulate, involves primarily Western European countries. As Figure 1-13 suggests, Western Europe is the very heart of Europe's coreland.

Counting the microstate of Liechtenstein, there are eight states in Western Europe: dominant Germany and France, the three Benelux countries, and three landlocked mountain states (Fig. 1-14). France is territorially the region's largest state, but Germany is Europe's most populous. And no European country has more neighbors than Germany. So let us focus first on this, Europe's largest and most productive economy.

Reunited Germany

Twice during the twentieth century Germany plunged Europe and the world into war, until, in 1945, the defeated and devastated German state was divided into two parts, West and East (see the delimitation in red on Fig. 1-15). Its eastern boundaries were also changed, leaving the industrial district of Silesia in newly defined Poland, that of Saxony in communist-ruled and Soviet-controlled East Germany, and the Ruhr in West Germany (see Fig. 1-14). Aware that these were the industrial centers that had enabled Nazi Germany to seek world domination through war, the victorious allies laid out this new boundary framework to make sure this would not happen again.

In the aftermath of World War II, Soviet and Allied administration of East and West Germany differed. Soviet rule in East Germany was established on the Russian-communist model and, given the extreme hardships the USSR had suffered at German hands during the war, harshly punitive. The American-led authority in West Germany was less strict and aimed more at rehabilitation. When the Marshall Plan was instituted, West Germany was included, and its economy recovered rapidly. Meanwhile, West Germany was reorganized politically into a modern federal state on democratic foundations.

MAJOR CITIES OF THE REALM

City	Population* (in millions)
Amsterdam, Netherlands	1.2
Athens, Greece	3.2
Barcelona, Spain	5.0
Berlin, Germany	3.4
Brussels, Belgium	1.0
Frankfurt, Germany	3.8
London, UK	8.6
Lyon, France	1.4
Madrid, Spain	5.9
Milan, Italy	4.1
Paris, France	9.8
Prague, Czech Republic	1.2
Rome, Italy	3.3
Stuttgart, Germany	2.7
Vienna, Austria	2.3
Warsaw, Poland	1.7

*Based on 2008 estimates.

FIGURE 1-13   © H. J. de Blij, P. O. Muller, and John Wiley & Sons, Inc.

Revival

West Germany's economy thrived. Between 1949 and 1964 its GNI tripled while industrial output rose 60 percent. It absorbed millions of German-speaking refugees from Eastern Europe (and many escapees from communist East Germany as well). Since unemployment was virtually nonexistent, hundreds of thousands of Turkish Kurds and other foreign workers arrived to take jobs Germans could not fill or did not want.

Simultaneously, West Germany's political leaders participated enthusiastically in the OEEC and in the negotiations that led to the six-member Common Market. Geog-

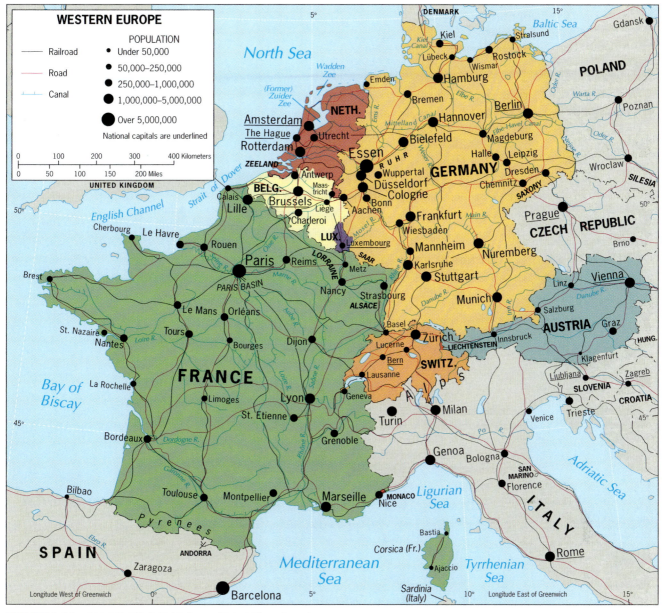

FIGURE 1-14 © H. J. de Blij, P. O. Muller, and John Wiley & Sons, Inc.

raphy worked in West Germany's favor: it had common borders with all but one of the EEC member-states. Its transport infrastructure, rapidly rebuilt, was second to none in the realm. More than compensating for its loss of Saxony and Silesia were the expanding Ruhr (in the hinterland of the Dutch port of Rotterdam) and the newly emerging industrial complexes centered on Hamburg in the north, Frankfurt (the leading financial hub as well) in the center, and Stuttgart in the south. West Germany exported huge quantities of iron, steel, motor vehicles, machinery, textiles, and farm products.

Setback and Shock

No economy grows without setbacks and slowdowns, however, and Germany experienced such problems in the 1970s and 1980s, when energy shortages, declining competitiveness on world markets, lagging modernization (notably in the aging Ruhr), and social dilemmas involving rising unemployment, an aging population, high taxation, and a backlash against foreign resident workers roiled West German society. And then, quite suddenly, the collapse of the communist Soviet Union opened the door to reunification with East Germany.

STATES (*LÄNDER*) OF REUNIFIED GERMANY

GDP PER CAPITA, IN EUROS, 2002

▮ (red)	Over 26,000
▮ (orange)	20,000–26,000
▮ (yellow)	Below 20,000
——	Railroads
——	Roads

City population
- • Under 50,000
- • 50,000–250,000
- ● 250,000–1,000,000
- ● 1,000,000–5,000,000
- ● Over 5,000,000

0   50   100 Kilometers
0   25   50 Miles
National capitals are underlined

FIGURE 1-15   © H. J. de Blij, P. O. Muller, and John Wiley & Sons, Inc.

Germany Restored

In 1990, West Germany had a population of about 62 million and East Germany 17 million. Communist misrule in the East had yielded outdated factories, crumbling infrastructures, polluted environments, drab cities, inefficient farming, and inadequate legal and other institutions. Reunification was more a rescue than a merger, and the cost to West Germany was enormous. When the West German government imposed sales-tax increases and an income-tax surcharge on its citizens, many Westerners doubted the wisdom of reunification. It was pro-

jected that it would take decades to reconstruct Virginia-sized East Germany: ten years later, exports from the former East still contributed only about 7 percent of the national total. Regional disparity would afflict Germany for a very long time to come.

The Federal Republic

Before reunification, West Germany functioned as a federal state consisting of ten States or *Länder* (Fig. 1-15). East Germany had been divided under communist rule into

15 districts including East Berlin. Upon reunification, East Germany was reorganized into six new States based on traditional provinces within its borders. Figure 1-15 makes a key point: regional disparity in terms of income per person remains a serious problem between the former East and West. Note that five of former East Germany's six States (Berlin being the sole exception) are in the lowest income category, while most of the ten former West German States rank in the two higher income categories. In the first decade of this century, Germany's economy was stagnant, raising unemployment and slowing former East Germany's recovery. Nonetheless, the gap continues to narrow, and with 82.1 million inhabitants including over 7 million foreigners and more than 4 million ethnic Germans born outside the country, Germany is again exerting its dominance over a mainland Europe in which it has no peer.

France

German dominance in the European Union is a constant concern in the other leading Western European country. The French and the Germans have been rivals in Europe for centuries. France (population: 61.7 million) is an old state, by most measures the oldest in Western Europe. Germany is a young country, created in 1871 after a loose association of German-speaking states had fought a successful war against . . . the French.

AMONG THE REALM'S GREAT CITIES . . . Paris

IF THE GREATNESS of a city were to be measured solely by its number of inhabitants, Paris (9.8 million) would not even rank in the world's top 20. But if greatness is measured by a city's historic heritage, cultural content, and international influence, Paris has no peer. Old Paris, near the Île de la Cité that housed the original village where Paris began and carries the eight-century-old Notre Dame Cathedral, contains an unparalleled assemblage of architectural and artistic landmarks old and new. The Arc de Triomphe, erected by Napoleon in 1806 (though not completed until 1836), commemorates the emperor's victories and stands as a monument to French neoclassical architecture, overlooking one of the world's most famous streets, the Champs Elysées, which leads to the grandest of city squares, the Place de la Concorde, and on to the magnificent palace-turned-museum, the Louvre.

Even the Eiffel Tower, built for the 1889 International Exposition over the objections of Parisians who regarded it as ugly and unsafe, became a treasure. From its beautiful Seine River bridges to its palaces and parks, Paris embodies French culture and tradition. It is perhaps the ultimate primate city in the world.

As the capital of a globe-girdling empire, Paris was the hearth from which radiated the cultural forces of Francophone assimilation, transforming much of North, West, and Equatorial Africa, Madagascar, Indochina, and many smaller colonies into societies on the French model. Distant cities such as Dakar, Abidjan, Brazzaville, and Saigon acquired a Parisian atmosphere. France, meanwhile, spent heavily to keep Paris, especially Old Paris, well maintained—not just as a relic of history, but as a functioning, vibrant center, an example to which other cities can aspire.

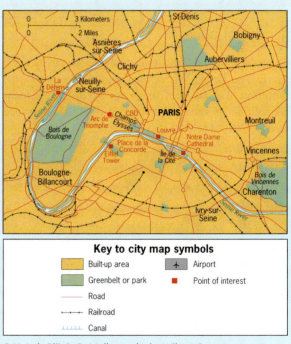

© H. J. de Blij, P. O. Muller, and John Wiley & Sons, Inc.

Today, Old Paris is ringed by a new and different Paris. Stand on top of the Arc de Triomphe and turn around from the Champs Elysées, and the tree-lined avenue gives way to La Défense, an ultramodern high-rise complex that is one of Europe's leading business districts (see photo, p. 61). But from atop the Eiffel Tower you can see as far as 80 kilometers (50 mi) and discern a Paris visitors rarely experience: grimy, aging industrial quarters, and poor, crowded neighborhoods where discontent and unemployment fester—and where Muslim immigrants cluster in a world apart from the splendor of the old city.

Territorially, France is much larger than Germany, and the map suggests that France has a superior relative location, with coastlines on the Mediterranean Sea, the Atlantic Ocean, and, at Calais, even a window on the North Sea. But France does not have any good natural harbors, and oceangoing ships cannot navigate its rivers and other waterways far inland. France has no equivalent to Rotterdam either internally or externally.

The map of Western Europe (Fig. 1-14) reveals a significant demographic contrast between France and Germany. France has one dominant city, Paris, at the heart of the Paris Basin, France's core area. No other city in France comes close to Paris in terms of population or centrality: Paris has 9.8 million residents, whereas its closest rival, Lyon, has only 1.4 million. Germany has no city to match Paris, but it does have a number of cities with populations between 1 and 5 million. And as Table G-1 shows, Germany is much more highly urbanized overall than France.

Paris: Site and Situation

Why should Paris, without major raw materials nearby, have grown so large? Whenever geographers investigate the evolution of a city, they focus on two important loca-

23 tional qualities: its **site** (the physical attributes of the place

it occupies) and its **situation** (its location relative to surrounding areas of productive capacity, other cities and towns, barriers to access and movement, and other aspects of the greater regional framework in which it lies). **24**

The site of the original settlement at Paris lay on an island in the Seine River, a defensible place where the river was often crossed. This island, the *Île de la Cité*, was a Roman outpost 2000 years ago; for centuries its security ensured continuity. Eventually the island became overcrowded, and the city expanded along the banks of the river (Fig. 1-16A).

Soon the settlement's advantageous situation stimulated its growth and prosperity. Its fertile agricultural hinterland thrived, and, as an enlarging market, Paris's focality increased steadily. The Seine River is joined near Paris by several navigable tributaries (the Oise, Marne, and Yonne). When canals extended these waterways even farther, Paris was linked to the Loire Valley, the Rhône-Saône Basin, Lorraine (an industrial area in the northeast), and the northern border with Belgium. When Napoleon reorganized France and built a radial system of roads—followed later by railroads—that focused on Paris from all parts of the country, the city's primacy was assured (Fig. 1-16B). The only disadvantage in Paris's situation lies in its seaward access: oceangoing ships can sail up the Seine River only as far as Rouen.

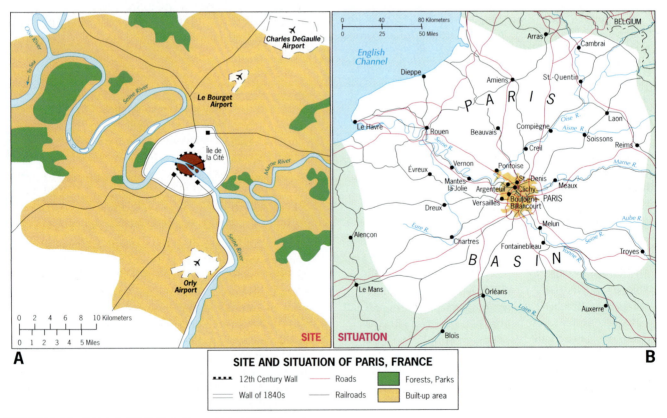

SITE AND SITUATION OF PARIS, FRANCE

- ◼◼◼◼ 12th Century Wall
- ══ Wall of 1840s
- —— Roads
- —— Railroads
- ▮ Forests, Parks
- ▮ Built-up area

FIGURE 1-16 © H. J. de Blij, P. O. Muller, and John Wiley & Sons, Inc.

Modern France

Paris, in accordance with Weber's agglomeration principle, grew into one of Europe's greatest cities. French industrial development was less spectacular, but northern French agriculture remained Europe's most productive and varied, exploiting the country's wide range of soils and climates and enjoying state subsidies and protections. Today France's economic geography is marked by new high-tech industries. It is a leading producer of high-speed trains, aircraft, fiber-optic communications systems, and space-related technologies. It also is the world leader in nuclear power, which currently supplies more than 75 percent of its electricity and thereby reduces its dependence on foreign oil imports.

Napoleon's Legacy

When Napoleon reorganized France in the early 1800s, he broke up the country's large traditional subregions and established more than 80 small departments (additions and subdivisions later increased this number to 96). Each *département* had representation in Paris, but the power was concentrated in the capital, not in the individual départements. France became a highly centralized state and remained so for nearly two centuries (see inset map, Fig. 1-17). Only the island département of Corsica produced a rebel movement, whose violent opposition to French rule continued for decades and even touched the mainland. In 2003 the voters in Corsica rejected an offer of special status for their island, including limited autonomy. They wanted more, and trouble lies ahead.

Decentralizing the State

Today, France is decentralizing. A new subnational framework of 22 historically significant provinces, groupings of départements called *regions* (Fig. 1-17), has been established to accommodate the devolutionary forces felt throughout Europe and, indeed, throughout the world. These regions, though still represented in the Paris government, have substantial autonomy in such areas as taxation, borrowing, and development spending. The cities that anchor them benefit because they are the seats of governing regional councils that can attract investment, not only within France but also from abroad.

Lyon, France's second city and headquarters of the region named Rhône-Alpes, has become a focus for growth industries and multinational firms. This region is evolving into a self-standing economic powerhouse that is becoming a driving force in the European economy; indeed, it is one of the Four Motors of Europe (see p. 70) with its own international business connections to countries as far away as China and Chile.

France has one of the world's most productive and diversified economies, based in one of humanity's richest cultures and vigorously promoted and protected (notably its heavily subsidized agricultural sector). And although France and Germany agree on many aspects of EU integration, they tend to differ on important issues. Old, historically centralized France is less eager than young, federal Germany to push political integration in supranational Europe.

Benelux

Three small countries are crowded into the northwest corner of Western Europe—the Netherlands, Belgium, and Luxembourg—and are collectively referred to by their first syllables (**Be-Ne-Lux**). Their total population, just under 28 million, reminds us that this is one of the most densely peopled corners of our planet. They are also a highly productive trio: both the Netherlands and Belgium rank among the top 20 economies of the world, and tiny Luxembourg has the world's highest per-capita GNI (see Table G-1). The seafaring Dutch had a thriving agricultural economy and a rich colonial empire; the Belgians forged ahead during the Industrial Revolution; and Luxembourg came into its own after the Second World War as a financial center for the evolving European Union and, indeed, the world.

The Netherlands, one of Europe's oldest democracies and a constitutional monarchy today, has for centuries been expanding its living space—not by warring with its neighbors but by wresting land from the sea. Its greatest project so far, the draining and reclaiming of almost the entire Zuider Zee (Southern Sea), began in 1932 and continues. In the southwestern province of Zeeland, islands are being connected by dikes and the water is being pumped out, adding still more *polders* (as the Dutch call inhabited land claimed from the sea lying behind dikes and below sea level) to the national territory. Among the technologies in which the Dutch lead the world, not surprisingly, is the engineering of coastal systems to control the sea and protect against storms.

The regional geography of this highly urbanized country (16.5 million) is noted for the *Randstad*, a roughly triangular urban core area anchored by Amsterdam, the constitutional capital, Rotterdam, Europe's largest port, and The Hague, the seat of government. This **conurbation**, as geographers call large urban areas when two or more cities merge spatially, now forms a ring-shaped complex that surrounds a still-rural center (in Dutch, *rand* means edge or margin; *stad* means city).

The economic geography of The Netherlands, like that of other Western European states, is heavily dominated by

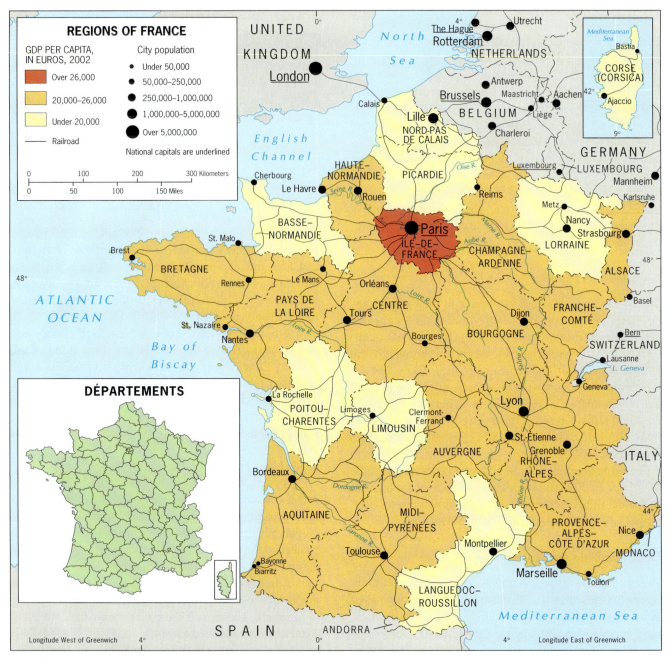

FIGURE 1-17 © H. J. de Blij, P. O. Muller, and John Wiley & Sons, Inc.

services, finance, and trade; manufacturing contributes about 14 percent of the value of the GDP annually and farming, once a mainstay, less than 3 percent. Amsterdam's airport, Schiphol, is regularly recognized as Europe's best; Rotterdam, one of the world's busiest ports, is also rated as the most efficient.

Belgium also has a thriving economy and a major port, Antwerp. But Belgium is a much younger state than The Netherlands, becoming independent within its present borders as recently as 1830. With 10.5 million people, Belgium's regional geography is dominated by a cultural fault line that cuts diagonally across the country, separating a Flemish-speaking majority (58 percent) centered on Flanders in the northwest from a French-speaking minority in southeastern Wallonia (31 percent). Brussels, the mainly French-speaking capital, lies like a cultural island in the Flemish-speaking sector, but Brussels also is one of Belgium's greatest assets: this is the headquarters, in some ways the functional capital, of the European Union. Still, devolution is a looming problem for Belgium, with political parties espousing Flemish separatism roiling the social landscape.

Luxembourg, one of Europe's many ministates, lies landlocked between Germany, Belgium, and France, with a Grand Duke as head of state (its official name is the Grand Duchy of Luxembourg), a territory of only 2600 square kilometers (1000 sq mi), and a population of just half a million. Luxembourg has translated sovereignty, relative location, and stability into a haven for financial, service, and information-technology industries. In 2007 there were more than 160 banks in this tiny country and nearly 14,000 holding companies (corporations that hold controlling stock in other businesses in Europe and worldwide). With its unmatched per-capita income (U.S. $64,000 in 2006) Luxembourg is in some ways the greatest beneficiary of the advent of the European Union. And in no country in Europe is support for the EU stronger than it is here.

The Alpine States

Switzerland, Austria, and the *microstate* of Liechtenstein on their border share an absence of coasts and the mountainous topography of the Alps—and little else (Fig. 1-14). Austria speaks one language; the Swiss speak German in their north, French in the west, Italian in the southeast, and even a bit of Rhaeto-Romansch in the remote central highlands (Fig. 1-8). Austria has a large primate city; multicultural Switzerland does not. Austria has a substantial range of domestic raw materials; Switzerland does not. Austria is twice the size of Switzerland and has a larger population, but far more trade crosses the Swiss Alps between Western and Mediterranean Europe than crosses Austria.

Switzerland, not Austria, is in most ways the leading state in Alpine Western Europe (Table G-1). Mountainous terrain and **landlocked location** can constitute crucial barriers to economic development, tending to inhibit the dissemination of ideas and innovations, obstruct circulation, constrain farming, and divide cultures. That is why Switzerland is such an important lesson in human geography. Through the skillful maximization of their opportunities (including the transfer needs of their neighbors), the Swiss have transformed their seemingly restrictive environment into a prosperous state. They used the waters cascading from their mountains to generate hydroelectric power to develop highly specialized industries. Swiss farmers perfected ways to optimize the productivity of mountain pastures and valley soils. Swiss leaders converted their country's isolation into stability, security, and neutrality, making it a world banking giant, a global magnet for money. Zürich, in the German sector, is the financial center; Geneva, in the French sector, is one of the world's most international cities. The Swiss

feel that they do not need to join the EU, and they have not done so.

Austria, which joined the EU in 1995, is a remnant of the Austro-Hungarian Empire and has a historical geography that is far more reminiscent of unstable Eastern Europe than that of Switzerland. Even Austria's physical geography seems to demand that the country look eastward: it is at its widest, lowest, and most productive in the east, where the Danube links it to Hungary, its old ally in the anti-Muslim wars of the past.

Vienna, by far the Alpine subregion's largest city, also lies on the country's eastern perimeter. One of the world's most expressive primate cities with magnificent architecture and monumental art, Vienna today is Western Europe's easternmost city, but Vienna's relative location changed dramatically with EU enlargement in 2004 and 2007. Peripheral to the EU and a vanguard of the West until 2004, Vienna found itself in a far more central position when the EU border moved eastward. But many Austrians had their doubts about their Eastern European neighbors becoming EU members of potentially equal standing, and in Austria public support for the EU fell to the lowest level of any member-state.

● THE BRITISH ISLES

Off the coast of mainland Western Europe lie two major islands, surrounded by a constellation of tiny ones, that constitute the British Isles, a discrete region of the European realm (Fig. 1-18). The larger of the two major islands, which also lies nearest to the mainland (a mere 34 kilometers, or 21 miles, at the closest point), is the island called *Britain*; its smaller neighbor to the west is *Ireland*.

States and Peoples

The names attached to these islands and the countries they encompass are the source of some confusion. They still are called the British Isles, even though British dominance over most of Ireland ended in 1921. The nation-state that occupies Britain and the northeastern corner of Ireland is officially called the United Kingdom of Great Britain and Northern Ireland—United Kingdom for short and UK by abbreviation. But this country often is referred to simply as Britain, and its people are known as the British. The nation-state of Ireland officially is the Republic of Ireland (*Eire* in Irish Gaelic), but it does not include the whole island of Ireland.

How convenient it would be if physical and political geography coincided! Unfortunately, the two do not.

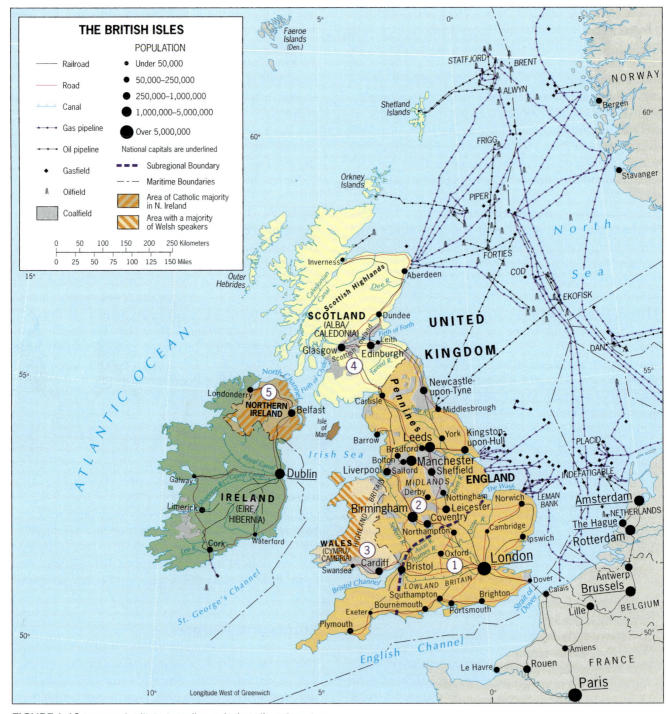

FIGURE 1-18 © H. J. de Blij, P. O. Muller, and John Wiley & Sons, Inc.

During the long British occupation of Ireland, which is overwhelmingly Catholic, many Protestants from northern Britain settled in northeastern Ireland. In 1921, when British domination ended, the Irish were set free—except in that corner in the north, where London kept control to protect the area's Protestant settlers. That is why the country to this day is officially known as the United Kingdom of Great Britain and Northern Ireland.

Roots of Devolution

Northern Ireland (Fig. 1-18) was home not only to Protestants from Britain, but also to a substantial population of Irish Catholics who found themselves on the wrong side of the border when Ireland was liberated. Ever since, conflict has intermittently engulfed Northern Ireland and spilled over into Britain and even, in the form of terrorism, into Western Europe. It has been and remains one of Europe's costliest struggles.

Although all of Britain lies in the United Kingdom, political divisions exist here as well. England is the largest of these units, the center of power from which the rest of the region was originally brought under unified control. The English conquered Wales in the Middle Ages, and Scotland's link to England, cemented when a Scottish king ascended the English throne in 1603, was ratified by the Act of Union of 1707. Thus England, Wales, Scotland, and Northern Ireland became the United Kingdom.

A Discrete Region

The British Isles form a distinct region of Europe for several reasons. Britain's insularity provided centuries of security from turbulent Europe, protecting the evolving British nation as it achieved a system of parliamentary government that had no peer in the Western world. Having united the Welsh, Scots, and Irish, the British set out to forge what would become the world's largest colonial empire. An era of mercantilism and domestic manufacturing (the latter based on water power from streams flowing off the Pennines, Britain's mountain backbone) foreshadowed the momentous Industrial Revolution, which transformed Britain—and much of the world. British cities became synonyms for specialized products as the smokestacks of factories rose like forests over the urban scene. London on the Thames River anchored an English core area that mushroomed into the headquarters of a global political, financial, and cultural empire. As recently as World War II, the narrow English Channel ensured the United Kingdom's impregnability against German invasion, giving the British time to

---

AMONG THE REALM'S GREAT CITIES . . . London

SAIL WESTWARD UP the meandering Thames River toward the heart of London, and be prepared to be disappointed. London does not overpower or overwhelm with spectacular skylines or beckoning beauty. It is, rather, an amalgam of towns—Chelsea, Chiswick, Dulwich, Hampstead, Islington—each with its own social character and urban landscape. Some of these towns come into view from the same river the Romans sailed 2000 years ago: Silvertown and its waterfront urban renewal; Greenwich with its famed Observatory; Thamesmead, the model modern commuter community. Others somehow retain their identity in the vast metropolis that remains (Fig. 1-9), in many ways Europe's most civilized and cosmopolitan city. And each contributes to the whole in its own way: every part of London, it seems, has its memories of empire, its memorials to heroes, its monuments to wartime courage.

Along the banks of the Thames, London displays the heritage of state and empire: the Tower and the Tower Bridge, the Houses of Parliament (officially known as the Palace of Westminster), the Royal Festival Hall. Step ashore, and you find London to be a memorable mix of the historic and (often architecturally ugly) modern, of the obsolete and the efficient, of the poor and the prosperous. Public transportation, by world standards, is excellent, even though facilities are aging rapidly; traffic, however, often is chaotic, gridlocked by narrow streets. Recreational and cultural amenities are second to none. London seems to stand with one foot in the twenty-first century and the other in the nineteenth. Its airports are ultramodern. But when the Eurostar TGV train from Paris emerges from the Channel Tunnel, it has to slow down for the final hour in its three-hour

© H. J. de Blij, P. O. Muller, and John Wiley & Sons, Inc.

run because the tracks into London's Waterloo Station still have not been upgraded to carry it at its normal high speed.

London remains one of the world's most livable metropolises for its size (8.6 million) largely because of farsighted urban planning that created and maintained, around the central city, a so-called Greenbelt set aside for recreation, farming, and other nonresidential, noncommercial uses (Fig. 1-9). Although London's growth eroded this Greenbelt, in places leaving only "green wedges," the design preserved crucial open space in and around the city, channeling suburbanization toward a zone at least 40 kilometers (25 mi) from the center. The map reveals the design's continuing impact.

organize their war machine. When the United Kingdom emerged from that conflict as a leading power among the victorious allies, it seemed that its superpower role in the postwar era was assured.

End of Empire

Two unanticipated developments changed that prospect: the worldwide collapse of colonial empires and the rapid resurgence of mainland Europe. Always ambivalent about the EC and EU, and with its first membership application vetoed by the French in 1963, Britain (admitted in 1973) has worked to restrain moves toward tighter integration. When most member-states adopted the new euro in favor of their national currencies, the British kept their pound sterling and delayed their participation in the EMU. As for a federalized Europe, to Britain this is out of the question. In this as in other respects, Britain's historic, insular standoffishness continues.

The United Kingdom

As we noted earlier, the British Isles as a region consists of two political entities: the United Kingdom and Ireland. The UK, with an area about the size of Oregon and a population of just over 60 million, is by European standards quite a large country. Based on a combination of physiographic, historical, cultural, economic, and political criteria, the United Kingdom can be divided into five subregions (numbered in Fig. 1-18):

1. *Southern England.* Centered on the gigantic London metropolitan area, this is the UK's most affluent subregion, with one-third of the country's population. Financial, communications, engineering, and energy-related industries cluster in this economically and politically dominant area. London exemplifies the momentum of long-term agglomeration; today this subregion benefits anew not only from its superior links to the European mainland but also (with New York and Tokyo) as one of the three leading "world cities" on the global scene.

2. *Northern England.* As the map suggests, this subregion should really be called Northern, Central, and Western England. Its center, appropriately called the Midlands, was the focus of the Industrial Revolution and the spectacular rise of Manchester, the source; Liverpool, the great port; Birmingham, unmatched industrial complex; and many other manufacturing cities and towns. But here obsolescence has overtaken what was once ultramodern, and the North now suffers from Rustbelt conditions and high unemployment

among immigrants from South Asia, Africa, and the Caribbean.

3. *Wales.* This nearly rectangular, rugged territory was a refuge for ancient Celtic peoples, and in its western counties more than half the inhabitants still speak Welsh. Because of the high-quality coal reserves in its southern tier, Wales too was engulfed by the Industrial Revolution, and Cardiff, the capital, was once the world's leading coal exporter. But the fortunes of Wales also declined, and many Welsh emigrated. Among the 3 million who remained, however, the flame of Welsh nationalism survived, and in 1997 the voters approved the establishment of a Welsh Assembly to administer public services in Wales, a first devolutionary step.

4. *Scotland.* Nearly twice as large as the Netherlands and with a population about the size of Denmark's, Scotland is a major component of the United Kingdom. As Figure 1-18 suggests, most of Scotland's more than 5 million people live in the Scottish Lowlands anchored by Edinburgh, the capital, in the east and Glasgow in the west. Attracted there by the labor demands of the Industrial Revolution (coal and iron reserves lay in the area), the Scots developed a world-class shipbuilding industry. Decline and obsolescence were followed by high-tech development, notably in the hinterland of Glasgow, and Scottish participation in the exploitation of oil and gas reserves under the North Sea (Fig. 1-18), which transformed the eastern ports of Aberdeen and Leith (Edinburgh). But many Scots feel that they are disadvantaged within the UK and should play a major role in the EU. Therefore, when the British government put the option of a Scottish parliament before the voters in 1997, 74 percent approved. Many Scots still hope that sovereignty lies in their future, and in local elections in 2007 the independence-minded Scottish National Party won more seats in the parliament than any of the other parties.

5. *Northern Ireland.* Prospects of devolution in Scotland pale before the devastation caused by political and sectarian conflict in Northern Ireland. With a population of 1.8 million occupying the northeastern one-sixth of the island of Ireland, this area represents the troubled legacy of British colonial rule. A declining majority, now about 54 percent of the people in Northern Ireland, trace their ancestry to Scotland or England and are Protestants; a growing minority, currently around 45 percent, are Roman Catholics, who share their Catholicism with virtually the entire population of the Irish Republic on the other side of the border. Although Figure 1-18 suggests that there are majority areas of Protestants and Catholics in North-

WILLIAMSON
SQUARE
ENVIRONMENTAL
IMPROVEMENTS

COMPLETION OF PAVING WORKS

THIS PROJECT IS BEING PART-FINANCED
BY THE EUROPEAN COMMUNITY
European Regional Development Fund

FROM THE FIELD NOTES

"From a ferry on the Mersey River, the skyline of downtown Liverpool seemed to mirror the troubles of this city, the nucleus of a metropolitan area with almost one million inhabitants. The dominant buildings in the CBD remain those built during Liverpool's heyday a century ago, when Liverpool's docks outnumbered even London's, and the city thrived on its trade with America and Africa as Merseyside's manufacturing industries prospered. After World War II, Liverpool declined as its port functions dwindled, its industries weakened, and unemployment soared; this view reveals no modern, glass-sided buildings towering over the older cityscape, symbols of investment and confidence in the future. Once the apex of what was called the *Liverpool Triangle* linking this port to West Africa and eastern America, the Merseyside area now lies remote from the centers of British and European action. Some investment has been made in waterfront redevelopment, but the city's stagnation is evident throughout its historic center (where memories of the famed rock group, the Beatles, have developed into a minor industry). 'There may be ambivalence about the European Union in other parts of Britain,' said my colleague of the Department of Geography at the University of Liverpool as we walked the streets, 'but not here. At least we can see more tangible results of EU investment, for which Liverpool is eligible.' But what will lift Merseyside out of its doldrums?"   © H. J. de Blij.

Concept Caching

www.conceptcaching.com

ern Ireland, no clear separation exists; mostly they live in clusters throughout the territory, including walled-off neighborhoods in the major cities of Belfast and Londonderry (photo p. 84). Partition is no solution to a conflict that has raged for over three decades at a cost of thousands of lives; Catholics accuse London as well as the local Protestant-dominated administration of discrimination, whereas Protestants accuse Catholics of seeking union with the Republic of Ireland. The persistent mediation efforts of former Prime Minister Tony Blair were crucial in the creation of a Northern Ireland Assembly to which powers will be devolved from London. After its first launch failed in 2003, forcing the British government to resume direct rule, a second try succeeded in May 2007 when all parties agreed to cooperate, and the Assembly began functioning again.

Republic of Ireland

What Northern Ireland is missing through its conflicts is shown by the Irish Republic itself: a booming service-based economy, the fastest-growing in all of Europe for a time around the turn of this century, when Ireland was dubbed the ***Celtic Tiger***. Burgeoning cities and towns (and rising real-estate prices), mushrooming industrial parks, bustling traffic, and construction everywhere reflect this new era. For the first time ever, workers of Irish descent returned from foreign places to take jobs at home; non-Irish immigrants also arrived, posing some new social problems for a closely knit, long-isolated society.

The Republic of Ireland fought itself free from British colonial rule just three generations ago. Its cool, moist climate had earlier led to the adoption of the American potato as the staple crop, but excessive rain and a blight

"The Troubles" between Protestants and Catholics, pro- and anti-British factions that have torn Northern Ireland apart for decades at a cost of more 3000 lives, are etched in the cultural landscape. The so-called "Peace Wall" across West Belfast, shown here separating Catholic and Protestant neighborhoods, is a tragic monument to the failure of accommodation and compromise, a physical manifestation of the emotional divide that still runs deep— despite the political compromise engineered by former Prime Minister Tony Blair and his negotiators. In May 2007, a Northern Ireland Assembly began functioning again, after hardline Catholic and Protestant political parties were persuaded to join in the effort. But what an analyst said to the author back in 2005 still is the view across the cityscape you see here: "Good fences make good neighbors . . . the Peace Wall will have to stay up for a few more decades yet." © Barry Chin/The Boston Globe/Redux.

in the late 1840s caused famine and cost over 1 million lives. Another 2 million Irish emigrated.

Independence in 1921 did not bring economic prosperity, and Ireland stagnated until the 1990s. Then, European telecommunications service industries saw labor and locational opportunities in Ireland, and soon Ireland was Europe's leading call center for the realm's rapidly expanding toll-free telephone market. Other service industries followed suit, benefiting from Ireland's well-educated but not highly paid labor pool. Changing economic conditions worldwide and in Europe slowed Ireland's economy in the early 2000s, but growth has now resumed and the country continues to be an EU success story.

● NORTHERN (NORDIC) EUROPE

North of Europe's Western European core area lies a disconnected group of six countries that exemplify core-periphery contrasts. Northern Europe is a region of difficult environments: generally cold climates, poorly developed soils, limited mineral resources, long distances. Together, the six countries in this region—Sweden, Norway, Denmark, Finland, Estonia, and Iceland—contain just over 26 million inhabitants, which is a lower total population than that of Benelux and only one-seventh that of Western Europe. The overall land area, on the other hand, is almost the size of the entire European core. Here in peripheral **Norden**, as the people call their northerly domain, national core areas lie in the south: note the location of the capitals of Helsinki, Stockholm, and Oslo at approximately the same latitude (Fig. 1-19).

Northern Europe's remoteness, isolation, and environmental severity also have had positive effects for this region. The countries of the Scandinavian Peninsula lay removed from the wars of mainland Europe (although Norway was overrun by Nazi Germany during World War II). The three major languages—Danish, Swedish, and Norwegian—are mutually intelligible, which creates one of the criteria delimiting this region. Another regional criterion is the overwhelming adherence to the same Lutheran church in each of the Scandinavian countries (Norway, Sweden, and Denmark), Iceland, and Finland. Furthermore, democratic and representative governments emerged early, and individual rights and social welfare have long been carefully protected. Women participate more fully in government and politics here than in any other region of the world.

Sweden is the largest Nordic country in terms of both population and territory. Most Swedes live south of 60° North latitude (which passes through Uppsala), in what is climatically the most moderate part of the country (Fig. 1-19). Here lie the capital, core area, and, as Figure 1-7 shows, the main industrial districts; here, too, are the main agricultural areas that benefit from the lower relief, better soils, and milder climate.

Sweden long exported raw or semifinished materials to industrial countries, but today the Swedes are making finished products themselves, including automobiles, electronics, stainless steel, furniture, and glassware. Much of this production is based on local resources, including a major iron ore reserve at Kiruna in the far north (there is a steel mill at Luleå). Swedish manufacturing, in contrast to that of several Western European countries, is based in dozens of small and medium-sized towns specializing in

FIGURE 1-19   © H. J. de Blij, P. O. Muller, and John Wiley & Sons, Inc.

particular products. Energy-poor Sweden was a pioneer in the development of nuclear power, but a national debate over the risks involved has reversed that course.

Norway does not need a nuclear power industry to supply its energy needs. It has found its economic opportunities on, in, and beneath the sea. Norway's fishing industry, now augmented by highly efficient fish farms, long

has been a cornerstone of the economy, and its merchant marine spans the world. But since the 1970s, Norway's economic life has been transformed by the bounty of oil and natural gas discovered in its sector of the North Sea.

With its limited patches of cultivable soil, high relief, extensive forests, frigid north, and spectacularly fjorded coastline, Norway has nothing to compare to Sweden's agricultural or industrial development. Its cities, from the capital Oslo and the North Sea port of Bergen to the historic national focus of Trondheim as well as Arctic Hammerfest, lie on the coast and have difficult overland connections. The isolated northern province of Finnmark has even become the scene of an autonomy movement among the reindeer-herding indigenous Saami (Fig. 1-12). Norway has been described as a necklace, its beads linked by the thinnest of strands. But this has not constrained national development. Norway in 2005 had the second-lowest unemployment rate in Europe (after tiny Luxembourg). In terms of income per capita, Norway is the fourth-richest country in the world.

Norwegians have a strong national consciousness and a spirit of independence. In 1994, when Sweden and Finland voted to join the European Union, the Norwegians again said no. They did not want to trade their economic independence for the regulations of a larger, even possibly safer, Europe.

Denmark, territorially small by Scandinavian standards, has a population of 5.4 million, second largest in the Nordic region after Sweden. It consists of the Jutland Peninsula and several islands to the east at the gateway to the Baltic Sea; it is on one of these islands, Sjaelland, that the capital of Copenhagen is located. Copenhagen, the "Singapore of the Baltic," has long been a port that collects, stores, and transships large quantities of goods. This **27 break-of-bulk** function exists because many oceangoing vessels cannot enter the shallow Baltic Sea, making the **28** city an **entrepôt** where transfer facilities and activities prevail. The completion of the Øresund bridge-tunnel link to southern Sweden enhanced Copenhagen's situation in 2000 (Fig. 1-19).

Denmark remains a kingdom, and in centuries past Danish influence spread far beyond its present confines. Remnants of that period now challenge Denmark's governance. Greenland came under Danish rule after union with Norway (1380) and remained a Danish domain when that union ended (1814). In 1953, Greenland's status changed from colony to province, and in 1979 the 60,000 inhabitants were given home rule with an Inuit name: *Kalaallit Nunaat*. They promptly exercised their rights by withdrawing from the European Union, of which they had become a part when Denmark joined. Another restive dependency is the Faroe Islands, located between Scotland and Iceland. These 17 small islands and their 45,000 inhabitants were awarded self-government in 1948, complete with their own flag and currency, but even this was not enough to defuse demands for total independence. A referendum in mid-2001 confirmed that even Denmark is not immune from Europe's devolutionary forces (Fig. 1-12).

Finland, territorially almost as large as Germany, has only 5.3 million residents, most of them concentrated in the triangle formed by the capital, Helsinki, the textile-producing center, Tampere, and the shipbuilding center, Turku (Fig. 1-19). A land of evergreen forests and glacial lakes, Finland has an economy that has long been sustained by wood and wood product exports. But the Finns, being a skillful and productive people, have developed a diversified economy in which the manufacture of precision machinery and telecommunications equipment (prominently including cell phones) as well as the growing of staple crops are key.

As in Norway and Sweden, environmental challenges and relative location have created Nordic cultural landscapes in Finland, but the Finns are not a Scandinavian people; their linguistic and historic links are instead with the Estonians across the Gulf of Finland. As we will see in Chapter 2, ethnic groups speaking Finno-Ugric languages are widely dispersed across what is today western Russia.

Estonia, northernmost of the three "Baltic states," is part of Nordic Europe by virtue of its ethnic and linguistic ties to Finland. But during the period of Soviet control from 1940 to 1991, Estonia's demographic structure changed drastically: today about 25 percent of its 1.3 million inhabitants are Russians, most of whom came there as colonizers.

After a difficult period of adjustment, Estonia today is forging ahead of its Baltic neighbors (see Table G-1) and catching up with its Nordic counterparts. Busy traffic links Tallinn, the capital, with Helsinki, and a new free-trade zone at Muuga Harbor facilitates commerce with Russia. But more important for Estonia's future was its entry into the European Union in 2004.

Iceland, the volcanic, glacier-studded island in the frigid waters of the North Atlantic just south of the Arctic Circle, is the sixth country in this region. Inhabited by people with Scandinavian ancestries (population: 288,000), Iceland and its small neighboring archipelago, the Westermann Islands, are of special scientific interest because they lie on the Mid-Atlantic Ridge, where the Eurasian and North American tectonic plates of the Earth's crust are diverging and new land can be seen forming (see Fig. G-4).

Iceland's population is almost totally urban, and the capital, Reykjavik, contains about half the country's inhabitants. The nation's economic geography is almost entirely oriented to the surrounding waters, whose seafood harvests give Iceland one of the world's highest standards of living—but at the risk of overfishing. Disputes over fishing grounds and fish quotas have intensified in recent decades; the Icelanders argue that, unlike the Norwegians or the British, they have little or no alternative economic opportunity.

MEDITERRANEAN EUROPE

South of Europe's core lie the six countries that constitute the Mediterranean region: Italy, Spain, Portugal, Greece, and the island countries of Cyprus and Malta (Fig. 1-20). Like Northern Europe, this is a region of peninsulas, and again it is discontinuous. It lies separated from the European core, and core-periphery contrasts in some areas are quite sharp. A degree of continuity, dating from Greco-Roman times, marks the region's languages and religion, lifeways, and cultural landscapes. As Figure 1-2 shows, natural environments in this region are dominated by a climatic regime that bears its very name—Mediterranean. Dry, hot summers are the norm, so that moisture is often in short supply during the growing season, and specially adapted plants mark local agriculture.

In terms of raw materials, Southern (Mediterranean) Europe is not nearly as well endowed as the European core, as reflected by the map of industrial complexes (Fig. 1-7). Only northern Italy and northern Spain (the former through massive imports of coal and iron ore) have become part of the core area. Furthermore, the region has been largely deforested, and the limited and

AMONG THE REALM'S GREAT CITIES . . . Rome

FROM A HIGH vantage point, Rome seems to consist of an endless sea of tiled roofs, above which rise numerous white, ochre, and gray domes of various sizes; in the distance, the urban perimeter is marked by high-rises fading in the urban haze. This historic city lives amid its past as perhaps no other as busy traffic encircles the Colosseum, the Forum, the Pantheon, and other legacies of Europe's greatest empire.

Founded about 3000 years ago at an island crossing point on the Tiber River about 25 kilometers (15 mi) from the sea, Rome had a high, defensible site. A millennium later, with a population some scholars estimate as high as 1 million, it was the capital of a Roman domain that extended from Britain to the head of the Persian Gulf and from the shores of the Black Sea to North Africa. Rome's emperors endowed the city with magnificent, marble-faced, columned public buildings, baths, stadiums, obelisks, arches, and statuary; when Rome became a Christian city, the domes of churches and chapels added to its luster.

It is almost inconceivable that such a city could collapse, but that is what happened after the center of Roman power shifted eastward to Constantinople (now Istanbul). By the end of the sixth century, Rome probably had fewer than 50,000 inhabitants, and in the thirteenth, a mere 30,000. Papal rule and a Renaissance revival lay ahead, but in 1870, when Rome became the capital of newly united Italy, it still had a population of only 200,000.

Now began a growth cycle that eclipsed all previous records. As Italy's political, religious, and cultural focus (though not an industrial center to match), Rome grew to 1 million by 1930, to 2 million by 1960, and subsequently to 3.3 million where it has leveled off today. The religious enclave of Vatican City, Roman Catholicism's headquarters, makes Rome a twin capital; the Vatican functions

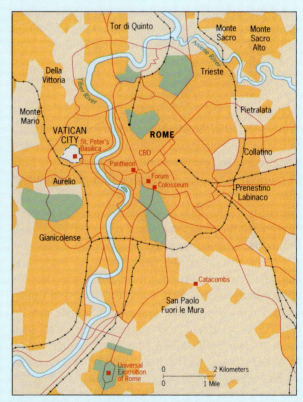

© H. J. de Blij, P. O. Muller, and John Wiley & Sons, Inc.

as an independent entity and has a global influence that Italy cannot equal.

Rome today remains a city whose economy is dominated by service industries: national and local government, finance and banking, insurance, retailing, and tourism employ three-quarters of the labor force. The new city sprawls far beyond the old, walled, traffic-choked center where the Roman past and the Italian future come face to face.

highly seasonal water supply constrains Southern Europe's hydroelectric opportunities.

A key contrast between Northern and Southern Europe lies in their populations. Although territorially smaller than Northern Europe, Mediterranean Europe has nearly five times as many people (128 million in 2008). Population distribution continues to reflect the agricultural bases of the preindustrial era, with large concentrations in coastal lowlands and fertile river basins, although the growth of major industrial centers, notably in northern Italy and northern Spain, has superimposed a new mosaic. Still, urbanization in Southern Europe is far below that in Western or Northern Europe, or in the British Isles (Portugal's 55 percent ranks among the lower levels in the realm). Other data indicate that living standards in Mediterranean Europe also lag behind (Table G-1). All this is changing, of course, with the EU's expansion into Eastern Europe. As of 2004, Southern Europe no longer was the lowest-ranking EU region in terms of GNI, living standards, or urbanization.

Italy

Centrally located in the Mediterranean region, most populous of the Mediterranean states, best connected to the European core, and economically most advanced is Italy (59.0 million), a charter member of Europe's Common Market.

Administratively, Italy is organized into 20 regions, many with historic roots dating back centuries (Fig. 1-21). Several of these regions have become powerful economic entities centered on major cities, such as Lombardy (Milan) and Piedmont (Turin); others are historic hearths of Italian culture, including Tuscany (Florence) and Veneto (Venice). These regions in the northern half of Italy stand in strong social, economic, and political contrast to such southern regions as Calabria (the "toe" of the Italian "boot") and Italy's two major Mediterranean islands, Sicily and Sardinia. Not surprisingly, Italy is often described as two countries—a progressive north and a stagnant south or *Mezzogiorno*.

North of the Ancona Line

North and south are bound by the ancient headquarters, Rome, which lies astride the narrow transition zone between Italy's contrasting halves. This zone is referred to as the *Ancona Line*, after the city on the Adriatic coast where the zone reaches the other side of the peninsula (Fig. 1-21). Whereas Rome remains Italy's capital and cultural focus, the functional core area of Italy has shifted

FIGURE 1-20 © H. J. de Blij, P. O. Muller, and John Wiley & Sons, Inc.

northward into Lombardy in the basin of the Po River. Here lies Southern Europe's leading manufacturing complex, in which a large, skilled labor force and ample hydroelectric power from Alpine and Appennine slopes combine with a host of imported raw materials to produce a wide range of machinery and precision equipment. The Milan–Turin–Genoa triangle exports appliances, instruments, automobiles, ships, and many specialized products. Meanwhile, the Po Basin, lying on the margins of the region's dominant Mediterranean climatic regime, enjoys a more even pattern of rainfall distribution throughout the year, making it a productive agricultural zone as well.

Metropolitan Milan embodies the new, modern Italy. Not only is Milan (at 4.1 million) Italy's largest city and leading manufacturing center—making Lombardy one

of Europe's Four Motors—but it also is the country's financial and service-industry headquarters. Today the Milan area, a cornerstone of the European core, has just 7 percent of Italy's population but accounts for one-third of the entire country's national income.

The Mezzogiorno

As in Germany, the lowest-income regions of Italy lie concentrated in one part of the country: the Mezzogiorno in the south. But Italy's north-south disparity continues to grow, which has led to taxpayers' revolts over the subsidies the state pays to the poorer southern regions (a separatist movement has even espoused an independent *Padania* in the north). In truth, the south receives the bulk of Italy's illegal immigrants, whether from Africa across the

Mediterranean or from the states of former Yugoslavia and Albania across the Adriatic, and it is these workers, willing to work for low wages, who move north to take jobs in the factories of Milan and Turin. Italian southerners tend to stay where they were born. Sicily and the Mezzogiorno underscore the problems of a periphery.

Iberia

At the western end of Southern Europe lies the Iberian Peninsula, separated from France and Western Europe by the rugged Pyrenees and from North Africa by the narrow Strait of Gibraltar (Fig. 1-20). Spain (population: 45.7 million) occupies most of this compact Mediterranean landmass. Portugal lies in its southwestern corner.

FIGURE 1-21 © H. J. de Blij, P. O. Muller, and John Wiley & Sons, Inc.

Imperial Romans, Muslim Moors, and Catholic kings left their imprints on Iberia, notably the boundary between Spain and Portugal dating from the twelfth century. The golden age of colonialism was followed by dictatorial rule and economic stagnation.

Today both Spain and Portugal are democracies, and their economies are doing well, helped by EU subsidies. As Figure 1-22 shows, Spain followed the leads of Germany and France and decentralized its administrative structure, creating 17 regions called Autonomous Communities (ACs). Every AC has its own parliament and administration that control planning, public works, cul-

tural affairs, education, environmental matters, and even, to some extent, international commerce. Each AC can negotiate its own degree of autonomy with the central government in Madrid. This new system, however, has not been enough to defuse Spain's most problematic devolutionary issue, which involves the Basques in the north-central, France-adjoining AC mapped as Basque Country (Fig. 1-22, especially the top inset map).

Particularly noteworthy is *Catalonia*, the triangular AC in Spain's northeastern corner, also adjacent to France. Centered on prosperous, productive Barcelona, Catalonia is Spain's leading industrial area, an AC with a popu-

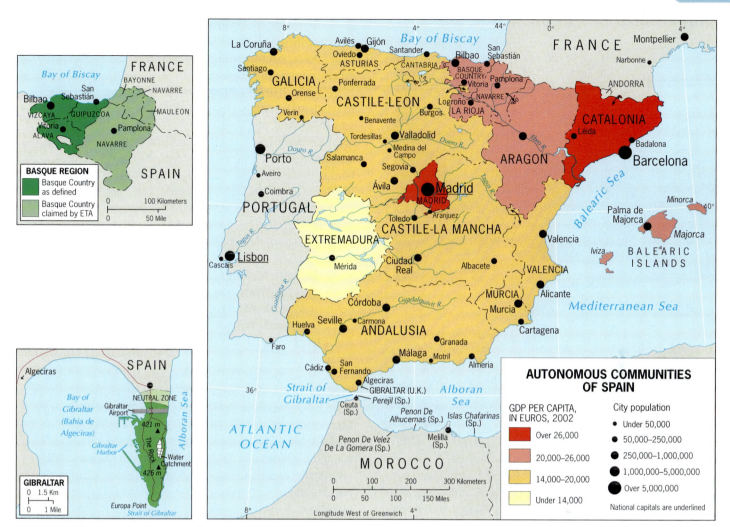

FIGURE 1-22   © H. J. de Blij, P. O. Muller, and John Wiley & Sons, Inc.

lation of 7.2 million. It is imbued with a fierce nationalism, endowed with its own language and culture, and economically it is one of the Four Motors of Europe. Note that while most of Spain's industrial raw materials lie in the northwest, its major industrial development has taken place in the northeast, where innovations and skills propel a high-technology-driven regional economy. In recent years, Catalonia—with 6 percent of Spain's territory and 16 percent of its population—has annually produced 25 percent of all Spanish exports and nearly 40 percent of its industrial exports. Such economic strength translates into political power, and in Spain the issue of Catalonian separatism is never far from the surface.

Here are two items that keep Catalans angry: first, the fact that Spain's first high-speed rail line, built with EU funds, runs south from Madrid to poor, agricultural Andalusia and not to rich, industrial Barcelona; and second, that the Madrid government is contemplating the southward diversion of river water from the Rio Ebro through dams and pipelines.

As Figure 1-22 shows, Spain's capital and largest city, Madrid, lies near the geographic center of the state. It also lies within an economic-geographic transition zone. In terms of people's annual income, Spain's north is far more affluent than its south, a direct result of the country's distribution of resources, climate (the south suffers from drought and poorer soils), and overall development opportunities. The most prosperous ACs are Barcelona-centered Catalonia and Madrid; between them, the contiguous group of ACs extending from the Basque Country to Aragon rank next. In the ACs of northern Spain, economies tend to do better in the east, where tourism and winegrowing (especially in La Rioja, a famous name in Spanish wine) are among the mainstays, than in the west, where industrial obsolescence, dwindling raw-material sources, the decline of the fishing industry, and emigration plague local economies. Spain's southernmost ACs have long been the least developed economically.

These economic contrasts have political consequences. In recent years Catalonia, with its economic strength and

FROM THE FIELD NOTES

"Catalonian nationalism is visible both obviously and subtly in Barcelona's urban landscape. Walking toward the Catalonian Parliament, I noticed that the flags of Spain (left) and Catalonia (right) flew from slightly diverging flagpoles above the entrance to the historic building. Is there a message here?"   © H. J. de Blij.

Concept Caching

www.conceptcaching.com

political clout, has pressed the Madrid government to approve a statute that would acknowledge Catalonia as a nation (photo above), give Barcelona its own tax-raising powers, replace Spain's constitutional court with Catalonia's regional high court as the AC's ultimate judicial authority, and compel all officials in Catalonia, national or regional, to speak Catalan. While Spain's government debated (and watered down) these demands, other ACs, though opposed to Catalonia's campaign, began thinking of ways to enhance their own "national" ambitions. So even as Spain yields some of its sovereignty as the price of participating in the EU, it faces centrifugal forces arising from its own, internal economic and social regionalism.

Gibraltar.   Although Spain and the United Kingdom are both EU members committed to cooperative resolution of territorial problems, the two countries are embroiled in a dispute over a sliver of land at the southern tip of Iberia: legendary Gibraltar (see lower inset map, Fig. 1-22). "The Rock" was ceded by Spain to the British in perpetuity in 1713 (though not the neck of the peninsula linking it to the mainland, which the British later occupied) and has been a British colony ever since. The 30,000-odd residents are used to British institutions, legal rules, and schools. Successive Spanish governments have demanded that Gibraltar be returned to Madrid's rule, but the colony's residents are against it. And under their 1969 constitution, Gibraltarians have the right to vote on any transfer of sovereignty.

Now the British and Spanish governments are trying to reach an agreement under which they would share the administration of Gibraltar for an indefinite time; both are EU members, so little would have to change. The advantages would be many: economic development is being slowed by the ongoing dispute, border checks would be lifted, EU benefits would flow. But Gibraltarians have their doubts and want to put the issue to a referendum. Spain will not accept the idea of a referendum, and the matter is far from resolved.

Ceuta and Melilla.   Spain's refusal to allow, and abide by, a referendum in Gibraltar stands in remarkable contrast to its demand for a referendum in its own outposts, two small ***exclaves*** on the coast of Morocco, Ceuta and Melilla (Fig. 1-22). Morocco has been demanding the return of these two small cities, but Spain has refused on the grounds that the local residents do not want this. The matter has long simmered quietly but came to international attention in 2002 when a small detachment of Moroccan soldiers seized the island of Perejil, an uninhabited, Spanish possession off the Moroccan coast also wanted by Morocco. In the diplomatic tensions that followed, the entire question of Spain's holdings in North Africa (and its anti-Moroccan stance in the larger matter of Western Sahara) exposed some contradictions Madrid had preferred to conceal.

The state of Portugal (population: 10.6 million), a comparatively poor country that has benefited enormously from its admission to the EU, occupies the southwestern corner of the Iberian Peninsula. Since one rule of EU membership is that the richer members assist the poorer ones, Portugal shows the results in a massive renovation project in the capital, Lisbon, as well as in the modernization of surface transport routes.

Unlike Spain, which has major population clusters on its interior plateau as well as its coastal lowlands, the Portuguese are concentrated along and near the Atlantic coast. Lisbon and the second city, Porto, are coastal cities; the best farmlands lie in the moister western and northern zones of the country. But the farms are small and inefficient, and although Portugal remains dominantly rural, it must import as much as half of its foodstuffs. Exporting textiles, wines, corks, and fish, and running up an annual deficit, the indebted Portuguese economy remains a far cry from those of other European countries of similar dimensions.

FIGURE 1-23   © H. J. de Blij, P. O. Muller, and John Wiley & Sons, Inc.

Greece

The eastern segment of the region that we define as Southern Europe is dominated by Greece, an outlier of both Mediterranean Europe and (until 2007) the European Union. Greece's land boundaries are with Turkey, Bulgaria, Macedonia, and Albania; as Figure 1-23 reveals, it also owns islands just offshore from mainland Turkey. Altogether, the Greek archipelago numbers some 2000 islands, ranging in size from Crete (8260 square kilometers [3190 sq mi]) to small specks of land in the Cyclades. In addition, Greeks represent the great majority on the now-divided island of Cyprus.

Ancient Greece was a cradle of Western civilization, and later it was absorbed by the expanding Roman Empire. For some 350 years beginning in the mid-fifteenth century, Greece was under the sway of the Ottoman Turks. Greece regained independence in 1827, but not until nearly a century later, through a series of Balkan wars, did it acquire its present boundaries. During World War II, Nazi Germany occupied and ravaged the country, and in the postwar period the Greeks have quarreled with the Turks, the Albanians, and the newly independent Macedonians. Today, Greece finds itself between the Muslim world of Southwest Asia and the Muslim communities of Eastern Europe.

Implications of EU Expansion

Volatile as Greece's surroundings are, Greece itself is a country on the move, an EU success story to rival Ireland's. Political upheavals in the 1970s and economic stagnation in the 1980s are all but forgotten in the new century: now Greece, its economy booming, is described as the locomotive for the Balkans, a beacon for the EU in a crucial part of the world. Recent infrastructure improvements have focused on metropolitan Athens, where nearly one-third of the population is concentrated, and include new subways, a new beltway, and a new airport.

Greece will be strongly affected by the accession of the ten EU members in 2004 and by the EU's admission of Romania and its neighbor, Bulgaria, in 2007. Until 2007, Greece lay separated from the contiguous EU by nonmember-states, but when Romania and Bulgaria joined the EU, Greece became part of the conterminous Union. The effect of what will eventually become a far more porous border with Bulgaria may still be distant, but more immediately, Greece will lose much of its EU subsidy because the poorer new members will start receiving EU financial support. Meanwhile, Greece's democratic institutions still need strengthening, its educational institutions need modernizing (street riots greeted the introduction of private university education, an EU regulation), and corruption remains a serious problem. And while Greece is on better terms today with its fractious neighbors and invests in development projects in Macedonia and Bulgaria (for example, an oil pipeline from Thessaloniki to a Greek-owned refinery in Skopje, Macedonia), it will take skillful diplomacy to navigate the shoals of historic discord.

Urban and Rural Greece

Modern Greece is a nation of 11.1 million centered on historic Athens, one of the realm's great cities. With its port of Piraeus, metropolitan Athens contains about 30 percent of the Greek population, making it one of Europe's most congested and polluted urban areas. Athens is the quintessential primate city; the monumental architecture of ancient Greece still dominates its cultural landscape. The Acropolis and other prominent landmarks attract a steady stream of visitors; tourism is one of Greece's leading sources of foreign revenues, and Athens is only the beginning of what the country has to offer.

Deforestation, soil erosion, and variable rainfall make farming difficult in much of Greece, but the country remains strongly agrarian. It is self-sufficient in staple foods, and

AMONG THE REALM'S GREAT CITIES . . . Athens

TAKE A MAP of Greece. Draw a line to encompass its land area as well as all the islands. Now find the geographic center of this terrestrial and maritime territory, and you will find a major city nearby. That city is Athens, capital of modern Greece, culture hearth of ancient Greece, source of Western civilization.

Athens today is only part of a greater metropolis that sprawls across, and extends beyond, a mountain-encircled, arid basin that opens southward to the Bay of Phaleron, an inlet of the Aegean Sea. Here lies Piraeus, Greece's largest port, linked by rail and road to the adjacent capital. Other towns, from Keratsinion in the west to Agios Dimitrios in the east, form part of an urban area that not only lies at, but is, the heart of modern Greece. With a population of 3.2 million, most of the country's major industries, its densest transport network, and a multitude of service functions ranging from government to tourism, Greater Athens is a well-defined national core area.

Arrive at the new Eleftherios Venizelos Airport east of Athens and drive into the city, and you can navigate on the famed Acropolis, the 150-meter (500 ft) high hill that was ancient Athens' sanctuary and citadel, crowned by one of humanity's greatest historic treasures, the Parthenon, temple to the goddess Athena. Only one among many relics of the grandeur of Athens 25 centuries ago, it is now engulfed by mainly low-rise urbanization and all too often obscured by smog generated by factories and vehicles and trapped in the basin. Conservationists decry the damage air pollution has inflicted on the city's remaining historic structures, but in truth the greatest destruction was wrought by the Ottoman Turks, who plundered the country during their occupation

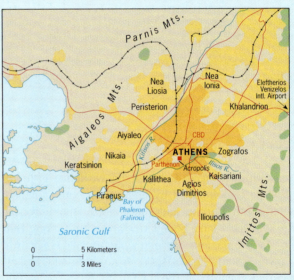

© H. J. de Blij, P. O. Muller, and John Wiley & Sons, Inc.

of it, and the British and Germans, who carted away priceless treasures to museums and markets in Western Europe.

Travel westward from the Middle East or Turkey, and Athens will present itself as the first European city on your route. Travel eastward from the heart of Europe, however, and your impression of its cultural landscape may differ: behind the modern avenues and shopping streets are bazaar-like alleys and markets where the atmosphere is unlike that of any other in this geographic realm. And beyond the margins of Greater Athens lies a Greece that lags far behind this bustling, productive coreland.

farm products continue to figure strongly among exports. But other sectors of the economy, including manufacturing (textiles) and the service industries, are growing rapidly. The challenge for Greece will be to maintain its growth after the momentous EU expansions of 2004 and 2007.

Cyprus and Malta

Cyprus lies in the far northeastern corner of the Mediterranean Sea, much closer to Turkey than to Greece (Fig. 1-23), but is peopled dominantly by Greeks rather than Turks. In 1571, the Turks conquered Cyprus, then ruled by Venice, and controlled it until 1878 when the British took over. Most of the island's Turks arrived during the Ottoman period; the Greeks have been there longest.

When the British were ready to give Cyprus independence after World War II, the 80-percent-Greek majority mostly preferred union with Greece. Ethnic conflict followed, but in 1960 the British granted Cyprus independence under a constitution that prescribed majority rule but guaranteed minority rights.

This fragile order broke down in 1974, and civil war engulfed the island. Turkey sent in troops and massive dislocation followed, resulting in the partition of Cyprus into northern Turkish and southern Greek sectors (Fig. 1-20, inset map). In 1983, the 40 percent of Cyprus under Turkish control, with about 100,000 inhabitants (and some 30,000 Turkish soldiers), declared itself the independent Turkish Republic of Northern Cyprus. Only Turkey recognizes this ministate (which now contains a population of about 200,000); the international community recognizes the government on the Greek side as legitimate. This legitimacy, however, is questionable in light of the Greek side's rejection of the United Nations plan to allow both the Greek and the Turkish side of the island to join the EU in 2004. The Turkish-Cypriot voters accepted it, yet they were left out when the Greek side of the island was admitted to the EU in 2004. Resentment was high on the Turkish side, and in Turkey itself as well—just as discussions on Turkey's own admission to the EU were getting under way. It was—and remains—a reminder that Cyprus's "Green Line" separating the Greek and Turkish communities constitutes not just a regional border but a boundary between geographic realms.

The Mediterranean region contains one other ministate, Malta, located south of Sicily. Malta is a small archipelago of three inhabited and two uninhabited islands with a population of just under 400,000 (Fig. 1-23, inset). An ancient crossroads and culturally rich with Arab, Phoenician, Italian, and British infusions, Malta became a British dependency and served British shipping and its military. It suffered terribly during World War II bombings, but despite limited natural resources recovered strongly during the postwar period. Today Malta has a booming tourist industry and a relatively high standard of living, and is a new member of the European Union.

● EASTERN EUROPE

In the past, Eastern Europe was one of those regional designations that had various interpretations. The map on page 72 suggests that it incorporated all of Europe east of Germany, Austria, and Italy, north of Greece, and south of Finland. The Soviet communist domination of Eastern Europe (1945–1990) behind an ideological and strategic "Iron Curtain" served to reinforce the division between "West" and "East."

Today, with the Soviet occupation mostly a distant memory and the European Union extending from the British Isles to the Black Sea, it might seem that the geographic notion of an Eastern Europe is defunct. But not so fast. Take a look at the statistics in Table G-1, and you will see that, with some notable but still minimal exceptions, there remain significant contrasts between West and East. True, Estonia has taken its place as one of Nordic (no longer Eastern) Europe's economic success stories. And in the south, Slovenia is another winner. But these small countries in favorable locations are less representative of the region than poorly governed Poland, deeply indebted Hungary, or corrupt Slovakia—not to mention barely postcommunist Romania and economically troubled Bulgaria (with a per-capita GNI one-quarter of Western Europe's).

Eastern Europe's tumultuous history played itself out on a physical stage of immense diversity, its landscapes ranging from open plains and wide river basins to strategic mountains and crucial corridors. Epic battles fought centuries ago remain fresh in the minds of many people living here today; pivotal past migrations are celebrated as though they happened yesterday. Nowhere in Europe is the cultural geography as complex. Illyrians, Slavs, Turks, Magyars, and other peoples converged on this region from near and far. Ethnic and cultural differences kept them in chronic conflict.

Geographers call this region a **shatter belt**, a zone of [29] persistent splintering and fracturing. Geographic terminology uses several expressions to describe the breakup of established order, and these tend to have their roots in this part of the world. One of them is **balkanization**. The southern [30] half of Eastern Europe is referred to as the Balkans or Balkan Peninsula, after the name of a mountain range in Bulgaria. Balkanization denotes the recurrent division and fragmentation of this part of Eastern Europe, and it is now applied to any place where such processes take place. A more recent term is *ethnic cleansing*—the forcible ouster of entire populations from their homelands by a stronger power bent on taking their territories. The term may be new, but the process is as old as Eastern Europe itself.

Cultural Legacies

Each episode in the historical geography of Eastern Europe has left its legacy in the cultural landscape. Twenty centuries ago the Roman Empire ruled much of it (Romania is a cartographic reminder of this period); during the past half-century, the Soviet Empire controlled almost all of it. In the intervening two millennia, Christian Orthodox church doctrines spread from the southeast, and Roman Catholicism advanced from the northwest. Turkish (Ottoman) Muslims invaded and created an empire that reached the environs of Vienna. By the time the Austro-Hungarian Empire ousted the Turks, millions of Eastern Europeans had been converted to Islam. Albania today

FIGURE 1-24   © H. J. de Blij, P. O. Muller, and John Wiley & Sons, Inc.

remains a predominantly Muslim country. In the twentieth century, Eastern Europe became a battleground between superpowers, and the complicated map reflects the results through 1991 (Fig. 1-24).

The subsequent collapse of the Soviet Empire freed the European countries that had been under Soviet rule, and with only one exception—Belarus—these countries turned their gaze from Moscow to the west. Meanwhile, the European Union had been expanding eastward, and membership in the EU became an overriding goal for the majority of the liberated states of Eastern Europe. This generated a new map of the region (Fig. 1-25), integrat-

ing most of Eastern Europe into the EU and creating, as we saw earlier, a new layout of insiders and outsiders. The Iron Curtain that long divided Germany is just a dim memory, and of the Soviet realm that once bordered Poland and Romania, Russia is the heir, so that the post-Soviet realm boundary (refer back to Fig. G-2, p. 7) now coincides with the Russian-European border. As noted, only Belarus survives as an almost totally unreformed Soviet-style state, and Mensk (Minsk) continues to look toward Moscow rather than Brussels. While Russia continues to meddle destructively in Moldova and plays its powerful energy card against other neighbors, it cannot

FIGURE 1-25   © H. J. de Blij, P. O. Muller, and John Wiley & Sons, Inc.

turn the social, economic, and political tides that are transforming Eastern Europe.

The Geographic Framework

Eastern Europe continues to change, and not just through members joining the European Union. You would think that the map of this old part of the world would not change dramatically, but in fact a new country appeared on the map in 2006 and another one may be in the offing. The new country is Montenegro, which voted to split from what used to be called Serbia-Montenegro, and joined the United Nations as the 192nd member-state. The state-in-waiting is dominantly Muslim Kosovo, another part of Serbia, under UN administration in mid-2007 and awaiting key decisions on its future. A problem loomed here: there was no doubt that the great majority of the population wanted independence, but there was a Serb minority that was totally opposed, and in 2006 the Serbian parliament declared Kosovo an "unalterable" part of the state. At press time, Kosovo's future remained clouded.

One way to approach the complex geography of Eastern Europe is to group the countries with reference to their EU membership (red type) and their relative location, which is what the numbers below signify. Counting Montenegro (but not Kosovo), there are 18 states:

1. Countries Facing the Baltic Sea (**3** + 1)
2. The Landlocked Center (**3**)
3. Countries Facing the Black Sea (**2** + 2)
4. Countries Facing the Adriatic Sea (**1** + 6)

Countries Facing the Baltic Sea

Poland dominates this subregion, which also includes Lithuania and Latvia and, by virtue of relative location, Belarus (although Belarus does not possess a Baltic Sea coast). Wedged between coastal Poland and Lithuania is the Russian **exclave** of *Kaliningrad*, which is not func-[31] tionally a part of Eastern Europe (see Chapter 2).

Figure 1-24 displays Poland's most recent boundary shifts: in the aftermath of World War II, the whole country shifted westward, losing land to the (then) USSR in the east and gaining it at the expense of Germany to the west. Situated on the North European Lowland, Poland traditionally was an agrarian country, but during the Soviet-communist period Silesia (once part of Germany) became its industrial heartland, and Katowice, Wroclaw, and Krakow grew into major industrial cities amid some of the world's worst environmental pollution. The Soviets invested far less in agriculture, collectivizing farms without modernizing the technology and leaving post-Soviet farming in abysmal condition.

By many measures, Poland is the most important state that joined the EU in 2004: with nearly 40 million people, it has half the total population of the ten new entrants; territorially it is larger than all the others combined; and its economy is the biggest by far.

Latvia, centered on the Baltic port of Riga, is Eastern Europe's northernmost country and was tightly integrated into the Soviet system during its long domination. Still today, only 59 percent of the population of 2.3 million is Latvian, and nearly 30 percent is Russian. After

FROM THE FIELD NOTES

"Turbulent history and prosperous past are etched on the cultural landscape of the old Hanseatic city of Gdansk, Poland. Despite major wartime destruction, much of the old architecture survives, and restoration has revived not only Gdansk but also its twin (port) city, Gdynia. Gdansk was the initial stage for the rebellion in the 1980s of the Solidarity labor union against Poland's communist regime, and after that regime was toppled, the country went through difficult economic times. Today Poland, a full member of the European Union, is making rapid progress; its historic cities, from Gdansk in the north to Krakow in the south, attract throngs of foreign visitors and the tourist industry is booming. Note: don't let that Dutch tricolor mislead you. Attracting foreign visitors is one way to raise revenues, and the historic old city on the Baltic is becoming a tourist draw."
© Sigmund Malinowski.

www.conceptcaching.com

independence there was ethnic tension between these Baltic and Slav sectors, but the prospect of EU membership required the end of discriminatory practices. Latvians concentrated instead on the economy, which had been left in dreadful shape, with the result that it qualified for 2004 admission. Consider this: 20 years ago, virtually all of Latvia's trade was with the Soviet Union. Today its principal trading partners are Germany, the United Kingdom, and Sweden. Russia figures in only one category: Latvia's import of oil and gas.

Lithuania (3.4 million) has a residual Russian minority of only about 7 percent, but relations with its giant neighbor are worse than Latvia's—this despite Lithuania's greater dependence on Russia as a trading partner. One reason for this bad relationship has to do with neighboring Kaliningrad, Russia's Baltic Sea exclave (Fig. 1-25). When Kaliningrad became a Russian territory after World War II, Lithuania was left with only about 80 kilometers (50 mi) of Baltic coastline and a small port that was not even connected by rail to the interior capital of Vilnius. Lithuania in 2005 called for the demilitarization of Kaliningrad as a matter of national security. Despite these problems, Lithuania's economy during 2003 had the highest growth rate in Europe, spurred by foreign investment and by profits from its oil refinery at Mazeikiai, facilitating EU admission in 2004.

Belarus has no Baltic coast, but it borders all three of the coastal states in this subregion. About 80 percent of its nearly 10 million people are Belarussians ("White" Russians), a West Slavic people. Only some 11 percent are (East Slavic) Russians. A small Polish minority complains of mistreatment and discrimination. Devastated during World War II, Belarus became one of Moscow's most loyal satellites, and the Soviets made Mensk (Minsk), the capital, into a large industrial center. But in the post-Soviet era, Belarus has lagged badly, and its government has retained powers reminiscent of the communist era. Unlike its neighbors, Belarus has no interest in joining the EU. On the contrary, its overtures have been toward Moscow: it seeks to rejoin Russia in some formal way.

The Landlocked Center

Three states comprise this subregion: the Czech Republic, Slovakia, and Hungary. Until 1993, the first two formed the state of Czechoslovakia, but their "velvet divorce" broke it up. The Czechs got the better of the deal; Slovakia is this subregion's poorest country by far.

The Czech Republic (10.3 million) centers on Bohemia, the mountain-enclosed core area that contains the historic capital, Prague. This province always has been

cosmopolitan and Western in its exposure, outlook, development, and linkages. Prague lies in the basin of the Elbe River, Bohemia's traditional outlet through northern Germany to the North Sea. It is a classic primate city, its cultural landscape faithful to Czech traditions; but it also is an industrial center. The encircling mountains contain many valleys with small towns that specialize, Swiss-style, in fabricating high-quality goods. In Eastern Europe the Czechs always were the leaders in technology and engineering; even during the communist period their products found markets in foreign countries near and far.

As Figure 1-25 shows, the Czech Republic incorporates an eastern province named Moravia, linked to Poland's Silesia through the Moravian Gate between the Sudeten and Carpathian Mountains. The Soviet industrialization drive made Ostrava and Brno important manufacturing cities but left behind obsolescence and environmental problems.

The separation of the Czech and Slovak societies created a virtual nation-state in the Czech Republic, with only a small minority of several hundred thousand Roma, or Gypsies, among the Czechs. The treatment of this minority became a human rights issue in the discussions leading to Czech membership in the EU, effective 2004.

As Figure 1-26 shows, Slovakia inherited Czechoslovakia's largest minority, the ethnic Hungarian community, which now constitutes about 10 percent of the country's 5.4 million inhabitants and is concentrated in the southern zone along the Danube River. The country's division resulted from ethnic and cultural factors, but the Slovaks also did not share the Czechs' enthusiasm for post-Soviet economic reforms. The Hungarian minority found itself at odds with the new government in Bratislava, but both communities saw the advantages of EU membership and found ways to improve their relationships. Although the economy performed well in 2005 and 2006, the political situation in the country deteriorated after the 2004 EU accession in what appeared to be a reversion to the immediate post-breakup period, when newly independent Slovakia was poorly served by one of the region's most corrupt governments. In 2007, Slovakia's prospects remained uncertain.

Hungary also approaches the status of nation-state, but there are Hungarians (Magyars) not only in Slovakia but also in Serbia, Austria, and Romania (Fig. 1-26), all remnants of a time when Hungary ruled much of this region. Hungarian governments have repeatedly expressed support for these ethnic cohorts in neighboring countries, a practice referred to as **irredentism**. The term derives from a nineteenth-century campaign by Italy to incorporate an Italian-speaking area of Austria, calling it *Italia Irredenta* (Unredeemed Italy). In the case of Hungary, its recent manifestation was the so-called status law giving persons

32

ETHNIC MOSAIC OF EASTERN EUROPE

Slavic		Non-Slavic
Poles	Macedonians	Magyars
Czechs	Serbs	Albanians
Slovaks	Montenegrins	Romanians
Slovenes	Belarussians	Turks
Croats	Ukrainians	Pomaks
Muslims		Latvians
Bulgars		Lithuanians
Russians		No group over 50%

FIGURE 1-26 © H. J. de Blij, P. O. Muller, and John Wiley & Sons, Inc.

of Hungarian ancestry living in neighboring countries certain rights (including work permits) in the "motherland." The governments of those neighboring countries did not like this "law" applying to its citizens, and neither did the EU leadership. It has been modified, but public opinion in Hungary still supports the notion.

The Hungarians moved into the middle Danube River Basin more than a thousand years ago from an Asian source; they have neither Slavic nor Germanic roots. They converted their fertile lowland into a thriving state while retaining their cultural and linguistic identity, eventually forging an imperial power in the region. The twin-cities capital astride the Danube, Buda and Pest (better known as Budapest), is a primate city nearly ten times the size of the next largest town in Hungary—reflecting the continuing rural character of the country.

With a population of 10 million and a considerable and varied resource base, Hungary should have good economic prospects, which was one factor in its admission to EU membership in 2004. However, economic mismanagement and political intrigue combined to make Hungary, in 2007, Europe's most deeply indebted country.

Immediately after midnight January 1, 2007, Romania joined the European Union, but this crowd in Bucharest couldn't wait. It was the evening of December 31, 2006 when this photo was taken as the EU flag was hoisted while Romania was still in the final countdown to a momentous event in its long and troubled history. © Landov LLC.

Countries Facing the Black Sea

As Figure 1-25 shows, three major European states have Black Sea coasts, and one small country comes close. Two of these four joined the European Union in January 2007: Romania (see photo above) and Bulgaria. The two others, Ukraine and Moldova, would like to but are not likely to get the opportunity anytime soon.

Just how Romania managed to persuade EU leaders to endorse its 2007 admission remains a question for many Europeans. As Table G-1 indicates, Romania has some of Europe's worst social indicators; its economy is weak, its incomes are low, its political system has not been sufficiently upgraded from communist times (many "apparatchiks" have acquired state assets under the guise of "privatization" and are controlling the political process), and political infighting and corruption are endemic.

But Romania is an important country, located in the lower basin of the Danube River and occupying much of the heart of Eastern Europe. With 21.5 million inhabitants and a pivotal situation on the Black Sea, Romania is a bridge between central Europe and the realm's southeastern periphery, where EU member Greece and would-be member Turkey face each other across land and water.

Romania's drab and decaying capital of Bucharest (once known as the Paris of the Balkans) and its surrounding core area lie in the interior, linked by rail to the Black Sea port of Constanta. The country's once-productive oilfields have been depleted, about one-third of the labor force works in agriculture, poverty is widespread in the country-side as well as the towns, and unemployment is high. Many talented Romanians continue to leave the country in search of opportunities elsewhere. Romania has been described as the basket case of the Balkans, but that did not derail its EU membership drive.

Across the Danube lies Romania's neighbor, Bulgaria, southernmost of this subregion's countries. The rugged Balkan Mountains form Bulgaria's physiographic backbone, separating the Danube and Maritsa basins. As the map shows, Bulgaria has five neighbors, several of which are in political turmoil.

The Bulgarian state appeared in 1878, when the Russian czar's armies drove the Turks out of this area. The Slavic Bulgars, who form 84 percent of the population of 7.6 million, were (unlike the Romanians) loyal allies of Moscow during the Soviet period. But they did not treat their Turkish minority, about 10 percent of the population, very kindly, closing mosques, prohibiting use of the Turkish language, and forcing Turkish families to adopt Slavic names. Conditions for the remaining Turks improved somewhat after the end of the Soviet period. They improved further as Bulgaria became a candidate and then a member of the European Union, which required judicial and other social reforms.

Bulgaria has a Black Sea coast and an outlet, the port of Varna, but the main advantage it derives from its coastal location is the tourist trade its beaches generate. The capital, Sofia, lies not on the coast but near the opposite border with Serbia, and in its core-area hinterland some foreign investment is changing the economic landscape—but slowly. Bulgaria's GNI is on a par with Romania's, although the telltale agricultural sector is much smaller here.

Bulgarians, too, are emigrating in droves—and this worries the countries of Western Europe, where an uncontrolled influx of immigrants with newly won access to their job markets would create serious problems. In October 2006 the United Kingdom announced that it would place severe restrictions on workers from Romania and Bulgaria—a surprising reversal for a country that has long championed openness in the EU job market. It was a signal that, by admitting these two states, the EU's leaders had crossed a line.

Ukraine is Eastern Europe's most populous country (46.1 million); territorially, it is the largest state in the entire realm. Its capital, Kiev (Kyyiv), is a major historic, cultural, and political focus. Briefly independent before the communist takeover in Russia, Ukraine regained its sovereignty as a much-changed country in 1991. Once a land of farmers tilling its famously fertile soils, Ukraine emerged from the Soviet period with a huge industrial complex in its east—and with a large (17 percent) Russian minority. Ukraine's boundaries also changed during the Soviet era. In 1954, a Soviet dictator capriciously transferred the entire Crimea Peninsula, including its Russian inhabitants, to Ukraine as a reward for its productivity.

The Dnieper River forms a useful geographic reference to comprehend Ukraine's spatial division (Fig. 1-25). To its west lies agrarian, rural, mainly Roman Catholic Ukraine; in its great southern bend and eastward lies industrial, urban, Russified (and Russian Orthodox) Ukraine. Soviet planners made eastern Ukraine a communist Ruhr based on local coal and iron ore, making the Donets Basin (*Donbas* for short) a key industrial complex. Meanwhile, the Russian Soviet Republic supplied Ukraine with oil and gas.

As Figure 1-26 shows, Ukraine, with the exception of its eastern and urban-concentrated Russian minority, has an ethnically homogeneous population by East European standards. Ukraine is a critically important country for Europe's future, but it suffers from numerous problems ranging from political mismanagement and corruption to a faltering economy and rising crime. Yet this country has access to international shipping lanes, a large resource base, massive farm production, educated and skilled labor, and a large domestic market.

In 2004, its problems were evident when Ukraine became a political battleground with geostrategic overtones. In a key election, the pro-Russian East and the pro-European West each had a presidential candidate. Moscow overtly supported the eastern candidate; Europe clearly preferred his opponent. When the votes were tallied, the map showed the political fault line dividing the country (Fig. 1-27); but the victorious pro-Russian candidate provoked an "Orange Revolution" when electoral fraud swiftly came to light. That peaceful protest led to a second election a few weeks later in which the result was reversed—but the dividing line on the map

remained unchanged. The new president immediately sought to reassure Russia, even as his government made it clear that its aspirations are to join the EU at some future time. Russia's leader, however, retaliated in the winter of 2006 by drastically raising the price of natural gas with which it supplies Ukraine, and then, when Ukraine sought to negotiate, by curtailing the supply—affecting not only Ukraine, but also European consumers depending on the same pipelines. Once again, Ukraine's weakness and vulnerability cost it dearly.

Moldova, Ukraine's small and impoverished neighbor (which is by many measures Europe's poorest country) was a Romanian province seized by the Soviets in 1940 and made into a landlocked "Soviet Socialist Republic." A half-century later, along with other such "republics," Moldova gained independence when the Soviet Union collapsed. Romanians remain in the majority among its 4 million people, but most of the Russians and Ukrainians (each about 13 percent) have moved across the Dniester River to an industrialized strip of land between that river and the Ukrainian border, proclaiming there a "Republic of Transdniestria" (see Fig. 1-12, p. 69). But such separatist efforts form only one of Moldova's many problems. Its economy, dominated by farming, is in decline; an estimated 40 percent of the population works outside its borders because unemployment in Moldova is as high as 30 percent; smuggling and illegal arms trafficking are rife. These misfortunes translate into weakness, and Russia's malign influence in the form of support for Transdniestria's separatists keeps the country in turmoil. In fact, in 2007 the elected president of Moldova was persuaded by Russian pressure to recognize the regime of breakaway Transdniestria as legitimate.

Countries Facing the Adriatic Sea

As recently as 1990, only two Eastern European countries fronted the Adriatic Sea: Yugoslavia and Albania (Fig. 1-24). Albania survives, but the former Yugoslavia has splintered into six countries—and its disintegration may not be over.

We turn first to the *former Yugoslavia* ("Land of the South Slavs"), a country extending from Austria to Greece, thrown together in 1918 after World War I, containing 7 major and 17 smaller ethnic and cultural groups (Fig. 1-24). The Slovenes and Croats in the north were Roman Catholics; the Serbs in the south adhered to the Serbian Orthodox church. Several million Muslims also formed part of the cultural mosaic. Two alphabets were in use in separate regions of the country. At first the Royal House of Serbia dominated Yugoslavia; after 1945, a communist dictatorship personified by World War II hero Marshal Tito held

UKRAINE'S ELECTORAL GEOGRAPHY

Majority supporting Western-oriented candidate

Majority supporting candidate favored by Russia

FIGURE 1-27   © H. J. de Blij, P. O. Muller, and John Wiley & Sons, Inc.

Yugoslavia together. But in the early 1990s when the communist system collapsed in Eastern Europe and the Soviet Union disintegrated, so did the Yugoslav state.

Communist social planners laid the groundwork for the disaster that befell Yugoslavia. They divided the country into six internal "republics" based on the Soviet model, each dominated (except Bosnia) by one major group. Since all these "republics" inevitably incorporated minorities, the state guaranteed their rights, albeit through autocratic means.

When the communist system collapsed, individual "republics" proclaimed their independence—that is, the majorities in these entities did so. What remained of the machine of state, still dominated by the Serbs, tried to halt this disintegration. That effort soon failed, and new countries named Slovenia, Croatia, Bosnia, Macedonia, and, by subtraction, Serbia appeared on the map. Minorities in these new states, however, objected and, in Croatia and Bosnia, rose against those who were advocating statehood. The result was a catastrophic conflict just when grandiose Euro-unification schemes were under way. Europe, which twice during the twentieth century had plunged the world into war and had vowed that genocide would "never again" cloud its horizons, failed this first test of its declarations. EU members, led by Germany, quickly accorded official recognition to the post-Yugoslavian states, even before the concerns of the frightened minorities could be considered. Then, European powers stood by as more than 250,000 people were killed (perhaps a million more were injured), ethnic cleansing and mass executions depopulated entire areas, refugees streamed into neighboring countries, and historic treasures were demolished.

As we will discover, the devolutionary process continues in the twenty-first century. Today, the still-evolving map emerging from the former Yugoslavia consists of six countries: Slovenia, Croatia, Bosnia, Macedonia, Serbia, and Montenegro (Fig. 1-25). In addition, devolutionary pressures within Serbia have wrested the province of Kosovo from Serbian control.

Slovenia was the first "republic" to secede and proclaim independence. Ethnically the most homogeneous but territorially small with only 2 million of the former Yugoslavia's 23 million citizens, Slovenia lay farthest from the Serbian power core and seized its opportunity. This Alpine-mountain state, with the most productive economy among the republics even before independence, today has this subregion's highest per-capita GNI by far. It qualified easily to join the EU in 2004.

Croatia's crescent-shaped territory, with prongs along the Hungarian border and along the Adriatic coast, only partially reflects the distribution of Croats in the former Yugoslavia (Fig. 1-26). About 90 percent of Croatia's 4.4 million people are Croats, but another 800,000 live in a broad zone within southern Bosnia. Croatia's Serb minority, about 12 percent when independence came, has dwindled under Croatian pressure to less than 5 percent. Recent democratic reforms have improved Croatia's prospects.

Bosnia was the cauldron of calamity. Multicultural, effectively landlocked, and situated between the Croatian state to the west and the Serbian stronghold to the east, Bosnia fell victim to disastrous conflict among Serbs, Croats, and Bosniaks (now the official name for Bosnia's Muslims, who constitute about 50 percent of the population). In 1995, a U.S. diplomatic initiative resulted in a truce that partitioned Bosnia as shown in Figure 1-28, making this country (of just under 4 million today) a fragile federation.

Macedonia was the southernmost "republic" of former Yugoslavia and became a state with 2 million inhabitants. More than 60 percent of these inhabitants are Macedonian Slavs, about 25 percent are Muslim Albanians, and the remainder are Turks, Serbs, and Roma (Gypsies). Small, landlocked, poor, and powerless, Macedonia first faced the ire of Greece, which argued that the name "Macedonia" was Greek property, then confronted a massive flow of refugees entering from war-torn Kosovo, and most recently coped with an Albanian autonomy movement in its northwestern corner, where this minority is concentrated. Accommodation of some of the Muslims' demands, coupled with political concessions, has reduced the tensions.

Serbia is the name of what is left of the larger domain once ruled by the Serbs, who were dominant in the former Yugoslavia. Serb minorities still remain in Bosnia, Croatia, and other former Yugoslav "republics," but the Serbian sphere of influence continues to shrink. Centered on the historic capital of Belgrade on the Danube River, *Serbia* still is the largest of the post-Yugoslav states, but the country faces still more territorial losses. While its official 2008 population totals 9.5 million, Serbia incorporates a Hungarian minority of about 400,000 in its northern province of *Vojvodina* plus nearly 2 million Muslims in its southern territory of *Kosovo*. Kosovo is already under United Nations administration and may be next to leave the Serbian fold—which would leave only about 7.5 million citizens within Serbia's borders.

Meanwhile Serbia's hopes to join the EU are thwarted by the lingering effects of its recent history of misdeeds during the collapse of the former Yugoslavia. A record of mass murder and ethnic cleansing sent its former ruler, Slobodan Milosevic, to The Hague for trial; the current government has been unable to arrest and extradite other offenders, assumed to be hiding among Serb minorities outside Serbia. The EU will not begin admission procedures until those culprits have been arrested. Meanwhile Serbia remains in limbo, its economy and social conditions beset by uncertainty.

FIGURE 1-28 © H. J. de Blij, P. O. Muller, and John Wiley & Sons, Inc.

Montenegro, with a mere 630,000 inhabitants (about one-third of them Serbs, mostly clustered in the tiny territory's northeast near the Serbian border), voted in 2006 to separate from the country then called Serbia-Montenegro to become Europe's 40th state. Once an independent kingdom, Montenegro was forced into the Serb-dominated sphere after World War I and stayed with Serbia when Yugoslavia disintegrated. As a result, it suffered from the sanctions imposed on Serbia-Montenegro, its economy collapsing and a black market taking over. Montenegro has spectacular Adriatic shorelines and scenic mountains, and tourists are expected to return now that the political issue is settled.

Although the still-nominally Serbian province of *Kosovo* in 2007 continued to be administered by the UN, independence from Serbia may be in the offing here as well. In 1999, NATO took over this Serb-ruled but overwhelmingly Albanian-Muslim territory (with a population of nearly 2 million today) after ending Serbian control in a brief but damaging military campaign; NATO subsequently turned over Kosovo's administration to the United Nations. Since then, the small remaining Serbian minorities in Kosovo (see Fig. 1-28) have not fared well, and the political situa-

tion among the Muslim majority is still maturing. In Belgrade, the prospect of "losing" still another part of the once-Serbian domain, especially one in which historic battles between Muslims and Christians established Christian dominance, constitutes yet another source of tension and dismay. In mid-2007, immediate hopes for independence were sidetracked by Russian opposition in the UN Security Council; but negotiations were set to continue, raising prospects for a resolution by 2008.

It is appropriate to save our discussion of Albania for last because even in this turbulent region Albania is unusual. It is the only dominantly Muslim state in Europe; some 70 percent of its 3.3 million people adhere to Islam. Albania also rivals Moldova as the poorest country in Europe, ranking lowest on several indices of well-being. It is tied for the fastest rate of population growth in the realm. Most Albanians subsist on livestock herding and farming on the one-fifth of this mountainous, earthquake-prone country that can be used for agriculture. Thousands of Albanians have tried to reach Italy across the Adriatic Sea.

Kosovar and Macedonian Muslims look to Albania for support, but the government headquartered in Tirane cannot embark on irredentist campaigns. Albania has its own cultural divide between the poverty-stricken Gegs in the north and the somewhat-better-off Tosks in the south. National unity is its primary goal.

With 590 million inhabitants in 40 countries, including some of the world's highest-income economies, a politically stable and economically integrated Europe would be a superpower in the twenty-first century. But Europe's political geography is anything but stable, as devolutionary forces and cultural conflict continue to trouble the realm. Moreover, economic integration, despite the momentous EU expansions of 2004 and 2007, still involves just two-thirds of the realm's countries, a process that will become more difficult as the European Union confronts applications from marginally qualifying states in the east. Europe always has been a realm of revolutionary change, and it remains so today.

What You Can Do

RECOMMENDATION: Make your point in the press! Most newspapers publish letters to the editor, and often that page is the most interesting of all. Writing a letter to the editor, however, is not as easy as it might appear. When you have to make your argument in just 150 to 200 words (sometimes a few dozen more, depending on editorial policy), this has a way of focusing the mind. But if you have a geographically informed opinion on a matter of current concern, or if you have spotted an error on a map or in a story, the editor would like to hear from you (submission instructions are found on the newspaper's editorial page and website). It takes practice, and it is best to start with a local or regional newspaper, but the effort may pay off in several unexpected ways because even a local newspaper today attracts a global audience through its website and other electronic links. And you will be amazed how practice in accurate, concise writing under space limitations and for a wide audience will improve your exam-writing skills.

GEOGRAPHIC CONNECTIONS

1 You are a member of the staff of a multinational corporation doing business in several European countries. Your management has been used to dealing with governments in the capitals of a number of countries including the Netherlands (The Hague), Poland (Warsaw), and Hungary (Budapest). Now the company has announced plans to expand operations in Spain at Barcelona, in Belgium at Antwerp, in Ukraine at Donetsk, and in the United Kingdom at Edinburgh. You have been asked to prepare a "hurdles and pitfalls" report as part of the preparation for this initiative. What advice will you provide based on your knowledge of Europe's cultural geography and devolutionary pressures?

2 The European Union expanded from 15 to 27 member-states during the middle years of this decade, and the voters in every one of the joining countries endorsed their governments' desire for membership. But even as other countries hope to enter the EU in the foreseeable future, certain states have actually declined membership. What countries are these, and why would they stay out of the EU? Under what circumstances might a member-state decide to leave the EU? Can you relate these considerations to the decision by 3 of the 15 pre-2004 members not to join the EMU and to reject the use of the euro?

3 The largest and most populous country in Europe not to be a member of the EU is largely Christian Ukraine, and in 2007 Ukraine was not even in discussions aimed at EU membership. Yet Turkey, not in Europe and almost entirely Muslim, is in the early stages of negotiations toward eventual membership. What cultural-geographic and geopolitical factors explain this remarkable situation, and in what ways does this illustrate the ever-greater difficulties confronted by the EU as it makes an ever-greater impact in and on its periphery?

Geographic Literature on Europe: The key introductory works on this realm were authored by Berentsen, Gottmann, Jordan-Bychkov & Bychkova Jordan, McDonald, Ostergren & Rice, and Shaw. These works, together with a comprehensive listing of noteworthy books and recent articles on the geography of Europe, can be found in the *References and Further Readings* section of this book's website at **www.wiley.com/college/deblij.**

2

- **RUSSIAN CORE AND PERIPHERIES**
- **EASTERN FRONTIER**
- **SIBERIA**
- **RUSSIAN FAR EAST**

CONCEPTS, IDEAS, AND TERMS

1. **Near Abroad**
2. **Oligarch**
3. **Climatology**
4. **Continentality**
5. **Weather**
6. **Tundra**
7. **Taiga**
8. **Permafrost**
9. **Forward capital**
10. **Colonialism**
11. **Imperialism**
12. **Russification**
13. **Federation**
14. **Collectivization**
15. **Command economy**
16. **Unitary state system**
17. **Distance decay**
18. **Population decline**
19. **Heartland theory**
20. **Core area**

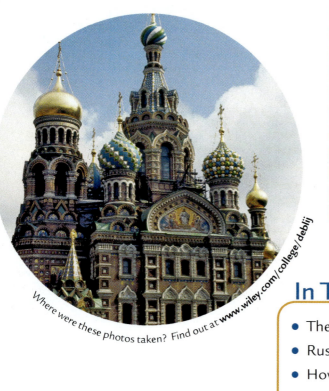

Where were these photos taken? Find out at www.wiley.com/college/deblij

In This Chapter

- The geography of Russian nationalism
- Russia's colonies and the Soviet Empire
- How Russia copes with distance and isolation
- Moscow and the War on Terror
- Where Russia meets China
- Trouble in Transcaucasia

Photos: © H. J. de Blij

FIGURE 2-1 *Map:* © H. J. de Blij, P. O. Muller, and John Wiley & Sons, Inc.

TAKE ANOTHER LOOK at Figure G-2 (p. 7) and you might wonder whether Russia should be counted as one of the world's geographic realms. Does this high-latitude country qualify as one of the dozen great building blocks of the so-called New World Order now in the making? Or is it merely another region of Europe, just an exceptionally large one? Well, here we have a geographic problem. If Russia is part of Europe, then Europe extends all the way across Eurasia to the Pacific Ocean. If it isn't, then Russia is either part of some other realm or it stands alone. Have a look at this issue on a globe (two-dimensional map projections have a way of distorting reality, and Russia in Figure G-2 looks a lot smaller than it really is).

In fact, Russia is territorially so large that the area of this single country exceeds that of East Asia (including China) and South Asia (including India) combined. Russia is almost three times the size of Europe, nearly as large as all of South America, and more than twice the size of continental Australia. From this perspective, there is no doubt that Russia is a world realm.

But Russia does not have a population to match its enormous territorial dimensions. East Asia and South Asia combined have a still-growing population of more than 3 billion; Russia has a shrinking population of just 141 million. That is less than the total of Brazil or Indonesia or Pakistan or Bangladesh! How can Russia be regarded as a major force in the political, economic, and social reordering of the world in the twenty-first century?

That is not just a geographic question. It vexes Russians as well. During the twentieth century Russia rose to global power, victorious in World War II, in control of a colonial empire that extended from the Baltic countries to Kazakhstan, empowered by massive armies and nuclear weapons, possessor of vast energy resources, a legitimate contender for world domination. Russia was the cornerstone of the Soviet Union, its communist ideology the key issue of the Cold War that followed World War II.

When the Soviet Union collapsed in 1991 and Russia lost its empire, chaos ensued. As millions of Russians streamed back to Russia from the former empire, Russia sought to keep some form of control over events in its **Near Abroad**, the so-called republics that previously fell under communist rule (and where millions of Russians today remain under the jurisdiction of non-Russian governments [Table 2-1]). At home, Russians struggled to reform their government toward some semblance of democracy even as minorities demanded rights ranging from autonomy to outright independence. A war broke out in the south, where Muslim rebels in Chechnya, one of Russia's internal republics, resisted continued Russian rule. The Chechens responded to Russian intervention by mounting a terrorist campaign that took hundreds of lives in the capital, Moscow, alone. Meanwhile, rapacious opportunists took advantage of Russia's weak economic controls to acquire huge components of the energy industry under the guise of "privatization." These newly rich **oligarchs** then tried to use their wealth to gain political power as well. Russia's first elected president, the ineffectual Boris Yeltsin, presided over a state in dangerous turmoil. In the world at large, Russia's status was at a low ebb.

Russian citizens, surveys indicated, yearned for a stronger government, better security, greater stability, and

TABLE 2-1

Social Characteristics of the Other 14 Ex-Soviet Republics

Name	Population (Millions)	Percent Russians	Religion	Official Language
Armenia	3.0	1	Armenian Orthodox (Christian) 79%	Armenian
Azerbaijan	8.7	3	Shi'ite Muslim 93%	Azeri
Belarus	9.6	11	Belarussian Orthodox 49%, Roman Catholic 13%	Belarussian
Estonia	1.3	26	Estonian Orthodox 17%, Estonian Lutheran 17%	Estonian
Georgia	4.4	6	Georgian, Russian Orthodox 84%; Sunni Muslim 11%	Georgian
Kazakhstan	15.5	30	Sunni Muslim 47%; Russian Orthodox 8%	Kazakh; Russian
Kyrgyzstan	5.3	13	Sunni Muslim 75%	Kyrgyz; Russian
Latvia	2.3	29	Christian Churches 67%	Latvian
Lithuania	3.4	7	Roman Catholic 79%	Lithuanian
Moldova	4.0	13	Eastern Orthodox 46%	Moldovan
Tajikistan	7.3	1	Sunni Muslim 85%	Tajik
Turkmenistan	5.5	6	Sunni Muslim 89%	Turkmen
Ukraine	46.1	17	Ukrainian Orthodox 38%	Ukrainian
Uzbekistan	27.0	6	Sunni Muslim 83%	Uzbek

MAJOR GEOGRAPHIC QUALITIES OF

Russia

1. Russia is the largest territorial state in the world. Its area is nearly twice as large as that of the next-ranking country (Canada).

2. Russia is the northernmost large and populous country in the world; much of it is cold and/or dry. Extensive rugged mountain zones separate Russia from warmer subtropical air, and the country lies open to Arctic air masses.

3. Russia was one of the world's major colonial powers. Under the czars, the Russians forged the world's largest contiguous empire; the Soviet rulers who succeeded the czars took over and expanded this empire.

4. For so large an area, Russia's population of under 141 million is comparatively small. The population remains heavily concentrated in the westernmost one-fifth of the country.

5. Development in Russia is concentrated west of the Ural Mountains; here lie the major cities, leading industrial regions, densest transport networks, and most productive farming areas. National integration and economic development east of the Urals extend mainly along a narrow corridor that stretches from the southern Urals region to the southern Far East around Vladivostok.

6. Russia is a multicultural state with a complex domestic political geography. Twenty-one internal Republics, originally based on ethnic clusters, function as politico-geographical entities.

7. Its large territorial size notwithstanding, Russia suffers from land encirclement within Eurasia; it has few good and suitably located ports.

8. Regions long part of the Russian and Soviet empires are realigning themselves in the postcommunist era. Eastern Europe and the heavily Muslim Southwest Asia realm are encroaching on Russia's imperial borders.

9. The failure of the Soviet communist system left Russia in economic disarray. Many of the long-term components described in this chapter (food-producing areas, railroad links, pipeline connections) broke down in the transition to the postcommunist order.

10. Russia long has been a source of raw materials but not a manufacturer of export products, except weaponry. Few Russian (or Soviet) automobiles, televisions, cameras, or other consumer goods reach world markets.

more effective delivery of social services ranging from education to pensions. Stronger government is what they got when Vladimir Putin succeeded Yeltsin as Russia's second elected president. Putin arrested oligarchs (some fled to other countries, but others wound up in jail), clamped down on the governments of Russia's regions, pursued Chechnya's rebels into the mountains, muzzled the once-freewheeling media, and vowed to the nation that he would restore Russia's powerful influence in world affairs. He managed to get Russia invited to join the so-called G-7, a group of the world's leading economies, making this the G-8 and raising more than a few eyebrows, because Russia's economy hardly qualified. Russia began to intervene forcefully in the Near Abroad with efforts to influence national elections (notably Ukraine's, see p. 102) and domestic affairs in such places as neighboring Georgia as well as nonneighbor Moldova. And Russia obstructed U.S. and European designs ranging from politics in the Balkans to nuclear development in Iran.

As you study the geography of this gigantic country, it is well to remember that Russia also remains one of the world's nuclear powers with the capacity to back up its renewed quest for global influence with military might. As we will see, post-Soviet Russia is a country troubled in many ways even as its leaders seek to resurrect its global geopolitical stature.

Defining the Realm

Although the name we give to this world realm is Russia, the geographic situation actually is not so simple (Fig. 2-1). The name is appropriate, because Russia is the overwhelming component of this realm. But, as we noted in the Introduction, geographic realms usually are not bounded by sharply defined political or physical features: they tend to display zones of transition on their margins. And so it is with Russia. In Chapter 1 we noted that Russia's western neighbor, Belarus, satisfies neither European nor Russian criteria. It was part of the Soviet Union and currently remains a throwback to the days of authoritarian politics and stagnant economics. In its western border areas, however, there is a European minority (mostly Polish) and some evidence of European influences. Still, in 2008 Belarus's convergence with Russia (rather than Europe) continued. That is why we map Belarus as one of the transition zones marking the "Russian" realm.

A second transitional area lies in the so-called Transcaucasus, the generally mountainous area between the Black and the Caspian seas just to Russia's south. As we will see in the regional discussion, the Russian realm suffers from internal as well as external problems in this zone. The internal problems focus on Chechnya, where Muslim rebels continue to resist Russian rule. The external issues focus on Georgia, one of the three former Soviet colonies in this area and the one with the largest Russian minority. While there is no doubt that these three countries lie to various degrees in the Russian orbit, things are changing in Transcaucasia and here, too, we must recognize a transition zone.

In a later chapter we will encounter yet another transition zone, one more relic of the Soviet past. During communist rule, millions of Russians moved to the Soviet "republics" outside Russia. But in one such republic, Kazakhstan, many of them moved into a zone right across the border from Russia and proceeded to "Russify" that area so completely that its cultural landscape became Russian. Many of those Russians and their descendants have now moved back to Russia, but northern Kazakhstan still carries its Russian imprint and constitutes another one of the realm's peripheral transition zones.

And still another transition zone may be developing, this one internal. As we will discover, Russia's population is shrinking, and this decline is especially rapid in Russia's Far East, in its Pacific margin where Russia borders China and North Korea. With North Korea in disarray and refugees crossing the border, and with Chinese workers and tradespeople filtering into Russia from China's economically lagging Northeast, Russia's distant frontier is taking on new demographic and social characteristics. The immigration question is fiercely debated in Moscow, but the forces of change are already altering the cultural landscape of Russia's remote eastern corner.

An Internal Empire

And within its borders, Russia still is an empire. It was the Soviet Empire that disintegrated, not the Russian one: about 20 percent of Russia's more than 140 million inhabitants are non-Russians, including numerous minorities large and small whose presence is reflected by the administrative map. The Chechens are but one of those minorities, many of which look back upon a history of mistreatment but most of which have accommodated themselves to the reality of Russian dominance. Russia's failure in Chechnya should not obscure what was in fact a remarkable success of the Yeltsin administration: avoiding the collapse of administrative order and the secession of minority-based

In the 1960s visitors to the (then) Soviet Union, one of the authors of this book among them, were shocked by the fate of virtually all of the country's Christian churches, whose condition ranged from bad repair to outright ruin. Architecturally ornate and historic church buildings with superb murals and exquisite glasswork stood empty, subject to the elements and vandalism. Doors and furniture had been taken for firewood; some churches served as barns, others as storage. In his atheist zeal, Josef Stalin ordered the demolition of one of Moscow's religious treasures, the Christ-the-Savior Cathedral (other Moscow cathedrals fared better and were turned into museums). This photograph symbolizes the reversal of fortunes following the collapse of the Soviet Union: religion is again a part of Russian life, the Russian Orthodox Church again dominates the religious map, and a reconstructed Cathedral of the Savior again stands where it did before the communist era.   © Harf Zimmermann.

"republics" throughout the country. For a time, it seemed that Tatarstan, the historic home of Muslim Tatars, would declare independence along with other minorities long unhappy under Soviet-Russian rule. But Yeltsin's government gave these peoples ways to express their demands and to negotiate their terms, and in the end the Russian map stayed intact—except in the Transcaucasus. Still, one of the risks inherent in a return to "strong" government is a reversal of this good fortune. Chechnya's rebellion and Muslim Chechens' terrorist actions have already made life much more difficult for other minorities on the streets of Moscow and other cities, where many ethnic Russians equate minority appearance and dress with rebellion and risk. Alienating Russia's other minorities risks the future of the state itself.

ROOTS OF THE RUSSIAN REALM

The name **Russia** evokes cultural-geographic images of a stormy past: terrifying czars, conquering Cossacks, Byzantine bishops, rousing revolutionaries, clashing cultures. Russians repulsed the Tatar (Mongol) hordes, forged a powerful state, colonized a vast contiguous empire, defeated Napoleon, adopted communism, and, when the communist system failed, lost most of their imperial domain. Today, Russia seeks a new place in the globalizing world.

A Vibrant Culture

Precommunist Russia may be described as a culture of extremes. It was a culture of strong nationalism, resistance to change, and despotic rule. Enormous wealth was concentrated in a small elite. Powerful rulers and bejeweled aristocrats perpetuated their privileges at the expense of millions of peasants and serfs who lived in dreadful poverty. The Industrial Revolution arrived late in Russia, and a middle class was slow to develop. Yet the Russian nation gained the loyalty of many of its citizens, and its writers and artists were among the world's greatest. Authors such as Tolstoy and Dostoyevsky chronicled the plight of the poor; composers celebrated the indomitable Russian people and, as Tchaikovsky did in his *1812 Overture*, commemorated their victories over foreign foes.

Empire of the Czars

Under the czars, Russia grew from nation into empire. The czars' insatiable demands for wealth, territory, and power sent Russian armies across the plains of Siberia, through

The Soviet Union, 1924–1991

FOR 67 YEARS Russia was the cornerstone of the Soviet Union, the so-called Union of Soviet Socialist Republics (USSR). The Soviet Union was the product of the Revolution of 1917, when more than a decade of rebellion against the rule of Czar Nicholas II led to his abdication. Russian revolutionary groups were called *soviets* ("councils"), and they had been active since the first workers' uprising in 1905. In that year, thousands of Russian workers marched on the czar's palace in St. Petersburg in protest, and soldiers fired on them. Hundreds were killed and wounded. Russia descended into chaos.

A coalition of military and professional men forced the czar's abdication in 1917. Then Russia was ruled briefly by a provisional government. In November 1917, the country held its first democratic election—and, as it turned out, its last for more than 70 years.

The provisional government allowed the exiled activists in the Bolshevik camp to return to Russia (there were divisions among the revolutionaries): Lenin from Switzerland, Trotsky from New York, and Stalin from internal exile in Siberia. In the ensuing political struggle, Lenin's Bolsheviks gained control over the revolutionary soviets, and this ushered in the era of communism. In 1924, the new communist empire was formally renamed the Union of Soviet Socialist Republics, or Soviet Union for short.

Lenin the organizer died in 1924 and was succeeded by Stalin the tyrant, and many of the peoples under Moscow's control suffered unimaginably. In pursuit of communist reconstruction, Stalin and his adherents starved millions of Ukrainian peasants to death, forcibly relocated entire ethnic groups (including a people known as the Chechens), and exterminated "uncooperative" or "disloyal" peoples. The full extent of these horrors may never be known.

On December 25, 1991, the inevitable occurred: the Soviet Union ceased to exist, its economy a shambles, its political system shattered, the communist experiment a failure. The last Soviet president, Mikhail Gorbachev, resigned, and the Soviet hammer-and-sickle flag flying atop the Kremlin was lowered for the last time and immediately replaced by the white, red, and blue Russian tricolor. A new and turbulent era had begun, but Soviet structures and systems will long cast their shadows over transforming Russia.

the deserts of interior Asia, and into the mountains along Russia's rim. Russian pioneers ventured even farther, entering Alaska, traveling down the Pacific coast of North America, and planting the Russian flag near San Francisco in 1812. But as Russia's empire expanded, its internal weaknesses gnawed at the power of the czars. Peasants rebelled. Unpaid (and poorly fed) armies mutinied. When the czars tried to initiate reforms, the aristocracy objected. The empire at the beginning of the twentieth century was ripe for revolution, which began in 1905 (see box titled "The Soviet Union, 1924–1991").

Communist Victory

The last czar, Nicholas II, was overthrown in 1917, and civil war followed. The victorious communists led by V. I. Lenin soon swept away much of the Russia of the past. The Russian flag disappeared. The czar and his family were executed. The old capital of Russia, St. Petersburg, was renamed Leningrad in honor of the revolutionary leader. Moscow, in the interior of the country, was chosen as the new capital for a country with a new name, the *Soviet Union*. Eventually, this Union consisted of 15 political entities, each a Soviet Socialist Republic (SSR). Russia was just one of these republics, and the name *Russia* disappeared from the international map.

But on the Soviet map, Russia was the giant, the dominant republic, the Slavic center. Not for nothing was the communist revolution known as the Russian Revolution. The other republics of the Soviet Union were for minorities the czars had colonized or for countries that fell under Soviet sway later, but none could begin to match Russia. The Soviet Empire was the legacy of czarist expansionism, and the new communist rulers were Russian first and foremost (with the exception of Josef Stalin, who was a Georgian).

Seven Fateful Decades

Officially, the Union of Soviet Socialist Republics endured from 1924 to 1991. When the USSR collapsed, the republics were set free and world maps had to be redrawn. Now Russia and its former colonies had to adjust to the realities of a changing world order. It proved to be a difficult, sometimes desperate journey.

RUSSIA'S PHYSICAL GEOGRAPHY

Russia's physiography is dominated by vast plains and plateaus rimmed by rugged mountains (Fig. 2-2). Only the Ural Mountains break an expanse of low relief that

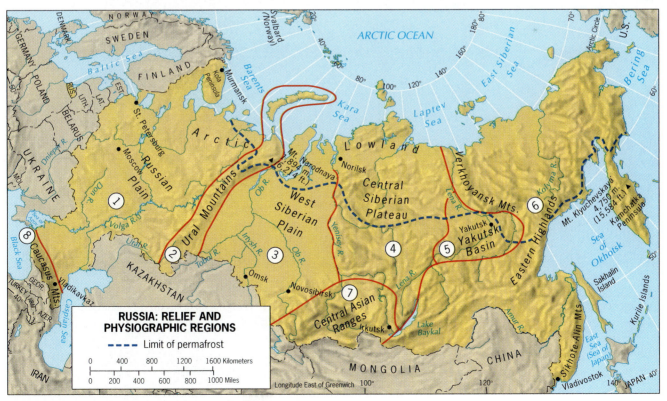

FIGURE 2-2 © H. J. de Blij, P. O. Muller, and John Wiley & Sons, Inc.

extends from the Polish border to eastern Siberia. In the western part of this huge plain, where the northern pine forests meet the midlatitude grasslands, the Slavs established their domain.

Harsh Environments

The historical geography of Russia is the story of Slavic expansion from its populous western heartland across interior Eurasia to the east, and into the mountains and deserts of the south. This eastward march was hampered not only by vast distances but also by harsh natural conditions. As the northernmost populous country on Earth, Russia has virtually no natural barriers against the onslaught of Arctic air. Moscow lies farther north than Edmonton, Canada, and St. Petersburg lies at latitude 60° North—the latitude of the southern tip of Greenland. Winters are long, dark, and bitterly cold in most of Russia; summers are short and growing seasons limited. Many a Siberian frontier outpost was doomed by cold, snow, and hunger.

It is therefore useful to view Russia's past, present, and future in the context of its **climatology**. This field of geography investigates not only the distribution of climatic conditions over the Earth's surface but also the processes that generate this spatial arrangement. The Earth's atmosphere traps heat received as radiation from the Sun, but this greenhouse effect varies over planetary space—and over time. As we noted in the introductory chapter, much of what is today Russia was in the grip of a glaciation until the onset of the warmer Holocene. But even today, with a natural warming cycle in progress augmented by human activity, Russia still suffers from severe cold and associated drought. If the enhanced global warming cycle continues, Russia (in contrast to low-lying countries faced by rising sea levels) may benefit. But that will take generations.

Currently, precipitation totals, even in western Russia, range from modest to minimal because the warm, moist air carried across Europe from the North Atlantic Ocean loses much of its warmth and moisture by the time it reaches Russia. Figures G-7, G-8, and 2-3 reveal the consequences. Russia's climatic **continentality** (inland climatic environment remote from moderating and moistening maritime influences) is expressed by its prevailing *Dfb* and *Dfc* conditions. Compare the Russian map to that of North America (Fig. G-8), and you note that, except for a small corner off the Black Sea, Russia's climatic conditions resemble those of the U.S. Upper Midwest and Canada. Along its entire north, Russia has a zone of *E* climates, the most frigid on the planet. In these Arctic latitudes originate the polar air masses that dominate its environments.

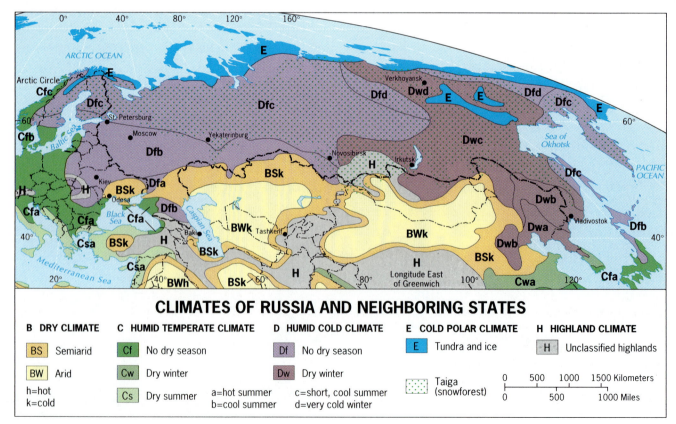

CLIMATES OF RUSSIA AND NEIGHBORING STATES

B DRY CLIMATE
BS Semiarid
BW Arid
h=hot
k=cold

C HUMID TEMPERATE CLIMATE
Cf No dry season
Cw Dry winter
Cs Dry summer
a=hot summer
b=cool summer

D HUMID COLD CLIMATE
Df No dry season
Dw Dry winter
c=short, cool summer
d=very cold winter

E COLD POLAR CLIMATE
E Tundra and ice
Taiga (snowforest)

H HIGHLAND CLIMATE
H Unclassified highlands

0   500   1000   1500 Kilometers
0   500   1000 Miles

FIGURE 2-3   © H. J. de Blij, P. O. Muller, and John Wiley & Sons, Inc.

Climates and Peoples

By studying Russia's climates we can begin to understand what the map of population distribution (Fig. G-9) shows. The overwhelming majority of the country's more than 140 million people are concentrated in the west and southwest, where climatic conditions were least difficult at a time when farming was the mainstay for most of the people. The map is a legacy that will mark Russia's living space for generations to come.

Environment and Politics

Climate and weather (there is a distinction: *climate* is a long-term average, whereas **weather** refers to existing atmospheric conditions at a given place and time) have always challenged Russia's farmers. Conditions are most favorable in the west, but even there temperature extremes, variable and undependable rainfall, and short growing seasons make farming difficult. During the Soviet period, fertile and productive Ukraine supplied much of Russia's food needs, but even then Russia often had to import grain. Soviet rulers wanted to reduce their country's dependence on imported food, and their communist planners built major irrigation projects to increase crop yields in the colonized republics of Central Asia. As we will see in Chapter 7, some of these attempts to overcome nature's limitations spelled disaster for the local people.

Physiographic Regions

To assess the physiography of this vast country, refer again to Figure 2-2. Note how mountains and deserts encircle Russia: the Caucasus in the southwest ⑧; the Central Asian Ranges in the center ⑦; the Eastern Highlands facing the Pacific from the Bering Sea to the East Sea (Sea of Japan) ⑥. The Kamchatka Peninsula has a string of active volcanoes in one of the world's most earthquake-prone zones (Fig. G-5). Warm subtropical air thus has little opportunity to penetrate Russia, while cold Arctic air sweeps southward without impediment. Russia's Arctic north is a gently sloping lowland broken only by the Urals and Eastern Highlands.

Russia's vast and complex physical stage can be divided into eight physiographic regions, each of which, at a larger scale, can be subdivided into smaller units. In the Siberian region, one criterion for such subdivision would be the vegetation. The Russian language has given us two terms to describe this vegetation: **tundra**, the treeless plain along the Arctic shore where mosses, lichens, and some grasses survive, and **taiga**, the mostly coniferous forests that begin to the south of where the tundra ends, and extend over vast reaches of Siberia, which means the "sleeping land" in Russian.

The Russian Plain

The Russian Plain ① is the eastward continuation of the North European Lowland, and here the Russian state formed its first core area. Travel north from Moscow at its heart, and the countryside soon is covered by needleleaf forests like those of Canada; to the south lie the grain fields of southern Russia and, beyond, those of Ukraine. Note the Kola Peninsula and Barents Sea in the far north: warm water from the North Atlantic comes around northern Norway and keeps the port of Murmansk ice free most of the year. The Russian Plain is bounded on the east by the Ural Mountains ②; though not a high range, it is topographically prominent because it separates two extensive plains. The main range of the Urals is more than 3200 kilometers (2000 mi) long and reaches from the shores of the Kara Sea to the border with Kazakhstan. It is not a barrier to east-west transportation, and its southern end is densely populated. Here the Urals yield minerals and fossil fuels.

Siberia

East of the Urals lies Siberia. The West Siberian Plain ③ has been described as the world's largest unbroken lowland; this is the vast basin of the Ob and Irtysh rivers. Over the last 1600 kilometers (1000 mi) of its course to the Arctic Ocean, the Ob falls less than 100 meters (330 ft). In Figure 2-2, note the dashed line that extends from west of the northern Urals to the Bering Sea and includes the Arctic Lowland. North of the line, water in the ground is permanently frozen; this **permafrost** creates another obstacle to permanent settlement. Looking again at the West Siberian Plain, we see that the north is permafrost-ridden and the central zone is marshy. The far south, however, has such major cities as Omsk and Novosibirsk, beyond the reach of permafrost conditions.

East of the West Siberian Plain the country begins to rise, first into the Central Siberian Plateau ④, another sparsely settled, remote, permafrost-affected region. Here winters are long and cold, and summers are short; the area remains barely touched by human activity. Beyond the Yakutsk Basin ⑤, the terrain becomes mountainous and the relief high. The Eastern Highlands ⑥ are a jumbled mass of ranges and ridges, precipitous valleys, and volcanic mountains. Lake Baykal lies in a trough that is over 1500 meters (5000 ft) deep—the deepest rift lake in the world (see Chapter 6). On the Kamchatka Peninsula, volcanic Mount Klyuchevskaya reaches 4750 meters (15,584 ft).

Encircling Mountains

The northern part of region ⑥ is Russia's most inhospitable zone, but southward along the Pacific coast the climate is less severe. Nonetheless, this is a true frontier region. The forests provide opportunities for lumbering, a fur trade exists, and there are gold and diamond deposits.

Mountains also mark the southern margins of much of Russia. The Central Asian Ranges ⑦, from the Kazakh border in the west to Lake Baykal in the east, contain many glaciers whose annual meltwaters send alluvium-laden streams to enrich farmlands at lower elevations. The Caucasus ⑧, in the land corridor between the Black and Caspian seas, form an extension of Europe's Alpine Mountains and exhibit a similarly high *relief* (range of elevations). Here Russia's southern border is sharply defined by *topography* (surface configuration).

As our physiographic map suggests, the more habitable terrain in Russia becomes latitudinally narrower from west to east. Beyond the southern Urals, the zone of settlement in Russia becomes discontinuous (in Soviet times it extended into northern Kazakhstan, where a large Russian population still lives). Isolated towns did develop in Russia's vast eastern reaches, even in Siberia, but the tenuous ribbon of settlement does not widen again until it reaches the country's Far Eastern Pacific Rim.

EVOLUTION OF THE RUSSIAN STATE

Four centuries ago, when European kingdoms were sending fleets to distant shores to search for riches and capture colonies, there was little indication that the largest of all empires would one day center on a city in the forests halfway between Sweden and the Black Sea. The plains of Eurasia south of the taiga had seen waves of migrants sweep from east to west: Scythians, Sarmatians, Goths, Huns, and others came, fought, settled, and were absorbed or driven off. Eventually, the Slavs, heirs to these Eurasian infusions, emerged as the dominant culture in what is today Ukraine, south of Russia and north of the Black Sea.

From this base on fertile, productive soils, the Slavs expanded their domain, making Kiev (Kyyiv) their southern headquarters and establishing Novgorod on Lake Ilmen in the north. Each was the center of a state known as a *Rus*, and both formed key links on a trade route between the German-speaking Hanseatic ports of the Baltic Sea and the trading centers of the Mediterranean. During the eleventh and twelfth centuries, the Kievan Rus and the Novgorod Rus united to form a large and comparatively prosperous state astride forest and *steppe* (short-grass prairie).

The Mongol Invasion

Prosperity attracts attention, and far to the east, north of China, the Mongol Empire had been building under Genghis Khan. In the thirteenth century, the Mongol (Tatar) "hordes" rode their horses into the Kievan Rus, and the state fell. By mid-century, Slavs were fleeing into the forests, where the Mongol horsemen had less success than on the open plains of Ukraine. Here the Slavs reorganized, setting up several new Russes. Still threatened by the Tatars, the ruling princes paid tribute to the Mongols in exchange for peace.

Moscow, deep in the forest on the Moscow River, was one of these Russes. Its site was remote and defensible. Over time, Moscow established trade and political links with Novgorod and became the focus of a growing area of Slavic settlement. When the Mongols, worried about Moscow's expanding influence, attacked near the end of the fourteenth century and were repulsed, Moscow emerged as the unchallenged center of the Slavic Russes. Soon its ruler, now called a Grand Duke, took control of Novgorod, and the stage was set for further growth.

Grand Duchy of Muscovy

Moscow's geographic advantages again influenced events. Its centrality, defensibility, links to Slavic settlements including Novgorod, and open frontiers to the north and west, where no enemies threatened, gave Moscow the potential to expand at the cost of its old enemies, the Mongols. Many of these invaders had settled in the basin of the Volga River and elsewhere, where the city of Kazan had become a major center of Islam, the Tatars' dominant religion. During the rule of Ivan the Terrible (1547–1584), the Grand Duchy of Muscovy became a major military power and imperial state. Its rulers called themselves czars and claimed to be the heirs of the Byzantine emperors. Now began the expansion of the Russian domain (Fig. 2-4), marked by the defeat of the Tatars at Kazan and the destruction of hundreds of mosques.

The Cossacks

This eastward expansion of Russia was spearheaded by a relatively small group of seminomadic peoples who came to be known as Cossacks and whose original home was in present-day Ukraine. Opportunists and pioneers, they sought the riches of the eastern frontier, chiefly fur-bearing animals, as early as the sixteenth century. By the mid-seventeenth century they reached the Pacific Ocean, defeating Tatars in their path and consolidating their gains

GROWTH OF THE RUSSIAN EMPIRE

Grand Duchy of Moscow 1462, Russia 1533

Territory Gained
- 1533–1598
- 1598–1689
- 1689–1725
- 1725–1801
- 1801–1945

Western Border
- 1864
- 1920

ALASKA Permanent settlements were established in 1784. Territory sold to United States in 1867. Fort Ross, California built by Russians in 1812. Relinquished in 1840.

FINLAND gained by Russia from Sweden 1809. Independent since 1918.

GRAND DUCHY OF WARSAW Gained by Russia under the Vienna Settlement (1815). Lost in 1918 on the formation of an independent Poland.

ESTONIA incorporated into Russia 1721. Independent 1918–1940; 1991–.

PECHENGA (PETSAMO) Area ceded by Finland to Russia in 1940.

KALININGRAD OBLAST under Soviet administration 1945–1991. Now Russian.

BELORUSSIA (BELARUS) Russian 1795–1920. Reincorporated into the Soviet Union in 1939. Independence in 1991.

SAKHALIN under joint Russo-Japanese control 1854–1875. Became Russian in 1875. Southern part ceded to Japan in 1905. Reincorporated 1945.

Incorporated into Ukraine from Czechoslovakia in 1945.

MOLDAVIA (BESSARABIA) Russian 1812–1918. Incorporated into Romania from 1918 to 1940. The territory then passed back to the USSR. Independence as Moldova in 1991.

KURILE ISLANDS Divided between Russia and Japan 1854. Passed to Japan in 1875. Incorporated into USSR 1945.

ARDAHAN AND KARS Changed hands between Russia and Turkey several times in the 19th century. Annexed by Russia in 1878 and returned to Turkey in 1921.

THE KANATE OF BUKHARA Became a Russian vassal in 1868 and then THE KHANATE OF KHIVA 1873. They were merged into the Soviet system in 1920.

TUVA made protectorate in 1911. Joined the USSR in 1944.

THE KWANYUNG TERRITORY Leased to Russia 1898–1905 and 1945–1955.

FIGURE 2-4   © H. J. de Blij, P. O. Muller, and John Wiley & Sons, Inc.

by constructing *ostrogs* (strategic fortified waystations) along river courses. Before the eastward expansion halted in 1812 (Fig. 2-4), the Russians had moved across the Bering Strait to Alaska and down the western coast of North America into what is now northern California (see box titled "Russians in North America").

Czar Peter the Great

When Peter the Great became czar (he ruled from 1682 to 1725), Moscow already lay at the center of a great empire—great, at least, in terms of the territories it controlled. The Islamic threat had been ended with the defeat of the Tatars. The influence of the Russian Orthodox Church was represented by its distinctive religious architecture and powerful bishops. Peter consolidated Russia's gains and hoped to make a modern, European-style state out of his loosely knit country. He built St. Petersburg as a **forward capital** on the doorstep of Swedish-held Finland, fortified it with major military installations, and made it Russia's leading port.

Czar Peter the Great, an extraordinary leader, was in many ways the founder of modern Russia. In his desire to

remake Russia—to pull it from the forests of the interior to the waters of the west, to open it to outside influences, and to relocate its population—he left no stone unturned. Prominent merchant families were forced to move from other cities to St. Petersburg. Ships and wagons entering the city had to bring building stones as an entry toll. The czar himself, aware that to become a major power Russia had to be strong at sea as well as on land, went to the Netherlands to work as a laborer in the famed Dutch shipyards to learn the most efficient method for building ships. Meanwhile, the czar's forces continued to conquer people and territory: Estonia was incorporated in 1721, widening Russia's window to the west, and major expansion soon occurred south of the city of Tomsk (Fig. 2-4).

Czarina Catherine the Great

Under Czarina Catherine the Great, who ruled from 1760 to 1796, Russia's empire in the Black Sea area grew at the expense of the Ottoman Turks. The Crimea Peninsula, the port city of Odesa (Odessa), and the whole northern coastal zone of the Black Sea fell under Russian control. Also during this

Russians in North America

THE FIRST WHITE settlers in Alaska were Russians, not Western Europeans, and they came across Siberia and the Bering Strait, not across the Atlantic and North America. Russian hunters of the sea otter, valued for its high-priced pelt, established their first Alaskan settlement at Kodiak Island in 1784. Moving south along the North American coast, the Russians founded additional villages and forts to protect their tenuous holdings until they reached as far as the area just north of San Francisco Bay, where they built Fort Ross in 1812.

But the Russian settlements were isolated and vulnerable. European fur traders began to put pressure on their Russian competitors, and St. Petersburg found the distant settlements a burden and a risk. In any case, American, British, and Canadian hunters were decimating the sea otter population, and profits declined. When U.S. Secretary of State William Seward offered to purchase Russia's North American holdings in 1867, St. Petersburg quickly agreed—for $7.2 million. Thus Alaska, including its lengthy southward coastal extension, became U.S. territory and, in 1959, the forty-ninth State. Although Seward was ridiculed for his decision—Alaska was called "Seward's Folly" and "Seward's Icebox"—he was vindicated when gold was discovered there in the 1890s. The twentieth century further underscored the wisdom of Seward's action, strategically as well as economically. At Prudhoe Bay off Alaska's northern Arctic slope, large oil reserves were tapped in the 1970s and are still being exploited. And like Siberia, Alaska probably contains yet-unknown riches.

period, the Russians made a fateful move: they penetrated the area between the Black and Caspian seas, the mountainous Caucasus with dozens of ethnic and cultural groups, many of which were Islamized. The cities of Tbilisi (now in Georgia), Baki (Baku) in Azerbaijan, and Yerevan (Armenia) were captured. Eventually, the Russian push toward an Indian Ocean outlet was halted by the British, who held sway in Persia (modern Iran), and also by the Turks.

Meanwhile, Russian colonists entered Alaska, founding their first American settlement at Kodiak Island in 1784. As they moved southward, numerous forts were built to protect their tenuous holdings against indigenous peoples. Eventually, they reached nearly as far south as San Francisco Bay where, in the fateful year 1812, they erected Fort Ross.

Catherine the Great had made Russia a colonial power, but the Russians eventually gave up on their North American outposts. The sea-otter pelts that had attracted the early pioneers were running out, European and white American hunters were cutting into the profits, and Native American resistance was growing. When U.S. Secretary of State William Seward offered to purchase Russia's Alaskan holdings in 1867, the Russian government quickly agreed—for $7.2 million. Thus Alaska and its Russian-held panhandle became U.S. territory and, ultimately, the forty-ninth State.

A Russian Empire

Although Russia withdrew from North America, Russian expansionism during the nineteenth century continued in Eurasia. While extending their empire southward, the Rus-sians also took on the Poles, old enemies to the west, and succeeded in taking most of what is today the Polish state, including the capital of Warsaw. To the northwest, Russia took over Finland from the Swedes in 1809.

Penetration of Muslim Asia

During most of the nineteenth century, however, the Russians were preoccupied with Central Asia—the region between the Caspian Sea and western China—where Tashkent and Samarqand (Samarkand) came under St. Petersburg's control (Fig. 2-4). The Russians here were still bothered by raids of nomadic horsemen, and they sought to establish their authority over the Central Asian steppe country as far as the edges of the high mountains that lay to the south. Thus Russia gained many Muslim subjects, for this was Islamic Asia they were penetrating. Under czarist rule, these people retained some autonomy.

Confronting China and Japan

Much farther to the east, a combination of Japanese expansionism and a decline of Chinese influence led Russia to annex from China several provinces east of the Amur River. Soon thereafter, in 1860, the port of Vladivostok on the Pacific was founded.

Now began the events that were to lead to the first involuntary halt to the Russian drive for territory. As Figure 2-4 shows, the most direct route from western Russia to the port of Vladivostok lay across northeastern China, the territory then still called Manchuria. The Russians had begun construction of the Trans-Siberian Railroad in 1892, and

FROM THE FIELD NOTES

"Not only the city of St. Petersburg itself, but also its surrounding suburbs display the architectural and artistic splendor of czarist Russia. The czars built opulent palaces in these outlying districts (then some distance from the built-up center), among which the Catherine Palace, begun in 1717 and completed in 1723 followed by several expansions, was especially majestic. During my first visit in 1994 the palace, parts of which had been deliberately destroyed by the Germans during World War II, was still being restored; large black and white photographs in the hallways showed what the Nazis had done and chronicled the progress of the repairs during communist and post-communist years. A return visit in 2000 revealed the wealth of sculptural decoration on the magnificent exterior (left) and the interior detail of a set of rooms called the 'golden suite' of which the ballroom (right) exemplifies eighteenth-century Russian Baroque at its height."   © H. J. de Blij.

www.conceptcaching.com

they wanted China to permit the track to cross Manchuria. But the Chinese resisted. Then, taking advantage of the Boxer Rebellion in China in 1900 (see Chapter 9), Russian forces occupied Manchuria so that railroad construction might proceed.

This move, however, threatened Japanese interests in this area, and the Japanese confronted the Russians in the Russo-Japanese War of 1904–1905. Not only was Russia defeated and forced out of Manchuria: Japan even took possession of the southern part of Sakhalin Island, which they named Karafuto and retained until the end of World War II.

THE COLONIAL LEGACY

Thus Russia, like Britain, France, and other European powers, expanded through **colonialism**. Yet whereas the other European powers expanded overseas, Russian influence traveled overland into Central Asia, Siberia, China, and the Pacific coastlands of the Far East. What emerged was not the greatest empire but the largest territorially contiguous empire in the world. At the time of the Japanese war, the Russian czar controlled more than 22 million square kilometers (8.5 million sq mi), just a tiny fraction less than the area of the Soviet Union after the 1917 Revolution. Thus the communist empire, to a large extent, was the legacy of St. Petersburg and European Russia, not the product of Moscow and the socialist revolution.

The czars embarked on their imperial conquests in part because of Russia's relative location: Russia always lacked warm-water ports. Had the Revolution not intervened, their southward push might have reached the Persian Gulf or even the Mediterranean Sea. Czar Peter the Great envisaged a Russia open to trading with the entire world; he developed St. Petersburg on the Baltic Sea into Russia's leading port. But in truth, Russia's historical geography is one of remoteness from the mainstreams of change and progress, as well as one of self-imposed isolation.

An Imperial, Multinational State

Centuries of Russian expansionism did not confine itself to empty land or unclaimed frontiers. The Russian state became an imperial power that annexed and incorporated many nationalities and cultures. This was done by employing force of arms, by overthrowing uncooperative rulers, by annexing territory, and by stoking the fires of ethnic conflict. By the time the ruthless Russian regime began to face revolution among its own people, czarist Russia was a hearth of **imperialism**, and its empire contained peoples representing more than 100 nationalities. The winners in the ensuing revolutionary struggle—the communists who forged the Soviet Union—did not liberate these subjugated peoples. Rather, they changed the empire's framework, binding the peoples colonized by the czars into a new system that would in theory give them autonomy and identity. In practice, it doomed those peoples to bondage and, in some cases, extinction.

When the Soviet system failed and the Soviet Socialist Republics became independent states, Russia was left without the empire that had taken centuries to build and consolidate—and that contained crucial agricultural and mineral resources. No longer did Moscow control the farms of Ukraine and the oil and natural gas reserves of Central Asia. But look again at Figure 2-4 and you will see that, even without its European and Central Asian colonies, Russia remains an empire. Russia lost the "republics" on its periphery, but Moscow still rules over a domain that extends from the borders of Finland to North Korea. Inside that domain Russians are in the overwhelming majority, but many subjugated nationalities, from Tatars to Yakuts, still inhabit ancestral homelands. Accommodating these many indigenous peoples is one of the challenges facing the Russian Federation today.

In the 1990s, Russia began to reorganize in the aftermath of the collapse of the Soviet Union. This reorganization cannot be understood without reference to the seven decades of Soviet communist rule that went before. We turn next to this crucial topic.

FROM THE FIELD NOTES

"June 25, 1964. We left Moscow this morning after our first week in the Soviet Union, on the road to Kiev. Much has been seen and learned after we arrived in Leningrad from Helsinki (also by bus) last Sunday. Some interesting contradictions: in a country that launched *Sputnik* nearly a decade ago, nothing mechanical seems to work, from elevators to vehicles. The arts are vibrant, from the ballet in the Kirov Theatre to the Moscow Symphony. But much of what is historic here seems to be in disrepair, and in every city and town you pass, rows and rows of drab, gray apartment buildings stand amid uncleared rubble, without green spaces, playgrounds, trees. My colleagues here talk about the good fortune of the residents of these buildings: as part of the planning of the 'socialist city,' their apartments are always close to their place of work, so that they can walk or ride the bus a short distance rather than having to commute for hours. Our bus broke down as we entered the old part of Tula, and soon we learned that it would be several hours before we could expect to reboard. We had stopped right next to a row of what must have been attractive shops and apartments a half-century ago; everything decorative seems to have been removed, and nothing has been done to paint, replace rotting wood, or fix roofs. I walked up the street a mile or so, and there was the 'new' Tula—the Soviet imprint on the architecture of the place. This scene could have been photographed on the outskirts of Leningrad, Moscow, or any other urban center we had seen. What will life be like in the Soviet city of the distant future?"   © H. J. de Blij.

www.conceptcaching.com

THE SOVIET LEGACY

The era of communism may have ended in the Soviet Empire, but its effects on Russia's political and economic geography will long remain. Seventy years of centralized planning and implementation cannot be erased overnight; regional reorganization toward a market economy cannot be accomplished in a day.

While the world of capitalism celebrates the failure of the communist system in the former Soviet realm, it should remember why communism found such fertile ground in the Russia of the 1910s and 1920s. In those days Russia was infamous for the wretched serfdom of its peasants, the cruel exploitation of its workers, the excesses of its nobility, and the ostentatious palaces and riches of the czars. Ripples from the Western European Industrial Revolution introduced a new age of misery for those laboring in factories. There were workers' strikes and ugly retributions, but when the czars finally tried to better the lot of the poor, it was too little too late. There was no democracy, and the people had no way to express or channel their grievances. Europe's democratic revolution passed Russia by, and its economic revolution touched the czars' domain only slightly. Most Russians, and tens of millions of non-Russians under the czars' control, faced exploitation, corruption, starvation, and harsh subjugation. When the people began to rebel in 1905, there was no hint of what lay in store; even after the full-scale Revolution of 1917, Russia's political future hung in the balance.

The Russian Revolution was no unified uprising. There were factions and cliques; the Bolsheviks ("Majority") took their ideological lead from Lenin, while the Mensheviks ("Minority") saw a different, more liberal future for their country. The so-called Red army factions fought against the Whites, while both battled the forces of the czar. The country stopped functioning; the people suffered terrible deprivations in the countryside as well as the cities. Most Russians (and other nationalities within the empire as well) were ready for radical change.

That change came when the Revolution succeeded and the Bolsheviks bested the Mensheviks, most of whom were exiled. In 1918, the capital was moved from Petrograd (as St. Petersburg had been renamed in 1914, to remove its German appellation) to Moscow. This was a symbolic move, the opposite of the forward-capital principle: Moscow lay deep in the Russian interior, not even on a major navigable waterway (let alone a coast), amid the same forests that much earlier had protected the Russians from their enemies. The new Soviet Union would look inward, and the communist system would achieve with Soviet resources and labor the goals that had for so long eluded the country. The chief political and economic architect of this effort was the revolutionary leader who prevailed in the power struggle: V. I. Lenin (born Vladimir Ilyich Ulyanov).

The Political Framework

Russia's great expansion had brought many nationalities under czarist control; now the revolutionary government sought to organize this heterogeneous ethnic mosaic into a smoothly functioning state. The czars had conquered, but

Here is another view of Tula, four decades after the photographs on the left were taken, and it provides some clues to the question posed at the end of the caption. Tula lies about 160 kilometers (100 mi) south of Moscow and was Russia's second city during the Middle Ages, endowed with an impressive kremlin (fortress) and thriving as a commercial center. Tula's twenty-first-century townscape is a mixture of historic buildings (built between 1514 and 1521), drab Soviet-era tenements (unchanged since 1964), and a few more modern apartment buildings. Old habits persist: where there is some fertile soil, you are likely to see vegetable plots, not flower gardens.   © Mauro Galligani/Contrasto/Redux Pictures.

they had done little to bring Russian culture to the peoples they ruled. The Georgians, Armenians, Tatars, and residents of the Muslim states of Central Asia were among dozens of individual cultural, linguistic, and religious groups that had not been "Russified." In 1917, however, the Russians themselves constituted only about one-half of the population of the entire realm. Thus it was impossible to establish a Russian state instantly over this vast political region, and these diverse national groups had to be accommodated.

The question of the nationalities became a major issue in the young Soviet state after 1917. Lenin, who brought the philosophy of Karl Marx to Russia, talked from the beginning about the "right of self-determination for the nationalities." The first response by many of Russia's subject peoples was to proclaim independent republics, as they did in Ukraine, Georgia, Armenia, Azerbaijan, and even in Central Asia. But Lenin had no intention of permitting the Russian state to break up. In 1923, when his blueprint for the new Soviet Union went into effect, the last of these briefly independent units was fully absorbed into the sphere of the Moscow regime. Ukraine, for example, declared itself independent in 1917 and managed to sustain this initiative until 1919. But in that year the Bolsheviks set up a provisional government in Kiev, the Ukrainian capital, thereby ensuring the incorporation of the country into Lenin's Soviet framework.

The Communist System

The Bolsheviks' political framework for the Soviet Union was based on the ethnic identities of its many incorporated peoples. Given the size and cultural complexity of the empire, it was impossible to allocate territory of equal political standing to all the nationalities; the communists controlled the destinies of well over 100 peoples, both large nations and small isolated groups. It was decided to divide the vast realm into Soviet Socialist Republics (SSRs), each of which was delimited to correspond broadly to one of the major nationalities. At the time, Russians constituted about half of the developing Soviet Union's population, and, as Figure 2-5 shows, they also were (and still are) the most widely dispersed ethnic group in the realm. The Russian Republic, therefore, was by far the largest designated SSR, comprising just under 77 percent of total Soviet territory.

Within the SSRs, smaller minorities were assigned political units of lesser rank. These were called Autonomous Soviet Socialist Republics (ASSRs), which in effect were republics within republics; other areas were designated Autonomous Regions or other nationality-based units. It was a complicated, cumbersome, often poorly designed framework, but in 1924 it was launched officially under the banner of the Union of Soviet Socialist Republics (USSR).

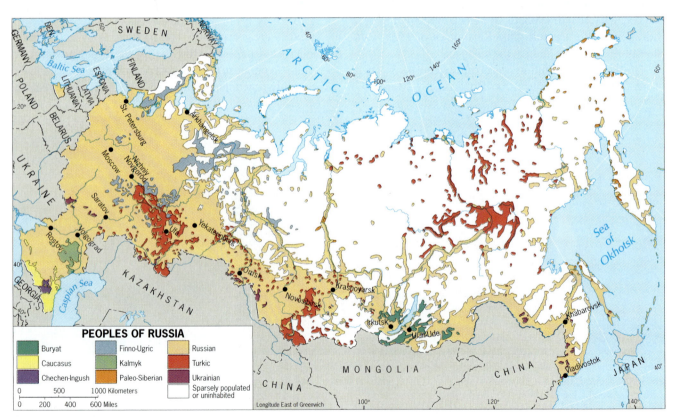

PEOPLES OF RUSSIA

Buryat	Finno-Ugric	Russian
Caucasus	Kalmyk	Turkic
Chechen-Ingush	Paleo-Siberian	Ukrainian
		Sparsely populated or uninhabited

0   500   1000 Kilometers
0   200   400   600 Miles

Longitude East of Greenwich

FIGURE 2-5   © H. J. de Blij, P. O. Muller, and John Wiley & Sons, Inc.

A Soviet Empire

Eventually, the Soviet Union came to consist of 15 SSRs (shown in Fig. 2-6), including not only the original republics of 1924 but also such later acquisitions as Moldova (formerly Moldavia in Romania), Estonia, Latvia, and Lithuania. The internal political layout often was changed, sometimes at the whim of the communist empire's dictators. But no communist apartheid-like system of segregation could accommodate the shifting multinational mosaic of the Soviet realm. The republics quarreled among themselves over boundaries and territory. Demographic changes, migrations, war, and economic factors soon made much of the layout of the 1920s obsolete. Moreover, the communist planners made it Soviet policy to relocate entire peoples from their homelands in order to better fit the grand design, and to reward or punish—sometimes capriciously. The overall effect, however, was to move minority peoples eastward and to **12** replace them with Russians. This **Russification** of the Soviet Empire produced substantial ethnic Russian minorities in all the non-Russian republics.

Phantom Federation

The Soviet planners called their system a **federation**. We **13** focus in more detail on this geographic concept in Chapter 11, but we note here that *federalism* involves the sharing of power between a country's central government and its political subdivisions (provinces, States, or, in the Soviet case, "Socialist Republics"). Study the map of the former Soviet Union (Fig. 2-6), and an interesting geographic corollary emerges: every one of the 15 Soviet Republics had a boundary with a non-Soviet neighbor. Not one was spatially locked within the others. This seemed to give geographic substance to the notion that any Republic was free to leave the USSR if it so desired. Reality, of course, was different. Moscow's control over the republics made the Soviet Union a federation in theory only.

The centerpiece of the tightly controlled Soviet "federation" was the Russian Republic. With half the vast state's population, the capital city, the realm's core area, and over three-quarters of the Soviet Union's territory, Russia was the empire's nucleus. In other republics, "Soviet" often was simply equated with "Russian"—it was the reality

FIGURE 2-6   © H. J. de Blij, P. O. Muller, and John Wiley & Sons, Inc.

with which the lesser republics lived. Russians came to the other republics to teach (Russian was taught in the colonial schools), to organize (and often dominate) the local Communist Party, and to implement Moscow's economic decisions. This was colonialism, but somehow the communist disguise—how could socialists, as the communists called themselves, be colonialists?—and the contiguous spatial nature of the empire made it appear to the rest of the world as something else. Indeed, on the world stage the Soviet Union became a champion of oppressed peoples, a force in the decolonization process. It was an astonishing contradiction that would, in time, be fully exposed.

The Soviet Economic Framework

The geopolitical changes that resulted from the founding of the Soviet Union were accompanied by a gigantic economic experiment: the conversion of the empire from a czarist autocracy with a capitalist veneer to communism. From the early 1920s onward, the country's economy would be centrally planned—the communist leadership in Moscow would make all decisions regarding economic planning and development. Soviet planners had two principal objectives: (1) to accelerate industrialization and (2) to **collectivize** agriculture. For the first time ever on such a scale, and for the first time in accordance with Marxist-Leninist principles, an entire country was organized to work toward national goals prescribed by a central government.

Land and the State

The Soviet planners believed that agriculture could be made more productive by organizing it into huge state-run enterprises. The holdings of large landowners were expropriated, private farms were taken away from the farmers, and the land was consolidated into collective farms. Initially, all such land was meant to be part of a *sovkhoz*, literally a grain-and-meat factory in which agricultural efficiency, through maximum mechanization and minimum labor requirements, would be at its peak. But many farmers opposed the Soviets and tried to sabotage the program in various ways, hoping to retain their land.

The farmers and peasants who obstructed the communists' grand design suffered a dreadful fate. In the 1930s, for instance, Stalin confiscated Ukraine's agricultural output and then ordered part of the border between the Russian and Ukrainian republics sealed—thereby leading to a famine that killed millions of farmers and their families. In the Soviet Union under communist totalitarianism, the ends justified the means, and untold hardship came to mil-

lions who had already suffered under the czars. In his book *Lenin's Tomb*, David Remnick estimates that between 30 and 60 million people lost their lives from imposed starvation, political purges, Siberian exile, and forced relocation. It was an incalculable human tragedy, but the secretive character of Soviet officialdom made it possible to hide it from the world.

The Soviet planners hoped that collectivized and mechanized farming would free hundreds of thousands of workers to labor in factories. Industrialization was the prime objective of the regime, and here the results were superior. Productivity rose rapidly, and when World War II engulfed the empire in 1941, the Soviet manufacturing sector was able to produce the equipment and weapons needed to repel the German invaders.

A Command Economy

Yet even in this context, the Soviet grand design entailed liabilities for the future. The USSR practiced a **command economy**, in which state planners assigned the production of particular manufactures to particular places, often disregarding the rules of economic geography. For example, the manufacture of railroad cars might be assigned (as indeed it was) to a factory in Latvia. No other factory anywhere else would be permitted to produce this equipment—even if supplies of raw materials would make it cheaper to build them near, say, Volgograd 1900 kilometers (1200 mi) away. Despite an expanded and improved transport network (Fig. 2-7), such practices made manufacturing in the USSR extremely expensive, and the absence of competition made managers complacent and workers less productive.

Of course, the Soviet planners never imagined that their experiment would fail and that a market-driven economy would replace their command economy. When that happened, the transition was predictably difficult; indeed, it is far from over, and it is putting severe strains on the now more democratic state.

RUSSIA'S CHANGING POLITICAL GEOGRAPHY

When the USSR dissolved in 1991, Russia's former empire devolved into 14 independent countries, and Russia itself was a changed nation. Russians now made up about 83 percent of the population of under 150 million, a far higher proportion than in the days of the Soviet Union. But numerous minority peoples remained under Moscow's new flag, and millions of Russians found themselves under new governments in the former Republics.

Soviet planners had created a complicated administrative structure for their "Russian Soviet Federative Socialist Republic," and Russia's postcommunist leaders had to use this framework to make their country function. In 1992, most of Russia's internal "republics," autonomous regions, oblasts, and krays (all of them components of the administrative hierarchy) signed a document known as the Russian Federation Treaty, committing them to cooperate in the new federal system. At first a few units refused to sign, including Tatarstan, scene of Ivan the Terrible's brutal conquest more than four centuries ago, and a republic in the Caucasus periphery, then known as Chechenya-Ingushetia, where Muslim rebels waged a campaign for independence. As the map shows, Chechnya-Ingushetiya split into two separate republics whose names at the time were spelled Chechenya and Ingushetia (Fig. 2-8). Eventually, only Chechnya refused to sign the Russian Federation Treaty, and subsequent Russian military intervention led to a prolonged and violent conflict, with disastrous consequences for Chechnya's people and infrastructure (the capital, Groznyy, was completely destroyed). The Chechnya war continues today and is a disaster for Russia's government as well (see the Issue Box titled "Chechnya").

The Federal Framework of Russia

The spatial framework of the still-evolving Russian Federation is as complex as that of the Russian Federative Socialist Republic of communist times. As the twenty-first century opened, the Federation consisted of 89 entities: 2 Autonomous Federal Cities, 21 Republics, 11 Autonomous Regions (Okrugs), 49 Provinces (Oblasts), and 6 Territories (Krays). Moscow and St. Petersburg are the two Autonomous Federal Cities. The 21 Republics, recognized to accommodate substantial ethnic minorities in the population, lie in several clusters (Fig. 2-8).

Russia's post–1991 governments have faced complicated administrative problems. A multinational, multicultural state that had been accustomed to authoritarian rule and government control over virtually everything—from factory production to everyday life—now had to be governed in a new way. Democratization of the political system, transition to a market economy, sale of state-owned industries (privatization), and liberation of the press and other media must all take place in an orderly way, or the country would, literally, fall apart.

Unitary and Federal Options

Russia's leaders knew that their options were limited. They could continue to hold as much power as possible at the center, making decisions in Moscow that would apply to all the Republics, Regions, and other subdivisions of the state. Such a **unitary state system**, with its centralized government and administration, marked authoritarian kingdoms of the past and serves totalitarian dictatorships of the present. Or they could share power with the Republics and Regions, allowing elected regional leaders to come to Moscow to represent the interests of their people. This is the federal system Russia chose as the only way to accommodate the country's economic and cultural diversity.

In a *federal system*, the national government usually is responsible for matters such as defense, foreign policy, and foreign trade. The Regions (or provinces, States, or other subdivisions) retain authority over affairs ranging from education to transportation. A federal system does not create unity out of diversity, but it does allow diverse components of the state to coexist, their common interests represented by the national government and their regional interests by their local administrations. Some countries owe their survival as coherent states to their federal frameworks. India and Australia are cases in point.

But to maintain a generally acceptable balance of power between the center and the Regions (or States) is difficult. Disputes over "States' rights" continue to roil the American political scene more than two centuries after the Constitution was adopted. Russia, which also owes its continuity to a grand federal experiment, is barely more than 15 years old. Power shifts continue, and the framework of the Russian Federation will undoubtedly change over time.

Problems of Size and Distance

An additional problem arises from Russia's sheer size. Geographers refer to the principle of **distance decay** to explain how increasing distances between places tend to reduce interactions among them. Because Russia is the world's largest country, distance is a significant factor in the relationships between the capital and outlying areas. Furthermore, Moscow lies in the far west of the giant country, half a world away from the shores of the Pacific. Not surprisingly, one of the most obstreperous Regions has been remote Primorskiy, the Region of Vladivostok.

Still another problem is reflected in Figure 2-8: the enormous size variation among the Republics (and to a lesser extent, the Regions). Whereas the (territorially) smallest Republics are concentrated in the Russian core area, the largest lie far to the east, where Sakha is nearly a thousand times as large as Ingushetiya. On the other hand, the populations of the huge eastern Republics are tiny compared to those of the smaller ones in the core. Such diversity spells administrative difficulty.

FIGURE 2-7 © H. J. de Blij, P. O. Muller, and John Wiley & Sons, Inc.

RUSSIA AND ITS REGIONS

Ethnic Republics and Autonomous Regions

Russian regions

National capital is underlined

0   200   400   600   800 Kilometers

0   100   200   300   400   500 Miles

RUSSIAN ADMINISTRATIVE UNITS
(named after their capitals)

1. Astrakhan	19. Penza
2. Belgorod	20. Pskov
3. Bryansk	21. Rostov
4. Chelyabinsk	22. Ryazan
5. Ivanovo	23. St. Petersburg
6. Kaluga	24. Samara
7. Kemerovo	25. Saratov
8. Kostroma	26. Smolensk
9. Krasnodar	27. Stavropol
10. Kurgan	28. Tambov
11. Kursk	29. Tula
12. Lipetsk	30. Tver
13. Mariy-el	31. Ulyanovsk
14. Moscow	32. Vladimir
15. Novosibirsk	33. Volgograd
16. Omsk	34. Voronezh
17. Orenburg	35. Yaroslavl
18. Orel	

Longitude East of Greenwich

FIGURE 2-8 © H. J. de Blij, P. O. Muller, and John Wiley & Sons, Inc.

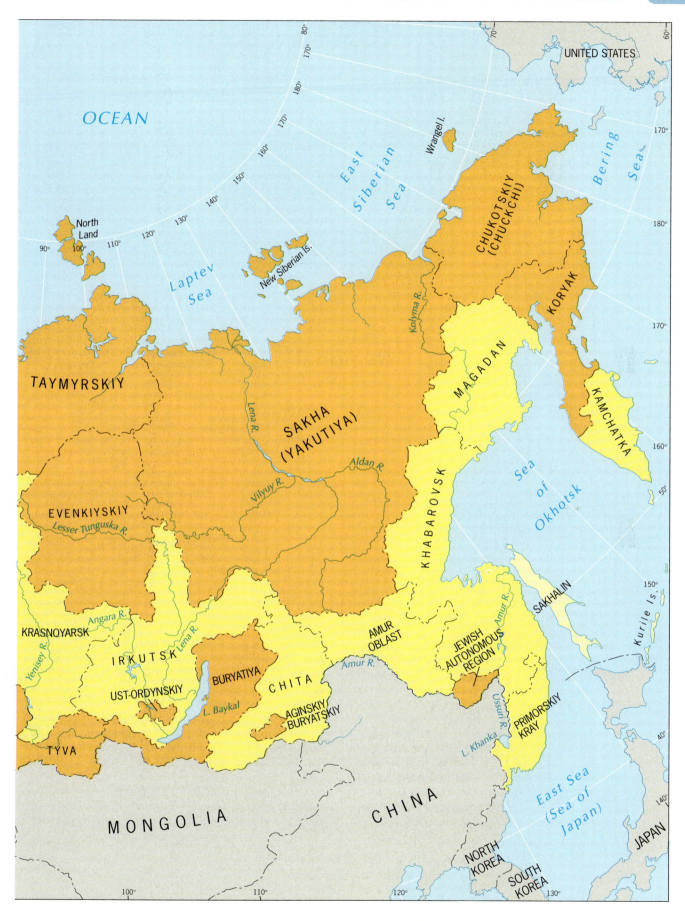

OCEAN

East
Siberian
Sea

Wrangel I.

UNITED STATES

Bering
Sea

North
Land

Laptev
Sea

New Siberian Is.

CHUKOTSKIY
(CHUCKCHI)

KORYAK

TAYMYRSKIY

Kolyma R.

MAGADAN

KAMCHATKA

Lena R.

SAKHA
(YAKUTIYA)

Aldan R.

Sea
of
Okhotsk

EVENKIYSKIY

Vilyuy R.

Lesser Tunguska R.

KHABAROVSK

SAKHALIN

Angara R.

Kurile Is.

KRASNOYARSK

Yenisey R.

Lena R.

IRKUTSK

AMUR
OBLAST

Amur R.

JEWISH
AUTONOMOUS
REGION

BURYATIYA

UST-ORDYNSKIY

Amur R.

CHITA

PRIMORSKIY
KRAY

Ussuri R.

L. Baykal

AGINSKIY
BURYATSKIY

East Sea
(Sea of
Japan)

TYVA

L. Khanka

MONGOLIA

CHINA

NORTH
KOREA

SOUTH
KOREA

JAPAN

Chechnya

IN SUPPORT OF RUSSIAN CONTROL OVER CHECHNYA

"As a policeman here in Moscow, I strongly support the government and the army in their efforts to establish control over the criminal elements trying to take control of Chechnya. Let us remember what happened there. When the Russian government back in 1991 had to take charge of our country and its many components, there was lots of opposition. The Tatars (also Muslims, like the Chechens) talked about establishing an independent state in their republic. The Kalmyks, the Udmurts, the Bashkirs, the Chuvash, and many others proclaimed that they wanted everything ranging from autonomy to independence. There were even Russians demanding it in places like Yekaterinburg and Primorskiy. But the government under President Yeltsin negotiated all these claims and gave those people reasons to want to stay under the Russian flag. Except the Chechens. Nothing was going to satisfy them.

"And the Chechens were even given their own republic after the changeover from Soviet to Russian administration. In 1991 they still shared their territory with the Ingush, who are also Muslims, in the so-called Chechen-Ingush Republic. So what did the Chechens do? They installed their separatist leader as ruler of the republic and started fighting with the Ingush minority. To help solve this crisis, Russia's government divided the republic's territory into two, the larger, richer part including the capital and the oilfields for the Chechens and the remainder for the Ingush.

"But it wasn't good enough for the Chechens. They attacked the Russians living in Chechnya, causing the army to move in to protect Russians and Russian interests. They broke the truce that followed. Then they mounted a full-scale war from their mountain hideouts and caused us hideous casualties. They did not care that their capital was totally destroyed; these Muslim warlords fought among themselves even as they fought us. Next they started using terror to get their way. Remember the autumn of 1999, when they blew up apartment buildings in Moscow, causing more than 300 dead? And how about what they did in Dagestan, where they took over a bunch of villages and declared an 'independent Chechen republic'? Not to mention that hospital full of doctors, nurses, and patients, whom they took hostage, eventually killing hundreds of innocent victims. These are people to whom we should entrust the government of an independent country on our borders? Never.

"So don't criticize us when it comes to our strategies to deal with these barbarians. Foreign governments say that the Russian army does terrible things to captured Chechens, but what about the treatment of Russian soldiers by Chechen rebels? War is brutal, and this is a war for the soil of Mother Russia.

"If we give up in Chechnya, other minorities will get ideas about independence too, and that would be the end of the nation. There is nothing to negotiate. Russia must and will prevail."

Regional ISSUE

WHY CHECHNYA DESERVES INDEPENDENCE

"My grandparents were born in what is today the Russian colony of Chechnya, and they died a horrible death somewhere in Soviet Central Asia. I am here to avenge their deaths and to punish the Russians for what they did to my people.

"Russians seem to think that we are fighting for independence just because we want it. Why aren't we like all those other minorities that have come to terms with their lives under Moscow's heel? Well, in our case there is more to it. We fought the Russian imperialists to a standstill less than two centuries ago, when the czars' armies colonized Islamic peoples from the Caucasus to Central Asia. Yes, we were eventually defeated, but those Russians never really penetrated our mountain hideouts. Then came the Soviets, who thought they did us a big favor by creating one of those 'Autonomous Republics' for us along with the poor Ingush. But look at the map. They combined our traditional homeland with a stretch of flatland to the north, which was full of Russian farmers and oilmen. Do you think Groznyy was a Chechen town? Think again. Or look for a mosque on those photographs of the 'good old days.'

"But we might have put up with it all except for what happened during the war between the Soviets and the Germans, World War II. Some of us Chechens were happy to see the Germans do to the Soviets what the Soviets had done to us, but Josef Stalin accused us all of collaborating with the enemy. In a few short months he packed all 500,000 of my people on trains and sent them to exile in Kazakhstan. Thousands died on the trains and were simply thrown onto the tracks. Many more, probably about 125,000, perished in the desert. I don't know where or when my grandparents died, but my parents survived.

"We got our homeland back because the Soviet dictator Khrushchev reinstated our Republic in 1957, and the survivors straggled back. But nobody ever forgot what was done to us. We're not talking of a few hundred hostages here. We're talking about the killing of a quarter of a nation. Still, we tried to restart our society under those atheist Soviets, keeping our Islamic practices quiet and private. But then the Soviet Empire collapsed and we got our chance. In 1991 we declared our independence from the new Russia, and at first it seemed that the Russians would be sensible and approve.

"But before long the Russians changed their minds and tried to intervene in our internal affairs. We managed to defeat the Russian army once again, and the war devastated Groznyy, but the Russians would not even consider independence. We got assistance from Muslims elsewhere and money from many sources, but we can see that Russia will not give us what we want. So we turn to any means we can to wound the Russian bear, and we will chase it out of Chechnya with the help of Allah."

But perhaps the most serious current problem for Russia is the growing social disharmony between Moscow and the subnational political entities. Almost wherever one goes in Russia today, Moscow is disliked and often berated by angry locals. The capital is seen as the privileged playground for those who have benefited most from the post-Soviet transition—bureaucrats and hangers-on whose economic policies have driven down standards of living, whose greed and corruption have hurt the economy, whose actions in Chechnya have been a disaster, who have allowed foreigners (prominently the United States) to erode Russian power and prestige, who fail to pay the wages of workers toiling in industries still owned by the state, and who do not represent the Russian people. Complaints about the capital are not uncommon in free countries, but in Russia the mistrust between capital and subordinate areas has become so serious that it constitutes a key challenge for the government.

Federal Administrative Districts

In 2000, the Putin administration moved to diminish the influence of the Regions by creating a new spatial framework (Fig. 2-9) that combines the 89 Regions, Republics, and other entities into seven new administrative units—not to enhance their influence in Moscow, but to increase Moscow's authority over them. As the map shows, each of these new *Federal Districts* has a capital, elevating such cities as Rostov and Novosibirsk to a status secondary to Moscow. In a related move, the Russian president proposed, and the Parliament approved, that Regional governors be appointed rather than elected—thereby concentrating still more power in Moscow. Russia's federal system appears to be moving in a unitary direction.

CHANGING SOCIAL GEOGRAPHIES

None of this political maneuvering reflects in any way what is happening in the daily lives of most of Russia's citizens. Since the end of communist rule, Russia's social geography has changed quite drastically, for the better for most but for the worse for others. These changes include:

1. ***Revival of Religion.*** The communist regime was avowedly atheist, and the end of Soviet rule revived the long-suppressed Russian Orthodox Church. In the cultural landscape, this change is reflected by the reconstruction of cathedrals and churches as well as the reappearance of public religious events and ceremonies.

FIGURE 2-9 © H. J. de Blij, P. O. Muller, and John Wiley & Sons, Inc.

2. *Failure of the Pension System.* During the Soviet era, millions of Russians depended on state pensions for their retirement. Political and financial turmoil during the post-Soviet years caused the system to falter, impoverishing many Russian families and widening the gap between rich and poor. Street people and the destitute are now seen in numbers unknown during the communist era, a significant change in Russia's social landscape.

3. *Rise of the Oligarchs.* During the chaotic 1990s, the new Russian government sold state industries to private entrepreneurs (often friends of the new leadership) at low prices in return for political and other favors. The newly rich oligarchs went on a binge of conspicuous consumption, further widening the gap between rich and poor. Some took their money and went abroad; others stayed and took to political meddling. After 2000, the government began a campaign to crack down on these newly rich, and some were imprisoned.

4. *Crime and Corruption.* Organized crime exploded after the fall of Soviet rule, in part because of the end of communism's iron-fisted rule and also because law enforcement in the new Russia could not cope with the new epidemic. Crime syndicates thrived and extended their networks abroad, aided by endemic corruption in Russia itself. Street crime eroded Russians' quality of life.

5. *Experimental Freedoms.* Russian citizens began to experience freedoms unimaginable under communist rule, and surveys confirm that a large majority preferred the new circumstances, despite the problems arising from the transition. Greater freedom of the press and other media; unobstructed street protests against governmental regulations; political campaigns and rallies and other forms of public participation signaled a new era in what some called Russia's "liberal revolution." But during the later years of the Putin administration, such liberties came under attack and others were misused. The media came under government pressure and individual reporters were intimidated, even murdered. Democratic reforms were reversed. Race relations, especially in the large cities, deteriorated as minorities found themselves threatened and inadequately protected by law enforcement agencies. Russia's new democracy was still in difficult transition nearly two decades after the collapse of the communist system.

RUSSIA'S DEMOGRAPHIC DILEMMA

All this pales, however, against what some geographers refer to as Russia's demographic disaster: its rapid decline in population coupled with the deteriorating health of its citizens. When the Soviet Union collapsed in 1991, Russia's population totaled about 149 million. By 2008, this number was down to 140.6 million—despite the immigration of several million ethnic Russians from the former Soviet "republics" beyond Russia's borders. Since the end of communist rule, Russia has seen over 10 million more deaths than births. Severe social dislocation is the cause.

Undoubtedly, the transition from Soviet rule is one cause: uncertainty tends to cause families to have fewer children, and abortion is widespread in Russia. But the birth rate has stabilized at around 10 per thousand; it is Russia's death rate that has skyrocketed, now exceeding more than 16 per thousand. This produces an annual loss of population of over 0.5 percent, or more than three-quarters of a million per year.

Russian males are the ones most affected. Average male life expectancy dropped from 71 in 1991 to 59 in 2006; it was at its lowest in the Irkutsk Region, where it was a mere 53. (Female life expectancy also declined, but markedly less, to 72.) Males are more likely to be affected by alcoholism, heavy smoking, suicide (Russia's rate is 15 times Europe's), accidents, and murder. On average, a Russian male is nine times more likely to die a violent or an accidental death than his European counterpart. Fewer than half of today's male Russian teenagers will survive to age 60.

Another cause of the high death rate among males, and the declining life expectancy even among females, is AIDS, whose impact is difficult to gauge because of Russia's high levels of tuberculosis and the questionable quality of official reporting. Estimates of the total number of Russians infected with AIDS range from about 1 million to nearly 4 million, although the official number in 2006 was given as 340,000, and only a fraction of those infected were receiving treatment.

When we encounter the realm worst affected by HIV/AIDS in the world today, Subsaharan Africa, we will see that Russia's figures—even at the high end of these estimates—are still far below those of Africa. In Africa, too, life expectancies are declining markedly. Yet Russia's population is dropping while Africa's continues to rise (see Table G-1). The key difference lies in the birth rates, which in Subsaharan Africa are still high enough (40 per thousand) to replace those who die. That is not the case in Russia (10 per thousand).

If this **population decline** continues, Russia will have only about 130 million citizens by 2025 and barely 100 million by 2050, raising doubts about the future of the state itself. Look again at that map of the Federal Districts (Fig. 2-9) and consider this: between 1990 and 2004, the Far East District lost 17 percent of its population, Siberia

5 percent, the Northwest 9 percent, and the South 12 percent. Only the Central District, the one around Moscow, had a loss on a par with European countries (0.2 percent).

What is the answer? Russia's leaders hope that improvements in the social circumstances of the average Russian will reduce the rate of population loss. Campaigns against alcoholism and careless lifestyles are under way. In 2006, the Russian government started spending some of its plentiful oil revenues on a program that awards well over 100 U.S. dollars per month to mothers who decide to have a second child, plus a one-time award of nearly U.S. $10,000 to be put toward family expenses such as children's education or a mortgage. Such schemes, however, have a history of causing a quick "baby rush" but little long-term impact on the problem. Immigration is another option, but it is alarming that the decline just discussed has taken place even as several million Russians returned home from the Near Abroad. As for immigration by other peoples, such as Chinese and North Koreans, the barriers against it remain strong and the reception of those who do manage to enter Russia is generally not welcoming. Russians in the western heartland are not eager to see their Far Eastern frontier transformed into an extension of East Asia. Russians in the cities of the west exhibit considerable **xenophobia** (fear or anger toward foreigners and strangers) in their response to the arrival of migrants from Transcaucasia, even to minorities from within Russia itself. Indeed, many Russian citizens were not especially welcoming to ethnic Russians arriving from the former Soviet Republics either. These are huge obstacles in the search for a demographic revival.

When the Soviet Union collapsed in 1991, Russia inherited only about half of the empire's population but more than 75 percent of its territory. Among the things Russia lacked was a true network of large, region-anchoring cities served by an efficient interconnecting transport system stimulating growth and development. Russia today has just two cities with populations exceeding 2 million, the same number as Australia, which contains less than one-sixth of Russia's population. To once again become the world power Russia's leaders envisage, Russia must reverse its demographic decline, but at present no solution is in sight.

RUSSIA'S PROSPECTS

When Peter the Great began to reorient his country toward Europe and the outside world, he envisaged a Russia with warm-water ports, a nation no longer encircled by Swedes, Lithuanians, Poles, and Turks, and a force in European affairs. Core-periphery relationships always have been crucial to Russia, and they remain so today.

Heartland and Rimland

Russia's relative location has long been the subject of study and conjecture by geographers. Just over a century ago, the British geographer Sir Halford Mackinder (1861–1947) argued in a still-discussed article entitled "The Geographical Pivot of History" that western Russia and Eastern Europe enjoyed a combination of natural protection and resource wealth that would someday propel its occupants to world power. This protected core, he reasoned, overshadowed the exposed periphery. Eventually, this pivotal interior region of Eurasia, which he later called the **heartland**, would become a stage for world domination.

Mackinder's **heartland theory** was published when `19` Russia was a weak, economically backward society ripe for revolution, but Mackinder stuck to his guns. When Russia, in control of Eastern Europe and with a vast colonial empire, emerged from World War II as a superpower, Mackinder's conjectures seemed prophetic.

But not all geographers agreed. Probably the first scholar to use the term **rimland** (today **Pacific Rim** is part of everyday language) was Nicholas Spykman, who in 1944 countered Mackinder by calculating that Eurasia's periphery, not its core, held the key to global power. Spykman foresaw the rise of rimland states as superpowers and viewed Japan's emergence to wealth and power as just the beginning of that process. In that perspective, Russia's twentieth-century superpower status was just a temporary phenomenon.

Russia and the World

The expansion of the European Union since 2003 has reached Russia's borders. Former components of the Soviet Empire now are members of NATO. How will Russia's relationship with Europe evolve, and what are the prospects for Russia in the wider world?

To Europeans, Russia has always been the enigmatic colossus to the east, alternately bungling and threatening but never in tandem with the West. When the Berlin Wall came down and the Soviet banner folded, there were hopeful predictions of a "European Russia" finally joining its erstwhile ideological adversaries in a regional partnership that would extend from the Atlantic to the Pacific. Such forecasts were based on real as well as idealized evidence: Russians' ethnic ties with Eastern Europe; the revival of

Christian churches after decades of communist atheism; the sprouting of Russian democracy; the budding of a market economy; the long-term dependence of Europe on Russian energy supplies.

Reality, however, was rather different. Russian democracy remained tinged by an authoritarian streak that scared investors. The media continued to face government interference. The Russian Orthodox Church suppressed efforts by other churches to become reestablished. Russian leaders objected vigorously, then acquiesced as NATO expanded toward its borders. Russia was unhelpful during efforts to mitigate the collapse of Yugoslavia. And Russia's treatment of minorities and migrants seemed to violate EU standards. All this occurred against the background of a military nuclear arsenal still capable of destroying the world in an afternoon.

Russia and Europe

Nevertheless, Russia has more in common with Europe than it has with any other realm of the world, and the "Europeanization of Russia," whatever form it takes, is likely to be a hallmark of the century just started. The process was given an inadvertent boost by al-Qaeda's 9/11 attack on New York and Washington, D.C., when Russian, European, and American leaders realized that they faced a common enemy far more capable than they had estimated. One result was a softening of Western criticism of Russian efforts to suppress Islamic dissidents in Chechnya and elsewhere in the Federation. Over the longer term, however, Russia's well-being, and its ability to reverse its social deterioration discussed above, will depend in large measure on its integration into Europe and its adoption of European norms.

Unresolved Problems

Russia's enormous territory endows it with numerous neighbors, and its imperial history leaves it with many border problems. In Central Asia, millions of Russians still reside in the former Soviet "republics" including neighboring Kazakhstan (see Chapter 7). In East Asia, China and Russia have settled a number of the boundary disputes shown in Figure 2-10, but for the future the larger question is likely to involve China's "lost" territories east of the Amur and Ussuri rivers, taken by Russia more than a century ago.

Other significant external problems Russia confronts include: (1) the ownership of oil and gas reserves in the Caspian Basin, where maritime boundaries in the Caspian Sea have not yet been satisfactorily delimited; (2) settlement of the Kurile Islands issue with Japan, a legacy of World War II still not settled (see Chapter 9); (3) relations with its Transcaucasian neighbor, Georgia;

BORDERLANDS OF THE SENSITIVE EAST

Areas claimed by China at some time since 1946 — Treaty definitions of disputed boundaries — Railroad

0 200 400 600 800 Kilometers
0 200 400 Miles

FIGURE 2-10 © H. J. de Blij, P. O. Muller, and John Wiley & Sons, Inc.

and (4) efforts by the authoritarian regime of Belarus to reunite in some formal way with the Russian state. But perhaps Russia's greatest challenge of all, in the international arena, is to sustain its residual position as a credible force in world affairs.

POINTS TO PONDER

● Organized crime with international links, local gangsterism and street crime, and corruption in government afflict Russia, but polls show that a majority of Russians still feel that life is better than it was under communism.

● The fate of about 1 million ethnic Russians stranded outside contiguous Russia's borders in Europe, including in Kaliningrad, is arousing passions in Russia.

● Embattled Chechnya's parliament in 2006 voted to rename the capital, Groznyy (which means "dreadful"), to Akhmad-Kala after an assassinated president. But the Kremlin would have to approve.

● Russia has encouraged separatism in neighboring Georgia's Abkhazia region, going so far as to issue Russian passports to Abkhazian residents.

Regions of the Realm

So vast is Russia, so varied its physiography, and so diverse its cultural landscape that regionalization requires a small-scale perspective and a high level of generalization. Figure 2-11 outlines a four-region framework: the Russian Core and its Peripheries west of the Urals, the Eastern Frontier, Siberia, and the Far East. As we will see, each of these massive regions contains major subregions.

● RUSSIAN CORE AND PERIPHERIES

20 The heartland of a state is its **core area**. Here much of the population is concentrated, and here lie its leading cities, major industries, densest transport networks, most intensively cultivated lands, and other key components of the country. Core areas of long standing strongly reflect the imprints of culture and history. The Russian core area, broadly defined, extends from the western border of the Russian realm to the Ural Mountains in the east (Fig. 2-11). This is the Russia of Moscow and St. Petersburg, of the Volga River and its industrial cities.

Central Industrial Region

At the heart of the Russian Core lies the Central Industrial Region (Fig. 2-12). The precise definition of this subregion varies, for all regional definitions are subject to debate.

MAJOR CITIES OF THE REALM

City	Population (in millions)
Baki (Baku), Azerbaijan	1.9
Chelyabinsk, Russia	1.0
Kazan, Russia	1.1
Moscow, Russia	10.9
Nizhniy Novgorod, Russia	1.3
Novosibirsk, Russia	1.4
St. Petersburg, Russia	5.3
Tbilisi, Georgia	1.0
Vladivostok, Russia	0.6
Volgograd, Russia	1.0
Yekaterinburg, Russia	1.3
Yerevan, Armenia	1.1

*Based on 2008 estimates.

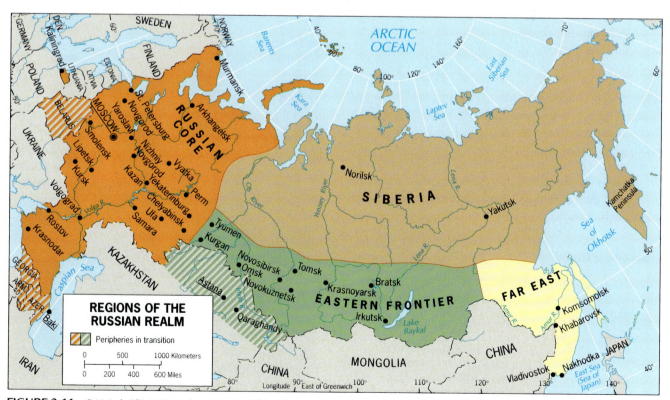

FIGURE 2-11   © H. J. de Blij, P. O. Muller, and John Wiley & Sons, Inc.

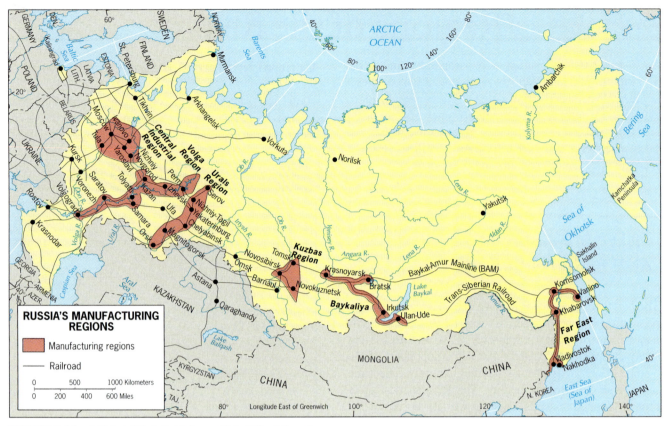

FIGURE 2-12 © H. J. de Blij, P. O. Muller, and John Wiley & Sons, Inc.

Some geographers prefer to call this the Moscow Region, thereby emphasizing that for over 400 kilometers (250 mi) in all directions from the capital, everything is oriented toward this historic focus of the state. As both Figures 2-1 and 2-7 show, Moscow has maintained its decisive centrality: roads and railroads converge in all directions from Ukraine in the south; from Mensk (Belarus) and the rest of Eastern Europe in the west; from St. Petersburg and the Baltic coast in the northwest; from Nizhniy Novgorod (formerly Gorkiy) and the Urals in the east; from the cities and waterways of the Volga Basin in the southeast (a canal links Moscow to the Volga, Russia's most important navigable river); and even from the subarctic northern periphery that faces the Barents Sea.

Two Great Cities

Moscow (population: 10.9 million) is the focus of an area that includes some 50 million inhabitants (more than one-

Old and new in Moscow: St. Basil's Cathedral, perhaps Russia's most famous historic structure and left intact by communist zealots who destroyed much of the country's architectural heritage, and a cell-phone advertisement in one of that industry's most lucrative markets. © Corbis.

AMONG THE REALM'S GREAT CITIES . . . Moscow

IN THE VASTNESS of Russia's territory, Moscow, capital of the Federation, seems to lie far from the center, close to its western margin. But in terms of Russia's population distribution, Moscow's centrality is second to none among the country's cities. Moscow (10.9 million) lies at the heart of Russia's primary core area and at the focus of its circulation systems.

On the banks of the meandering Moscow River, Moscow's skyline of onion-shaped church domes and modern buildings rises from the forested flat Russian Plain like a giant oasis in a verdant setting. Archeological evidence points to ancient settlement of the site, but Moscow enters recorded history only in the middle of the twelfth century. Forest and river provided defenses against Tatar raids, and when a Muscovy force defeated a Tatar army in the late fourteenth century, Moscow's primacy was assured. A huge brick Kremlin (citadel; fortress), with walls 2 kilometers in length and with 18 towers, was built to ensure the city's security. From this base Ivan the Terrible expanded Muscovy's domain and laid the foundations for Russia's vast empire.

The triangular Kremlin and the enormous square in front of it (Red Square of revolutionary times), flanked by St. Basil's Cathedral (photo at left) and overlooking the Moscow River, is the center of old Moscow and is still the heart of the city. From here, avenues and streets radiate in all directions to the Garden Ring and beyond. Until the 1970s, the Moscow Circular Motorway encircled most of the built-up area, but today the metropolis sprawls far outside this beltway. For all its size, Moscow never developed a world-class downtown skyline. Communist policy was to create a "socialist city" in which neighborhoods—consisting

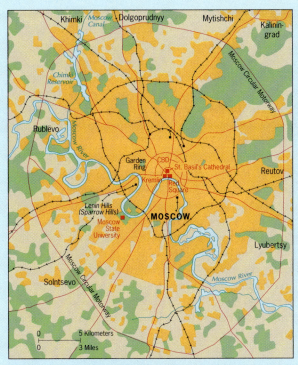

© H. J. de Blij, P. O. Muller, and John Wiley & Sons, Inc.

of apartment buildings, workplaces, schools, hospitals, shops, and other amenities—would be clustered so as to obviate long commutes into a high-rise city center.

Moscow may not be known for its architectural appeal, but the city contains noteworthy historic as well as modern structures, including the Cathedral of the Archangel and the towers of Moscow State University.

third of the country's total population), many of them concentrated in such major cities as Nizhniy Novgorod, the automobile-producing "Soviet Detroit"; Yaroslavl, the tire-producing center; Ivanovo, the heart of the textile industry; and Tula, the mining and metallurgical center where lignite (brown coal) deposits are worked (Fig. 2-12).

St. Petersburg (the former Leningrad) remains Russia's second city, with a population of 5.3 million. Under the czars, St. Petersburg was the focus of Russian political and cultural life, and Moscow was a distant second city. Today, however, St. Petersburg has none of Moscow's locational advantages, at least not with respect to the domestic market. It lies well outside the Central Industrial Region near the northwestern corner of the country, 650 kilometers (400 mi) from Moscow. Neither is it better off than Moscow in terms of resources: fuels, metals, and foodstuffs must all be brought in, mostly from far away. The former Soviet

emphasis on self-sufficiency even reduced St. Petersburg's asset of being on the Baltic coast because some raw materials could have been imported much more cheaply across the Baltic Sea from foreign sources than from domestic sites in distant Central Asia (only bauxite deposits lie nearby, at Tikhvin).

Yet St. Petersburg was at the vanguard of the Industrial Revolution in Russia, and its specialization and skills have remained important. Today, the city and its immediate environs contribute about 10 percent of the country's manufacturing, much of it through fabricating high-quality machinery.

Facing the Barents Sea

North of the latitude of St. Petersburg, the region we have named the Russian Core takes on Siberian properties.

AMONG THE REALM'S GREAT CITIES . . . St. Petersburg

CZAR PETER THE Great and his architects transformed the islands of the Neva Delta, at the head of the Gulf of Finland, into the Venice of the North, its palaces, churches, waterfront façades, bridges, and monuments giving St. Petersburg a European look unlike that of any other city in Russia. Having driven out the Swedes, Peter laid the foundations of the Peter and Paul Fortress on the bank of the wide Neva River in 1703, and the city was declared the capital of Russia in 1712.

Peter's "window on Europe," St. Petersburg (named after the saint, not the czar) was to become a capital to match Paris, Rome, and London. During the eighteenth century, St. Petersburg acquired a magnificent skyline dominated by tall, thin spires and ornate cupolas, and graced by baroque and classical architecture. The Imperial Winter Palace and the adjoining Hermitage Museum at the heart of the city are among a host of surviving architectural treasures.

Revolution and war lay in St. Petersburg's future. In 1917, the Russian Revolution began in the city (then named Petrograd), and following the communist victory it lost its functions as a capital to Moscow and its name to Lenin. As Leningrad, it suffered through the 872-day Nazi siege during World War II, holding out heroically through endless bombardment and starvation that took nearly 1 million lives and severely damaged many of its buildings.

The communist period witnessed the neglect and destruction of some of Leningrad's most beautiful churches, the intrusion and addition of crude monuments and bleak apartment complexes, and the rapid industrialization and growth of the city (which now totals 5.3 million). Immediately after the collapse of the Soviet Union, it re-

© H. J. de Blij, P. O. Muller, and John Wiley & Sons, Inc.

gained its original name and a new era opened. The Russian Orthodox Church revived, building restoration went forward, and tourism boomed. But the transformation from communist to capitalist ways has been accompanied by social problems that include crime and poverty.

However, the Russian presence is much stronger in this remote northern periphery than it is in Siberia proper. Two substantial cities, Murmansk and Arkhangelsk, are road- and rail-connected outposts in the shadow of the Arctic Circle.

Murmansk lies on the Kola Peninsula not far from the border with Finland (Fig. 2-7). In its hinterland lie a variety of mineral deposits, but Murmansk is particularly important as a naval base. During World War II, Allied ships brought supplies to Murmansk; the city's remoteness shielded it from German occupation. After the war, it became a base for nuclear submarines. This city is also an important fishing port and a container facility for cargo ships.

Arkhangelsk is located near the mouth of the Northern Dvina River where it reaches an arm of the White Sea (Fig. 2-7). Ivan the Terrible chose its site during Mus-

covy's early expansion; the czar wanted to make this the key port on a route to maritime Europe. Yet Arkhangelsk, mainly a port for lumber shipments, has a shorter ice-free season than Murmansk, whose port can be kept open with the help of the warm North Atlantic Drift ocean current (and by icebreakers if necessary).

Nothing in Siberia east of the Urals rivals either of these cities—yet. But their existence and growth prove that Siberian barriers to settlement can be overcome.

Povolzhye: The Volga Region

A second region lying within the Russian Core is the *Povolzhye*, the Russian name for an area that extends along the middle and lower valley of the Volga River. It would be appropriate to call this the Volga Region, for that greatest of

Russia's rivers is its lifeline and most of the cities that lie in the Povolzhye are on its banks (Fig. 2-12). In the 1950s, a canal was completed to link the lower Volga with the lower Don River (and thereby the Black Sea).

The Volga River was an important historic route in old Russia, but for a long time neighboring regions overshadowed it. The Moscow area and Ukraine were far ahead in industry and agriculture. The Industrial Revolution came late in the nineteenth century to the Moscow Region and did not have much effect in the Povolzhye. Its major function remained the transit of foodstuffs and raw materials to and from other regions.

Changing Times

This transport function is still important, but the Povolzhye has changed. First, World War II brought furious development because the Volga River, located east of Ukraine, was far from the German armies that invaded from the west. Second, in the postwar era the Volga-Urals Region for some time was the largest known source of petroleum and natural gas in the entire Soviet Union. From near Volgograd (formerly Stalingrad) in the southwest to Perm on the Urals' flank in the northeast lies a belt of major oilfields (Fig. 2-13).

Third, the transport system has been greatly expanded. The Volga-Don Canal directly connects the Volga waterway to the Black Sea; the Moscow Canal extends the northern navigability of this river system into the heart of the Central Industrial Region; and the Mariinsk canals provide a link to the Baltic Sea. Today, the Volga Region's population exceeds 25 million, and the cities of Samara (formerly Kuybyshev), Volgograd, Kazan, and Saratov all have populations between 1.0 and 1.3 million. Manufacturing has also expanded into the middle Volga Basin, emphasizing more specialized engineering industries. The huge Fiat-built auto assembly plant in Tolyatti, for example, is one of the world's largest of its kind.

The Internal Southern Periphery

We turn now to a subregion that has generated some of the most difficult politico-geographical problems the new Russia has faced since the collapse of the Soviet Union. Between the Black Sea to the west and the Caspian Sea to the east, and with the Caucasus Mountains to the south, Moscow seeks to stabilize and assert its authority over a tier of minority "republics," holdovers from the Soviet era (Fig. 2-14).

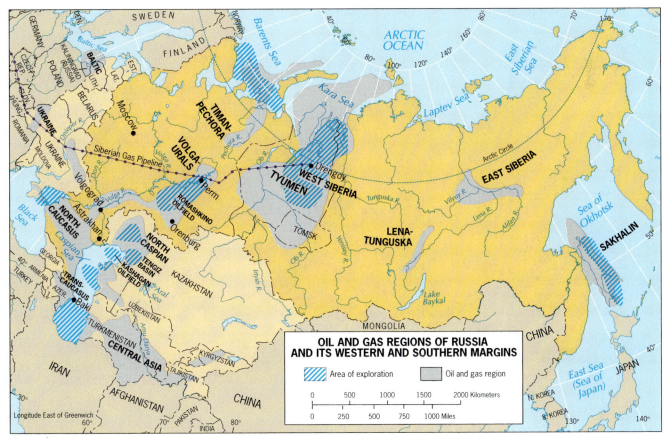

FIGURE 2-13   © H. J. de Blij, P. O. Muller, and John Wiley & Sons, Inc.

FIGURE 2-14   © H. J. de Blij, P. O. Muller, and John Wiley & Sons, Inc.

Islam and Oil

A combination of geographic factors causes the problems Moscow confronts in this peripheral zone. Here the Russians (and later the Soviets) met and stalled the advance of Islam. Here the cultural mosaic is as intricate as anywhere in the Federation: Dagestan, the southernmost Republic fronting the Caspian Sea, has 2.7 million inhabitants comprising some 30 ethnic groups speaking about 80 languages. And beneath the land as well as under the waters of the Caspian Sea lie oil and gas reserves that make this a crucial corner of the vast country. Indeed, Russia has a stake even in the oilfields belonging to neighboring Azerbaijan: as the map shows, the Russians control the pipelines leading from those reserves across Russian territory to the Black Sea oil terminal at Novorossiysk. They watch with great concern as new pipelines are completed from Azerbaijan that traverse Georgia and Turkey to reach alternate outlets.

Disaster in Chechnya

The presence of oil and oil installations endow this Southern Periphery with far greater significance in the Russian Federation than would otherwise be the case. Note again the route of the pipeline to Novorossiysk: it crosses the Republic called Chechnya whose now-devastated capital, Groznyy, was a major oil-industry center and service hub during the Soviet era (Fig. 2-14, including inset).

But Chechnya also contains a sizeable Muslim population, and fiercely independent Muslim Chechens used Caucasus mountain hideouts to resist Russian colonization during the nineteenth century. Accused of collaboration with the Nazis during World War II, on Stalin's orders the Soviets exiled the entire Chechen population to Central Asia, with much loss of life. Rehabilitated and allowed to return by Premier Nikita Khrushchev in the 1950s, the Chechens seized their opportunity in 1991 after the collapse of the Soviet Union and fought the Russian army to a stalemate. But Moscow never granted Chechnya independence (one quarter of the population of 1.2 million was Russian), and an uneasy standoff continued. Meanwhile, Chechen hit-and-run attacks on targets in neighboring Republics continued, and in 1999 three apartment buildings in Moscow were bombed, resulting in 230 deaths. Russian authorities blamed Chechnyan terrorists for these bombings and ordered a full-scale attack on Chechnya's Muslim holdouts. Groznyy, already severely damaged in the earlier conflict, was totally devastated and the rebels were driven into the mountains, but the Russian armed forces, still taking substantial losses, were unable to establish unchallenged control of all of the Republic.

The cycle of violence resumed in 2002 when a band of 41 terrorists took more than 700 performers and theatergoers hostage in a Moscow auditorium, demanding an immediate end to Russian military action in their homeland and the withdrawal of all Russian forces from Chechnya. After a three-day standoff, a bungled rescue effort led to the deaths of nearly 130 of the hostages and all of the terrorists. In 2004 an attack by Islamic militants from Chechnya and neighboring (also partly Muslim) Ingushetiya on a school in Beslan, a small town in North Ossetia (see Fig. 2-14), killed as many as 350, nearly half of them schoolchildren. And only a week before the Beslan attack, suicide bombers had struck simultaneously to down two airliners shortly after taking off from a Moscow airport. Whatever Moscow has tried in Chechnya, from accommodation to manipulation to repression, has not worked. Thus a "republic" the size of Connecticut with a population of barely over 1 million threatens security and stability in what is territorially the largest state in the world with a population of more than 140 million. It is a mirror not only on Russia but on the world of the present.

It is important to familiarize yourself with the layout of Russia's Internal Southern Periphery, because the problems here not only afflict Russia but spill over into its neighbors. The southern Republics, colored pink in Figure 2-14, all present their particular (sometimes intertwined) problems. Half the population in Kalmykia, north of Dagestan, is Buddhist, and local leaders presented Moscow with a dilemma when they decided in 2004 to invite Tibet's Dalai Lama to visit (China, which controls Tibet and calls it Xizang, does not recognize the Dalai Lama's position and denigrates his role). Also bordering the Caspian Sea is Dagestan which, as we noted, contains more than two dozen ethnic groups. Ingushetiya, which was once part of the Chechnyan republic, has a Muslim majority but has not joined the Chechnyan campaign although individual Islamic fighters have done so. North Ossetia is called **North** Ossetia because its kinspeople live outside Russia in **South** Ossetia—which is part of neighboring Georgia. Russian leaders argue that terrorists pass through South Ossetia into North Ossetia and on into Chechnya, and that the government in Georgia is not doing enough to prevent this movement. The Muslim Balkar people of Kabardino-Balkariya, like the Chechens, were accused by Stalin of collaboration with the Nazis and were forcibly exiled with great loss of life; today they constitute only about 12 percent of the Republic's population. The same thing happened to the Muslim Karachay of Karachayevo-Cherkessiya: many were deported, and today they constitute a minority in their Republic, where Russians and Christian Cherkess form the majority. Only in the separate Republic of Adygeya is there no significant Muslim element: Cherkess form the non-Russian population sector here.

This summation of the regional geography of the Internal Southern Periphery helps explain why Russia faces such severe difficulties here. Even after a solution to the

Chechnyan problem is found, the challenge will continue: stability in this physiographically and culturally fractured subregion has always been elusive.

The Urals Region

The Ural Mountains form the eastern limit of the Russian Core. They are not particularly high; in the north they consist of a single range, but southward they broaden into a hilly zone. Nowhere are they an obstacle to east-west transportation. An enormous storehouse of metallic mineral resources located in and near the Urals has made this area a natural place for industrial development. Today, the Urals Region, well connected to the Volga and Central Industrial Region, extends from Serov in the north to Orsk in the south (Fig. 2-12).

The Central Industrial, Volga, and Urals regions form the anchors of the Russian core area. For decades they have been spatially expanding toward one another, their interactions ever more intensive. These subregions of the Russian Core stand in sharp contrast to the comparatively less developed, forested, Arctic north and the remote upland of the Southern Periphery lying to the southwest between the Black and Caspian seas. Thus even within this Russian coreland, frontiers still await growth and development.

As noted in Chapter 1, because the Ural Mountains form a prominent north-south divide between Russia's western heartland and its eastern expanses, geographers sometimes use this to separate a "European" Russia from its presumably non-European, "Asian" east. (The National Geographic Society on its maps and in its atlases even draws a green line along the crest of the Urals marking this transition.) But geographically this makes no sense. Take the train or drive eastward across the southern end of the Urals, and you will see no changes in cultural landscapes, no geographic evidence on which to base such a partition. Indeed, as we will see, Russia remains Russia all the way to the end of its transcontinental railroad at Vladivostok on the Pacific coast.

● THE EASTERN FRONTIER

From the eastern flanks of the Ural Mountains to the headwaters of the Amur River, and from the latitude of Tyumen to the northern zone of neighboring Kazakhstan, lies Russia's vast Eastern Frontier Region, product of a gigantic experiment in the eastward extension of the Russian Core (Fig. 2-11). As the map of cities and surface communications suggests (Fig. 2-7), this Eastern Frontier is more densely peopled and more fully developed in the west than in the east. At the longitude of Lake Baykal, settlement has become linear, marked by ribbons and clusters along the east-west railroads. Two subregions dominate the geography: the Kuznetsk Basin in the west and the Lake Baykal area in the east.

The Kuznetsk Basin (Kuzbas)

Some 1500 kilometers (950 mi) east of the Urals lies another of Russia's primary regions of heavy manufacturing resulting from the communist period's national planning: the Kuznetsk Basin, or *Kuzbas* (Fig. 2-12). In the 1930s, it was opened up as a supplier of raw materials (especially coal) to the Urals, but that function became less important as local industrialization accelerated. The original plan was to move coal from the Kuzbas west to the Urals and allow the returning trains to carry iron ore east to the coalfields. However, good-quality iron ore deposits were subsequently discovered near the Kuznetsk Basin itself. As the new resource-based Kuzbas industries grew, so did its urban centers. The leading city, located just outside the region, is Novosibirsk, which stands at the intersection of the Trans-Siberian Railroad and the Ob River as the symbol of Russian enterprise in the vast eastern interior. To the northeast lies Tomsk, one of the oldest Russian towns in all of Siberia, founded in the seventeenth century and now caught up in the modern development of the Kuzbas Region. Southeast of Novosibirsk lies Novokuznetsk, a city that produces steel for the region's machine and metalworking plants and aluminum products from Urals bauxite.

The Lake Baykal Area (Baykaliya)

East of the Kuzbas, development becomes more insular, and distance becomes a stronger adversary. North of the Tyva Republic and eastward around Lake Baykal, larger and smaller settlements cluster along the two railroads to the Pacific coast (Fig. 2-12). West of the lake, these rail corridors lie in the headwater zone of the Yenisey River and its tributaries. A number of dams and hydroelectric projects serve the valley of the Angara River, particularly the city of Bratsk. Mining, lumbering, and some farming sustain life here, but isolation dominates it. The city of Irkutsk, near the southern end of Lake Baykal, is the principal service center for a vast Siberian region to the north and for a lengthy east-west stretch of southeastern Russia.

Beyond Lake Baykal, the Eastern Frontier really lives up to its name: this is southern Russia's most rugged, remote, forbidding country. Settlements are rare, many being mere camps. The Buryat Republic (Fig. 2-8) is part of this zone; the territory bordering it to the east was taken from China by the czars and may become an issue in the future. Where the Russian-Chinese boundary turns southward, along the Amur River, the region called the Eastern Frontier ends and Russia's Far East begins.

SIBERIA

Before we assess the potential of Russia's Pacific Rim, we should remember that the ribbons of settlement just discussed hug the southern perimeter of this giant country, avoiding the vast Siberian region to the north (Fig. 2-11). Siberia extends from the Ural Mountains to the Kamchatka Peninsula—a vast, bleak, frigid, forbidding land. Larger than the conterminous United States but inhabited by only an estimated 15 million people, Siberia quintessentially symbolizes the Russian environmental plight: vast distances, cold temperatures worsened by strong Arctic winds, difficult terrain, poor soils, and limited options for survival.

But Siberia also has resources. From the days of the first Russian explorers and Cossack adventurers, Siberia's riches have beckoned. Gold, diamonds, and other precious minerals were found there, and later metallic ores including iron and bauxite were discovered. Still more recently, the Siberian interior proved to contain sizeable quantities of oil and natural gas (Fig. 2-13) and began to contribute significantly to Russia's energy supply. As the physiographic map (Fig. 2-2) shows, major rivers—the Ob, Yenisey, and Lena—flow gently northward across Siberia and the Arctic Lowland into the Arctic Ocean. Hydroelectric power development in the basins of these rivers has generated electricity used to extract and refine local ores, and run the lumber mills that have been set up to exploit the vast Siberian forests.

The Future

The human geography of Siberia is fragmented, and much of the region is virtually uninhabited (Fig. 2-5). Ribbons of Russian settlement have developed; the Yenisey River, for instance, can be traced on this map of Soviet peoples (a series of small settlements north of Krasnoyarsk), and the upper Lena Valley is similarly fringed by ethnic Russian settlement. Yet hundreds of miles of empty territory separate these ribbons and other islands of habitation.

The political geography of eastern Siberia is marked by the growing identity of Sakha (the Yakut Republic). As additional resources are discovered here (including oil and natural gas), this Republic, centered on the capital, Yakutsk, will become more important.

Siberia, Russia's freezer, is stocked with goods that may become mainstays of future national development. Already, precious metals and mineral fuels are bolstering the Russian economy. In time, we may expect Siberian resources to play a growing role in the economic development of the Eastern Frontier and the Russian Far East as well. One step in that process was already taken during Soviet times: the completion of the BAM (Baykal-Amur Mainline) Railroad in the 1980s. This route, lying north of and parallel to the old Trans-Siberian Railroad, extends 3540 kilometers (2200 mi) eastward from Tayshet (near the important center of Krasnoyarsk) directly to the Far East city of Komsomolsk (Fig. 2-7). In the post-Soviet era, the BAM Railroad has been beset by equipment breakdowns and workers' strikes. Nonetheless, it is a key element of the infrastructure that will serve the Eastern Frontier's economic growth in the twenty-first century.

An unusual photograph from the energy-rich island of Sakhalin: workers laying part of a pipeline that will transport natural gas from one of Sakhalin's offshore deposits to the export terminal. State-owned Russian companies and foreign oil and gas corporations are both involved here, but there are problems over ownership and rights. These have made the news; what has not is the impact of Sakhalin's energy boom on local minorities, including the Evenki reindeer herders and the Nivkhi fishers and hunters, whose livelihoods are affected by the environmental impact of projects like this one.   © Zuma Press.

THE RUSSIAN FAR EAST

Imagine this: a country with 8000 kilometers (5000 mi) of Pacific coastline, two major ports, interior cities nearby, huge reserves of resources ranging from minerals to fuels to timber, directly across from one of the world's largest economies—all this at a time when the Asian Pacific Rim was the world's fastest-growing economic region. Would not that country have burgeoning cities, busy harbors, growing industries, and expanding trade?

In the Russian Far East (Fig. 2-11), the answer is—no. Activity in the port of Vladivostok is a shadow of what it was during the Soviet era, when it was the communists' key naval base. The nearby container terminal at Nakhodka suffers from breakdowns and inefficiencies. The railroad to western Russia carries just a fraction of the trade it did dur-

ing the 1970s and 1980s. Cross-border trade with China is minimal. Trade with Japan is inconsequential. The region's cities are grimy, drab, moribund. Utilities are shut off for hours at a time because of fuel shortages and system breakdowns. Outdated factories are closed down, their workers dismissed. Political relations with Moscow are poor. There is potential here, but little of it has been realized.

Mainland and Island

As a region, the Russian Far East consists of two parts: the mainland area extending from Vladivostok to the Stanovoy Range and the large island of Sakhalin (Figs. 2-1, 2-15). This is cold country: icebreakers have to keep the ports of Vladivostok and Nakhodka open throughout the winter. Winters here are long and bitterly cold; summers are brief

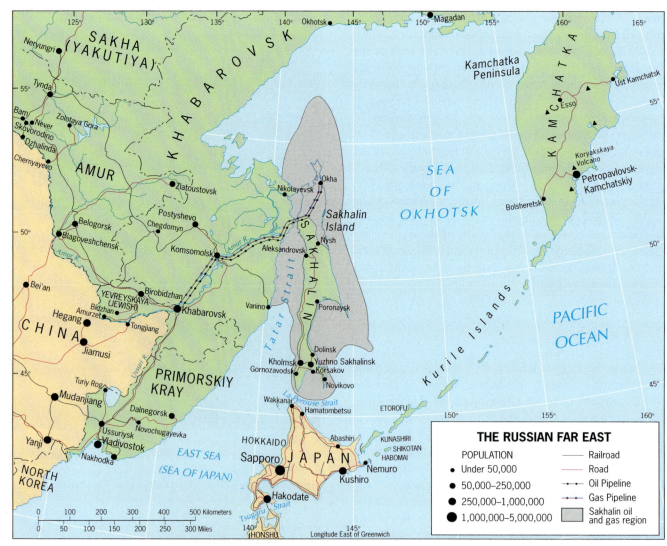

FIGURE 2-15 © H. J. de Blij, P. O. Muller, and John Wiley & Sons, Inc.

and cool. Although the population is small (about 7 million), food must be imported because not much can be grown. Most of the region is rugged, forested, and remote. Vladivostok, Khabarovsk, and Komsomolsk are the only cities of any size. Nakhodka and the newer railroad terminal at Vanino are smaller towns; the population of the whole island of Sakhalin is about 600,000 (on an island the size of Caribbean Hispaniola [18.2 million]). Offshore lie productive fishing grounds. Most important of all are the huge oil and gas reserves recently discovered on and around Sakhalin: a new era may be dawning as Russian-partnered foreign companies move in to exploit them (see photo p. 143).

Post-Soviet Malaise

The Soviet regime rewarded people willing to move to this region with housing and subsidies. The communists, like the czars before them, realized the importance of this frontier (Vladivostok means "We Own the East"), and they used every possible incentive to develop it and link it ever closer to Russia's distant western core. Freight rates on the Trans-Siberian Railroad, for example, were about 10 percent of their real costs; the trains were always loaded in both directions. Vladivostok was a military base and a city closed to foreigners, and Moscow invested heavily in its infrastructure. Komsomolsk in the north and Khabarovsk near the region's center were endowed with state-owned industries using local resources: iron ore from Komsomolsk, oil from Sakhalin, timber from the ubiquitous forests. Steel, chemical, and furniture industries sent their products westward by train and received food and other consumer goods from the Russian heartland.

For several reasons, the post-Soviet transition has been especially difficult here in the Far East. The new economic order has canceled the region's communist-era advantages: the Trans-Siberian Railroad now must charge the real cost of hauling products from this Pacific-fronting outpost to the Russian Core. State-subsidized industries must compete

FROM THE FIELD NOTES

"Standing in an elevated doorway in the center of Vladivostok, you can see some of the vestiges of the Soviet period: the omnipresent, once-dominant, GUM department store behind the blue seal on the left, and the communist hammer-and-sickle on top of the defunct hotel on the right. But in the much more colorful garb of the people, and the private cars in the street, you see reflections of the new era. Once a city closed to foreigners, Vladivostok is now open to visitors—and has become a major point of entry for contraband goods."   © George W. Moore.

www.conceptcaching.com

The Kamchatka Peninsula constitutes one of the Pacific Rim's most geologically active zones, with more than 125 volcanoes of which more than 20 are active. Sudden massive eruptions can endanger airliners on the northern Pacific route between North America and East Asia, which skirts Kamchatka. The region's largest city, Petropavlovsk-Kamchatskiy, lies at the foot of the 3450-meter (11,400-ft) high Koryakskaya Volcano near the Pacific coast in southeastern Kamchatka. Soviet-era apartment buildings dominate the cultural landscape of this bleak fishing port founded during Russia's eighteenth-century eastward expansion. Note that, in this *Dfc* climate, the countryside is much greener (with deciduous as well as coniferous vegetation) than it is toward the north and the Siberian interior. © Mark Newman/Photo Researchers.

on market principles, their subsidies having ended. The decline of Russia's armed forces has hit Vladivostok hard (photo p. 145). The fleet lies rusting in port; service industries have lost their military markets; the shipbuilding industry has no government contracts. Coal miners in the Bureya River Valley (a tributary of the Amur) go unpaid for months and go on strike; coal-fired power plants do not receive fuel shipments, and cities and towns go dark.

Locals put much of the blame for their region's failure on Moscow, and with reason. As Figure 2-8 shows, the Far East contains only five administrative regions: Primorskiy Kray, Khabarovsk Kray, Amur Oblast, Sakhalin Oblast, and Yevreyskaya, originally the Jewish Autonomous Region. This does not add up to much political clout, which is the way Moscow appears to want it. Compare Figures 2-9 and 2-15 and you will note that the city of Khabarovsk, not the Primorskiy Kray capital of Vladivostok, has been made the headquarters of the Far Eastern Region as defined by Moscow.

For all its stagnation, Russia's Far East will figure prominently in Russian (and probably world) affairs. Here Russia meets China on land and Japan at sea (an unresolved issue between Russia and Japan involves several of the Kurile islands that separate the Sea of Okhotsk from the Pacific Ocean). Here lie vast resources ranging from Sakhalin's fuels to Siberia's lumber, Sakha's gold to Khabarovsk's metals. Here Russia has a foothold on the Pacific Rim, a window on the ocean on whose shores the world is being transformed.

TRANSCAUCASIA: THE EXTERNAL SOUTHERN PERIPHERY

Beyond Russia's southern tier of internal republics (discussed on pp. 139–142) lies a crucial periphery, legally outside the Russian sphere but closely tied to it nevertheless. This is ***Transcaucasia***, historically a battleground for Christians and Muslims, Armenians and Turks, Russians and Persians. Today it is an external subregion containing three former Soviet Socialist Republics: landlocked Armenia, coastal Georgia on the Black Sea, and Azerbaijan on the land-encircled Caspian Sea. Although these three countries are independent states today, they are so closely bound up with Russia that they remain, functionally, part of the Russian realm (Fig. 2-14).

Armenia

As Figure 2-1 shows, landlocked Armenia (population: 3.0 million) occupies some of the most rugged and mountainous terrain in the earthquake-prone Transcaucasus. The Armenians are an embattled people who adopted Christianity 17 centuries ago and for more than a millennium sought to secure their ancient homeland here on the margins of the Muslim world. During World War I, the Ottoman Turks massacred much of the Christian Armenian minority and drove the survivors from eastern Anatolia (Turkey) and what is now Iraq into the Transcaucasus. At the end of that war in 1918, an independent Armenia arose, but its autonomy lasted only two years. In 1920, Armenia was taken over by the Soviets; in 1936, it became one of the 15 constituent republics of the Soviet Union. The collapse of the Soviet Empire gave Armenia what it had lost three generations earlier: independence.

Embattled Exclave

Or so it seemed. Soon afterward, the Armenians found themselves at war with neighboring Azerbaijan over the fate of some 150,000 Armenians living in Nagorno-Karabakh, a pocket of territory surrounded by Azerbaijan. Such a separated territory is called an ***exclave***, and this one had been created by Soviet sociopolitical planners who, while acknowledging the cultural (Christian) distinctiveness of this cluster of Armenians, nevertheless gave (Muslim) Azerbaijan jurisdiction over it.

That was a recipe for trouble: in the ensuing conflict, Armenian troops entered Azerbaijan and gained control over the exclave, even ousting Azerbaijanis from the zone between the main body of Armenia and Nagorno-Karabakh (Fig. 2-14, vertical-striped zone). The international community, however, has not recognized Armenia's occupation, and officially the territory remains a part of Azerbaijan. In 2008, this issue is still unresolved.

Georgia

Of the three former Soviet Republics in Transcaucasia, only Georgia has a Black Sea coast and thus an outlet to the wider world. Smaller than South Carolina, Georgia is a country of high mountains and fertile valleys. Its social and political geographies are complicated. The population of 4.4 million is more than 70 percent Georgian but also includes Armenians (8 percent), Russians (6 percent), Ossetians (3 percent), and Abkhazians (2 percent). The Georgian Orthodox Church dominates the religious community, but about 10 percent of the people are Muslims, most of them concentrated in Ajaria in the southwest.

Unlike Armenia and Azerbaijan, Georgia has no exclaves, but its political geography is problematic nonetheless (Fig. 2-14). Within Georgia's borders lie three minority-based autonomous entities: the Abkhazian and Ajarian Autonomous Republics, and the South Ossetian Autonomous Region.

Sakartvelos, as the Georgians call their country, has a long and turbulent history. Tbilisi, the capital for 15 centuries, lay at the core of an empire around the turn of the thirteenth century, but the Mongol invasion ended that era. Next, the Christian Georgians found themselves in the path of wars between Islamic Turks and Persians. Turning northward for protection, the Georgians were annexed in 1800 by the Russians, who were looking for warm-water ports. Like other peoples overpowered by the czars, the Georgians took advantage of the Russian Revolution to reassert their independence; but the Soviets reincorporated Georgia in 1921 and proclaimed a Georgian Soviet Socialist Republic in 1936. Josef Stalin, the communist dictator who succeeded Lenin, was a Georgian.

Georgia is renowned for its scenic beauty, warm and favorable climates, agriculture (especially tea), timber, manganese, and other products. Georgian wines, tobacco, and citrus fruits are much in demand. The diversified economy could support a viable state, but has yet to prosper.

Georgia and Russia

Unfortunately, Georgia's political geography is loaded with centrifugal forces. Georgia declared its independence in 1991, but in the Autonomous Region of South Ossetia, a movement for union with (Russian) North Ossetia created havoc. Abkhazia proclaimed independence. The price of all this instability was Russian intervention, and just when things calmed down and Russian military bases were closed, Moscow charged that Georgia was not doing enough to deter Muslim extremists operating in the Caucasus. Today, Georgia remains a dysfunctional state.

Azerbaijan

Azerbaijan is the name of an independent state and of a province in neighboring Iran. The *Azeris* (short for Azerbaijanis) on both sides of the border have the same ancestry: they are a Turkish people divided between the (then) Russian and Persian empires by a treaty signed in 1828. By that time, the Azeris had become Shi'ite Muslims, and when the Soviet communists laid out their grand design for the USSR, they awarded the Azeris their own republic. On the Persian side, the Azeris were assimilated into the Persian Empire, and their domain became a province. Today, the former Soviet Socialist Republic is the independent state of Azerbaijan (population: 8.7 million), and the 10 million Azeris to the south live in the Iranian province.

During the brief transition to independence and at the height of their war with the Armenians, the dominantly Muslim Azeris tended to look southward, toward Iran. But geographic realities dictate a more practical orientation. Azerbaijan possesses huge reserves of oil and natural gas; under the Soviets it was one of Moscow's chief regional sources of fuels. The center of the oil industry is Baki (Baku), the capital on the shore of the Caspian Sea—but the Caspian Sea is a lake. To export its oil, Azerbaijan needs pipelines, but those of Soviet vintage link Baki to Russia's Black Sea terminal of Novorossiysk.

Playing the Energy Card

In the post-Soviet period, Azerbaijan has played its energy card to national advantage. Its authoritarian government is sometimes compared to that of Belarus, but unlike Belarus it has tried to establish some independence from Moscow. Oil and gas have made this easier: Washington, which supposedly promotes democracy, does not press the issue when it negotiates for Azerbaijan's oil, and America has become its most important customer. During Soviet and immediate post-Soviet times, oil from Baki continued to flow exclusively through Novorossiysk (Fig. 2-14), but the United States negotiated the construction of an alternate route across Georgia and Turkey to the Mediterranean oil terminal at Ceyhan, thus avoiding Russian transit altogether. Oil began to flow through this outlet in 2006. Soviet state companies once extracted Azerbaijan's oil, but now American, French, British, and Japanese oil companies are developing the country's vast Caspian reserves.

Azerbaijan's nearly 9 million inhabitants have not benefited greatly from their country's energy riches. Given its Muslim roots and relative location, the time may come when this country turns toward the Islamic world. For the present, it is caught in the worldwide web of energy demand and supply.

What You Can Do

RECOMMENDATION: Do you have a good-sized political globe and/or a large wall map of the world in your home or room? If not, we recommend that you do so (we are aware of how crowded desktop and table space can become in a room, especially a shared one). There even are self-standing globes you can place on the floor. We predict that you will be referring to this useful addition to your scholarly arsenal more than you can imagine, and you will find the map equally useful. National Geographic Maps, the Society's cartographic division, produces world maps on a sensible projection (see Appendix A) at several scales, so you can select one that will fit on your wall. In general, the bigger the better, and you will soon be challenging guests to find places you have discovered in this course.

GEOGRAPHIC CONNECTIONS

1 The American Geographical Society in New York, the country's oldest geographical association, presents travelers the opportunity not only to visit remote parts of the world, but to do so accompanied by a professional geographer who specializes in the region or area being visited. Other organizations also appoint geographers as lecturers and guides on ships, planes, and even trains and buses. Imagine that you have been asked to accompany a group of your fellow students taking a riverboat down the Volga and Don rivers in July from Moscow to Rostov near the Sea of Azov, an arm of the Black Sea. In your first briefing, what will you tell them about the weather they should expect, about the minority "republics" of which they will get a glimpse, the crops and land uses they will encounter, the cities and towns they will see, and how they will get from the Volga to the Don? What might be the most interesting side trips along this itinerary?

2 President Vladimir Putin of Russia, who succeeded President Boris Yeltsin, soon after his reelection proclaimed his intention to restore Russia to superpower status, to recapture much of the global might the Soviet Union once possessed. How will Russia's leaders use their vast country's natural resources to advance this plan? Is any Soviet-era legacy likely to be helpful in achieving it? In what ways will this plan affect future relations between the European Union and Russia? Might the boundaries of the Russian realm as mapped in this chapter change as a result, and if so, where and how?

3 Russia has experienced the impact of Islamic terrorism principally in two locales: the Moscow urban area and the Interior Southern Periphery. But Muslim minorities are far more widely distributed throughout this vast country beyond those two areas. Why is Islamic terrorism in Russia so prevalent in these locales and virtually (though not totally) absent elsewhere? What role does physiography play in the persistence of terrorism in the Interior Southern Periphery?

Geographic Literature on Russia: The key introductory works on this realm were authored by Bater, Remnick (both titles), Shaw (both titles), and Symons. These works, together with a comprehensive listing of noteworthy books and recent articles on Russia's geography, can be found in the *References and Further Readings* section of this book's website at **www.wiley.com/college/deblij**.

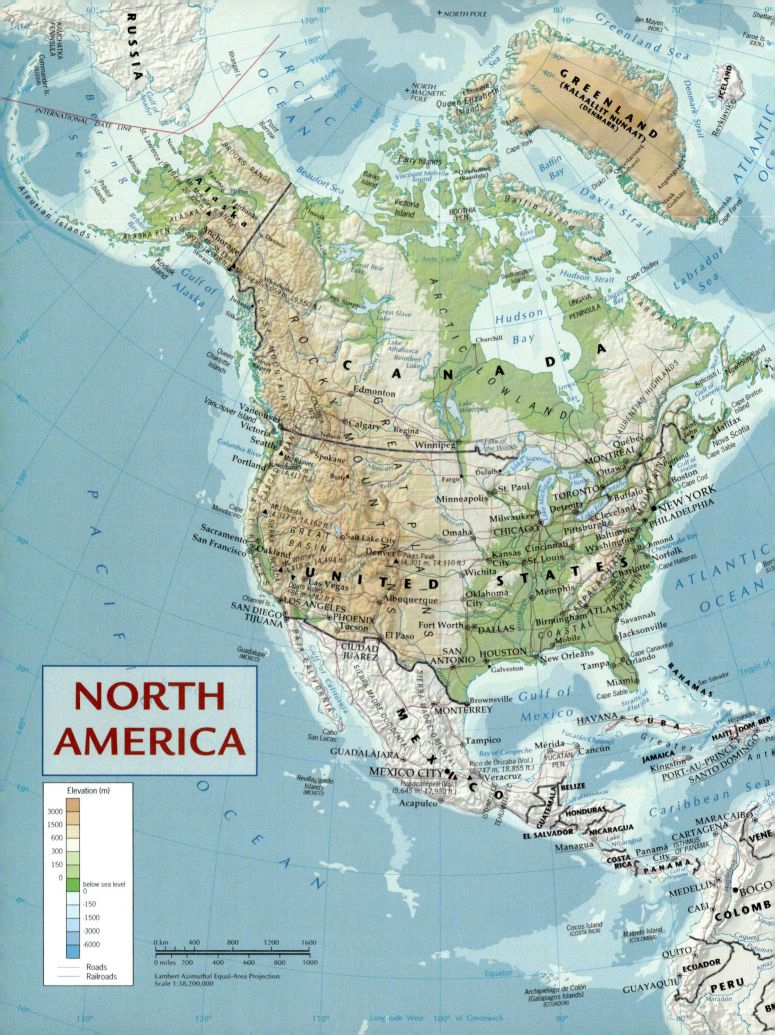

3

CONCEPTS, IDEAS, AND TERMS

1. Cultural pluralism
2. Physiographic province
3. Rain shadow effect
4. Sunbelt
5. Migration
6. Push/pull factors
7. American Manufacturing Belt
8. Ghetto
9. Outer city
10. Suburban downtown
11. Urban realms model
12. Mosaic culture
13. Productive activities
14. Fossil fuels
15. Economies of scale
16. Postindustrialism
17. Technopole
18. Ecumene
19. Regional state
20. Pacific Rim

REGIONS

- NORTH AMERICAN CORE
- MARITIME NORTHEAST
- FRENCH CANADA
- CONTINENTAL INTERIOR
- SOUTH
- SOUTHWEST
- WESTERN FRONTIER
- NORTHERN FRONTIER
- PACIFIC HINGE

Where were these photos taken? Find out at **www.wiley.com/college/deblij**

In This Chapter

- Spectacular scenery, violent weather
- North America before the European invasion
- What makes Americans and Canadians move so often
- Immigration and the changing map of the United States
- Canada and Quebec: Still together, but forever?
- China's impact on the Pacific Hinge

Photos: © H. J. de Blij

FIGURE 3-1 © H. J. de Blij, P. O. Muller, and John Wiley & Sons, Inc.

NORTH AMERICA AS a geographic name means different things to different people. To physical geographers, North America refers to the landmass that extends from the shores of the Arctic Ocean to Panama. As Figure G-4 shows, the North American continent is the dominant feature of the North American tectonic plate, which reaches almost—but not quite—to Panama. To human geographers, North America (sometimes inappropriately called "Anglo" America) signifies a geographic realm consisting of two large countries with much in common, Canada and the United States. We therefore recognize three realms in the Americas. The first, covered here in Chapter 3, is *North America*, which is propelled by the United States. The second and third are treated, respectively, in Chapters 4 and 5: *Middle America* (dominated by Mexico) and *South America* (anchored by Brazil).

Defining the Realm

Defined in the context of human geography, North America is constituted by two of the world's most highly advanced countries by virtually every measure of human development. Blessed by an almost endless range of natural resources and bonded by trade as well as culture, Canada and the United States are locked in a mutually productive embrace that is reflected by the statistics. In an average year of the recent past, more than 80 percent of Canadian exports went to the United States and about two-thirds of Canada's imports came from its southern neighbor. For the United States, Canada is its leading export market and its number one source of imports.

Both countries rank among the world's most highly urbanized, and in each about 80 percent of the population is concentrated in cities that have evolved into metropolitan agglomerations. The old core cities are now surrounded (and often engulfed) by rings of suburban development, which draw an ever-greater proportion of the metropolitan population and activity base into their newly urbanized environments. Nothing symbolizes the North American city as strongly as the skyscrapered panoramas of New York, Chicago, and Toronto—or the vast, beltway-connected suburban expanses of Los Angeles, Washington, and Houston.

North Americans also are the most mobile people in the world. Commuters stream into and out of suburban business centers and central-city downtowns by the millions each working day (see photos below); most of them drive cars, whose numbers have multiplied more than six

The photo at the left shows the heart of the central business district (CBD) of Houston, Texas, the fourth-largest U.S. city and one of the leading centers in the nation's energy industry. The central city's skyscraper-dominated skyline originated during the era of industrial urbanization that peaked in the mid-twentieth century. . . . But we now live in the twenty-first century, and the postindustrial transformation of urban America is all but complete. The landscape of the latter is symbolized by the low-rise world headquarters of IBM, completed in 1997, the centerpiece of a forested suburban office campus located in Armonk, New York about 30 miles north of Times Square. This is actually IBM's second headquarters facility on the site, the company having first moved out of Manhattan to this suburban Westchester County community more than 50 years ago. These scenes represent the most far-reaching change in the history of the American city, shaped by the forces of the past four decades that turned it inside-out as well as replacing its single-centered with a multicentered spatial structure. *left*: © Alan Schein Photography/Corbis Images. *right*: © Kohn Pedersen Fox Associates.

times faster than the human population since 1970. Moreover, each year nearly one out of every six individuals changes his or her residence.

GEOGRAPHIC AND SOCIAL CONTRASTS

Although Canada and the United States share many historical, cultural, and economic qualities, they also differ in significant ways, as can be seen on the map. The United States, somewhat smaller territorially than Canada, occupies the heart of the North American continent and, as a result, encompasses a greater environmental range. The U.S. population is dispersed across most of the country, forming major concentrations along both the (north-south-trending) Atlantic and Pacific coasts. The overwhelming majority of Canadians, however, live in an interrupted east-west corri-

dor that lies across southern Canada, mainly within 300 kilometers (200 mi) of the U.S. border. The United States also encompasses North America's northwestern extension, Alaska. (Offshore Hawai'i, however, belongs in the Pacific Realm.)

Demographic contrasts also exist between the two countries because the U.S. population is nearly ten times larger than Canada's. In fact, the United States is the world's third most populous country after China and India (Canada ranks 36th), having surpassed 300 million in 2006 on its way to 303 million two years later. A snapshot of the United States at this very recent milestone shows that each 100 million was added more rapidly: the first took more than 125 years, the second 52 years, and the third 39 years (the addition of the next 100 million is likely to continue the trend and is projected to require 37 years). This relatively high growth rate is unusual for such an advantaged country, with about 60 percent of its annual growth accounted for by net natural population increase and the

MAJOR GEOGRAPHIC QUALITIES OF
North America

1. North America encompasses two of the world's biggest states territorially (Canada is the second largest in size; the United States is third).

2. Both Canada and the United States are federal states, but their systems differ. Canada's is adapted from the British parliamentary system and is divided into ten provinces and three territories. The United States separates its executive and legislative branches of government, and it consists of 50 States, the Commonwealth of Puerto Rico, and a number of island territories under U.S. jurisdiction in the Caribbean Sea and the Pacific Ocean.

3. Both Canada and the United States are plural societies. Although ethnicity is increasingly important, Canada's pluralism is most strongly expressed in regional bilingualism. In the United States, divisions occur largely along racial and income lines.

4. A substantial number of Quebec's French-speaking citizens supports a movement that seeks independence for the province. The movement's high-water mark was reached in the 1995 referendum in which (minority) non-French-speakers were the difference in the narrow defeat of separation. Prospects for a break-up of the Canadian state have diminished since 2000 but have not disappeared.

5. North America's population, not large by international standards, is one of the world's most highly urbanized and mobile. Largely propelled by a continuing wave of immigration, the realm's population total is expected to grow by more than 40 percent over the next half-century.

6. By world standards, this is a rich realm where high incomes and high rates of consumption prevail. North America possesses a highly diversified resource base, but nonrenewable fuel and mineral deposits are consumed prodigiously.

7. North America is home to one of the world's great manufacturing complexes. The realm's industrialization generated its unparalleled urban growth, but today both of its countries are experiencing the emergence of a new postindustrial society and economy.

8. The two countries heavily depend on each other for supplies of critical raw materials (e.g., Canada is a leading source of U.S. energy imports) and have long been each other's chief trading partners. Today, the North American Free Trade Agreement (NAFTA), which also includes Mexico, is linking all three economies ever more tightly as the remaining barriers to international trade and investment flow are dismantled.

9. North Americans are the world's most mobile people. Although plagued by recurrent congestion problems, the realm's networks of highways, commercial air routes, and cutting-edge telecommunications are still the most efficient on Earth.

remainder by immigration. Many of the key population changes that have occurred since the 200-million milestone was passed in 1967 have important geographical dimensions and will be discussed in this chapter. They include the continuing shift of people both westward and southward; the rise of the suburbs to a dominant position; the proliferation and widening dispersal of foreign-born residents, whose numbers are now at an all-time high; and the heightened flow of immigrants.

Comparatively small as its population of 33 million may be, however, Canada has varied cultures and traditions ranging from multilingualism (French has equal status with English) to ethnic diversity that, as we will note later, have strong spatial expression. Thus Canada shares with the United States a characteristic feature of this realm: the prevalence of **cultural pluralism**. In the United States, this **1** is exhibited by a range of ethnic backgrounds from European and African to Asian and Native American. And while the civil rights movement and other initiatives have eliminated *de jure* (legal) segregation in the United States, ethnic clustering and *de facto* (real-world) self-segregation continue to mark the social landscape. In part, this is due to a troubling aspect of this wealthiest of all realms: poverty and deprivation still afflict a substantial minority as income gaps between the successful rich and the struggling poor continue to widen.

Immigration Issues

Both countries have been and continue to be shaped by immigration. We just noted that 40 percent of the annual population growth of the United States results from immigration; and in Canada, net yearly immigration per 1000 population is more than double the U.S. rate. Not surprisingly, in the global ranking of urban areas according to foreign-born residents as a percentage of total population, five of the top eight are located in North America—three in the United States (Miami, Los Angeles, and New York) and two in Canada (Toronto and Vancouver).

Immigration to the United States has become an increasingly important and divisive issue since 2000, and came to a head during 2006 in a major public debate about those who have crossed the southern border illegally. The number of illegal immigrants inside the country today is believed to be at least 12 million, and the influx continues (a yearly average of 650,000 are estimated to have entered the United States between 1999 and 2007, 80 percent of them Latinos). Conceding that illegals strain social services and schools, their supporters point out that they work hard for low wages and productively perform vital jobs not easily filled by employers. Opponents insist that federal legislation is needed to improve border security (which began under presidential order in 2006) and strengthen en-

forcement of immigration laws to penalize employers of illegals and deport all undocumented immigrants.

North of the border, the Canadians are having a different debate about their immigrants. The country now faces a critical labor shortage, particularly in its westernmost provinces where employers are scrambling to find enough workers to sustain an oil boom in Alberta and the burgeoning Asian trade in neighboring British Columbia. As we shall see later in the chapter, the main issues center around adjusting the legal immigration process to keep it compatible with Canada's changing employment and demographic needs.

NORTH AMERICA'S PHYSICAL GEOGRAPHY

Before we examine the human geography of the United States and Canada more closely, we need to consider the physical setting in which they are rooted. The North American continent extends from the Arctic Ocean to Panama (Fig. 3-1), but we will confine ourselves here to the territory north of Mexico—a geographic realm that still stretches from the near-tropical latitudes of southern Florida and Texas to subpolar Alaska and Canada's farflung northern periphery. The remainder of the North American continent comprises a separate realm, *Middle America*, which is covered in Chapter 4.

Physiography

North America's physiography is characterized by its clear, well-defined division into physically homogeneous regions called **physiographic provinces**. Each region is **2** marked by considerable uniformity in relief, climate, vegetation, soils, and other environmental conditions, resulting in a scenic sameness that comes readily to mind. For example, we identify such regions when we refer to the Rocky Mountains, the Great Plains, and the Appalachian Highlands. However, not all the physiographic provinces of North America are so easily delineated.

Figure 3-2 maps the complete layout of the continent's physiography and also includes a cross-sectional terrain profile along the 40th parallel. The most obvious aspect of this map of North America's physiographic provinces is the north-south alignment of the continent's great mountain backbone, the Rocky Mountains, whose rugged topography dominates the western segment of the continent from Alaska to New Mexico. The major feature of eastern North America is another, much lower chain of mountain ranges called the Appalachian Highlands, which also trend approximately north-south and extend from

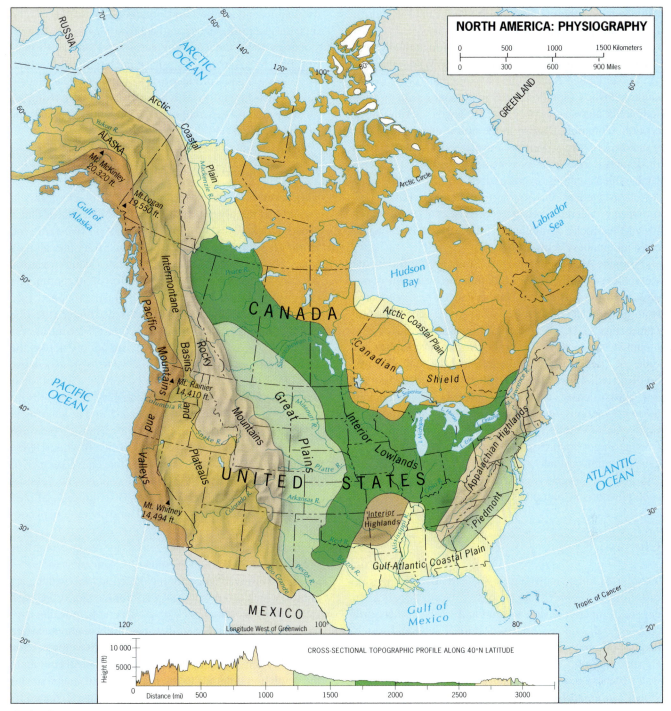

NORTH AMERICA: PHYSIOGRAPHY

CROSS-SECTIONAL TOPOGRAPHIC PROFILE ALONG 40°N LATITUDE

FIGURE 3-2   © H. J. de Blij, P. O. Muller, and John Wiley & Sons, Inc.

Canada's Atlantic Provinces to Alabama. The orientation of the Rockies and Appalachians is important because, unlike Europe's Alps, they do not form a topographic barrier to polar or tropical air masses flowing southward or northward, respectively, across the continent's interior.

Between the Rocky Mountains and the Appalachians lie North America's vast interior plains, which extend from the Mackenzie Delta on the Arctic Ocean to the Gulf of Mexico. We can subdivide these into several provinces: (1) the

great Canadian Shield, which is the geologic core area containing North America's oldest rocks; (2) the Interior Lowlands, covered largely by glacial debris laid down by ice, meltwater, and wind during the Pleistocene glaciation; and (3) the Great Plains, the extensive sedimentary surface that rises gently westward toward the Rocky Mountains. Along the southern margin, these interior plainlands merge into the Gulf-Atlantic Coastal Plain, which extends from southern Texas along the seaward margin of the Appalachian

Highlands and the neighboring Piedmont until it ends at Long Island just to the east of New York City.

On the western side of the Rocky Mountains lies the zone of Intermontane Basins and Plateaus. Within the conterminous United States, this physiographic province includes: (1) the Colorado Plateau in the south, with its thick sediments and spectacular Grand Canyon; (2) the lava-covered Columbia Plateau in the north, which forms the watershed of the Columbia River; and (3) the central Basin-and-Range country (Great Basin) of Nevada and Utah, which contains several extinct lakes from the glacial period as well as the surviving Great Salt Lake. This province is called *intermontane* because of its position between the Rocky Mountains to the east and the Pacific coast mountain system to the west.

From the Alaskan Peninsula to Southern California, the west coast of North America is dominated by an almost unbroken corridor of high mountain ranges that originated from the contact between the North American and Pacific Plates (Fig. G-4). The major components of this coastal mountain belt include California's Sierra Nevada, the Cascades of Oregon and Washington, and the long chain of highland massifs that line the British Columbia and southern Alaska coasts. Three broad valleys—which contain dense populations—are the only noteworthy interruptions: California's Central (San Joaquin-Sacramento) Valley; the Cowlitz-Puget Sound lowland of Washington State, which extends southward into western Oregon's Willamette Valley; and the lower Fraser Valley, which slices through southern British Columbia's coast range.

Climate

The world climate map (Fig. G-8) clearly depicts the various climatic regimes and regions of North America. In general, temperature varies latitudinally—the farther north one goes, the cooler it gets. Regional land-and-water-heating differentials, however, distort this broad pattern. Because land surfaces heat and cool far more rapidly than water bodies, yearly temperature ranges are much larger where **continentality** (interior remoteness from the sea) is greatest.

Western Rain Shadow

Precipitation generally tends to decline toward the west (except for the Pacific coastal strip itself) as a result of the **rain shadow effect**. This occurs because Pacific air masses, driven by prevailing winds, carry their moisture onshore but soon collide with the Sierra Nevada-Cascades wall, forcing them to rise—and cool—in order to crest those mountain ranges. Such cooling is accompanied

by major condensation and precipitation, so that by the time these air masses descend and warm along the eastern slopes to begin their journey across the continent's interior, they have already deposited much of their moisture. Thus the mountains produce a downwind "shadow" of dryness, which is reinforced whenever eastward-moving air must surmount other ranges farther inland, especially the massive Rockies.

Arid and Humid America

This semiarid—and in places truly arid—environment extends so deeply into the central United States that a broad division can be made between Arid (western) and Humid (eastern) America, which face each other along a fuzzy boundary that is best viewed as a wide transition zone. Although the separating criterion of 50 centimeters (20 in) of annual precipitation is easily mapped (see Fig. G-7), that generally north-south **isohyet** (the line connecting all places receiving exactly 50 centimeters [20 in] per year) can and does swing widely across the drought-prone Great Plains from year to year because highly variable warm-season rains from the Gulf of Mexico come and go in unpredictable fashion. Drought is indeed a constant environmental hazard, and much of Arid America today remains in the grip of one of the most severe dry periods ever recorded. Since the late 1990s, major parts of the western United States have been challenged by the shrinkage of water supplies, the dessication of soils, and the proliferation of wildfires—all at a time when population growth and land development continue to mushroom.

Precipitation in Humid America is far more regular. The prevailing westerly winds (blowing from west to east—winds are always named for the direction *from* which they come), which usually come up dry for the large zone west of the 100th meridian, pick up considerable moisture over the Interior Lowlands and distribute it throughout eastern North America. A large number of storms develop here on the highly active weather front between tropical Gulf air to the south and polar air to the north. Even if major storms do not materialize, local weather disturbances created by sharply contrasting temperature differences are always a danger. There are more tornadoes (nature's most violent weather) in the central United States each year than anywhere else on Earth. And in winter, the northern half of this region receives large amounts of snow, particularly around the Great Lakes. Moreover, along the southeastern Atlantic and Gulf coasts, hurricanes spawned in nearby tropical waters pose a major seasonal threat; as Hurricane Katrina showed in 2005 (see photo pair, p. 157), these tropical cyclones are capable of inflicting catastrophic devastation on vulnerable low-lying areas.

On August 29, 2005, Category-4 Hurricane Katrina cut a 230,000-square-kilometer (90,000-sq-mi) path of destruction along the Louisiana and Mississippi Gulf Coast centered around landfall just west of the mouth of the Mississippi River. Most devastated was the city of New Orleans, 80 percent of which lay underwater in the immediate aftermath of the storm. The pair of aerial photos above shows central New Orleans under normal conditions in the spring of 2004 (left) and two days after the tropical cyclone struck (right). Katrina was by far the most costly natural disaster in U.S. history, and its toll was truly staggering: more than 1800 lives lost, at least 200,000 homes destroyed, nearly a million people displaced, and more than $25 billion in insured property damage (a total equal to the GDP of Luxembourg). Despite $110 billion in federal aid during the year following the disaster, rebuilding in 2007 was spotty and only just beginning. Nowhere can the impact of Katrina be measured more profoundly than in this profile of New Orleans on the first anniversary of the storm. What had been the nation's 31st-largest city saw its population shockingly reduced by more than half from 444,000 to 191,000. Only 60 percent of its residential units had electric power, just 39 percent of its schools had reopened, 4 of its 7 hospitals were closed down, a mere 23 percent of daycare facilities were functioning, and only 17 percent of the city's buses and trolleys were operational on 49 percent of the pre-Katrina route network. The Crescent City's tourist economy fared somewhat better, but with 15 percent of the hotels and 56 percent of the restaurants still out of commission, the Big Easy continues to face a long uphill struggle to complete its restoration.   © Digital Globe/Eurimage/Photo Researchers, Inc.

Figure G-8 shows the absence of humid temperate (*C*) climates from Canada (except along the narrow Pacific coastal zone) and the prevalence of cold in Canadian environments. East of the Rocky Mountains, Canada's most *moderate* climates correspond to the *coldest* of the United States. Nonetheless, southern Canada shares the environmental conditions that mark the Upper Midwest and Great Lakes areas of the United States, so that agricultural productivity in the Prairie Provinces and in Ontario is substantial. Canada is a leading food exporter (chiefly wheat), as is the United States, despite its comparatively short growing season.

Soils and Vegetation

The broad environmental partitioning into Humid and Arid America is also reflected in the distribution of the realm's soils and vegetation. For farming purposes there is usually sufficient soil moisture to support crops where annual precipitation exceeds the critical 50 centimeters (20 in); where the yearly total is less, soils may still be fertile (especially in the Great Plains), but irrigation is often necessary to achieve their agricultural potential. As for vegetation, the Humid/Arid America dichotomy is again a valid generalization: the natural vegetation of areas receiving more than 50 centimeters of water annually is *forest*, whereas the drier climates give rise to a *grassland* cover.

Hydrography (Surface Water)

Surface water patterns in North America are dominated by the two major drainage systems that lie between the Rockies and the Appalachians: (1) the five Great Lakes (Superior, Michigan, Huron, Erie, and Ontario) that drain into the St. Lawrence River, and (2) the mighty Mississippi-Missouri river network, fed by such major tributaries as the Ohio, Tennessee, and Arkansas rivers. Both are products of the last episode of Pleistocene glaciation, and together they amount to nothing less than the finest natural inland waterway system in the world. Human intervention has further enhanced this network of navigability, mainly through the building of canals that link the two systems as well as the St. Lawrence Seaway.

Elsewhere, the northern east coast of the continent is well served by a number of short rivers leading inland from the Atlantic. In fact, many of the major northeastern seaboard cities of the United States—such as Washington, D.C., Baltimore, and Philadelphia—are located at the waterfalls that marked the limit to tidewater navigation (hence their designation as **Fall Line cities**). Rivers in the Southeast and west of the Rockies at first offered little practical value because of their orientation and the difficulty of navigating them. In the far west, however, the Colorado and Columbia rivers have become supremely important as suppliers of drinking and irrigation water as well as hydroelectric power.

INDIGENOUS NORTH AMERICA

When the first Europeans set foot on North American soil, the continent was occupied by millions of people whose ancestors had reached the Americas from Asia, via Alaska and probably also across the Pacific, more than 13,000 years before (and possibly as long as 30,000 years ago). In search of Asia, the Europeans misnamed them "Indians," but the historic affinities of these earliest Americans were with the peoples of eastern and northeastern Asia, not India. In North America these **Native Americans** or **First Nations**—as they are now called in the United States and Canada, respectively—had organized themselves into hundreds of nations with a rich mosaic of languages and a great diversity of cultures (Fig. 3-3). Farmers grew crops that the Europeans had never seen; other nations depended chiefly on fishing, herding, hunting, or some combination of these. Elaborate houses, efficient watercraft, effective weaponry, decorative clothing, and wide-ranging art forms distinguished the aboriginal nations. Certain nations had formulated sophisticated health and medical practices; ceremonial life was complex and highly developed; and political institutions were mature and elaborate.

The eastern nations were the first to bear the brunt of the European invasion. By the end of the eighteenth century, ruthless, land-hungry settlers had driven most of the Native American peoples living along the Atlantic and Gulf coasts from their homes and lands, beginning a westward push that was to devastate indigenous society. The U.S. Congress in 1789 proclaimed that "Indian . . . land and property shall never be taken from them without their consent," but in fact this is just what happened. One of the sorriest episodes in American history involved the removal of the eastern Cherokee, Chickasaw, Choctaw, Creek, and Seminole from their homelands in forced marches a thousand miles westward to Oklahoma. One-fourth of the entire Cherokee population died along the way from exposure, starvation, and disease, and the others fared little better. Again, Congress approved treaties that would at least protect the native peoples of the Plains (Fig. 3-3) and those farther to the west, but after the mid-nineteenth century the white settlers ignored those guarantees as well. A half-century of war left what remained of North America's nations with about 4 percent of U.S. territory in the form of mostly impoverished reservations.

In what is today Canada, too, the comparatively small First Nations population was overwhelmed by the numbers and power of European settlers, and decimated by the diseases they introduced. Efforts at restitution and recognition of First Nations rights, however, have gone farther in Canada than in the United States.

As we noted earlier, two large countries, with geographic similarities as well as differences, constitute the North American realm. Canada and the United States share physiographic provinces as well as border-straddling economic and cultural subregions (for example, large-scale grain cultivation in the Great Plains). It would be possible to base the regionalization of this realm on contrasts between the two states; but as we will see in the second half of this chapter, quite a different set of regions emerges from our geographic analysis. To fully appreciate this regional framework, we must first examine in some detail the changing human geography of the United States and Canada individually.

THE UNITED STATES

As we note throughout this book, the administrative components of states are assuming ever-greater importance in the scheme of things. In Europe, all the major states have reorganized their political-economic systems, forged regions from smaller units, and devolved power to the provinces. As a result, it is increasingly important not just to know Spain but also Catalonia, not just France but also Rhône-Alpes, not just Italy but also Lombardy. So it is in the United States. We should have a clear mental map of the functional layout of this federation, which is why it may be helpful to take a few moments to review Figure 3-4.

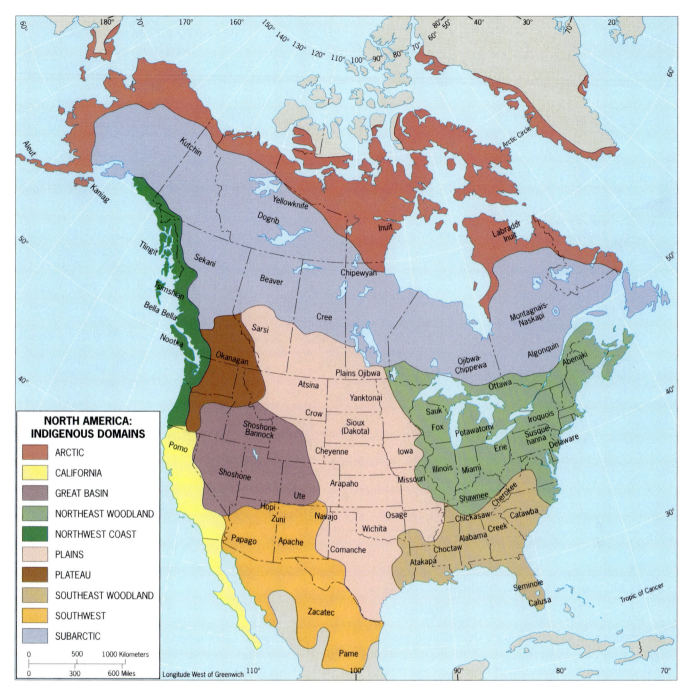

FIGURE 3-3   © H. J. de Blij, P. O. Muller, and John Wiley & Sons, Inc.

Population in Time and Space

The current population distribution of the United States is shown in Figure 3-5. It is important to note that this map is the latest "still" in a motion picture, one that has been unreeling for four centuries since the founding of the first permanent European settlement on the northeastern coast. Slowly at first, then with accelerating speed after 1800, as one major transportation breakthrough followed another,

Americans (and Canadians) took charge of their remarkable continent and pushed the settlement frontier westward to the Pacific. The swiftness of this expansion was dramatic, but North Americans have long been the world's most mobile people. In fact, migrations continue to redistribute population in the United States today, perhaps the most significant being the persistent drift of people and livelihoods toward the South and West (the so-called **Sunbelt**).

4

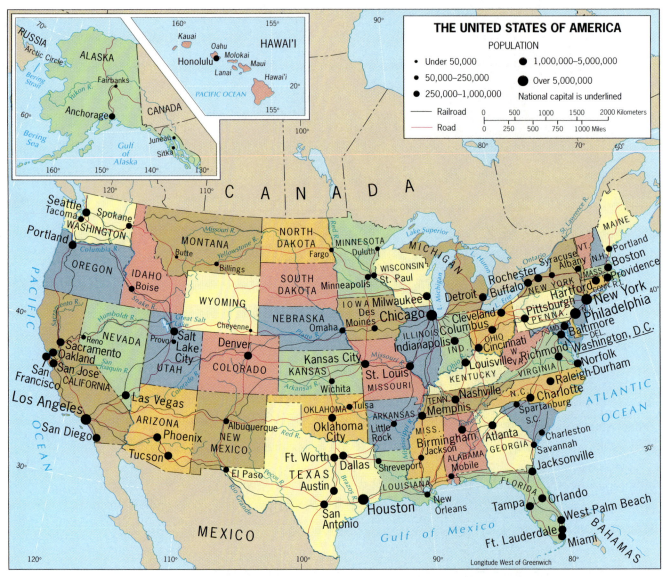

FIGURE 3-4 © H. J. de Blij, P. O. Muller, and John Wiley & Sons, Inc.

To understand the contemporary population map, we need to review the major forces that have shaped, and continue to shape, the distribution of Americans and their activities. Since its earliest days, the United States has attracted a steady influx of immigrants who were rapidly assimilated into the societal mainstream (see box titled "The Migration Process"). Within the country, people have sorted themselves out to maximize their proximity to existing economic opportunities, and they have shown little resistance to relocating as the nation's evolving economic geography has successively favored different sets of places over time.

During the past century these transformations spawned a number of major migrations: (1) the still-continuing westward shift but now with that southerly, Sunbelt deflection; (2) the rapid growth of metropolitan areas, first

triggered by the late-nineteenth-century Industrial Revolution, which since the 1960s has been largely rechanneled from the central cities to the suburban ring; (3) the movement of African Americans from the rural South to the urban North, which over the last few decades has become a stronger return flow, particularly among middle-income blacks; and (4) the strong influx of immigrants from outside the United States, mostly European before 1960 but now dominated by a new wave from Middle America and eastern Asia that is directed, respectively, toward the States near the southern border and the urban areas of the Pacific coast.

Let us now look more closely at the historical geography of these changing population patterns, considering first the initial rural influence and then the decisive impacts of industrial urbanization.

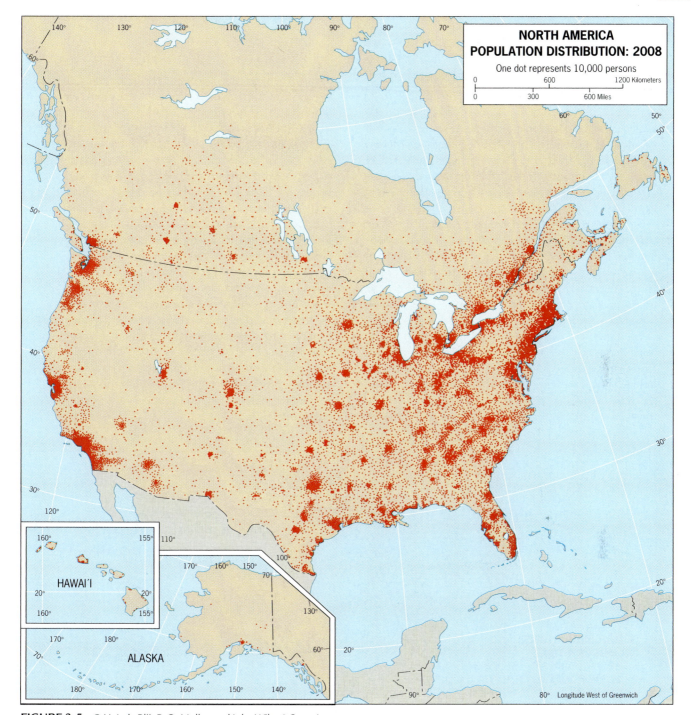

FIGURE 3-5 © H. J. de Blij, P. O. Muller, and John Wiley & Sons, Inc.

Colonial and Nineteenth-Century Development

The current spatial distribution of the U.S. population is rooted in the colonial era of the seventeenth and eighteenth centuries that was dominated by England and France. The French sought mainly to organize a lucrative fur-trading network, while the English established settlements along the coast of what is today the northeastern U.S. seaboard (the oldest, Virginia's Jamestown, was founded just over 400 years ago). These British colonies quickly became differentiated in their local economies, a diversity that was to endure and later shape American cultural geography. The northern colony of New England (Massachusetts Bay and environs) specialized in commerce; the southern Chesapeake Bay colony (Tidewater Virginia and Maryland) emphasized the plantation farming of tobacco; the Middle Atlantic area lying in between (southeastern New York,

The Migration Process

THE UNITED STATES as well as Canada is the product of **migration** (a change in residential location intended to be permanent). After tens of thousands of years of native settlement, European explorers reached the shores of North America beginning with Columbus in 1492 (and probably centuries before that). The first permanent colonies were established in the early 1600s, and from them evolved the modern United States of America. The Europeanization of North America doomed the continent's aboriginal societies, but this was only one of many areas around the world where local cultures and foreign invaders came face to face. Between 1835 and 1935, perhaps as many as 75 million Europeans departed for distant shores—most of them bound for the Americas (Fig. 3-6). Some sought religious freedom; others escaped poverty and famine; still others simply hoped for a better life.

Studies of the *migration decision* indicate that migration flows vary in size with: (1) the perceived degree of difference between one's home, or source, and the destination; (2) the effectiveness of the information flow, that is, the news about the destination that migrants sent to those who stayed behind waiting to decide; and (3) the distance between the source and the destination (shorter moves attract many more migrants than longer ones). More than a century ago, the British social scientist Ernst Georg Ravenstein studied the migration process, and many of his conclusions remain valid today. For example, every migration stream from source to destination produces a counter-stream of returning migrants who cannot adjust, are unsuccessful, or are otherwise persuaded to go back home. Studies of population movements also conclude that several factors are at work in the migration process. **Push factors** motivate people to move away; **pull factors** attract them to new destinations. To those early Europeans, the United States was a new frontier, a place where one might acquire a piece of land and some livestock. The opportunities were reported to be unlimited.

That perception has never changed, and immigration continues to significantly shape the human-geographic complexion of the United States. Today's in-migrants account for at least 40 percent of the country's annual population growth. Important details about this latest surge of the international migration process are discussed on pages 169–170.

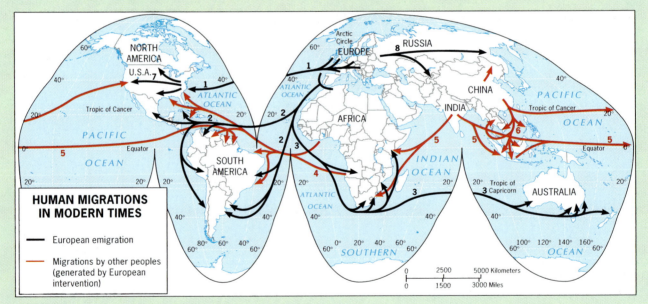

HUMAN MIGRATIONS IN MODERN TIMES

— European emigration

— Migrations by other peoples (generated by European intervention)

FIGURE 3-6 © H. J. de Blij, P. O. Muller, and John Wiley & Sons, Inc.

New Jersey, eastern Pennsylvania) was home to a number of smaller, independent-farmer colonies.

These neighboring colonies soon thrived and yearned to expand, but the British government responded by closing the inland frontier and tightening economic controls. By 1783 this move had led to colonial unification, British defeat in the Revolutionary War, and independence for the newly formed United States of America. The western frontier of the fledgling nation now swung open, and the zone north of the Ohio River was promptly settled following the discovery that the soils and climate of the Interior Lowlands were more favorable for farming than those of the Atlantic Coastal Plain and Piedmont. This triggered the rapid growth of trans-Appalachian agriculture and the widening of seaboard-interior trading ties. The new inter-regional complementarities signified that U.S. spatial organization was assuming national-scale proportions.

Culture Hearths and Frontier Expansion. By the time the westward-moving frontier swept across the Mississippi Valley in the 1820s, the three former seaboard colonies (**A**, **B**, and **C** in Fig. 3-7) had become separate ***culture hearths***—primary source areas and innovation centers from which migrants carried cultural traditions into the central United States (as the arrows in Fig. 3-7 indicate). The northern half of this vast interior space soon became well unified as its infrastructure steadily improved following the introduction of the railroad in the 1830s. The American South, however, did not wish to integrate itself economically with the North, preferring to export tobacco and cotton from its plantations to overseas markets; its insistence on preserving slavery to support this system soon led the South into secession, disastrous Civil War (1861–1865), and a dismal aftermath that took a full century to overcome.

The second half of the nineteenth century saw the frontier cross the western United States, and by 1869 agriculturally booming California was linked to the rest of the nation by transcontinental railroad (these same steel tracks also opened up the bypassed, semiarid Great Plains). When the American frontier closed in the 1890s, today's rural settlement pattern was firmly in place, anchored to a set of national-scale agricultural regions. By then, however, the exodus of rural Americans toward the burgeoning cities had begun in response to the Industrial Revolution that had taken hold after 1870.

Post–1900 Industrial Urbanization

The Industrial Revolution occurred almost a century later in the United States than in Europe, but when it finally did cross the Atlantic in the 1870s, it took hold so successfully and advanced so robustly that only 50 years later America was surpassing Europe as the world's mightiest industrial power. The impact of industrial urbanization occurred simultaneously at two levels of spatial generalization. At the national or ***macroscale***, a system of new cities swiftly emerged, specializing in the collection, processing, and

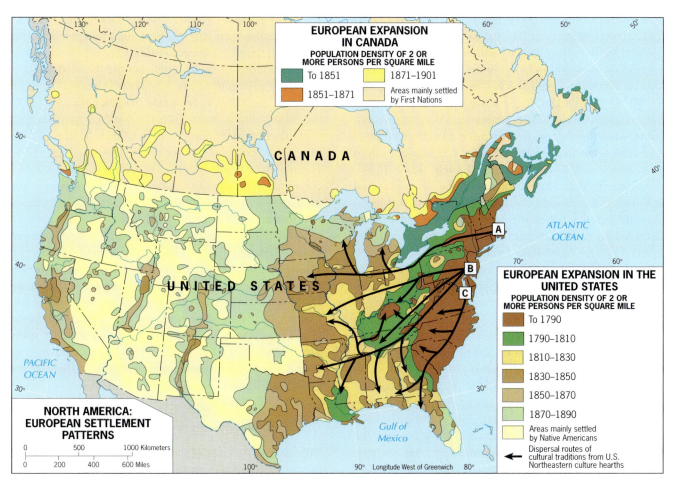

FIGURE 3-7   © H. J. de Blij, P. O. Muller, and John Wiley & Sons, Inc.

distribution of raw materials and manufactured goods, linked together by an efficient web of railroad lines. Within that urban network, at the local or *microscale*, individual cities prospered in their new roles as manufacturing centers, generating an internal structure that still forms the geographic framework of most of the central cities of America's large metropolitan areas. We now examine the urban trend at both of these scales.

Five-Stage Evolution of the U.S. Urban System.

The rise of the national urban system in the late nineteenth century was based on the traditional external role of cities: providing goods and services for their hinterlands in exchange for raw materials. Because people (both as laborers and as consumers), commercial activities, investment capital, and transport facilities were already agglomerated in existing (preindustrial) cities, the emerging industrialization movement tended to favor such locations. Their growing incomes, in turn, permitted industrially intensifying cities to invest in a bigger local infrastructure of private and public services as well as housing—and thereby convert each round of industrial expansion into a new stage of urban development.

The national urban system had actually been in the process of formation long before its flowering after 1870. Its evolution over the past two centuries has been studied by John Borchert, who identified five epochs of metropolitan evolution based on transportation technology and industrial energy. (1) *The Sail-Wagon Epoch* (1790–1830), marked by primitive overland and waterway circulation; the leading cities were the northeastern ports, which were more heavily oriented to the European overseas trade than to their barely accessible western hinterlands. (2) *The Iron Horse Epoch* (1830–1870), dominated by the arrival and spread of the steam-powered railroad; a nationwide transport system had now been forged, and the national urban system began to take shape, with New York emerging as the primate city by 1850. (3) *The Steel-Rail Epoch* (1870–1920), which spanned the Industrial Revolution and saw the full establishment of the national metropolitan system; the new forces shaping growth were the increasing scale of manufacturing, the rise of the steel and auto industries in Midwestern cities, and the introduction of steel rails that enabled trains to travel faster and haul heavier cargoes. (4) *The Auto-Air-Amenity Epoch* (1920–1970), which encompassed the later stage of U.S. industrial urbanization and the maturation of the national urban hierarchy; its key elements were the automobile and the airplane, the expansion of white-collar services jobs, and the growing locational pull of *amenities* (pleasant environments) that increasingly stimulated the urbanization of the suburbs and selected Sunbelt locales. (5) *The Satellite-Electronic-Jet Propulsion Epoch* (1970–), shaped by the newest advancements in information management, computer technologies, global communications, and intercontinental travel; it favors globally oriented metropolises, particularly those along the Pacific and Atlantic coasts that function as international gateways.

The American Manufacturing Belt.

Industrialization and the accompanying growth of the urban system reconfigured the realm's economic landscape. The most notable regional transformation was the emergence of the North American Core, or **American Manufacturing Belt**, **7** which contained the lion's share of industrial activity in both the United States and Canada. As Figure 3-8 shows, the geographic form of the Core Region—which includes southern Ontario—was a near-rectangle whose four corners were Boston, Milwaukee, St. Louis, and Baltimore. However, because manufacturing is such a spatially concentrated activity, less than 1 percent of the territory of the Belt is devoted to industrial land use: most of its factories are tightly clustered into a dozen districts centered on the cities mapped in Figure 3-8.

Megalopolitan Growth.

At the subregional scale, as transportation breakthroughs permitted progressive urban decentralization and *megalopolitan growth*, the expanding peripheries of major cities soon coalesced to form a number of conurbations. The most important of these by far is the *Atlantic Seaboard Megalopolis* (Fig. 3-9), the 1000-kilometer (600-mi) urbanized northeastern coastal strip extending from southern Maine to Virginia that contains metropolitan Boston, New York, Philadelphia, Baltimore, and Washington. This was the economic heartland of the Core; the seat of U.S. government, business, and culture; and the trans-Atlantic trading interface between much of North America and Europe. Six other primary conurbations have also emerged: *Lower Great Lakes* (Chicago–Detroit–Cleveland–Pittsburgh), *Piedmont* (Atlanta–Charlotte–Raleigh/Durham), *Florida* (Jacksonville–Tampa–Orlando–Miami), *Texas* (Houston–Dallas/Ft. Worth–San Antonio), *California* (San Diego–Los Angeles–San Francisco), and the *Pacific Northwest* (Portland–Seattle–Vancouver). Note that the last spills across the border into Canada, which has also spawned its own nationally predominant conurbation—*Main Street* (Windsor–Toronto–Montreal–Quebec City).

Changing Internal Structure of the U.S. Metropolis

The internal structure of the metropolis reflected the same mixture of forces that shaped the national urban system, es-

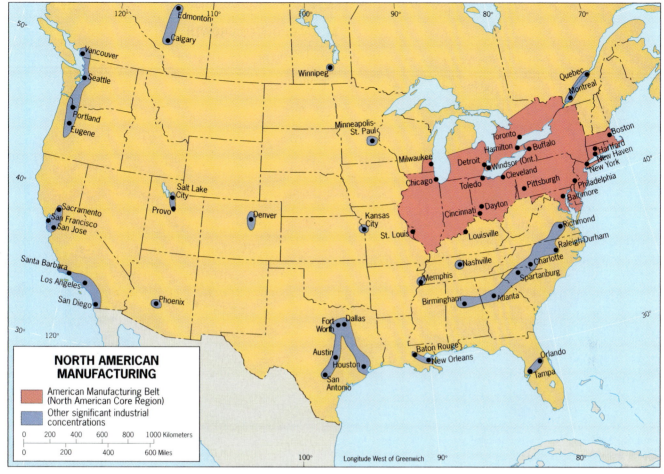

FIGURE 3-8   © H. J. de Blij, P. O. Muller, and John Wiley & Sons, Inc.

pecially transportation technology. Rails—in this case, lighter street-rail lines—once again shaped spatial organization as horse-drawn trolleys were succeeded by electric streetcars in the late nineteenth century. The mass introduction of the automobile after World War I changed all that, and America steadily turned from building compact cities to the widely dispersed metropolises of the post–World War II highway era. By 1970, the new intraurban expressway network had equalized location costs throughout the metropolis, setting the stage for suburbia to swiftly transform itself from a residential preserve into a complete outer city with amenities and new prestige that proved highly attractive to the business world. As the newly urbanized suburbs increasingly captured major economic activities, many large cities saw their status diminish to that of coequal. Their once thriving central business districts (CBDs) were now all but reduced to serving the less affluent populations that increasingly dominated the central city's neighborhoods.

This growth process was conceptualized into a four-stage model by John Adams, who identified the four **eras of intraurban structural evolution** (see Fig. 3-10). Stage I,

prior to 1888, was the ***Walking-Horsecar Era***, which produced a compact pedestrian city in which everything had to be within a 30-minute walk, a layout only slightly augmented when horse-drawn trolleys began to operate after 1850. The 1888 invention of the electric traction motor launched Stage II, the ***Electric Streetcar Era*** (1888–1920); higher speeds enabled the 30-minute travel radius and the urbanized area to expand considerably along new outlying trolley corridors; in the older core city, the CBD, industrial, and residential land uses differentiated into their modern form. Stage III, the ***Recreational Automobile Era*** (1920–1945), was marked by the initial impact of cars and highways that steadily improved the accessibility of the outer metropolitan ring, thereby launching a wave of mass suburbanization that further extended the urban frontier. During this era, the still-dominant central city experienced its economic peak and the partitioning of its residential space into neighborhoods sharply defined by income, ethnicity, and race. Stage IV, the ***Freeway Era*** (under way since 1945), saw the full impact of automobiles, with the metropolis turning inside-out as expressways pushed

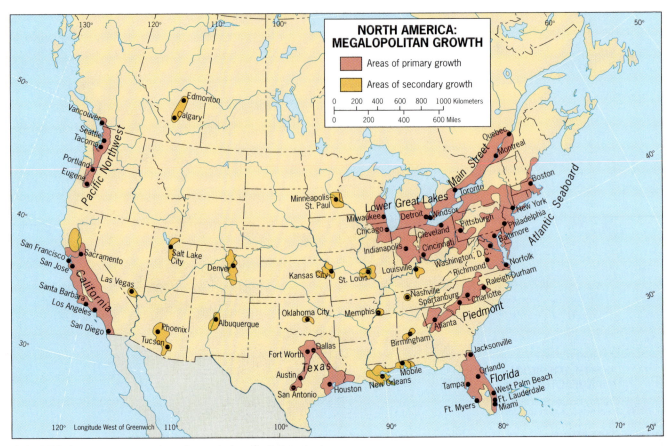

FIGURE 3-9   © H. J. de Blij, P. O. Muller, and John Wiley & Sons, Inc.

suburban development more than 50 kilometers (30 mi) from the CBD.

The social geography of the evolving industrial metropolis has been marked by the development of a residential mosaic that exhibited the congregating of ever-more-specialized groups. The electric streetcar, which introduced "mass" transit that every urbanite could afford, allowed the heterogeneous, immigrant-dominated city population to sort itself into ethnically uniform neighborhoods. When the United States sharply curtailed foreign immigration in the 1920s, industrial managers discovered the large African American population of the rural South—increasingly unemployed there as cotton-related agriculture declined—and began to recruit these workers by the thousands for the factories of Manufacturing Belt cities. This influx had an immediate impact on the social geography of the industrial city because whites were unwilling to share their living space with the racially different newcomers. The result was the involuntary segregation of these newest migrants, who were channeled into geographically separate, all-black areas. By the 1950s, these mostly inner-city areas became large expanding **ghettos**, speeding the departure of many white central-city communities and reinforcing the trend toward a racially divided urban society.

Urbanization of the Suburbs.   The huge suburban component of the intraurban residential mosaic is not home just to the more affluent residents of the metropolis: over the past three decades it received so massive an infusion of nonresidential activities that it was transformed into a full-fledged **outer city**. As its ties to the central city loosened, the outer city's growing independence was accelerated by the rise of major new suburban nuclei (particularly near key freeway interchanges) to serve the new local economies. These multipurpose activity nodes often developed around large regional shopping centers, whose prestigious images attracted scores of industrial parks, office campuses and high-rises, hotels, restaurants, entertainment facilities, and even major league sports stadiums and arenas, which together formed burgeoning new **suburban downtowns** that are an automobile-age version of the CBD (see photo, p. 168). As suburban downtowns flourish, they attract tens of thousands of local residents to organize their lives around them—offering workplaces, shopping, leisure activities, and all the other elements of a complete urban environment.

These newest spatial elements of the contemporary metropolis are assembled in the model displayed in Figure 3-11. The rise of the outer city has produced a ***multicentered***

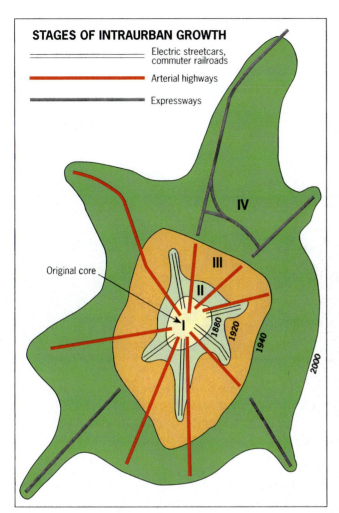

STAGES OF INTRAURBAN GROWTH

———— Electric streetcars, commuter railroads
———— Arterial highways
———— Expressways

Original core

IV
III
II
I
1880
1920
1940
2000

FIGURE 3-10   © H. J. de Blij, P. O. Muller, and John Wiley & Sons, Inc.

metropolis consisting of the traditional CBD as well as a set of increasingly coequal suburban downtowns, with each activity center serving a discrete and self-sufficient surrounding area. James Vance defined these tributary areas as ***urban realms***, recognizing in his studies that each such realm maintains a separate, distinct economic, social, and political significance and strength. Figure 3-12 applies the **urban realms model** to Los Angeles; we could easily draw a similar regionalization scheme for other large U.S. metropolises.

The position of the central city within the new multi-nodal metropolis of realms is eroding. No longer the dominant metropolitanwide center for goods and services, the CBD increasingly serves the less affluent residents of the innermost realm and those working downtown. As manufacturing employment declined precipitously, many large cities adapted successfully by shifting toward service industries. Accompanying this switch is downtown commercial revitalization, but in many cities for each shining new skyscraper that goes up several old commercial buildings are abandoned. Residential reinvestment has also occurred

in many downtown-area neighborhoods but usually requires the displacement of established lower-income residents, an emotional issue that has sparked many conflicts. Beyond the CBD zone, the vast inner city remains the problem-ridden domain of low- and moderate-income people, with most forced to reside in ghettos.

Cultural Geography

In the United States, over the past two centuries, the contributions of a wide spectrum of immigrant groups have shaped, and continue to shape, a rich and varied cultural complex. Great numbers of these newcomers were willing to set aside their original cultural baggage in favor of assimilation into the emerging culture of their adopted homeland, which itself was a hybrid nurtured by constant infusions of new influences. For most upwardly mobile immigrants, this plunge into the much-touted "melting pot" promised a ticket for acceptance into mainstream American society.

American Cultural Bases

As the hybrid culture of the United States matured, it built on a set of powerful values and beliefs: (1) love of newness; (2) desire to be near nature; (3) freedom to move; (4) individualism; (5) societal acceptance; (6) aggressive pursuit of goals; and (7) a firm sense of destiny. Brian Berry has discerned these cultural traits in the behavior of people throughout the evolution of urban America. A "rural ideal" has prevailed throughout U.S. history and is still expressed in a strong bias against residing in cities. When industrialization made urban living

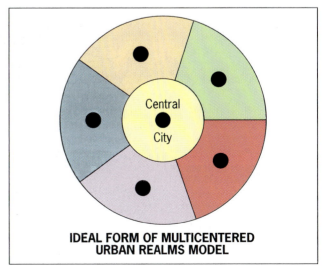

Central City

IDEAL FORM OF MULTICENTERED URBAN REALMS MODEL

FIGURE 3-11   © H. J. de Blij, P. O. Muller, and John Wiley & Sons, Inc.

FROM THE FIELD NOTES

"Monitoring the urbanization of U.S. suburbs for the past three decades has brought us to Tyson's Corner, Virginia on many a field trip and data-gathering foray. It is now hard to recall from this recent view that only 40 years ago this place was merely a near-rural crossroads. But as nearby Washington, D.C. steadily decentralized, *Tyson's* capitalized on its unparalleled regional accessibility (its Capital Beltway location at the intersection with the radial Dulles Airport Toll Road) to attract a seemingly endless parade of high-level retail facilities, office complexes, and a plethora of supporting commercial services. We would rank it today among the three largest and most influential suburban downtowns (or 'edge cities') in North America."
© Rob Crandall/
The Image Works.

Concept Caching

www.conceptcaching.com

unavoidable, those able to afford it soon moved to the emerging suburbs (*newness*) where a form of country life (*close to nature*) was possible in a semiurban setting. The fragmented metropolitan residential mosaic, composed of myriad slightly different neighborhoods, encouraged frequent *unencumbered mobility* as middle-class life revolved around the *individual* nuclear family's *aggressive pursuit* of its aspirations for *acceptance into the next higher stratum of society*. To most Americans these accomplishments confirmed that they could attain their goals through hard work and perseverance and that they had the ability to realize their *destiny* by achieving the "American Dream" of homeownership, affluence, and total satisfaction.

Language and Religion

Although linguistic variations play a far more important role in Canada, nearly one-fifth of the U.S. population speaks a language other than English at home. Differences in English usage are also evident at the subnational level in the United States, where regional variations (*dialects*) can still be noted despite the recent trend toward a truly national society. The South immediately comes to mind as a region that still possesses a distinctive accent.

North America's Christian-dominated kaleidoscope of religious faiths contains important spatial variations. Many major Protestant denominations are clustered in particular regions, with Southern Baptists localized in the southeastern quadrant of the United States, Lutherans in the Upper Midwest and northern Great Plains, and

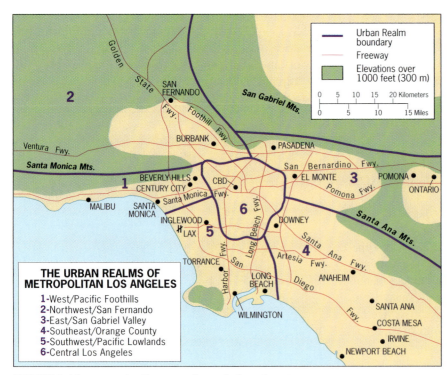

THE URBAN REALMS OF METROPOLITAN LOS ANGELES

1-West/Pacific Foothills
2-Northwest/San Fernando
3-East/San Gabriel Valley
4-Southeast/Orange County
5-Southwest/Pacific Lowlands
6-Central Los Angeles

FIGURE 3-12   © H. J. de Blij, P. O. Muller, and John Wiley & Sons, Inc.

Mormons west of the Rockies but focused on Utah. Roman Catholics are most visibly concentrated in Manufacturing Belt metropolises, New England, and the Mexican borderland zone. Judaism is the nation's most highly agglomerated major religious group, whose largest congregations are clustered in the suburbs of Megalopolis, Southern California, South Florida, and the Midwest.

Ethnicity and Geography

Ethnicity (national ancestry) has always played a key role in American cultural geography. Today, whites of European background no longer dominate the increasingly diverse U.S. ethnic tapestry, with ethnics of color and non-European origin comprising more than one-third of the population—a proportion that demographers predict will rise to 50 percent by mid-century. In the late 1990s, Hispanic Americans surpassed African Americans to become the nation's largest minority, and today they account for 15 percent of the U.S. total.

The spatial distribution of the four largest ethnic minorities is mapped in Figure 3-13. Hispanics are regionally clustered, with over 60 percent residing in California, Texas, New Mexico, and Florida; nonetheless, population geographers expect this rapidly expanding minority to continue to disperse across the country, especially in the West, the interior South, and along the eastern seaboard. African Americans are regionally concentrated as well, most notably in the South, reflecting the legacy of slavery and the plantation economy; blacks are also a major presence in the central cities of the Manufacturing Belt and west coast (most metropolitan areas do not register prominently on these small-scale maps). Asian Americans are the most agglomerated of the leading ethnic minorities, their distribution dominated by urban-based clusters along the Pacific coast (entry points for the large number who are recent immigrants). The main concentrations of Native Americans are found in the West, where they largely occupy tribal lands on reservations ceded by the federal government; many also reside in communities widely scattered across the nation.

Persistent Immigration

The changing ethnic complexion of the United States has long been influenced by immigration. That trend continues strongly, and during the first half of this decade an average

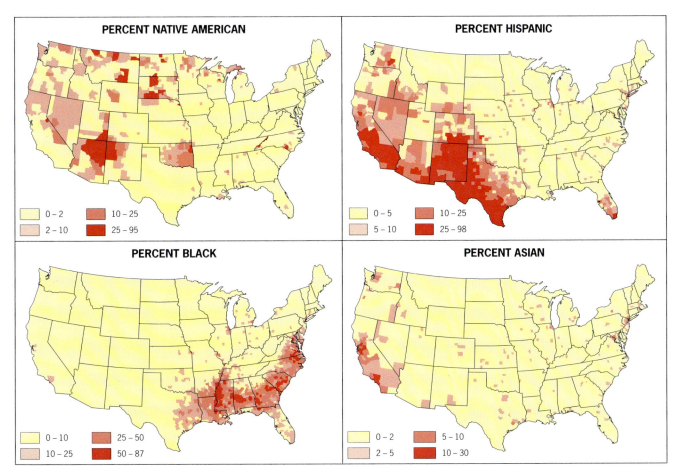

FIGURE 3-13   © H. J. de Blij, P. O. Muller, and John Wiley & Sons, Inc.

People have walled and fenced themselves off since the earliest days, as the Great Wall of China and Hadrian's Wall (built by the Romans in northern Britain) remind us. But no wall—whether the Berlin Wall during the Cold War or the Israeli "Security Barrier" of today—has ever halted all migration. Neither will migration be stopped by a fortified fence along the Mexican border from the Pacific coast to the Gulf of Mexico. However, long stretches of the U.S.–Mexico boundary already look like this, and more construction will take place. This photo shows a ground-level view of the improvised wall that slashes westward, between Tijuana, Mexico and the southernmost suburbs of San Diego, California, as it makes its way toward the Pacific shore on the horizon. The poverty-stricken landscape of this part of Tijuana extends almost to the rusty barrier itself, while development on the U.S. side has kept its distance from this well-patrolled section of the border.   © Diane Cook & Ken Jenshel.

of 957,000 legal immigrants (plus another 650,000 illegal immigrants) annually entered the country. The source areas, however, have changed dramatically over the past half-century. During the 1950s, just over 50 percent came from Europe, 25 percent from Middle and South America, 15 percent from Canada, and 5 percent from Asia. Today 41 percent come from Middle and South America, 32 percent from Asia, and only about 18 percent from Europe and Canada.

Given this ongoing influx of immigrants, it is not surprising that in 2006 the Census Bureau reported that more than 36 million residents (12 percent of the national population) had been born in other countries—the highest total in U.S. history. A profile of the approximately 8 million of these newest Americans who entered the United States since 1999 reveals the following: only 6 percent have become naturalized citizens; more than half are Hispanic; 90 percent speak languages other than English at home (vs. 19 percent for all Americans); their households are larger and their incomes are lower than the U.S. average; 45 percent work in construction, manufacturing, and food and accommodation services (vs. 27 percent of all Americans employed in these sectors); and their leading countries of origin are Mexico (which alone accounts for fully 35 percent), the Philippines, India, China, and Vietnam. As noted earlier in the chapter, this latest wave of immigration has now triggered a major public debate whose viewpoints are underscored in the Issue Box titled "Immigrants: How Many Can North America Accommodate?"

Although still strongly directed at urban areas, the geography of immigrant destinations is changing. The once-dominant Manufacturing Belt States have slipped to less than 25 percent, with most of their newcomers now heading for either metropolitan New York or Chicago. Not unexpectedly, with the upsurge in the growth of the Hispanic and Asian American populations (Fig. 3-13), the Sunbelt States increasingly began to take up the slack in the late twentieth century, particularly California, Texas, and Florida. Today, however, these mega-States are leveling off as immigrants flock to such new growth areas as the Carolinas, Georgia, and Nevada—a dispersal likely to continue because an additional 15 States in the middle of the country have experienced a doubling of their immigrant populations since 1990.

The Emerging Mosaic Culture

American cultural geography continues to evolve. What is now taking place is a new fragmentation into the emerging nationwide **mosaic culture**, an increasingly heterogeneous complex of separate, uniform "tiles" that cater to more specialized groups than ever before. No longer based solely on such broad divisions as income, race, and ethnicity, today's residential communities of interest are also forming along the dimensions of age, occupational status, and especially lifestyle. Their success and steady proliferation reflect an obvious satisfaction on the part of most Americans. Yet such balkanization—fueled by people choosing to interact only with others exactly like themselves—threatens the survival of important democratic values that have prevailed throughout the evolution of U.S. society.

12

Immigrants: How Many Can North America Accommodate?

IMMIGRATION BRINGS BENEFITS— THE MORE THE MERRIER!

"The United States and Canada are nations of immigrants. What would have happened if our forebears had closed the door to America after they arrived and stopped the Irish, the Italians, the Eastern Europeans, and so many other nationalities from entering this country? Now we're arguing over Latinos, Asians, Russians, Muslims, you name it. Fact is, newcomers have always been viewed negatively by most of those who came before them. When Irish Catholics began arriving in the 1830s, the Protestants already here accused them of assigning their loyalty to some Italian pope rather than to their new country, but Irish Catholics soon proved to be pretty good Americans. Sound familiar? Muslims can be very good Americans too. It just takes time, longer for some immigrant groups than others. But don't you see that America's immigrants have always been the engine of growth? They become part of the world's most dynamic economy and make it more dynamic still.

"My ancestors came from Holland in the 1800s, and the head of the family was an architect from Rotterdam. I work here in western Michigan as an urban planner. People who want to limit immigration seem to think that only the least-educated workers flood into the United States and Canada, depriving the less-skilled among us of jobs and causing hardship for citizens. But in fact America attracts skilled and highly educated as well as unskilled immigrants, and they all make contributions. The highly educated foreigners, including doctors and technologically skilled workers, are quickly absorbed into the workforce; you're very likely to have been treated by a physician from India or a dentist from South Africa. The unskilled workers take jobs we're not willing to perform at the wages offered. Things have changed! A few decades ago, American youngsters on summer break flooded the job market in search of temporary employment in hotels, department stores, and restaurants. Now they're vacationing in Europe or trekking in Costa Rica, and the managers of those establishments bring in temporary workers from Jamaica and Romania.

"And our own population is aging, which is why we need the infusion of younger people immigration brings with it. We don't want to become like Japan or some European countries, where they won't have the younger working people to pay the taxes needed to support the social security system. I agree with opponents of immigration on only one point: what we need is legal immigration, so that the new arrivals will get housed and schooled, and illegal immigration must be curbed. Otherwise, we need more, not fewer, immigrants."

LIMIT IMMIGRATION NOW

"The percentage of recent immigrants in the U.S. population is the highest it has been in 70 years, and in Canada in 60 years. America is adding the population of San Diego every year, over and above the natural increase, and not counting illegal immigration. This can't go on. By 2010, 15 percent of the U.S. population will consist of recent immigrants. One-third of them will not have a high-school diploma. They will need housing, education, medical treatment, and other social services that put a huge strain on the budgets of the States they enter. The jobs they're looking for often aren't there, and then they start displacing working Americans by accepting lower wages. It's easy for the elite to pontificate about how great immigration is for the American melting pot, but they're not the ones affected on a daily basis. Immigration is a problem for the working people. We see company jobs disappearing across the border to Mexico, and at the same time we have Mexicans arriving here by the hundreds of thousands, legally and illegally, and more jobs are taken away.

"And don't talk to me about how immigration now will pay social security bills later. I know a thing or two about this because I'm an accountant here in Los Angeles, and I can calculate as well as the next guy in Washington. Those fiscal planners seem to forget that immigrants grow older and will need social security too. And as for that supposed slowdown in the aging of our population because immigrants are so young and have so many children, over the past 20 years the average age in the United States has dropped by four months. So much for that nonsense. What's needed is a revamping of the tax structure, so those fat cats who rob corporations and then let them go under will at least have paid their fair share into the national kitty. There'll be plenty of money to fund social services for the aged. We don't need unskilled immigrants to pay those bills.

"And I'm against this notion of amnesty for illegal immigrants being talked about these days. All that would do is to attract more people to try to make it across our borders. I heard the president of Mexico propose opening the U.S.-Mexican border the way they're opening borders in the European Union. Can you imagine what would happen? What our two countries really need is a policy that deters illegal movement across that border, which will save lives as well as jobs, and a system that will confine immigration to legal channels. These days, that's not just a social or economic issue; it's a security matter as well."

Regional ISSUE

Vote your opinion at www.wiley.com/college/deblij

The Changing Geography of Economic Activity

The economic geography of the United States today is the product of all of the foregoing, as bountiful environmental, human, and technological resources have cumulatively blended together to create one of the world's most advanced economies. Perhaps the greatest triumph was overcoming the tyranny of distance, so that people and activities could be organized into a continentwide spatial economy that took maximum advantage of agricultural, industrial, and urban development opportunities. Yet, despite these past achievements, American economic geography in the early twenty-first century is once again in the throes of restructuring as the transition is completed from industrial to postindustrial society.

Major Components of the Spatial Economy

Economic geography is mainly (though not exclusively) **13** concerned with the locational analysis of **productive activities**. Four major sets may be identified:

- **Primary activity:** the extractive sector of the economy in which workers and the environment come into direct contact, especially in *mining* and *agriculture*.
- **Secondary activity:** the *manufacturing* sector in which raw materials are transformed into finished industrial products.
- **Tertiary activity:** the *services* sector, including a wide range of activities from retailing to finance to education to routine office-based jobs.
- **Quaternary activity:** today's dominant sector, involving the collection, processing, and manipulation of *information*; a subset, sometimes referred to as **quinary activity**, is the managerial activity associated with decision-making in large organizations.

Historically, each of these activities has successively dominated the American labor force for a time over the past 200 years, with the quaternary sector now dominant. Agriculture dominated until the late nineteenth century, giving way to manufacturing by 1900. The steady growth of services after 1920 finally surpassed manufacturing in the 1950s but now shares a dwindling portion of the limelight with the still-rising quaternary sector. The approximate breakdown by major sector of employment in the U.S. labor force today is agriculture, 2 percent; manufacturing, 16 percent; services, 17 percent; and quaternary, 65 percent (with about 10 percent in the quinary sector). We now proceed to review these major productive components of the spatial economy in the following coverage of resource use, agriculture, manufacturing, and the postindustrial revolution.

Resource Use

The United States (and Canada) was blessed with abundant deposits of mineral and energy resources. Fortunately, these resources were usually concentrated in sufficient quantities to make long-term extraction an economically feasible proposition, and most of the richest raw material sites are still the scene of major drilling or mining operations. Moreover, the continental and offshore mineral/fuel storehouse may yet contain significant undiscovered resources for future exploitation.

Mineral Resources. North America's rich mineral deposits are localized in three zones: the Canadian Shield north of the Great Lakes, the Appalachian Highlands, and scattered areas throughout the mountain ranges of the West. The Shield's most noteworthy minerals are iron ore, nickel, gold, uranium, and copper. Besides vast deposits of soft (bituminous) coal, the Appalachian region also contains hard (anthracite) coal in northeastern Pennsylvania and iron ore in central Alabama. The western mountain zone contains significant deposits of coal, copper, lead, zinc, molybdenum, uranium, silver, and gold.

Fossil Fuel Energy Resources. The realm's most strategically important resources are its petroleum (oil), natural gas, and coal supplies—the **fossil fuels**, so named **14** because they were formed by the geologic compression and transformation of plant and tiny animal organisms that lived hundreds of millions of years ago. These energy supplies are mapped in Figure 3-14, which reveals abundant deposits and far-flung distribution networks.

The leading *oil*-production areas of the United States are located along and offshore from the Texas–Louisiana Gulf Coast; in the Midcontinent district, extending through western Texas–Oklahoma–eastern Kansas; and along Alaska's central North Slope facing the Arctic Ocean. (Canada's major oilfields lie in a wide crescent curving southeastward from northern Alberta to southern Manitoba and beneath the waters around Newfoundland in the extreme east.) The distribution of *natural gas* deposits generally resembles the geography of oilfields because petroleum and energy gas are usually found in similar geologic formations (the floors of ancient shallow seas); natural gas is now considered the most glamorous fossil fuel because of its growing dominance in the generation of North America's electrical power supply. The realm's *coal* reserves rank among the greatest on Earth, and the U.S. portion alone contains at least a 400-year supply; the main producing coalfields are found in Appalachia, the northern U.S. Great Plains/southern Alberta, and southern Illinois/western Kentucky.

FIGURE 3-14   © H. J. de Blij, P. O. Muller, and John Wiley & Sons, Inc.

Agriculture

Despite the post–1900 emphasis on developing the nonprimary sectors of the spatial economy, agriculture remains an important element in America's human geography. Because it is the most space-consuming economic activity, vast expanses of the U.S. (and Canadian) landscape are clothed with fields of grain. In addition, great herds of livestock are sustained by pastures and fodder crops because this wealthy realm can afford the luxury of feeding animals from its farmlands to meet huge demands for red meat in its diet. The growing use of high-technology mechanization in farming has steadily increased both the volume and value of total agricultural production. It also has been accompanied by a sharp reduction in the number of those actively engaged in agriculture, and today only about 1 percent of the U.S. population still lives on farms.

As a food provider, the United States continues to be the world's colossus. At mid-decade it produced 40 percent of the global supply of corn, one-third of all soybeans, and more than one-eighth of the world's wheat. Nonetheless, while U.S. food exports have increased by more than half since 1990, globalization has introduced new challenges as well as new opportunities. Chief among these is rising competition in the international marketplace from faster-developing commercial agricultural powerhouses such as Brazil and China. This has led to falling commodity prices, and forces American farmers and agribusinesses to constantly restructure production in order to keep pace.

An important geographical expression of this ongoing change is the restructuring of U.S. agricultural regions. For most of the twentieth century farming regions were organized within the macrospatial framework of the *von Thünen model* (see pp. 53–54). As in Europe (Fig. 1-6), the early-nineteenth-century, original-scale model of town and hinterland had expanded outward (driven by improving transport technology) until it encompassed the entire continent by 1900. The "supercity" anchoring this macro-Thünian system of decreasingly intensive production zones was the northeastern Megalopolis, well on its way toward coalescence and already the dominant food market and transportation focus of the entire country. But as the twentieth century gave way to the twenty-first, globalization forces, which required U.S. producers to become ever more competitive on the international market, were speeding the transformation of the Thünian regional system.

Although the new U.S. agricultural geography is still in its formative stage, geographer John Fraser Hart has already identified its three key features: (1) a Midwestern heartland, increasingly devoted to cash-grain farming; (2) a constellation of livestock-raising zones along the heartland's wide periphery; and (3) a coastal "rimland" of specialized crop-producing areas. These elements are seen in greater detail in Figure 3-15, whose legend and accompanying dot maps should be studied carefully in conjunction with the main map of *farm resource regions*. This nine-region framework was recently developed by the U.S. Department of Agriculture for planning purposes, and to date is the most comprehensive attempt to assemble the components of the reorganizing regional geography of agricultural activity.

Manufacturing

The geography of North America's industrial production has long been dominated by the Manufacturing Belt (Fig. 3-8). The emergence of this region was propelled by (1) superior access to the Megalopolis national market that formed its eastern edge and (2) proximity to industrial re-

sources, particularly iron ore and coal for the pivotal steel industry that arose in its western half. We noted earlier that manufacturers had a strong locational affinity for cities, and the internal structure of the Belt became organized around a dozen urban-industrial districts interlinked by a dense transportation network. As these industrial centers expanded, they swiftly achieved **economies of scale**, savings accruing from large-scale production in which the cost of manufacturing a single item was further reduced as individual companies mechanized assembly lines, specialized their workforces, and purchased raw materials in massive quantities.

This efficient production pattern served the nation well throughout the remainder of the industrial age. Because of the effects of *historical inertia*—the need to continue using hugely expensive manufacturing facilities for their full, multiple-decade lifetimes to cover initial investments—the Manufacturing Belt is not going to disappear anytime soon. But as its aging facilities are retired, a process well under way, the distribution of American industry will change. As transportation costs equalize among U.S. regions, as energy costs now favor the south-central oil- and gas-producing States, as high-technology manufacturing advances reduce the need for less skilled labor, and as locational decision making intensifies its attachment to noneconomic factors, industrial management has increasingly demonstrated its willingness to relocate to regions it perceives as more desirable in the South and West.

Parts of the Manufacturing Belt have been resisting this trend, particularly the industrial Midwest, whose prospects have greatly improved since the "Rustbelt" days of the 1970s and 1980s. Shedding that label in recent years, many Midwestern manufacturers are successfully reinventing their operations by rooting out inefficiencies, investing in cutting-edge factories and technologies, and not only beating back foreign competition in the United States but significantly expanding their exports. Nonetheless, there is a downside to this progress because as high-tech manufacturing proliferates, older factory workers discover that the demand for their lower-tech skills shrinks accordingly.

As the U.S. manufacturing map continues to change, the economic lives of countless communities are affected. If the past 15 years were a sign of things of come, entrepreneurial restructuring will also play an increasingly important role because corporate mergers and acquisitions often result in the geographic reshuffling of thousands of jobs. Another recent trend likely to intensify is the economic warfare that occurs when employers seek to locate major new facilities and deliberately pit cities, counties, and even States against one another in order to get the best deal. Often exhorted by politicians seeking reelection, communities offer ever greater incentive packages that

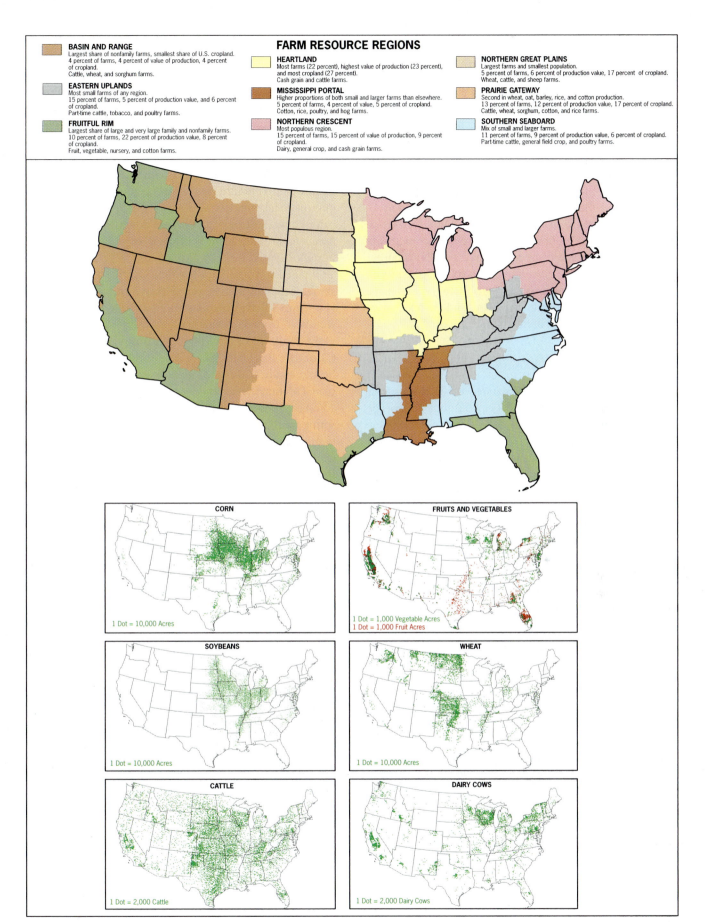

FARM RESOURCE REGIONS

BASIN AND RANGE
Largest share of nonfamily farms, smallest share of U.S. cropland.
4 percent of farms, 4 percent of value of production, 4 percent of cropland.
Cattle, wheat, and sorghum farms.

EASTERN UPLANDS
Most small farms of any region.
15 percent of farms, 5 percent of production value, and 6 percent of cropland.
Part-time cattle, tobacco, and poultry farms.

FRUITFUL RIM
Largest share of large and very large family and nonfamily farms.
10 percent of farms, 22 percent of production value, 8 percent of cropland.
Fruit, vegetable, nursery, and cotton farms.

HEARTLAND
Most farms (22 percent), highest value of production (23 percent), and most cropland (27 percent).
Cash grain and cattle farms.

MISSISSIPPI PORTAL
Higher proportions of both small and larger farms than elsewhere.
5 percent of farms, 4 percent of value, 5 percent of cropland.
Cotton, rice, poultry, and hog farms.

NORTHERN CRESCENT
Most populous region.
15 percent of farms, 15 percent of value of production, 9 percent of cropland.
Dairy, general crop, and cash grain farms.

NORTHERN GREAT PLAINS
Largest farms and smallest population.
5 percent of farms, 6 percent of production value, 17 percent of cropland.
Wheat, cattle, and sheep farms.

PRAIRIE GATEWAY
Second in wheat, oat, barley, rice, and cotton production.
13 percent of farms, 12 percent of production value, 17 percent of cropland.
Cattle, wheat, sorghum, cotton, and rice farms.

SOUTHERN SEABOARD
Mix of small amd larger farms.
11 percent of farms, 9 percent of production value, 6 percent of cropland.
Part-time cattle, general field crop, and poultry farms.

CORN
1 Dot = 10,000 Acres

FRUITS AND VEGETABLES
1 Dot = 1,000 Vegetable Acres
1 Dot = 1,000 Fruit Acres

SOYBEANS
1 Dot = 10,000 Acres

WHEAT
1 Dot = 10,000 Acres

CATTLE
1 Dot = 2,000 Cattle

DAIRY COWS
1 Dot = 2,000 Dairy Cows

FIGURE 3-15   © H. J. de Blij, P. O. Muller, and John Wiley & Sons, Inc.

include tax breaks, labor retraining, and infrastructural improvements. However, many such competitions can fail to produce a decent rate of return. For example, when Alabama recently went to extreme lengths to lure Mercedes-Benz to build a new assembly plant near Tuscaloosa, it wound up costing the State government $200,000 per expected job. (The cost to Tennessee's government to attract the Saturn automaking complex a few years earlier had been one-tenth that figure.) Moreover, these business strategies can be worked in reverse; not long ago, New York City offered a $60 million incentive package to keep two Wall Street brokerage firms from moving to the suburbs, and Disneyland got its city and State to spend hundreds of millions on local road improvements after the company threatened to cancel expansion plans and perhaps even move the theme park out of Southern California.

The Postindustrial Revolution

16 The signs of **postindustrialism** are visible throughout the United States (and much of Canada) today; they are popularly grouped under such labels as "the computer age" or "the New Economy." The term *postindustrial* by itself, of course, tells us mainly what the theme of the American economy no longer is. Yet many social scientists also use the term to refer to a set of societal traits that signal an historic break with the past (for example, work experiences now focus mainly on person-to-person interaction rather than person-product or person-environment contact). Many of the urban spatial expressions of the ongoing societal transformation have already been highlighted; here we focus on some broader economic-geographical patterns.

High-technology, white-collar, office-based activities are the leading growth industries of the postindustrial economy. Most are relatively footloose and are therefore responsive to such noneconomic locational forces as geographic prestige, local amenities, and proximity to recreational opportunities. Northern California's **Silicon Valley**—the world's leading center for computer research and development and the headquarters of the U.S. microprocessor industry—epitomizes the blend of locational qualities that attract a critical mass of high-tech companies to a given locality. These include: (1) a world-class research university (Stanford); (2) technological know-how in the form of a large pool of highly educated, highly skilled labor; (3) close proximity to a cosmopolitan urban center (San Francisco); (4) abundant venture capital; (5) a local economic climate and entrepreneurial culture that supports risk-taking and forgives failure by both smaller and big companies; (6) a locally based network of global business linkages; and (7) a high-amenity environment in the form of top-quality housing, pleasant weather, scenic countrysides, and year-round recreational opportunities.

The development of Silicon Valley is so significant to the new postindustrial era that regional-planning scholars Manuel Castells and Peter Hall have conceptualized it as the first technopole. **Technopoles** are planned techno-industrial complexes that innovate, promote, and manufacture the hardware and software products of the new informational economy. On the landscape of the outer suburban city, where almost all of these complexes are located, the signature of a technopole is a low-density cluster of ultramodern, low-rise buildings laid out as a campus—an image as symbolic of today's spatial economy as the smoke-belching factory was of the industrial age a century ago. From Silicon Valley, technopoles have spread in all directions—from San Diego in southwesternmost California to the Route 128 corridor around Boston and from North Carolina's Research Triangle to the lakeside suburbs of Seattle; many of these technopoles will be noted in the regional section that concludes this chapter. These high-tech complexes are also becoming a global phenomenon as they spring up in other geographic realms. They often arise as joint ventures between government and the private sector; among the local names they go by are science city (Japan; Russia), technopolis (France; Japan; Brazil), science park (Taiwan)—and even Silicon Glen (Scotland).

17

One of the hallmarks of the postindustrial revolution is the emergence of the Internet, which has already been called the defining technology of the twenty-first century. The recent growth of the Internet has been driven by the need to transmit ever greater quantities of data at ever higher speeds. Fiber-optic cables are best able to meet these needs, and have given rise to a network of backbones or linkages that connect information-rich centers across the planet. The leading component of this global infrastructure is its United States network, which handles up to 75 percent of the world's Internet traffic and is dominated by a set of major backbone cities and their interconnections (Fig. 3-16). Interestingly, the three leading metropolitan areas in terms of connected backbone capacity are Washington, D.C. (federal government operations coupled with a booming technopole in suburban northern Virginia), Chicago (central location on the network, which facilitates wide access to the continental interior), and San Francisco (preeminent influence of Silicon Valley). Thus, at least in its initial stage of development, the hierarchy of backbone capacity does not match the urban population hierarchy. The primate city, New York, continues to rank in the second tier of Internet backbone cities together with another major technopole (Dallas's "Silicon Prairie") and a leading regional hub (Atlanta, which serves the South).

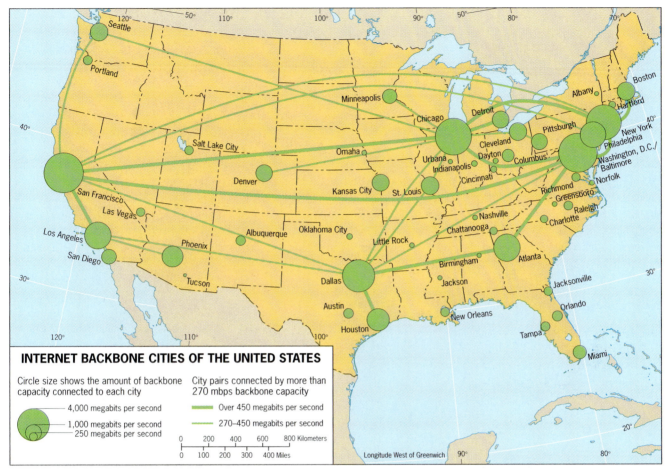

INTERNET BACKBONE CITIES OF THE UNITED STATES

Circle size shows the amount of backbone capacity connected to each city

- 4,000 megabits per second
- 1,000 megabits per second
- 250 megabits per second

City pairs connected by more than 270 mbps backbone capacity

— Over 450 megabits per second
— 270–450 megabits per second

0   200   400   600   800 Kilometers
0   100   200   300   400 Miles

Longitude West of Greenwich

FIGURE 3-16   © H. J. de Blij, P. O. Muller, and John Wiley & Sons, Inc.

CANADA

Like the United States, Canada is a federal state, but it is organized differently. Canada is divided administratively into ten provinces and three territories (Fig. 3-17). The provinces—where 99.7 percent of all Canadians live—range in territorial size from tiny, Delaware-sized Prince Edward Island to sprawling Quebec, more than twice the area of Texas. Beginning in the east, the four Atlantic Provinces are Nova Scotia, New Brunswick, Prince Edward Island, and Newfoundland and Labrador. To their west lie Quebec and Ontario, Canada's two biggest provinces. Most of western Canada (which is what the Canadians call everything west of Lake Superior) is covered by the three Prairie Provinces—Manitoba, Saskatchewan, and Alberta. In the far west, beyond the Canadian Rockies and facing the Pacific, lies the tenth province, British Columbia.

The three territories—Yukon, the Northwest Territories, and Nunavut—together occupy a massive area half the size of Australia but are inhabited by only about 100,000 people. Nunavut is the newest addition to Canada's political map and deserves special mention. Created in 1999, this new territory is the outcome of a major aboriginal land claim agreement between the ***Inuit*** people (formerly called Eskimos) and the federal government, and encompasses all of Canada's eastern Arctic. Nunavut covers fully one-fifth of Canada's land, an area larger than that of any other province or territory. Since 80 percent of its 31,000 residents are Inuit, Nunavut—which means "our land"—is the first territory to have a substantial degree of native self-government (in addition to outright ownership of an area half the size of Texas).

As for the capital city, unlike the United States and a number of other federations, Ottawa is not anchored to a federal territory that lies separate from the country's subnational political units. Rather than establishing a District of Columbia-type administrative entity, the Canadians were content to locate their capital astride the Ottawa River, which marks the boundary between Ontario and Quebec. Even though Parliament and other leading capital-city functions are situated on the Ontario side, many federal-government operations have long been based in Hull and nearby communities on the Quebec side of the provincial border.

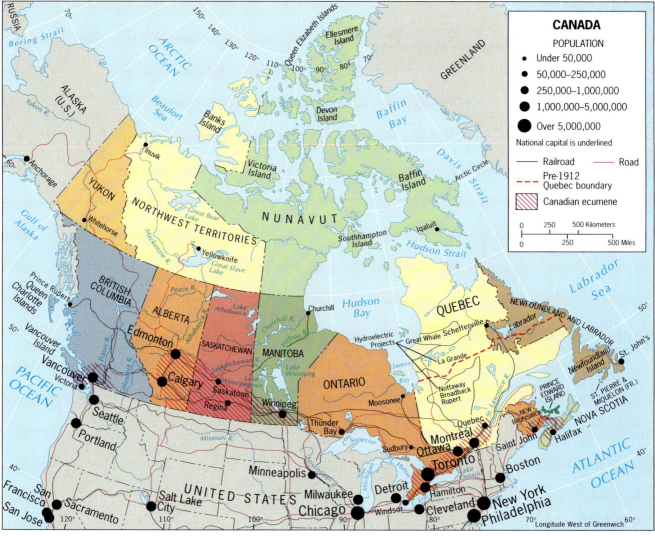

FIGURE 3-17 © H. J. de Blij, P. O. Muller, and John Wiley & Sons, Inc.

In population size (in descending order), Ontario (12.8 million) and Quebec (7.8 million) are again the leaders; British Columbia ranks third with 4.4 million; next come the three Prairie Provinces with a combined total of 5.5 million; the Atlantic Provinces are the four smallest, together containing 2.3 million. The sparsely populated territories in the far north contain barely 100,000 residents. Canada's total population of 32.9 million is only slightly larger than one-tenth the size of the U.S. population. Spatially, as noted at the outset of the chapter, Canadians overwhelmingly reside within 300 kilometers (ca. 200 mi) of the U.S. border.

Though comparatively small, Canada's population is divided by culture and tradition, and this division has a pronounced regional expression. Just under 60 percent of Canada's citizens speak English as their mother tongue, 23 percent speak French, and the remaining 17 percent other languages; about 18 percent of the population speak both English and French (of these, 41 percent of those who speak French as a first language are bilingual, but only 10 percent of the English speakers). The spatial clustering of more than 85 percent of the country's **Francophones** (French-speakers) in Quebec accentuates this Canadian social division along ethnic and linguistic lines (Fig. 3-18). More than 80 percent of this province's population is French Canadian, and Quebec is the historic, traditional, and emotional focus of French culture in Canada. Over the past few decades a strong nationalist movement emerged in Quebec to demand outright separation from the rest of Canada. That movement crested in 1995, when independence was narrowly defeated in a landmark referendum. Even though support for secession has waned since then, this ethnolinguistic division continues to be the cornerstone of Canada's cultural pluralism, and the issue could still pose a threat to national unity in the future.

PERCENTAGE OF CANADA'S POPULATION IN 1996 WHOSE MOTHER TONGUE WAS FRENCH
90–100%
60–90%
20–60%
5–20%
2.5–5%
0–2.5%
0 250 500 Kilometers
0 250 500 Miles
Arctic Circle
Longitude West of Greenwich

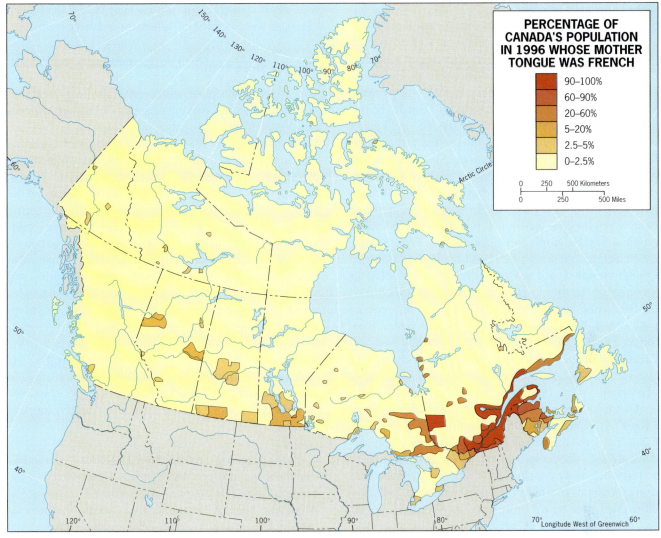

FIGURE 3-18 © H. J. de Blij, P. O. Muller, and John Wiley & Sons, Inc.

Population in Time and Space

The map showing the distribution of Canada's population (Fig. 3-5, p. 161) reveals that only about one-eighth of this
18 enormous country can be classified as its **ecumene**—the inhabitable zone of permanent settlement. As Figure 3-17 indicates, the Canadian ecumene is dominated by a discontinuous strip of population clusters that lines the southern border. We can identify four such clusters on the map, the most populous by far being Main Street (home to more than six out of every ten Canadians). As noted earlier (Fig. 3-9), ***Main Street*** is the conurbation that stretches across southernmost Quebec and Ontario, from Quebec City on the lower St. Lawrence River southwest through Montreal and Toronto to Windsor on the Detroit River. The three lesser clusters are: (1) the Saint John–Halifax crescent in central New Brunswick and Nova Scotia; (2) the prairies of southern Alberta, Saskatchewan, and Manitoba; and (3) the southwestern corner of British Columbia, focused on Canada's third-largest metropolis, Vancouver.

Pre-Twentieth-Century Canada

Canada's population map evolved more slowly than that of the United States. As Figure 3-7 shows, compared to European expansion in the United States, penetration of the Canadian interior lagged several decades behind. In 1850, for example, when the eastern United States and much of the Pacific coast had already been settled, Canada's frontier had only reached Lake Huron. In terms of political geography, Canada did not unify before the last third of the nineteenth century—and then only reluctantly because of fears the United States was about to expand in a northerly direction. Moreover, the Canadian push to the west was delayed by major physical obstacles, including the barren Canadian Shield north and west of the Great Lakes, and the

rugged highlands separating the Great Plains from the Pacific (which lacked convenient mountain passes).

The evolution of modern Canada—as well as its contemporary cultural geography—is also deeply rooted in the bicultural division discussed above. Its origin lies in the fact that it was the French, not the British, who were the first European colonizers of present-day Canada, beginning in the 1530s. *New France*, during the seventeenth century, grew to encompass the St. Lawrence Basin, the Great Lakes region, and the Mississippi Valley. A series of wars between the English and French began in the 1680s, ending with France's defeat and the cession of New France to Britain in 1763. By the time London took control of its new possession, the French had made considerable progress in their North American domain. French laws, the French land-tenure system, and the Roman Catholic Church prevailed, and substantial settlements (including Montreal on the St. Lawrence) had been established. The British, anxious to avoid a war of suppression and preoccupied with problems in their other American colonies, gave former French Quebec—the territory extending from the Great Lakes to the mouth of the St. Lawrence—the right to retain its legal and land-tenure systems as well as freedom of religion.

After the American War for Independence, London was left with a region it called British North America (the name *Canada*—derived from an aboriginal word meaning settlement—was not yet in use), but whose cultural imprint still was decidedly French. The Revolutionary War drove many thousands of English refugees northward, and soon difficulties arose between them and the French in British North America. In 1791, heeding appeals by these new settlers, the British Parliament divided Quebec into two provinces: Upper Canada, the region upstream from Montreal centered on the north shore of Lake Ontario, and Lower Canada, the valley of the St. Lawrence. Upper and Lower Canada became, respectively, the provinces of Ontario and Quebec (Fig. 3-17). By parliamentary plan, Ontario would become English-speaking and Quebec would remain French-speaking.

This earliest cultural division did not work well, and in 1840 the British Parliament tried again. This time it reunited the two provinces through the Act of Union under which Upper and Lower Canada would have equal representation in the provincial legislature. That, too, failed, and efforts to find a better system finally led to the 1867 British North America Act that established the Canadian federation (consisting initially of Upper and Lower Canada, New Brunswick, and Nova Scotia, and later, between 1870 and 1999, to be joined by the other provinces and territories). Under the 1867 Act, Ontario and Quebec were again separated, but this time Quebec was given important guarantees:

the French civil code was left unchanged, and the French language was protected in Parliament and in the courts.

Canada Since 1900

By the close of the nineteenth century, the fledgling Canadian federation was making major strides toward regional development and the spatial integration of a continentwide economy. In 1886, the transcontinental Canadian Pacific Railway had finally been completed to Vancouver, a condition of British Columbia's entry into the federation 15 years earlier. This new lifeline soon spawned the settlement of not only the far west but also the fertile Prairie Provinces, whose wheat-raising economy expanded steadily as immigrants from the east and abroad poured in during the early twentieth century. Industrialization also began to stir, and by 1920 Canadian manufacturing had surpassed agriculture as the leading source of national income. As noted earlier (Fig. 3-8), the dominant zone of industrial activity is the Toronto–Hamilton–Windsor corridor of southern Ontario (which also functions as one of the 12 districts of the American Manufacturing Belt). That has been the case since the Industrial Revolution took hold in Canada during World War I (1914–1918), when the country heavily supported Britain's war effort on the European mainland.

Three Stages of Metropolitan Evolution. As industrial intensification proceeded, it was accompanied by the expected parallel maturation of the national urban system. Along the lines of Borchert's five epochs of American metropolitan evolution, Maurice Yeates has constructed a similar multistage model of urban-system development that divides the telescoped Canadian experience into three eras. The initial ***Frontier-Staples Era*** (prior to 1935) encompasses the century-long transition from a frontier-mercantile economy to one oriented to staples (production of raw materials and agricultural goods for export), with increasing manufacturing activity in the budding industrial heartland. By 1930, Montreal and Toronto, reflecting their different cultural constituencies, had emerged as the two leading cities atop the national urban hierarchy (thus Canada has no single primate city).

Next came the ***Era of Industrial Capitalism*** (1935–1975), during which Canada achieved U.S.-style prosperity. However, most of this development—involving the massive growth of manufacturing, tertiary activities, and urbanization—took place after 1950 because the Great Depression lasted until World War II, which was followed by a lengthy recession. A major stimulus was the investment of U.S. corporations in Canadian branch-plant construction, especially in the automobile industry in Ontario near the automakers' Detroit-area headquarters. In western Canada, the rapid growth of oil and natural gas production fueled Alberta's urban development, but new agricultural technologies reduced farm labor needs and sparked a rural outmigration in neighboring Saskatchewan and Manitoba. The postwar period also saw the ascent of Main Street, which on less than 2 percent of Canada's land quickly came to contain more than 60 percent of its people, contributed two-thirds of its national income, and claimed nearly 75 percent of its manufacturing jobs.

The third stage, ongoing since 1975, is the ***Era of Global Capitalism***, signifying the rise of additional foreign investment from the Asian Pacific Rim and Europe. This, of course, is also the era of transformation into a postindustrial economy and society, and in the process Canada is experiencing many of the same upheavals as the United States. Most development today is occurring in the form of new suburbanization. This represents a departure from the recent past because the pre-1990 Canadian metropolis had experienced far less automobile-generated deconcentration than the United States, with the central city retaining much of its middle-income population and economic base. But many large Canadian cities are now turning inside-out, and the new intraurban geography is increasingly symbolized by the suburban downtowns that anchor the ultramodern business complexes along the Highway 401 freeway north of Toronto and Alberta's West Edmonton mega-mall, the world's biggest shopping center.

Cultural/Political Geography

The historic cleavage between Canada's French- and English-speakers has resurfaced over the past four decades to dominate the country's cultural and political geography. By the time the Canadian federation observed its centennial in 1967, it had become evident that Quebecers regarded themselves as second-class citizens; they believed that bilingualism meant that French-speakers had to learn English but not vice versa; and they perceived that Quebec was not getting its fair share of the country's wealth. Since the 1960s, the intensity of ethnic feelings in Quebec has risen in surges despite the federal government's efforts to satisfy the province's demands. During the 1970s, while a separatist political party came to power in Quebec, a new federal constitution was drawn up in Ottawa. In 1980, Quebec's voters solidly rejected independence when given that choice in a referendum. But the new constitution did *not* satisfy the Quebecers, and throughout the 1980s and early 1990s the Ottawa government struggled unsuccessfully to devise a plan, acceptable to all the provinces, that would keep Quebec in the Canadian federation.

This thriving high-rise CBD anchors a metropolis that is located where the Great Plains meet the Rocky Mountains. No, this is *not* Denver, Colorado—it is Alberta's Calgary. Calgary started out as a fort on the western edge of Canada's Great Plains; the settlement got a boost when the Canadian Pacific Railway reached it in 1883. Later a second railroad, followed by the Trans-Canada Highway, further improved its cross-continental linkages. But Calgary's biggest shot in the arm was the discovery of oil and gas fields in its hinterland, which put petroleum refining and petrochemical industries ahead of flour milling and meatpacking as economic mainstays. Today Calgary also functions as the leading service center for much of the western Prairie Provinces, with a high-tech sector spearheading its latest growth boom. © Wally Bauman/ Alamy Images.

Quebec's 1995 Referendum and Its Aftermath

By 1995, with Canada's interest in constitutional reform exhausted, a second referendum on Quebec's sovereignty could no longer be put off. With the reenergized separatist Parti Québécois again leading the way, the Francophone-dominated electorate very nearly approved independence (which attracted an astonishing 49.4 percent of the vote), an outcome now regarded by Canadians as their country's "near-death experience." Subsequently, despite calls by many separatists for a follow-up vote that might turn narrow defeat into victory, plans for a third referendum were shelved because opinion polls in Quebec have shown an overall decline in public support for secession. Several reasons lie behind that shift in attitude: (1) the implementation of provincial laws that firmly established the use of French and the primacy of Québécois culture; (2) a substantial increase in the bilinguality of Quebec's Anglophones (English speakers), which has calmed Francophone fears concerning assimilation into Canada's English-speaking mainstream; (3) the arrival of a new wave of immigrants (some Francophone, some Anglophone) from Asia, Africa,

AMONG THE REALM'S GREAT CITIES . . . Toronto

TORONTO, CAPITAL of Ontario and Canada's largest metropolis (5.5 million), is the historic heart of English-speaking Canada. The landscape of much of its center is dominated by exquisite Victorian-era architecture and surrounds a healthy downtown that is one of Canada's leading economic centers. Landmarks abound in this CBD, including the SkyDome stadium, the famous City Hall with its facing pair of curved high-rises, and mast-like CN Tower, which at 553 meters (1815 ft) is the world's tallest free-standing structure. Toronto also is a major port and industrial complex, whose facilities line the shore of Lake Ontario.

Livability is one of the first labels Torontonians apply to their city, which has retained more of its middle class than central cities of its size in the United States. *Diversity* is another leading characteristic because this is North America's richest urban ethnic mosaic. Among the largest of more than a dozen thriving ethnic communities are those dominated by Italians, Portuguese, Chinese, Greeks, and Ukrainians; overall, Toronto now includes residents from about 170 countries who speak more than 100 languages; and the immigrant inflow continues strongly, with those born outside Canada constituting nearly half of the city's population (in all the world only two other million-plus cities exhibit a higher percentage). *Vibrancy* is yet another hallmark of this remarkable city, whose resiliency was tested but not broken by the SARS-virus outbreak of 2003.

Toronto has worked well in recent decades, thanks to a metropolitan government structure that fostered central

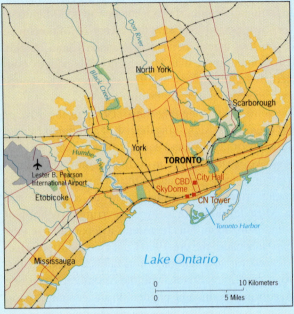

© H. J. de Blij, P. O. Muller, and John Wiley & Sons, Inc.

city–suburban cooperation. But that relationship is increasingly stressed as the outer city gains a critical mass of population, economic activity, and political clout. Toronto has responded with plans to become a "city-state," but first must overcome problems of municipal finances and aging infrastructure.

Eastern Europe, and the Caribbean, who have accelerated the weakening of language barriers by settling in both French-speaking and English-speaking neighborhoods; and (4) the landslide defeat of the ruling separatist Parti Québécois in the provincial election of 2003.

Although the issue of Quebec's separation from the Canadian federation has receded in this decade, it has not been buried and could resurface. Anticipating that possibility, the federal government has armed itself by adopting the Clarity Act, which mandates that any future referendum on independence must be based on a clearly worded question that passes with the support of a "clear majority" (i.e., well over 50 percent) before Ottawa would open negotiations on separation.

The most insightful analyses of public opinion in the province today suggest that what Quebecers really want is a better relationship with the rest of Canada that falls short of independence. There is also considerable evidence that the Québécois regard themselves as a *distinct nation* and

wish to be treated accordingly. That is precisely the recognition they received under a new Parliamentary resolution passed at the end of 2006—but with the important added caveat . . . *within a united Canada*. This was followed by a resounding defeat of the Parti Québécois in the provincial election of March 2007, and the emergence of a new political movement whose guiding vision replaced independence with "autonomy"—a Quebec that would remain part of Canada but with (as-yet-undefined) increased powers. Even if the supporters of separatism never again come close to attaining their goal, they can nonetheless claim several successes that include (1) the achievement of Francophone cultural and economic dominance in Quebec; (2) the strengthening position of the French language throughout the Canadian federation; and (3) an upsurge of federal spending in the province that now surpasses the collection of Ottawa-bound tax revenues, a monetary inflow that is likely to increase as the country's energy windfall continues to propel rising national income.

The Ascendancy of Indigenous Peoples

On the wider Canadian scene, the culture-based Quebec crisis has stirred the ethnic feelings of the country's 1.4 million native (First Nations and Inuit) peoples. Their assertions have also received a sympathetic hearing in Ottawa, and in recent years breakthroughs have been achieved with the creation of Nunavut and a treaty for limited tribal self-government in northern British Columbia. Nonetheless, this fastest-growing segment of the Canadian population is plagued by serious social and economic problems, and after seven years of self-rule Nunavut's development has been minimal. Elsewhere, confrontations over aboriginal land claims in both populated and frontier areas continue to proliferate, but governments and private landowners are notoriously slow to resolve such issues.

A foremost concern among these peoples is that their aboriginal rights be protected by the federal government against the provinces. This is especially true for the First Nations of Quebec's northern frontier, the Cree, whose historic domain covers more than half of the province of Quebec as it appears on current maps. Administration of the Cree was assigned to Quebec's government in 1912, a responsibility that a move toward independence might well invalidate. In any case, the Cree would also likely be empowered to seek independence. This would leave the French-speaking remnant of Quebec with only about 45 percent of the province's present territory. As Figure 3-17 shows, the territory of the Cree is no unproductive wilderness: it contains vital facilities of the James Bay Hydroelectric Project, a massive scheme of dikes, dams, and artificial lakes that has transformed much of northwestern Quebec and generates electrical power for a huge market within and outside the province. In 2002, after its attempts to block construction of the Project were supported by the federal courts, the Cree negotiated a groundbreaking treaty with the Quebec government whereby this First Nation dropped its opposition in return for a portion of the income earned from electricity sales as well as the granting of power to control its own economic and community development.

Intensifying Regionalism

Finally, the events of the past three decades have profoundly impacted Canada's national political landscape, and not only in Quebec. Regionalism has also intensified in the increasingly affluent west, whose leaders opposed federal concessions to Quebec and insisted that the equal treatment of all ten provinces is a basic principle that precludes the designation of special status for anyone. Recent federal elections have clearly revealed the emerging fault lines that surround Quebec and set the western provinces (British Columbia, Alberta, Saskatchewan, and Manitoba) off from the rest of the country. Even the remaining Ontario-led

center and the eastern bloc of Atlantic Provinces voted divergently, and today the politico-geographical hypothesis of "Four Canadas" may be on its way to becoming reality. Thus, with its national unity under pressure, Canada today confronts the coalescing forces of *devolution* that threaten to transform the new fault lines into permanent fractures.

Economic Geography

As in the United States, the growth of Canada's spatial economy has been supported by a diversified, high-quality *resource base*. We noted earlier that the Canadian Shield is endowed with rich mineral deposits and that oil and natural gas are extracted in sizeable quantities in Alberta to the southwest of the Shield. Canada has long been a leading *agricultural* producer and exporter, especially of wheat and other grains from its breadbasket in the Prairie Provinces; new technologies keep productivity high, but labor requirements continue to diminish and farm workers now account for less than 2 percent of the national workforce. Postindustrialization has caused substantial employment decline in the *manufacturing sector*, with Southern Ontario's industrial heartland being the most adversely affected area. On the other hand, Canada's robust *tertiary and quaternary sectors* (which today employ more than 70 percent of the total workforce) are creating a host of new economic opportunities. Fortunately, Southern Ontario is benefiting from this postindustrial transformation because the country's leading high-technology, research-and-development complex has grown up around Waterloo and Guelph, a pair of university towns about 100 kilometers (60 mi) west of Toronto.

Canada–United States Trade

Canada's economic future is also going to be strongly affected by the continuing development of its trading relationships. The landmark United States–Canada Free Trade Agreement, signed in 1989, initiated a now-completed phasing out of all tariffs and investment restrictions between the two countries, whose annual cross-border flow of goods and services is the largest in the international trade arena. (In the mid-2000s more than four-fifths of Canada's exports went to the United States, from which it also derived about two-thirds of its imports.) In 1994 these economic linkages were further tightened by the implementation of the *North American Free Trade Agreement (NAFTA)*, which consolidated the gains of the 1989 pact and opened major new opportunities for both countries by adding Mexico to the trading partnership (see box titled "North American Free Trade Agreement [NAFTA]").

Because these free-trade agreements increasingly impact the Canadian spatial economy, they likely will continue to weaken domestic east-west linkages and strengthen international north-south ties. Since many local cross-border linkages built on geographical and historical commonalities are already well developed, they can be expected to intensify in the future: the Atlantic Provinces with neighboring New England; Quebec with New York State; Ontario with Michigan and adjacent Midwestern States; the Prairie Provinces with the Upper Midwest; and British Columbia with the (U.S.) Pacific Northwest. Such functional reorientations, of course, constitute yet another set of powerful devolutionary forces confronting the Ottawa government because most of these potential economic fault lines coincide with those that politically demarcate the "Four Canadas."

Emerging Regional States

The rising importance of this framework of transnational regions straddling the U.S.-Canadian border recalls Kenichi Ohmae's regional state concept introduced in Chapter 1. A **19** **regional state** is a "natural economic zone" that defies old

POINTS TO PONDER

● The United States does not have an official language. Should it?

● Global warming may create an ice-free passage around northern Canada. Goodbye Panama Canal?

● Quebec's off-again, on-again independence movement is off again as of the March 2007 provincial-election aftermath.

● Three hundred million people in the U.S.A. as of October 2006. Get ready for 400 million in 2043.

borders, and it is shaped by the global economy of which it is a part; its leaders deal directly with foreign partners and negotiate the best terms they can with the national governments under which they operate. Writing about Canada, Ohmae identified both a Pacific Northwest (the

North American Free Trade Agreement (NAFTA)

THE ECONOMIES OF Canada, the United States, and Mexico formalized their interactions under the North American Free Trade Agreement (NAFTA), which took effect in 1994. This economic alliance has forged the world's second-largest trading bloc, a (U.S.) $8-trillion-plus market of 448 million consumers (not far behind the European Union's 494 million). Between now and its completion date in 2009, NAFTA will further integrate its constituent economies through a number of steps. Thousands of individual tariffs, quotas, and import licenses have been or are scheduled to be rolled back to eliminate most trade barriers for agricultural and manufactured goods as well as tertiary and quaternary services. Moreover, restrictions on the flow of investment capital across international borders are being lifted, and regulations concerning foreign ownership of productive facilities continue to be liberalized.

The architects of NAFTA cite the opportunities the maturing pact will bring. Business investment will surge as unimpeded exports flow throughout the vast new marketplace. At the same time, intensified competition will give consumers access to a wider variety of quality products at reduced prices. Canada—which joined NAFTA mainly to safeguard the concessions it received in its 1989 Free Trade Agreement with the United States—faced the most skeptical citizenry at the outset, but those doubts evaporated as Canadian exports (and new jobs) under NAFTA have risen

substantially over the past decade. Trade with the United States, already heightened under the bilateral 1989 agreement, has propelled much of this growth, especially the southward export of goods whose tariffs had been lowered.

This progress notwithstanding, NAFTA's evolution continues to be criticized. Organized labor objects to the redistribution and losses of jobs in a number of industries, dislocations which it perceives stem largely from the implementation of the trade agreement. Environmental activists complain that the pact's architects and component governments have not adequately addressed their concerns. Social scientists and public policy planners argue that NAFTA is ineffective at reducing regional income disparities and uneven development within its trade area. And, despite its stimulation of trade and investment, most residents of Canada, the United States, and Mexico continue to view NAFTA as no great success.

As NAFTA realizes its goals, it will require each member country to concentrate on the production of those goods and services for which it has a comparative advantage (in research and development, technological skills, managerial know-how, and the like). As workers move from less productive to more productive industries, the geography of employment will be restructured—a process that would be happening anyway as the three countries continue to adapt to the forces shaping continental and global economic change.

Seattle–Vancouver axis) and a Great Lakes regional state (the remarkably intertwined Ontario–Michigan industrial complex), and warned that the manner in which Ottawa's leaders dealt with these new economic entities would be critical to the survival of the Canadian state. Today his warning is more meaningful than ever because the Pacific Northwest and Great Lakes have been joined by most of the other transnational areas listed in the previous paragraph. Indeed, these growing international interactions are increasingly evident in the overall regional configuration of North America, to which we now turn in the second part of this chapter.

Regions of the Realm

The ongoing transformation of North America's human geography is fully reflected in its internal regional organization. As new forces uproot and redistribute people and activities, old locational rules no longer apply. However, the varied character of the realm's physical, cultural, and economic landscapes ensures that meaningful regional differences will persist. We will now examine the current areal arrangement of the United States and Canada within a framework of nine regions. These regions are mapped in Figure 3-19 and should be studied carefully because they differ from the set of popularly perceived regions (discussed in the box titled "The Informal Geography of Popular Regions") many readers are already familiar with.

● THE NORTH AMERICAN CORE

The Core Region (Fig. 3-19)—synonymous with the American Manufacturing Belt—was introduced earlier in this chapter. Serving as the historic workshop for the linked spatial economies of the United States and Canada, this region was the unquestioned leader and centerpiece during the century between the Civil War and the close of the industrial age (1865–1970). The rise of postindustrialism over the past four decades has diminished that linchpin regional role, and strengthening challengers to the south and west will continue to siphon key functions away from the Core as our new century unfolds. But make no mistake: this is still the geographic heart of North America. Here one finds the largest city and federal capital of both countries. Here, too, are the leading financial markets, corporate headquarters, media centers, cultural facilities, research complexes, and the busiest airports and intercity expressways. Moreover, both the U.S. and Canadian portions of the Core still contain more than one-third of their respective national populations.

Although manufacturing remains highly important within the transformed American economy, productivity and obsolescence problems during the late twentieth century increased production costs in the Core Region, thereby erasing many of its historical competitive advantages over newly emerging industrial areas in the southern and western United States. In parts of the Core (especially its eastern portion), employment dropped sharply in the smokestack industries as factories closed or relocated, accelerating the economic decline of numerous mill towns, blue-collar suburbs, and inner central cities. But other areas fought back, none more successfully than the Midwest portion of the Manufacturing Belt, which reinvented itself by aggressively pursuing the high-tech upgrading of its aged industrial base and becoming far more competitive through expanded connections in the global marketplace.

The reconcentration of the pivotal automobile industry since the 1980s is a prime example of the Midwest's thriving new manufacturing geography, which foreign as well as domestic carmakers have shaped. A key role was played by Japanese-owned Toyota, which successfully challenged

MAJOR CITIES OF THE REALM

City	Population (in millions)*
Atlanta	5.0
Boston	6.1
Chicago	9.9
Dallas–Ft. Worth	6.2
Denver	3.1
Detroit	5.6
Houston	5.5
Los Angeles	17.8
Montreal, Canada	3.7
New York	22.3
Ottawa, Canada	1.2
Philadelphia	6.4
San Diego	3.2
San Francisco	7.7
Seattle	4.0
Toronto, Canada	5.5
Vancouver, Canada	2.3
Washington, D.C.	5.5

*Based on 2008 estimates.

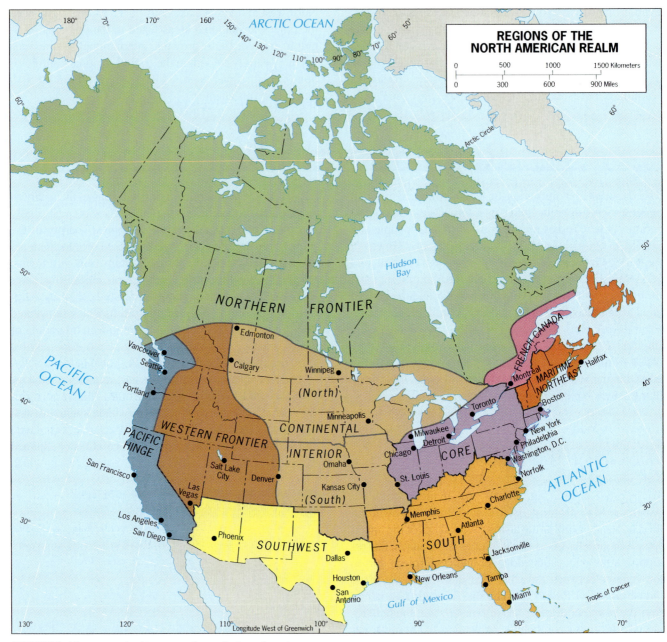

FIGURE 3-19 © H. J. de Blij, P. O. Muller, and John Wiley & Sons, Inc.

the "Big Three" (GM, Ford, and then Daimler-Chrysler) on its home turf. Toyota's investments were directed at the Core's less-developed southwestern border zone, where an 800-kilometer (500-mi)-long corridor of Interstate-64 between St. Louis and northern West Virginia became the axis of a self-sufficient, state-of-the-art automaking complex centered on the Lexington, Kentucky area. Meanwhile, and with diminishing success since 2000, Detroit's Big Three struggled to keep pace by modernizing fabrication technologies and dispersing certain operations away from their traditional base in Michigan to locations elsewhere in the Midwest and as far away as Mexico.

Postindustrial Hot Spots

Postindustrial development has also spawned a number of new (or recycled) growth centers in the Core. The major metropolitan complexes of the northeastern Megalopolis—already the scene of much quaternary and quinary economic activity—are adjusting fairly well. The Boston area, well endowed with research facilities, continues to attract innovative high-tech businesses, mostly toward the Route 128 freeway corridor that girdles the central city (North America's pioneering technopole of the 1950s). New York City remains the national leader in finance and advertising,

The Informal Geography of Popular Regions

AS IN OTHER world realms, popularly perceived regions have emerged in North America, and it is useful to briefly compare them to those delimited by geographers using the regional criteria introduced in the opening chapter. In the United States there is a long history of this kind of informal regionalization, and popular or *vernacular* regions such as New England, the South, and the West have been widely used by the media and the public for more than two centuries. Most are imprecisely defined and marked by fuzzy boundaries, as we noted for the Midwest on page 5.

Popular regions exist at many scales. At the microregional end of the spectrum we have city neighborhoods. In recent times, however, far more attention has been focused on broad divisions at a much higher level of generalization. During the late twentieth century, social commentators made a big deal about the purported Sunbelt/Frostbelt partitioning of the conterminous United States. But this overly simplistic conceptualization was not convincingly supported by demographic, economic-development, or climatic evidence. Thus claims concerning the existence of a continent-spanning, monolithic, uniformly booming tier of more than a dozen "Sunbelt" States (separated from the rest of the country along an east-west line roughly following 37°N latitude), have always been regarded skeptically by scholars. Although the term *Sunbelt* has been disappearing from news reports and magazine articles since 2000, the popular media today champion even more simplistic and unsatisfactory bifurcations of the United States along such dimensions as "the coasts versus the interior" or "red" States of the Republican Party versus "blue" States of the Democratic Party.

and is home to significant elements of the broadcast and print media. But Manhattan must now increasingly share its economic leadership with its huge outer suburban city: for example, a new upsurge in direct seaborne trade with China is sparking a revival of New York-area port functions; however, most of this cargo flow is being funneled through state-of-the-art container-handling facilities in Ports Newark and Elizabeth on the New Jersey side of the harbor.

The Core Region metropolis that has gained the most from postindustrialization is Washington, D.C. As the information-producing and managerial sectors have blossomed, and as the U.S. federal government began to outsource operations to the private sector in the 1980s, Washington and its surrounding outer cities in the affluent Maryland and Virginia suburbs (interconnected by the 105-kilometer [66-mi] Beltway encircling the capital city) have amassed an enormous complex of office, research, high-tech, lobbying, and consulting firms. The most important development has occurred along the outer-suburban expressway corridor between Tyson's Corner on the Beltway (see photo, p. 168) and Dulles International Airport, recently upgraded to a world-class facility. Since the mid-1990s, a large number of major telecommunications and Internet companies have been attracted to the Dulles area, one of the country's leading fiber-optic cable hubs (see Fig. 3-16) thanks to its local cluster of federal intelligence agencies as well as the corporate headquarters of America Online and Sprint Nextel. And since 2001, this area has also become the nexus of the new and still-expanding homeland-security industry. Today this growth is fanning out from here all across

northern Virginia's sector of the Washington metropolis, which is well on its way to becoming one of the realm's foremost technopoles and job machines. No wonder that many are now calling this activity complex the "technological capital" of the United States because more technical workers are employed here than in either Silicon Valley or suburban Boston's Route 128 agglomeration. And the economic transformation is being felt throughout metropolitan Washington as its shrinking total of about 300,000 federal government workers is steadily being outnumbered by the more than 750,000 employed by computer-service, telecom, Internet, aerospace, and biotech companies.

● THE MARITIME NORTHEAST

The Maritime Northeast consists of upper New England and the neighboring Atlantic Provinces of easternmost Canada (Fig. 3-19). New England, one of the realm's historic culture hearths, has retained a strong regional identity for almost 400 years. The urbanized southern half of New England has been the northeastern anchor of the Core since the mid-nineteenth century—where it must be regionally classified. However, the six New England States (Maine, New Hampshire, Vermont, Rhode Island, Massachusetts, and Connecticut) still share many common characteristics. Besides this overlap with the Manufacturing Belt (and Megalopolis) in its south, the Maritime Northeast region also extends northeastward across the Canadian border to encompass almost all of the four Atlantic Provinces of New

AMONG THE REALM'S GREAT CITIES . . . Chicago

CHICAGO LIES NEAR the southern end of Lake Michigan not far from the geographic center of the conterminous United States. Its centrality and crossroads location were evident to its earliest indigenous settlers, who developed the site as a portage where canoes could be hauled from the lake to the headwaters of a nearby stream that led to the Mississippi River. Centuries later, when the modern U.S. spatial economy emerged, Chicago became the leading hub on the continental transport network. As a freight-rail node, it still reigns supreme; and even though most long-distance passengers switched to jet planes a generation ago, Chicago remains the quintessential North American hub city, with O'Hare International consistently ranking among the world's busiest airports.

Poet Carl Sandburg described Chicago as "the city of big shoulders," underscoring its personality and prowess as the leading U.S. manufacturing center during the century following the Industrial Revolution of the 1870s and 1880s. Among the myriad products that emanated from Chicago's industrial crucible were the first steel-frame skyscraper, elevated railway, refrigerated boxcar, cooking range, electric iron, nuclear reactor, and window envelope. Chicago also became a major commercial center, spawning a skyscraper-dominated central business district second only to New York's in size and influence.

Chicago today (metro population: 9.9 million) is trying hard to make the transition to the postindustrial age. Its diversified economy has proven to be a precious asset in this quest, and the city is determined to reinvent itself as a first-order service center. The new forces it must contend with, however, are symbolized in the fate of the emptying Sears Tower, the city's (and the nation's) tallest building: not long ago Sears & Roebuck itself abandoned the structure for a new headquarters campus along Interstate-90 in the far northwestern suburb of Hoffman Estates. This shift acknowledges the rise of Chicagoland's

© H. J. de Blij, P. O. Muller, and John Wiley & Sons, Inc.

thriving outer city, one of the country's largest and most successful. Around O'Hare, a huge and still-expanding suburban office/industrial-park complex has truly become Chicago's "Second City," underscoring that major airports themselves will increasingly function as urban growth magnets in the ever more globalized spatial economy of the twenty-first century.

Brunswick, Nova Scotia, Prince Edward Island, and Newfoundland-and-Labrador.

A long association based on economic and cultural similarities has tied northern New England to Atlantic Canada. Both have a strong maritime orientation, are rural in character, possess difficult environments with limited land resources, and were historically bypassed in favor of more fertile inland areas. Thus economic growth in upper New England has always lagged behind that of the rest of the realm. Development has centered on primary activities, mainly fishing the (once) rich offshore banks of the nearby North Atlantic, forestry in the uplands, and farming in the few fertile valleys available. Recreation and tourism have boosted the regional economy in recent times, with scenic

coasts and mountains attracting millions from the neighboring Core Region. The growth of skiing has also helped, extending the tourist season through the harsh winter months.

Upper New England

Since 1980, New England's roller-coaster economy has experienced spectacular prosperity, hard times, and, most recently, steady recovery. Weaning itself from its overreliance on the activities that brought on the troubles of the 1990s, New England is developing a more diversified economic base built around high-tech manufacturing, financial services, education, health care, and biotechnology.

AMONG THE REALM'S GREAT CITIES . . . New York

NEW YORK IS much more than the largest city of the North American realm. It is one of the most famous places on Earth; it is the hemisphere's gateway to Europe and the rest of the Old World; it is a tourist mecca; it is the seat of the United Nations; it is one of the globe's most important financial centers—a true "world city" in every sense of that overused term. Unfortunately, the terrorists who planned and executed the attacks of 9/11 were only too well aware of New York's symbolic importance on the international stage.

New York City consists of five boroughs, centered by the island of Manhattan, which contains the CBD south of 59th Street (the southern border of Central Park). Here is a skyscrapered landscape unequaled anywhere, studded with more fabled landmarks, streets, squares, and commercial facilities than any other cities except London and Paris. This is also the cultural and media capital of the United States, which means that the city's influence constantly radiates across the planet thanks to New York-based television networks, newspapers and magazines, book publishers, fashion and design leaders, and artistic trendsetters.

At the metropolitan scale, New York forms the center of a vast urban region, 250 kilometers (150 mi) square (population: 22.3 million), that sprawls across parts of three States in the heart of Megalopolis. That outer city has become a giant in its own right with its population of 14 million, massive business complexes, and flourishing suburban downtowns. Thus, despite its global connections, New York is caught in the currents that affect U.S. central cities today. Its ghettos of disadvantaged populations grow steadily; its aging port facilities and industrial base continue to decline; its corporate community is increasingly stressed by the high cost of doing business in ultra-expensive Manhattan.

These challenges notwithstanding, New York today is seeking to reinforce its position as a great, sustainable,

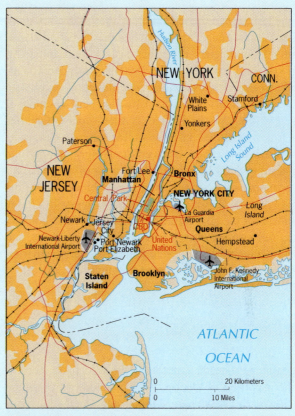

© H. J. de Blij, P. O. Muller, and John Wiley & Sons, Inc.

twenty-first-century city that will continue to lure the hypermobile, super-affluent global elite. Its bold new plans are geared to meet its biggest problems head-on: a major increase in affordable housing for a growing population, across-the-board upgrading of a disintegrating infrastructure, and a sizeable expansion of parks and other outdoor recreational spaces.

Nonetheless, the economic revival has not taken hold in much of northern New England. The most visible benefits of the recovery are concentrated in the zone closest to metropolitan Boston, not just its immediate suburbs focused on the again-booming Route 128 beltway corridor, but a wide swath of fringe areas that spill over into the southern parts of Maine and particularly New Hampshire. New England's continued development, however, is not assured because the region's costs of doing business are higher than those in the South and Southwest. Nevertheless, New England has for now stemmed the outflow of its highly skilled workforce, amassed sufficient venture capital to fuel continuing regional growth, and demonstrated a remarkable ability to adapt to changing economic conditions.

Atlantic Canada

The Atlantic Provinces have also endured hard economic times through the recent past. Most adversely affected was the groundfish industry as offshore stocks of flounder, haddock, and cod became severely depleted through overfishing. Here, too, disagreements over Canada's maritime fishing boundaries led to the seizure of vessels and heightened tensions with other nations, including the United States. New opportunities, never easy to come by, are most promising in the provinces peripheral to the more heavily populated Nova Scotia and New Brunswick. In tiny Prince Edward Island, tourism has been boosted by the opening of a 13-kilometer (8-mi) bridge across Northumberland

Strait, the island's first direct link to the mainland since it joined the Canadian federation in 1873.

Economic opportunities are even brighter in remote Newfoundland and Labrador, Canada's poorest province and the last to enter the federation in 1949. Major offshore oil deposits (see Fig. 3-14), centered on the Grand Banks about 300 kilometers (200 mi) east of the capital, St. John's, were discovered in the 1990s just as government-mandated fisheries restrictions were inflicting disaster on the seafood industry. The construction and opening of seabed drilling platforms and coastal support facilities soon followed, additional oil as well as natural gas deposits were discovered closer to the coast and onshore, and the long-term outlook became promising for oil to transform the economy of Newfoundland and Labrador along the lines of Norway's over the past quarter-century. But today prospects for the near-future achievement of this goal are fading because big energy strikes have been elusive, and the extraction of oil and gas in this harsh marine environment has proven to be far costlier than first envisioned.

● FRENCH CANADA

Francophone Canada constitutes the effectively settled, southern portion of Quebec, which straddles the central and lower St. Lawrence Valley from where that river crosses the Ontario–Quebec border just upstream from Montreal to its mouth in the Gulf of St. Lawrence. Also included is a sizeable concentration of French-speakers, known as the Acadians, who reside just beyond Quebec's provincial boundary in neighboring New Brunswick (Fig. 3-19). The Old World charm of Quebec's cities is matched by an equally unique rural settlement landscape introduced by the French: narrow rectangular farms, known as ***long lots***, are laid out in sequence perpendicular to the St. Lawrence, other rivers, and the roads that parallel them, thereby allowing each farm access to an adjacent routeway.

The economy of French Canada, however, is no longer rural (although dairying remains a leading agricultural pursuit) and exhibits urbanization rates similar to those of the rest of the country. Industrialization is widespread, supported by cheap hydroelectric energy generated at huge dams in northern Quebec, but relatively little of the region's manufacturing could be classified as high-tech. Tertiary and postindustrial commercial activities are concentrated around Montreal, and tourism and recreation are also important to the regional economy. But the health of these sectors—now improving as the independence movement has subsided—is tied to the resolution of Quebec's political status within Canada.

This uncertainty reminds us that, over the past 40 years, Quebec's heightened nationalism has had a major impact on the province that extends well beyond the economic sphere. During the 1980s, the provincial government enacted new laws that strengthened the French language and culture throughout Quebec. With English domination ended, the French Canadians then channeled their energies into developing Quebec's economy. A byproduct of that effort was the rapid urbanization of young families and a loosening of their ties to Roman Catholicism. The outcome was

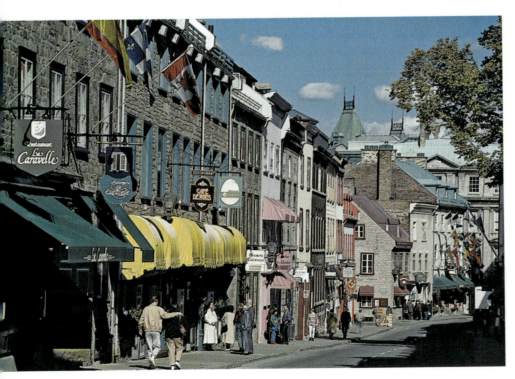

An unmistakably French cultural landscape in North America: the Rue Saint-Louis in Quebec City, the capital of Quebec Province. Not an English sign in sight in this center of French/Québécois culture. The green spires in the background belong to Le Chateau Frontenac, a grand hotel built on the site of the old Fort St. Louis (the Count of Frontenac was a prominent governor of what was then known as New France). French-speaking, Roman Catholic Quebec lies at the core of French Canada. © SUPERSTOCK.

AMONG THE REALM'S GREAT CITIES . . .   Montreal

MONTREAL LIES ON a large triangular island in the St. Lawrence River, its historic core laid out along the base of what Montrealers call "the mountain"—flat-topped Mount Royal, from which the city takes its name. Culturally, Montreal (3.7 million) is the largest French-speaking city in the world after Paris, and therein lie its opportunities and challenges. Without a doubt, this is still one of the realm's most cosmopolitan cities, a strong European flavor pervading its stately CBD, its lively neighborhoods, and its bustling street life. To escape the long, harsh winters here, much of downtown Montreal has become an underground city with miles of handsome passageways lined with shops, restaurants, and cinemas, all of it served by one of the world's most modern and attractive subway systems.

Yet, despite the sparkle and continental ambience, Montreal has not been a happy place in recent times. With the escalation of tensions between Francophone and Anglophone Canada over the past four decades, Quebec's largest metropolis became a crucible of confrontation. For most of that period, the ethnolinguistic division was most visible along the Boulevard St. Laurent, the longstanding boundary between the French-speakers of Montreal's East End (who constitute about 75 percent of the city's population) and the English-speakers who clustered in the West End (the remnant of a larger Anglophone community that was reduced by 40 percent when the French-supremacy language laws were implemented). These days, however, things are improving as linguistically mixed communities proliferate throughout Montreal, driven by an influx of immigrants from outside North America who settle in both English- and French-speaking neighborhoods. At the same time, polarization continues to abate as stores, newspapers, and universities cater to

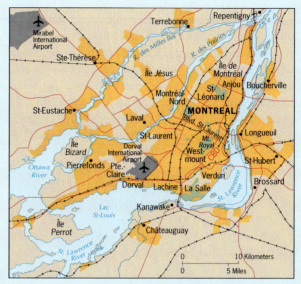

© H. J. de Blij, P. O. Muller, and John Wiley & Sons, Inc.

increasingly bilingual clienteles, and intermarriage between Anglophones and Francophones is on the rise.

Montreal's economy has also been affected. During the 1990s, hundreds of Anglophone firms (and thousands of their workers) fled to Ontario and points west, leaving in their wake a commercial real estate tailspin, rising unemployment, and a noticeable decline in the city's prosperity. Today, with the secession issue no longer on the front burner in Quebec, Montreal is steadily progressing in its comeback. Tourists have returned in droves; new foreign-trading ties, especially with the United States, are reviving the business community; and vibrant suburbs are leading the metropolis to reinvent itself as a hub for information technology, telecommunications, and biopharmaceutical companies.

---

a sharp decline in the provincial birth rate, which the Quebec government tried to offset by encouraging (with cash payments) large families and Francophone immigration (mostly from former colonies of France around the world).

Acadian New Brunswick

As noted earlier, French Canada also includes Acadia, Canada's largest cluster of Francophones outside Quebec. Here in northernmost New Brunswick, the approximately 270,000 French-speakers constitute 35 percent of the province's population (Fig. 3-18). The Acadians, however, not only shun the notion of independence for themselves but actively promote all efforts to keep Quebec within the Canadian federation. Unlike the Quebecers, the Acadians

have devoted their energies in recent years to accommodation with the Anglophone community. Today, a new relationship (known as *cohabitation*) has been worked out between the two groups, based as much on mutual respect for each other's languages as on the strict equality of a new, government-mandated system of bilingualism.

THE CONTINENTAL INTERIOR

The Continental Interior extends across the center of both the conterminous United States and the southern tier of Canada (Fig. 3-19). With few exceptions—most notably in the region's northeastern corner where the barren but mineral-rich Canadian Shield meets the Great

Lakes—agriculture is the primary land use. As Figure 3-15 indicates, meat and grain production dominate farming from the Mississippi Valley in the east to the region's western border along the base of the Rocky Mountains. Because the eastern half of the Continental Interior lies in Humid America and closer to the national food market on the northeastern seaboard, mixed crop-and-livestock farming (based on corn) won out over less competitive wheat raising, which was relegated to the fertile but semiarid environment of the central and western Great Plains on the dry side of the 100th meridian. The latter area also contains Canada's agricultural heartland north of the 49th-parallel border. Although these fertile Prairie Provinces are less subject to serious drought than the U.S. high plains to the south, they are situated at a more northerly latitude and thus have a shorter growing season.

Agricultural Supremacy

The distribution of corn and wheat farming is mapped in Figure 3-15, and so is the production of soybeans—the increasingly important grain crop that is raised in conjunction with corn (see box titled "Geography of the Mighty Soybean"). Note that the edges of the most productive zones of these leading Continental Interior crops accord with several of the region's boundaries. As for their cores, note that corn and soybeans are focused on Iowa and neighboring Illinois (the center of the historic Corn Belt), while wheat is concentrated in the drier Great Plains to the west.

The wheat-producing zone has historically been centered on two different Plains States (North Dakota and Kansas), a separation that is still the basis for subdividing the Continental Interior region, as the dashed line in Figure 3-19 indicates. The North subregion encompasses the Upper Midwest, the northern Great Plains, and the cultivable southern portions of Canada's three Prairie Provinces; spring wheat dominates here, so named because the crop is planted in the spring and harvested in late summer or early autumn. The South subregion specializes in the raising of winter wheat, which is sown in the fall and harvested by late spring; during its infancy winter wheat can withstand the milder winter temperatures south of 40°N latitude, and the great advantage of this growth cycle is that the crop is harvested before the hot, dry summer can take hold in this semiarid

Geography of the Mighty Soybean

OVER THE PAST century, no crop in this realm has expanded faster than the soybean. This hardy legume was one of the earliest plants domesticated, and for nearly 5000 years the Chinese have revered it as a grain vital to life itself. The advantages of soybeans are numerous. They are the world's cheapest and most efficient source of protein, with a level of concentration double that of meat or fish. They are easy to cultivate, adapting to a wide range of environmental conditions, and they are less vulnerable to pests and diseases than other temperate-climate food crops. Economically, crop yields are comparatively high, and the demand for soybeans has steadily risen across the world in recent decades.

The dramatic increase of soybean production in the United States has been propelled not only by these advantages but also by this crop's particular suitability to the historic Corn Belt (the Heartland mapped in Fig. 3-15). It was quickly discovered that soybeans were an ideal crop to rotate with corn because they recondition the soil with nitrogen and other nutrients needed by corn. In terms of cultivation, soybeans also proved highly profitable to grow on the large, flat fields of the Heartland agricultural region (which continues to account for more than 70 percent of all U.S. soybean production) where their farm-machinery requirements were identical to those already in place for raising corn. Moreover, clever marketing constantly widened the versatility of soybeans beyond their use as a nutritious livestock feed, and

today they are used in a variety of human foods (ranging from vegetable oil to a plethora of low-calorie products) as well as such industrial goods as fertilizers, paints, and insect sprays.

All of the above also contributed to steadily rising foreign demand, and in many recent years soybeans have ranked as the leading U.S. export commodity. Other producing countries have recognized this opportunity as well, and intensifying global competition since 1990 has resulted in a decline of America's share of the world market. The United States today produces roughly 40 percent of the world's annual soybean output (down from more than 80 percent in 1980, and 50 percent as recently as the late 1990s). As we note in Chapter 5, most of its competition—totaling about 45 percent of the global supply—comes from Brazil, Argentina, and Paraguay in mid-latitude South America. (Brazilian exports, in fact, surpassed those of the United States in 2006, a lead that is expected to widen in the years ahead.) China, the largest single market country and soybean importer, is also seeking to expand its own acreage in a bid to become a major producer, but the unavailability of sufficient agricultural land has kept its production to just under 10 percent of the world supply. Soybeans will continue to play a key role in North American and international agriculture as the twenty-first century unfolds, buoyed in no small part by this grain's important potential in the struggle to ease the global food shortage.

environment. Nonetheless, winter wheat is declining today for a number of reasons. More profitable corn and soybean production increasingly penetrate the wheatlands from the east with new strains that have been genetically modified to adapt to the drier climate. At the same time, wheat exporting has become less lucrative as foreign competitors multiply, keeping world prices low and reducing the U.S. share of the global market. And government subsidies for wheat farming have fallen behind those for corn and soybeans, which have risen as new opportunities open for producing biofuels (discussed below). As a result, Kansas is no longer synonymous with wheat: corn has now pulled ahead (by about 25 percent in 2006) to become the Sunflower State's leading crop.

Throughout the Continental Interior, economic activity is markedly oriented toward farming. Its leading metropolises—Kansas City, Minneapolis–St. Paul, Winnipeg, Omaha, and even Denver—are major processing and marketing points for pork and beef packing, flour milling, and soybean, sunflower, and canola oil production (all of which are increasingly exported). Although family farms are still distributed widely across the rural landscape, the unrelenting, aggressive incursion of large-scale corporate farming threatens that way of life.

The Ethanol Factor

Anyone following the headlines of the past three years is aware of the rising importance of ethanol in the farmlands of the central United States. The United States is now the world's foremost producer of this gasoline substitute, and as corn has become the leading source of ethanol and other biofuels, the agricultural landscape of the Continental Interior is increasingly impacted. Because the more than 100 modest-sized ethanol refineries that have been built are widely dispersed across the region, the production of these alternative fuels is not dominated by large, centralized, agribusiness corporations. Thus the perception is taking hold that local control of this booming industry will prevail, and ethanol can become the cornerstone of a new prosperity that will revitalize much of rural America.

But just how realistic is this rosy scenario? Despite substantial government subsidies for corn farmers, currently strong retail demand, and handsome profits, biofuel production is a decidedly risky longer-term proposition. The projected output of ethanol in 2007 would only be sufficient to replace about 4 percent of U.S. gasoline consumption; and even if the *entire* corn crop could be converted to ethanol, it would only account for 12 percent of total gasoline consumption (and thus would minimally enhance national energy independence). Far more serious is that so much of the corn crop (about 50 percent in 2007) is already being used to produce biofuels that the United States is approaching food shortages in this commodity—signified by rising corn prices since 2005, especially for cattle feed. Moreover, the economics of these biofuels are proving to be problematic, because corn is a much less efficient source of ethanol than the sugarcane that has been so successful in reducing Brazil's gasoline usage. In the United States, not only is more energy required to produce a gallon of ethanol versus a gallon of gasoline, but that gallon of ethanol yields one-third less mileage than the same quantity of gasoline. And finally, the benefits of ethanol-based development may be overstated: in many of the small towns in which biorefineries have operated for the past few years, employment gains have been modest and have not reversed the outflow of population.

As of mid-2007, 119 ethanol refineries were operating in the United States, with about another hundred under construction or expansion. This plant is located in Liddonia, Missouri (population: approximately 600), at the southern edge of the largest cluster of refineries that is focused on Iowa, southern Minnesota, and the eastern Dakotas. Ethanol is basically alcohol distilled from corn mash and was first utilized as an additive to gasoline to reduce vehicle emissions. But with the recent surge in oil prices, it is increasingly used to run cars and trucks as a fuel blend that contains up to 85% ethanol/15% gasoline. The text points out the risks and potential problems involved, but they are not yet constraining the development of this new energy industry. Ethanol refining is particularly attractive to small towns on railroad lines in less accessible farming areas, where converting local corn to more valuable biofuel is also a more cost-effective way of shipping their product to distant markets.   © Peter Newcomb/Redux Pictures.

Depopulation and Regional Change

The dream of ethanol-based salvation notwithstanding, individual agricultural opportunities are declining in the Continental Interior. This is particularly evident in the less intensively farmed Great Plains in the region's western half, where the exodus of younger people is accelerating. Left behind is an ever more elderly and poorer population, thinly dispersed across a constellation of stressed rural communities, struggling to survive. The Great Plains depopulation trend, in fact, is rapidly plummeting toward demographic collapse: fewer than 12 million residents remained on the Plains in 2005, a population total equal to only about half that of the New York metropolitan area. As Great Plains society unravels, a policy debate intensifies as to whether or not the best solution is planned abandonment and the return of at least a portion of these grasslands to their natural state.

Despite the key role that agriculture plays in the regional economy, the nonfarming sectors of the Continental Interior continue to develop and diversify, especially in and around major metropolitan centers. Minnesota has been the most successful in promoting such growth, capitalizing on the reputations of that State's stable economy, highly educated workforce, track record as a product innovator, and "north woods" amenities. And in the urban areas of the eastern Great Plains, led by Omaha, Nebraska, telemarketing has become a major pursuit.

Burgeoning Alberta

The Continental Interior is also home to one of the most spectacular growth booms in Canadian history: the energy-propelled development of the region's northwestern corner, focused on southern Alberta's Calgary–Edmonton corridor. Alberta's known petroleum reserves have recently been confirmed to be the world's second-largest, ranking directly behind those of Saudi Arabia. The province's oil production is increasingly dominated by the mining of its massive deposits of tar sands (discussed on page 199), which are centered on Fort McMurray in northeastern Alberta and alone account for over 90 percent of Canada's total reserves.

With Alberta now establishing itself as the "Kuwait of North America," it is so awash in prosperity that it has been able to completely retire its provincial debt. Not surprisingly, it has also become a magnet for businesses and highly skilled workers, making it Canada's fastest-growing province with a population gain of nearly 30 percent since the mid-1990s. With secure and expanding economic connections to the Canadian east and the huge U.S. market to the south, Alberta today is steadily turning its attention westward toward the mighty Asian-Pacific arena. Trading linkages to China have become important and are growing steadily; most notably, Chinese investment is expected to help build a major oil pipeline across the Rockies to the Pacific coast by 2010.

● THE SOUTH

The American South occupies the realm's southeastern corner, extending from the lush Bluegrass Basin of northern Kentucky to the swampy bayous of Louisiana's Gulf Coast, and from the knobby hills of West Virginia to the flat sandy islands off the southernmost extremity of peninsular Florida (Fig. 3-19). Of the realm's nine regions, none has undergone more overall change during the past half-century. Choosing to pursue its own sectional interests practically from the advent of nationhood, the South experienced an even deeper economic and cultural isolation from the rest of the United States during the long and bitter aftermath of the ruinous Civil War. For over a century it languished in economic stagnation, but the 1970s finally witnessed a reassessment of the nation's perception of the region that launched a still-ongoing wave of growth and change unprecedented in Southern history.

Propelled by the forces that created the Sunbelt phenomenon, people and activities streamed into the urban South. Cities such as Atlanta, Charlotte, Miami, and Tampa became booming metropolises practically overnight, and conurbations swiftly formed in such places as southern and central Florida, the Carolina Piedmont, and the western Gulf Coast from Houston (just across the Texas border) through New Orleans to Mobile, Alabama. This "bulldozer revolution" was matched in the most favored rural areas by an agricultural renaissance that stressed such higher-value commodities as soybeans, winter wheat, poultry, and wood products. On the social front, institutionalized racial segregation was dismantled more than three decades ago.

Uneven Development

Yet for all the growth that has taken place since 1970, the South remains a region beset by many economic problems because the geography of its development has been decidedly uneven. Although several metropolitan and certain farming areas have benefited, many others containing sizeable populations have not, and the juxtaposition of progress and backwardness is frequently encountered in the Southern landscape. Not surprisingly, the gap between rich and poor is wider here than in any other U.S. region. This economic disparity also still carries racial dimensions: compared to whites, blacks exhibit higher poverty rates and rank well behind in median family income and achievement of middle-class status.

Within this checkerboard spatial framework of development, the South's new affluence is largely concentrated in its central-city CBDs and burgeoning suburban rings. At the other end of the income spectrum, its far more widely dispersed struggling areas include less favored agricultural zones (such as the chronically depressed bottomlands that line the lower Mississippi) and myriad rural towns whose jobs are tied mainly to small-scale plants in locally stagnating industries such as food processing and garmentmaking. And at the regional scale, yet another geographic pattern of inequality emerges: except for centrally located Atlanta and the Interstate-85 "Boombelt" corridor extending from there northeastward into central North Carolina, a significant proportion of the South's recent economic growth has taken place on its periphery in the Washington suburbs of northern Virginia, on the western Gulf Coast, and along Florida's central (Interstate-4) and southerncoast corridors.

Social and Economic Transformation

The transformation of the South is also changing the complexion of its demographic mosaic. Racially and economically, the composition of the region's population continues to be reshaped by the in-migration of whites and blacks from the North, with both groups exhibiting higher income levels than their resident Southern counterparts. Ethnically, the steadily increasing presence of Hispanics and Asians is intensifying the South's cultural heterogeneity. The region's Hispanic population growth has been particularly strong, more than quadrupling since 1990 as largely Mexican migrants (many of them illegals) have streamed in to fill agricultural and unskilled industrial jobs others would not take. (With most of this influx aimed at small towns, social tensions are escalating as long-time residents feel overwhelmed by the changes these newcomers have brought to daily life.) In age structure, too, the region is diversifying as an influx of retirees continues, not just in Florida but also in Georgia, Virginia, and the Carolinas. Finally, we should remember that the massive construction of new urban landscapes since 1980 has provided ample opportunity for creating fragmented residential environments that typify today's mosaic culture. Thus the filling of the tiles of the Southern population mosaic by an ever more diverse array of social and lifestyle groups is another signal that this once outcast region has all but completed its convergence with the rest of the country.

This 30-year transition period has ended with the South not only fully absorbed into the *national* economy and society, but also trying to forge a role on the *international* stage. During the 1990s, European and Japanese manufacturers were active investors in the South, but industrial jobs are now proving hard to hang onto as factories increasingly relocate outside the United States to take advantage of cheaper foreign labor. Thus the overall impact of globalization has so far been modest, and the South today is the least successful exporter of any U.S. region. As for the future, in terms of relative location the most promising new foreign business opportunities to develop in the NAFTA era lie to the south in Middle and South America. South Florida has been in the vanguard here, with Miami appropriately billing itself as "the gateway to Latin America." Lately, other cities of the South have indicated an interest in undercutting this monopoly, most notably Atlanta (expanded nonstop air routes), Memphis (global distribution hub for both Federal Express and United Parcel Service), and New Orleans (seaport enlargement).

● THE SOUTHWEST

Before the late twentieth century, most geographers did not even identify a distinct Southwestern region, classifying the U.S.-Mexican borderland zone into separate southward extensions of the Continental Interior and the mountain/plateau country to its west. Today, however, this steadily developing area—constituted by the States of Texas, New Mexico, and Arizona—has earned its place on the regional map of North America (Fig. 3-19). The still-emerging Southwest is also unique in the United States because it is a bicultural regional complex. Atop the crest of the Sunbelt wave, in-migrating Anglo-Americans joined a rapidly expanding Mexican American population anchored to the sizeable long-time resident Hispanic population, which traces its roots to the Spanish colonial territory that once extended from Texas's Gulf Coast to central California. In fact, when we add the large Native American population, it is not inappropriate to recognize the Southwest as a ***tricultural*** region.

Recent rapid development in Texas, Arizona, and New Mexico is built on a three-pronged foundation: (1) availability of huge amounts of *electricity* to power air conditioners through the long, brutally hot summers; (2) sufficient *water* to supply everything from swimming pools to irrigated crops, so that large numbers may reside in this dry environment; and (3) the *automobile*, so that affluent newcomers may spread themselves out at much-desired low densities. The first and third of these have been rather easily attained because the eastern half of the Southwest is abundantly endowed with oil and natural gas (Fig. 3-14). But the future of water supplies, especially in this decade of record drought throughout the western United States, is far more problematical. For instance, Phoenix and Tucson could not keep growing at their current swift pace without the massive new Central Arizona Project (CAP) Canal, which transports Colorado River water across hundreds of miles of desert.

Three Mushrooming States

The economic development of the Southwest in recent years has been led by *Texas*, which during the 1990s weaned itself away from dependence on its oil and gas industries (whose share of the State's annual income is now only about one-quarter of the 1980 level). The Lone Star State's mix of economic activities has been successfully restructured and diversified, and today the eastern Texas triangle formed by Dallas–Ft. Worth, Houston, and San Antonio has become one of the world's most productive postindustrial complexes, specializing in business-information and health-care services and high-technology manufacturing. Austin, which lies near the heart of the triangle, has become a leading technopole for computer development and the production of semiconductors and PCs. To the north at the triangle's apex, the Dallas–Ft. Worth (DFW) Metroplex is home to an expanding cluster of high-tech industries focused on the *Telecom Corridor* technopole just north of Dallas (the world's largest concen-

tration of telecommunications firms). As Figure 3-20 shows, the geography of employment is increasingly oriented to booming suburban freeway corridors, a spatial pattern intensifying as well in metropolitan Austin, Houston, and San Antonio. The DFW Metroplex is also the main Texas command post for international trade and NAFTA-spawned development, and is the northern anchor of a new transnational growth corridor that extends for more than 800 kilometers (500 mi) southward from Dallas through Austin and San Antonio to Monterrey 250 kilometers (150 mi) inside Mexico.

New Mexico in the region's center may still lag behind its neighbors, but it is now stirring and intent on catching up to Texas and Arizona. Propelled by income from its natural gas and oil production, the State's development has been accelerating since 2000. Abetted by a comparatively low cost of living, modest land prices and tax rates, and widespread environmental amenities, New Mexico has had no trouble in luring a rising influx of affluent migrants and relocating businesses. Benefiting most has been the Upper Rio Grande Valley centered on Albuquerque and Santa Fe–Los Alamos, whose leading growth sectors include research and development, alternative energy, and financial services.

The western flank of the Southwest—whose development is concentrated mostly in southern *Arizona*—is unique. Equaled only by parts of southern Israel's Negev, this is the most technologically transformed desert environment on Earth. But while the Negev is devoted to agriculture, Arizona's far more intensive development is marked by vast urbanized landscapes focused on two of the largest western U.S. cities, Phoenix and Tucson. During the 1990s, metropolitan Phoenix was expanding into the surrounding desert at the rate of an acre per hour, and it still grows at a robust yearly rate of around 3 percent, driven by relocating Californians. Although Arizona's conservative political and business leaders are loath to restrict development, such expansion cannot be sustained for much longer. Retrievable water supplies are nearing their limit, and strains on the highway and other infrastructural networks are mounting steadily.

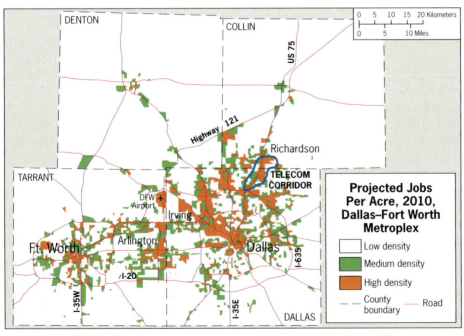

FIGURE 3-20 © H. J. de Blij, P. O. Muller, and John Wiley & Sons, Inc.

● THE WESTERN FRONTIER

To the north of the western half of the Southwest lies the realm's newest region: the swiftly growing, economically booming Western Frontier. In its east-west extent, this mountain-studded plateau region stretches between (and includes) the Rocky Mountains and the Sierra Nevada–Cascades chain; its longer north-south axis reaches from northern Arizona's Grand Canyon country to the edge of the subarctic snowforest west of the Canadian Rockies (Fig. 3-19). The region's heartland encompasses Utah, Nevada, Idaho, and western Colorado, and it is here that the old intermountain West is most actively reinventing itself. Until recently, such an opportunity would have been unthinkable because of this region's remoteness, dryness, and sparse population. But the Western Frontier's geographic prospects are changing significantly as the postindustrial national economy matures, advances in communications and transportation technologies eliminate interregional differences in the costs of doing business, and amenities become the leading locational preference of employers. Undoubtedly, proximity to the increasingly expensive and overcrowded Pacific coast is also a reinforcing factor: about 2 million disenchanted Californians have moved into the Western Frontier since 1995, lured by its sunny climate, wide-open spaces, spectacular scenery, proliferating jobs, and reduced cost of living.

The label *Western Frontier* further emphasizes that the transformation under way is being driven by the inflow of people and activities from outside the region. Indeed, the populations of the four States of its heartland all rank among the fastest-growing in the United States since 1990, with the more than 3 million migrants who arrived during the past decade already constituting around 30 percent of the Western Frontier's population. In the economic-geographic sphere, the region's identity is being reforged as traditional resource-based industries (mining, timber, livestock grazing) decline, while in-migrating and new indigenous companies create thousands of high-tech manufacturing and specialized services jobs. The spatial pattern of this development is strongly focused on the region's mushrooming urban areas and creates an unusual frontier: instead of a single line of advancing settlement, a dispersed set of population clusters is simultaneously expanding around large- and middle-sized cities, high-amenity zones, and interstate freeway corridors.

Las Vegas-Led Boom

The hottest growth area of all in the Western Frontier region—and the fastest-growing large metropolis in the United States—is the Las Vegas Valley of southernmost Nevada. This boom was triggered by Las Vegas's burgeoning recreation industry (which now attracts about 40 million visitors annually), but it is also the product of

There is something quite remarkable about the ongoing urbanization of arid valleys in the high desert of the Western Frontier, and nowhere is this phenomenon more starkly on view than in the Las Vegas Valley in Nevada's southernmost extremity. This particular municipality is Henderson, not only Las Vegas's largest suburb but also the fastest-growing urban settlement in the entire United States between 1990 and 2000. Many affluent retirees have also been attracted to this immediate area, and together with their younger counterparts have shaped a classic mosaic-culture urban fabric dominated by a low-density checkerboard of self-contained, socially homogeneous communities. Architectural critics and environmentalists, of course, would have no trouble in declaring the filling of this landscape to be one of the most blatant examples of urban sprawl on view anywhere in North America.   © Mike Yamashita/Woodfin Camp & Associates.

new economic development dominated by the influx of high-tech and professional services firms that specialize in computers, finance, and health care. Business entrepreneurs have flocked here, and the new employment opportunities sparked a local population explosion (photo p. 197), spearheaded by those relocating Californians. Relatively cheap land, modest housing prices, low taxes, and the sunny climate further fueled this migration.

Other urban areas are also experiencing accelerated growth, particularly the region's two largest metropolises, centered on Denver and Salt Lake City (which have spawned conurbations extending more than 80 kilometers [50 mi] north and south of the central city). In both cases the leading amenity is the Rocky Mountains, which soar majestically above the urbanizing flatlands and foothills. Metropolitan Denver has expanded rapidly along (and into) central Colorado's Front Range, building a diverse new, globally oriented economic base dominated by telecommunications, computer software, and financial services. On a somewhat smaller scale, Salt Lake City has followed in Denver's footsteps on the Rockies' western flank and now centers a thriving conurbation that lines northern Utah's Wasatch Front from Ogden in the north to Provo in the south. In fact, the budding technopole of *Software Valley*, the 65-kilometer (40-mi) corridor linking Salt Lake City and Provo, contains one of the world's largest concentrations of companies specializing in the production of word-processing and computer-simulation software.

Penetration of the Back Country

As much as 90 percent of the Western Frontier's development is confined to its widely dispersed urban areas, but the effects of this rapid growth are rippling deeply into the region's vast rural zones. Some of the problems—such as the steadily rising incidence of wildfires as zones bordering forests become ever more populated—are caused by the spillover effects of urbanization on adjacent areas. But many others result from the collision of lifestyles as higher-income, recreation-seeking city dwellers invade a countryside whose long-time, nonaffluent residents are struggling to preserve a way of life based on declining activities such as mining and extensive agriculture. In the valleys of the Colorado Rockies, ranchers are further squeezed by the upwardly spiraling land values set off by the massive growth of ski resorts—and usually are forced to sell and relocate their operations to less favorable land in even more remote areas. In the small towns of northern Idaho and western Montana, social tensions are inflamed as local right-wing militias encounter more liberal transplanted Californians. And seemingly everywhere, the

plans of developers are blocked by environmentalists. The rapid emergence of a new preservation ethic, which increasingly threatens environmentally insensitive land and resource developers, is part of a larger antigrowth movement that is certain to intensify. Throughout this region today, the buzzwords are "managing growth without restricting economic opportunity"—which usually translates into bringing opposing interests together to work out compromises.

Whatever our new century brings, the growth boom of the 1990s that forged the realm's newest region has forever altered the economic-geographic and demographic complexion of much of the old intermountain West. Now reborn as the Western Frontier, this no-longer-remote region is putting the often disastrous boom/bust cycles of its mining- and grazing-dominated past behind it and is making major strides toward establishing itself as a globally connected, technological powerhouse on the inland doorstep of North America's burgeoning sector of the Pacific Rim (see box on p. 201).

● THE NORTHERN FRONTIER

The northern half of the realm, lying poleward of roughly 52°N latitude, constitutes the Northern Frontier region (Fig. 3-19). This is by far North America's biggest regional subdivision, covering almost 90 percent of Canada and all of the U.S. State of Alaska. Indeed, if this huge territory were a separate country, it would rank as the world's fifth-largest in size. The latitudinal extent of this region leaves no doubt as to why it is designated as *northern*. In fact, after northern Russia (whose dimensions of continentality are even more pronounced), this is our planet's second greatest expanse of harsh cold environment affecting an inhabited area. As with the Arctic fringes of Eurasia, very few physical barriers in northern North America ameliorate the masses of supercold air, which throughout the long winter plunge directly southward from the ice-covered Arctic Ocean. Not surprisingly, most of the Northern Frontier's people and activities are concentrated in the milder, heavily forested southern half of the region.

Unquestionably, this region is also a *frontier*—and far more classically so than the Western Frontier just discussed. The latter's newly activated, dispersed, urbanized development pattern and dramatic population influx are largely attributable to its improved economic-geographic position astride a national axis, within which significant east-west interaction was already taking place. The Northern Frontier, on the other hand, represents a thrust into undeveloped country in an entirely new direction, one marked by the leisurely, long-term advance of a line across parts of Canada's near- and middle-northern periphery that absorbs the areas it crosses into the national spatial economy. Boom and bust cycles since the late nineteenth century have controlled the rate of movement of this development line, which frequently has ebbed backward as much as it has flowed forward. In general, the still sparsely populated Northern Frontier is mostly a region of recent European settlement based on the exploitation of newly discovered resources. As we shall see, however, First Nations and other indigenous peoples, who have always regarded this region as a homeland, are today playing a growing role in the affairs of the Northern Frontier and will have much to say about shaping its future.

Canada's Stirring Northland

As noted earlier, the Canadian Shield, which covers the eastern half of the Northern Frontier (Fig. 3-2), is one of the world's richest storehouses of mineral resources. Most of the Shield's developing areas are focused on mining complexes that extract such high-demand metallic ores as nickel, uranium, copper, gold, silver, lead, and zinc. The leading zones of production are located along the Shield's southern margins, but since 1980 the search for additional deposits, the expansion of road and rail networks, and the opening of new mines have all advanced steadily northward. The Yukon and Northwest Territories have proven bountiful, with gold- and diamond-mining (Canada now ranks among the world's top five producers) paving the way; and at the opposite, eastern edge of the Shield, near Voisey's Bay on the central coast of Labrador, the largest body of high-grade nickel ores ever discovered is being opened up for full-scale production.

Besides metallic minerals, the Northern Frontier is endowed with an abundance of other resources, which include fossil fuels, hydroelectric power opportunities, timber, and fisheries. These are dominated by the substantial oil and gas reserves that are concentrated in a wide zone east of the Canadian Rockies between the Yukon and the U.S. border (Fig. 3-14). Most spectacular are the Athabasca Tar Sands around Fort McMurray in northeastern Alberta. We noted earlier that they contain more crude oil than all of Saudi Arabia (as well as 90+ percent of Canada's total supply), but their petroleum is deposited as a tar-like substance—known as bitumen—in near-surface sand formations. Extraction involves stripmining these deposits, separating bitumen from sand, and converting it into liquid crude oil—an expensive and highly polluting process that consumes massive quantities of local (natural gas) energy and water. However, the recent advance of world oil prices beyond U.S. $30 a barrel, combined with new technological break-

As the price of a barrel of oil (and a gallon of gasoline) rises, it becomes profitable to derive oil from sources other than liquid reserves. The Canadian province of Alberta contains vast deposits of "oil sands" in which the petroleum is mixed with sand, requiring a relatively expensive and complicated process to extract it. The quantity of oil locked in these Athabasca Tar Sands is estimated to constitute one of the world's largest reserves, and this huge open-pit mining project is under way around the town of Fort McMurray to recover it. Those who assume that Canada will sell this oil to the United States should be aware that the Chinese are offering to fund construction of a pipeline across the mountains to Canada's Pacific coast; the rising cost of oil has political as well as economic consequences. © Jim Wark.

throughs that include better pollution control, enables this energy supply to be exploited ever more profitably. Output is now rising sharply (see photo above), and over the next decade the Tar Sands are expected to become one of the world's leading oil-producing zones.

The productive activities of the Northern Frontier are linked together within a far-flung network of mines, oil- and gasfields, pulp mills, and hydropower stations that have spawned hundreds of settlements and thousands of miles of interconnecting transport and communications facilities. Though dominant, this commercial economy constitutes just one component of a dual regional economy. The other component, as Robert Bone points out in his book *The Geography of the Canadian North*, consists of the native economy that "evolved from a subsistence hunting economy to one in which wage employment, transfer payments, and trapping provide cash income while hunting provides country food." This native economy, which adheres to the traditional values of the indigenous peoples (First Nations), has long been tied to the commercial economy as a source of wages. It has also been heavily land-based, and in recent years First Nations throughout the region have become more assertive and have filed legal actions to recover lands that Europeans took in the past without negotiating treaties. Both the courts and the federal government have been sympathetic, and the creation of Nunavut in 1999 (see map, p. 178) has been the most dramatic breakthrough to date. It now seems clear that First Nations will increasingly be consulted on future resource development in their traditional homelands; however, because vast tracts of northern lands are likely to be involved, this represents a potential threat to national unity should related issues of self-government become enmeshed in the process.

Alaska's Energy Storehouse

The Northern Frontier also contains Alaska, whose geographic opportunities and challenges differ somewhat from those of the Canadian North. Besides being the largest U.S. State, Alaska earns a sizeable income from oil production on its northeastern Arctic slope, and the south-central coast around Anchorage increasingly benefits from its near-midpoint location on the Pacific Rim air route between the "Lower 48" States and East Asia. These advantages have propelled a growth spurt of more than 100 percent since 1970, and today Alaska's total population of just over 700,000 accounts for nearly one-third of the Northern Frontier's entire population on less than one-sixth of its land area. Here, too, metallic minerals have long drawn migrants northward. But, unlike Canada to the east, Alaska's late-twentieth-century boom was triggered by the opening of its vast oilfields on the North Slope, one of the hemisphere's leading energy sources since the 1300-kilometer (800-mi) Trans-Alaska Pipeline began operating in 1977 (Fig. 3-14). Dwindling supplies at the main field at Prudhoe Bay are now compelling producers to turn to the huge additional reserves of the Arctic slope, but strong opposi-

tion from nature preservationist groups has slowed their plans. To the east lies the Arctic National Wildlife Refuge, which the Bush administration wants to open for drilling. To the west lies the National Petroleum Reserve-Alaska, which the federal government has partially opened to oil leasing by drillers, who insist that their newly gained sensitivity paired with the latest technologies will minimize the impact of production on the tundra's fragile ecosystems. Another major project that faces a similar battle is the proposal to construct a natural-gas pipeline to parallel the oil pipeline (huge supplies of liquified gas are a byproduct of oil drilling). As noted earlier, demand for clean-burning natural gas by electricity producers is rising all across North America.

● THE PACIFIC HINGE

The Pacific coastlands of the conterminous United States and southwesternmost Canada, which constitute the realm's ninth region (Fig. 3-19), have been a powerful lure to migrants since the Oregon Trail was pioneered more than 160 years ago. Unlike the remainder of western North America south of 50°N latitude, the strip of land between the Sierra Nevada–Cascade mountain wall and the sea receives adequate moisture. It also possesses a far more hospitable environment, with generally delightful weather south of San Francisco, highly productive farmlands in California's Central Valley, and such scenic glories as the Big Sur coast, Washington State's Olympic Mountains, and the spectacular waters surrounding San Francisco, San Diego, Seattle, and Vancouver. Most of the major development here took place during the post–World War II era, accommodating enormous population and economic growth along this outermost edge of the conterminous United States. But with the arrival of the twenty-first century, it is clear that, in terms of its economic geography, the West Coast no longer represents an end but rather a beginning—a gateway to an abundance of growing opportunities that in recent times have blossomed on many of the distant shores encircling the Pacific Basin. That is why we use the term *hinge*, because now this region increasingly forms an interface between North America and the booming, still-emerging *Pacific Rim* (see box titled "The Pacific Rim Connection").

Challenges and Opportunities in Maturing California

More than 60 years of unrelenting growth have taken their toll, and whereas the West Coast States and British Columbia are now savoring the prospects of their Pacific Rim lo-

The Pacific Rim Connection

TODAY WE CONTINUE to witness the rise of a far-flung region born of a string of economic miracles on the shores of the Pacific Ocean. It is a still-discontinuous region, led by its foci on the Pacific's western margins: Japan, coastal China, South Korea, Taiwan, Thailand, Malaysia, and Singapore. The regional term **Pacific Rim** has come into use to describe this dramatic development, which over the past three decades has redrawn the map of the Pacific periphery—not only in East and Southeast Asia but also in the United States and Canada, and even in such Southern Hemisphere locales as Australia and South America's Chile.

The Pacific Rim also is a superb example of a functional region, with economic activity in the form of capital flows, raw-material movements, and trade linkages generating urbanization, industrialization, and labor migration. Within this process, human landscapes from Sydney to Santiago are being transformed inside a 32,000-kilometer (20,000-mi) corridor that girdles the globe's largest body of water. In later chapters, wherever the Pacific Rim intersects Pacific-bordering geographic realms, we will continue to discuss this important regional development and its spatial impacts.

cation, they must also confront the less pleasant consequences of regional maturity. This is especially true in California, where the massive development of America's most populated and multiethnic State (almost 40 million in 2008, 20 percent larger than all of Canada) has been overwhelmingly concentrated in the teeming conurbation ex-

tending south from San Francisco through San Jose, the San Joaquin Valley, the Los Angeles Basin, and the southwestern coast into San Diego at the Mexican border. Environmental hazards bedevil this entire corridor, including inland droughts, coastal-zone flooding, mudslides, brushfires, and particularly earthquakes—with the ominous San

AMONG THE REALM'S GREAT CITIES . . . Los Angeles

IT IS HARD to think of another city that has been in more headlines in the past several years than Los Angeles. Much of this publicity has been negative. But despite its earthquakes, mudslides, brushfires, pollution alerts, showcase trials, and riots, L.A.'s glamorous image has hardly been dented. This is compelling testimony to the city's resilience and vibrancy, and to its unchallenged position as the Western world's entertainment capital.

The plane-window view during the descent into Los Angeles International Airport, almost always across the heart of the metropolis, gives a good feel for the immensity of this urban landscape. It not only fills the huge natural amphitheater known as the Los Angeles Basin, but it also oozes into adjoining coastal strips, mountain-fringed valleys and foothills, and even the margins of the Mojave Desert more than 90 kilometers (60 mi) inland. This quintessential spread city, of course, could only have materialized in the automobile age. In fact, most of it was built rapidly over the past 70 years, propelled by the swift expansion of a high-speed freeway network unsurpassed anywhere in metropolitan America. In the process, Greater Los Angeles became so widely dispersed that today it has reorganized within a geographic framework of six sprawling urban realms (see Fig. 3-12).

The metropolis as a whole is home to 17.8 million Southern Californians, constitutes North America's second largest urban agglomeration, and forms the southern anchor of the huge California conurbation that lines the Pacific coast from San Francisco southward to the Mexi-

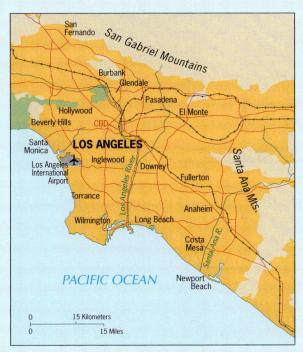

© H. J. de Blij, P. O. Muller, and John Wiley & Sons, Inc.

can border (Fig. 3-9). It also is the Pacific Hinge's leading trade, manufacturing, and financial center. In the global arena, Los Angeles is the eastern Pacific Rim's biggest city and the origin of the largest number of transoceanic flights and sailings to Asia.

FROM THE FIELD NOTES

"Minutes after takeoff from the Seattle-Tacoma International Airport on the way to Tokyo we got a superb view of central Seattle from the ferry terminals on Elliott Bay to Interstate-5 linking California to Canada and beyond. Like Sydney, Australia, Seattle has suburbs across the water reached most easily by boat; this is one of America's most scenic and vibrant cities. Seattle has transformed itself repeatedly through economic cycles reflected by the urban landscape: the dominance of precision-equipment manufacture, long dominated by Boeing Aircraft, is ending in favor of a new era of computer technology (Microsoft has its headquarters in a suburb) and related industries of the 'New Economy.'" (There is something, though, that dates this photograph . . . what is it?)   © H. J. de Blij.

www.conceptcaching.com

fense expenditures. By the early 2000s, a (still-continuing) turnaround was underway as Southern California successfully diversified its economy and developed such new growth sectors as the entertainment industry, foreign trade, and the nation's largest single metropolitan manufacturing complex—whose products range from high-tech medical equipment to the clothing churned out by the low-tech factories of a burgeoning, immigrant-staffed garment industry that continues to outpace New York City's.

The northern portion of the region encompasses Northern California and the Pacific Northwest, the latter extending northward from Oregon's Willamette Valley through the Cowlitz-Puget Sound lowland of western Washington State into the adjoining coastal zone of British Columbia (and offshore Vancouver Island) beyond the Canadian border. Since 1990, Northern California, focused on the San Francisco Bay Area, has experienced a smoother path than the southern part of the State. Here an unprecedented period of economic expansion continues, led by ***Silicon Valley***. Within that technopole, no community stood to gain more from this boom than Palo Alto, which is situated at the largest switching point on the Internet that handles more than half of its global traffic (see Fig. 3-16). As Internet commerce grows explosively, companies flock to Palo Alto to take advantage of its unparalleled access to cyberspace. This is further facilitated by the city's state-of-the-art local fiber-optic network, and what is taking shape here is nothing less than an infrastructural hub that will become one of the cornerstones of the so-called New Economy.

Growth and Change in the Pacific Northwest

Continuing up the coast, the Pacific Northwest is also deeply involved in the pioneering and production of computer technology, especially in the technopole centered on suburban Seattle's Redmond (Microsoft's headquarters). Originally built on timber and fishing—primary activities that still survive here despite human pressures on the environment—this subregion found its impetus for industrialization in the massive Columbia River dam projects of the 1930s and 1950s, which generated the cheap hydroelectricity that attracted aluminum and aircraft manufacturers. Although Boeing is still a major employer, the recent influx of affluent newcomers is fueling the growth of software and Internet companies that is spearheading Seattle's transformation into the prototype metropolis of the New Economy. Nonetheless, what appear to be short-term economic problems have arisen in the early twenty-first century as a result of fluctuations in the global computer and aviation industries.

North of the Canadian border, the Pacific Hinge's northernmost subregion stands out from the others not only be-

Andreas Fault practically the axis of megalopolitan coalescence. To all this, humans have added their own abuses of California's fragile habitat, from overuse of water supplies (requiring vast aqueduct systems to import water from hundreds of kilometers away) to the chronic air pollution of most of its metropolitan areas.

These challenges notwithstanding, Californians over the past half-century have built one of the realm's most productive economic machines. Indeed, if the Golden State were an independent country, its economy would today rank as the world's sixth largest. The road to success, however, has often proven to be a bumpy one, as was the case in Southern California after 1990. This subregion began the struggling 1990s as national recession combined with the unexpected end of the Cold War to deliver a serious blow to the Los Angeles-area economy, which at the time was heavily dependent on its aerospace industry driven by federal government de-

cause of its British Columbia location but also because its economic core, Vancouver, has become the most Asianized metropolis in North America. Vancouver lies closer to East Asia on the air and sea routes than the cities to its south, and it got a head start in forging trade and investment linkages to the western Pacific Rim thanks to its large and growing Asian population. Today, ethnic Chinese residents constitute more than 20 percent of the metropolitan population, whose total Asian component now approaches a remarkable 40 percent. Although a new east-west hybrid culture is trying to establish itself, frictions intensified during the late 1990s as wealthy Asians (many of them expatriates of now-Chinese Hong Kong) took over neighborhoods and replaced traditional homes with much larger houses, surrounded them with walls and fences where none had existed, and cut down most of the trees. More serious is the growing belief by the majority Anglos that the newcomers do not really want to assimilate, preferring instead to avoid the use of English and live, shop, and transact business exclusively within their own communities.

With North America ever more tightly enmeshed in the global economy, nowhere are the realm's international linkages more apparent today than in the Pacific Hinge. From the northern end of this region to the Mexican border, its geographic advantages as a gateway to the Pacific Rim are providing opportunities undreamed of barely a decade ago. Asianizing British Columbia now exports close to 50 percent of its goods to the countries of the western Pacific Rim; China alone has sparked a resurgence of the province's mining industries, importing ever increasing quantities of coal and metallic ores. The rest of the Vancouver–Seattle–Portland corridor is following suit, led by its high-technology industries. In California, the foreign trade sector of the State's gigantic economy has tripled since 1990, and today more than one-sixth of its jobs are export-dependent. Both Northern and Southern California are expected to thrive as these trends continue, with metropolitan Los Angeles becoming the financial, manufacturing, and trading capital of the eastern Pacific Rim as well as its transport hub (L.A. overtook New York in the 1990s as the leading U.S. seaport in the value of goods passing through it). Thus it is abundantly clear today that the mid-latitude West Coast no longer serves as North America's back door, but forms the hinges of a new front door to the Pacific arena thrown wide open to foster international interactions of every kind.

What You Can Do

RECOMMENDATION: Get involved! Help a Geographic Alliance! Because "geographic illiteracy" still prevails in the United States, the National Geographic Society in the 1980s launched a project to help teachers prepare themselves for geography instruction. Teachers from every State in the Union were invited to participate in seminars at NGS headquarters in Washington, D.C., and these teachers subsequently presented similar seminars for colleagues in their home States. Thus a State-based network of Geographic Alliances was created and still functions today. These Geographic Alliances are always looking for assistance, resources, ideas, and other contributions. It would be an interesting research project to discover whether geography is taught in the schools in the town where your college is located, and whether the teachers teaching it were part of the Alliance network.

GEOGRAPHIC CONNECTIONS

1 Imagine that you are the chief executive of a successful young company that manufactures software for the newest high-speed computers. Where would you locate your company and why? Establish a "short list" of three possible sites, and choose one by a process of elimination in which your decision is built upon the most important locational variables for your plant and its highly skilled workforce.

2 Immigration has long been a potent force in shaping the population mosaic of North America. Why is the "geography of demography" in the United States today so different from that of stagnant Europe? How does that differ-

ence express itself regionally? Does the immigration process affect your daily life, and how might it affect your competitive position when you are ready to enter the job market following completion of your education?

3 Enhance your experience in considering these questions by surfing the Internet and consulting appropriate websites. As an extra endeavor, ponder these same questions as they would apply to the contemporary Canadian scene, keeping an eye out for similarities and differences between the two countries of this realm.

Geographic Literature on North America: The key introductory works on this realm were authored by Birdsall et al., Boal & Royle, Bone (both titles), Getis & Getis, Hardwick et al., Hudson, McGillivray, and McKnight. These works, together with a comprehensive listing of noteworthy books and recent articles on the geography of the United States and Canada, can be found in the *References and Further Readings* section of this book's website at **www.wiley.com/college/deblij.**

4

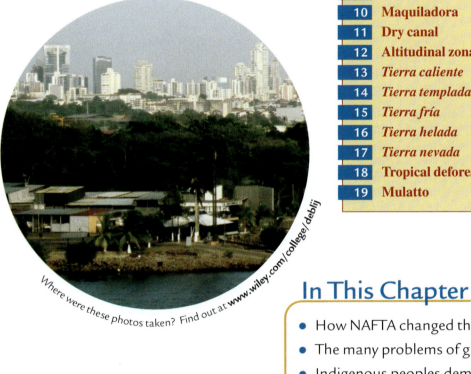

Where were these photos taken? Find out at **www.wiley.com/college/deblij**

Photos: © H. J. de Blij

FIGURE 4-1 *Map:* © H. J. de Blij, P. O. Muller, and John Wiley & Sons, Inc.

In This Chapter

- How NAFTA changed the economic geography of Mexico
- The many problems of gigantic Mexico City
- Indigenous peoples demand recognition and rights
- Projected Panama Canal expansion fuels boom in Panama City
- Cuba: Future star of the Caribbean
- The debate over Puerto Rico's status continues

L OOK AT A world map, and it is obvious that the Americas comprise two landmasses: North America extending from Alaska to Panama and South America from Colombia to Argentina. But here we are reminded that continents and geographic realms do not usually coincide. In Chapter 3 we discussed a North American realm that is bounded by the U.S.-Mexican border and the Gulf of Mexico. Between North America and South America lies a small but important geographic realm known as Middle America. Consisting of a mainland corridor and numerous Caribbean islands, Middle America is a highly fragmented realm.

From Figure 4-1 it is clear that Middle America is much wider longitudinally than it is long latitudinally. The distance from Baja California to Barbados is about 6000 kilometers (3800 mi), but from the latitude of Tijuana to Panama City is only half that distance. In terms of total area, Middle America is the second-smallest of the world's geographic realms (see Table G-1). As the map shows, the dominant state of this realm is Mexico, larger than all its other countries and territories combined.

What Middle America lacks in size it makes up in physiographic and cultural diversity. This is a realm of soaring volcanoes and spectacular shorelines, of tropical forests and barren deserts, of windswept plateaus and scenic islands. It holds the architectural and technological legacies of ancient indigenous civilizations. Today it is a mosaic of immigrant cultures from Africa, Europe, and elsewhere, richly reflected in music and the visual arts. Material poverty, however, is endemic: Haiti, on one of the Caribbean islands, is the poorest country in the Americas. Nicaragua, on the mainland, is almost as badly off. As we will find, a combination of factors has produced a distinctive but troubled realm between North and South America.

MAJOR GEOGRAPHIC QUALITIES OF
Middle America

1. Middle America is a fragmented realm that consists of all the mainland countries from Mexico to Panama and all the islands of the Caribbean Basin to the east.

2. Middle America's mainland constitutes a crucial barrier between Atlantic and Pacific waters. In physiographic terms, this is a land bridge that connects the continental landmasses of North and South America.

3. Middle America is a realm of intense cultural and political fragmentation. The political geography defies unification efforts, but countries and regions are beginning to work together to solve mutual problems.

4. Middle America's cultural geography is complex. African influences dominate the Caribbean, whereas Spanish and Amerindian traditions survive on the mainland.

5. The realm contains the Americas' least-developed territories. New economic opportunities may help alleviate Middle America's endemic poverty.

6. In terms of area, population, and economic potential, Mexico dominates the realm.

7. Mexico is reforming its economy and has experienced major industrial growth. Its hopes for continuing this development are tied to overcoming its remaining economic problems and to expanding trade with the United States and Canada under the North American Free Trade Agreement (NAFTA).

Defining the Realm

Sometimes you will see Middle and South America referred to, in combination, as "Latin" America, alluding to their prevailing Spanish-Portuguese heritage. This is an inappropriate regional name, just as "Anglo" America, a term once commonly used for North America, was also improper. Such culturally based terminologies reflect historic power and dominance, and they tend to make outsiders out of those people they do not represent. In North America, the term "Anglo" (as a geographic appellation) was offensive to many Native Americans, to African Americans, to Hispanics, to Quebecers, and to others. In Middle (and South) America, millions of people of indigenous American,

African, Asian, and European ancestries do not fit under the "Latin" rubric. You will not find the cultural landscape very "Latin" in the Bahamas, Barbados, Jamaica, Belize, or large parts of Guatemala and Mexico. So let us adopt the geographic neutrality of North, Middle, and South.

But is Middle America sufficiently different from either North or South America to merit regional distinction? Certainly, in this age of globalization and migration, border areas are becoming transition zones, as is happening along the U.S.-Mexican boundary. Still, consider this: North America encompasses just 2 states, and the entire continent of South America only 13 (even including France's dependency on the northeast coast). But far smaller Middle America, as we define it, incorporates more than three dozen political entities, including several dependencies (or quasi-dependencies) of the Netherlands, the United Kingdom, and France, but not counting constituent territories of the United States. Therefore, unlike South America, Middle America is a multilingual patchwork of independent states, territories in political transition, and residual colonial dependencies, with strong continuing ties to the United States and non-Iberian Europe. Middle America is defined in large measure by its vivid cultural-geographic pluralism.

The Regions

As Figure 4-2 indicates, Middle America contains four distinct regions. Dominant *Mexico* occupies the largest part of the collective territory, with much of its northern border

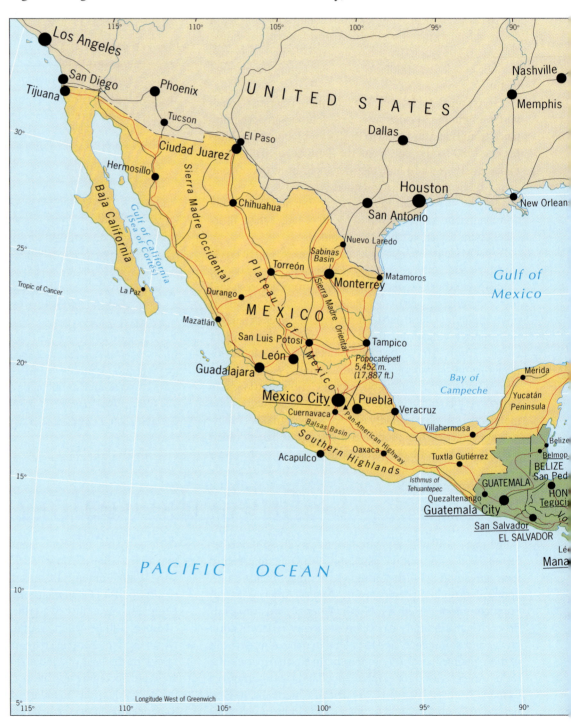

FIGURE 4-2   © H. J. de Blij, P. O. Muller, and John Wiley & Sons, Inc.

defined by the Rio Bravo (Rio Grande) from El Paso to the Gulf of Mexico. To the southeast, Mexico yields to the region called *Central America*, consisting of the seven republics of Guatemala, Belize, Honduras, El Salvador, Nicaragua, Costa Rica, and Panama. Sometimes the entire realm is referred to as Central America, but only these seven countries constitute this region. The *Greater Antilles* is the regional name referring to the four large islands in the northern sector of the Caribbean Sea: Cuba, Jamaica, His-

paniola, and Puerto Rico; two countries, Haiti and the Dominican Republic, share the island of Hispaniola. And the *Lesser Antilles* form an extensive arc of smaller islands from the Virgin Islands off Puerto Rico to the Netherlands Antilles near the northwestern coast of Venezuela. In the Caribbean islands large and small, moist winds sweep in from the east, watering windward (wind-facing) coasts while leaving leeward (wind-protected) areas dry. On the map, the "Leeward" and "Windward" Islands actually are

not "dry" and "wet," these terms having navigational rather than environmental implications.

PHYSIOGRAPHY

As Figure 4-1 shows, fragmented Middle America is a realm of high relief, studded with active volcanoes. Figure G-4 reminds us why: the Caribbean, North American, South American, and Cocos tectonic plates converge here, creating dangerous landscapes subject to earthquakes and landslides (see photo p. 211). Add to this the realm's exposure to Atlantic hurricanes, and it amounts to some of the highest-risk real estate on Earth.

A Land Bridge

The funnel-shaped mainland, a 4800-kilometer (3000-mi) connection between North and South America, is wide enough in the north to contain two major mountain chains and a vast interior plateau, but narrows to a slim 65-kilometer (40-mi) ribbon of land in Panama. Here this strip of land, or *isthmus*, bends eastward so that Panama's orientation is east-west. Thus mainland Middle America is what physical geographers call a **land bridge**, an isthmian link between continents.

If you examine a globe, you can see other present and former land bridges: Egypt's Sinai Peninsula between Asia and Africa, the (now-broken) Bering land bridge between northeasternmost Asia and Alaska, and the shallow waters between New Guinea and Australia. Such land bridges, though temporary features in geologic time, have played crucial roles in the dispersal of animals and humans across the planet. But even though mainland Middle America forms a land bridge, its internal fragmentation has always inhibited movement. Mountain ranges, swampy coastlands, and dense rainforests make contact and interaction difficult.

Island Chains

As shown in Figure 4-1, the approximately 7000 islands of the Caribbean Sea stretch in a lengthy arc from Cuba and the Bahamas east and south to Trinidad, with numerous outliers outside (such as Barbados) and inside (e.g., the Cayman Islands) the main chain. As we noted earlier, the four large islands—Cuba, Hispaniola (containing Haiti and the Dominican Republic), Puerto Rico, and Jamaica—are called the *Greater Antilles*, and all the remaining smaller islands constitute the *Lesser Antilles*.

The entire Antillean **archipelago** (island chain) consists of the crests and tops of mountain chains that rise from the floor of the Caribbean, the result of collisions between the Caribbean Plate and its neighbors (Fig. G-4). Some of these crests are relatively stable, but elsewhere they contain active volcanoes, and almost everywhere in this realm earthquakes are an ever-present danger—in the islands as well as on the mainland.

LEGACY OF MESOAMERICA

Mainland Middle America was the scene of the emergence of a major ancient civilization. Here lay one of the world's true **culture hearths** (see Fig. 7-3), a source area from which new ideas radiated and whose population could expand and make significant material and intellectual progress. Agricultural specialization, urbanization, and transport networks developed, and writing, science, art, and other spheres of achievement saw major advances. Anthropologists refer to the Middle American culture hearth as *Mesoamerica*, which extended southeast from the vicinity of present-day Mexico City to central Nicaragua. Its development is especially remarkable because it occurred in very different geographic environments, each presenting obstacles that had to be overcome in order to unify and integrate large areas. First, in the low-lying tropical plains of what is now northern Guatemala, Belize, and Mexico's Yucatán Peninsula, and perhaps simultaneously in Guatemala's highlands to the south, the Maya civilization arose more than 3000 years ago. Later, far to the northwest on the high plateau in central Mexico, the Aztecs founded a major civilization centered on the largest city ever to exist in pre-Columbian times.

The Lowland Maya

The Maya civilization is the only major culture hearth in the world that arose in the lowland tropics. Its great cities, with their stone pyramids and massive temples (photo p. 212), still yield archeological information today. Maya culture reached its zenith from the third to the tenth centuries AD.

The Maya civilization, anchored by a series of city-states, unified an area larger than any of the modern Middle American countries except Mexico. Its population probably totaled between 2 and 3 million; certain Maya languages are still used in the area to this day. The Maya city-states were marked by dynastic rule that functioned alongside a powerful religious hierarchy, and the great cities that now lie in ruins were primarily ceremonial centers. We also know that Maya culture produced skilled

In terms of natural hazards, Middle America is the most dangerous realm of all. Hurricanes, volcanic eruptions, earthquakes, and landslides threaten virtually every corner of this part of the world. In January 2001, two of these hazards lethally combined to devastate the Las Colinas neighborhood of Santa Tecla, a suburb of the city of San Salvador, capital of the Central American country of El Salvador. A severe earthquake (magnitude 7.6 on the Richter Scale) rocked much of that country, causing only moderate damage overall, but a few places took the brunt of the seismic forces unleashed by the temblor. One was the geologically unstable hillside above Las Colinas, which in a matter of seconds produced this massive landslide that buried hundreds of homes and claimed more than half of the 687 Salvadorians who perished in the quake that day.   © Lopez/ El Diario de Hoy/Gamma-Presse, Inc.

artists and scientists, and that these people achieved a great deal in agriculture and trade. They grew cotton, created a rudimentary textile industry, and exported cotton cloth by seagoing canoes to other parts of Middle America in return for valuable raw materials. They domesticated the turkey and grew cacao, developed writing systems, and studied astronomy.

The Highland Aztecs

In what is today the intermontane highland zone of Mexico, significant cultural developments were also taking place. Here, just north of present-day Mexico City, lay Teotihuacán, the first true urban center in the Western Hemisphere, which prospered for nearly seven centuries after its founding around the beginning of the Christian era.

A Once and Future Core Area

The Aztec state, the pinnacle of organization and power in pre-Columbian Middle America, is thought to have originated in the early fourteenth century with the founding of a settlement on an island in a lake that lay in the *Valley of Mexico* (the area surrounding what is now Mexico City). This urban complex, a functioning city as well as a ceremonial center, named Tenochtitlán, was soon to become the greatest city in the Americas and the capital of a large powerful state. The Aztecs soon gained control over the entire Valley of Mexico, a pivotal 50-by-65-kilometer (30-by-40-mi) mountain-encircled basin that is still the heart of the modern state of Mexico. Both elevation and interior location moderate its climate; for a tropical area, it is quite dry and very cool. The basin's lakes formed a valuable means of internal communication, and the Aztecs built canals to connect several of them. This fostered a busy canoe traffic, bringing agricultural produce to the cities, and tribute was paid by their many subjects to the headquarters of the ruling nobility.

A Regional Empire

With its heartland consolidated, the new Aztec state soon began to conquer territories to the east and south. The Aztecs' expansion of their empire was driven by their desire to subjugate peoples and towns in order to extract taxes and tribute. As Aztec influence spread throughout Middle America, the state grew ever richer, its population mushroomed, and its cities thrived and expanded.

The Aztecs produced a wide range of impressive accomplishments, although they were better borrowers and refiners than they were innovators. They developed irrigation systems, and they built elaborate walls to terrace slopes where soil erosion threatened. Indeed, the greatest contributions of Mesoamerica's Amerindians surely came from the agricultural sphere. Corn (maize), the sweet potato, various kinds of beans, the tomato, squash, cacao beans (the raw material of chocolate), and tobacco are just a few of the crops that grew in Mesoamerica when the Europeans first made contact.

FROM THE FIELD NOTES

"We spent Monday and Tuesday upriver at Lamanai, a huge, still mostly overgrown Maya site deep in the forest of Belize. On Wednesday we drove from Belize City to Altun Ha, which represents a very different picture. Settled around 200 BC, Altun Ha flourished as a Classic Period center between AD 300 and 900, when it was a thriving trade and redistribution center for the Caribbean merchant canoe traffic and served as an entrepôt for the interior land trails, some of them leading all the way to Teotihuacán. Altun Ha has an area of about 6.5 square kilometers (2.5 sq mi), with the main structures, one of which is shown here, arranged around two plazas at its core. I climbed to the top of this one to get a perspective, and sat down to have my sandwich lunch, imagining what this place must have looked like as a bustling trade and ceremonial center when the Roman Empire still thrived, but a more urgent matter intruded. A five-inch tarantula emerged from a wide crack in the sun-baked platform, and I noticed it only when it was about two feet away, apparently attracted by the crumbs and a small piece of salami. A somewhat hurried departure put an end to my historical-geographical ruminations."
© H. J. de Blij.

Concept Caching

www.conceptcaching.com

COLLISION OF CULTURES

We in the Western world all too often believe that history began when the Europeans arrived in some area of the world and that the Europeans brought such superior power to the other continents that whatever existed there previously had little significance. Middle America confirms this misperception: the great, feared Aztec state apparently fell before a relatively small band of Spanish invaders in an incredibly short period of time (1519–1521). But let us not lose sight of a few realities. At first, the Aztecs believed the Spaniards were "White Gods" whose arrival had been predicted by Aztec prophecy. Hernán Cortés, for all his 508

soldiers, did not singlehandedly overthrow this powerful empire: he ignited a rebellion by Amerindian peoples who had fallen under Aztec domination and had seen their relatives carried off for human sacrifice to Aztec gods. Led by Cortés with his horses and guns, these peoples rose against their Aztec oppressors and joined the band of Spaniards headed toward Tenochtitlán (where thousands of them would die in combat against Aztec defenders).

Effects of the Conquest

Spain's defeat of Middle America's dominant indigenous state opened the door to Spanish penetration and supremacy. Throughout the realm, the confrontation between Hispanic and native cultures spelled disaster for the Amerindians: a catastrophic decline in population (perhaps as high as 90 percent), rapid deforestation, pressure on vegetation from grazing animals, substitution of Spanish wheat for maize (corn) on cropland, and the concentration of Amerindians into newly built towns.

The Spaniards were ruthless colonizers but not more so than other European powers that subjugated other cultures. True, the Spaniards first enslaved the Amerindians and were determined to destroy the strength of indigenous society. But biology accomplished what ruthlessness could not have achieved in so short a time: diseases introduced by the Spaniards and slaves imported from Africa killed millions of Amerindians.

The Rural Impact

Middle America's cultural landscape—its great cities, its terraced fields, its dispersed aboriginal villages—was thus drastically modified. Unlike the Amerindians, who had used stone as their main building material, the Spaniards employed great quantities of wood and used charcoal for heating, cooking, and smelting metal. The onslaught on the forests was immediate, and rings of deforestation swiftly expanded around the colonizers' towns. The Spaniards also introduced large numbers of cattle and sheep, and people and livestock now had to compete for available food (requiring the opening of vast areas of marginal land that further disrupted the region's food-production balance). Moreover, the Spaniards introduced their own crops (notably wheat) and farming equipment, and soon large fields of wheat began to encroach upon the small plots of corn that the natives cultivated.

The New Urban Settlements

The Spaniards' most far-reaching cultural changes derived from their traditions as town dwellers. To facilitate domination, the Amerindians were moved off their land

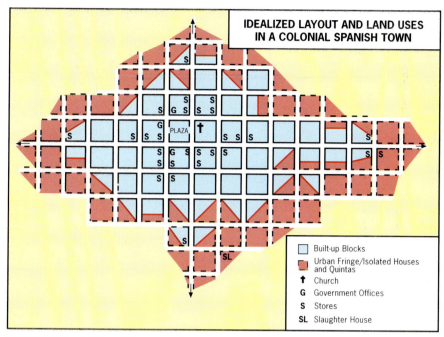

FIGURE 4-3 © H. J. de Blij, P. O. Muller, and John Wiley & Sons, Inc.

Amerindians could go out each day and work in the fields. Packed tightly into these towns and villages, they came face to face with Spanish culture. Here they learned the Europeans' Roman Catholic religion and Spanish language, and they paid their taxes and tribute to a new master. Nonetheless, the nucleated indigenous village survived under colonial (and later postcolonial) administration and is still a key feature of remote Amerindian areas in southeastern Mexico and inner Guatemala, where to this day native languages prevail over Spanish (Fig. 4-4).

Church and State

Once the indigenous population was conquered and resettled, the Spaniards were able to pursue another primary goal in their New World territory: the exploitation of its wealth for their own benefit. Lucrative trade, commercial agriculture, livestock ranching, and especially mining were avenues to affluence. And wherever the Spaniards ruled—in towns, farms, mines, or indigenous villages—the Roman Catholic Church was the supreme cultural force transforming Amerindian society. With Jesuits and soldiers working together to advance the frontiers of New Spain, in the wake of conquest it was the church that controlled, pacified, organized, and acculturated the Amerindian peoples.

into nucleated villages and towns that the Spaniards established and laid out. In these settlements, the Spaniards could exercise the kind of rule and administration to which they were accustomed (Fig. 4-3). The internal focus of each Spanish town was the central *plaza* or market square, around which both the local church and government buildings were located. The surrounding street pattern was deliberately laid out in gridiron form, so that any insurrections by the resettled Amerindians could be contained by having a small military force seal off the affected blocks and then root out the troublemakers. Each town was located near what was thought to be good agricultural land (which was often not so good), so that the

MAINLAND AND RIMLAND

In Middle America outside Mexico, only Panama, with its twin attractions of interoceanic transit and gold deposits, became an early focus of Spanish activity. From there, following the Pacific side of the isthmus, Spanish influence radiated northwestward through Central America and into Mexico. The major arena of international competition in Middle America, however, lay not on the Pacific side but on the islands and coasts of the Caribbean Sea. Here the British gained a foothold on the mainland, controlling a narrow coastal strip that extended southeast from Yucatán to what is now Costa Rica. As the colonial-era map (Fig. 4-5) shows, in the Caribbean the Spaniards faced not only the British but also the French and Dutch, all interested in the lucrative sugar trade, all searching for instant wealth, and all seeking to expand their empires.

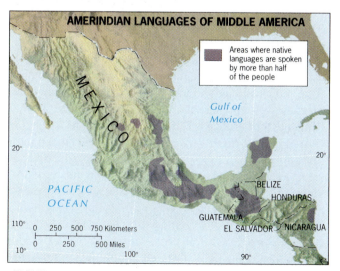

FIGURE 4-4 © H. J. de Blij, P. O. Muller, and John Wiley & Sons, Inc.

FIGURE 4-5 © H. J. de Blij, P. O. Muller, and John Wiley & Sons, Inc.

Much later, after centuries of European colonial rivalry in the Caribbean Basin, the United States entered the picture and made its influence felt in the coastal areas of the mainland, not through colonial conquest but through the introduction of widespread, large-scale, banana plantation agriculture. The effects of these plantations were as far-reaching as the impact of colonialism on the Caribbean islands. Because the diseases the Europeans had introduced were most rampant in these hot, humid lowlands (as well as the Caribbean islands to the east), the Amerindian population that survived was too small to provide a sufficient workforce. This labor shortage was remedied through the trans-Atlantic slave trade from Africa that transformed the Caribbean Basin's demography (see map in Chapter 6, p. 304).

Lingering Regional Contrasts

These contrasts between the Middle American highlands on the one hand and the coastal areas and Caribbean islands on the other were conceptualized by John Augelli into the **Mainland-Rimland framework** (Fig. 4-6). Augelli recognized (1) a Euro-Amerindian **Mainland**, which consisted of continental Middle America from Mexico to Panama, excluding the Caribbean coastal belt from mid-Yucatán southeastward; and (2) a Euro-African **Rimland**, which included this coastal zone as well as the islands of the Caribbean. The terms *Euro-Amerindian* and *Euro-African* underscore the cultural heritage of each region. On the Mainland, European (Spanish) and Amerindian influences are paramount and also include **mestizo** sectors where the two ancestries mixed. In the Rimland, the heritage is European and African.

As Figure 4-6 shows, the Mainland is subdivided into several areas based on the strength of the Amerindian legacy. The Rimland is also subdivided, with the most obvious division the one between the mainland-coastal plantation zone and the islands. Note, too, that the islands themselves are classified according to their cultural heritage (Figs. 4-5, 4-6).

Supplementing these contrasts are regional differences in outlook and orientation. The Rimland was an area of sugar and banana plantations, of high accessibility, of sea-

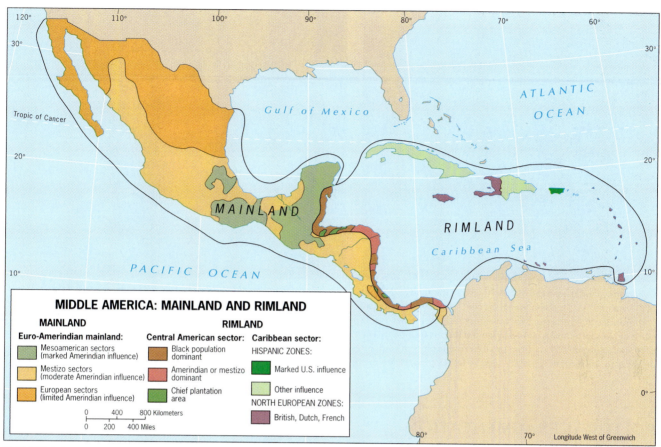

FIGURE 4-6 © H. J. de Blij, P. O. Muller, and John Wiley & Sons, Inc.

ward exposure, and of maximum cultural contact and mixture. The Mainland, being farther removed from these contacts, was an area of greater isolation. The Rimland was the region of the great *plantation*, and its commercial economy was therefore susceptible to fluctuating world markets and tied to overseas investment capital. The Mainland was the region of the *hacienda*, which was more self-sufficient and less dependent on external markets.

The Hacienda

This contrast between plantation and hacienda land tenure in itself constitutes strong evidence for the Rimland-Mainland division. The hacienda was a Spanish institution, but the modern plantation, Augelli argued, was the concept of Europeans of more northerly origin. In the **6 hacienda**, Spanish landowners possessed a domain whose productivity they might never push to its limits: the very possession of such a vast estate brought with it social prestige and a comfortable lifestyle. Native workers lived on the land—which may once have been their land—and had plots where they could grow their own subsistence crops. All this is written as though it is mostly in the past, but

the legacy of the hacienda system, with its inefficient use of land and labor, still exists throughout mainland Middle America (see photos p. 216).

The Plantation

The **plantation** was conceived as something entirely different from the hacienda. Robert West and John Augelli list five characteristics of Middle American plantations that illustrate the differences between hacienda and plantation: (1) plantations are located in the humid tropical coastal lowlands; (2) plantations produce for export almost exclusively—usually a single crop; (3) capital and skills are often imported so that foreign ownership and an outflow of profits occur; (4) labor is seasonal—needed in large numbers mainly during the harvest period—and such labor has been imported because of the scarcity of Amerindian workers; and (5) with its "factory-in-the-field" operation, the plantation is more efficient in its use of land and labor than the hacienda. The objective was not self-sufficiency but profit, and wealth rather than social prestige is a dominant motive for the plantation's establishment and operation.

Contrasting land uses in the Middle American Rimland and Mainland give rise to some very different rural cultural landscapes. Huge stretches of the realm's best land continue to be controlled by (often absentee) landowners whose haciendas yield export or luxury crops, or foreign corporations that raise fruits for transport and sale on their home markets. The banana plantation shown here (*left*) lies near the Caribbean coast of Honduras and is owned by United Brands Company. The vast fields of banana plants stand in strong contrast to the lone peasant who ekes out a bare subsistence from small cultivable plots of land, often in high-relief countryside where grazing some goats or other livestock is the only way to use most of the land.   *Left:* © J. P. Courau/D. Donne Bryant Stock Photography; *Right:* © Jean-Gerard Sidaner/Photo Researchers.

POLITICAL FRAGMENTATION

Today continental Middle America is fragmented into eight countries, all but one of which have Hispanic origins. The largest of them all is Mexico—the giant of Middle America—whose 1,960,000 square kilometers (756,000 sq mi) constitute more than 70 percent of the realm's entire land area (the Caribbean region included) and whose 112 million people outnumber those of all the other countries and islands of Middle America combined.

The cultural variety in Caribbean Middle America is much greater. Here Hispanic-influenced Cuba dominates: its area is almost as large as that of all the other islands put together, and its population of 11.4 million is well ahead of the next-ranking country—the Dominican Republic (9.3 million), also of Spanish heritage. As we noted, however, the Caribbean is hardly an arena of exclusive Hispanic cultural heritage: for example, Cuba's southern neighbor, Jamaica (population 2.8 million, mostly of African ancestry), has a legacy of British involvement, while to the east in Haiti (8.9 million, overwhelmingly of African ancestry) the strongest imprints have been African and French. The Lesser Antilles also exhibit great cultural diversity. There are the (once Danish) U.S. Virgin Islands; French Guadeloupe and Martinique; a group of British-influenced islands, including Barbados, St. Lucia, and Trinidad and Tobago; and the Dutch St. Maarten (shared with the French) and A-B-C islands of the Netherlands Antilles—Aruba,

Bonaire, and Curaçao—off the northwestern Venezuelan coast.

Independence

Independence movements stirred Middle America at an early stage. On the mainland, revolts against Spanish authority (beginning in 1810) achieved independence for Mexico by 1821 and for the Central American republics by

POINTS TO PONDER

- Existing and projected border fences between the United States and Mexico are sparking anger on the Mexican side, with public opinion toward America plummeting.

- The Panama Canal is slated for a major renovation. Watch for China to play a major role in the process.

- Puerto Rico is part of the United States, but nearly 50 percent of all Puerto Ricans live below the federal poverty line.

- The Netherlands Antilles were dissolved as a Caribbean political entity in 2007, its five islands accorded new status as parts of the Kingdom of the Netherlands.

the end of the 1820s. The United States, concerned over European designs in the realm, proclaimed the Monroe Doctrine in 1823 to deter any European power from re-asserting its authority in the newly independent republics or from further expanding its existing domains. By the end of the nineteenth century, the United States itself had become a major force in Middle America. The Spanish-American War of 1898 made Cuba independent and put Puerto Rico under the U.S. flag; soon afterward, the Americans were in Panama constructing the Panama Canal. Meanwhile, with U.S. corporations driving a boom based on huge banana plantations, the Central American republics had become colonies of the United States in all but name.

Independence came to the Caribbean Basin in fits and starts. Afro-Caribbean Jamaica as well as Trinidad and Tobago, where the British had brought a large South Asian population, attained full sovereignty from the United Kingdom in 1962; other British islands (among them Barbados, St. Vincent, and Dominica) became independent later. France, however, retains Martinique and Guadeloupe as overseas *départements* of the French Republic, and the Dutch islands are at various stages of autonomy.

Regions of the Realm

Middle America consists of four geographic regions: (1) Mexico, the giant of the realm in every respect; (2) Central America, the string of seven small republics occupying the land bridge to South America; (3) the four large islands that constitute the Greater Antilles of the Caribbean; and (4) the numerous islands of the Caribbean's Lesser Antilles (Fig. 4-2).

● MEXICO

By virtue of its physical size, large population, cultural identity, resource base, economy, and relative location, the state of Mexico by itself constitutes a geographic region in this multifaceted realm. The U.S.-Mexican boundary on the map crosses the continent from the Pacific to the Gulf, but Mexican cultural influences penetrate deeply into the southwestern States and American impacts reach far into Mexico. To Mexicans the border is a reminder of territory lost to the United States in historic conflicts; to Americans it is a symbol of economic contrasts and illegal immigration. Along the Mexican side of this border the effects of NAFTA (the North American Free Trade Agreement among Canada, the United States, and Mexico [see p. 184]) are transforming Mexico's economic geography.

Physiography

The physiography of Mexico is reminiscent of that of the western United States, although its environments are more tropical. Figure 4-7 shows several prominent features: the elongated Baja (Lower) California Peninsula in the northwest, the far eastern Yucatán Peninsula, and the Isthmus of Tehuantepec in the southeast where the Mexican landmass tapers to its narrowest extent. Here in the southeast, Mexico most resembles Central America physiographically; a mountain backbone forms the isthmus, curves southeast into Guatemala, and extends northwest toward Mexico City. Shortly before reaching the capital, this mountain range divides into two chains, the Sierra Madre Occidental in the west and the Sierra Madre Oriental in the east (Figs. 4-1, 4-7). These diverging ranges frame the funnel-shaped Mexican heartland, the center of which consists of the rugged, extensive Plateau of Mexico (the Valley of Mexico lies near its southeastern end). As Figure G-8 reveals, Mexico's climates are marked by dryness, particularly in the broad, mountain-flanked north. Most of the better-watered

MAJOR CITIES OF THE REALM	
City	Population* (in millions)
Ciudad Juárez, Mexico	1.8
Guadalajara, Mexico	4.1
Guatemala City, Guatemala	1.1
Havana, Cuba	2.2
Managua, Nicaragua	1.3
Mexico City, Mexico	28.2
Monterrey, Mexico	3.8
Panama City, Panama	1.3
Port-au-Prince, Haiti	2.4
Puebla, Mexico	1.8
San José, Costa Rica	1.3
San Juan, Puerto Rico	2.7
San Salvador, El Salvador	1.6
Santo Domingo, Dominican Rep.	2.2
Tegucigalpa, Honduras	1.0
Tijuana, Mexico	1.9

*Based on 2008 estimates.

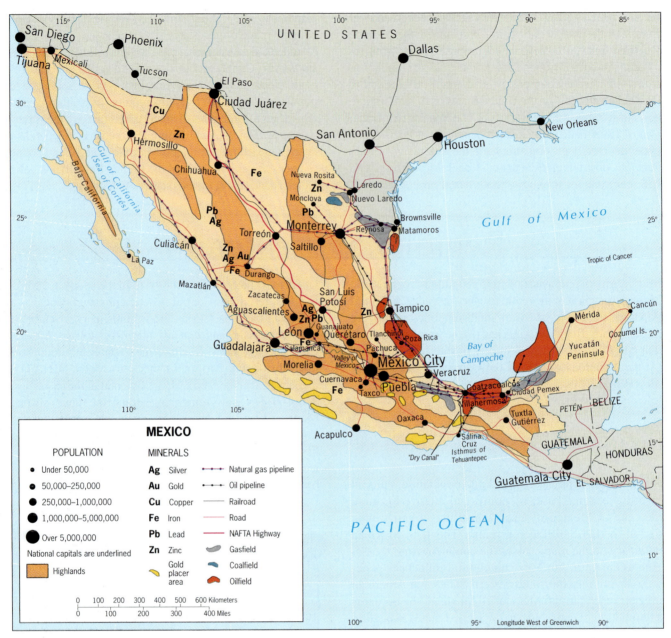

FIGURE 4-7   © H. J. de Blij, P. O. Muller, and John Wiley & Sons, Inc.

areas lie in the southern half of the country where the major population concentrations have developed.

Population Patterns

Mexico's population grew rapidly during the last three decades of the twentieth century, doubling in just 28 years; but demographers have recently noted a sharp drop in fertility, and they are predicting that Mexico's population (currently 112 million) will stop growing altogether by about 2050. This will have enormous implications for the country's economy, and it will reduce the cross-border migration that is of so much concern at present.

Where the People Live

The distribution of population across Mexico's 31 internal States is shown in Figure 4-8. The largest concentration, containing more than half the Mexican people, extends across the densely populated "waist" of the country from Veracruz State on the eastern Gulf coast to Jalisco State on the Pacific. The center of this corridor is dominated by the most populous State, Mexico (**3** on the map), at whose heart lies the Federal District of Mexico City (**9**). In the dry and rugged terrain to the north of this central corridor lie Mexico's least-populated States. Southern Mexico also exhibits a sparsely peopled periphery in the hot and humid lowlands of the Yucatán Peninsula, but

FIGURE 4-8   © H. J. de Blij, P. O. Muller, and John Wiley & Sons, Inc.

here most of the highlands of the continental spine contain sizeable populations.

Pull and Push

Another major feature of Mexico's population map is urbanization, driven by the ***pull*** of the cities (with their perceived opportunities for upward mobility) in tandem with the ***push*** of the economically stagnant countryside. Today, 75 percent of the Mexican people reside in towns and cities, a surprisingly high proportion for a developing country. Undoubtedly, these numbers are affected by the explosive recent growth of Mexico City, which now totals 28 million (making it the largest urban concentration on Earth) and is home to 25 percent of the national population. Among the other leading cities are Guadalajara, Puebla, and León in the central population corridor, and Monterrey, Ciudad Juarez, and Tijuana in the northern U.S. border zone (Fig. 4-8). Urbanization rates at the other end of Mexico, however, are at their lowest in those remote uplands where Amerindian society has been least touched by modernization.

A Mix of Cultures

Nationally, the Amerindian imprint on Mexican culture remains strong. Today, 60 percent of all Mexicans are mestizos, 22 percent are predominantly Amerindian, and about 8 percent are full-blooded Amerindians; most of the remaining 10 percent are Europeans. Certainly the Mexican Amerindian has been Europeanized, but the Amerindianization of modern Mexican society is so powerful that it would be inappropriate here to speak of one-way, European-dominated **acculturation**. Instead, what took place **8** in Mexico is **transculturation**—the two-way exchange of **9** culture traits between societies in close contact. In the southeastern periphery (Fig. 4-4), several hundred thousand Mexicans still speak only an Amerindian language, and millions more still use these languages daily even though they also speak Mexican Spanish. The latter has been strongly shaped by Amerindian influences, as have Mexican modes of dress, foods and cuisine, artistic and architectural styles, and folkways. This fusion of heritages, which makes Mexico unique, is the product of an upheaval that began to reshape the country nearly a century ago.

AMONG THE REALM'S GREAT CITIES . . . Mexico City

MIDDLE AMERICA HAS only one great metropolis: Mexico City. With 28.2 million inhabitants, Mexico City is home to just over one-fourth of Mexico's population and grows by about 300,000 each year. Even more significantly, Mexico City has surpassed Tokyo, Japan (current population: 26.7 million) to become the world's largest urban agglomeration.

Lakes and canals marked this site when the Aztecs built their city of Tenochtitlán here seven centuries ago. The conquering Spaniards made it their headquarters, and following independence the Mexicans made it their capital. Centrally positioned and well connected to the rest of the country, Mexico City, hub of the national core area, became the quintessential primate city.

Vivid social contrasts mark the cityscape. Historic plazas, magnificent palaces, churches, villas, superb museums, ultramodern skyscrapers, and luxury shops fill the city center. Beyond lies a zone of comfortable middle-class and struggling, but stable, working-class neighborhoods. Outside this belt, however, lies a ring of more than 500 slums and countless, even poorer *ciudades perdidas*—the "lost cities" where newly arrived peasants live in miserable poverty and squalor. (These squatter settlements contain no less than one-third of the metropolitan area's population.) Mexico City's more affluent residents have also been plagued by problems in recent years as the country's social and political order came close to unraveling. Rampant crime remains a serious concern, much of it associated with corrupt police as control of the country's long-time ruling party (the PRI) weakened and collapsed.

Environmental crises parallel the social problems. Local surface waters have long since dried up, and groundwater supplies are approaching depletion; to meet demand, the metropolis must now import a third of its water by pipeline from across the mountains (with about 40 percent lost through leakages in the city's crumbling waterpipe network). Air pollution here is probably the world's worst as more than 4 million motor vehicles and 40,000 factories churn out smog that in Mexico City's

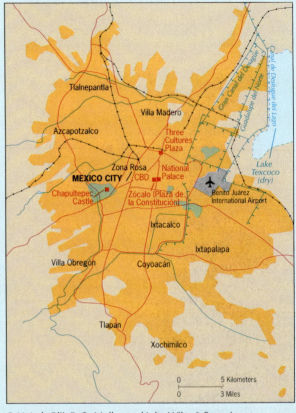

© H. J. de Blij, P. O. Muller, and John Wiley & Sons, Inc.

thin, high-altitude air sometimes reaches 100 times the acceptably safe level. And add to all this a set of geologic hazards: severe land subsidence as underground water supplies are overdrawn; the ever-present threat of earthquakes that can wreak havoc on the city's unstable surface (the last big one occurred in 1985); and the risk of volcanic activity as nearby Mount Popocatépetl may be ending centuries of dormancy.

In spite of it all, the great city continues to beckon, and every year more than 100,000 of the desperate and the dislocated arrive with hope—and little else.

Revolution and Its Aftermath

Modern Mexico was forged in a revolution that began in 1910 and set into motion events that are still unfolding today. At its heart, this revolution was about the redistribution of land, an issue that had not been resolved after Mexico freed itself from Spanish colonial control in the early nineteenth century. As late as 1900, more than 8000 haciendas blanketed virtually all of Mexico's good farmland, and about 95 percent of all rural families owned no land whatsoever and toiled as *peones* (landless, constantly indebted serfs) on the haciendas. The triumphant revolution produced a new constitution in 1917 that launched a program of expropriation and parceling out of the haciendas to rural communities.

Land Reform

Since 1917, more than half the cultivated land of Mexico has been redistributed, mostly to peasant communities consisting of 20 families or more. On such farmlands (known as *ejidos*) the government holds title to the land, and use rights are parceled out to villages and then individuals for cultivation. Most of these *ejidos* lie in central and southern Mexico, where Amerindian agricultural traditions survived, but the newly created landholdings were too finely fragmented to provide a real development opportunity. Despite an understandable temporary decline in agricultural productivity during the transition, the miracle is that land reform was carried off without major dislocation. Although considerable poverty persisted in the countryside, it was recognized that Mexico, alone among Middle America's countries with large Amerindian populations, had made significant strides toward solving the land question. Just how much farther Mexico has to go, however, became evident during the 1990s when the issue resurfaced at the heart of a rebellion that broke out in southeasternmost Chiapas State.

The Problem of Chiapas

Chiapas, the poorest of Mexico's 31 States, borders Guatemala to the east and has much in common with its Central American neighbor. Amerindian (Maya) peoples such as the Tzeltal and the Tzotzil eke out a precarious subsistence on small plots of land in the rainforest that covers the hills; the better valley soils have for centuries been held by large landowners whose estates remained unaffected by the postrevolutionary land reform that transformed so much of Mexico elsewhere. The Mexican government in the 1980s pledged to do something about this, but nothing much happened, and so in January 1994 a group of Maya peasant farmers calling themselves the Zapatista National Liberation Army (ZNLA) launched a guerrilla war by attacking several Chiapas towns. By taking the name of a legendary leader of the 1910 Revolution (Emiliano Zapata), and by timing their insurgency to coincide with the birth of NAFTA, the ZNLA achieved maximum impact and publicity.

Indeed, the ZNLA had considerable public support in Mexico, where the public was impatient with a long-dominant, high-handed ruling party, where doubts about NAFTA were widespread, and where there was considerable sympathy not only for the Chiapas rebels, but also for the plight of Mexico's other Amerindian peoples. Mexico has at least 20 widely dispersed Amerindian nations, ranging from the Nahuatl (1.2 million) in the country's center to the Yucatán Maya (700,000+) in the east and from the Zapotec of Oaxaca (400,000+), the southern State bordering Chiapas, to the Tarahumara (ca. 50,000) in the northern State of Chihuahua. These peoples, who choose to preserve pre-Hispanic cultural traditions, often suffer economically. But their plight is no longer ignored. They constitute more than 9 million of Mexico's 112 million people, too large a constituency to neglect.

Regions of Mexico

Physiographic, demographic, economic, historical, and cultural criteria combine to reveal a regionally diverse Mexico extending from the lengthy ridge of Baja California to the tropical lowlands of the Yucatán Peninsula, and from the economic frenzy of the NAFTA North to the Amerindian traditionalism of the Chiapas southeast (Fig. 4-9). In the Core Area, anchored by Mexico City, and the West, centered on Guadalajara, lies the transition zone from the more Hispanic-mestizo north to the more Amerindian-infused mestizo south. East of the Core Area lies the Gulf Coast, once dominated by major irrigation projects and huge livestock-raising schemes but now the mainland center of Mexico's petroleum industry. The dry, scrub-vegetated Balsas Lowland separates the Core Area from the rugged, Pacific-fronting Southern Highlands, where Acapulco's luxury hotels stand in stark contrast to the Amerindian villages and *ejidos* of the interior, scene of major land reform in the wake of the 1910 revolution.

The dry, vast north stands in sharp contrast to these southern regions: only in the Northwest was there significant sedentary Amerindian settlement when the Spanish arrived. Huge haciendas, major irrigation projects, and some large cities separated by great distances mark this area, now galvanized by the impact of NAFTA. The NAFTA North region is still formative and discontinuous but is changing northern Mexico significantly. This is true even in Yucatán, where Mérida and environs are strongly affected by NAFTA development. To go from comparatively well-off northern Yucatán to poverty-stricken southern Chiapas is to see the whole range of Mexico's regional geography.

The Changing Geography of Economic Activity

During the last two decades of the twentieth century and in the first years of the twenty-first, Mexico's economic geography has changed, and in some respects progressed—though not without setbacks. During the early 1990s, the implementation of NAFTA led to an economic boom as Mexico became part of a free-trade zone and market comprising over 400 million people. This boom transformed urban landscapes along the 3115-kilometer (1936-mi) border between Mexico and the United States, but it could not, of course, close the economic gap between the two sides (see photo p. 222).

Today, Mexico is in progressive transition in many spheres: its democratic institutions are strengthening; its economy is more robust than it was at the time of NAFTA's inception; its once-rapid population growth is declining; its social fabric (notably relations with Amerindian minorities)

FIGURE 4-9 © Robert C. West & John P. Augelli, *Middle America: Its Lands and Peoples*, 3rd edition, p. 340, © 1989. Adapted by permission of Pearson Education, Inc., Upper Saddle River, N.J.

The border between the United States and Mexico occasionally provides some stunning contrasts. Here, looking westward, we observe the opposing economic geographies that mark the border landscape of the Imperial Valley east of the urban area formed by Mexicali, Mexico and Calexico, California. The larger town of Mexicali (on the left) sprawls eastward along the border, with a cluster of maquiladora assembly plants, visible in the left foreground, surrounded by high-density, poor-quality housing. On the U.S. side, lush irrigated croplands blanket the otherwise dry countryside, fed by the All-American Canal that taps the waters of the Colorado River before they cross into Mexico. At this point, the canal swings northward to bypass border-hugging Calexico (right rear) before rejoining the international boundary beyond the town's western edge.   © Alex McLean/Landslides.

is improving. The familiar problems of a country in transition, such as unchecked urbanization, inadequate infrastructure, corruption, and violent crime, will continue to afflict Mexico for decades to come. But, as the following discussion confirms, Mexico is a far stronger economy and society today than it was just one generation ago.

Agriculture

Although traditional subsistence agriculture and the output of the inefficient *ejidos* have not changed a great deal in the poorer areas of rural Mexico, larger-scale commercial agriculture has diversified during the past three decades and made major gains with respect to both domestic and export markets. The country's arid northern tier has led the way as major irrigation projects have been built on streams flowing down from the interior highlands. Along the booming northwest coast of the mainland, which lies within a day's drive of Southern California, mechanized large-scale cotton production now supplies an increasingly profitable export trade. Here, too, wheat and winter vegetables are grown, with fruit and vegetable cultivation attracting foreign investors.

But many Mexican small farmers are having a tough time of it, for two reasons. First, cheap corn from the United States, grown by American farmers heavily subsidized by the government, floods Mexico's markets, so that local farmers cannot make a profit. Second, the United States tried to create barriers against the import of low-priced Mexican produce such as avocados and tomatoes—violating the very rules NAFTA is supposed to stand for.

Energy and Industry

Until after 1990, Mexican industry was typical of a developing economy with state control over income-generating natural resources, among which oil and natural gas were (and are) the most productive. The government set up industries (such as the country's first steel plant, opened in Monterrey in 1903) without allowing competition. Employment in agriculture far exceeded that in manufacturing or mining, despite Mexico's substantial reserves of silver and other metals, high-quality coal (especially in Coahuila State), and an additional inventory ranging from antimony to zinc. Inefficiency prevailed, wages were low, inflation was rife, and Mexico's public debt (see p. 30) was one of the highest in the world. The Mexican economy suffered from ups and downs directly related to world oil prices; when the price of oil dropped, federal expenditures had to be cut and deficits soared. In 2007, oil still accounted for more than one-third of total government revenues, a dangerous situation in a volatile world (the distribution of Mexican oil and gas reserves is mapped in Fig. 4-7).

During the 1980s, successive Mexican governments began trying to address these problems as part of the preparations for the momentous inception of NAFTA. Some of the inefficient mining and manufacturing operations were sold to private investors, and Mexico's financial systems (banking, tax collection) were strengthened. But with the inception of NAFTA in 1994, Mexico's industrial geography changed dramatically. Under the new agreement, factories based in Mexico could assemble imported, duty-free raw materials and components into finished products, which were then exported back to the United States market. Logically, these factories, called **maquiladoras**, would locate as close to the U.S. border **10** as possible. As a result, manufacturing employment in the cities and towns along that border, from Tijuana in the west to Matamoros in the east, expanded rapidly. After only seven years of NAFTA's existence, there were 4000 factories with more than 1.2 million workers in the border zone and in northern Yucatán (where Mérida was part of the process), accounting for nearly one-third of Mexico's industrial jobs and 45 percent of its total exports.

NAFTA's impact on Mexico was far-reaching. Mexico benefited from foreign investment, job creation, tax receipts, and technology transfer. But Mexican employees worked long hours for low wages with few benefits and lived in the most basic shacks and slum dwellings encircling the burgeoning towns of the region we call the "NAFTA North" (Fig. 4-9). And there was no job security. Ten years after NAFTA's start, hundreds of American and other foreign corporations that had moved their factories to northern Mexico decided to relocate once again—to East and Southeast Asia where wages were even lower than the U.S. $2.00 per hour average paid by the maquiladoras. While factories assembling heavy and bulky items such as vehicles and refrigerators were still better off right across the border, others producing lighter and smaller goods such as electronic equipment and cameras moved to China, Vietnam, and other countries where wages were less than half of Mexico's. As a result, thousands of Mexican workers found themselves unemployed—and many crossed the border into the United States.

How can Mexico counter this trend? Here is one indicator: while maquiladora jobs were lost to Asia in textiles, jobs were being added in electronics, and Mexican company managers were in short supply. Education in high-tech and management fields is part of the answer. In the Northeast, the city of Monterrey in the high-income State of Nuevo Léon has a substantial international business community and modern industrial facilities that have attracted major multinational companies. The *Tecnológico* campus near Monterrey's airport, a degree-granting college supported by both government and private enterprise, lies at the heart of a network of more than 30 campuses throughout Mexico offering degrees in information technology, engineering, and administration. In the West, near Guadalajara, in the still-less-prosperous State of Jalisco, you can discern the beginnings of a "Silicon Valley of Mexico." In such places, the outlines of the landscape of the global economy are evident. Here lies hope for Mexico's future.

Geography of Inequality

The statistics of economic geography often hide the implications of the data for the people they purportedly represent. There is no doubt about it: NAFTA has enriched Mexico in general and northern Mexico in particular, and almost all the northern States bordering the United States have per-capita incomes above the national average. In fact, northern Mexico has historically been better off than the south—and NAFTA has widened the gap. We tend to think of NAFTA in terms of the industries it represents, but NAFTA also had an impact on Mexico's farmers. One stipulation of the agreement was that Mexico would lower and then drop its tariff barriers against U.S. and Canadian farm produce, including Mexico's staple, corn. But the United States generously subsidizes its corn farmers, who as a result can market corn in Mexico for prices even lower than Mexican farmers have to charge in order to make a small profit. In the process, hundreds of thousands of Mexican farmers are being put out of business. No wonder the U.S. government finds itself having to build fences to keep illegal immigrants out.

Mexico's north-south divide is starkly evident from the economic data (Fig. 4-10). In very general terms, the annual per-capita income in the northern States exceeds U.S.

$10,000; in the southern States it falls below $5000. Economic growth in the northern States has averaged more than 4 percent in recent years; in southern States it is less than 2 percent. Far more people are poor in the south than in the north. Mexico's infrastructure, already inadequate, serves the south far less well than the north. Whatever the index—literacy, electricity use, water availability—the south lags.

These still-widening contrasts were thrown into sharp relief in 2006, when Mexico's presidential election was contested by three candidates, of whom the two leading ones were a conservative and a populist. When the ballots were counted, there was a near-tie, a result so close that the populist at first refused to concede. But on the map, it was clear that the north was won by the conservative candidate and the south by the populist (Fig. 4-10, inset map). Mexico's electoral map thus showed that the country's regional economic disparities were having serious social consequences (see photo below).

Mexico's Future

Mexico has accomplished much since the mid-1990s: it has strengthened its democratic institutions, brought greater stability to the economy, absorbed the impacts of NAFTA, and improved relations with its Amerindian minorities. Mexico's rate of population growth continues to decline, and while millions of Mexicans have emigrated to the United States since NAFTA's inception, fraying social fabrics, even this has a positive side: remittances from these workers in recent years are estimated to amount to between U.S. $25 and $30 billion, which is much greater than "official" foreign investment.

Mexico's 2006 presidential election not only divided the country geographically (see Fig. 4-10) but produced a result (between the two front-runners) so close that both sides claimed victory. For months, supporters of the eventual loser blockaded parts of the capital, and the leftist candidate himself established a "provisional" government even as recounts showed that the conservative-party candidate, Felipe Calderón, had narrowly won. Even after the matter was settled, disorder continued in polarized Mexico, notably in Oaxaca City, capital of Oaxaca State (the map suggests why this was a particularly tense part of the country). Scenes like this one, with federal police lobbing tear gas canisters to disperse a crowd involved in a clash up the street, were common. © AP/Wide World Photos.

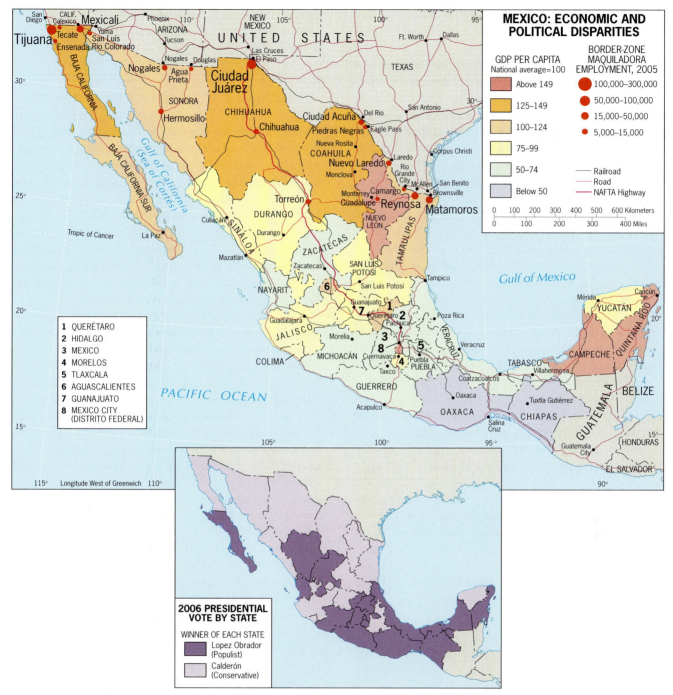

FIGURE 4-10   © H. J. de Blij, P. O. Muller, and John Wiley & Sons, Inc.

But, as noted above, the impact of globalization has fractured Mexico's economic and political geography, reflecting success in the north and failure in the south and deepening historic regional disparities. Northern Mexico is becoming more like North America; southern Mexico resembles Central America. The future of the country depends on the government's efforts to close this gap and on its plans to spread the positive effects of NAFTA from north to south. One such effort, a federal antipoverty pro-

gram, requires families to keep their children in school, and it is having good effect. Another, the improvement of infrastructure to reduce the isolation of southern communities, is less successful. This cannot be done by government alone, but Mexico's history of reliance on government (inefficient and corrupt as it has long been) is not easily overcome. Take a look at Figure 4-1: the dominance of Mexico City, the seat of power, is reflected by the national transport system. All roads (such as they are) seem to go to and through Mexico City, creating a costly bottleneck that disadvantages the movement of southern products on their way to northern markets.

Mexico's future is inextricably bound up with the United States; when the American economy stalls, Mexico's suffers. NAFTA's brief history has not yet witnessed a serious economic recession in the United States, but the mild ones did serve as a warning. Meanwhile, Mexico's proximity to the American market has other consequences. The United States is the world's richest single market for illegal narcotics, and drug cartels that once operated from Colombia are now based in the maquiladora cities, contributing to the rising crime wave that afflicts Mexico today.

Mexico is a country in tumultuous transition, a land of many opportunities in need of modernization and regional integration. Its tourist industry—from the great Maya cities of the Yucatán Peninsula to the "Mexican Riviera" on the Pacific coast—could contribute far more to the economy than it presently does, another consequence of inadequate infrastructure. Its energy industries are burdened by insufficient investment in research and renovation. In the maquiladora industries, the drive should be toward higher-value manufacturing, reducing vulnerability to Asian competition. At times the Mexican government reverts to its old penchant for megaprojects, such as the notion of a so-called **dry canal** across the narrowest part of the country from the Pacific port of Salina Cruz in Oaxaca State to the Mexican Gulf port of Coatzacoalcos in Veracruz State (Fig. 4-7). This would be built to compete with the Panama Canal, but there are less spectacular and more effective ways to spend money—for example, on improving access roads, electricity supply, and irrigation systems for small farmers in the impoverished south. Mexico has made significant progress in the past two decades, but old problems continue to slow the pace.

● THE CENTRAL AMERICAN REPUBLICS

Crowded onto the narrow segment of the Middle American land bridge between Mexico and the South American continent are the seven countries of Central America (Fig. 4-11). Territorially, they are all quite small; their population sizes range from Guatemala's 13.7 million down to Belize's 315,000. Physiographically, the land bridge here consists of a highland belt flanked by coastal lowlands on both the Caribbean and Pacific sides (Fig. 4-1). These highlands are studded with volcanoes, and local areas of fertile volcanic soils are scattered throughout them. From earliest times, the region's inhabitants have been concentrated in this upland zone, where tropical temperatures are moderated by elevation and rainfall is sufficient to support a variety of crops.

Altitudinal Zonation of Environments

Continental Middle America and the western margin of South America are areas of high relief and strong environmental contrasts. Even though settlers have always favored temperate intermontane basins and valleys, people also cluster in hot tropical lowlands as well as high plateaus just below the snow line in South America's Andes Mountains. In each of these zones, distinct local climates, soils, vegetation, crops, domestic animals, and modes of life prevail. Such **altitudinal zones** (diagrammed in Fig. 4-12) are **12** known by specific names as if they were regions with distinguishing properties—as, in reality, they are.

The lowest of these vertical zones, from sea level to 750 meters (2500 ft), is known as the *tierra caliente*, the "hot **13** land" of the coastal plains and low-lying interior basins where tropical agriculture predominates. Above this zone lie the tropical highlands containing Middle and South America's largest population clusters, the *tierra templada* **14** of temperate land reaching up to about 1800 meters (6000 ft). Temperatures here are cooler; prominent among the commercial crops is coffee, while corn (maize) and wheat are the staple grains. Still higher, from about 1800 to 3600 meters (6000 to nearly 12,000 ft), is the *tierra fría*, the **15** cold country of the higher Andes where hardy crops such as potatoes and barley are mainstays. Above the tree line, which marks the upper limit of the *tierra fría*, lies the *tierra helada*; this fourth altitudinal zone, extending from **16** about 3600 to 4500 meters (12,000 to 15,000 ft), is so cold and barren that it can support only the grazing of sheep and other hardy livestock. The highest zone of all is the *tierra nevada*, a zone of permanent snow and ice associ- **17** ated with the loftiest Andean peaks. As we will see, the varied human geography of Middle and western South America is closely related to these diverse environments.

Population Patterns

Even at its comparatively small scale, Figure G-9 indicates that Central America's population tends to concentrate in the uplands of the *tierra templada* and that population densities are generally greater toward the Pacific than toward the Caribbean side. The most significant exception is El Salvador, whose political boundaries confine its people mostly to its tropical *tierra caliente*, tempered here by the somewhat cooler Pacific offshore. On the opposite side of the isthmus, Belize has the typically sparse population of

FIGURE 4-11   © H. J. de Blij, P. O. Muller, and John Wiley & Sons, Inc.

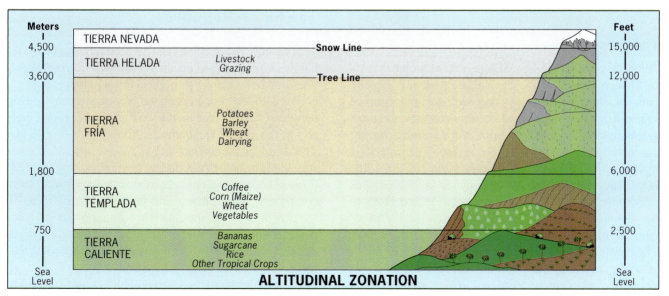

FIGURE 4-12   © H. J. de Blij, P. O. Muller, and John Wiley & Sons, Inc.

the hot, wet Caribbean coastlands and their infertile soils. Panama is the only other exception to the rule, but for different reasons. Economic development has focused on the Panama Canal and the coasts, although new settlement is now moving onto the mountain slopes of the Pacific side.

Central America, we noted earlier, actually begins within Mexico, in Chiapas and in Yucatán, and the region's republics face many of the same problems as less-developed parts of Mexico. Population pressure is one of these problems. A population explosion began in the mid-twentieth century, increasing the region's human inhabitants from 9 million to almost 43 million in 2008. Unlike Mexico, Central America's population growth is not yet slowing down significantly except in Costa Rica and Panama, and population geographers talk of a demographic catastrophe in the making.

Emergence from a Turbulent Era

Devastating inequities, repressive governments, external interference, and the frequent unleashing of armed forces have destabilized Central America for much of its modern history. The roots of these upheavals are old and deep, and today the region continues its struggle to emerge from a period of turmoil that lasted through the 1980s into the mid-1990s.

Central America is not a large region, but because of its physiography it contains many isolated, comparatively inaccessible locales. Conflicts between Amerindian population clusters and mestizo groups are endemic to the region, and contrasts between the privileged and the poor are especially harsh. Dictatorial rule by local elites followed authoritarian rule by Spanish colonizers.

The Gang Problem

Almost everywhere in Central America today (and in parts of Mexico as well) a growing social problem is causing concern: the rising tide of gang violence. International criminal gangs of young and unemployed men, the most infamous of which currently is the *Mara Salvatrucha*, are committing random acts of violence (such as an attack on civilians in San Pedro Sula, Honduras, in late 2004 that cost 28 lives) and are provoking paramilitary death squads into combating them, resulting in increased murder rates in the region's cities. In El Salvador, for example, the homicide rate rose from 37 per 100,000 in 2002 to 48 in 2006. The problem reflects not only chronic unemployment, but also migration (including the return of deported criminals), corruption in law enforcement, and living conditions in the poorest neighborhoods.

The Promise of CAFTA

Political leaders and economic planners in the United States have long advocated a "free trade" agreement to boost the economies of Central American countries. In 2005, this campaign resulted in the U.S. Congress's approval of the Central American Free Trade Agreement (CAFTA) linking the huge U.S. economy to five small Central American economies (Guatemala, El Salvador, Honduras, Nicaragua, and Costa Rica) and one Caribbean economy (the Dominican Republic). This is not the first time such an integrative effort has been made in Central America: the Central American Common Market, created in 1960, fell apart within a decade.

But today the region is more stable, the United States has stopped supporting dictators, democratic governments are holding their own, and trade between the republics and the United States, while still modest, is expanding. It is true, as critics of CAFTA argue, that many of its free-trade provisions are already in place, the result of earlier bilateral agreements, and that other tariffs will be reduced gradually over a period of two decades. But CAFTA will do more than enhance trade: it will signal to the world that the region has turned a corner. It will encourage investment, force governments to be more open and less corrupt, expand trade among these five Central American republics themselves, promote economic diversification and thus reduce excessive dependence on farm products. It will take a long time for Central America's vital statistics (see Table G-1) to improve significantly, and CAFTA will have some negative as well as many positive consequences. But overall, its advantages outweigh its liabilities.

The Seven Republics

Guatemala, the westernmost of Central America's republics, has more land neighbors than any other. Straight-line boundaries across the tropical forest mark much of the border with Mexico, creating the box-like region of Petén between Chiapas State on the west and Belize on the east; also to the east lie Honduras and El Salvador (Fig. 4-11). This heart of the ancient Maya Empire, which remains strongly infused by Amerindian culture and tradition, has just a small window on the Caribbean but a longer Pacific coastline. Guatemala was still part of Mexico when the Mexicans threw off the Spanish yoke, and though independent from Spain after 1821, it did not become a separate republic until 1838. Mestizos, not the Amerindian majority, secured the country's independence.

Most populous of the seven republics with 13.7 million inhabitants (mestizos are in the majority with 59 percent,

Amerindians 41 percent), Guatemala has seen much conflict. Repressive regimes made deals with U.S. and other foreign economic interests that stimulated development, but at a high social cost. Over the past half-century, military regimes have dominated political life. The deepening split between the wretchedly poor Amerindians and the better-off mestizos, who here call themselves *ladinos*, generated a civil war that started in 1960 and claimed more than 200,000 lives as well as 50,000 "disappearances" before it ended in 1996. An overwhelming number of the victims were of Mayan descent; the mestizos continue to control the government, army, and land-tenure system.

Guatemala's tragedy is that its economic geography has considerable potential but has long been shackled by the unending internal conflicts that have kept the income of 80 percent of the population below the poverty line. The country's mineral wealth includes nickel in the highlands and oil in the lower-lying north. Agriculturally, soils are fertile and moisture is ample over highland areas large enough to produce a wide range of crops including excellent coffee.

Belize, strictly speaking, is not a Central American republic in the same tradition as the other six. Until 1981, this country, a wedge of land between northern Guatemala, Mexico's Yucatán Peninsula, and the Caribbean, was a dependency of the United Kingdom known as British Honduras. Slightly larger than Massachusetts and with a minuscule population of only 315,000 (many of African descent), Belize has been more reminiscent of a Caribbean island than of a continental Middle American state. Today, all that is changing as the demographic complexion of Belize is being reshaped. Thousands of residents of African descent have recently emigrated (many went to the United States) and were replaced by tens of thousands of Spanish-speaking immigrants. Most of the latter are refugees from strife in nearby Guatemala, El Salvador, and Honduras, and their proportion of the Belizean population has risen from 33 to nearly 50 percent since 1980. Within the next few years the newcomers will be in the majority, Spanish will become the *lingua franca*, and Belize's cultural geography will exhibit an expansion of the Mainland at the expense of the Rimland.

The Belizean transformation extends to the economic sphere as well. No longer just an exporter of sugar and bananas, Belize is producing new commercial crops, and its seafood-processing and clothing industries have become major revenue earners. Also important is tourism, which annually lures more than 150,000 vacationers to the country's Mayan ruins, resorts, and newly legalized casinos; a growing speciality is ecotourism, based on the natural attractions of the country's near-pristine environment. Belize is also known as a center for *offshore banking*—a financial haven for foreign companies and individuals who want to avoid paying taxes in their home countries.

Honduras is a country on hold as it struggles to rebuild its battered infrastructure and economy. In 1998 Hurricane Mitch struck Honduras, and the consequences were catastrophic as massive floods and mudslides were unleashed across the country, killing 9200 people, demolishing more than 150,000 homes, destroying 21,000 miles of roadway and 335 bridges, and rendering 2 million homeless. Also devastated was the critical agricultural sector that employed two-thirds of Honduras's labor force, accounted for nearly a third of its gross domestic product, and earned more than 70 percent of its foreign revenues.

With 7.8 million inhabitants, about 90 percent mestizo, bedeviled Honduras still has years to go even to restore what was already the third-poorest economy in the Americas (after Haiti and Nicaragua). Agriculture, livestock, forestry, and limited mining formed the mainstays of the pre-1998 economy, with the familiar Central American products—bananas, coffee, shellfish, and apparel—earning most of the external income.

Honduras, in direct contrast to Guatemala, has a lengthy Caribbean coastline and a small window on the Pacific (Fig. 4-11). The country also occupies a critical place in the political geography of Central America, flanked as it is by Nicaragua, El Salvador, and Guatemala—all continuing to grapple with the aftermath of years of internal conflict and, most recently, natural disaster. The road back to economic viability is an arduous one, but once traversed will still leave four out of five Hondurans deeply mired in poverty and the country with little overall improvement in its development prospects.

El Salvador is Central America's smallest country territorially, smaller even than Belize, but with a population about 25 times as large (7.3 million) it is the most densely peopled. With Belize, it is one of only two continental republics that lack coastlines on both the Caribbean and Pacific sides (Fig. 4-11). El Salvador adjoins the Pacific in a narrow coastal plain backed by a chain of volcanic mountains, behind which lies the country's heartland. Unlike neighboring Guatemala, El Salvador has a quite homogeneous population (90 percent mestizo and just 1 percent Amerindian). Yet ethnic homogeneity has not translated into social or economic equality or even opportunity. Whereas other Central American countries were called banana republics, El Salvador was a coffee republic, and the coffee was produced on the huge landholdings of a few landowners and on the backs of a subjugated peasant labor force. The military supported this system and repeatedly suppressed violent and desperate peasant uprisings.

From 1980 to 1992, El Salvador was torn by a devastating civil war that was worsened by outside arms supplies from the United States (supporting the government) and Nicaragua (aiding the Marxist rebel forces). But ever since the negotiated end to that war, efforts have been under way to prevent a recurrence because El Salvador is having difficulty overcoming its legacy of searing inequality. The civil war did have one positive result: affluent citizens who left the country and did well in the United States and elsewhere send substantial funds back home, which now provide the largest single source of foreign revenues. This has helped stimulate such industries as apparel and footwear manufacturing, as well as food processing. But a major stumbling block to revitalization of the agricultural sector has again been land reform, and El Salvador's future still hangs in the balance.

Nicaragua is best approached by reexamining the map (Fig. 4-11), which underscores the country's pivotal position in the heart of Central America. The Pacific coast follows a southeasterly direction, but the Caribbean coast is oriented north-south so that Nicaragua forms a triangle of land with its lakeside capital, Managua, located in a valley on the mountainous, earthquake-prone, Pacific side (the country's core area has always been located here). The Caribbean side, where the uplands yield to a coastal plain of rainforest, savanna, and swampland, has for centuries been home to Amerindian peoples such as the Miskito, who have been remote from the focus of national life.

Until the end of the 1970s, Nicaragua was the typical Central American republic, ruled by a dictatorial government and exploited by a wealthy land-owning minority, its export agriculture dominated by huge plantations owned by foreign corporations. It was a situation ripe for insurgency, and in 1979 leftist rebels overthrew the government. But the new regime quickly produced its own excesses, resulting in civil war through most of the 1980s, a conflict in which the United States got involved by arming the rebel **Contras**, who were trying to undermine the left-wing Sandinista regime. This strife ended in 1990, and since then more democratic governments have been voted into office. In 2006, Nicaraguans elected a former Sandinista leader, Daniel Ortega, back into office, to the consternation of the American government.

Nicaragua's economy has been a leading casualty of this turmoil, and for the past two decades it has ranked as continental Middle America's poorest. Hurricane Mitch struck here too, devastating the country's farms and driving tens of thousands into the impoverished towns just at a time when the agricultural sector was recovering and land reform promised a better life for some 200,000 peasant families.

Nicaragua's options are limited: none of the billions of aid dollars headed for Bosnia, Iraq, and Subsaharan Africa will be matched here. For years there has been talk of a *dry canal* transit role for this country, but other potential overland routes are more promising. Meanwhile, Nicaragua's population explosion continues (2008 total: 5.9 million), dooming hopes for a rise in standards of living.

Costa Rica underscores what was said about Middle America's endless variety and diversity because it differs significantly from its neighbors and from the norms of Central America as well. Bordered by two volatile countries (Nicaragua to the north and Panama to the east), Costa Rica is a nation with an old democratic tradition and, in this cauldron, no standing army for the past half-century! Although the country's Hispanic imprint is similar to that found elsewhere on the Mainland, its early independence, its good fortune to lie remote from regional strife, and its leisurely pace of settlement allowed Costa Rica the luxury of concentrating on its economic development. Perhaps most important, internal political stability has prevailed over much of the past 175 years.

Like its neighbors, Costa Rica is divided into environmental zones that parallel the coasts. The most densely settled is the central highland zone, lying in the cooler *tierra templada*, whose heartland is the *Valle Central* (Central Valley), a fertile basin that contains the country's main coffee-growing area and the leading population cluster focused on San José (Fig. 4-11)—the most cosmopolitan urban center between Mexico City and the primate cities of northern South America. To the east of the highlands are the hot and rainy Caribbean lowlands, a sparsely populated segment of Rimland where many plantations have been abandoned and replaced by subsistence farming. Between 1930 and 1960, the U.S.-based United Fruit Company shifted most of the country's banana plantations from this crop-disease-ridden coastal plain to Costa Rica's third zone—the plains and gentle slopes of the Pacific coastlands. This move gave the Pacific zone a major boost in economic growth, and it is now an area of diversifying and expanding commercial agriculture.

The long-term development of Costa Rica's economy has given it the region's highest standard of living, literacy rate, and life expectancy. Agriculture continues to dominate (with bananas, coffee, tropical fruits, and seafood the leading exports), and tourism is expanding steadily. Costa Rica is widely known for its superb scenery and for its efforts to protect what is left of its diverse tropical flora and fauna. But rainforest destruction is the regionwide price of human population growth and woodland exploitation, and even here more than 80 percent of the original forest has vanished (see box titled "Tropical Deforestation").

Still, the country's veneer of development cannot mask serious problems. In terms of social structure, about one-quarter of its population of 4.4 million is trapped in an unending cycle of poverty, and the huge gap between the poor and the affluent is constantly widening. With volatile neighbors and an economy that remains insufficiently diversified against risk, Costa Rica is only one step ahead of its regional partners.

Tropical Deforestation

18

BEFORE THE EUROPEANS arrived, two-thirds of continental Middle America was covered by tropical rainforests. The clearing and destruction of this precious woodland resource to make way for expanding settlement frontiers and the exploitation of new economic opportunities began in the sixteenth century during the Spanish colonial era, and the practice has continued systematically ever since. In recent decades, however, the pace of **tropical deforestation** in Central America has accelerated alarmingly, and since 1950 almost 90 percent of the region's forests have been decimated. Today, about 1.2 million hectares (3 million acres) of Central American and Mexican woodland disappear each year, an area equivalent to one-third the size of Belgium. El Salvador has already lost 99 percent of its forests, and most of the six other republics will soon reach that stage.

The causes of tropical deforestation are related to the persistent economic and demographic problems of disadvantaged countries. In Central America, the leading cause has been the need to clear rural lands for cattle pasture as many countries, especially Costa Rica, became meat producers and exporters. The price of this environmental degradation has been enormous, although some gains have been recorded. Because tropical soils are so nutrient-poor, newly deforested areas can function as pastures for only a few years at most. These fields are then abandoned for other freshly cut lands and quickly become the ravaged landscape seen in the photo in this box. Without the protection of trees, local soil erosion and flooding immediately become problems, affecting still-productive nearby areas (a sequence of events that reached catastrophic dimensions all across Honduras and northern Nicaragua when Hurricane Mitch struck in 1998). A second cause of deforestation is the rapid logging of tropical woodlands as the timber industry increasingly turns from the exhausted forests of the mid-latitudes to harvest the rich tree resources of the equatorial zones, responding to accelerating global demands for housing, paper, and furniture. The third major contributing factor is related to the region's population explosion: as more and more peasants are required to extract a subsistence from inferior lands, they have no choice but to cut down the remaining forest for both firewood and additional crop-raising space, and their intrusion prevents the trees from regenerating (Haiti is the realm's extreme example of this denudation process).

Although deforestation is a depressing event, tropical pastoralists, farmers, and timber producers do not consider it to be life-threatening, and perhaps it even seems to offer some short-term economic advantages. Why, then, should there be such an outcry from the scientific community? And why should the World Resources Institute call this "the world's most pressing land-use problem"? The answer is that unless immediate large-scale action is taken, by the middle of this century the world's tropical rainforests will be reduced to two disappearing patches—the western Amazon Basin of northern South America and the middle Congo Basin of Equatorial Africa.

The tropical forest, therefore, must be a very important part of our natural world, and indeed it is. Biologically, the rainforest is by far the richest, most diversified arena of life on our planet: even though it covers only about (a shrinking) 3 percent of the Earth's land area, it contains about three-quarters of all plant and animal species. Its loss would cause not only the extinction of millions of species, but also what ecologist Norman Myers calls "the death of birth" because the evolutionary process that produces new species would be terminated. Because tropical rainforests already yield countless valuable medicinal, food, and industrial products, many potential disease-combatting drugs or new crop varieties to feed undernourished millions will be irretrievably lost if they are allowed to disappear.

This scene in Costa Rica shows how badly the land can be scarred in the wake of deforestation. Without roots to bind the soil, tropical rains swiftly erode the unprotected topsoil. © Peter Poulides.

Panama owes its existence to the idea of a canal connecting the Atlantic and Pacific oceans to avoid the lengthy circumnavigation of South America. In the 1880s, when Panama was still an extension of neighboring Colombia, a French company tried and failed to build such a waterway here. By the turn of the twentieth century, U.S. interest in a Panama canal rose sharply, and the United States in 1903 proposed a treaty that would permit a renewed effort at construction across Colombia's Panamanian isthmus. When the Colombian Senate refused to go along, Panamanians rebelled and the United States supported this uprising by preventing Colombian forces from intervening. The Panamanians, at the behest of the United States, declared their independence from Colombia, and the new republic immediately granted the United States rights to the Canal Zone, averaging about 15 kilometers (10 mi) in width and just over 80 kilometers (50 mi) in length.

Soon canal construction commenced, and this time the project succeeded as American technology and medical advances triumphed over a formidable set of obstacles. The Panama Canal (see inset map, Fig. 4-11) was opened in 1914, a symbol of U.S. power and influence in Middle America. The Canal Zone was held by the United States under a treaty that granted it "all the rights, powers, and authority" in the area "as if it were the sovereign of the territory." Such language might suggest that the United States held rights over the Canal Zone in perpetuity, but the treaty nowhere stated specifically that Panama permanently yielded its own sovereignty in that transit corridor. In the 1970s, as the canal was transferring more than 14,000 ships per year (that number is now only slightly lower, but the cargo tonnage is up significantly) and generating hundreds of millions of dollars in tolls, Panama sought to ter-

minate U.S. control in the Canal Zone. Delicate negotiations began. In 1977, an agreement was reached on a staged withdrawal by the United States from the territory, first from the Canal Zone and then from the Panama Canal itself (a process completed on December 31, 1999).

Panama today reflects some of the usual geographic features of the Central American republics. Its population of 3.4 million is nearly 70 percent mestizo and also contains substantial Amerindian, white, and black minorities. Spanish is the official language, but English is also widely used. Ribbon-like and oriented east-west, Panama's topography is mountainous and hilly. Eastern Panama, especially Darien Province adjoining Colombia, is densely forested, and here is the only remaining gap in the intercontinental Pan American Highway. Most of the rural population lives in the uplands west of the canal; there, Panama produces bananas, shrimps and other seafood, sugarcane, coffee, and rice. Much of the urban population is concentrated in the vicinity of the waterway, anchored by the cities at each end of the canal.

Near the northern end of the Panama Canal lies the city of Colón, site of the Colón Free Zone, a huge trading entrepôt designed to transfer and distribute goods bound for South America. It is augmented by the Manzanillo International Terminal, an ultramodern port facility capable of transshipping more than 1000 containers a day. By 2002 China had become the third-largest user of the Canal (after the United States and Japan) and accounted for more than 20 percent of the cargo entering the Colón Free Zone. Near the southern end lies Panama City, the "Miami of the Caribbean" because of its waterfront location and skyscrapered skyline (photo p. 206). The capital is the financial center that handles the funds generated by the Canal,

FROM THE FIELD NOTES

"The Panama Canal remains an engineering marvel 90 years after it opened in August 1914. The parallel lock chambers each are 1000 feet long and 110 feet wide, permitting vessels as large as the *Queen Elizabeth II* to cross the isthmus. Ships are raised by a series of locks to Gatún Lake, 85 feet above sea level. We watched as tugs helped guide the *QEII* into the Gatún Locks, a series of three locks leading to Gatún Lake, on the Atlantic side. A container ship behind the *QEII* is sailing up the dredged channel leading from the Limón Bay entrance. The lock gates are 65 feet wide and 7 feet thick, and range in height from 47 to 82 feet. The motors that move them are recessed in the walls of the lock chambers. Once inside the locks, the ships are pulled by powerful locomotives called mules that ride on rails that ascend and descend the system. It was still early morning, and a major fire, probably a forest fire, was burning near the city of Colón, where land clearing was in progress. This was the beginning of one of the most fascinating days ever."   © H. J. de Blij

www.conceptcaching.com

but its high-rise profile also reflects the proximity of Colombia's illicit drug industry and associated money-laundering and corruption. Panamanians' pervasive poverty presents a stark contrast to the city's modern image.

The Panama Canal has political as well as economic implications for Panama. For many years, Panama has recognized Taiwan as an independent entity and has sponsored resolutions to get Taiwan readmitted to the United Nations (where it was ousted in favor of China in 1971). In return, Taiwan has spent hundreds of millions of dollars in Panama in the form of investment and direct aid. But now China's growing presence in the Canal and the Colón Free Zone is causing a conundrum. Panama has awarded a contract to a Hong Kong-based (and thus Chinese) firm to operate and modernize the ports at both ends of the Canal, resulting in investments approaching U.S. $500 million. Not surprisingly, China has pressed Panama for a switch in recognition, using this commitment as leverage.

Meanwhile, many huge ships now sailing the oceans are too large to transit the Canal, and the Panama Canal Authority embarked on a plan to modernize the waterway by building new locks that will be able to handle vessels twice as large as the current maximum. Not only will this reverse the decline in the share of world cargo passing through the Canal, but it will also increase the Canal's efficiency. That will make potential competitors (such as Mexico with its dry-canal plan and Nicaragua with a scheme that would exploit Lake Managua in a mixed land-water venture) less eager to make huge investments in such projects.

In October 2006 Panamanian voters overwhelmingly approved the modernization plan, which will involve building a new set of locks that will be 40 percent longer and 60 percent wider than the existing ones (see diagram above). The cost will be financed by charging ships more for passage and by international loans (China is sure to play its role), and the project is due to be completed in time for the Canal's centenary in 2014. To what extent it will help ordinary Panamanians escape poverty (which afflicts 40 percent of the population) remains uncertain.

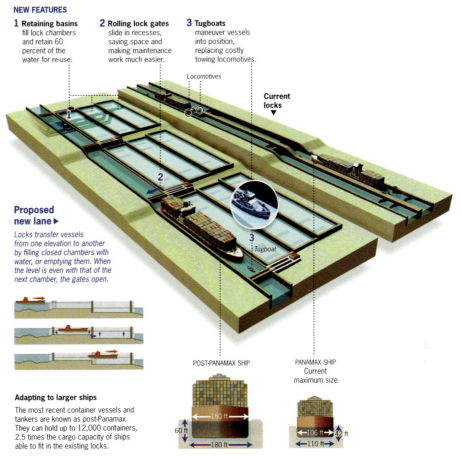

NEW FEATURES

1 Retaining basins fill lock chambers and retain 60 percent of the water for re-use.

2 Rolling lock gates slide in recesses, saving space and making maintenance work much easier.

3 Tugboats maneuver vessels into position, replacing costly towing locomotives.

Locomotives

Current locks ▼

Proposed new lane ▶

Locks transfer vessels from one elevation to another by filling closed chambers with water, or emptying them. When the level is even with that of the next chamber, the gates open.

Tugboat

POST-PANAMAX SHIP

PANAMAX SHIP
Current maximum size

150 ft
60 ft
180 ft

106 ft   42 ft
110 ft

Adapting to larger ships
The most recent container vessels and tankers are known as post-Panamax. They can hold up to 12,000 containers, 2.5 times the cargo capacity of ships able to fit in the existing locks.

Diagram showing the planned expansion and upgrading of the Panama Canal's lock system in order to accommodate ships that currently exceed the waterway's ("Panamax") capacity. The target date for completing the project is August 15, 2014, the one hundredth anniversary of the first interocean transit of the Canal by the passenger-cargo vessel, *S. S. Ancon*.   © NG Image Collection.

THE CARIBBEAN BASIN

As Figure 4-2 reveals, the Caribbean Basin, Middle America's island region, consists of a broad arc of numerous islands extending from the western tip of Cuba to the southern coast of Trinidad. The four larger islands or *Greater Antilles* (Cuba, Hispaniola, Jamaica, and Puerto Rico) are clustered in the western half of this arc. The smaller islands or *Lesser Antilles* extend to the east in a crescent-shaped zone from the Virgin Islands to Trinidad and Tobago. (Breaking this tectonic plate-related regularity are the Bahamas and the Turks and Caicos, north of the Greater Antilles, and numerous other islands too small to appear on a map at the scale of Fig. 4-2.)

On these islands, whose combined land area constitutes only 9 percent of Middle America, lie 33 states and several other entities. (Europe's colonial flags have not totally disappeared from this region, and the U.S. flag flies over Puerto Rico.) The populations of these states and territories, however, comprise 21 percent of the entire geographic realm, making this the most densely peopled part of the Americas.

Economic and Social Patterns

The Caribbean is also one of the poorest regions of the world. Environmental, demographic, political, and economic circumstances combine to impede development in the region at almost every turn. The Caribbean Basin is scenically beautiful, but it is a difficult place to make a living, and poverty is the norm.

It was not always thus. Untold wealth flowed to the European colonists who invaded this region, enslaved and eventually obliterated the indigenous Amerindian (Carib) population, and brought to these islands in bondage the Africans to work on the sugar plantations that made them rich. By the time competition from elsewhere ended the near-monopoly Caribbean sugar had enjoyed on European markets, the seeds of disaster had been sowed.

As the sugar trade collapsed, millions were pushed into a life of subsistence, malnutrition, and even hunger. Population growth now created ever-greater pressure on the land, but Caribbean islanders had few emigration options in their fragmented, mountainous habitats. The American market for sugar allowed some island countries to revive their sugar exports, including the Dominican Republic and Jamaica (Cuba's exports went to the Soviet Union, but when the USSR disintegrated Cuba's sugar industry was doomed). Some agricultural diversification occurred in the Lesser Antilles where bananas, spices, and sea-island cotton replaced sugar, but farming is a risky profession in the Caribbean. Foreign markets are undependable, and producers are trapped in a disadvantageous international economic system they cannot change.

Not surprisingly, many farmers and their families simply abandon their land and leave for towns and cities, hoping to do better there. Although the Caribbean region is less urbanized, overall, than other parts of the Americas, it is far more urbanized than, say, Africa or the Pacific Realm. Nearly two-thirds of the Caribbean's population now live in cities such as Santo Domingo in the Dominican Republic, Havana in Cuba, Port-au-Prince in Haiti, and San Juan in Puerto Rico. Many of the region's cities reflect the poverty of the citizens who were driven to seek refuge there. Port-au-Prince has some of the world's worst slums (see photo below), and desolate squatter settlements ring cities such as Kingston, Jamaica and smaller towns.

Ethnicity and Advantage

The human geography of the Caribbean region still carries imprints of the cultures of Subsaharan Africa. And the legacy of European domination also lingers. The historical geography of Cuba, Hispaniola, and Puerto Rico is suffused with Hispanic culture; Haiti and Jamaica carry stronger African legacies. But the reality of this ethnic diversity is that European lineages still hold the advantage. Hispanics tend to be in the best positions in the Greater Antilles; people who have mixed European-African ancestries, and who are described as **mulatto**, rank next. The largest part of this social pyramid is also the least advantaged: the Afro-Caribbean majority. In virtually all societies of the Caribbean, the minorities hold disproportionate power and exert overriding influence. In Haiti, the mulatto minority accounts for barely 5 percent of the population but

Labadie, Haiti: Cruise ships from one of the richest countries in the world visit the shores of the Western Hemisphere's poorest. Because tourism provides jobs at coastal resorts favored by the visitors and generates revenues for local governments, the industry is welcomed. But most of the thousands of passengers sailing on ships like this never see the malnourished children, open sewage ditches, or fetid urban slums where UN peacekeeping forces struggle to keep order. At right, an appalling example of these most desperate of human environments is this hillside shacktown overlooking the capital, Port-au-Prince, where the most rudimentary structures are precariously crammed into every available space—on a steep slope fully exposed to the hazards of rainy-season downpours and intermittent tropical storms and hurricanes. *Left:* © Ruth Fremson/*The New York Times*/Redux Pictures. *Right:* © Charles Trainor Jr./*Miami Herald* Staff.

has long held most of the power. In the adjacent Dominican Republic, the pyramid of power puts Hispanics (16 percent) at the top, the mixed sector (73 percent) in the middle, and the Afro-Caribbean minority (11 percent) at the bottom. Historic advantage has a way of perpetuating itself.

The composition of the population of the islands is further complicated by the presence of Asians from both China and India. During the nineteenth century, the emancipation of slaves and ensuing local labor shortages brought some far-reaching solutions. Some 100,000 Chinese emigrated to Cuba as indentured laborers, and Jamaica, Guadeloupe, and especially Trinidad saw nearly 250,000 South Asians arrive for similar purposes. To the African-modified forms of English and French heard in the Caribbean, therefore, can be added several Asian languages. The ethnic and cultural variety of the plural societies of Caribbean America is indeed endless.

Tourism: Promising Alternative?

Given the Caribbean Basin's limited economic options, does the tourist industry offer better opportunities? Opinions on this question are divided (see the Issue Box entitled "The Role of the Tourist Industry in Middle American Economies"). The resort areas, scenic treasures, and historic locales of Caribbean America attract well over 20 million visitors annually, with about half of these tourists traveling on Florida-based cruise ships. Certainly, Caribbean tourism is a prospective money-maker for many islands. In Jamaica alone, this industry now accounts for about one-sixth of the gross domestic product and employs more than one-third of the labor force.

But Caribbean tourism also has serious drawbacks. The invasion of poor communities by affluent tourists contributes to rising local resentment, which is further fueled by the glaring contrasts of shiny new hotels towering over substandard housing and luxury liners gliding past poverty-stricken villages. At the same time, tourism can have the effect of debasing local culture, which often is adapted to suit the visitors' tastes at hotel-staged "culture" shows. And while tourism does generate income in the Caribbean, the intervention of island governments and multinational corporations removes opportunities from local entrepreneurs in favor of large operators and major resorts.

The Greater Antilles

The four islands of the Greater Antilles contain five political entities: Cuba, Haiti, the Dominican Republic, Jamaica, and Puerto Rico (Fig. 4-2). Haiti and the Dominican Republic share the island of Hispaniola.

Cuba, the largest Caribbean island-state in terms of both territory (111,000 square kilometers/43,000 sq mi) and population (11.4 million), lies only 145 kilometers (90 mi) from the southern tip of Florida (Fig. 4-13). Havana, the now-dilapidated capital, lies almost directly across from the Florida Keys on the northwest coast of the elongated island. Cuba was a Spanish possession until the late 1890s when, with American help in the Spanish-American War,

FIGURE 4-13   © H. J. de Blij, P. O. Muller, and John Wiley & Sons, Inc.

The Role of the Tourist Industry in Middle American Economies

IN SUPPORT OF THE TOURIST INDUSTRY

"As the general manager of a small hotel in St. Maarten, on the Dutch side of the island, I can tell you that without tourists, we would be in deep trouble economically. Mass tourism, plain and simple, has come to the rescue in the Caribbean and in other countries of Middle America. Look at the numbers. Here it's the only industry that is growing, and it already is the largest worldwide industry. For some of the smaller countries of the Caribbean, this is not just the leading industry but the only one producing external revenues. Whether it's hotel patrons or cruise-ship passengers, tourists spend money, create jobs, fill airplanes that give us a link to the outside world, require infrastructure that's good not just for them but for locals too. We've got better roads, better telephone service, more items in our stores. All this comes from tourism. I employ 24 people, most of whom would be looking for nonexistent jobs if it weren't for the tourist industry.

"And it isn't just us here in the Caribbean. Look at Belize. I just read that tourism there, based on their coral reefs, Mayan ruins, and inland waterways, brought in U.S. $100 million last year, in a country with a total population of only about 300,000! That sure beats sugar and bananas. In Jamaica, they tell me, one in every three workers has a job in tourism. And the truth is, there's still plenty of room for the tourist industry to expand in Middle America. Those Americans and Europeans can close off their markets against our products, but they can't stop their citizens from getting away from their awful weather by coming to this tropical paradise.

"Here's another good thing about tourism. It's a clean industry. It digs no mine shafts, doesn't pollute the atmosphere, doesn't cause diseases, doesn't poison villagers, isn't subject to graft and corruption the way some other industries are.

"Last but not least, tourism is educational. Travel heightens knowledge and awareness. There's always a minority of tourists who just come to lie on the beach or spend all their time in some cruise-ship bar, but most of the travelers we see in my hotel are interested in the place they're visiting. They want to know why this island is divided between the Dutch and the French, they ask about coral reefs and volcanoes, and some even want to practice their French on the other side of the border (don't worry, no formalities, just drive across and start talking). Tourism's the best thing that happened to this part of the world, and other parts too, and I hope we'll never see a slowdown."

CRITICAL OF THE TOURIST INDUSTRY

"You won't get much support for tourism from some of us teaching at this college in Puerto Rico, no matter how important some economists say tourism is for the Caribbean. Yes, tourism is an important source of income for some countries, like Kenya with its wildlife and Nepal with its mountains, but for many countries that income from tourism does not constitute a real and fundamental benefit to the local economies. Much of it may in fact result from the diversion to tourist consumption of scarce commodities such as food, clean water, and electricity. More of it has to be reinvested in the construction of airport, cruise-port, overland transport, and other tourist-serving amenities. And as for items in demand by tourists, have you noticed that places with many tourists are also places where prices are high?

"Sure, our government people like tourism. Some of them have a stake in those gleaming hotels where they can share the pleasures of the wealthy. But what those glass-enclosed towers represent is globalization, powerful multinational corporations colluding with the government to limit the opportunities of local entrepreneurs. Planeloads and busloads of tourists come through on prearranged (and prepaid) tour promotions that isolate those visitors from local society.

"You geographers talk about cultural landscapes. Well, picture this: luxury liners sailing past poverty-stricken villages, luxury hotels towering over muddy slums, restaurants serving caviar when, down the street, children suffer from malnutrition. If the tourist industry offered real prospects for economic progress in poorer countries, such circumstances might be viewed as the temporary, unfortunate byproducts of the upward struggle. Unfortunately, the evidence indicates otherwise. Name me a tourism-dependent economy where the gap between the rich and poor has narrowed.

"As for the educational effect of tourism, spare me the argument. Have you sat through any of those 'culture' shows staged by the big hotels? What you see there is the debasing of local culture as it adapts to visitors' tastes. Ask hotel workers how they really feel about their jobs, and you'll hear many say that they find their work dehumanizing because expatriate managers demand displays of friendliness and servitude that locals find insulting to sustain.

"I've heard it said that tourism doesn't pollute. Well, the Alaskans certainly don't agree—they sued a major cruise line on that issue and won. Not very long ago, cruise-ship crews routinely threw garbage-filled plastic bags overboard. That seems to have stopped, but I'm sure you've heard of the trash left by mountain-climbers in Nepal, the damage done by off-road vehicles in the wildlife parks of Kenya, the coral reefs injured by divers off Florida. Tourism is here to stay, but it is no panacea."

Regional **ISSUE**

Vote your opinion at www.wiley.com/college/deblij

it attained independence. Fifty years later, a U.S.-backed dictator was in control, and by the 1950s Havana had become an American playground. The island was ripe for revolution, and in 1959 Fidel Castro's insurgents gained control, thereby making Cuba a communist dictatorship and a Soviet client. Castro's rule survived the collapse of the Soviet Empire despite the loss of subsidies and sugar markets on which it had long relied.

As the map shows, sugar was Cuba's economic mainstay for many years; the plantations, once the property of rich landowners, extend all across the territory. But as this map also shows, sugarcane is losing its position as the leading Cuban foreign exchange earner. Mills are being closed down, and the canefields are being cleared for other crops and for pastures. Cuba has other economic opportunities, however, especially in its highlands. There are three mountainous areas, of which the southeastern chain, the Sierra Maestra, is the highest and most extensive. These highlands create considerable environmental diversity as reflected by extensive, timber-producing tropical forests and varied soils on which crops ranging from tobacco to subtropical and tropical fruits are grown. Rice and beans are the staples, but Cuba cannot meet its needs and so must import food. The savannas of the center and west support livestock. Although Cuba has only limited mineral reserves, its nickel deposits are extensive and have been mined for a century.

Early in the twenty-first century, Cuba found a crucial new supporter in Venezuela's leader, Hugo Chávez. Cuba has no domestic petroleum reserves, but Venezuela is oil-rich and, since 2003, has been providing all of the fuel that Cuba needs. In return, Castro sent 30,000 health workers and other professionals to Venezuela, where they aid the poor.

Poverty, crumbling infrastructure, crowded slums, and unemployment mark the Cuban cultural landscape, but Cuba's regime still has support among the general population. During the Castro period, much was done to bring the Afro-Cuban population into the mainstream through education and health provision. Cubans point to Guatemala, Nicaragua, and El Salvador and ask whether those countries are better off than they are under Castro. But in the United States, the view is different: Cuba, exiles and locals agree, could be the shining star of the Caribbean, its people free, its tourist economy booming, its products flowing to American markets. "The future Ireland of the Caribbean," suggested a Cuban geographer recently. Much will have to change for that prediction to come true.

Jamaica lies across the deep Cayman Trench from southern Cuba, and a cultural gulf separates these two countries as well. Jamaica, a former British dependency, has an almost entirely Afro-Caribbean population. As a member of the British Commonwealth, Jamaica still recognizes the British monarch as the chief of state, represented by a governor-general. The effective head of government in this democratic country, however, is the prime minister. English remains the official language here, and British customs still linger.

Smaller than Connecticut and with 2.8 million people, Jamaica has experienced a steadily declining GNI over the past few decades despite its relatively slow population growth. Tourism has become the largest source of income, but the markets for bauxite (aluminum ore), of which Jamaica is a major exporter, have dwindled. And like other Caribbean countries, Jamaica has trouble making money from its sugar exports. Jamaican farmers also produce crops ranging from bananas to tobacco, but the country faces the disadvantages on world markets common to those in the periphery. Meanwhile, Jamaica must import all of its oil and much of its food because the densely populated coastal flatlands suffer from overuse and shrinking harvests.

The capital, Kingston on the south coast, reflects Jamaica's economic struggle. Almost none of the hundreds of thousands of tourists who visit the country's beaches, explore its Cockpit Country of (karst) limestone towers and caverns, or populate the cruise ships calling at Montego Bay or other points along the north coast get even a glimpse of what life is like for the ordinary Jamaican.

Haiti, the poorest state in the Western Hemisphere by virtually every measure, occupies the western part of the island of Hispaniola, directly across the Windward Passage from eastern Cuba (Fig. 4-14). Arawaks, not Caribs, formed the dominant indigenous population here, but the Spanish colonists who first took Hispaniola killed many, worked most of the survivors to death on their plantations, and left the others to die of the diseases they brought with them. French pirates established themselves in coastal coves along Hispaniola's west coast even as Spanish activity focused on the east, and just before the eighteenth century opened France formalized the colonial status of "Saint Dominique." During the following century, French colonists established vast plantations in the valleys of the northern mountains and in the central plain, laid out elaborate irrigation systems, built dams, and brought in a large number of Africans in bondage to work the fields. Prosperity made the colonists rich, but the African workers suffered terribly. They rebelled and, in a momentous victory over their European oppressors, established the independent republic of Haiti (resurrecting the original Arawak name for it) in 1804.

Strife among Haitian groups, American intervention, mismanagement, and dictatorship doomed the fortunes of the republic. In the 1990s, the United States attempted to help move Haiti toward more representative government, but the country was in economic and social collapse. By 2003, Haiti's GNI per capita had fallen to less than half that

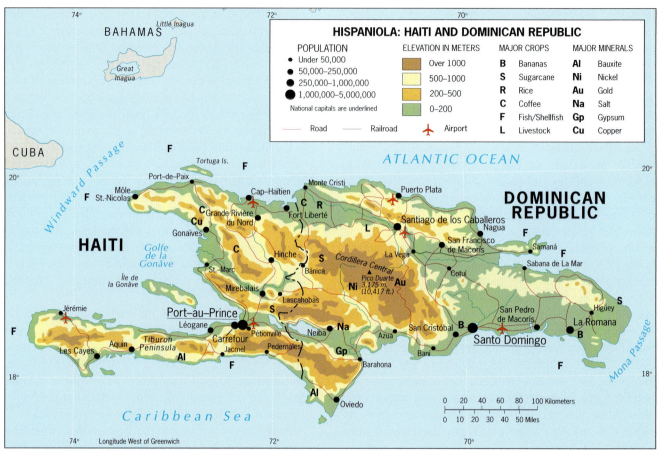

FIGURE 4-14 © H. J. de Blij, P. O. Muller, and John Wiley & Sons, Inc.

of Jamaica, a level lower than that of many poor African countries; foreign aid makes possible most of the country's limited public expenditures. Health conditions within the population of 8.9 million are dreadful: malnutrition is common, AIDS is rampant, diseases ranging from malaria to tuberculosis are rife, but hospital beds and doctors are in short supply. Conditions in and around the capital, Port-au-Prince, are among the worst in the world—less than 1000 kilometers (600 mi) from the United States.

The Dominican Republic has a larger share of the island of Hispaniola than Haiti (Fig. 4-14) in terms of both territory and population. Fly along the north-south border between the two countries, and you see a crucial difference: to the west, Haiti's hills and plains are treeless and gulleyed, its soils eroded and its streams silt-laden. To the east, forests drape the countryside and streams run clear.

Indigenous Caribs inhabited this eastern part of Hispaniola, and when the European colonists arrived on the island, they were in the process of driving the Arawaks westward. Spanish colonists made this a prosperous colony, but then Mexico and Peru diverted Spanish attention from Hispaniola and the territory was ceded to France. But Hispanic culture lingered, and after the revolution in Haiti the French gave eastern Hispaniola back to Spain. After the

Dominican Republic declared its independence in 1821, Haitian forces invaded it and occupied the republic until 1844, creating an historic animosity that persists today.

The mountainous Dominican Republic has a wide range of natural environments and a far stronger resource base than Haiti. Nickel, gold, and silver have long been exported along with sugar, tobacco, coffee, and cocoa, but tourism (the great opportunity lost to Haiti) is the leading industry. A long period of dictatorial rule punctuated by revolutions and U.S. military intervention ended in 1978 with the first peaceful transfer of power following a democratic election.

Political stability brought the Dominican Republic rich rewards, and during the late 1990s the economy, based on manufacturing, high-tech industries, and remittances from Dominicans abroad as well as tourism, grew at an average 7 percent per year. But in the early 2000s the economy imploded, not only because of the downturn in the world economy but also because of bank fraud and corruption in government. Suddenly the Dominican peso collapsed, inflation skyrocketed, jobs were lost, and blackouts prevailed. As the people protested, lives were lost and the self-enriched elite blamed foreign financial institutions that were unwilling to lend the government more money. Once again the hopes of ordinary citizens were dashed by greed and corruption among those in power.

Puerto Rico is the largest U.S. domain in Middle America, the easternmost and smallest island of the Greater Antilles (Fig. 4-15). This 9000-square-kilometer (3500-sq-mi) island, with a population of 3.9 million, is larger than Delaware and more populous than Oregon. It fell to the United States more than a century ago during the Spanish-American War of 1898. Since the Puerto Ricans had been struggling for some time to free themselves from Spanish control, this transfer of power was, in their view, only a change from one colonial power to another. As a result, the first half-century of U.S. administration was difficult, and it was not until 1948 that Puerto Ricans were permitted to elect their own governor.

When the island's voters approved the creation of a Commonwealth in a 1952 referendum, Washington, D.C., and San Juan, the two seats of government, entered into a complicated arrangement. Puerto Ricans are U.S. citizens but pay no federal taxes on local incomes. The Puerto Rican Federal Relations Act governs the island under the terms of its own constitution and awards it considerable autonomy. Puerto Rico also receives a large annual subsidy from Washington, totaling about U.S. $9 billion per year at the end of the 1990s.

Despite these apparent advantages in the poverty-mired Caribbean, Puerto Rico has not thrived under U.S. administration. Long dependent on a single-crop economy (sugar), the island's industrialization during the 1950s and 1960s was based on its comparatively cheap labor, tax breaks for corporations, political stability, and special access to the U.S. market. As a result, pharmaceuticals, electronic equipment, and apparel top today's list of exports, not sugar or bananas. But this industrialization failed to stem a tide of emigration that carried more than 1 million Puerto Ricans to the New York City area alone. The same wages that favored corporations kept many Puerto Ricans poor or unemployed. By some estimates, unemployment on the island stands at about 45 percent today. Most receive federal support. Another 30 percent work in the public sector, that is, in government. Puerto Rico's welfare-state condition discourages initiative, so that many people who could be working do not because their federal subsidy would decrease accordingly. Economists report that workers from the Dominican Republic even come to Puerto Rico to take low-wage jobs Puerto Ricans do not want.

Although Puerto Ricans are vocal in demanding political change, it is perhaps not surprising that successive referendums during the 1990s resulted in retention of the status quo—continuation of Commonwealth status rather than either Statehood or independence. The issue will no doubt continue to pose a formidable challenge to American statecraft in the years ahead.

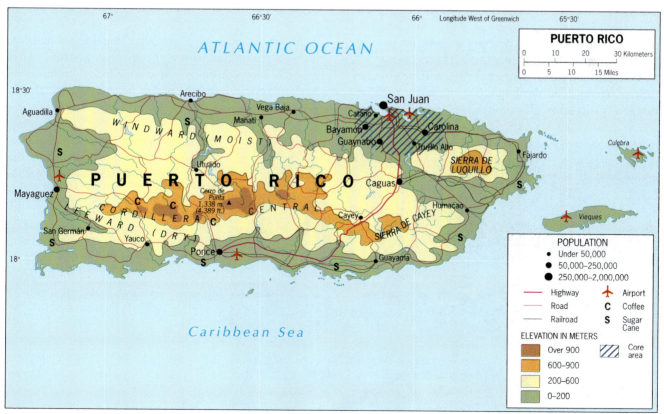

FIGURE 4-15 © H. J. de Blij, P. O. Muller, and John Wiley & Sons, Inc.

The Lesser Antilles

As Figure 4-2 shows, the Greater Antilles are flanked by two clusters of islands: the extensive Bahamas-Turks/Caicos archipelago to the north and the Lesser Antilles to the east and south. The Bahamas, the former British colony that is now the closest Caribbean neighbor to the United States, alone consists of nearly 3000 coral islands, most of them rocky, barren, and uninhabited, but about 700 carrying vegetation, of which some 30 are inhabited. Centrally positioned New Providence Island carries most of the country's 330,000 inhabitants and contains the capital, Nassau, a leading tourist attraction.

The Lesser Antilles are grouped geographically into the Leeward Islands and the Windward Islands, a (climatologically incorrect) reference to the prevailing airflows in this tropical area. The Leeward Islands extend from the U.S. Virgin Islands to the French dependencies of Guadeloupe and Martinique, and the Windward Islands from St. Lucia to the Netherlands Antilles off the Venezuelan coast (Fig. 4-2). It would be impractical to detail the individual geographic characteristics of each of the Lesser Antilles, but we should note that these countries and territories share the environmental risks of this region: earthquakes, volcanic eruptions, and hurricanes; that they face, to varying degrees, similar economic problems in the form of limited domestic resources, overpopulation, soil deterioration, land fragmentation, and market limitations; that tourism has become the leading industry for many; that political status ranges from complete sovereignty to continuing dependency; and that cultural diversity is strong not only between islands but within them as well. In economic terms, the GNI figures given (where available) in Table G-1 may look encouraging, but these numbers conceal the social reality in virtually every entity: the gap between the fortunate few who are well-off and the great majority who are poor remains enormous.

Under such circumstances one looks for hopeful signs, and one of these signs comes from Trinidad and Tobago, the two-island republic at the southern end of the Lesser Antilles. This country (population: 1.3 million) has embarked on a natural-gas-driven industrialization boom that could turn it into an economic tiger (see photo below).

Trinidad has long been an oil producer, but lower world prices and dwindling supplies in the 1990s forced a re-examination of its natural gas deposits to help counter the downturn. That quickly resulted in the discovery of major new supplies, and this cheap and abundant fuel has sparked a local gas-production boom as well as an influx of energy, chemical, and steel companies from Western Europe, Canada, and even India. (Meanwhile, Trinidad has become the largest supplier of liquefied natural gas for the United States.) Many of the new industrial facilities have agglomerated at the ultramodern Point Lisas Industrial Estate outside the capital, Port of Spain, and they have propelled Trinidad to become the world's leading exporter of ammonia and methanol. Natural gas is also an efficient fuel for the manufacturing of metals, and steelmakers as well as aluminum refiners have been attracted to locate here. With Trinidad lying only a few kilometers from the Venezuelan coast of South America, it is also a sea-lane crossroads that is well connected to the vast, near-coastal supplies of iron ore and bauxite that are mined in nearby countries, particularly the Brazilian Amazon.

Middle America is a physically, culturally, and economically fragmented and diverse geographic realm that defies generalization. Given its strong Amerindian presence, its North American infusions, and its lingering Western European traditions, this certainly is not "Latin" America.

The capital city of Trinidad and Tobago may have a historic colonial name (Port of Spain), but after nearly three centuries of Spanish rule, the British took control here. English became the *lingua franca*, democratic government followed independence in 1962, and natural gas reserves propelled a thriving economy. As can be seen here, the harbor of Port of Spain is bursting at the seams as a car carrier delivers automobiles from Japan, and containers are stacked high on the docks. © Boutin/Sipa Press.

What You Can Do

RECOMMENDATION: Take a "serious" vacation! Hundreds of thousands of hard-working students take off for warmer weather between semesters or during spring break. Most go to relax in Florida or another "Sunbelt" State. But Middle America is nearby, and you could combine some sun and warmth with a valuable geographic field experience—if you prepare yourself and are ready for some challenges. Take some good, large-scale maps and a notebook and, when you spend a week in Puerto Rico, ask the people you meet how they feel about their island's political status. Visit Martinique and learn on the spot what happened to the "Paris of the Caribbean," St. Pierre, just over a century ago: the evidence is still visible. Go to Curaçao and find out how the island's economy survived the decline of its oil-refinery era. Or if you're in the West, cross the Mexican border and get your own take on the impact of NAFTA. The opportunities are right on our doorstep, and when you use your geographic perspective, you'll see the world in new ways of lasting value.

GEOGRAPHIC CONNECTIONS

1 When we studied the expansion of the European Union (EU), the issue of Turkey's application for membership loomed large (Chapter 1). Turkish leaders want admission and are working to meet the conditions of membership, but Turkey still fails to meet these criteria in important respects. European Union members' views range across the spectrum, from enthusiastic to ambivalent to negative. Now imagine a North American Union of which Canada and the United States are members, and which Mexico wishes to join (NAFTA is a far less comprehensive union than the EU, in effect just a first step toward it). Use your geographic insights to analyze the prospects of such a merger, and predict reactions in Canada and the United States. In what ways would Mexico qualify for membership, and in what respects would it not? Do you see any similarities between current Turkey–Europe and Mexico–North America relationships? In what ways are the two situations crucially different?

2 For many years, Caribbean islands such as St. Lucia have been exporting bananas to markets in North America and Europe, with these markets protected by preferential trade agreements. Now those agreements are ending as trade barriers fall all over the world, and the single-crop economies of the Caribbean are in trouble. What are the options for the farmers of St. Lucia and its neighbors? Relate your answer to core-periphery contrasts in the modern world (which can be reviewed on pp. 28–29).

Geographic Literature on Middle America: The key introductory works on this realm were authored by Blouet & Blouet, Clawson, James & Minkel, Kent, Preston, and West & Augelli et al. These works, together with a comprehensive listing of noteworthy books and recent articles on the geography of Mexico, Central America, and the Caribbean, can be found in the *References and Further Readings* section of this book's website at **www.wiley.com/college/deblij**.

5

Where were these photos taken? Find out at **www.wiley.com/college/deblij**

In This Chapter

- The growing power of indigenous peoples: "Latin" America no more
- U.S. initiatives in economics and politics less welcome today
- Tentative efforts toward economic integration: China lends a hand
- Chile: Star of the realm
- The poor performance of rich Argentina
- Brazil: Superpower in the making?

Photos: © H. J. de Blij

FIGURE 5-1   © H. J. de Blij, P. O. Muller, and John Wiley & Sons, Inc.

O F ALL THE continents, South America has the most familiar shape—a giant triangle connected by mainland Middle America's tenuous land bridge to its neighbor in the north. South America also lies not only south but mostly east of its northern counterpart. Lima, the capital of Peru—one of the continent's westernmost cities—lies farther east than Miami, Florida. Thus South America juts out much more prominently into the Atlantic Ocean toward Southern Europe and Africa than does North America. But lying so far eastward means that South America's western flank faces a much wider Pacific Ocean, with the distance from Peru to Australia nearly twice that from California to Japan.

As if to reaffirm South America's northward and eastward orientation, the western margins of the continent are rimmed by one of the world's longest and highest mountain ranges, the Andes, a gigantic wall that extends unbroken from Tierra del Fuego near the continent's southern tip in Chile to northeastern Venezuela in the far north (Fig. 5-1). The other major physiographic feature of South America dominates its central north—the Amazon Basin; this vast humid-tropical amphitheater is drained by the mighty Amazon, which is fed by several major tributaries. Much of the remainder of the continent can be classified as plateau, with the most important components being the Brazilian Highlands that cover most of Brazil southeast of the Amazon Basin, the Guiana Highlands located north of the lower Amazon Basin, and the cold Patagonian Plateau that blankets the southern third of Argentina. Figure 5-1 also reveals two other noteworthy river basins beyond Amazonia: the Paraná-Paraguay Basin of south-central South America, and the Orinoco Basin in the far north that drains interior Colombia and Venezuela.

MAJOR GEOGRAPHIC QUALITIES OF
South America

1. South America's physiography is dominated by the Andes Mountains in the west and the Amazon Basin in the central north. Much of the remainder is plateau country.

2. Almost half of the realm's area as well as its population are concentrated in one country—Brazil.

3. South America's population remains concentrated along the continent's periphery. Most of the interior is sparsely peopled, but sections of it are now undergoing significant development.

4. Interconnections among the states of the realm are improving rapidly. Economic integration has become a major force, particularly in southern South America.

5. Regional economic contrasts and disparities, both in the realm as a whole and within individual countries, are strong.

6. Cultural pluralism exists in almost all of the realm's countries and is often expressed regionally.

7. Rapid urban growth continues to mark much of the South American realm, and the urbanization level overall is today on a par with the levels in the United States and Western Europe.

Defining the Realm

South America is a realm in dramatic transition, and it is not clear where this transition will lead. During much of the twentieth century, South American countries were in frequent political turmoil. Dictatorial regimes ruled from one end of the realm to the other; unstable governments fell with damaging frequency. Widespread poverty, harsh regional disparities, poor internal surface connections, limited international contact, and economic stagnation prevailed.

Toward the end of the first decade of the twenty-first century, there were signs of improvement. Democratic governments had replaced authoritarian regimes. South America's countries were becoming more interconnected. New transport routes crossed international borders, and new settlement frontiers were opened. High energy and commodity prices fueled economic growth in several countries.

But achieving economic growth and democratic statecraft rarely come without setbacks, and over the past decade South America has seen reversals as well as gains. The realm's giant, Brazil, is embarked on a program of land reform, a campaign against poverty, and an effort to maintain financial rigor that have all run up against endemic corruption in government. The economy of Argentina is just recovering after an implosion that shook the country to its core. Whereas Chile is the realm's success story, emerging from the horrors of the Pinochet era as a stable and vibrant democracy with a thriving economy, neighboring Bolivia is in the grip of a social revolution arising in part from the realization of its energy riches. And on the north coast lies Venezuela, its oil reserves among the largest in the world (and by far the largest in the realm) and its political life dominated by a one-time coup leader whose closest ideological ally is Cuba's communist ruler and whose major adversary is the U.S. government.

Today, the United States hopes to foster democracy and encourage regional economic integration, but many South Americans remember past U.S. toleration of, and even support for, the realm's former dictators. Venezuela's populist dictator champions the poor and uses the decade's high oil prices to counter American influence, campaigning vigorously against the notion of a Free Trade Area of the Americas (FTAA) and warning South American governments against capitalist plots. Such advice finds a ready market because the great majority of South Americans remain mired in poverty. By some measures, the disparity between rich and poor is wider in this realm than in any other, and wealth is disproportionately concentrated in the hands of a small minority (the richest 20 percent of the realm's inhabitants control 70 percent, while the poorest 20 percent own 2 percent). The question of the day is whether South America can sustain its recovery against political, ideological, and economic odds.

THE HUMAN SEQUENCE

Although modern South America's largest populations are situated in the east and north, during the height of the Inca Empire the Andes Mountains contained the most densely peopled and best organized state on the continent. Although the origins of Inca civilization are still shrouded in mystery, it has become generally accepted that the Incas were descendants of ancient peoples who came to South America via the Middle American land bridge. Thus, for thousands of years before the Europeans arrived in the sixteenth century, indigenous Amerindian societies had been developing in South America.

The Inca Empire

About one thousand years ago, a number of regional cultures thrived in Andean valleys and basins and at places along the Pacific coast. By AD 1300, the Incas had established themselves in an elongated basin called an *altiplano* at Cuzco in the high Andes (Fig. 5-2). As the term implies, these *altiplanos* are high-elevation valleys between parallel Andean ranges, filled and floored by sediments from the adjacent mountains. The Incas exploited fertile soils and

FROM THE FIELD NOTES

"From this high vantage point I got a good perspective of a valley near Pisac in the Peruvian Andes, not far from Cuzco. This was part of the Incan domain when the Spaniards arrived to overthrow the empire, but the terraces you can see actually predate the Inca period. Human occupation in these rugged mountains is very old, and undoubtedly the physiography here changed over time. Today these slopes are arid and barren, and only a few hardy trees survive; the stream in the valley bottom is all the water in sight. But when the terrace builders transformed these slopes, the climate may have been more moist, the countryside greener."
© Courtesy Philip L. Keating.

www.conceptcaching.com

mountain meltwaters to achieve specialized agriculture that sustained an ever-growing population, and from their mountainous abode they extended their authority over the peoples of other *altiplanos* as well as others living along the Pacific coast.

FIGURE 5-2   © H. J. de Blij, P. O. Muller, and John Wiley & Sons, Inc.

The Inca State

The Inca were great military strategists, but even more impressive was their ability to integrate vanquished peoples into a stable and efficiently functioning state. They were expert road and bridge builders, colonizers, and administrators, and in a short period they unified an empire that extended from present-day Colombia to Chile (Fig. 5-2). In fact, the Incas were in a minority in their own far-flung state, becoming a ruling elite in a rigidly class-structured society. So centralized and controlled was this state that a takeover at the top was enough to gain power over the whole empire—as a small army of Spanish in-

vaders realized in the 1530s. Once the Inca system broke down, there was no resistance and the state collapsed, leaving behind a cultural landscape studded with magnificent structures of which Machu Picchu, in present-day Peru, is the most spectacular. It also left a legacy of contrasting social values between Hispanic and Amerindian citizens that remains a part of life in Andean South America to this day.

Modern Heirs

It is important to remember that South America's "Indian" population, including those whose ancestors were Incas or Inca subjects as well as those who were not, extends geographically in a wide arc from present-day Chile through Bolivia, Peru, and Ecuador to Colombia (Fig. 5-2). This matters because South America's long-downtrodden indigenous peoples are experiencing a social, political, and economic awakening. They are not alone in this—we noted the Zapatista movement based in southern Mexico's Chiapas State, and we will see that Amerindians in Brazil are resisting encroachment by farmers—but they have had greater impact in such countries as Ecuador and Bolivia, where they form a larger proportion of the population. Amerindian leaders are emerging to exercise political influence and to bring the plight of the realm's indigenous peoples to regional as well as international attention. They were conquered, decimated by foreign diseases, robbed of their best lands, subjected to forced labor, denied the right to grow traditional crops, socially discriminated against, and swindled out of their fair share of the revenues from resources in their (modern) countries. Today they are the poorest of the poor, but they are asserting themselves. As we will discover, there are even calls for secession and independence for Amerindian-dominated areas of Andean states. It may not come to this, but the "Bolivarian Revolution" that gave the poor of Venezuela their champion is more popular in Andean South America than the preferences of their own governments—such as free-trade agreements and globalization.

The Iberian Invaders

In South America as in Middle America, the location of indigenous peoples largely determined the direction of the thrusts of European invasion. The Incas, like Mexico's Maya and Aztec peoples, had accumulated gold and silver at their headquarters, possessed productive farmlands, and constituted a ready labor force. Not long after the defeat of the Aztecs in 1521, Francisco Pizarro sailed southward along the continent's northwestern coast, learned of the existence of the Inca Empire, and withdrew to Spain to organize its overthrow. He returned to the Peruvian coast in 1531 with 183 men and two dozen horses, and the events that followed are well known. In 1533, his party rode victorious into Cuzco.

At first, the Spaniards kept the Incan imperial structure intact by permitting the crowning of an emperor who was under their control. But soon the breakdown of the old order began. The new order that gradually emerged in western South America placed the indigenous peoples in serfdom to the Spaniards. Great haciendas were formed by **land alienation** (the takeover of indigenously held land by foreigners), taxes were instituted, and a forced-labor system was introduced to maximize the profits of exploitation.

Lima, the west-coast headquarters of the Spanish conquerors, soon became one of the richest cities in the world, its wealth based on the exploitation of vast Andean silver deposits. The city also served as the capital of the viceroyalty of Peru, as the Spanish authorities quickly integrated the new possession into their colonial empire (Fig. 5-2). Subsequently, when Colombia and Venezuela came under Spanish control and, later, when Spanish settlement expanded in what is now Argentina and Uruguay, two additional viceroyalties were added to the map: New Granada and La Plata.

Portuguese Brazil

Meanwhile, another vanguard of the Iberian invasion was penetrating the east-central part of the continent, the coastlands of present-day Brazil. This area had become a Portuguese sphere of influence because Spain and Portugal

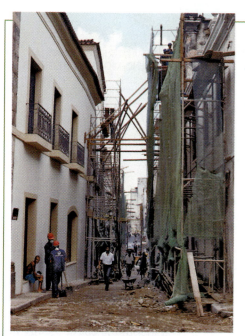

FROM THE FIELD NOTES

"Salvador is one of Brazil's oldest and most vibrant cities. Magnificent churches, public buildings, and mansions were built during the time when this was, by many measures, the most important city in the Southern Hemisphere. Long the capital of Brazil, Salvador was the point of entry for tens of thousands of Africans, and the fortune-making plantation economy, augmented by the whaling industry, concentrated enormous wealth here, some of which went into the construction of an opulent city center. But fortunes change, Salvador lost its political as well as economic advantages, and the city fell into disrepair. Walking the streets of the old town in 1982, I noted the state of decay of much of Salvador's architectural heritage and wondered about its survival: weathering in this tropical environment was destroying woodwork, façades, and roofs. But then the United Nations proclaimed Salvador's old town a World Heritage site, and massive restoration began (left). By the late 1990s, much of the district had been revived (right), and tourism's contribution to the local economy was on the rise."   © H. J. de Blij.

had signed a treaty in 1494 to recognize a north-south line 370 leagues west of the Cape Verde Islands as the boundary between their New World spheres of influence. This border ran approximately along the meridian of 50°W longitude, thereby cutting off a sizeable triangle of eastern South America for Portugal's exploitation (Fig. 5-2). But a brief look at the political map of South America (Fig. 5-3) shows that this treaty did not limit Portuguese colonial territory to the east of the 50th meridian. Instead, Brazil's boundaries were bent far inland to include almost the entire Amazon Basin, and the country came to be only slightly smaller in territorial size than all the other South American countries combined. This westward thrust was the result of Portuguese and Brazilian penetration, particularly by the *Paulistas*, the settlers of São Paulo who needed Amerindian slave labor to run their plantations.

The Africans

As Figure 5-2 shows, the Spaniards initially got very much the better of the territorial partitioning of South America—not just in land quality but also in the size of the aboriginal labor force. When the Portuguese began to develop their territory, they turned to the same lucrative activity that their Spanish rivals had pursued in the Caribbean—the plantation cultivation of sugar for the European market. And they, too, found their labor force in the same source region, as millions of Africans (nearly half of all who came to the Americas) were brought in bondage to the tropical Brazilian coast north of Rio de Janeiro. Not surprisingly, Brazil now has South America's largest black population, which is still heavily concentrated in the country's poverty-stricken northeastern States. With Brazilians of direct or mixed African ancestry today accounting for nearly half of the population of 192 million, the Africans decidedly constitute the third major immigration of foreign peoples into South America.

Longstanding Isolation

Despite their adjacent location on the same continent, their common language and cultural heritage, and their shared

FIGURE 5-3 © H. J. de Blij, P. O. Muller, and John Wiley & Sons, Inc.

national problems, the countries that arose out of South America's Spanish viceroyalties (along with Brazil) until quite recently existed in a considerable degree of isolation from one another. Distance and physiographic barriers reinforced this separation, and the realm's major population agglomerations still adjoin the coast, mainly the eastern and northern coasts (Fig. G-9). The viceroyalties existed primarily to extract riches and fill Spanish coffers. In Iberia there was little interest in developing the American lands for their own sake. Only after those who had made Spanish and Portuguese America their home and who had a stake there rebelled against Iberian authority did things begin to change, and then very slowly. Thus, South America was saddled with the values, economic outlook, and social attitudes of eighteenth-century Iberia—not the best tradition from which to begin the task of forging modern nation-states.

Independence

Certain isolating factors had their effect even during the wars for independence. Spanish military strength was always concentrated at Lima, and those territories that lay farthest from their center of power—Argentina and Chile—were the first to establish their independence from Spain (in 1816 and 1818, respectively). In the north Simón Bolívar led the burgeoning independence movement, and in 1824 two decisive military defeats there spelled the end of Spanish power in South America.

This joint struggle, however, did not produce unity because no fewer than nine countries emerged from the three former viceroyalties. It is not difficult to understand why this fragmentation took place. With the Andes intervening between Argentina and Chile and the Atacama Desert between Chile and Peru, overland distances seemed even greater than they really were, and these obstacles to contact proved quite effective. Hence, from their outset the new countries of South America began to grow apart amid friction and even wars. Only within the past two decades have the countries of this realm finally begun to recognize the mutual advantages of increasing cooperation and to make lasting efforts to steer their relationships in this direction.

CULTURAL FRAGMENTATION

South America may still be referred to in some writings as part of "Latin" America, but in Chapter 4 we noted the inappropriateness of this regional appellation. Tens of millions of South Americans have Amerindian, African, Asian, and even non-"Latin" European ancestries. English, Dutch, and French—not Spanish or Portuguese—are the official languages of three of South America's 13 political entities.

But there is no doubt about the dominant cultural imprint on South America. Amerindians were numerous, culturally advanced, and well-organized, but they were soon overpowered by Iberian and other invaders who took their best lands and divided up their realm, fighting among themselves in the process. As we noted above, the Portuguese, who claimed Brazil, proceeded to transform the realm's ethnic mosaic by bringing in several million Africans in bondage. The British and Dutch carried tens of thousands of South Asians to their possessions, and the Dutch in addition brought in substantial numbers of Indonesians. Much later, independent Brazil became a favorite destination for emigrating Japanese, and today Brazil has the largest Japanese community outside Japan. So much for "Latin" America.

South America, therefore, is a realm of **plural societies**, ▣3 states in which two or more cultural groups live in adjacent areas but rarely integrate. South America's cultural geography is a kaleidoscope of almost endless variety, whose entrenched divisions and more modern mixtures combine to create diverse economic landscapes as well.

Land Use and Agriculture

When you examine statistics on the economic geography of countries, it appears that agriculture no longer employs most of the people. Even in countries like Guatemala and Honduras, these data are likely to indicate that less than 40 percent of the labor force is engaged in farming. But make no mistake: many more people than these numbers suggest are engaged in some form of farming. What you are seeing here is the percentage of the labor force that works in **commercial agriculture**. This is farming for financial return, the field workers are paid, and the land (or corporate) owners make a profit. But millions more South Americans live directly off the land in various forms of **subsistence agriculture**, which may be life-sustaining ▣5 but not profitable. These *peones* (peasants) do not show up in the economic statistics and reports, because they are not counted as members of the "labor force."

As a result, the map of South America's agricultural systems is a complex mosaic of different uses of the land ranging from highly profitable commercial cropping (the growing of flowers for export by air to the United States and Europe is only the latest of these enterprises) to subsistence farming with some local marketing, mostly in the Andean highlands, and shifting cultivation as shown by the vast region colored green covering much of the Amazon Basin (Fig. 5-4). Note the vestiges of the past in the coast-oriented plantation agriculture in the tropical north; haciendas prevail in the temperate south where large-scale livestock ranching is a key industry. As the map confirms, cattle raising occurs across most of the continent from Argentina (where the Pampas is South America's most famous meat-producing area) to Venezuela and from Ecuador to Uruguay.

A string of small oases along Peru's desert coast produce specialty crops such as vegetables and fruits, an important industry made possible by water from the Andes Mountains. Go farther south along the Pacific coast, and the oases give way to unbroken desert in northern Chile—until you reach the streams that make it to the shore across the still-dry countryside. Soon you see vineyards along those streams, and as you enter Chile's Mediterranean climatic zone, farming diversifies into export crops, dairy products, and grains. As the map shows, on the Argentinian side of the Andes farming is so varied that no single product dominates.

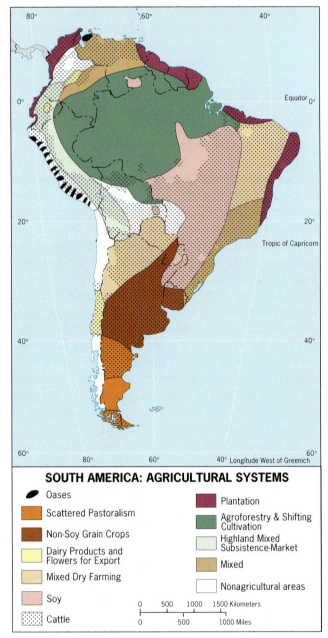

SOUTH AMERICA: AGRICULTURAL SYSTEMS

- ⬬ Oases
- Scattered Pastoralism
- Non-Soy Grain Crops
- Dairy Products and Flowers for Export
- Mixed Dry Farming
- Soy
- Cattle
- Plantation
- Agroforestry & Shifting Cultivation
- Highland Mixed Subsistence-Market
- Mixed
- Nonagricultural areas

0 500 1000 1500 Kilometers

0 500 1000 Miles

FIGURE 5-4 © H. J. de Blij, P. O. Muller, and John Wiley & Sons, Inc.

We should take note of one other dimension of the map: the vast region of soybean production extending from the northern corner of Uruguay all the way to the edge of the Amazon Basin. That signifies a major change: forest gave way to pasture and pastures to farming. Today Brazil is a world leader in the production of this versatile crop.

Cultural Landscapes

The countries, environments, and regional cultures of South America display enormous variety. To make sense of this diversity, several geographers have tried to map dominant regions and subregions of the realm based on

criteria ranging from economic mainstay to ethnic dimension. One early example of this kind was a map drawn by the cultural geographer John Augelli that showed the status of **South America's culture spheres**, as he called them, in the mid-twentieth century (Fig. 5-5A). It was a debatable delimitation, but the map did reveal contrasts between north and south that still prevail. Chile, Argentina, Uruguay, and southern Brazil retain cultural landscapes with stronger European imprints than Brazil and, for different reasons, Bolivia, Peru, and the countries to the north. Another merit of the Augelli map lies in its recognition that political and cultural borders in South America do not match. As one of our students on a field trip exclaimed, "you can go from one world to another without ever leaving Peru!" You can say the same about Brazil, Bolivia, or Ecuador.

Nevertheless, a map based on similar criteria today would look different (Fig. 5-5B). As we noted earlier, the cultural mix of South America is changing, and so are its economic activities. The area mapped in Figure 5-5A as "undifferentiated," by which Augelli meant the subsistence-dominated interior Amazon Basin and the far southwest, has shrunk (and the regional term is no longer acceptable—there is nothing "undifferentiated" about indigenous peoples living in forest habitats). Tropical-plantation agriculture continues, but does not dominate entire subregions as the earlier map suggests; economies have become far too diversified. And distant hinterlands have become more closely tied to national core areas. All this proves is that, in this fast-changing world, any cultural-geographic map represents a still-picture of a moment in time, not a permanent situation.

ECONOMIC INTEGRATION

As noted above, the separatism that has so long characterized international relations is giving way as South American countries discover the benefits of forging new partnerships with one another. With mutually advantageous trade the catalyst, a new continentwide spirit of cooperation is blossoming at every level. Periodic flareups of boundary disputes now rarely escalate into open conflict. Cross-border rail, road, and pipeline projects, stalled for years, are multiplying steadily. In southern South America, five formerly contentious nations are developing the *Hidrovia*, a system of river locks that is opening most of the Paraná-Paraguay Basin to barge transport. Investments today flow freely from one country to another, particularly in the agricultural sector.

Recognizing that free trade can solve many of the realm's economic-geographic problems, governments are pursuing several avenues of economic supranationalism.

FIGURE 5-5 © H. J. de Blij, P. O. Muller, and John Wiley & Sons, Inc.

In 2007, South America's republics were affiliating with the following major trading blocs:

Mercosur: Launched in 1995, this "Market South" comprises a free-trade zone and customs union linking Brazil, Argentina, Uruguay, Paraguay, and Venezuela (pending final approval at press time). Bolivia, Chile, and Mexico are associate members.

Andean Community: Formed as the Andean Pact in 1969 but restarted in 1995 as a customs union with common tariffs for imports, its members are Colombia, Peru, Ecuador, and Bolivia. Venezuela was a member until it withdrew in 2006.

South American Community of Nations: Founded in 2004 in Cuzco, Peru; intended as a forerunner to a union of all 12 South American states similar to the EU that would have a common currency, a South American passport, a parliament, and a common market. Its first agreement was to cooperate on the construction of a highway from Brazil across Peru to the Pacific coast, which would allow part of Brazil's growing trade with China to move overland and be handled by Peruvian ports (see Fig. 5-10).

7 **Free Trade Area of the Americas (FTAA):** Vigorously promoted by NAFTA proponents from North and

Middle America and energetically opposed by peasants and workers in South America, FTAA would create a free-trade area extending from Alaska to Tierra del Fuego and would thus incorporate NAFTA, Mercosur, and all other regional trading blocs. Recent political circumstances have not been conducive to this initiative.

URBANIZATION

As in most other less advantaged realms, South Americans are leaving the land and moving to the cities. This **urbanization** process intensified throughout the twentieth century, and it persists so strongly that South America's urban population percentage (80 percent, or 312 million of the continent's 389 million inhabitants) has now surpassed those of Europe and the United States. The urban population of South America has grown annually by about 5 percent since 1950, while the increase in rural areas was less than 2 percent. These numbers underscore not only the dimensions but also the durability of the **rural-to-urban migration** from the countryside **9** to the cities.

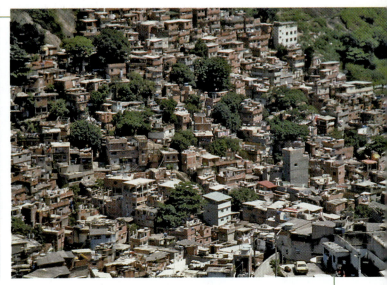

FROM THE FIELD NOTES

"Two unusual perspectives of Rio de Janeiro form a reminder that here the wealthy live near the water in luxury high-rises, such as these overlooking Ipanema Beach, while the poor have million-dollar views from their hillslope *favelas*, such as Rocinho."
© H. J. de Blij.

www.conceptcaching.com

Regional Patterns

The generalized spatial pattern of South America's urban transformation is displayed in Figure 5-6, which shows a *cartogram* of the continent's population.* Here we see not only the realm's countries in population-space relative to each other, but also the proportionate sizes of individual large cities within their total national populations.

Regionally, southern South America is most highly urbanized. Today in Argentina, Chile, and Uruguay, at least 87 percent of the people reside in towns and cities (statistically on a par with Europe's most urbanized countries). Ranking next on the urban scale is Brazil at 83 percent. Close behind, the next highest group of countries, averaging 82 percent urban, border the Caribbean in the north. Not surprisingly, the Amerindian, subsistence-dominated Andean countries constitute the realm's least urbanized zone (Peru, whose population is 74 percent urbanized, is that region's leader). Figure 5-6 also tells us a great deal about the relative positions of major metropolises in their countries. Three of them—Brazil's São Paulo and Rio de Janeiro, and Argentina's Buenos Aires—rank among the world's **megacities** (populations exceed 10 million). **10**

Causes and Challenges of Cityward Migration

In South America, as in Middle America, Africa, and Asia, people are attracted to the cities and driven from the poverty of the rural areas. Both ***push*** and ***pull*** factors are at work. Rural land reform has been slow in coming, and every year tens of thousands of farmers simply give up and leave, seeing little or no possibility for economic advancement. The cities lure them because they are perceived to provide opportunity—the chance to earn a regular wage. Visions of education for their children, better medical care, upward social mobility, and the excitement of life in a big city draw hordes to places such as São Paulo and Caracas.

But the actual move can be traumatic. Cities in developing countries are surrounded and often invaded by squalid slums, and this is where the urban immigrant most often finds a first—and frequently permanent—abode in a makeshift shack without even the most basic amenities and sanitary facilities. Unemployment is persistently high, often exceeding 25 percent of the available labor force. But still the people come, the overcrowding in the shantytowns worsens, and the threat of epidemic-scale disease (and other disasters) rises.

---

*A cartogram is a specially transformed map in which countries and cities are represented in proportion to their populations. Those containing large numbers are "blown up" in population-space, while those containing lesser numbers are "shrunk" in size accordingly. A similar map of world population distribution can be found on p. 20.

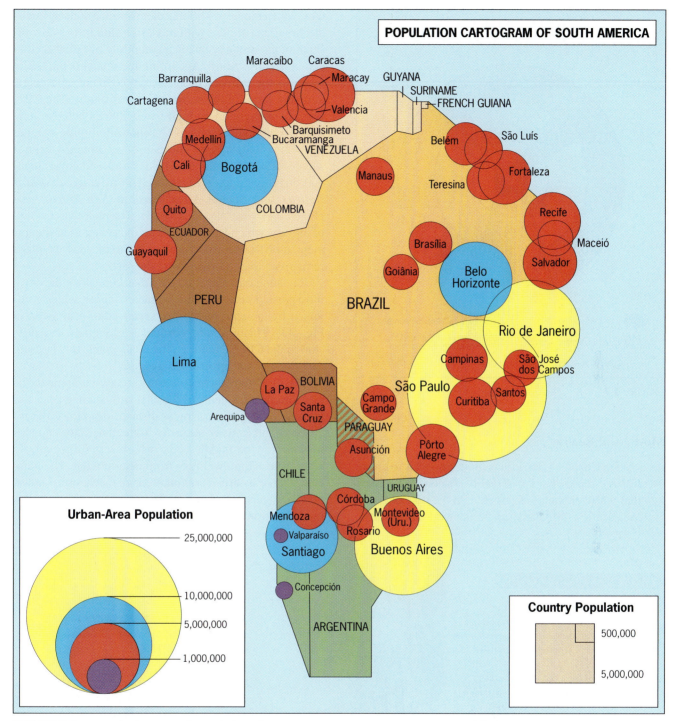

POPULATION CARTOGRAM OF SOUTH AMERICA

Urban-Area Population

— 25,000,000

— 10,000,000

— 5,000,000

— 1,000,000

Country Population

500,000

5,000,000

FIGURE 5-6   © H. J. de Blij, P. O. Muller, and John Wiley & Sons, Inc.

The "Latin" American City Model

The urban experience in the South and Middle American realms varies because of diverse historical, cultural, and economic influences. Nonetheless, there are many common threads that have prompted geographers to search for useful generalizations. One is the model of the intra-urban spatial structure of the **"Latin" American city** proposed by Ernst Griffin and Larry Ford (Fig. 5-7), which may have wider application to cities of developing countries in other geographic realms as well.

11

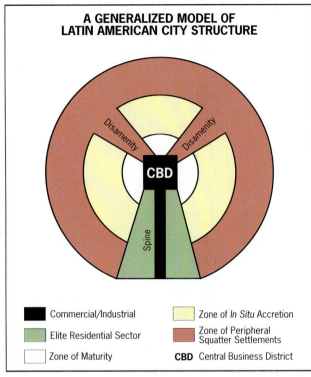

A GENERALIZED MODEL OF LATIN AMERICAN CITY STRUCTURE

Legend:
- ■ Commercial/Industrial
- ■ Elite Residential Sector
- □ Zone of Maturity
- □ Zone of *In Situ* Accretion
- ■ Zone of Peripheral Squatter Settlements
- **CBD** Central Business District

FIGURE 5-7 © H. J. de Blij, P. O. Muller, and John Wiley & Sons, Inc.

Model and Reality

As noted in Chapter 1, the idea behind a model is to create an idealized representation of reality, displaying as many key real-world elements as possible. In the case of South America's cities, the basic spatial framework of city struc-

ture, which blends traditional elements of South and Middle American culture with modernization forces now reshaping the urban scene, is a composite of radial sectors and concentric zones. Anchoring the model is the CBD, which is the primary business, employment, and entertainment focus of the surrounding metropolis. The CBD contains many modern high-rise buildings but also mirrors its colonial beginnings. As shown in Figure 4-3, the Spanish colonizers laid out their cities around a central square, or *plaza*, dominated by a church and government buildings. Santiago's *Plaza de Armas*, Bogotá's *Plaza Bolívar*, and Buenos Aires' *Plaza de San Martín* (see photo below) are classic examples. The plaza was the hub of the city, which later outgrew its old center as new commercial districts formed nearby; but to this day the plaza remains an important link with the past.

Radiating outward from the urban core along the city's most prestigious axis is the commercial *spine*, which is adjoined by the *elite residential sector* (shown in green in Fig. 5-7). This widening corridor is essentially an extension of the CBD, featuring offices, shopping, and housing for the upper and upper-middle classes.

The three remaining concentric zones are home to the less fortunate residents of the city, with income level and housing quality decreasing as distance from the CBD increases. The *zone of maturity* in the inner city contains housing for the middle class, who invest sufficiently to keep their aging dwellings from deteriorating. The adjacent *zone of* in situ *accretion* is one of much more modest housing interspersed with unkempt areas, representing a transition from inner-ring affluence to outer-ring poverty.

FROM THE FIELD NOTES

"At the heart of Buenos Aires lies the Plaza de San Martín, flanked by impressive buildings but, unusual for such squares in Iberian America, carpeted with extensive lawns shaded by century-old trees. Getting a perspective of the plaza was difficult until I realized that you could get to the top of the 'English Tower' across the avenue you see in the foreground. From there, one can observe the prominent location occupied by the monument to the approximately 700 Argentinian military casualties of the Falklands War of 1982 (center), where an eternal flame behind a brass map of the islands symbolizes Argentina's undiminished determination to wrest the islands from British control."
© H. J. de Blij.

www.conceptcaching.com

The outermost *zone of peripheral squatter settlements* is home to the millions of impoverished and unskilled workers who have recently migrated to the city. Here many newcomers earn their first cash income by becoming part of the **informal sector**, where workers are undocumented and money transactions are beyond the control of government. The settlements consist mostly of self-help housing, vast shantytowns known as *barrios* in Spanish-speaking South America and *favelas* in Brazil. Some of their entrepreneurial inhabitants succeed more than others, transforming parts of these shantytowns into beehives of activity that can propel resourceful workers toward a middle-class existence.

A final structural element of many South American cities forms an inward, wedgelike extension of the zone of peripheral squatter settlements and is known as the *disamenity sector*. It consists of undesirable land along highways, railway lines, river banks, and other low-lying areas; people are so poor that they are forced to live in the open. Thus this realm's cities present enormous contrasts between poverty and wealth, squalor and comfort—harsh contrasts all too frequently seen in the cityscape.

POINTS TO PONDER

- Mercosur, the South American common market, supposedly stands for open trade and democracy. Venezuela's (pending) accession raises questions about the organization's future direction.

- Colombian cocaine comes to North American consumers from staging points in remote parts of the country. On its way to Brazil, much of it is redistributed from interior French Guiana, where it may account for one-fifth of the territorial economy.

- President Evo Morales of Bolivia is campaigning for a 10-kilometer (6-mi) stretch of Chilean Pacific coastline to end his country's landlocked situation. His ally, Hugo Chávez of Venezuela, proclaims that he would "like to swim on a Bolivian beach."

- Watch for Argentina's president to revive Argentina's claim to the (British) Falkland Islands. Hugo Chávez has told the United Kingdom that the islands belong to Argentina and should be "turned over."

Regions of the Realm

South America divides geographically into four rather clearly defined regions (Fig. 5-3):

1. *The North* consists of five entities that display a combination of Caribbean and South American features: Colombia, Venezuela, and those that represent three historic colonial footholds by Britain (Guyana), the Netherlands (Suriname), and France (French Guiana).

2. *The West* is formed by four republics that share a strong Amerindian cultural heritage as well as powerful influences resulting from their Andean physiography: Ecuador, Peru, Bolivia, and, transitionally, Paraguay.

3. *The South*, often called the "Southern Cone," includes three countries that actually conform to the much-misused regional term "Latin" America: Argentina, Chile, and Uruguay (all with strong European imprints and little remaining Amerindian influence) plus aspects of Paraguay.

4. *Brazil* by itself accounts for just under half the continent in terms of territory as well as population. Here the dominant Iberian influence is Portuguese, not Spanish; and here Africans, not Amerindians, form a significant element in demography and culture. Brazil in some ways is a bridge between the Americas and Africa, and since we turn to Subsaharan Africa next, we discuss it last in this chapter.

MAJOR CITIES OF THE REALM

City	Population* (in millions)
Asunción, Paraguay	2.1
Belo Horizonte, Brazil	5.8
Brasília, Brazil	3.8
Bogotá, Colombia	8.2
Buenos Aires, Argentina	12.9
Caracas, Venezuela	3.0
Guayaquil, Ecuador	2.6
La Paz, Bolivia	1.7
Lima, Peru	7.5
Manaus, Brazil	1.8
Montevideo, Uruguay	1.3
Quito, Ecuador	1.6
Rio de Janeiro, Brazil	12.0
Santiago, Chile	5.9
São Paulo, Brazil	25.8

*Based on 2008 estimates.

THE NORTH: FACING THE CARIBBEAN

As Figure 5-5A reminds us, the countries of South America's northern tier have something in common besides their coastal location: each has a coastal tropical-plantation zone on the Caribbean colonial model. Especially in the three Guianas, early European plantation development entailed the forced immigration of African laborers and eventually the absorption of this element into the population matrix. Far fewer Africans were brought to South America's northern shores than to Brazil's Atlantic coasts, and tens of thousands of South Asians also arrived as contract laborers and stayed as settlers, so the overall situation here is not comparable to Brazil's. And it is also distinctly different from that of the rest of South America.

Today, Guyana, Suriname, and French Guiana still display the coastal orientation and plantation dependency with which the colonial period endowed them, although logging their tropical forests is penetrating and ravaging the interior. In Venezuela and Colombia, however, farming, ranching, and mining drew the population inland, overtaking the coastal-plantation economy and creating diversified economies.

Figure 5-8 shows that not only Venezuela but also Colombia is Caribbean in its orientation: in Venezuela, coastal oil and natural gas reserves have replaced plantations as the economic mainstay, and Colombia's Andean valleys open toward the north where roads, railways, and pipelines lead toward Caribbean ports such as Cartagena and Barranquilla. Colombia's Pacific coast is hardly a factor in the national economy despite the outlet at Buenaventura. But, as we will see, the North's locational advantages are countered by physical, economic, and political obstacles that continue to prevent the realization of their full potential.

Colombia

Imagine a country more than twice the size of France, not at all burdened by overpopulation, with a physical geography so varied that it can produce crops ranging from the temperate to the tropical, possessing world-class reserves

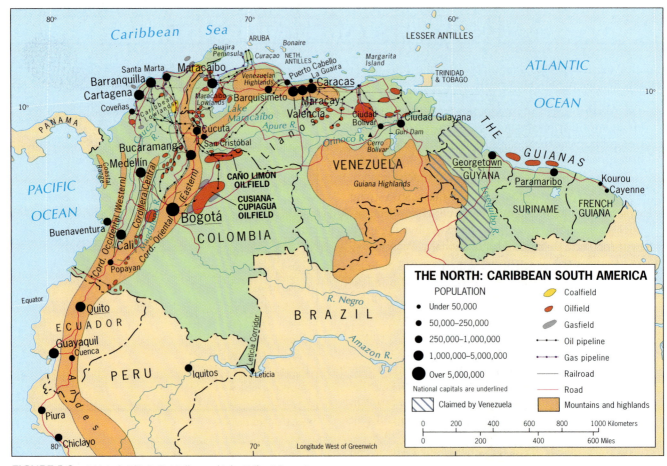

FIGURE 5-8   © H. J. de Blij, P. O. Muller, and John Wiley & Sons, Inc.

of oil and natural gas, plus nearly 2000 kilometers (1200 mi) of coastline on Atlantic and Pacific waters. Situated in the northwestern corner of South America, closer than any of its neighbors to the markets of the north, with a single language and one dominant religion, would such a country not assuredly thrive amidst the burgeoning economic geography of its hemisphere?

The answer is no. For a combination of reasons, historic as well as current, Colombia is a troubled state. Colombia's cultural uniformity never did produce social cohesion: the same physical geography that diversifies its environments also separates its population clusters, which are so tenuously interconnected that the country, larger than Texas and California combined, had only about 400 kilometers (250 mi) of paved four-lane highway in 2007. In recent decades, Colombia's proximity to U.S. markets has become a curse as well as a blessing: at the root of Colombia's current problems lies its illicit drug industry, generating another phase in a seemingly endless cycle of violence.

History of Conflict

Colombia's current disorder is not its first. In the past, civil wars between conservatives and liberals (on Roman Catholic religious issues) developed into conflicts pitting rich against poor, elites against workers. In Colombia today, people still refer to the last of these wars as *La Violencia*, a decade of strife beginning in 1948 during which as many as 200,000 people died. When, in the late 1950s, the two sides—the Conservative and Liberal parties—called a truce and agreed on constitutional reform to preclude a recurrence, Colombia at last seemed able to realize its economic potential. Exports of coffee, for which the country was famous, soared. Oil exports from reserves in the northern part of the country increased. Minerals, fruits, and cut flowers flowed to foreign markets, half by value to the United States, a quarter to Europe.

But in the 1970s, disaster struck. In remote parts of the country, groups opposed to the power-sharing monopoly between the political parties began a campaign of terrorism, damaging the developing infrastructure and destroying confidence in the future. Cattle ranchers on haciendas in remote northern and eastern areas of the country began forming paramilitaries to protect their interests and to resist government "interference" in the form of regulation, inspection, and taxation. Simultaneously, the U.S. market for narcotics expanded rapidly, and many Colombians got involved in the drug trade. Powerful and wealthy drug cartels formed in major cities such as Medellín and Cali, with networks that influenced all facets of Colombian life from the peasantry to the politicians. The fabric of Colombian society unraveled.

Looking over the CBD of Bogotá, Colombia in a southeasterly direction, where high relief curtails building, you are reminded of the formidable and divisive topography of this large country. The airport (Eldorado International!) lies behind us to the northwest, and the larger and less prosperous suburbs occupy the hilly area to the west (to the right of this view). With more than 8 million inhabitants, metropolitan Bogotá is larger than Colombia's next seven urban areas combined and is the country's primate city—but terrain, distance, and persistent insurgencies erode the capital's supremacy.   © Lonely Planet Images/Getty Images, Inc.

People and Resources

As Figure 5-8 shows, western Colombia is mountainous while the east lies in the Orinoco and Amazon basins. Most of the country's 48.2 million people (the largest population in all of Spanish-speaking South America) live in the western and northern parts of Colombia. Physiography has contributed to the concentration of this population into more than a dozen separate clusters. Some of them are located

The Geography of Cocaine

ANY GEOGRAPHICAL DISCUSSION of northwestern South America today must take note of one of its most widespread activities: the production of illegal narcotics. Of the enormous flow of illicit drugs that enter the United States each year, the most widely used substance undoubtedly is cocaine—all of which comes from South America, mainly Bolivia, Peru, and Colombia (the sources of at least 75 percent of the total world supply). Within these three countries, cocaine annually brings in billions of (U.S.) dollars and "employs" tens of thousands of workers, constituting an industry that functions as a powerful economic force. Those who operate the industry have accumulated considerable power through bribery of politicians, kidnappings and other forms of intimidation, threats of terrorism, and alliances with guerrilla groups in outlying zones beyond governmental control. The cocaine industry itself is structured within a tightly organized network of territories that encompass the various stages of this drug's production.

The first stage of cocaine production is the extraction of coca paste from the coca plant, a raw-material-oriented activity that is located near the areas where the plant is grown. The coca plant was domesticated in the Andes by the Incas centuries ago; millions of their descendants still chew coca leaves for stimulation and brew them into coca tea, the leading beverage of highland South America. The main zone of coca-plant cultivation is along the eastern slopes of the Andes and in adjacent tropical lowlands in Bolivia, Peru, and Colombia. Today, five areas dominate in the growing of coca leaves for narcotic production (see Fig. 5-10): Bolivia's Chaparé district in the marginal Amazon lowlands northeast of the city of Cochabamba; the Yungas Highlands north of the Bolivian capital, La Paz; north-central Peru's Huallaga and neighboring valleys; south-central Peru's Apurimac Valley, east of Huancayo; and the green-colored zones (in Fig. 5-9) around the guerrilla-controlled territories of southern Colombia, which are now extending southeastward along the Putumayo River that marks the Peru–Colombia border. These areas, which can produce as many as nine crops per year, are especially conducive to high leaf yields thanks to favorable local climatic and soil environments that also allow plants to develop immunity to many diseases and insect ravages. Despite government efforts to aerially spray herbicides and manually eradicate plants in the field, all of these areas continue to thrive as coca cultivation has become a full-fledged cash crop. Operations in guerrilla-controlled portions of Colombia have reached the scale of plantations, inducing thousands of local subsistence farmers (sometimes at gunpoint) to join the more lucrative ranks of the field workforce. In Peru and Bolivia, the specialized coca-cultivation zones have lured an even larger peasant-farmer population to abandon the less profitable production of food crops (coca's per-acre income is now seven times greater than cocoa's), thereby further reducing the capability of these nutrition-poor countries to feed themselves.

As (U.S.-supported) government efforts to eliminate coca-raising have intensified in recent years, the geography of production has made certain adjustments. The most aggressive pursuit occurred in Colombia, where total acreage may have been halved since 2000, and the net effect has been the displacement of coca cultivation to nearby countries. This shift, however, actually appears to have strengthened rather than weakened the industry: *The Economist* recently reported that not only are coca-leaf prices—and overall productivity—rising steadily, but also that street prices for cocaine were falling worldwide. Geographically, the fastest-growing areas outside Colombia are in the rainforested montaña of Peru, particularly in the Putumayo Valley along the Colombian border in the far north and in the Bolivian borderland at the opposite end of the country (Fig. 5-10).

The coca leaves harvested in the source areas of the eastern Andes and adjoining interior lowlands make their way to local collection centers, located at the convergence of rivers and trails, where coca paste is extracted and prepared. This begins the second stage of production, which involves the refining of that coca paste (about 40 percent pure cocaine) into cocaine hydrochloride (more than 90 percent pure), a lethal concentrate that is diluted with substances such as sugar or flour before being sold on the streets to consumers. Cocaine refining requires sophisticated chemicals, carefully controlled processes, and a labor force skilled in their supervision, and here Colombia has predominated. Most of this activity takes place in ultramodern processing centers located in the rebel-held territory of the lowland central-south and east, beyond the reach of the Bogotá government. Interior Colombia also possesses the geographic advantage of intermediate location, lying between the source areas to the south (as well as locally) and the U.S. market to the north, a short hop for smugglers by air or sea across the Caribbean. Nonetheless, Colombia's pursuit of the drug industry has made inroads here since 2000, and the latest evidence suggests that cocaine refining is rapidly expanding south of the border in Peru—where the government has not intervened and well-financed Mexican drug mobsters are successfully rerouting the region's output through their country.

The final stage of production entails the distribution of cocaine to the marketplace, which depends on an efficient, clandestine transportation network that leads to the United States (and increasingly to Europe). Private planes operating out of remote airstrips are a favorite method, but much cocaine now travels overland to be smuggled by sea through Pacific seaports, and increasingly via land routes through Central America and Mexico all the way to the U.S. border.

When it comes to energy resources, Venezuela is in the lead in South America, but Colombia has formidable reserves as well, the most extensive of which lie in the central north on the inland side of the Andes' easternmost range. This example of the industry's environmental impact shows what is reputedly the largest oilfield in South America, based on reserves in the Cusiana river basin, about 250 kilometers (150 mi) east of Bogotá in the heart of the country. © Gamma Presse, Inc.

in the Caribbean Lowlands; others lie in the Magdalena, Cauca, and lesser valleys between the parallel ranges of the Andes (including Medellín and Cali); and still others are found in basins within the mountains themselves, the most important being the cluster around the capital city of Bogotá (elevation: 2700 meters [8500 ft]; see photo p. 257). Colombia's huge coffee crop—second only to Brazil's on the continent and in the world—comes from the numerous *templada* zones on the Andean slopes. But the more sparsely populated east and south have become production centers for the coca crop that sustains the country's illicit narcotics industry (see box titled "The Geography of Cocaine").

Colombia's belt of oil and natural gas reserves flanks the border with Venezuela from Lake Maracaibo south to the Caño Limón oilfield, then continues onward along the Andean front through the country's largest petroleum deposits (the Cusiana-Cupiagua oilfield; see photo above), and terminates at the newly discovered Guando oilfield in the upper Magdalena Valley southwest of Bogotá (Fig. 5-8). Pipelines connect the oilfields to the Caribbean ports of Cartagena, Barranquilla, Santa Marta, and Coveñas. Beneath that coastal lowland, Colombia also has substantial reserves of coal, a product that now ranks high on the list of exports. With its fuel, mineral, and agricultural productivity, Colombia's economy might have been one of South America's most prosperous. But the rise of the narcotics industry, coupled with the legacy of *La Violencia*, has dealt the country a devastating blow.

Cocaine's Cost

The era of narcotics brought to Colombia a new version of violence, a period of convulsion and anarchy that escalated from shocking acts of core-area terrorism to virtual warfare

in the countryside. Rival armies controlled production in different parts of the country; the Colombian armed forces could not defeat them. Meanwhile, the market price of cocaine was supported by the largest consumer, the United States, even as Washington sent money, equipment, and advisors to help Bogotá in its "eradication" campaign. Rebels sabotaged pipelines and took prominent Colombians hostage; civilians by the thousands died in the crossfire among the belligerents.

Threats of the Insurgent State

Colombia's nightmare worsened in the 1990s, when the two main rebel forces, the so-called United Self-Defense Forces of Colombia (AUC) and the Revolutionary Armed Forces of Colombia (FARC), became powerful enough to create permanent domains in the north and south of the country, respectively, forcing the civilian populations in these domains to live under their authority and even negotiating their territorial status with the legitimate government in the capital. At the time, the map of Colombia showed a Switzerland-sized enclave in southern Colombia controlled by FARC and called *Farclandia*, and a smaller but similar area near the Panamanian border in the remote northwest, ruled by the AUC (Fig. 5-9). Both of these enclaves operated on drug money as the rebels, on confiscated real estate, established their own coca-paste collection networks, drug laboratories, production units, and airfields to move the product to market—all the while maintaining command posts and border controls.

What happened in Colombia was not unique, although the combination of circumstances—historical, physiographic, agricultural—was. Many countries were and are affected by rebel movements, and most of them manage

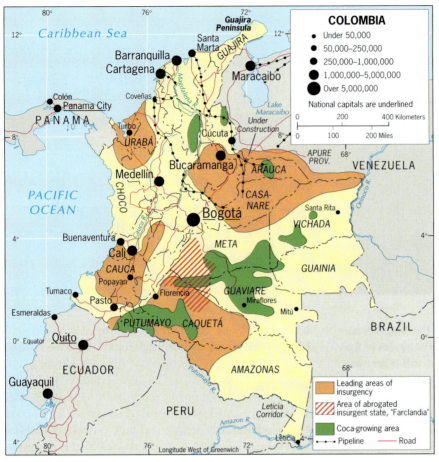

FIGURE 5-9 © H. J. de Blij, P. O. Muller, and John Wiley & Sons, Inc.

try. The geographer Robert McColl, who translated the political construct of equilibrium into the geographic concept of the insurgent state, identified Cuba and South Vietnam as countries in which insurgent states had been established before their rebel leaders took over the government. The insurgent state obviously cannot last indefinitely; no government can allow it, and no rebel movement is ever satisfied without a final resolution. Thus in the third and decisive stage of the process, referred to as *counteroffensive*, the government and the rebels, usually following the breakdown of negotiations, engage in a final struggle that will decide the future of the country.

A Nation in Crisis

When insurgent states arose in Colombia, it appeared that the nation would devolve into a **failed state**, a country whose institutions have collapsed and in which anarchy prevails. (Afghanistan during the rule of the fanatic Taliban was a failed state [see pp. 397–398], and terrorists quickly established bases there.) The government's willingness

to overcome the challenge. Sometimes, however, one or more rebel movements succeed in wresting part of the national territory from the control of the central government, **14** and when this happens an **insurgent state** may be in the making. In the case of Colombia, the first stage in this process occurred during *La Violencia*, when the country was sufficiently destabilized to enter a period political scientists refer to as *contention*. During this phase, a rebel movement reveals itself to be strong enough to challenge the government and its forces, inflicting casualties, attacking infrastructure, and damaging the economy. Most of the time, perhaps with outside help, governments eventually defeat the rebels. But when a rebel movement gains effective control over a substantial area of the country and its population, as happened in Colombia, a second, more ominous stage referred to as *equilibrium* is reached. At this point the possibility arises that the rebels will achieve a stalemate, forcing the government to acknowledge its inability to protect its citizens under rebel control and to make concessions such as the recognition of rebel authority. Now the insurgent state has emerged, and some insurgent states achieve secession and international recognition—or their regimes take over the whole coun-

to negotiate with the revolutionaries and its agreement to recognize rebel control in large parts of the country appeared to signal a prolonged prelude to an ineffective counteroffensive. But the election in 2002 of President Alvaro Uribe changed the picture. Uribe reversed the government's course: he changed tactics by withdrawing recognition of the rebel enclaves, mounted a major military offensive against the regional revolutionary movements, and then offered amnesty to those paramilitaries willing to turn over their weapons. This demobilization program had some success—both in Urabá Province in the north, where the AUC had held sway, and in the vast swath of south-central Colombia, where FARC had established its insurgent state—and large numbers of rebels gave up the battle and laid down their arms.

President Uribe was elected to a second term in 2006, reflecting Colombians' approval of his policies as enacted by the government. But there were concerns over the implications of the demobilization plan: just by laying down their arms, paramilitary "soldiers" could escape punishment for criminal acts committed while serving their masters. And the process exposed ugly relationships between paramilitary commanders and members of the government, revealing an even deeper penetration of the administration than had been assumed. In 2007, the outcome of the government's program was still uncertain, and the political geography of Colombia hung in the balance.

Venezuela

A long and tortuous boundary separates Colombia from Venezuela, its neighbor to the east. Much of what is important in Venezuela is concentrated in the northern and western parts of the country, where the Venezuelan Highlands form the eastern spur of the north end of the Andes system. Most of Venezuela's 27.9 million people are concentrated in these uplands, which include the capital of Caracas, its early rival Valencia, and the commercial/industrial centers of Barquisimeto and Maracay.

The Venezuelan Highlands are flanked by the Maracaibo Lowlands and Lake Maracaibo to the northwest and by a vast plainland of savanna country, known as the **Llanos**, in the Orinoco Basin to the south and east (Fig. 5-8). The Maracaibo Lowlands, once a disease-infested, sparsely peopled coastland, today constitute one of the world's leading oil-producing areas. Much of the oil is drawn from reserves that lie beneath the shallow waters of the lake itself. Actually, Lake Maracaibo is a misnomer, for the "lake" is open to the ocean and is in fact a gulf with a very narrow entry. The country's third-largest city, Maracaibo, is the focus of the petroleum industry that transformed the Venezuelan economy in the 1970s; however, as we shall see, since then oil has been more of a curse than a blessing.

The *Llanos* on the southern side of the Venezuelan Highlands and the Guiana Highlands in the country's southeast, like much of Brazil's interior, are in an early stage of development. The 300- to 650-kilometer (180- to 400-mi)-long *Llanos* slope gently from the base of the Andean spur to the Orinoco River. Their mixture of savanna grasses and scrub woodland support cattle grazing on higher ground, but widespread wet-season flooding of the more fertile lower-lying areas has thus far inhibited the plainland's commercial farming potential (much of the development of the *Llanos* to date has been limited to the exploitation of its substantial oil reserves). Crop-raising conditions are more favorable in the *tierra templada* areas (see discussion of the altitudinal zonation concept on p. 226) of the Guiana Highlands. Economic integration of this more remote interior zone with the rest of Venezuela has been spearheaded by the discovery of rich iron ores on the northern flanks of the Guiana Highlands southwest of Ciudad Guayana. Local railroads now connect with the Orinoco, and from there ores are shipped directly to foreign markets.

Oil and Politics

Despite these opportunities, Venezuela since 1998 has been in upheaval as longstanding economic and social problems finally intensified to the point where the electorate decided to push the country into a new era of radical political change. For nearly two decades since the euphoric 1970s, oil had not bettered the lives of most Venezuelans. A major reason was that the government unwisely acquired the habit of living off oil profits, forcing the country to suffer the consequences of the long global oil depression that began in the early 1980s. Now Venezuela found itself heavily burdened by a huge foreign debt it had incurred in borrowing against future oil revenues that were not materializing fast enough. By the mid-1990s, the government was required to sharply devalue the currency, and a political crisis ensued that resulted in a severe recession and widespread social unrest. With more and more Venezuelans enraged at the way their oil-rich country was approaching bankruptcy without making progress toward the more equitable distribution of the national wealth, voters resoundingly turned in an extreme direction in the 1998 presidential election. Expressing their disgust with Venezuela's ruling elite of both political parties, they elected Hugo Chávez, a former colonel who in 1992 had led a failed military coup. Reaffirming their decision in 2000, Venezuelans gave nearly 60 percent of their votes to Chávez, apparently providing him with a mandate to act as strongman on behalf of the urban poor and the increasingly penurious middle class.

Chávez has indeed pursued this course since entering office in 1999, sweeping aside Congress and the Supreme Court, supervising the rewriting of the Venezuelan constitution in his own image, and proclaiming himself the leader of a "peaceful leftist revolution" that will transform the country. Although he professes that social equality ranks highest on his agenda, Chávez has stirred up racial divisions by actively promoting mestizos (68 percent of the population) over those of European background (21 percent). And in the international arena, Chávez sparks controversy at every turn: angering the government of neighboring Colombia by expressing his "neutrality" in its confrontation with cocaine-producing insurgent forces; unsettling another neighbor, Guyana, by aggressively reviving a century-old territorial claim to the western zone of that country (the striped area in Fig. 5-8); embracing Fidel Castro and providing communist Cuba with oil in exchange for the assignment to Venezuela of thousands of doctors and technicians; and obstructing the globalization plans of the United States throughout South America, using the windfall income from high-priced oil to reward allies in this effort with substantial subsidies. In 2006, Chávez was reelected with an overwhelming majority over an opponent from the oil-rich western State of Zulia, where oil and cattle dominate the economy—the "Texas of Venezuela." Opposition to Chávez is so strong in this corner of the country (the State encircles Lake Maracaibo) that a pro-autonomy movement has arisen there, a devolutionary response to the current government's populist policies.

Chávez has positioned himself as the champion of Venezuela's (and South America's) poor, trumpeting his "Bolivarian Revolution" as a local alternative to what he describes as the insidious encroachment of U.S. imperialism. He has taken Venezuela to center stage in the ideological contest in this realm, casting doubt on the virtues of democracy and free enterprise, castigating elites for not doing enough to reduce the gap between rich and poor, and urging the downtrodden to assert themselves. His message resonates across the continent.

Guyana

The eastern flank of Caribbean South America is the most "Caribbean" of the region, and its cultural landscapes often resemble the Caribbean rather than South America. Here British, Dutch, and French colonial powers acquired possessions and established plantations, brought in African and Asian workers, and created economies similar to those of their Middle American domains. Today these three entities (two of which are independent states and one still a *département* of France) have small populations—also more reminiscent of Caribbean islands than of their South American neighbors.

With 740,000 people, Guyana still has more inhabitants than Suriname and French Guiana combined. When Guyana became independent in 1966, its British rulers left behind an ethnically and culturally divided population in which people with South Asian (Indian) ancestry make up about 50 percent and those with African heritage (including African-European ancestry) 39 percent. This makes for contentious politics, given the religious mix, which is approximately 57 percent Christian and 37 percent Hindu and Muslim. Guyana remains dominantly rural, and plantation crops continue to figure strongly among exports (gold from the interior is the most valuable single product). Oil may soon become a factor in the economy, though, because a recently discovered reserve that lies offshore from Suriname extends westward beneath Guyana's waters. Still, Guyana is among the realm's poorest and least urbanized countries, and it is strongly affected by its neighbor's narcotics industry. Its thinly populated interior, beyond the reach of anti-drug campaigns, has become a staging area for drug distribution to Brazil, North America, and even Europe. A recent official report suggests that drug money amounts to as much as one-fifth of Guyana's total economy.

Guyana's economic weakness translates into vulnerability, and its government for many years has had reason to worry about Venezuela's territorial intentions. At various times since 1966, Venezuela has laid claim to a large part of the country (see Fig. 5-8), and while the Chávez regime has not pressed these demands, previous Venezuelan reassurances have been followed by renewed assertions of ownership.

Suriname

Suriname actually progressed more rapidly than Guyana after it became independent in 1975, but persistent political instability soon ensued. The Dutch colonists brought South Asians, Indonesians, Africans, and even some Chinese to their colony, making for a fractious nation. More than 100,000 residents—about one-quarter of the entire population—emigrated to the Netherlands, and were it not for support from its former colonial ruler Suriname would have collapsed. Still, the rice farms laid out by the Dutch give Suriname self-sufficiency and even allow for some exporting, and again plantation crops continue to figure among the exports. Suriname's leading income producer, however, is its bauxite (aluminum ore) mined in a zone across the middle of the country. And the recent oil finds (Fig. 5-8) may provide important revenues in years ahead.

Suriname, whose population totals 480,000, is one of South America's poorest countries, and its prospects remain bleak. Along with Guyana, Suriname is involved in an environmental controversy centering on its luxuriant tropical forests. Timber companies from Asia's Pacific Rim offer large rewards for the right to cut down the magnificent hardwood trees, but conservationists are trying to slow the destruction by buying up concessions before they can be opened to logging.

Suriname's cultural geography is enlivened by its languages. Dutch may be the official tongue, but other than by officials and in schools it is not heard much. A mixture of Dutch and English, *Sranan Tongo*, serves as a kind of common tongue, but you can also hear Amerindian, Hindi, Chinese, Indonesian, and even some French Creole in the streets. Suriname may have a Dutch past, but today it is a member of Caricom, the English-speaking Caribbean economic organization.

French Guiana

This easternmost outpost of Caribbean South America is a dependency—mainland South America's only one. This territory is an anomaly in other ways as well. Consider this: it is not much smaller than South Korea (population: nearly

50 million) with a population of just over 200,000. Its status is an *Overseas Département* of France, and its official language is French. Nearly half the population resides in the urban area of the capital, Cayenne.

In 2007 there still was no prospect of independence for this severely underdeveloped relic of the former French Empire. Gold remains the most valuable export, and the small fishing industry sends some exports to France. But what really matters here in French Guiana is the European Space Agency's launch complex at Kourou on the coast, which accounts for more than half of the territory's entire economic activity. From plantation farming to spaceport . . . this must be the ultimate story of globalization.

● THE WEST: ANDEAN SOUTH AMERICA

The second regional grouping of South American states—the Andean West (Fig. 5-10)—is dominated physiographically by the great Andes Mountains and historically by Amerindian peoples. This region encompasses Peru, Ecuador, Bolivia, and transitional Paraguay, the last with one foot in the West and the other in the South (Figs. 5-5A and 5-5B). Bolivia and Paraguay also constitute South America's only two landlocked countries.

Spanish conquerors overpowered the Amerindian nations, but they did not reduce them to small minorities as happened to so many indigenous peoples in other parts of the world. Today, more than 55 percent of the people in Peru, the region's most populous country, are Amerindian; in Bolivia the Amerindian majority (62 percent) is even larger. More than 30 percent of Ecuador's population identifies itself as Amerindian, and in Paraguay the ethnic mix, not regionally clustered as in the other three countries, is overwhelmingly weighted toward Amerindian ancestry.

As Table G-1 reports, this is South America's poorest region economically, with lower incomes, higher numbers of subsistence farmers, and fewer opportunities for job-seekers. There is no oil-rich Venezuela or agriculturally bountiful Argentina in the South America's West. For a very long time, the urbane lives of the land-owning elite have been worlds away from the hard-scrabble existence of the landless peonage (the word *peon* is an old Spanish term for an indebted day laborer). But today, this region, like the realm as a whole, is stirring, and oil and natural gas are part of the story. In Bolivia, the first elected president of Amerindian ancestry is trying to gain control over an energy industry not used to his aggressive tactics. In Ecuador's 2006 elections, a populist gained the presidency on a promise to divert more of the country's oil revenues toward domestic needs. In Peru, where an energy

era seems to be opening as new reserves are discovered, the government is under pressure to limit foreign involvement and to put domestic needs and rights first.

In short, this is a crucial region in significant transition, where U.S. ambitions and intentions collide with domestic initiatives and where globalizers and locals find themselves face to face.

Peru

Peru straddles the Andean spine for more than 1600 kilometers (1000 mi) and is the largest of the region's four republics in both territory and population (29.1 million). Physiographically and culturally, Peru divides into three subregions: (1) the desert coast, the European-mestizo region; (2) the Andean highlands or *Sierra*, the Amerindian region; and (3) the eastern slopes, the sparsely populated Amerindian-mestizo interior (Fig. 5-10).

Three Subregions

Lima and its port, Callao, lie at the center of the desert *coastal strip*, and it is symptomatic of the cultural division prevailing in Peru that for nearly 500 years the capital city has been positioned on the periphery, not in a central location in a basin of the Andes. From an economic point of view, however, the Spaniards' choice of a headquarters on the Pacific coast proved to be sound, for the coastal subregion has become commercially the most productive part of the country. A thriving fishing industry contributes significantly to the export trade; so do the products of irrigated agriculture in some 40 oases distributed all along the arid coast, which include fruits such as citrus, olives, and avocadoes, and vegetables such as asparagus (a big money-maker) and lettuce.

The *Sierra* (Andean) subregion occupies about one-third of the country and is the ancestral home of the largest component in the total population, the Quechua-speakers who had been subjugated by the Inca rulers when the Spanish conquerors arrived. Their survival during the harsh colonial regime was made possible by their adaptation to the high-altitude environments they inhabited, but their social fabric was ripped apart by communalization, forced cropping, religious persecution, and outward migration to towns and haciendas where many became serfs. Another Amerindian people, the Aymara, live in the area of Lake Titicaca, making up about 7 percent of Peru's population but larger numbers in neighboring Bolivia.

Although these Amerindian people make up more than half of the population of Peru, their political influence remains slight—even during the recent term of President Alejandro Toledo, who had strong support in the Andean

FIGURE 5-10 © H. J. de Blij, P. O. Muller, and John Wiley & Sons, Inc.

districts. Nor is the Andean subregion a major factor in Peru's commercial economy—except, of course, for its mineral storehouse, which yields copper, zinc, and lead from mining centers, the largest of which is Cerro de Pasco. In the high valleys and intermontane basins, the Amerindian population is clustered either in isolated villages, around which people practice a precarious subsistence agriculture, or in the more favorably located and fertile areas where they are tenants, peons on white- or mestizo-owned haciendas. Most of these people never receive an adequate daily caloric intake or balanced diet of any sort. The wheat produced around Huancayo, for example, is mainly exported and would in any case be too expensive for the Amerindians themselves to buy. Potatoes, barley, and corn are among the subsistence crops grown here in the *tierra fría* zone, and in the *tierra helada* of the higher

AMONG THE REALM'S GREAT CITIES . . . Lima

EVEN IN A realm marked by an abundance of primate cities, Lima (7.5 million) stands out. Here reside 26 percent of the Peruvian population—who produce over 70 percent of the country's gross national income, 90 percent of its collected taxes, and 98 percent of its private investments. Economically, at least, Lima is Peru.

Lima began as a modest oasis in a narrow coastal desert squeezed between the cold waters of the Pacific and the soaring heights of the nearby Andes. Near here the Spanish conquistadors discovered one of the best natural harbors on the western shoreline of South America, where they founded the port of Callao; but they built their city on a site 11 kilometers (7 mi) inland where soils and water supplies proved more beneficial. This city was named Lima and quickly became the Spaniards' headquarters for all their South American territories.

Since independence, Lima has continued to dominate Peruvian national life. The city's population growth was manageable through the end of the 1970s, but the past quarter-century has witnessed a disastrous doubling in its size. As usual, the worst problems are localized in the peripheral squatter shantytowns that now house close to one-half of the metropolitan population. Lima's CBD,

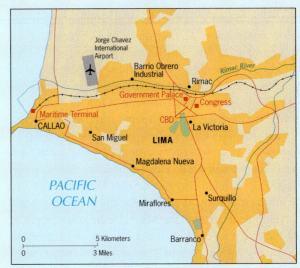

© H. J. de Blij, P. O. Muller, and John Wiley & Sons, Inc.

however, is worlds removed from this squalor, its historic landscape dotted with five-star hotels, new office towers, and prestigious shops as foreign investments have poured in since 1990.

---

basins the Amerindians graze their llamas, alpacas, cattle, and sheep.

Of Peru's three subregions, the ***Oriente***, or East—the inland slopes of the Andes and the Amazon-drained, rainforest-covered ***montaña***—is the most isolated. The focus of the eastern subregion, in fact, is Iquitos, a city that looks east rather than west and can be reached by oceangoing vessels sailing 3700 kilometers (2300 mi) up the Amazon River across northern Brazil. Iquitos grew rapidly during the Amazon wild-rubber boom of a century ago and then declined; now it is finally growing again and reflects Peruvian plans to open up the eastern interior. Petroleum was discovered west of Iquitos in the 1970s, and since 1977 oil has flowed through a pipeline built across the Andes to the Pacific port of Bayóvar. But what the *Oriente* needs most is a highway to the coast.

An Energy Age

Over the past decade, Peruvian planners have come to view their country as being on the verge of an energy age as exploration in the interior is confirming large reserves of oil and natural gas. Pipelines now carry gas from the recently discovered Camisea reserve (north of Cuzco) to a conver-

sion plant on the Paracas Peninsula south of Lima, from where an ocean-floor pipeline sends it to an offshore loading platform for tankers taking it to the U.S. market (Fig. 5-10). But environmentalists and supporters of indigenous rights are raising issues in the interior while political activists, following the example of Bolivia's leader, are arguing against the terms of trade. As the region's political landscape changes, its economic geography will grow more complicated, and the U.S. objective of an FTAA, the regional symbol of globalization, seems to be slipping beyond reach.

Insurgency to Stability

Peru, by the turn of this century, appeared to have put behind it the lengthy period of instability during which its government was threatened by well-organized guerrilla movements. The most serious threat, which for a time during the 1980s seemed on the verge of achieving an insurgent state in its Andean base, was the so-called Maoist-inspired ***Sendero Luminoso*** (Shining Path) movement. But forceful action under the later-disgraced president of Japanese ancestry, Alberto Fujimori, defeated the rebel movement and returned Peru to political stability and economic growth.

The social cost of the Fujimori regime, however, was high. Democratic principles were violated and corruption was endemic. The Amerindian population made no significant gains, and the wealth gap increased. Nevertheless, the country held together during the next administration despite low public support, violent clashes between Andean mining companies and Amerindian villagers, strikes by coca farmers demanding legalization, and other major challenges. Then, in 2006, a presidential election evolved into an ideological contest between a former president, supported by the traditional Hispanic and mestizo sector, and a candidate who had significant Amerindian support as well as the endorsement of Venezuela's Hugo Chávez and Bolivia's Evo Morales. The former president, Alan Garcia, was narrowly elected, suggesting that Peru was not yet ready for the transition already underway in its neighbors. But developments in other parts of Andean South America, and in the realm as a whole, make it likely that Peru, too, will come to confront the implications of its cultural geography. Amerindians are in the majority, they are restive, and they now have models of empowerment they have not previously seen. The question is what course Peru's transition will take.

Ecuador

On the map, Ecuador, smallest of the three Andean West republics, appears to be just a corner of Peru. But that would be a misrepresentation because Ecuador possesses a full range of regional contrasts (Fig. 5-10). It has a coastal belt; an Andean zone that may be narrow (under 250 kilometers or 150 miles) but by no means of lower elevation than elsewhere; and an *Oriente*—an eastern subregion that is as sparsely settled and as economically marginalized as that of Peru. As in Peru, nearly half of Ecuador's population (which totals 13.9 million) is concentrated in the Andean intermontane basins and valleys, and the most productive region is the coastal strip. Here, however, the similarities end.

Ecuador's Pacific coastal zone consists of a belt of hills interrupted by lowlands, of which the most important lies in the south between the hills and the Andes, drained by the Guayas River. Guayaquil—the country's largest city, main port, and leading commercial center—forms the focus of this subregion. Unlike Peru's coastal strip, Ecuador's is not a desert: it consists of fertile tropical plains not afflicted by excessive rainfall. Seafood (especially shrimp) is a leading product, and these lowlands support a thriving commercial agricultural economy built around bananas, cacao, cattle-raising, and coffee on the hillsides. Moreover, Ecuador's western subregion is also far less Europeanized than Peru's

because its white component of the national population is only about 7 percent.

A greater proportion of whites are engaged in administration and hacienda ownership in the central Andean zone, where most of the Ecuadorians who are Amerindian also reside—and, not surprisingly, where land-tenure reform is an explosive issue. The differing interests of the Guayaquil-dominated coastal lowland and the Andean-highland subregion focused on the capital (Quito) have long fostered a deep regional cleavage between the two. This schism has intensified in recent years, and autonomy and other devolutionary remedies are now being openly discussed in the coastlands.

In the rainforests of the *Oriente* subregion, oil production is expanding as a result of the discovery of additional reserves. Some analysts are predicting that interior Ecuador as well as Peru will prove to contain reserves comparable to those of Venezuela and Colombia, and that an "oil era" will soon transform their economies. As it is, oil already tops Ecuador's export list, and the industry's infrastructure is being modernized. This is essential because much ecological damage has already been done: the trans-Andean pipeline to the port of Esmeraldas, constructed in 1972, was the source of numerous oil spills, and toxic waste was dumped along its route in a series of dreadful environmental disasters. A second, more modern pipeline went into operation in 2003, but like Peru, Ecuador faced growing opposition from environmentalists and activists who, in 2005, shut it down for a week. During 2006 and 2007, Ecuador's already-fragile government had to deal with demands for greater financial rewards by people living in the producing zones and with disputes with foreign oil companies over taxes and ownership issues (one Canadian company gave up and sold its interests to a Chinese firm). With revenues from oil and gas exports exceeding 50 percent of all exports, Ecuador's leaders are hearing an increasingly familiar refrain from its people: demand more from the companies and give these funds to those who need it. But it is not as simple as that: oil and gas exploration and exploitation require huge investments, and foreign companies can afford to spend what the government's own state company, PetroEcuador, cannot. That leads to difficult choices, because the state needs income to cover its obligations. Clearly, energy riches are a double-edged sword.

There is more to Ecuador's economic geography than its energy sector, of course, but the country's other commercial opportunities (fish from its waters, fruits from the coast, flowers from the cooler slopes) do not reach their full potential because the country is in a difficult social transition. Ecuador's plural society is strained by a growing social movement, led by its indigenous peoples, that reflects

FROM THE FIELD NOTES

"I can't remember being hotter anyplace on Earth, not in Kinshasa, not in Singapore . . . not only are you near the equator here in steamy Guayaquil, but the city lies in a swampy, riverine lowland too far from the Pacific to benefit from any cooling breezes and too far from the Andes foothills to enjoy the benefits of suburban elevation. But Guayaquil is not the disease-ridden backwater it used to be. Its port is modern, its city-center waterfront on the Guayas River has been renovated, its international airport is the hub of a commercial center, and it has grown into a metropolis of 2.5 million. Ecuador's oil revenues have made much of this possible, but from the White Hill (a hill that serves as a cemetery, with the most elaborate vaults near its base and the poorest at the feet of the giant statue of Jesus that tops it) you can see that globalization has not quite arrived here. High-rise development remains limited, international banks, hotels and other businesses remain comparatively few, and the 'middle zone' encircling the city shows little evidence of prosperity (top). Beating the heat and glare is an everyday priority: whole streets have been covered by makeshift tarpaulins and more permanent awnings to protect shoppers (bottom). Talk to the locals, though, and you find that there is another daily concern: the people 'up there in the mountains who rule this country always put us in second place.' Take the 45-minute flight from Guayaquil to cool and comfortable Quito, the capital, and you're in another world, and you quickly forget Guayaquil's problems. That's just what the locals here say the politicians do."
© H. J. de Blij.

www.conceptcaching.com

Concept Caching

what is happening in the Andean region generally. It is grounded in the persistent exclusion of Amerindians in each of the country's three traditional societies and has been aggravated by Ecuador's ineffective governments, which have raised expectations and then failed to deliver on promises. Ongoing geographic change further fuels this simmering unrest because the poorest people are increasingly drawn into the urban areas—where they witness close-up how their second-class standing and lack of educational opportunities bar them from making progress. Amerindians want a stronger political voice, they demand land reform, and they insist that leaders elected on populist commitments keep their promises. In 2006, another such leader, Rafael Correa, was elected with Amerindian support. The question is whether he (or any other president) can reverse a course that threatens to destabilize an entire South American region.

Bolivia

Nowhere in this region are the problems faced by Ecuador and to a (hitherto) lesser extent by Peru more acute than they are in landlocked, volatile Bolivia. Before reading further, take a careful look at Bolivia's regional geography in both Figures 5-10 and 5-11. Bolivia is bounded by remote parts of Brazil and Argentina, mountainous Andean highlands and *altiplanos* (elongated, high-altitude basins) of Peru, and coveted coastal zones of northern Chile. As the maps show, the Andes in this area broaden into a vast

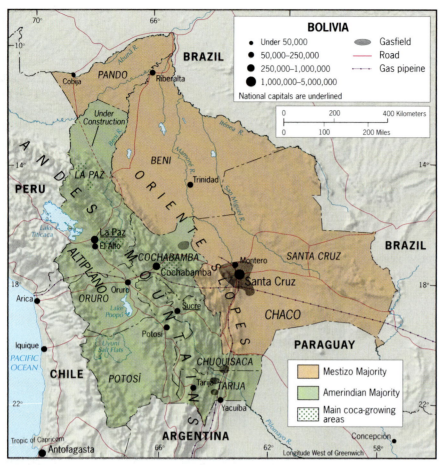

FIGURE 5-11 © H. J. de Blij, P. O. Muller, and John Wiley & Sons, Inc.

mountain complex some 700 kilometers (450 mi) wide. On the boundary between Peru and Bolivia, freshwater Lake Titicaca lies at 3700 meters (12,500 ft) above sea level and helps make the adjacent *Altiplano* livable by ameliorating the coldness in its vicinity, where the snow line lies just above the plateau surface. On the surrounding cultivable land, potatoes and grains have been raised for centuries dating back to pre-Inca times, and the Titicaca Basin still supports a major cluster of Aymara subsistence farmers. This portion of the *Altiplano* is the heart of modern Bolivia and also contains the capital city, La Paz.

The landlocked state of Bolivia is the product of the Hispanic impact, and the country's Amerindians (who still make up more than 60 percent of the national population of 9.5 million) no more escaped the loss of their land than did their Peruvian or Ecuadorian counterparts. What made the richest Europeans in Bolivia wealthy, however, was not land but minerals. The town of Potosí in the eastern cordillera became a legend for the immense deposits of silver in its vicinity; tin, zinc, copper, and several ferroalloys were also discovered there. Amerindian workers were forced to work in the mines under the most dreadful conditions.

Today oil and natural gas, exported to Argentina and Brazil, are major sources of foreign revenues, and zinc has replaced tin as the leading metal export. But Bolivia's economic prospects will always be impeded by the loss of its outlet to the Pacific Ocean during its war with Chile in the 1880s, despite its transit rights and dedicated port facilities at Antofagasta.

More critical than its economic limitations or its landlocked situation is Bolivia's social predicament. The government's history of mistreatment of indigenous people and the harsh exploitation of its labor force hangs heavily over a society where about 70 percent of the people, almost all Amerindian, live in dire poverty. In recent years, however, this underrepresented Aymara majority has been making an impact on national affairs. In 2003, violent opposition to a government plan to export natural gas to the United States via a new pipeline to the Chilean coast led to chaos and the government's resignation. In 2005, Bolivian voters elected their first Aymara (Amerindian-ancestry) president, Evo Morales, who proceeded to nationalize the country's natural-gas resources.

All this reflects not only the social cleavages in Bolivian society but also the country's sharp regional contrasts. In the east, centered on Santa Cruz Province and its capital, the hacienda system persists almost unchanged from colonial times, its profitable agriculture supporting a wealthy aristocracy. Now this eastern region proves to contain Bolivia's rich energy resources as well. While the government in La Paz struggles to meet growing Aymara demands for a share in the action, the prosperous east sees its objectives (among them that pipeline to export gas via the Chilean coast) thwarted. So there is talk of autonomy, even secession, in the east, but neither the haciendas nor the energy industry could do without the Amerindian labor force working here. Toward the end of this decade, Bolivia confronted several centrifugal forces that at times made it resemble a *failed state*. Its president was under the considerable influence of the populist leader and antiglobalization hero, President Hugo Chávez of Venezuela, whose involvement in Bolivian affairs was resented by many in the European and mestizo minority communities. The election of President Morales and his government defused the protests and blockades that in 2005 had brought Bolivia to the verge of civil war, but Morales's efforts to nationalize the energy industry, and talk of land expropriation, fueled separatism in Santa Cruz and other eastern *departments*, as provinces are called in Bolivia (Fig. 5-11). Santa Cruz normally generates more than 40 percent of Bolivia's tax revenues, but receives only about 20 percent of public expenditures. It is a measure of how far land reform has to go when it is reported that nearly 90 percent of Bolivia's agriculturally productive land is still owned by about 50,000 families, many of them from the white minority and much of it expropriated during military dictatorships.

By 2007, a combination of sustained high oil and gas prices and a slowdown in the reform program had stabilized the situation, but rising expectations among Amerindians and devolutionary pressures in the east continue to challenge this divided country in perilous transition.

Paraguay

Paraguay, Bolivia's landlocked neighbor to the southeast, is one of those transitional countries between regions and exhibits properties of each (see Fig. 5-3; additional examples on the world map [Fig. G-2] are Belarus, Kazakhstan, and Sudan). Certainly, Paraguay (population: 6.5 million) is not an "Andean" country: it has no highlands of consequence. Its well-watered eastern plains give way to the dry scrub of the Chaco in the west. Nor does it have clear, spatially entrenched ethnic divisions among Amerindians, mestizos, and others as do Bolivia and Peru. But Amerindian ancestry dominates the ethnic complexion of Paraguay, and Amerindian Guaraní is so widely spoken in the country that this is one of the world's most thoroughly bilingual societies. And continuing protests by landless peasants mirror those occurring elsewhere in South America's West. Looking south, there is little in Paraguay's economic geography to compare to Argentina, Uruguay, or Chile, as Table G-1 confirms. In a sense, Paraguay is a bridge between West and South, but not a heavily traveled one.

Paraguay is transitional in another way: as many as 300,000 Brazilians have crossed the border to settle in eastern Paraguay, where they have created a thriving commercial agricultural economy that produces soybeans, livestock, and other farm products exported to or through Brazil. Brazil, of course, is the giant in the Mercosur free-trade zone, but Paraguay is in a geographically difficult position, and often complains that Brazil does not live up to its regional-trade obligations and creates unacceptable difficulties for Paraguayan exporters. Meanwhile, politicians raise fears that growing Brazilian immigration is creating a foreign enclave within Paraguay, where people speak Portuguese (including in the local schools), the Brazilian rather than the Paraguayan flag flies over public buildings, and a Brazilian cultural landscape is evolving. Like Bolivia, Paraguay pays heavily for its landlocked weakness.

Paraguay's human geography reflects this. As Table G-1 shows, the country's low GNI resembles that of countries of the West, not the more advantaged South. This is also one of South America's least urbanized states, and poverty dominates the countryside as well as the slums encircling the capital, Asunción, and other towns (Fig. 5-10). In 2007, nearly two-thirds of the population lived at or below the official poverty level. Moreover, a history of dictatorial, corrupt, and unstable governments gives Paraguay one of the lowest ranks in South America on Transparency International's latest corruption index (Table G-1, column 17).

The Lawless Triple Frontier

These conditions create opportunities for lawlessness, and the chaotic area where Paraguay, Brazil, and Argentina converge, the so-called **Triple Frontier**, has raised security concerns for several reasons. A notable influx here of Arab Muslims, possible links with the terrorist organization Hizbollah during the attacks on Jewish targets in Argentina in the early 1990s, evidence of substantial money flows between this area and the Middle East, and the discovery of Triple Frontier maps in a Taliban safe house in Afghanistan after 9/11 all suggest that Paraguay's weaknesses may have more than just regional significance for the future.

● # THE SOUTH: MID-LATITUDE SOUTH AMERICA

South America's three southern countries—Argentina, Chile, and Uruguay—constitute a region sometimes referred to as the **Southern Cone** because of its tapered, ice-cream-cone shape (Fig. 5-12). As noted earlier, Paraguay has strong links to this region and in some ways forms part of it, although social contrasts between Paraguay and those in the Southern Cone remain sharp.

Since 1995, the countries of this region have been drawing closer together in an economic union named **Mercosur**, the hemisphere's second-largest trading bloc after NAFTA. Despite setbacks and disputes, Mercosur has expanded and today encompasses Argentina, Uruguay, Paraguay, Brazil, and (pending ratification) Venezuela, with Chile and Bolivia participating as associate members.

FIGURE 5-12   © H. J. de Blij, P. O. Muller, and John Wiley & Sons, Inc.

Argentina

The largest Southern Cone country by far is Argentina, whose territorial size ranks second only to Brazil in this geographic realm; its population of 39.8 million ranks third after Brazil and Colombia. Argentina exhibits a great deal of physical-environmental variety within its boun-

daries, and the vast majority of the Argentines are concentrated in the physiographic subregion known as the *Pampa* (a word meaning "plain"). Figure G-9 underscores the degree of clustering of Argentina's inhabitants on the land and in the cities of the Pampa. It also shows the relative emptiness of the other six subregions (mapped in Fig. 5-12): the scrub-forest *Chaco* in the northwest; the mountainous

AMONG THE REALM'S GREAT CITIES . . . Buenos Aires

ITS NAME MEANS "fair winds," which first attracted European mariners to the site of Buenos Aires alongside the broad estuary of the muddy Rio de la Plata. The shipping function has remained paramount, and to this day the city's residents are known throughout Argentina as the *porteños* (the "port dwellers"). Modern Buenos Aires was built on the back of the nearby Pampa's grain and beef industry. It is often likened to Chicago and the Corn Belt in the United States because both cities have thrived as interfaces between their immensely productive agricultural hinterlands and the rest of the world.

Buenos Aires (12.9 million) is yet another classic South American primate metropolis, housing nearly one-third of all Argentines, serving as the capital since 1880, and functioning as the country's economic core. Moreover, Buenos Aires is a cultural center of global standing, a monument-studded city that contains the world's widest boulevard (Avenida 9 de Julio).

During the half-century between 1890 and 1940, the city was known as the "Paris of the South" for its architecture, fashion leadership, book publishing, and performing arts activities (it still has the world's biggest opera house, the Teatro Colón). With the recent restoration of democracy, Buenos Aires is now trying to recapture its golden years. Besides reviving these cultural functions,

© H. J. de Blij, P. O. Muller, and John Wiley & Sons, Inc.

the city has added a new one: the leading base of the hemisphere's motion picture and television industry for Spanish-speaking audiences.

Andes in the west, along whose crestline lies the boundary with Chile; the arid plateaus of *Patagonia* south of the Rio Colorado; and the undulating transitional terrain of intermediate *Cuyo, Entre Rios* (also known as "Mesopotamia" because it lies between the Paraná and Uruguay rivers), and the *North*.

The Argentine Pampa is the product of the past 150 years. During the second half of the nineteenth century, when the great grasslands of the world were being opened up (including those of the interior United States, Russia, and Australia), the economy of the long-dormant Pampa began to emerge. The food needs of industrializing Europe grew by leaps and bounds, and the advances of the Industrial Revolution—railroads, more efficient ocean transport, refrigerated ships, and agricultural machinery—helped make large-scale commercial meat and grain production in the Pampa not only feasible but also highly profitable. Large haciendas were laid out and farmed by tenant workers; railroads radiated ever farther outward from the booming capital of Buenos Aires and brought the entire Pampa into production.

A Culture Urban and Urbane

Argentina once was one of the richest countries in the world. Its historic affluence is still reflected in its architecturally splendid cities whose plazas and avenues are flanked by ornate public buildings and private mansions. This is true not only of the capital, Buenos Aires, at the head of the Rio de la Plata estuary—it also applies to interior cities such as Mendoza and Córdoba. The cultural imprint is dominantly Spanish, but the cultural landscape was diversified by a massive influx of Italians and smaller but influential numbers of British, French, and German immigrants. A sizeable immigration from Lebanon resulted in the diffusion of Arab ancestry to more than 12 percent of the Argentinian population.

Argentina has long been one of the realm's most urbanized countries: 90 percent of its population is concentrated in cities and towns, a higher percentage even than Western Europe or the United States. Almost one-third of all Argentinians live in metropolitan Buenos Aires, by far the leading industrial complex where processing

Pampa products dominates. Córdoba has become the second-ranking industrial center and was chosen by foreign automobile manufacturers as the car-assembly center for the expanding Mercosur market. One in three Argentinian wage-earners is engaged in manufacturing, another indication of the country's economic progress. But what concentrates the urban populations is the processing of products from the vast, sparsely peopled interior: Tucumán (sugar), Mendoza (wines), Santa Fe (forest products), and Salta (livestock). Argentina's product range is enormous. There is even an oil reserve near Comodoro Rivadávia on the coast of Patagonia.

Economic Boom and Bust

Despite all these riches, Argentina's economic history is one of boom and bust. With just under 40 million inhabitants, a vast territory with diverse natural resources, adequate infrastructure, and good international linkages, Argentina should still be one of the world's wealthiest countries, as it once was. But political infighting and economic mismanagement have combined to ruin a vibrant and varied economy. What began as a severe recession toward the end of the 1990s became an economic collapse in the first years of this century.

To understand how Argentina finds itself in this situation, a map of its administrative structure is useful. On paper, Argentina is a federal state consisting of the Buenos Aires Federal District and 23 provinces (Fig. 5-13). As would be expected from what we have just learned, the urbanized provinces are populous while the mainly rural ones have smaller populations—but the gap between Buenos Aires Province (whose capital is La Plata), with nearly 15 million, and Tierra del Fuego (capital: Ushuaia), with barely over 100,000, is wide indeed. Several other provinces have under 500,000 inhabitants, so that the larger ones in addition to dominant Buenos Aires are also disproportionately influential in domestic politics, especially Córdoba and Santa Fe, both with capitals of the same name.

The never-ending problem for Argentina has been corrupt politics and associated mismanagement. Following an army coup in 1946, Juan Perón got himself elected president, bankrupted the country, and was succeeded by a military junta that plunged the country into its darkest days, culminating in the "Dirty War" of 1976–1983 which saw more than ten thousand Argentinians disappear without a trace. In 1982, this ruthless military clique launched an invasion of the British-held Malvinas (Falkland Islands), resulting in a major defeat for Argentina. By the time civilian government replaced the discredited junta, inflation was soaring and the national debt had become staggering (see the Issue Box entitled "Who Needs Democracy?"). Eco-

FIGURE 5-13 © H. J. de Blij, P. O. Muller, and John Wiley & Sons, Inc.

nomic revival during the 1990s was followed by another severe downturn that exposed the flaws in Argentina's fiscal system, including scandalously inefficient tax collection and unconditional federal handouts to the politically powerful provinces.

In 2003 a new administration took office led by President Néstor Kirchner, who hails from a small Patagonian province, not from one of the country's traditional power centers. He won on the Peronist Party platform, and then, true to Argentinian political tradition, two years later took on the political bosses in Buenos Aires Province by supporting the Senate candidacy of his wife against the spouse

Who Needs Democracy?

SOUTH AMERICA DOES!

"I'm old enough to remember the good as well as the bad old days here in Argentina. As a young man I worked for the national railroad company when that guy Perón came on the scene. I guess he learned his craft from the fascists in Italy, because he excelled in election by intimidation. He divided this country as never before, spending money on the workers and the poor while curbing freedoms and even abolishing freedoms guaranteed under our constitution. To tell you the truth, my salary actually went up a bit, and I even voted for him the second time around—largely, I think, because I wanted to do something for his beautiful wife Evita, who was loved by the whole country (except the military).

"Anyway, if you looked around South America in those days, we could have done worse. Wall to wall dictators. The military and police, secret and otherwise, put the fear of God into the populace, helped by the church. At least we had a guy who went through the motions of election. While those others divided the wealth among their cronies, Perón tried to spread it around. Even wealth we didn't have! Yes, he certainly put us into debt. And he had his cronies too.

"When they finally deposed Perón, I thought that we might see democratic elections here. (In retrospect, I don't know why I thought that—Spain and Portugal were ruled by two of the worst, Franco and Salazar; that's our 'Latin' heritage). Well, you already know how wrong I was. Instead of more democracy, we got military repression. We lived in terror, not just fear. You've probably heard of those nuns who asked questions about the fate of a baby whose mother had 'disappeared.' They loaded them into a helicopter, in their outfits, and threw them out over the Rio de la Plata. Later I heard some soldiers joke about 'the flying nuns.' Rumors of what could happen to you were rife. Everyone knew someone who knew someone who was a victim.

"Our cloud of military terror lifted after they made their mistake in the Malvinas, 25 years ago, and since then we've had a taste of real democracy. It hasn't been easy, but let me tell you, it's better than 'strongman' rule so many of us South Americans are familiar with. Most of all, there's openness. You can express your views without fear. Political parties can argue their positions without military intimidation. Corrupt public officials can be found out by reporters, and they can't send machine-gun-toting colonels to kill their pursuers. Yes, there are downsides—that openness applies to the economy too, and as we've discovered, outsiders can interfere in our financial affairs. But I'd rather have democracy and open disorder than dictatorship and isolated order, and I've known them both."

Regional ISSUE

SAY NO TO THE EXCESSES OF DEMOCRACY!

"You can't help noticing that the Great Democratic Revolution that was supposed to be sweeping Middle and South America is not exactly a success. Here in Brazil we've had democratic government of a sort since 1989, when our man Fernando Collor de Mello was allowed by the military to win an election. You'll remember what happened to him. He resigned three years later after being implicated in a corruption and influence-peddling scheme. I'm sure the poor benefited hugely from this return to democracy.

"South American countries have a history of being led by men who captured the imagination of the nation and whose vision forged the character of the state. These men knew that the state must serve the people. The state must ensure that its railroads and bus services are available at low cost to all citizens. The state must provide electricity, fuel, and clean water. The state must staff and maintain the schools. The state and the church are indivisible, and the networks of the church are in the service of the state. If the military is needed to maintain order in the interest of stability, so be it. Call them strongmen if you like, but they personified their nations and did so for centuries.

"Democracy is a luxury for rich countries. The Americans like to teach us about democracy, but as a geography high-school teacher here in Santa Catarina, let me ask you this: how can the U.S. call itself a democracy when, in their Senate, a few hundred thousand people in Wyoming have the same representation as over 30 million in California? Oh, I see. You call yourself a 'representative republic.' Well, then don't lecture us about democracy. Even Juan Perón was 'democratically' elected, you know. In fact, most of those allegedly dictatorial rulers you so harshly criticize would probably win at the ballot box.

"Anyway, look at what democracy has done for Argentina and for Venezuela before the Venezuelans had the good sense to throw the rascals out and elect Chávez. Look what it is doing for Peru, where democratically elected President Toledo was trying to sell off the electric company serving Arequipa before the citizens stopped him. That 'openness' you hear pro-democracy advocates talk about just means that outsiders can come in, put your country in debt, demand 'privatization' of public companies, buy up corporations and banks and haciendas, fire thousands of workers to increase profits and share prices back home, and make the poor even poorer. What we need in this part of the world is a strong hand, not only to guide the nation but to protect the state."

Vote your opinion at www.wiley.com/college/deblij

of the province's most powerful politician. Kirchner's wife won decidedly, cementing the president's power and allowing him to pursue goals he had long embraced: combating foreign economic intervention, controlling domestic companies, reviving Argentina's export economy by keeping its currency artificially low, and ending military immunity from prosecution (in response to growing public demands for justice following the military dictatorship).

By 2007, these initiatives were having good effect, although Argentina's future remained clouded. The growth of Argentina's realm-leading GNI remained high, but inflation was rising; foreign investors were reluctant to assume risks in a country whose government habitually ignored contract terms, and Argentina faced serious uncertainties over its energy supplies. The old boom-and-bust cycle may not have ended quite yet.

Chile

For 4000 kilometers (2500 mi) between the crestline of the Andes and the coastline of the Pacific lies the narrow strip of land that is the Republic of Chile (Fig. 5-14). On average just 150 kilometers (90 mi) wide (and only rarely over 250 kilometers or 150 mi in width), Chile is the world's quintessential example of what *elongation* means to the functioning of a state. Accentuated by its north-south orientation, this severe territorial attentuation not only results in Chile extending across numerous environmental zones; it has also contributed to the country's external political, internal administrative, and general economic problems. Nonetheless, throughout most of their modern history, the Chileans have made the best of this potentially disastrous centrifugal force: from the beginning, the sea has constituted an avenue of longitudinal communication; the Andes Mountains continue to form a barrier to encroachment from the east; and when confrontations loomed at the far ends of the country, Chile proved to be quite capable of coping with its northern rivals, Bolivia and Peru, as well as Argentina in the extreme south.

Three Subregions

As Figures 5-12 and 5-14 indicate, Chile is a three-subregion country. About 90 percent of its 16.8 million people are concentrated in what is called Middle Chile, where Santiago, capital and largest city, and Valparaíso, the chief port, are located. North of Middle Chile lies the Atacama Desert, wider, drier, and colder than the coastal desert of Peru. South of Middle Chile, the coast is broken by a plethora of fjords and islands, the topography is mountainous, and the climate—wet and cool near the Pacific—soon turns drier and colder against the Andean

FIGURE 5-14 © H. J. de Blij, P. O. Muller, and John Wiley & Sons, Inc.

interior. South of the latitude of Chiloé Island, there are few permanent overland routes and hardly any settlements. These three subregions are also clearly apparent on the realm's cultural map (Fig. 5-5), as well as the map of South America's agricultural systems (Fig. 5-4). In addition, a small Amerindian subsistence zone in northeasternmost Chile's Andes is shared with Argentina and Bolivia.

Some intraregional differences exist between northern and southern Middle Chile, the country's core area. Northern Middle Chile, the land of the hacienda and of Mediterranean climate with its dry summer season, is an area of (usually irrigated) crops that include wheat, corn, grapes, and other Mediterranean products. Livestock raising and fodder crops also take up much of the productive land, but continue to give way to the more efficient and profitable cultivation of fruits for export. Southern Middle Chile, into which immigrants from both the north and Europe (especially Germany) have pushed, is a better-watered area where raising cattle has predominated. But here, too, more lucrative fruits, vegetables, and grains are changing the area's agricultural specializations.

Prior to the 1990s, the arid Atacama region in the north accounted for more than half of Chile's foreign revenues. The Atacama Desert contains the world's largest exploitable deposits of nitrates, which was the country's economic mainstay before the discovery of methods of synthetic nitrate production a century ago. Subsequently, copper became the chief export (Chile again possesses the world's largest reserves, which in 2006 accounted for nearly half of all export revenues). It is found in several places, but the main concentration lies on the eastern margin of the Atacama near Chuquicamata, not far from the port of Antofagasta. During the closing years of this decade, commodity prices have been high and Chile's export revenues have soared, but such price increases tend to be followed by declines that underscore the risks involved—and the continuing need for economic diversification.

Political and Economic Transformation

Chile today is emerging from a development boom that transformed its economic geography during the 1990s, a growth spurt that established its reputation as South America's greatest success story. Following the withdrawal of its brutal military dictatorship in 1990, Chile embarked on a program of free-market economic reform that brought stable growth, lowered inflation and unemployment, reduced poverty, and attracted massive foreign investment. The last is of particular significance because these new international connections enabled the export-led Chilean economy to diversify and develop in some badly needed new directions. Copper remains the single leading export, but

FROM THE FIELD NOTES

"You could see this (what I thought was a) church from many kilometers away on the fieldtrip to Coquimbo, a coastal town about 300 kilometers (200 mi) north of Valparaíso, Chile, where the climatic transition to the desert north is evident all around. But it was this urban scene that taught me a lesson: I glanced at the tower and assumed that this was an unusually prominent Roman Catholic church looming over the townscape, as you see all over the realm. But my colleague advised me to look closer. 'See any crosses?' he asked. I didn't. 'Anything unusual about the architecture?' I noticed the arched courtyard. 'Remind you of anything?' Well, yes, but surely not here? Here indeed—a mosque serving a widely scattered but significant minority, visible from afar and symbolizing change in a realm where evangelical movements are by no means the only challenge to Roman Catholicism's historic domination. I had seen the growing presence of Islam in the greater Lima area days earlier, but that was to be expected in a city of its dimensions and international linkages. But here in relatively remote Coquimbo?" © H. J. de Blij.

www.conceptcaching.com

Concept Caching

many other mining ventures have been launched. In the agricultural sphere, fruit and vegetable production for export has soared because Chile's harvests coincide with the winter farming lull in the affluent countries of the Northern Hemisphere. Industrial expansion is occurring as well, though at a more leisurely pace, and new manufactures include a modest array of goods that range from basic chemicals to computer software.

Chile's increasingly globalized economy has propelled the country into a prominent role on the international economic scene. The United States, long Chile's leading trade partner, now is in second place compared to the East Asian Pacific Rim, where China takes the bulk of Chile's exports followed by Japan and South Korea (Argentina remains

Chile's leading source of imports, mostly energy, which is a potential problem given Argentina's own rising needs). Chile's regional commerce also is growing, and Chile has affiliated with Mercosur as an associate member.

A "Chilean Model"?

Chile's economic and democratic transformation is widely touted as a "Chilean model" to be followed by other Middle and South American countries. But Chile benefits from an unusual combination of opportunities, and some economic geographers argue that Chile has in fact not taken sufficient advantage of all of them. Exports should be more diversified, and Chile's dependence on high copper prices entails a major risk. The education system remains inadequate. Income inequality—that realm-wide affliction—has improved, but nearly 20 percent of Chileans still live in poverty. The plight of Chile's small number of indigenous people, brought to national attention by the Mapuche of the forested south, is of national concern. But compare Chile to virtually any other country in this realm: the Chilean model may not always apply, but the Chilean example certainly provides hope for the future.

Uruguay

Uruguay, unlike Argentina or Chile, is compact, small, and rather densely populated. This **buffer state** of old became a fairly prosperous agricultural country, in effect a smaller-scale Pampa (though possessing less favorable soils and topography). Figures 5-1 and G-8 show the similarity of physical conditions on the two sides of the Plata estuary. Montevideo, the coastal capital, contains almost 40 percent of the country's population of 3.3 million; from here, railroads and roads radiate outward into the productive agricultural interior (Fig. 5-12). In the immediate vicinity of Montevideo lies Uruguay's major farming area, which produces vegetables and fruits for the metropolis as well as wheat and fodder crops. Most of the rest of the country is used for grazing cattle and sheep, with beef products, wool and textile manufactures, and hides dominating the export trade. Tourism is another major economic activity as Argentines, Brazilians, and other visitors increasingly flock to the Atlantic beaches at Punta del Este and other thriving resort towns.

Concentricity and Conflict

16 In Chapter 1, we noted the work of **von Thünen** and his **model** layout of agricultural zones (pp. 53–54). Here in Uruguay we can discern a real-world example of that scheme (inset map, Fig. 5-12). Market gardens and dairy-

FROM THE FIELD NOTES

"In a small, admission-charge private museum I found a treasure of historic information about this place in Chile—its environmental challenges ranging from droughts to earthquakes, its growth as a center for irrigated agriculture augmented by mining in the hinterland, its religious importance as the site of many churches and convents. But what matters today is tourism and retirement. Not only is La Serena (just inland from Coquimbo) on the international visitor circuit, but it lies at the center of a growing cluster of second-home and retirement complexes, benefiting from its equable climate and attractive scenery, and from the Southern Cone's comparative economic prosperity. So La Serena has become an amusement park, its stores converted to boutique shops, its historic homes on visitor tours, its plaza throbbing with the music of competing boomboxes, its streets clogged by buses. An older resident sitting on a park bench had his doubts. 'All this started when they built this Pan-American Highway right past us,' he said. 'Now you can get here by sea, by road, even by air. Is this your first visit? You came too late.' No doubt about it: what I saw here wasn't what I read in those faded pages in the museum." © H. J. de Blij.

www.conceptcaching.com

ing cluster nearest the urban area of Montevideo, with increasingly extensive agriculture ringing this national market in concentric land-use zones.

Uruguay's 177,000 square kilometers (68,000 sq mi), about the size of Florida, evince a low physiologic density (see p. 413), and with the Montevideo urban area comprising 1.3 million people, there is plenty of agricultural potential here. But Uruguay's government seeks to diversify the economy, and one of its plans—the construction of two large cellulose (paper) factories on the Uruguayan side of the Uruguay River where it forms the border with Argentina—has caused a quarrel between the two Merco-

BRAZIL: GIANT OF SOUTH AMERICA **277**

sur neighbors that has exposed some serious rifts. Montevideo is the administrative headquarters of Mercosur, but, like Paraguay, Uruguay has long felt neglected and even obstructed by its much larger neighbors. When Argentinians across the river began demonstrating, blocked a bridge, and stopped taking vacations on Uruguay's beaches, they cost the Uruguayan economy hundreds of millions of dollars.

The Argentinians, who dislike large foreign investments anyway (the factories were to be built by Finnish and Spanish companies), argued that the paper mills would accelerate deforestation, cause river pollution, create acid rain, and damage the farming, fishing, and tourist industries. When the presidents of the two countries met to negotiate a solution and Uruguay's leader agreed to halt construction for a "study period," Urugayan public opinion showed strong opposition to the compromise.

This issue reveals how quickly nationalist feelings can overwhelm the need for international cooperation. It also shows that Mercosur's partners are far from united on economic matters, and indicates that the collaboration implied by Mercosur membership is still a distant reality.

● BRAZIL: GIANT OF SOUTH AMERICA

The next time you board an airplane in the United States, don't be surprised if the aircraft you enter was built in Brazil. When you go to the supermarket to buy provisions, take a look at their sources. Chances are some of them will come from Brazil. When you listen to your car radio, some of the best music you hear is likely to have originated in Brazil. The emergence of Brazil as the regional superpower of South America and an economic superpower in the world at large is going to be the story of the twenty-first century. In 2003 Brazil's newspapers carried quotes from national leaders who asked why Brazil, for all its territorial size, population numbers, and energy needs, did not have a nuclear-energy program. Why, asked one editorial, should countries like North Korea and Pakistan have nuclear weapons when Brazil does not even possess nuclear technology? Certainly not because Brazil lacks the capacity.

Why is Brazil so upward-bound today? In large measure it is because, after a long period of dictatorial rule by a minority elite that used the military to stay in power, Brazil embraced democratic government in 1989 (its first democratically elected president was ousted on charges of corruption three years later) and has not looked back since. The era of military coups and repeated quashings of civil liberties is over, and in Brazil's presidential election of 2002 the leader of the Workers Party, Luiz Inácio Lula da

Silva, popularly known as Lula, was elected in a runoff that marked a watershed in Brazilian democratic politics. (In a show of displeasure at the result, the United States did not send a single high-ranking representative to Lula's inauguration.) A champion of the poor and dispossessed, Lula came into office amid a rising tide of expectations but with the reality of a faltering economy confronting his government. In his first term, Lula skillfully managed Brazil's numerous countervailing forces, unleashing the country's regional and global potential. But toward its end his party faced bribery and corruption allegations that came close to the president's inner circle. Nonetheless, he was reelected to a second term in 2006 and continued his successful administration.

The Realm's Giant

By any measure, Brazil is South America's giant. It is so large that it has common boundaries with all the realm's other countries except Ecuador and Chile (Fig. 5-3). Its tropical and subtropical environments range from the equatorial rainforest of the Amazon Basin to the humid temperate climate of the far south. Territorially, Brazil occupies just under 50 percent of South America and is exceeded on the global stage only by Russia, Canada, the United States, and China. In population size, it also accounts for just under half the realm's total and is again the world's fifth-largest country (surpassed only by China, India, the United States, and Indonesia). The Brazilian economy, with its modern industrial base, is now the eleventh largest on Earth, and is likely to continue advancing in the international ranks and become a world force as the twenty-first century unfolds.

Population Patterns

Brazil's population of 192.4 million is as diverse as that of the United States. In a pattern quite familiar in the Americas, the indigenous inhabitants of the country were decimated following the European invasion (fewer than 200,000 Amerindians now survive deep in the Amazonian interior). Africans came in great numbers, too, and today there are about 12 million Afro-Brazilians in Brazil. Significantly, however, there was also much racial mixing, and 82 million Brazilians have combined European, African, and minor Amerindian ancestries. The remaining 98 million—now barely in the majority at 50.5 percent— are mainly of European origin, the descendants of immigrants from Portugal, Italy, Germany, and Eastern Europe. The complexion of the population was further diversified by the arrival of Lebanese and Syrian Muslims following

FROM THE FIELD NOTES

"Near the waterfront in Belém lie the now-deteriorating, once-elegant streets of the colonial city built at the mouth of the Amazon. Narrow, cobblestoned, flanked by tiled frontages and arched entrances, this area evinces the time of Dutch and Portuguese hegemony here. Mapping the functions and services here, we recorded the enormous diversity of activities ranging from carpentry shops to storefront restaurants and from bakeries to clothing stores. Dilapidated sidewalks were crowded with shoppers, workers, and people looking for jobs (some newly arrived, attracted by perceived employment opportunities in this growing city of 2.2 million). The diversity of population in this and other tropical South American cities reflects the varied background of the region's peoples and the wide hinterland from which these urban magnets have drawn their inhabitants."   © H. J. de Blij.

Concept Caching

www.conceptcaching.com

the collapse of the Ottoman Empire nearly a century ago. Estimates of the Muslim component in the Brazilian population suggest that about 10 million can claim to be of Arab descent, of whom about 1 million are nominally Muslims. But practicing Muslims are far fewer in number: in 2007 Brazil had less than 50 functioning mosques.

Another significant, though small, minority began arriving in Brazil in 1908: the Japanese, who today are concentrated in São Paulo State. The more than 1 million Japanese-Brazilians form the largest ethnic Japanese community outside Japan, and in their multicultural environ-

ment they have risen to the top ranks of Brazilian society as business leaders, urban professionals, farmers—and even as politicians in the city of São Paulo. Committed to their Brazilian homeland as they are, the Japanese community also retains its contacts with Japan, resulting in many a trade connection.

A National Culture

Brazilian society, to a greater degree than is true elsewhere in the Americas, has made progress in dealing with its racial divisions. To be sure, blacks are still the least advantaged among the country's major population groups, and community leaders continue to complain about discrimination. But ethnic mixing in Brazil is so pervasive that hardly any group is unaffected, and official census statistics about "blacks" and "Europeans" are meaningless. What the Brazilians do have is a true national culture, expressed in an historic adherence to the Catholic faith (now eroding under Protestant-evangelical and secular pressures), in the universal use of a modified form of Portuguese as the common language ("Brazilian"), and in a set of lifestyles in which soccer, "beach culture," distinctive music, and a growing national consciousness and pride are fundamental ingredients.

Changing Demography

Brazil's population grew rapidly during the world's twentieth-century population explosion. But over the past three decades, the rate of natural increase has slowed from nearly 3.0 percent to 1.4 percent, and the average number of children born to a Brazilian woman has been nearly halved from 4.4 in 1980 to 2.3 in 2006. These dramatic reductions occurred in the absence of any active family-planning policies by Brazil's federal or State governments and in direct contradiction to the teachings of the Catholic Church, reflecting religion's changing role in this former bastion of Roman Catholicism. Brazil's rapid urbanization, its prevalent economic uncertainties, and the widespread use of contraceptives are among factors influencing this significant change.

African Heritage

As we noted earlier, Afro-Brazilians still suffer disproportionately from the social ills of this relatively integrated nation. And yet Brazil's culture is infused with African themes, a quality that has marked it from the very beginning. Three centuries ago, the Afro-Brazilian sculptor and architect affectionately called Aleijadinho was Brazil's most famous artist. The world-famous com-

FIGURE 5-15 © H. J. de Blij, P. O. Muller, and John Wiley & Sons, Inc.

poser Heitor Villa-Lobos used numerous Afro-Brazilian folk themes in his music. So many Africans were brought in bondage to the city and hinterland of Salvador (Bahia State) from what is today Benin (formerly Dahomey) in West Africa that Bahia has become a veritable outpost of African culture. Candomblé, one of the Macumba religious sects arising from the arrival of African cultures in Brazil, is concentrated in Salvador and Bahia (Fig. 5-15).

Its presence there has led to a reconnection with modern Benin as Afro-Brazilians travel to West Africa to search for their roots and West Africans come to Bahia to experience the cultural landscape to which their ancestors gave rise.

In some parts of Bahia, the rural Northeast, and urban areas in Brazil's core you can almost imagine being in Africa. As a nation, Brazil is taking an increased interest in African linkages, not least because the Lusitanian world incorporates not only Portugal and Brazil but also Angola and Moçambique, where the Portuguese colonial experiment failed but where the Lusitanian imprint also survives.

Inequality in Brazil

For all its accomplishments in multiculturalism, Brazil remains a country of stark, appalling social inequalities (Fig. 5-16). Although such inequality is hard to measure precisely, South America is often cited as the geographic realm exhibiting the world's sharpest division between affluence and poverty. And in South America, Brazil is reputed to have the widest gap of all.

The Scourge of Poverty

Data that underscore this situation include the following: today the richest 10 percent of the population own two-thirds of all the land and control more than half of Brazil's wealth. The poorest 20 percent of the people live in the most squalid conditions prevailing anywhere on the planet, even including the megacities of Africa and Asia. According to UN estimates, in this age of adequate available (but not everywhere affordable) food, about half the population of Brazil suffers some form of malnutrition and, in the northeastern States, even hunger. Packs of young orphans and abandoned children roam the cities, sleeping where they can, stealing or robbing when they must. Some of the world's most magnificent central cities are ringed and sectored by wretched *favelas* where poverty, misery, and crime converge (see Fig. 5-7 and right-hand photo on p. 252).

Most Brazilian voters had long believed that these conditions were worsening, that what economic development was taking place benefited those already well off, and that powerful corporations and foreign investors were to blame for the spiral of rising and growing unemployment. When the latest official reports announced that the number of Brazilians afflicted by poverty had increased by 50 percent since 1980, the people demanded change and, as we will note, they achieved that in 2002.

Development Prospects

Brazil is richly endowed with mineral resources, including enormous iron and aluminum ore reserves, extensive tin and manganese deposits, and sizeable oil- and gasfields (Fig. 5-15). Other significant energy developments involve massive new hydroelectric facilities and the successful substitution of sugarcane-based alcohol (gasohol) for gasoline—allowing well over half of Brazil's cars to use this fuel instead of costly imported petroleum. Besides these natural endowments, Brazilian soils sustain a bountiful agricultural output that makes the country a global leader in the production and export of coffee, soybeans, and orange juice concentrate. Commercial agriculture, in fact, is now the fastest growing economic sector, driven by mechanization and the opening of a major new farming frontier in the fertile grasslands of southwestern Brazil.

Industrialization has propelled Brazil's rise as a global economic power. Much of the momentum for this continuing development was unleashed in the early 1990s after the government opened the country's long-protected industries to international competition and foreign investment. These new policies are proving to be effective because productivity has risen by over a third since 1990, as Brazilian manufacturers attain world-class quality. During the mid-1990s, the revenues from industrial exports surpassed those from agriculture. Commerce with Argentina made that member of *Merco-sul* (as the Portuguese-speaking Brazilians call Mercosur) one of Brazil's leading trade

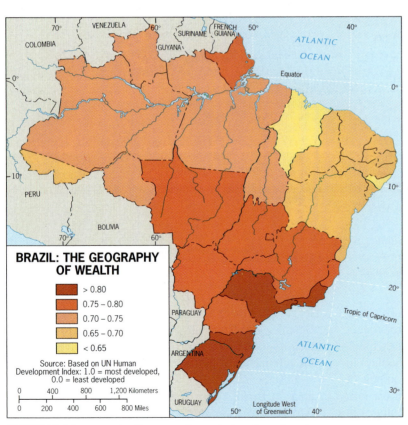

BRAZIL: THE GEOGRAPHY OF WEALTH

- > 0.80
- 0.75 – 0.80
- 0.70 – 0.75
- 0.65 – 0.70
- < 0.65

Source: Based on UN Human Development Index: 1.0 = most developed, 0.0 = least developed

0   400   800   1,200 Kilometers

0   200   400   600   800 Miles

FIGURE 5-16   © H. J. de Blij, P. O. Muller, and John Wiley & Sons, Inc.

partners. On the global stage, Brazil became a formidable presence in other ways. The country's enormous and easily accessible iron ore deposits, the relatively low wages of its workers, and the mechanized efficiency of its steelmakers enable Brazil to produce that commodity at half the cost of steel made in the United States. This causes American steel producers to demand government protection through tariffs, which goes against purported U.S. principles of free trade.

A New Era

Even though Brazil has become a major player on the world's economic stage, it has not escaped the cycles of boom and bust that affect transforming economies. Needing loans from international lending agencies to weather the downturns, Brazil's government had to agree to terms that included the privatization of public companies ranging from telephones to utilities. This often meant layoffs, rising prices to consumers, and even civil unrest. Brazil's currency, the *real*, was devalued in 1999, resulting in the outflow of billions of dollars withdrawn from Brazil's banks by panicked depositors. Strikes and marches against increased gas prices, highway tolls, and utility charges paralyzed the country for days. Newspapers reported scandals and corruption involving leading public figures and hundreds of millions of *reais*. In the political campaign that led up to the presidential election of 2002, these issues guaranteed the defeat of the right-of-center government held responsible for it all. In that election a leftist candidate, Luiz Inácio Lula da Silva, won in a landslide victory. He faced the daunting task of adhering to Brazil's existing financial commitments while steering a new course toward more fairness, openness, and honesty in government.

Continuing Challenges

The new left-of-center government in 2002 inherited Brazil's massive social problems without adequate resources to address them. The incoming administration focused on two key areas: land reform and poverty reduction. At the turn of this century, 1 percent of Brazilians owned 50 percent of all productive land. But the government was able to settle only about 60,000 landless families per year, far below what was needed. This resulted in protests and disruptions, which in turn scared off investors, affecting the jobless rate. By 2006, the government had found the resources to increase the rate of settlement, but the voters who elected President Lula da Silva had expected much more. And in the poverty arena, the administration began a *Fome Zero* (Zero Hunger) program, but in the process created a huge, corruption-riddled bureaucracy. Officially, Brazil counts some 40 million people who live on half the minimum wage of about U.S. $80 per month or less, and

the question was how to aid the poorest of the poor with modest government handouts. By 2006, the government's muddled plan had been replaced by a far more efficient subsidy program that was having significant effect in the poorest States. Thus the government had to operate the land reform and antipoverty programs while trying to keep the economy growing by maintaining the confidence of investors without whom such growth could not occur.

Brazil's Subregions

Brazil is a federal republic consisting of 26 States and the federal district of the capital, Brasília (Fig. 5-15). As in the United States, the smallest States lie in the northeast and the larger ones farther west; their populations range from about 450,000 in the northernmost, peripheral Amazon State of Roraima to more than 42 million in burgeoning São Paulo State. Although Brazil is about as large as the 48 contiguous United States, it does not exhibit a clear physiographic regionalism. Even the Amazon Basin, which covers almost 60 percent of the country, is not entirely a plain: between the tributaries of the great river lie low but extensive tablelands. Given this physiographic ambiguity, the six Brazilian subregions discussed next have no absolute or even generally accepted boundaries. In Figure 5-15, those boundaries have been drawn to coincide with the borders of States, making identifications easier.

The Northeast ① was Brazil's source area, its culture hearth. The plantation economy took root here at an early date, attracting Portuguese planters who soon imported the country's largest group of African slaves to work in the sugar fields. But the ample rainfall occurring along the coast soon gives way to lower and more variable patterns in the interior, which is home to about half of the region's 50-plus million people. This drier inland backcountry—called the *sertão*—is not only seriously overpopulated but also contains some of the worst poverty to be found anywhere in the Americas. The Northeast produces less than one-sixth of Brazil's gross domestic product, but its inhabitants constitute almost one-third of the national population. Given this staggering imbalance, it is not surprising that the region contains half of the country's poor, a literacy rate 20 percent below Brazil's mean, and an infant mortality rate twice the national average.

Much of the Northeast's misery is rooted in its unequal system of land tenure. Farms must be at least 100 hectares (250 acres) to be profitable in the hard-scrabble *sertão*, a size that only large landowners can afford. Moreover, the Northeast is plagued by a monumental environmental problem: the recurrence of devastating droughts at least partly attributable to **El Niño** (periodic sea-surface-warming events off the continent's northwestern coast).

AMONG THE REALM'S GREAT CITIES . . . Rio de Janeiro

SAY "SOUTH AMERICA" and the first image most people conjure up is Sugar Loaf Mountain, Rio de Janeiro's landmark sentinel that guards the entrance to beautiful Guanabara Bay. Nicknamed the "magnificent city" because of its breathtaking natural setting, Rio replaced Salvador as Brazil's capital in 1763 and held that position for almost two centuries until the federal government shifted its headquarters to interior Brasília in 1960. Rio de Janeiro's primacy suffered yet another blow in the late 1950s: São Paulo, its hated urban rival 400 kilometers (250 mi) to the southwest, surpassed Rio to become Brazil's largest city—a gap that has been widening ever since. Although these events triggered economic decline, Rio (12.0 million) remains a major *entrepôt*, air travel and tourist hub, and center of international business. As a focus of cultural life, however, Rio de Janeiro is unquestionably Brazil's leader, and the *cariocas* (as Rio's residents call themselves) continue to set the national pace with their entertainment industries, universities, museums, and libraries.

On the darker side, this city's reputation is increasingly being tarnished by the widening abyss between Rio's affluent and poor populations—symbolizing inequities that rank among the world's most extreme (see photos, p. 252). All great cities experience problems, and Rio de Janeiro is currently bedeviled by the drug use and crime waves emanating from its most desperate hillside *favelas* (slums) that continue to grow explosively.

Nonetheless, Rio's planners recently launched a major project (known as "Rio-City") to improve urban life for all residents. This ambitious scheme is designed to reshape

© H. J. de Blij, P. O. Muller, and John Wiley & Sons, Inc.

nearly two dozen of the aging city's neighborhoods, introduce an ultramodern crosstown expressway to relieve nightmarish traffic congestion, and—most importantly—bring electrical power, paved streets, and a sewage-disposal network to the beleaguered *favelas*.

Understandably, these conditions propel substantial emigration toward the coastal cities and, increasingly, out of the Northeast to the more prosperous Brazilian subregions that line the Atlantic seaboard to the south. To stem that human tide, Brazil's government has pursued the purchase of underutilized farmland for the settling of landless peasants; although tens of thousands of families have benefited, overall barely a dent has been made in the Northeast's massive rural poverty crisis.

The Northeast today is Brazil's great contradiction. In cities such as Recife and Salvador, hordes of peasants driven from the land constantly arrive to expand the surrounding shantytowns. As yet, few of the generalizations about emerging Brazil apply here, but there are some bright spots. A petrochemical complex has been built near Salvador, creating thousands of jobs and luring foreign investment. Irrigation projects have nurtured a num-

ber of productive new commercial agricultural ventures. Tourism is booming along the entire Northeast coast, whose thriving beachside resorts attract thousands of vacationing Europeans. Recife has spawned a budding software industry and a major medical complex. And Fortaleza is the center of new clothing and shoe industries that have already put the city on the global economic map.

The Southeast ② has been modern Brazil's *core area*, with its major cities and leading population clusters. Gold first drew many thousands of settlers, and other mineral finds also contributed to the influx—with Rio de Janeiro itself serving as the terminus of the "Gold Trail" and as the long-term capital of Brazil until 1960. Rio de Janeiro became the cultural capital as well, the country's most international center, *entrepôt*, and tourist hub. The third quarter of the twentieth century brought another mineral age to the Southeast, based on the iron ores around

Lafaiete carried to the steelmaking complex at Volta Redonda (Fig. 5-15).

The surrounding State of Minas Gerais (the name means "General Mines") formed the base from which industrial diversification in the Southeast has steadily expanded. The booming metallurgical center of Belo Horizonte paved the way and is now the endpoint of a rapidly developing, ultramodern manufacturing corridor that stretches 500 kilometers (300 mi) southwest to metropolitan São Paulo.

São Paulo State ③ is the leading industrial producer and primary focus of ongoing Brazilian development. This economic-geographic powerhouse accounts for nearly half of the country's gross domestic product, with an economy that today matches Argentina's in overall size. Not surprisingly, this subregion is growing phenomenally (it already contains 22 percent of Brazil's population) as a magnet for migrants, especially from the Northeast.

The wealth of São Paulo State was built on its coffee plantations (known as *fazendas*), and Brazil is still the world's leading producer. But coffee today has been eclipsed by other farm commodities. One of them is orange juice concentrate (here, too, Brazil leads the world). São Paulo State now produces more than double the annual output of Florida, thanks to a climate all but devoid of winter freezes, to ultramodern processing plants, and to a fleet of specially equipped tankers that ship the concentrate to foreign markets. Another leading pursuit is soybeans, in which Brazil ranks second among the world's producers.

Matching this agricultural prowess is the State's industrial strength. The revenues derived from the coffee plantations provided the necessary investment capital, ores from Minas Gerais supplied the vital raw materials, the nearby outport of Santos facilitated access to the ocean, and immigration from Europe, Japan, and other parts of Brazil contributed the increasingly skilled labor force. As the capacity of the domestic market grew, the advantages of central location and agglomeration secured São Paulo's primacy. This also resulted in metropolitan São Paulo becoming the country's—and South America's—leading industrial complex and megacity (population: 25.8 million).

The South ④ consists of three States, whose combined population exceeds 28 million: Paraná, Santa Catarina, and Rio Grande do Sul. Southernmost Brazil's excellent agricultural potential has long attracted large numbers of European immigrants. Here the newcomers introduced their advanced farming methods to several areas. Portuguese rice farmers clustered in the valleys of Rio Grande do Sul, where tobacco production has now propelled Brazil to become the world's leading exporter. The Germans, specialists in raising grain and cattle, occupied the somewhat higher areas to the north and in Santa Catarina. The Italians

São Paulo may not have an imposing skyline to match New York or a skyscraper to challenge the Sears Tower of Chicago, but what this Brazilian megacity does have is mass. São Paulo is more than just another city—it is a vast conurbation of numerous cities and towns containing more than 25 million inhabitants who constitute the third-largest human agglomeration on Earth. This view shows part of the razor-sharp edge of the CBD, a concrete jungle with little architectural distinction but vibrant urban cultures. In the foreground, juxtaposed against the affluence of downtown, the teeming inner city reflects the opposite end of the social spectrum—a chaotic jumble of *favelas* that mainly house the newest urban migrants, especially from Brazil's hard-pressed Northeast. São Paulo, demographers say, gives us a glimpse of the urban world of the future.   © AP/Wide World Photos.

selected the highest slopes, where they established thriving vineyards. All of these fertile lands proved highly productive, and with growing markets in the large urban areas to the north, this tristate subregion became Brazil's most affluent corner.

AMONG THE REALM'S GREAT CITIES . . . São Paulo

SÃO PAULO, WHICH lies on a plateau 50 kilometers (30 mi) inland from its Atlantic outport of Santos, possesses no obvious locational advantages. Yet here on this site we find the third largest metropolis on Earth, whose population has multiplied so uncontrollably that São Paulo has more than doubled in size (from 11 to just under 26 million) during the past three decades.

Founded in 1554 as a Jesuit mission, modern São Paulo was built on the nineteenth-century coffee boom. It has since grown steadily as both an agricultural processing center (soybeans, orange juice concentrate, and sugar besides coffee) and a manufacturing complex (accounting for about half of all of Brazil's industrial jobs). It has also become Brazil's primary focus of commercial and financial activity. Today São Paulo's bustling, high-rise CBD (photo, p. 283) is the very symbol of urban South America and attracts the continent's largest flow of foreign investment as well as the trade-related activities that befit the city's rise as the business capital of Mercosul (Mercosur).

Nonetheless, even for this biggest industrial metropolis of the Southern Hemisphere, the increasingly global tide of postindustrialism is rolling in and São Paulo is being forced to adapt. To avoid becoming a Detroit-style Rustbelt, the aging automobile-dominated manufacturing zone on the central city's southern fringes is today attracting new industries. Internet companies, in particular, have flocked here in such numbers that they already constitute Brazil's largest cluster of e-commerce activity—and prompt many to now call this area Silicon Village.

Elsewhere in São Paulo's vast urban constellation—whose suburbs now sprawl outward up to 100 kilometers (60 mi) from the CBD—additional opportunities are being exploited. In the outer northeastern sector, new research facilities as well as computer and telecommunications-equipment factories are springing up. And to the west of central São Paulo, lining the ring road that follows the Pinheiros River, is South America's largest suburban office complex replete with a skyline of ultramodern high-rises.

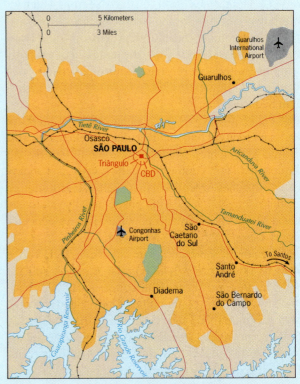

© H. J. de Blij, P. O. Muller, and John Wiley & Sons, Inc.

The huge recent growth of this megacity has also been accompanied by massive problems of overcrowding, pollution, and congestion. Traffic jams here are among the world's worst, with more than twice as many gridlocked motor vehicles on any given day than in Manhattan. Staggering poverty—on a par with Mexico City's—is the most pressing crisis as the ever-expanding belt of shantytowns tightens its grip on much of the metropolis. With its prodigious growth rate expected to persist through the foreseeable future, can anything prevent Greater São Paulo from approaching an unimaginable population of 50 million barely 20 years from today?

With the South firmly rooted in the European-commercial culture sphere (Fig. 5-5A), European-style standards of living match the diverse Old World heritage that is reflected in the towns and countryside (where German and Italian are spoken alongside Portuguese). This has led to hostility against non-European Brazilians, and many communities actively discourage poor, job-seeking migrants from the North by offering to pay return bus fares or even blocking their household-goods-laden vehicles. Moreover, extremist groups have arisen to openly espouse the secession of the South from Brazil.

Economic development in the South is not limited to the agricultural sector. Coal from Santa Catarina and Rio Grande do Sul is shipped north to the steel plants of Minas Gerais. Local manufacturing is growing as well, especially in Pôrto Alegre and Tubarão. During the 1990s, a major center of the computer software industry was established in Florianópolis, the island city and State capital just off

Santa Catarina's coast. Known as ***Tecnópolis***, this budding technopole continues to grow by capitalizing on its seaside amenities, skilled labor force, superior air-travel and global communications linkages, and government and private-sector incentives to support new companies. Only the sparsely populated interior portion of the South has lagged behind the rest of the region, despite the recent completion of massive Itaipu Dam in westernmost Paraná State.

The Interior ⑤ subregion—constituted by the States of Goiás, Mato Grosso, and Mato Grosso do Sul—is also known as the ***Central-West***. This is the region that Brazil's developers have long sought to make a part of the country's productive heartland, and in 1960 the new capital of Brasília was deliberately situated on its margins (Fig. 5-15).

By locating the new capital city in the wilderness 650 kilometers (400 mi) inland from its predecessor, Rio de Janeiro, the nation's leaders dramatically signaled the opening of Brazil's development thrust toward the west. Brasília is noteworthy in another regard because it represents what political geographers call a **forward capital**. A state will sometimes relocate its capital to a sensitive area, perhaps near a peripheral zone under dispute with an unfriendly neighbor, in part to confirm its determination to sustain its position in that contested zone. Brasília does not lie near a contested area, but Brazil's interior was an internal frontier to be conquered by a growing nation. Spearheading that drive, the new capital occupied a decidedly forward position.

Despite the subsequent growth of Brasília to 3.8 million inhabitants today (which includes a sizeable ring of peripheral squatter settlements), it was not until the 1990s that the Interior began its economic integration with the rest of Brazil. The catalyst was the exploitation of the vast ***cerrado***—the fertile savannas that blanket the Central-West and make it one of the world's most promising agricultural frontiers (at least two-thirds of its arable land still awaits development). As with the U.S. Great Plains, the flat terrain of the ***cerrado*** is one of its main advantages because it facilitates the large-scale mechanization of farming with a minimal labor force. Another advantage is rainfall, which is more prevalent than in the Great Plains or Argentine Pampa (Fig. G-7).

The leading crop is soybeans, whose output per hectare here exceeds even that of the U.S. Corn Belt. Other grains and cotton are also expanding across the farmscape of the ***cerrado***, but the current pace of regional development is inhibited by a serious accessibility problem. Unlike the Great Plains and Pampa, the growth of an efficient transportation network did not accompany the opening of this farming frontier. Thus the Interior's products must travel along poor roads and intermittent railroads to reach the markets and ports of the Atlantic seaboard. Today several projects are finally underway to alleviate these bottlenecks, including the privately financed ***Ferronorte*** railway that links Santos to the southeastern corner of Mato Grosso State, and an improved highway north to the Amazon River port city of Santarém.

The North ⑥ is Brazil's territorially largest and most rapidly developing subregion, which consists of the seven States of the Amazon Basin (Fig. 5-15). This was the scene of the great rubber boom a century ago, when the wild rubber trees in the ***selvas*** (tropical rainforests) produced huge profits and the central Amazon city of Manaus enjoyed a brief period of wealth and splendor. But the rubber boom ended in 1910, and for most of the seven decades that followed Amazonia was a stagnant hinterland lying remote from the centers of Brazilian settlement. All that changed dramatically during the 1980s as new development began to stir throughout this awakening region, which currently is the scene of the world's largest migration into virgin territory as more than 200,000 new settlers arrive each year. Most of this influx is occurring south of the Amazon River, in the tablelands between the major waterways and along the Basin's wide rim.

Development projects abound in the Amazonian North. One of the most durable is the ***Grande Carajás Project*** in southeastern Pará State, a huge multifaceted scheme centered on one of the world's largest known deposits of iron ore in the hills around Carajás (Fig. 5-15). In addition to a vast mining complex, other new construction here includes the Tucuruí Dam on the nearby Tocantins River and an 850-kilometer (535-mi) railroad to the Atlantic port of São Luis. This ambitious development project also emphasizes the exploitation of additional minerals, cattle raising, crop farming, and forestry. What is occurring is a manifestation of the **growth-pole concept**. A growth pole is a location where a set of activities, given a start, will expand and generate widening ripples of development in the surrounding area. In this case, the stimulated hinterland could one day cover one-sixth of all Amazonia.

Understandably, tens of thousands of settlers have descended on this part of the Amazon Basin. Those seeking business opportunities have been in the vanguard, but they have been followed by masses of lower-income laborers and peasant farmers in search of jobs and land ownership. The initial stage of this colossal enterprise has boosted the fortunes of many towns, particularly Manaus northwest of Carajás. Here, a thriving industrial complex (specializing in the production of electronic goods) has emerged in the free-trade zone adjoining the city thanks to the outstanding air-freight operations at Manaus's ultramodern airport. But many problems have also arisen as the tide of pioneers

This is what the Amazon's equatorial rainforest looks like from an orbiting satellite after the human onslaught in preparation for settlement. The colors on this Landsat image emphasize the destruction of the trees, with the dark green of the natural forest contrasted against the pale green and pinks of the leveled forest. The linear branching pattern of deforestation here in Rondônia State's Highway BR-364 corridor is explained in the text. But farming here is not likely to succeed for very long, and much of the cleared land is likely to be abandoned. Then the onslaught will resume to clear additional land—as an entire ecosystem comes ever closer to failing forever.
© NRSC LTS/Photo Researchers.

rolled across central Amazonia. One of the most tragic involved the Yanomami people, whose homeland in Roraima State was overrun by thousands of claim-stakers (in search of newly discovered gold) who triggered violent confrontations that ravaged the fragile aboriginal way of life.

Another leading development scheme, known as the **Polonoroeste Plan**, is located about 800 kilometers (500 miles) to the southwest of Grande Carajás in the 1500-mile-long Highway BR-364 corridor that parallels the Bolivian border and connects the western Brazilian towns of Cuiabá, Pôrto Velho, and Rio Branco (Fig. 5-15). Although the government had planned for the penetration of western Amazonia to proceed via the east-west Trans-Amazon Highway, the migrants of the 1980s and 1990s preferred to follow BR-364 and settle within the Basin's southwestern rim zone, mostly in Rondônia State. Agriculture has been the dominant activity here, but in the quest for land, bitter conflicts continue to break out between peasants and landholders as the Brazilian government pursues the volatile issue of land reform.

This region of Brazil is well known for the **deforestation** occurring there. Cutting the forest results directly from logging operations, but more of it is a matter of land occupation and use by settlers. The usual pattern of settlement in this part of Brazil goes like this. As main and branch high-ways are cut through the forest, settlers follow (often due to government-sponsored land-occupation schemes), and they move out laterally to clear land for farming. Subsistence crops are planted, but within a few years weed infestation and declining soil fertility make the plots unproductive. As soil fertility continues to decline, and with few markets for commercial crops, the settlers plant grasses and then sell their land to cattle ranchers. The peasant farmers then move on to newly opened areas, clear more land for planting, and the cycle repeats itself. Unfortunately, this not only entrenches low-grade land uses across widespread areas, but also entails the burning and clearing of vast stands of tropical woodland (see photo above). Since the 1980s, an area of rainforest almost the size of Ohio has been lost *annually* in Amazonia—accounting for over half of all tropical deforestation on the planet and worsening an environmental crisis of global proportions.

Brazil is the cornerstone of South America, the dominant economic force in Mercosur/Mercosul, the only dimensional counterweight to the United States in the Western Hemisphere, a maturing democracy, and an emerging global giant. The future of the entire South American realm depends on Brazil's stability and social as well as economic progress.

What Can You Do

RECOMMENDATION: Take up a Region! As you encounter the world's regions in this book, is there one that interests you more than others, one that arouses your curiosity so that you might want to learn more about it than this course can offer? Well, regional specialists are ever fewer and farther between in our increasingly computer-screen-dominated world, and regional knowledge could benefit you for a lifetime. We recommend that you follow your interest in a realm or a region that forms part of it—South America's "Southern Cone" or Southern Africa or Mainland Southeast Asia—and pursue it in the news, in your reading, in other courses. English-language newspapers (with corresponding websites) are published in almost every major city, from Jakarta to Buenos Aires to Beijing, and it's fascinating to see what concerns people there on a daily basis—the local news and the letters to the editor can give you insights you'll never find on the networks. Who knows, you may even travel there and start a specialization that could become a hobby, if not a professional sideline!

GEOGRAPHIC CONNECTIONS

1 South America is a realm of diverse cultural landscapes and numerous immigrant sources. Some countries that are relatively homogeneous culturally, such as Colombia and Venezuela, are as divided politically and economically as are other countries with strong and obvious cultural-geographic divisions such as Peru and Bolivia. In virtually all South American countries, such divisions have slowed development and hampered national integration. What do you see as the best solution(s) for the realm and its states as they seek to overcome these historic obstacles? What evidence is there on the continent of "what works" and what does not?

2 Brazil is the giant of South America, and it has the largest and most productive economy. But Brazil is unlike other South American states in a number of ways (for example, there is no megacity on a scale with São Paulo anywhere else in the realm) so that what locals call the "Brazilian model" of development may not work in other countries. Still, Brazil does share key qualities with its many neighbors, and its experiences can help these neighbors in their efforts to make progress. Discuss the important similarities and differences between Brazil and the rest of South America that have a bearing on the still-evolving relationships among countries in this realm.

Geographic Literature on South America: The key introductory works on this realm were authored by Blouet & Blouet, Bromley & Bromley, Caviedes & Knapp, Clawson, James & Minkel, Kent, Morris, and Preston. These works, together with a comprehensive listing of noteworthy books and recent articles on the geography of South America, can be found in the *References and Further Readings* section of this book's website at **www.wiley.com/college/deblij**.

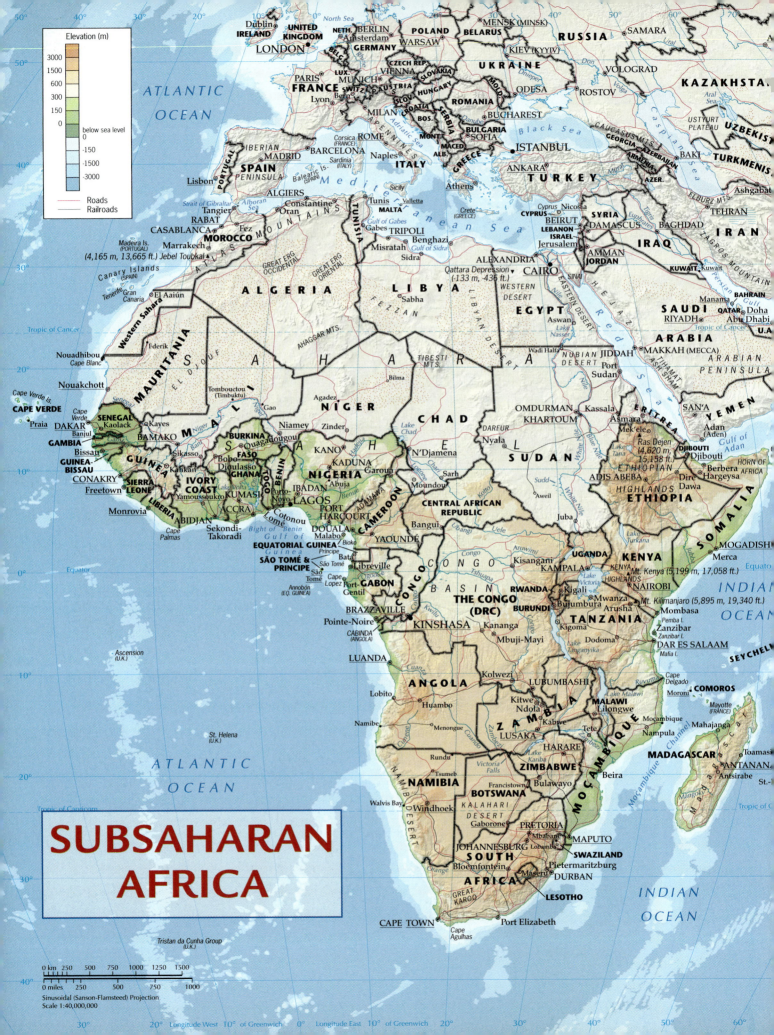

CONCEPTS, IDEAS, AND TERMS

1. Human evolution
2. Rift valley
3. Continental drift
4. Medical geography
5. Endemic
6. Epidemic
7. Pandemic
8. Land tenure
9. Land alienation
10. Green Revolution
11. State formation
12. Colonialism
13. Multilingualism
14. Apartheid
15. Separate development
16. Landlocked state
17. Exclave
18. Desertification
19. Periodic market
20. Islamic Front

REGIONS

- SOUTHERN AFRICA
- EAST AFRICA
- EQUATORIAL AFRICA
- WEST AFRICA
- AFRICAN TRANSITION ZONE

Where were these photos taken? Find out at www.wiley.com/college/deblij

In This Chapter

- Heart of the world, cradle of humanity
- Africa's rich and varied cultures
- Why Africa remains in the grip of poverty
- South Africa: Engine of its region, beacon for the realm
- Nigeria: Cracking cornerstone of West Africa
- The Islamic Front looms over the north

Photos: © H. J. de Blij

FIGURE 6-1 © H. J. de Blij, P. O. Muller, and John Wiley & Sons, Inc.

THE AFRICAN continent occupies a special place in the physical as well as the human world. You need a globe to confirm the former: the Earth has a Land Hemisphere and a Water Hemisphere, and Africa lies at the center of the Land Hemisphere, surrounded in all directions by the other landmasses. As for Africa's human significance, this is where the great saga of **human evolution** began. In Africa we formed our first communities, spoke our first words, created our first art and our first weapons. From Africa our ancestor hominids spread outward into Eurasia more than 2 million years ago. From Africa our own species emigrated, beginning perhaps 130,000 years ago, northward into present-day Europe and eastward via southern Asia into Australia and, much later, farther afield into the Americas. Disperse our forebears did, but we should remember that, at the source, we are all Africans.

Cradle and Cauldron

For millions of years, therefore, Africa served as the cradle for humanity's emergence. For tens of thousands of years, Africa was the source of human cultures. For thousands of years, Africa led the world in countless spheres ranging from tool manufacture to plant domestication.

But in this chapter we will encounter an Africa that has been struck by a series of disasters ranging from environmental deterioration to human dislocation on a scale unmatched anywhere in the world. When we assess Africa's misfortunes, though, we should remember that they have lasted hundreds, not thousands, or millions, of years. Africa's catastrophic interlude will end, and Africa's time and turn will come again.

The focus in this chapter will be on Africa south of the Sahara, for which the unsatisfactory but convenient name *Subsaharan Africa* has come into use to signify not physically "under" the great desert but directionally "below" it. The African continent contains two geographic realms: the African, extending from the southern margins of the Sahara to the Cape of Good Hope, and the western flank of the realm dominated by the Muslim faith and Islamic culture whose heartland lies in the Middle East and the Arabian Peninsula. The great desert forms a formidable barrier between the two, but the powerful influences of Islam crossed it centuries before the first Europeans set foot in West Africa. By that time, the African kingdoms in what is

---

MAJOR GEOGRAPHIC QUALITIES OF

Subsaharan Africa

1. Physiographically, Africa is a plateau continent without a linear mountain backbone, with a set of Great Lakes, variable rainfall, generally low-fertility soils, and mainly savanna and steppe vegetation.

2. Dozens of nations, hundreds of ethnic groups, and many smaller entities make up Subsaharan Africa's culturally rich and varied population.

3. Most of Subsaharan Africa's peoples depend on farming for their livelihood.

4. Health and nutritional conditions in Subsaharan Africa need improvement as the incidence of disease remains high and diets are often unbalanced. The AIDS pandemic began in Africa and has become a major health crisis in this realm.

5. Africa's boundary framework is a colonial legacy; many boundaries were drawn without adequate knowledge of or regard for the human and physical geography they divided.

6. The realm is rich in raw materials vital to industrialized countries, but much of Subsaharan Africa's population has little access to the goods and services of the world economy.

7. Patterns of raw-material exploitation and export routes set up in the colonial period still prevail in most of Subsaharan Africa. Interregional and international connections are poor.

8. During the Cold War, great-power competition magnified conflicts in several Subsaharan African countries, with results that will be felt for generations.

9. Severe dislocation affects many Subsaharan African countries, from Liberia to Rwanda. This realm has the largest refugee population in the world today.

10. Government mismanagement and poor leadership afflict the economies of many Subsaharan African countries.

known today as the Sahel had been converted, creating an Islamic foothold all along the northern periphery of the African realm (see Fig. G-2, p. 7). As we note later, this cultural and ideological penetration had momentous consequences for Subsaharan Africa.

Peril of Proximity

In the three previous chapters on the Americas, we made frequent reference to the forced migration of Africans to Brazil, the Caribbean region, and the United States. The slave trade was one of those African disasters alluded to above, and it was facilitated in part by what we may call the peril of proximity. The northeastern tip of Brazil, by far the largest single destination for the millions of Africans forced from their homes in bondage, lies about as far from the nearest West African coast as South Carolina lies from Venezuela, a short maritime intercontinental journey indeed (it is more than twice as far from West Africa to South Carolina). That proximity facilitated the forced migration of millions of West Africans to Brazil, which in turn contributed to the emergence of an African cultural diaspora in Brazil that is without equal in the New World. It is therefore logical to focus next on the African realm.

Defining the Realm

The African continent may be partitioned into two human-geographic realms, but the landmass is indivisible. Before we investigate the human geography of Subsaharan Africa, therefore, we should take note of the entire continent's unique physical geography (Fig. 6-1). We have already noted Africa's situation at the center of the planet's Land Hemisphere; moreover, no other landmass is positioned so squarely astride the equator, reaching almost as far to the south as to the north. This location has much to do with the distribution of Africa's climates, soils, vegetation, agricultural potential, and human population.

AFRICA'S PHYSIOGRAPHY

Africa accounts for about one-fifth of the Earth's entire land surface. The north coast of Tunisia lies 7700 kilometers (4800 mi) from the south coast of South Africa. Coastal Senegal, on the extreme western *Bulge* of Africa, lies 7200 kilometers (4500 mi) from the tip of the *Horn* in easternmost Somalia. These distances have environmental implications. Much of Africa is far from maritime sources of moisture. In addition, as Figure G-8 shows, large parts of the landmass lie in latitudes where global atmospheric circulation systems produce arid conditions. The Sahara in the north and the Kalahari in the south form part of this globe-girdling desert zone. Water supply is one of Africa's great problems.

Rifts and Rivers

Africa's topography reveals several properties that are not replicated on other landmasses. Alone among the continents, Africa does not have an Andes-like linear mountain backbone; neither the northern Atlas nor the southern Cape Ranges are in the same league. Where Africa does have high mountains, as in Ethiopia and South Africa, these are really deeply eroded plateaus or, as in East Africa, high, snowcapped volcanoes. Furthermore, Africa is one of only two continents containing a cluster of Great Lakes, and the only one whose lakes result from powerful tectonic forces in the Earth's crust. These lakes (with the exception of Lake Victoria) lie in deep trenches called **rift valleys**, which form when huge parallel [2] cracks or faults appear in the Earth's crust and the strips of crust between them sink, or are pushed down, to form great, steep-sided, linear valleys. In Figure 6-2 these rift valleys, which stretch almost 10,000 kilometers (6300 mi) from the Red Sea to Swaziland, are indicated by red lines.

Africa's rivers, too, are unusual: their upper courses often bear landward, seemingly unrelated to the coast toward which they eventually flow. Several rivers, such as the Nile and the Niger, have inland as well as coastal deltas. Major waterfalls, notably Victoria Falls on the Zambezi, or lengthy systems of cataracts, separate the upper from the lower courses.

Finally, Africa may be described as the "plateau continent." Except for some comparatively limited coastal plains, almost the entire continent lies above 300 meters (1000 ft) in elevation, and fully half of it lies over 800 meters (2500 ft) high. As Figure 6-2 shows, the plateau surface has sagged under the weight of accumulating sediments into a half dozen major basins (three of them in the Sahara). The margins of Africa's plateau are marked by escarpments, often steep and step-like. Most notable among these is the Great Escarpment of South Africa, marking the eastern edge of the Drakensberg Mountains.

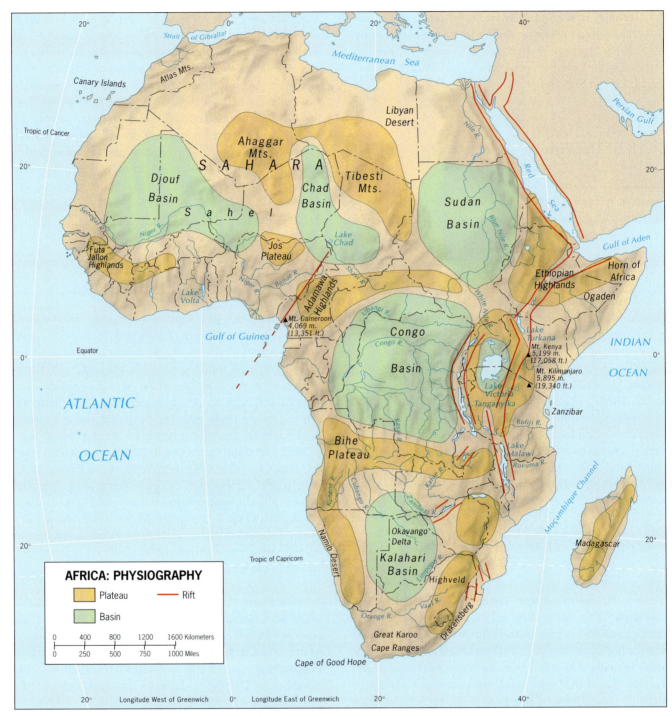

FIGURE 6-2 © H. J. de Blij, P. O. Muller, and John Wiley & Sons, Inc.

Continental Drift and Plate Tectonics

Africa's remarkable and unusual physiography was one piece of evidence that geographer Alfred Wegener used to construct his hypothesis of **continental drift**. The present continents, Wegener reasoned, lay assembled as one giant landmass called *Pangaea* not very long ago geologically (220 million years ago). The southern part of this super-continent was *Gondwana*, of which Africa formed the core (Fig. 6-3). When, about 200 million years ago, tectonic forces began to split Pangaea apart, Africa (and the other landmasses) acquired their present configurations. That

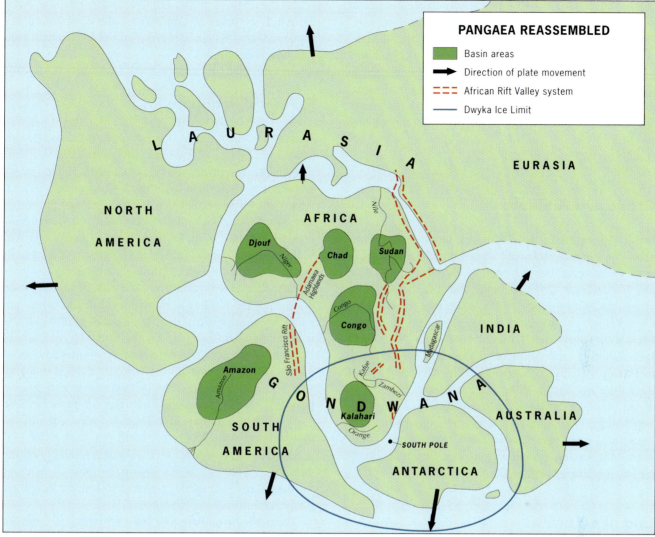

FIGURE 6-3   © H. J. de Blij, P. O. Muller, and John Wiley & Sons, Inc.

process, now known as *plate tectonics*, continues, marked by earthquakes and volcanic eruptions. By the time it started, however, Africa's land surface had begun to acquire some of the features that mark it today—and make it unique. The rift valleys, for example, demarcate the zones where plate movement continues—hence the linear shape of the Red Sea, where the Arabian Plate is separating from the African Plate (Fig. G-4). And yes, the rift valleys of East Africa probably mark the further fragmentation of the African Plate (some geophysicists have already referred to a "Somali Plate," which in the future will separate, Madagascar-like, from the rest of Africa).

So Africa's ring of escarpments, its rifts, its river systems, its interior basins, and its lack of Andes-like mountains, all relate to the continent's central location in Pangaea, all pieces of the puzzle that led to the plate-tectonics solution. Once again, geography was the key.

NATURAL ENVIRONMENTS

Only the southernmost tip of Subsaharan Africa lies outside the tropics. Although African elevations are comparatively high, they are not high enough to ward off the heat that comes with tropical location except in especially favored locales such as the Kenya Highlands and parts of Ethiopia. And, as we have noted, Africa's bulky shape means that much of the continent lies far from maritime moisture sources. Variable weather and frequent droughts are among Africa's environmental problems.

It is useful at this point to refer back to Figure G-8 on page 15. As that map shows, Africa's climatic regions are distributed almost symmetrically about the equator, though more so in the center of the landmass than in the east, where elevation changes the picture. The hot, rainy climate of the Congo Basin merges gradually, both northward and southward, into climates with distinct winter-dry seasons. "Winter," however, is marked more by drought than by cold. In parts of the area mapped *Aw* (savanna), the annual seasonal cycle produces two rainy seasons, often referred to locally as the "long rains" and the "short rains," separated by two "winter" dry periods. As you go farther north and south, away from the moist Congo Basin, the dry season(s) grow longer and the rainfall diminishes and becomes less and less dependable.

Crops and Animals

Most Africans make their living by farming, and many grow crops in marginal areas where rainfall variability can have catastrophic consequences. Compare Figures G-7 and G-8 and note the steep decline in annual rainfall from more than 200 centimeters (80 in) around the equator in the Congo Basin to a mere 10 centimeters (4 in) in parts of Chad to the north and Namibia to the south. Millions of farmers till the often unproductive soils of the *savanna*, and many mix livestock herding with crop-raising to reduce their risk in this difficult environment. But the savanna's wildlife carries diseases that also infect livestock, which makes herding a risky proposition, too. Even where the savanna gives way to the still drier *steppe*, human pressure continues to grow, and people as well as animals trample fragile, desert-margin ecologies.

End of an Era

Africa's shrinking rainforests and vast savannas form the world's last refuges for wildlife ranging from primates to wildebeests. Gorillas and chimpanzees survive in dwindling numbers in threatened forest habitats, while millions of herbivores range in great herds across the savanna plains where people compete with them for space. European colonizers, who introduced hunting as a "sport" (a practice that was not part of African cultural traditions) and who brought their capacities for mass destruction to animals as well as people in Africa, helped clear vast areas of wildlife and push species to near extinction. Later they laid out game reserves and other types of conservation areas, but these were not sufficiently large or well enough connected to allow herd animals to follow their seasonal and annual migration routes. The same climatic variability that affects farmers also affects wildlife, and when the rangelands wither, the animals seek better pastures. When the fences of a game reserve wall them off, they cannot survive. When there are no fences, the wildlife invades neighboring farm-

lands and destroys crops, and the farmers retaliate. After thousands of years of equilibrium, the competition between humans and animals in Africa has taken a new turn. It is the end of an era.

Farmers' Problems

And yet it would seem that there could be room for humans as well as wildlife in Subsaharan Africa. As Table G-1 confirms, all the countries of this realm *combined* have a population barely more than half that of China alone. Indeed, Africa does have some areas of good soils, ample water, and high productivity: the volcanic soils of Mount Kilimanjaro and those of the highlands around the Western Rift Valley, the soils in the Ethiopian Highlands, and those in the moister areas of higher-latitude South Africa and parts of West Africa yield good crops when social conditions do not disrupt the farming communities. But such areas are small in the vastness of Africa. Ominously, a 2006 study by the International Center for Soil Fertility and Agricultural Development reports that the overall situation is deteriorating, and that it is worst in East Africa. Soils in Subsaharan Africa are losing nutrients at the highest rate in the world through erosion, exhaustion, and lack of fertilizer. The Center cites an example: in southern Somalia, soils are losing 88 kilograms (200 lbs) of nutrients per hectare (2.5 acres) per year, compared to 9 kilograms (20 lbs) in overstressed Egypt. And according to the journal *Science*, "African farmers desperate for fresh soil are clearing fragile forestlands and wildlife habitat."

The Subsaharan Africa realm has nothing to compare to the huge alluvial basins of India and China, not even a lower Nile Valley and Delta, where comparatively small areas of fertile, irrigated soils can support tens of millions of people. Except for small, often experimental patches of rice and wheat, Africa is the land of corn (maize), millet, and root crops, far less able to provide high per-hectare yields. Africa's natural environment poses a formidable challenge to the millions who depend directly on it.

ENVIRONMENT AND HEALTH

From birth, Africans (especially rural people living in *A* climates) are exposed to a wide range of diseases spread by insects and other organisms. The study of human health in spatial context is the field of **medical geography**, and medical geographers employ modern methods of analysis (including geographic information systems) to track disease outbreaks, identify their sources, detect their carriers, and prevent their repetition. Alliances between doctors and geographers have already yielded significant results. Doctors understand how a disease ravages the body; geographers understand how climatic conditions such as wind direction or variations in river flow can affect the distribution and effectiveness of disease carriers. This collaboration helps protect vulnerable populations.

Tropical Africa, the source of many serious illnesses, is the focus of much of medical geography's work. Not only the carriers (*vectors*) of infectious diseases but also cultural traditions that facilitate transmission, such as sexual practices, food selection and preparation, and personal hygiene, play their role—and all can be mapped. Comparing medical, environmental, and cultural maps can lead to crucial evidence that helps combat the scourge.

In Africa today, hundreds of millions of people carry one or more maladies, often without knowing exactly what ails them. A disease that infects many people (the *hosts*) in a kind of equilibrium, without causing rapid and widespread deaths, is said to be **endemic** to the population. People affected may not die suddenly or dramatically, but their health deteriorates, energy levels fall, and the quality of life declines. In tropical Africa hepatitis, venereal diseases, and hookworm are among the public health threats in this category.

Epidemics and Pandemics

When a disease outbreak has local or regional dimensions, it is called **epidemic**. It may claim thousands, even tens of thousands of lives, but it remains confined to a certain area, perhaps one defined by the range of its vector. In tropical Africa, trypanosomiasis, the disease known as sleeping sickness and vectored by the tsetse fly, has regional proportions. The great herds of savanna wildlife form the *reservoir* of this disease, and the tsetse fly transmits it to livestock and people. It is endemic to the wildlife, but it kills cattle, so Africa's herders try to keep their animals in tsetse-free zones. African sleeping sickness appears to have originated in a West African source area during the fifteenth century, and it spread throughout much of tropical Africa (Fig. 6-4). Its epidemic range was limited by that of the tsetse fly: where there are no tsetse flies, there is no sleeping sickness. More than anything else, the tsetse fly has kept Subsaharan Africa's savannalands largely free of livestock and open to wildlife. Should a remedy be found, livestock would replace the great herds on the grasslands.

When a disease spreads worldwide, it is described as **pandemic**. Africa's and the world's most deadly vectored disease is malaria, transmitted by a mosquito and killer of as many as 1 million children each year. Whether malaria

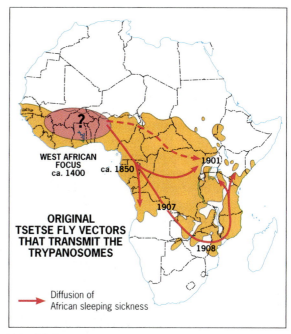

FIGURE 6-4 © H. J. de Blij, P. O. Muller, and John Wiley & Sons, Inc.

has an African origin is not known, but it is an ancient affliction. Hippocrates, the Greek physician of the fifth century BC, mentions it in his writings. Apes, monkeys, and several other species also suffer from it. Fever attacks, anemia, and enlargement of the spleen are its symptoms. Malaria has diffused around the world and prevails not only in tropical but also in temperate areas. Eradication campaigns against the mosquito vector have had some success, but always the carrier has come back with renewed vigor. At present, as many as 300 million people annually are recorded as having malaria, but the actual number probably is much higher because in much of tropical Africa malaria is simply an omnipresent malady affecting entire populations. The short life expectancies for tropical Africa reported in Table G-1 in part reflect infant and child mortality from malarial infection.

Another pandemic disease, also vectored by a mosquito and with African origins, is yellow fever. It crossed the Atlantic, reached South and Middle America, and even penetrated the United States during the nineteenth century; its defeat in Panama a century ago made possible the construction of the Panama Canal. In Africa, however, it remains a threat: Senegal had an outbreak during the 1960s that claimed more than 20,000 lives.

During the 1970s, Africa was afflicted by yet another malady that appears to have started in the equatorial forest, became an epidemic, and grew into a pandemic: AIDS (see box titled "AIDS in Subsaharan Africa"). Monkeys may have activated this disease in humans, but whatever the

AIDS in Subsaharan Africa

ACQUIRED IMMUNE DEFICIENCY Syndrome—AIDS—spread in Africa during the 1980s and became a pandemic during the 1990s. Its impact on Africa has been devastating. In 2006, more than 2.1 million people died of AIDS in Subsaharan Africa, 72 percent of the global toll.

Persons infected by HIV (human immunodeficiency virus) do not immediately or even soon display symptoms of AIDS. In the early stages, only a blood test will reveal infection, and then only by indicating that the body is mobilizing antibodies to fight HIV. People can carry the virus for years without being aware of it; during that period they can unwittingly transmit it to others. For that reason, official statistics of AIDS cases lag far behind the reservoir of those infected, especially in countries where populations cannot be thoroughly tested. Since AIDS was identified in the early 1980s, more than 25 million people have died of it.

By 2006, more than 40 million people were known to be infected by AIDS worldwide, and of these about 27 million lived in 34 tropical African countries according to United Nations sources. In the early 1990s, the worst-hit countries were in Equatorial Africa in what was called the AIDS Belt from (then) Zaïre to Kenya. But a decade later, the most severely afflicted countries lay in Southern Africa. In Zimbabwe, Botswana, Lesotho, and Swaziland, more than 25 percent of all persons 15–49 were infected with HIV; in South Africa, the region's most populous country, the proportion is 22 percent (6.1 million). These are the official numbers; medical geographers estimate that 20 to 25 percent of the entire population of several tropical African countries are infected.

As the number of deaths from AIDS rises, African countries' vital statistics are showing the demographic impact. In Botswana, life expectancy in 1994 was 60 years; by 2006 it had declined to 34. In South Africa it declined from 66 to 47. In Zimbabwe, population growth was 3.3 percent per year in the early 1980s; today it is 0.7 percent. Meanwhile, AIDS is killing parents at such a rate that by 2010, it is projected by the United Nations, it will be responsible for 20 million African orphans.

Geographer Peter Gould, in his classic book *The Slow Plague*, called Africa a "continent in catastrophe," pointing out that the official data on which these depressing reports are based understate the magnitude of the calamity. African governments, Gould wrote, conceal the real dimensions of the tragedy for various reasons: a sense of shame, concern that the real figures will drive investors out and keep tourists away, and the arrogance of military rulers used to denying reality. Many doctors in Africa, he added, say that AIDS cases are 80 to 90 percent underreported.

No part of tropical Africa has been spared. West African countries, too, are reporting growing numbers of AIDS cases. Ivory Coast, reporting that nearly 7 percent of adults are infected there, may provide the most accurate data. In Nigeria, where the full onslaught may not yet be known, AIDS incidence among adults is more than 5 percent (which, in that populous country, means that as many as 8 million people are ill with AIDS).

The misery the AIDS pandemic has created is beyond measure, but its economic impact is beyond doubt. Workers are sick on the job; but jobs are scarce and medical help is unavailable or unaffordable for many, so the workers stay on the job until they collapse. In a realm where unemployment is high, new workers immediately fill the vacancies, but they, too, are unwell. Corporations report skyrocketing AIDS-related costs. In Botswana, whose economy has been prospering due to increased diamond production, highly trained diamond sorters are being lost to AIDS, and training replacement workers is depressing profits. Medical expenses, multiplying death benefits, funeral payments, and recruiting and training costs are reducing corporate incomes from Kenya's sugar plantations to South Africa's gold mines. Thus the impact of AIDS is recorded in the continent's economic statistics. The pattern shown in Figure G-12, where Africa is concerned, is due in substantial part to the AIDS pandemic.

In countries of the global core, progress has been made in containing AIDS and prolonging the lives of those afflicted. This has taken AIDS off the medical center stage, but such a turnaround is not in prospect for Africa. The cost of the medications that combat AIDS is too high, and private drug companies will not dispense their products at prices affordable to African patients. Thus the best short-term strategy is for governments to attempt to alter public attitudes and behavior, notably by making condoms available as widely and freely as possible. Uganda, once one of tropical Africa's worst-afflicted countries, managed to slow the rate of AIDS dispersal by these methods—and through a vigorous advertising campaign on billboards, in newspapers, and on radio and television.

But what Africa needs is a coordinated attack on AIDS through the provision of medicines. In 2003 President George W. Bush announced an initiative to provide Subsaharan Africa with the financial means to combat the disease: a U.S. $10 billion fund to facilitate treatment as well as prevention. The urgency of the matter is underscored by estimates that, in the absence of a massive campaign, some of Africa's countries may see population declines of as much as 10 to 20 percent, a disaster comparable only to Europe's fourteenth-century bubonic plague and the collapse of Native American populations after the introduction of smallpox by the European invaders in the sixteenth. Once again environmental, cultural, and historical factors combine to afflict Africa's peoples.

source, Africa has been the most severely affected by it. In 2006, there were an estimated 41 million AIDS cases in the world, and at least 27 million of them were in Africa. Again, these official calculations probably conceal a much worse crisis, and the population growth rates for many African countries are declining as a result. During the early part of the twenty-first century, some countries will experience actual population decreases. Life expectancies are dropping to levels not seen for centuries.

Challenges to Improving Health Conditions

The great incubator of vectored and nonvectored diseases in humid equatorial Africa continues to threaten local populations. Sudden outbreaks of Ebola fever in Sudan during the 1970s, in The Congo* in the 1990s, and in Uganda in 2000–2001 projected the risks that many Africans live with onto the world's television screens. But Africa's woes are soon forgotten. Only the fear that an African epidemic might evolve into a global pandemic mobilizes world attention. Africans cope with the planet's most difficult environments, but they are least capable of combating its hazards because their resources are so limited.

Even when international efforts are made to assist Africa, the unintended results may be catastrophic. Among Africa's endemic diseases is schistosomiasis, also called bilharzia. The vector is a snail, and the transmitted parasites enter via body openings when people swim or wash in slow-moving water where the snails thrive (fast-moving streams are relatively safe). When development projects designed to help African farmers dammed rivers and streams and sent water into irrigation ditches, farm production rose—but so did the incidence of schistosomiasis, which causes internal bleeding and fatigue, although it is rarely fatal. Medical geographers' maps showed the spread of schistosomiasis around the dam projects, and soon the cause was clear: by slowing down the water flow, the engineers had created an ideal environment for the snail vector.

Poverty is a powerful barrier to improving health, and major interventions in the natural environment carry costs as well as benefits. Africa needs improved medical services, more effective child-immunization programs, mobile clinics for remote areas, and similar remedies. While two-thirds of the realm's people continue to live in rural areas, the natural environment will weigh heavily on a vulnerable population.

_____
*Two countries in Africa have the same short-form name: Congo. In this book, we use **The Congo** for the larger Democratic Republic of the Congo and **Congo** for the smaller Republic of Congo.

LAND AND FARMING

In their penetrating book, *Geography of Sub-Saharan Africa*, editor Samuel Aryeetey-Attoh and his colleagues focus on the issue of land tenure, crucial in Africa because most Africans are farmers. **Land tenure** refers to the way people own, occupy, and use land. African traditions of land tenure are different from those of Europe or the Americas. In most of Subsaharan Africa, communities, not individuals, customarily hold land. Occupants of the land have temporary, custodial rights to it and cannot sell it. Land may be held by large (extended) families, by a village community, or even by a traditional chief who holds the land in trust for the people. His subjects may house themselves on it and farm it, but in return they must follow his rules.

Stolen Lands

When the European colonizers took control of much of Subsaharan Africa, their land ownership practices clashed head-on with those of Africa. Africans believed that their land belonged to their ancestors, the living, and the yet-unborn; Europeans saw unclaimed space and felt justified in claiming it. What Africans called **land alienation**—the expropriation of (often the best) land by Europeans—changed the pattern of land tenure in Africa. By the time the Europeans withdrew, private land ownership was widely dispersed and could not be reversed. Postcolonial African states tried to deal with this legacy by nationalizing all the land and doing away with private ownership, reverting in theory to the role of the traditional chief. But this policy has not worked well. In the rural areas, the government in the capital is a remote authority often seen as unsympathetic to the plight of farmers. Governments keep the price of farm products low on urban markets, pleasing their supporters but frustrating farmers. Large landholdings once owned by Europeans seem now to be occupied by officials or those the government favors. The colonial period's approach to land tenure left Africa with a huge social problem.

Rapid population growth makes the problem worse. Traditional systems of land tenure, which involve subsistence farming in various forms ranging from shifting cultivation to pastoralism, work best when population is fairly stable. Land must be left fallow to recover from cultivation, and pastures must be kept free of livestock so the grasses can revive. A population explosion of the kind Africa experienced during the mid-twentieth century destroys this equilibrium. Soils cannot rest, and pastures are overgrazed. As land becomes degraded, yields decline.

Shifting agriculture in Africa's tropical forest zones still forms the mainstay for millions. Here in the Ituri Forest in the northeastern corner of The Congo, trees have been felled, the land cleared, and between stumps and rotting trunks the people farm cassava, yams, and bananas. The soil will sustain these crops for some years, but then yields will diminish and the land will be abandoned, so that another plot must be cleared. Women do the traditional farming in Africa; here three women from the village of Ngodi are weeding the field while two youngsters look on.   © Jose Azel/Aurora & Quanta Productions.

This is the situation in much of Subsaharan Africa, where two-thirds of the people depend on farming for their livelihood. That percentage is now declining, but it remains the highest for any realm in the global periphery. Subsaharan Africa is sometimes thought of as a storehouse of minerals and fuels and great underground riches, but farming is the key to its economy. And not just subsistence farming: agricultural exports also produce foreign exchange revenues needed to pay for essential imports. The European colonizers identified suitable environments and laid out plantations to grow cocoa, coffee, tea, and other luxury crops for the markets of the global core. Irrigation schemes enabled the export of cotton, groundnuts (peanuts), and sesame seeds. Today, national governments and some private owners continue to operate these projects.

Persistent Subsistence

But most African farmers are subsistence farmers who grow grain crops (corn, millet, sorghum) in the drier areas and root crops (yams, cassava, sweet potatoes) where moisture is ample. Shifting cultivation still occurs in remote parts of The Congo (photo above) and neighboring locales in Equatorial Africa, but this practice is declining as the forest shrinks. Others are pastoralists, driving their cattle and goats along migratory routes to find water and pasture, sometimes clashing with sedentary farmers whose fields they invade. Whatever their principal mode of farming, all African farmers today feel the effects of population pressure and must cope with dwindling space and damaged ecologies. By *intercropping*, planting several types of crops in one cleared field, some of which resemble those in the forest, shifting cultivators extend the life of their plots. In *compound farming*, usually near a market center, farmers combine the use of compound (village) and household waste as fertilizer with intensive care of the crops to produce vegetables, fruits, and root crops as well as eggs and chickens for urban consumers. Other subsistence farmers use *diversification* to cope with such problems; they combine their cultivation of subsistence crops with nonfarm work, including full-time jobs in nearby towns.

Notwithstanding all these adaptations, African farm production is declining for many reasons, including gov-

A Green Revolution for Africa?

10

THE **Green Revolution**—the development of more productive, higher-yielding types of grains—has narrowed the gap between world population and food production. Where people depend mainly on rice and wheat for their staples, the Green Revolution has pushed back the specter of hunger. The Green Revolution has had less impact in Subsaharan Africa, however. In part, this relates to the realm's high rate of population growth (which is substantially higher than that of India or China). Other reasons have to do with Africa's staples: rice and wheat support only a small part of the realm's population. Corn (maize) supports many more, along with sorghum, millet, and other grains. In moister areas, root crops, such as the yam and cassava, as well as the plantain (similar to the banana) supply most calories. These crops were not priorities in Green Revolution research.

Lately there have been a few signs of hope. Even as terrible famines struck several parts of the continent and people went hungry in places as widely scattered as Moçambique and Mali, scientists worked toward two goals: first, to develop strains of corn and other crops that would be more resistant to Africa's virulent crop diseases, and second, to increase the productivity of those strains. But these efforts faced serious problems. An average African hectare (2.5 acres) planted with corn, for example, yields only about half a ton of corn, whereas the world average is 1.3 tons. When a virus-resistant variety was developed and distributed throughout Subsaharan Africa in the 1980s, yields rose significantly where farmers could afford to buy the new strain as well as the fertilizers it requires. Nigeria saw a near-doubling of its production before it leveled off but population growth negated the advantage. Hardier types of root crops also raised yields in some areas.

But Africa needs more than this to reverse the trend toward food deficiency. (Nor is the Green Revolution an unqualified remedy: the poorest farmers, who need the help most, can least afford the more expensive higher-yielding seeds and also cannot pay for the pesticides that may be required.) It has been estimated that, for the realm as a whole, food production has fallen about 1 percent annually even as population grew by more than 2 percent. Lack of capital, inefficient farming methods, inadequate equipment, soil exhaustion, male dominance, apathy, and devastating droughts contributed to this decline. The seemingly endless series of civil conflicts (The Congo, Somalia, Rwanda, Ivory Coast, Chad) also reduced farm output. The Green Revolution may narrow the gap between production and need, but the battle for food sufficiency in Africa is far from won.

ernment mismanagement and corruption. When Nigeria became independent in 1960, its government embarked on costly industrialization, and agriculture was given a lower priority. In that year, Nigeria was the world's largest producer and leading exporter of palm oil; today, Nigeria must import palm oil to meet its domestic demand. Inadequate rural infrastructures create another obstacle. Roads from the rural areas to the market centers are in bad and worsening shape, often impassable during the rainy season and rutted when dry. Ineffective links between producers and consumers create economic losses. Storage facilities in the villages tend to be inadequate, so that untold quantities of grains, roots, and vegetables are lost to rot, pests, and theft. Electricity supply in rural areas also requires upgrading: without power, pumps cannot be run for water supply, and the processing of produce cannot be mechanized. Rural farmers have inadequate access to credit: banks would rather lend to urban homeowners than to distant villagers. This is especially true for women who, according to current estimates, produce 75 percent of the food in Subsaharan Africa.

Along the banks of Africa's rivers and on the shores of the Great Lakes fishing augments local food supplies, but Africa's fishing industries are still underdeveloped and contribute only modestly to overall nutrition. During the postcolonial period, Subsaharan Africa's population has more than doubled; food production has declined. Agricultural exports have also decreased, but food imports have mushroomed. The situation is especially clear in terms of per-capita output: in 2004 it was 15 percent below that in 1989. Even advances in biotechnology that closed the global gap between food production and demand largely passed Africa by (see box titled "A Green Revolution for Africa?")

AFRICA'S HISTORICAL GEOGRAPHY

Africa is the cradle of humanity. Archeological research has chronicled 7 million years of transition from australopithecenes to hominids to *Homo sapiens*. It is therefore ironic that we know comparatively little about Subsaharan Africa from 5000 to 500 years ago—that is, before the onset of European colonialism. This is partly due to the colonial period itself, during which African history was neglected, many African traditions and artifacts were destroyed, and many misconceptions about African cultures and institutions became entrenched. It is also a result of the absence of a written history over most of Africa south of the Sahara until the sixteenth century—and over a large part of it until much later than that. The best records are those of the savanna belt immediately south of the Sahara, where contact with North African peoples was greatest and where Islam achieved a major penetration.

The absence of a written record does not mean, as some scholars have suggested, that Africa does not have a history as such prior to the coming of Islam and Christianity. Nor does it mean that there were no rules of social behavior, no codes of law, no organized economies. Modern historians, encouraged by the intense interest shown by Africans generally, are now trying to reconstruct the African past, not only from the meager written record but also from folklore, poetry, art objects, buildings, and other such sources. Much has been lost forever, though. Almost nothing is known of the farming peoples who built well-laid terraces on the hill-

sides of northeastern Nigeria and East Africa or of the communities that laid irrigation canals and constructed stone-lined wells in Kenya; and very little is known about the people who, perhaps a thousand years ago, built the great walls of Zimbabwe (photos below). Porcelain and coins from China, beads from India, and other goods from distant sources have been found in Zimbabwe and other points in East and Southern Africa, but the trade routes within Africa itself—let alone the products that circulated and the people who handled them—still remain the subject of guesswork.

African Genesis

Africa on the eve of the colonial period was a continent in transition. For several centuries, the habitat in and near one of the continent's most culturally and economically

FROM THE FIELD NOTES

"Got up before dawn this date in the hope of being the first visitor to the Great Zimbabwe ruins, a place I have wanted to see ever since I learned about it in a historical geography class. Climbed the hill and watched the sun rise over the great elliptical 'temple' below, then explored the maze of structures of the so-called fortification on the hilltop above. Next, for an incredible 90 minutes I was alone in the interior of the great oval walls of the temple. What history this site has seen—it was settled for perhaps six centuries before the first stones were hewn and laid here around AD 750, and from the 11th to the 15th centuries this may have been a ceremonial or religious center of a vast empire whose citizens smelted gold and copper and traded them via Sofala and other ports on the Indian Ocean for goods from South Asia and even China. No cement was used: these walls are built with stones cut to fit closely together, and in the valley they rise over 30 feet (nearly 10 m) high. And what secrets Great Zimbabwe still conceals—where are the plans, the tools, the quarries? The remnants of an elaborate water-supply system suggest a practical function for this center, but a decorated conical tower and adjacent platform imply a religious role. Whatever the answers, history hangs heavily, almost tangibly, over this African site."
© H. J. de Blij

www.conceptcaching.com

Concept Caching

productive areas—West Africa—had been changing. For 2000 years, probably more, Africa had been innovating as well as adopting ideas from outside. In West Africa, cities were developing on an impressive scale; in central and Southern Africa, peoples were moving, readjusting, sometimes struggling with each other for territorial supremacy. The Romans had penetrated to southern Sudan, North African peoples were trading with West Africans, and Arab *dhows* were sailing the waters along the eastern coasts, bringing Asian goods in exchange for gold, copper, and a comparatively small number of slaves.

Consider the environmental situation in West Africa as it relates to the past. As Figures G-7 and G-8 indicate, the environmental regions in this part of the continent exhibit a decidedly east-west orientation. The *isohyets* (lines of equal rainfall totals) run parallel to the southern coast (Fig. G-7); the climatic regions, now positioned somewhat differently from where they were two millennia ago, still trend strongly east-west (Fig. G-8); the vegetation pattern also reflects this situation, with a coastal forest belt yielding to savanna (tall grass with scattered trees in the south, shorter grass in the north) that gives way, in turn, to steppe and desert.

We know that African cultures had been established in all these environmental settings for thousands of years. One of these, the Nok culture, endured for over eight centuries on the Benue Plateau (north of the Niger-Benue confluence in modern Nigeria) from about 500 BC to the third century AD. The Nok people made stone as well as iron tools, and they left behind a treasure of art in the form of clay figurines representing humans and animals. But we have no evidence that they traded with distant peoples. The opportunities created by environments and technologies still lay ahead.

Early Trade in West Africa

West Africa, over a north-south span of a few hundred kilometers, displayed an enormous contrast in environments, economic opportunities, modes of life, and products. The peoples of the tropical forest produced and needed goods that were different from the products and requirements of the peoples of the dry, distant north. For example, salt is a prized commodity in the forest, where the humidity precludes its formation, but it is plentiful in the desert and steppe. This enabled the desert peoples to sell salt to the forest peoples in exchange for ivory, spices, and dried foods. Thus there evolved a degree of ***regional complementarity*** between the peoples of the forest and those of the dryland. And the savanna peoples—those located in between—found themselves in a position to channel and handle the trade (which is always economically profitable).

The markets in which these goods were exchanged prospered and grew, and cities arose in the savanna belt of West Africa. One of these old cities, now an epitome of isolation, was once a thriving center of commerce and learning and one of the leading urban places in the world—Timbuktu. Others, predecessors as well as successors of Timbuktu, have declined, some of them into oblivion. Still other savanna cities, such as Kano in the northern part of Nigeria, remain important.

Early States

Strong and durable states arose in the West African culture hearth (see Fig. 7-3). The oldest state we know anything about is Ghana. Ancient Ghana was located to the northwest of the modern country of Ghana, and covered parts of present-day Mali, Mauritania, and adjacent territory. Ghana lay astride the upper Niger River and included gold-rich streams flowing off the Futa Jallon Highlands, where the Niger has its origins. For a thousand years, perhaps longer, old Ghana managed to weld various groups of people into a stable state. The country had a large capital city complete with markets, suburbs for foreign merchants, religious shrines, and, some distance from the city center, a fortified royal retreat. Taxes were collected from the citizens, and tribute was extracted from subjugated peoples on Ghana's periphery; tolls were levied on goods entering Ghana, and an army maintained control. Muslims from the northern drylands invaded Ghana in about AD 1062, when it may already have been in decline. Even so, the capital was protected for 14 years. However, the invaders had ruined the farmlands and destroyed the trade links with the north. Ghana could not survive. It finally broke into smaller units.

Eastward Shift

In the centuries that followed, the focus of politico-territorial organization in the West African culture hearth shifted almost continuously eastward—first to ancient Ghana's successor state of Mali, which was centered on Timbuktu and the middle Niger River Valley, and then to the state of Songhai, whose focus was Gao, a city on the Niger that still exists. This eastward movement may have been the result of the growing influence and power of Islam. Traditional religions prevailed in ancient Ghana, but Mali and its successor states sent huge, gold-laden pilgrimages to Mecca along the savanna corridor south of the Sahara, passing through present-day Khartoum and Cairo. Of the tens of thousands who participated in these pilgrimages, some remained behind. Today, many Sudanese trace their ancestry to the West Africa savanna kingdoms.

Beyond the West

West Africa's savanna region undoubtedly witnessed momentous cultural, technological, and economic developments, but other parts of Africa also progressed. Early states emerged in present-day Sudan, Eritrea, and Ethiopia. Influenced by innovations from the Egyptian culture hearth, these kingdoms were stable and durable: the oldest, Kush, lasted 23 centuries (Fig. 6-5). The Kushites built elaborate irrigation systems, forged iron tools, and built impressive structures as the ruins of their long-term capital and industrial center, Meroe, reveal. Nubia, to the southeast of Kush, was Christianized until the Muslim wave overtook it in the eighth century. And Axum was the richest market in northeastern Africa, a powerful kingdom that controlled Red Sea trade and endured for six centuries. Axum, too, was a Christian state that confronted Islam, but Axum's rulers deflected the Muslim advance and gave rise to the Christian dynasty that eventually shaped modern Ethiopia.

The process of **state formation** spread throughout **11** Africa and was still in progress when the first European contacts occurred in the late fifteenth century. Large and effectively organized states developed on the equatorial west coast (notably Kongo) and on the southern plateau from the southern part of The Congo to Zimbabwe. East Africa had several city-states, including Mogadishu, Kilwa, Mombasa, and Sofala.

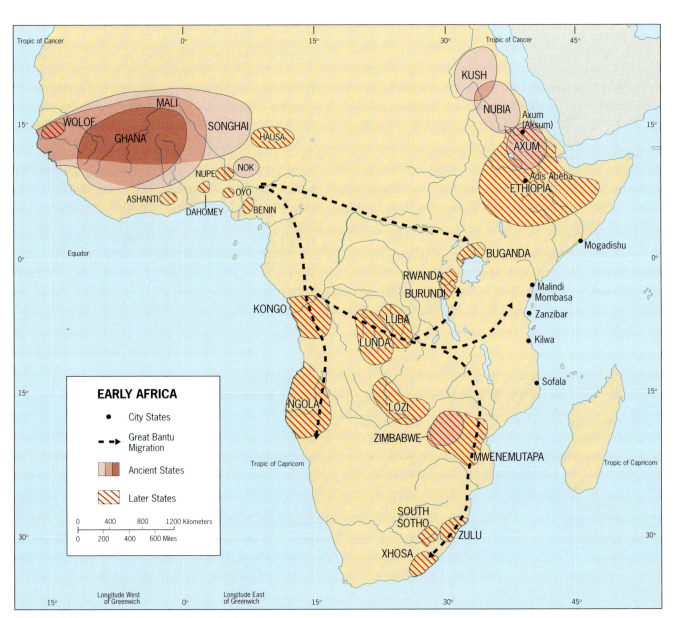

FIGURE 6-5 © H. J. de Blij, P. O. Muller, and John Wiley & Sons, Inc.

Bantu Migration

A crucial event affected virtually all of Equatorial, West, and Southern Africa: the great Bantu migration from present-day Nigeria–Cameroon southward and eastward across the continent. This migration appears to have occurred in waves starting as long as 5000 years ago, populating the Great Lakes area and penetrating South Africa, where it resulted in the formation of the powerful Zulu Empire in the nineteenth century (Fig. 6-5).

All this reminds us that, before European colonization, Africa was a realm of rich and varied cultures, diverse lifestyles, technological progress, and external trade (Fig. 6-6). But Europe's intervention would forever change its evolving political map.

The Colonial Transformation

European involvement in Subsaharan Africa began in the fifteenth century. It would interrupt the path of indigenous African development and irreversibly alter the entire cultural, economic, political, and social makeup of the continent. It started quietly in the late fifteenth century, with Portuguese ships groping their way along the west coast and rounding the Cape of Good Hope. Their goal was to find a sea route to the spices and riches of the Orient. Soon other European countries were sending their vessels to African waters, and a string of coastal stations and forts sprang up. In West Africa, the nearest part of the continent to European spheres in Middle and South America, the initial impact was strongest. At their coastal control points, the Europeans traded with African middlemen for the slaves who were needed to work New World plantations, for the gold that had been flowing northward across the desert, and for ivory and spices.

Coastward Reorientation

Suddenly, the centers of activity lay not with the cities of the savanna but in the foreign stations on the Atlantic coast. As the interior declined, the coastal peoples thrived. Small forest states gained unprecedented wealth, transferring and selling slaves captured in the interior to the European traders on the coast. Dahomey (now called Benin) and Benin (now part of neighboring Nigeria) were states built on the slave trade. When slavery eventually came under attack in Europe, those who had inherited the power and riches it had brought opposed abolition vigorously in both continents.

Horrors of the Slave Trade

Although slavery was not new to West Africa, the *kind* of slave raiding and trading the Europeans introduced certainly was. In the savanna, kings, chiefs, and prominent families traditionally took a few slaves, but the status of those slaves was unlike anything that lay in store for those who were shipped across the Atlantic. In fact, large-scale slave trading had been introduced in East Africa long before the Europeans brought it to West Africa. African middlemen from the coast raided the interior for able-bodied men and women and marched them in chains to the Arab markets on the coast (Zanzibar was a notorious market). There, packed in specially built *dhows*, they were carried off to Arabia, Persia, and India. When the European slave trade took hold in West Africa, however, its volume was far greater. Europeans, Arabs, and collaborating Africans ravaged the

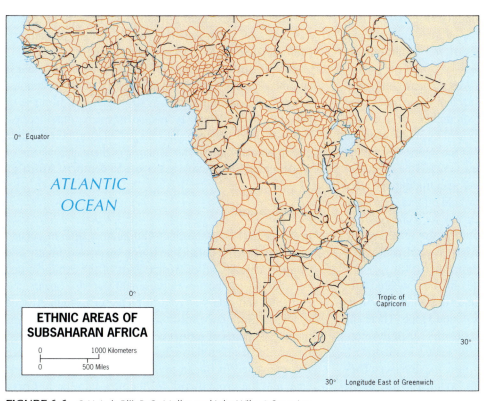

ATLANTIC OCEAN

0° Equator

Tropic of Capricorn

ETHNIC AREAS OF SUBSAHARAN AFRICA

0   1000 Kilometers
0   500 Miles

0°

30°

30°   Longitude East of Greenwich

FIGURE 6-6   © H. J. de Blij, P. O. Muller, and John Wiley & Sons, Inc.

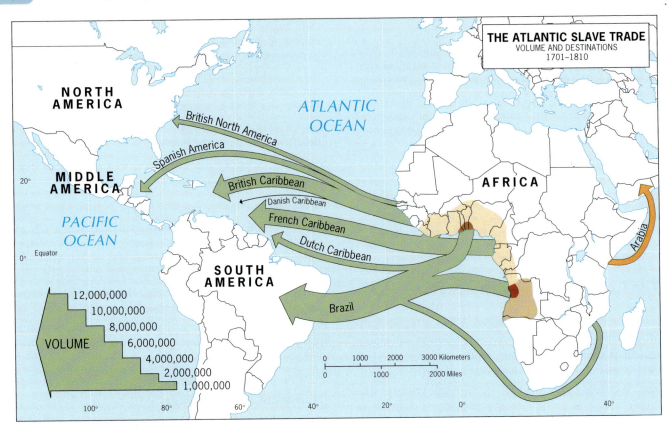

FIGURE 6-7   © H. J. de Blij, P. O. Muller, and John Wiley & Sons, Inc.

continent, forcing perhaps as many as 30 million persons away from their homelands in bondage (Fig. 6-7). Families were destroyed, as were whole villages and cultures; those who survived their exile suffered unfathomable misery.

The European presence on the West African coast completely reoriented its trade routes, for it initiated the decline of the interior savanna states and strengthened the coastal forest states. Moreover, the Europeans' insatiable demand for slaves ravaged the population of the interior. But it did not lead to any major European thrust toward the interior or produce colonies overnight. The African middlemen were well organized and strong, and they held off their European competitors, not just for decades but for centuries. Although the Europeans first appeared in the fifteenth century, they did not carve West Africa up until nearly 400 years later, and in many areas not until after 1900.

In all of Subsaharan Africa, the only area where European penetration was both early and substantial was at its southernmost end, at the Cape of Good Hope. There the Dutch founded Cape Town, a waystation to their developing empire in the East Indies. They settled in the town's hinterland and migrated deeper into the interior. They brought thousands of slaves from Southeast Asia to the Cape, and intermarriage was common. Today, people of mixed ancestry form the largest part of Cape Town's popu-

lation. Later the British took control not only of the Cape but also of the expanding frontier and brought in tens of thousands of South Asians to work on their plantations. Multiracial South Africa was in the making.

Elsewhere in the realm, the European presence remained confined almost entirely to the coastal trading stations, whose economic influence was strong. No real frontiers of penetration developed. Individual travelers, missionaries, explorers, and traders went into the interior, but nowhere else in Africa south of the Sahara was there an invasion of white settlers comparable to that of southernmost Africa's.

Colonization

In the second half of the nineteenth century, after more than four centuries of contact, the European powers finally laid claim to virtually all of Africa. Parts of the continent had been "explored," but now representatives of European governments and rulers arrived to create or expand African spheres of influence for their patrons. Competition was intense. Spheres of influence began to crowd each other. It was time for negotiation, and in late 1884 a conference was convened in Berlin to sort things out. This conference laid the groundwork for the now familiar politico-geographical map of Africa (see box titled "The Berlin Conference").

The Berlin Conference

IN NOVEMBER 1884, the imperial chancellor and architect of the German Empire, Otto von Bismarck, convened a conference of 14 states (including the United States) to settle the political partitioning of Africa. Bismarck wanted not only to expand German spheres of influence in Africa but also to play off Germany's colonial rivals against one another to the Germans' advantage. The major colonial contestants in Africa were: (1) the British, who held beachheads along the West, South, and East African coasts; (2) the French, whose main sphere of activity was in the area of the Senegal River and north of the Congo Basin; (3) the Portuguese, who now desired to extend their coastal stations in Angola and Moçambique deep into the interior; (4) King Leopold II of Belgium, who was amassing a personal domain in the Congo Basin; and (5) Germany itself, active in areas where the designs of other colonial powers might be obstructed, as in Togo (between British holdings), Cameroon (a wedge into French spheres), South West Africa (taken from under British noses in a swift strategic move), and East Africa (where German Tanganyika broke the British design for a solid block of territory from the Cape north to Cairo).

When the conference convened in Berlin, more than 80 percent of Africa was still under traditional African rule. Nonetheless, the colonial powers' representatives drew their boundary lines across the entire map. These lines were drawn through known as well as unknown regions, pieces of territory were haggled over, boundaries were erased and redrawn, and African real estate was exchanged among European governments. In the process, African peoples were divided, unified regions were ripped apart, hostile societies were thrown together, hinterlands were disrupted, and migration routes were closed off. Not all of this was felt immediately, of course, but these were some of the effects when the colonial powers began to consolidate their holdings and the boundaries on paper became barriers on the African landscape (Fig. 6-8).

The Berlin Conference was Africa's undoing in more ways than one. The colonial powers superimposed their domains on the African continent. By the time Africa regained its independence after the late 1950s, the realm had acquired a legacy of political fragmentation that could neither be eliminated nor made to operate satisfactorily. The African politico-geographical map is thus a permanent liability that resulted from three months of ignorant, greedy acquisitiveness during a period when Europe's search for minerals and markets had become insatiable.

Figure 6-8 shows the result. The French dominated most of West Africa, and the British East and Southern Africa. The Belgians acquired the vast territory that became The Congo. The Germans held four colonies, one in each of the realm's regions. The Portuguese held a small colony in West Africa and two large ones in Southern Africa (see the map dated 1910).

The European colonial powers shared one objective in their African colonies: exploitation. But in the way they governed their dependencies, they reflected their differences. Some colonial powers were themselves democracies (the United Kingdom and France); others were dictatorships (Portugal, Spain). The British established a system of indirect rule over much of their domain, leaving indigenous power structures in place and making local rulers representatives of the British Crown. This was unthinkable in the Portuguese colonies, where harsh, direct control was the rule. The French sought to create culturally assimilated elites that would represent French ideals in the colonies. In the Belgian Congo, however, King Leopold II, who had financed the expeditions that staked Belgium's claim in Berlin, embarked on a campaign of ruthless exploitation. His enforcers mobilized almost the entire Congolese population to gather rubber, kill elephants for their ivory, and build public works to improve export routes. For failing to meet production quotas, entire communities were massacred. Killing and maiming became routine in a colony in which horror was the only common denominator. After the impact of the slave trade, King Leopold's reign of terror was Africa's most severe demographic disaster. By the time it ended, after a growing outcry around the world, as many as 10 million Congolese had been murdered. In 1908 the Belgian government took over, and slowly its Congo began to mirror Belgium's own internal divisions: corporations, government administrators, and the Roman Catholic Church each pursued their sometimes competing interests. But no one thought to change the name of the colonial capital: it was Leopoldville until the Belgian Congo achieved independence in 1960.

Colonialism transformed Africa, but in its post-Berlin **12** form it lasted less than a century. In Ghana, for example, the Ashanti (Asante) Kingdom still was fighting the British into the early twentieth century; by 1957, Ghana was independent again. A few years from now, much of Subsaharan Africa will have been independent for half a century, and the colonial period is becoming an interlude rather than a paramount chapter in modern African history.

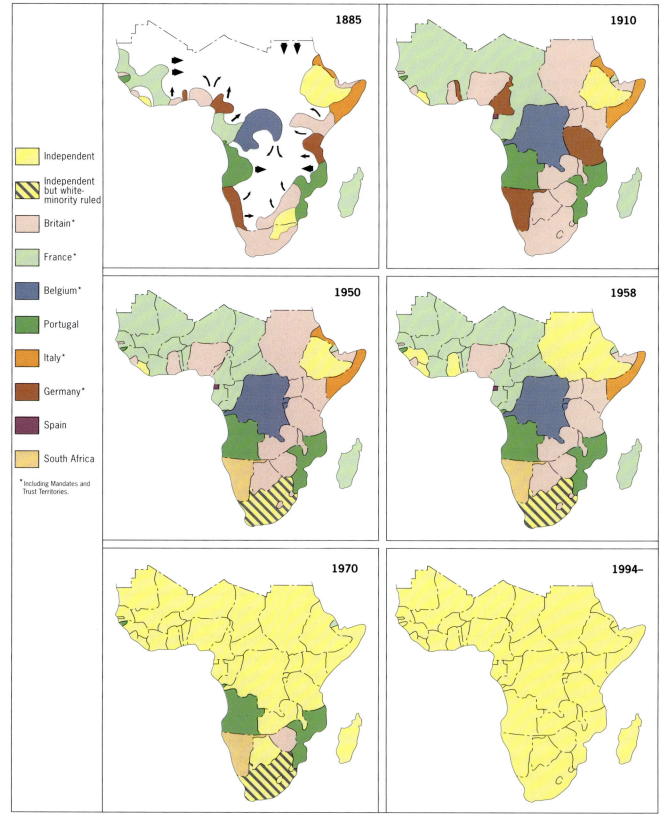

Independent

Independent but white-minority ruled

Britain*

France*

Belgium*

Portugal

Italy*

Germany*

Spain

South Africa

*Including Mandates and Trust Territories.

1885

1910

1950

1958

1970

1994–

COLONIZATION AND DECOLONIZATION SINCE 1885

FIGURE 6-8 © H. J. de Blij, P. O. Muller, and John Wiley & Sons, Inc.

Legacies

Nevertheless, some colonial legacies will remain on the map for centuries to come. The boundary framework has been modified in a few places, but it is a politico-geographical axiom that boundaries, once established, tend to become entrenched and immovable. The transport systems were laid out to facilitate the movement of raw materials from interior sources to coastal outlets. Internal circulation was only a secondary objective, and most African countries still are not well connected to each other. Moreover, the colonizers founded many of Subsaharan Africa's cities or built them on the sites of small towns or villages.

In many countries, the elites that gained advantage and prominence during the colonial period also have retained their preeminence. This has led to authoritarianism in some postcolonial African states (for details, see the regional discussion to come), and to violence and civil conflict in others. Military takeovers of national governments have been a byproduct of decolonization, and hopes for democracy in the struggle against colonialism have too often been dashed. Yet a growing number of African countries, including Ghana, Tanzania, Botswana, South Africa, and, most recently, Nigeria, have overcome the odds and achieved representative government under dauntingly difficult circumstances.

Although not strictly a colonial legacy, the impact of the Cold War (1945–1990) on African states should be noted. In three countries—Ethiopia, Somalia, and Angola—great-power competition, weapons, military advisers, and, in Angola, foreign armed forces magnified civil wars that would perhaps have been inevitable in any case. In other countries, ideological adherence to foreign dogmas led to political and social experiments (such as one-party Marxist regimes and costly farm collectivization) that cost Africa dearly. Today, debt-ridden Africa is again being told what to do, this time by foreign financial institutions. The cycle of poverty that followed colonialism exacts a high price from African societies. As for future development prospects, the disadvantages of peripheral location vis-à-vis the world's core areas continue to handicap Subsaharan Africa.

CULTURAL PATTERNS

We may tend to think of Africa in terms of its prominent countries and famous cities, its development problems and political dilemmas, but Africans themselves have another perspective. The colonial period created states and capitals, introduced foreign languages to serve as the *lingua franca*, and brought railroads and roads. The colonizers stimulated labor movements to the mines they opened, and they disrupted other migrations that had been part of African life for many centuries. But they did not change the ways of life of most of the people. No less than two-thirds of the realm's population still live in, and work near, Africa's hundreds of thousands of villages. They speak one of more than a thousand languages in use in the realm. The villagers' concerns are local; they focus on subsistence, health, and safety. They worry that the conflicts over regional power or political ideology will engulf them, as has happened to millions in Liberia, Sierra Leone, Ethiopia, Rwanda, The Congo, Moçambique, and Angola since the 1970s. Africa's largest peoples are major nations, such as the Yoruba of Nigeria and the Zulu of South Africa. Africa's smallest peoples number just a few thousand. As a geographic realm, Subsaharan Africa has the most complex cultural mosaic on Earth (see Fig. 6-6).

African Languages

Africa's linguistic geography is a key component of that cultural intricacy. Most of Subsaharan Africa's more than 1000 languages do not have a written tradition, making classification and mapping difficult. Scholars have attempted to delimit an African language map, and Figure 6-9 is a composite of their efforts. One feature is common to all language maps of Africa: the geographic realm begins approximately where the Afro-Asiatic language family (mapped in yellow in Fig. 6-9) ends, although the correlation is sharper in West Africa than to the east.

In Subsaharan Africa, the dominant language family is the Niger-Congo family, of which the Kordofanian subfamily is a small, historic northeastern outlier (Fig. 6-9), and the Niger-Congo languages carry the other subfamily's name. This subfamily extends across the realm from West to East and Southern Africa. The Bantu language forms the largest branch in this subfamily, but Niger-Congo languages in West Africa, such as Yoruba and Akan, also have millions of speakers. Another important language family is the Nilo-Saharan family, extending from Maasai in Kenya northwest to Teda in Chad. No other language families are of similar extent or importance: the Khoisan family, of ancient origins, now survives among the dwindling Khoi and San peoples of the Kalahari; the small white minority in South Africa speak Indo-European languages; and Malay–Polynesian languages prevail in Madagascar, which was peopled from Southeast Asia before Africans reached it.

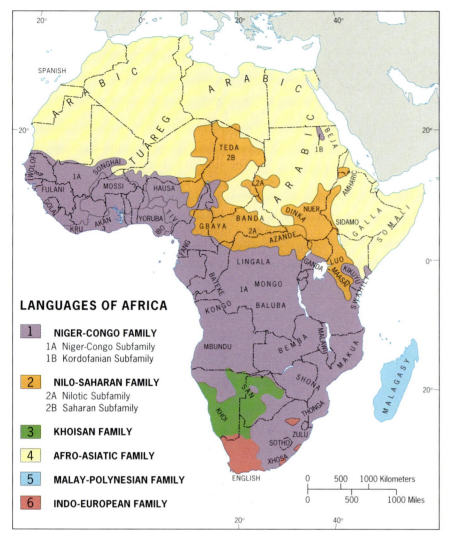

LANGUAGES OF AFRICA

1	NIGER-CONGO FAMILY
	1A Niger-Congo Subfamily
	1B Kordofanian Subfamily
2	NILO-SAHARAN FAMILY
	2A Nilotic Subfamily
	2B Saharan Subfamily
3	KHOISAN FAMILY
4	AFRO-ASIATIC FAMILY
5	MALAY-POLYNESIAN FAMILY
6	INDO-EUROPEAN FAMILY

FIGURE 6-9 © H. J. de Blij, P. O. Muller, and John Wiley & Sons, Inc.

The Most Widely Used Languages

About 40 African languages are spoken by 1 million people or more, and a half-dozen by about 10 million or more: Hausa (42 million), Yoruba (23 million), Ibo, Swahili, Lingala, and Zulu. Although English and French have become important *linguae francae* in multilingual countries such as Nigeria and Côte d'Ivoire (where officials even insist on spelling the name of their country—Ivory Coast—in the Francophone way), African languages also serve this purpose. Hausa is a common language across the West African savanna; Swahili is widely used in East Africa. And pidgin languages, mixtures of African and European tongues, are spreading along West Africa's coast. Millions of Pidgin English (called *Wes Kos*) speakers use this medium in Nigeria and Ghana.

Language and Culture

Multilingualism can be a powerful **13** centrifugal force in society, and African governments have tried with varying success to establish "national" alongside local languages. Nigeria, for example, made English its official language because none of its 250 languages, not even Hausa, had sufficient internal interregional use. But using a European, colonial language as an official medium invites criticism, and Nigeria remains divided on the issue. On the other hand, making a dominant local language official would invite negative reactions from ethnic minorities. Language remains a potent force in Africa's cultural life.

Religion in Africa

Africans had their own religious belief systems long before Christians and Muslims arrived to convert them. And for all of Subsaharan Africa's cultural diversity, Africans had a consistent view of their place in nature. Spiritual forces, according to African tradition, are manifest everywhere in the natural environment, not in a supreme deity that exists in some remote place. Thus gods and spirits affect people's daily lives, witnessing every move, rewarding the virtuous, and punishing (through injury or crop failure, for example) those who misbehave. Ancestral spirits can inflict misfortune on the living. They are everywhere: in the forest, rivers, mountains.

As with land tenure, the religious views of Africans clashed fundamentally with those of outsiders. Monotheistic Christianity first touched Africa in the northeast when Nubia and Axum were converted, and Ethiopia has been a Coptic Christian stronghold since the fourth century AD. But the Christian churches' real invasion did not begin until the onset of colonialism after the turn of the sixteenth century. Christianity's various denominations made inroads in different areas: Roman Catholicism in much of Equatorial Africa mainly at the behest of the Belgians, the Anglican Church in British colonies, and Presbyterians and others elsewhere. However, almost everywhere, Christianity's penetration led to a blending of traditional and Christian beliefs, so that much of Subsaharan Africa is nominally, though not exclusively, Christian. Go to a church in Gabon or Uganda or Zambia, and you may hear drums instead of church bells, sing African music rather than hymns, and see African carvings alongside the usual statuary.

Islam had a different arrival and impact. Long before the colonial invasion, Islam advanced out of Arabia, across the desert, and down the east coast. Muslim clerics converted the rulers of African states and commanded them to convert their subjects. They Islamized the savanna states and penetrated into present-day northern Nigeria, Ghana, and Ivory Coast. They encircled and isolated Ethiopia's Coptic Christians and Islamized the Somali people in Africa's Horn. They established beachheads on the Kenya coast and took over Zanzibar. On the map, the African Transition Zone defines the Islamic Front (see Fig. 6-10). In the field, Arabizing Islam and European Christianity competed for African minds, and Islam proved to be a far more pervasive force. From Senegal to Somalia, the population is virtually 100 percent Muslim, and Islam's rules dominate everyday life (see photo below). The Sunni *mullahs* would never allow the kind of marriage between traditional and Christian beliefs seen in much of formerly colonial Africa. This fundamental contradiction between Islamic dogma and

Christian accommodation creates a potential for conflict in countries where both religions have adherents.

MODERN MAP AND TRADITIONAL SOCIETY

The political map of Subsaharan Africa contains almost 50 states but no nation-states (apart from some microstates and ministates in the islands and in the south). Centrifugal forces are powerful, and outside interventions during the Cold War, when communist and anticommunist foreigners took sides in local civil wars, worsened conflict within African states. Colonialism's economic legacy was not much better. In tropical Africa, core areas, capitals, port cities, and transport systems were laid out to maximize profit and facilitate exploitation of minerals and soils; the colonial mosaic inhibited interregional communications except where cooperation enhanced efficiency. Colonial Zambia and Zimbabwe, for example (then called Northern and Southern Rhodesia), were landlocked and needed outlets, so railroads were built to Portuguese-owned ports. But such routes did little to create intra-African linkages. The modern map reveals the

The faithful kneel during Friday prayers at a mosque in Kano, northern Nigeria. The survival of Nigeria as a unified state is an African success story; the Nigerians have overcome strong centrifugal forces in a multi-ethnic country that is dominantly Muslim in the north, Christian in the south. In the 1990s, some Muslim clerics began calling for an Islamic Republic in Nigeria, and after the death of the dictator Abacha and the election of a non-Muslim president, the Islamic drive intensified. A number of Nigeria's northern States adopted Sharia (strict Islamic) law, which led to destructive riots between the majority Muslims and minority Christians who felt threatened by this turn of events. Can Nigeria avoid the fate of Sudan (see pp. 368–369)? © M. & E. Bernheim/Woodfin Camp & Associates.

results: in West Africa you can travel from the coast into the interior of all the coastal states along railways or adequate roads. But no high-standard roadway was ever built to link these coastal neighbors to each other.

Supranationalism

To overcome such disadvantages, African states must cooperate internationally, continentwide as well as regionally. The Organization of African Unity (OAU) was established for this purpose in 1963 and in 2001 was superseded by the African Union. In 1975 the Economic Community of West African States (ECOWAS) was founded by 15 countries to promote trade, transportation, industry, and social affairs in the region. And in the early 1990s another important step was taken when 12 countries joined in the Southern African Development Community (SADC), organized to facilitate regional commerce, intercountry transport networks, and political interaction.

Population and Urbanization

Subsaharan Africa is the second least urbanized world realm, but it is urbanizing at a fast pace. The percentage of urban dwellers today stands at 34 percent. This means that almost 260 million people now live in cities and towns, many of which were founded and developed by the colonial powers.

African cities became centers of embryonic national core areas, and of course they served as government headquarters. This *formal sector* of the city used to be the dominant one, with government control and regulations affecting civil service, business, industry, and workers. Today, however, African cities look different. From a distance, the skyline still resembles that of a modern center. But in the streets, on the sidewalks right below the shopwindows, there are hawkers, basket weavers, jewelry sellers, garment makers, wood carvers—a second economy, most of it beyond government control. This *informal sector* now dominates many African cities. It is peopled by the rural immigrants, who also work as servants, apprentices, construction workers, and in countless other menial jobs.

Millions of urban immigrants, however, cannot find work, at least not for months or even years at a time. They live in squalid circumstances, in desperate poverty, and governments cannot assist them. As a result, the squatter rings around (and also within) many of Africa's cities are unsafe—uncomfortable, unhealthy slums without adequate shelter, water supply, or basic sanitation. Garbage-strewn (no solid-waste removal here), muddy and insect-infested

during the rainy season, and stifling and smelly during the dry period, they are incubators of disease. Yet very few residents return to their villages. Every new day brings hope.

In our regional discussion we focus on some of Subsaharan Africa's cities, all of which, to varying degrees, are stressed by the rate of population influx. Despite the plight of the urban poor and the poverty of Africa's rural areas, some of Africa's capitals remain the strongholds of privileged elites who, dominant in governments, fail to address the needs of other ethnic groups. Discriminatory policies and artificially low food prices disadvantage farmers and create even greater urban-rural disparities than the colonial period saw. But today the prospect of democracy brings hope that Africa's rural majorities will be heard and heeded in the capitals.

ECONOMIC PROBLEMS

As the world map of economic development (Fig. G-12) shows, Subsaharan Africa as a geographic realm is the weakest link in the international economy. Proportionately more countries are poor and debt-ridden in Subsaharan Africa than anywhere else. Annual income per person is low, infrastructure is inadequate, linkages with the rest of the world are weak, farmers face obstacles to profit at home as well as abroad. Champions of globalization say that Africa's poverty is proof that a lack of globalizing interaction spells disadvantage for any country. Opponents of globalization argue that Africa's condition is proof that globalization is making the poor poorer and the already-rich richer.

From Mauritania to Madagascar, Africa's low-income economies reflect a combination of difficulties so daunting that overcoming them should be a global, not just an African, objective. If the world is a village of neighborhoods, as is so often said, this is the neighborhood that needs the most help—ranging from medical assistance against AIDS to fair market conditions for farmers. Lifting tens of millions of subsistence farmers from the vagaries of individual survival to the security of collective prosperity will take more than a Green Revolution. Africa's vast expanse of savanna soil may not be any less fertile than that of Brazil's booming *cerrado*, but if there is no capital to provide fertilizers and no incentive to produce, no secure financial system and no roads to markets, no government policies to secure fair domestic prices and no equal access to the international marketplace, Africa's key industry will not make progress. That is the crucial problem: farming will form Africa's dominant economic activity for decades to come. True, Africa continues to sell commodities ranging from asbestos to zinc to foreign buyers in a pattern established during the colonial era, but the gauge of Africa's well-being lies in its agriculture.

Figure G-12 may appear to contradict this prospect, but consider how few the exceptions are. Of Subsaharan Africa's nearly 50 countries, all but six have low-income economies. The only well-balanced economy among these six is South Africa, for reasons we discuss in the regional section of this chapter: South Africa combines industrial production, commodity sales, and commercial farming. Among the others, Botswana's prosperity is based on commodity (diamond) sales, Namibia's on metals, and Gabon's on oil and forest exploitation. But even the export of oil or metals cannot lift Nigeria, The Congo, Zimbabwe, or any other African economy out of its low-income status.

The remnants of colonial economic ties are mostly gone, and profits from commodities, from gold to manganese, have dropped as production costs rise and market prices fluctuate. But a new buyer has arrived on the African scene: China. The Chinese demand for metals, especially iron ore but also ferroalloys and copper, is burgeoning, and Africa has what China wants—except that Subsaharan Africa's railroad system, even in comparatively efficient South Africa, cannot get the cargoes to the coast in sufficient quantity or, in many areas, at all. China is offering to pay for infrastructure improvements, and there is hardly a country in

POINTS TO PONDER

● More than 1 million infants and children die of malaria each year in Africa, many times more than in the rest of the susceptible countries of the world combined.

● China is building a dam on Ethiopia's Tekeze River that will be the largest in Africa, higher even than China's Three Gorges project, to generate electricity and facilitate irrigation.

● Liberia in 2005 became the first country in Africa to be governed by a woman when President Ellen Johnson-Sirleaf was elected with 59 percent of the vote.

● The staggering death toll in the wars in The Congo—some 4 million since 1998—is the worst in the world, but international assistance, including UN peacekeeping efforts, has not been enough to end it. Despite a national election in 2006, the crisis is not over.

Africa today without Chinese business representatives negotiating with the government. If China's interest in commodities were to help link farmers as well as mines to cities and ports, Africa's economic condition might improve.

Regions of the Realm

On the face of it, Africa seems to be so massive, compact, and unbroken that any attempt to justify a contemporary regional breakdown is doomed to fail. No deeply penetrating bays or seas create peninsular fragments as they do in Europe. No major islands (other than Madagascar) provide the broad regional contrasts we see in Middle America. Nor does Africa really taper southward to the peninsular proportions of South America. And Africa is not cut by an Andean or a Himalayan mountain barrier. Given Africa's colonial fragmentation and cultural mosaic, is regionalization possible? Indeed it is.

Maps of environmental distributions, ethnic patterns, cultural landscapes, historic culture hearths, and other spatial data yield a four-region structure complicated by a fifth, overlapping zone as shown in Figure 6-10. Beginning in the south, we identify the following regions:

1. *Southern Africa*, extending from the southern tip of the continent to the northern borders of Angola, Zambia, Malawi, and Moçambique. Ten countries constitute this region, which extends beyond the tropics and whose giant is South Africa. The island state of Madagascar, with Southeast Asian influences, is neither Southern nor East African.

MAJOR CITIES OF THE REALM

City	Population* (in millions)
Abidjan, Ivory Coast	3.9
Accra, Ghana	2.2
Adis Abeba, Ethiopia	3.2
Cape Town, South Africa	3.2
Dakar, Senegal	2.4
Dar es Salaam, Tanzania	3.1
Durban, South Africa	2.8
Harare, Zimbabwe	1.6
Ibadan, Nigeria	2.7
Johannesburg, South Africa	3.5
Kinshasa, The Congo	6.9
Lagos, Nigeria	12.7
Lusaka, Zambia	1.4
Mombasa, Kenya	0.9
Nairobi, Kenya	3.2

*Based on 2008 estimates.

2. *East Africa*, where natural (equatorial) environments are moderated by elevation and where plateaus, lakes, and mountains, some carrying permanent snow, define the countryside. Six countries, including the highland part of Ethiopia, comprise this region.

3. *Equatorial Africa*, much of it defined by the basin of the Congo River, where elevations are lower than in East Africa, temperatures are higher and moisture more ample, and where most of Africa's surviving rainforests remain. Among the eight countries that form this region, The Congo dominates territorially and demographically, but others are much better off economically and politically.

4. *West Africa*, which includes the countries of the western coast and those on the margins of the Sahara in the interior, a populous region anchored in the southeast by Africa's demographic giant, Nigeria. Fifteen countries form this crucial African region.

5. *The African Transition Zone*, the complicating factor on the regional map of Africa. In Figure 6-10, note that this zone of increasing Islamic influence completely dominates some countries (e.g., Somalia in the east and Senegal in the west) while cutting across others, thereby creating Islamized northern areas and non-Islamic southern zones (Nigeria, Chad, Sudan).

FIGURE 6-10 © H. J. de Blij, P. O. Muller, and John Wiley & Sons, Inc.

SOUTHERN AFRICA

Southern Africa, as a geographic region, consists of all the countries and territories lying south of Equatorial Africa's The Congo and East Africa's Tanzania (Fig. 6-11). Thus defined, the region extends from Angola and Moçambique (on the Atlantic and Indian Ocean coasts, respectively) to South Africa and includes a half-dozen landlocked states. Also marking the northern limit of the region are Zambia and Malawi. Zambia is nearly cut in half by a long land extension from The Congo, and Malawi penetrates deeply into Moçambique. The colonial boundary framework, here as elsewhere, produced many liabilities.

FIGURE 6-11 © H. J. de Blij, P. O. Muller, and John Wiley & Sons, Inc.

Africa's Richest Region

Southern Africa constitutes a geographic region in both physiographic and human terms. Its northern zone marks the southern limit of the Congo Basin in a broad upland that stretches across Angola and into Zambia (the tan corridor extending eastward from the Bihe Plateau in Fig. 6-2). Lake Malawi is the southernmost of the East African rift-valley lakes; Southern Africa has none of East Africa's volcanic and earthquake activity. Most of the region is plateau country, and the Great Escarpment is much in evidence here. There are two pivotal river systems: the Zambezi (which forms the border between Zambia and Zimbabwe) and the Orange-Vaal (South African rivers that combine to demarcate southern Namibia from South Africa).

Southern Africa is the continent's richest region materially. A great zone of mineral deposits extends through the heart of the region from Zambia's Copperbelt through Zimbabwe's Great Dyke and South Africa's Bushveld Basin and Witwatersrand to the goldfields and diamond mines of the Orange Free State and northern Cape Province in the heart of South Africa. Ever since these minerals began to be exploited in colonial times, many migrant laborers have come to work in the mines.

Southern Africa's agricultural diversity matches its mineral wealth. Vineyards drape the slopes of South Africa's Cape Ranges; tea plantations hug the eastern escarpment slopes of Zimbabwe. Before civil war destroyed its economy, Angola was one of the world's leading coffee producers. South Africa's relatively high latitudes and its range of altitudes create environments for apple orchards, citrus groves, banana plantations, pineapple farms, and many other crops.

Despite this considerable wealth and potential, the countries of Southern Africa have not prospered. As Figure G-12 shows, many remain mired in the low-income category (Malawi, with a per-capita GNI of only $650, is one of the world's poorest states); South Africa and its western desert neighbors, Namibia and Botswana, are in the middle-income rank, and they have recently been joined by oil- and mineral-rich Angola. However, a period of rapid population growth, followed by the devastating onslaught of AIDS, civil conflict, political instability, incompetent government, widespread corruption, unfair competition on foreign markets, and environmental problems have constrained regional development.

Still, Southern Africa as a region is better off than any other in Subsaharan Africa, with half of its countries having risen above the lowest-income rank among world states. Some international cooperation including a regional association and a customs union are emerging. And South Africa, the realm's most important country by many measures, gives hope for a better future.

South Africa

The Republic of South Africa is the giant of Southern Africa, an African country at the center of world attention, a bright ray of hope not only for Africa but for all humankind.

Long in the grip of one of the world's most notorious racial policies (**apartheid**, or "apartness," and its derivative, **separate development**), South Africa today is shedding its past and is building a new future. That virtually all parties to the earlier debacle are now working cooperatively to restructure the country under a new flag, a new national anthem, and a new leadership was one of the great events of the late twentieth century. Now, nearly 15 years after Nelson Mandela became South Africa's first democratically elected president, and with his successor in his second and final term of office, South Africa has assumed its role as the economic engine of the region and a constructive force in the realm beyond.

South Africa stretches from the warm subtropics in the north to Antarctic-chilled waters in the south. With a land area in excess of 1.2 million square kilometers (470,000 sq mi) and a heterogeneous population of 47.8 million, South Africa is the dominant state in Southern Africa. It contains the bulk of the region's minerals, most of its good farmlands, its largest cities, best ports, most productive factories, and most developed transport networks. Mineral exports from Zambia and Zimbabwe move through South African ports. Workers from as far away as Malawi and as nearby as Lesotho work in South Africa's mines, factories, and fields.

An African—and Global—Magnet

South Africa's location has much to do with its human geography. On the African continent, peoples migrated southward—first the Khoisan-speakers and then the Bantu peoples—into the South African cul-de-sac. On the oceans, the Europeans arrived to claim the southernmost Cape as one of the most strategic places on Earth, the gateway from the Atlantic to the Indian Ocean, a waystation on the route to Asia's riches. The Dutch East India Company founded Cape Town as early as 1652, and soon the Hollanders began to bring Southeast Asians to the Cape to serve as domestics and laborers. When the British took over about 150 years later, Cape Town had a substantial population of mixed ancestry, the source of today's so-called Coloured sector of the country's citizenry.

The British also altered the demographic mosaic by bringing tens of thousands of indentured laborers from their South Asian domain to work on the sugar plantations of east-coast Natal. Most of these laborers stayed after their period of indenture was over, and today South Africa counts more than 1 million Indians among its people. Most are still concentrated in Natal and prominently so in metropolitan Durban.

British and Boers

As we noted earlier, South Africa had been occupied by Europeans long before the colonial "scramble for Africa" gained momentum. The Dutch, after the British took control of the Cape, trekked into the South African interior and, on the high plateau they called the *highveld*, founded their own republics. When diamonds and gold were discovered there, the British challenged the *Boers* (descendants of the Dutch settlers) for these prizes. In 1899–1902, the British and the Boers fought the Boer War. The British won, and British capitalists took control of South Africa's economic and political life. But the Boers negotiated a power-sharing arrangement and eventually achieved hegemony. Having long since shed their European links, they now called themselves *Afrikaners*, their word for Africans, and proceeded to erect the system known as apartheid.

Zulu and Xhosa

These foreign immigrations and struggles took place on lands that Africans had already entered and fought over. When the Europeans reached the Cape, Bantu nations were driving the weaker Khoi and San peoples into less hospitable territory or forcing them to work in bondage. One great contest was taking place in the east and southeast, below the Great Escarpment. The Xhosa nation was moving toward the Cape along this natural corridor. Behind them, in Natal (Fig. 6-11), the Zulu Empire became the region's most powerful entity in the nineteenth century. On the highveld, the North and South Sotho, the Tswana, and other peoples could not stem the tide of European aggrandizement, but their numbers ensured survival.

In the process, South Africa became Africa's most pluralistic and heterogeneous society. People had converged on the country from Western Europe, Southeast Asia, South Asia, and other parts of Africa itself. At the end of the twentieth century, Africans outnumbered non-Africans by just about 4 to 1 (Table 6-1).

The Ethnic Mosaic

Heterogeneity also marks the spatial demography of South Africa. Despite centuries of migration and (at the Cape) intermarriage, labor movement (to the mines, farms, and factories), and sustained urbanization, regionalism pervades the human mosaic. The Zulu nation still is largely concentrated in the province the Europeans called Natal. The Xhosa still cluster in the Eastern Cape, from the city of East

TABLE 6-1

Demographic Data for South Africa

Population Groups	2008 Estimated Population (in millions)
African nations	37.8
Zulu	11.4
Xhosa	8.4
Tswana	3.9
Sotho (N and S)	3.8
Others (6)	10.3
Whites	4.5
Afrikaners	2.8
English-speakers	1.6
Others	0.1
Mixed (Coloureds)	4.3
African/White	4.0
Malayan	0.3
South Asian	1.2
Hindus	0.8
Muslims	0.4
TOTAL	47.8

London to the Natal border and below the Great Escarpment. The Tswana still occupy ancestral lands along the Botswana border. Cape Town still is the core area of the Coloured population; Durban still has the strongest Indian imprint. Travel through South Africa, and you will recognize the diversity of rural cultural landscapes as they change from Swazi to Ndebele to Venda.

This historic regionalism was among the factors that led the Afrikaner government to institute its "separate development" scheme, but it could not stem the tide of urbanization. Millions of workers, job-seekers, and illegal migrants converged on the cities, creating vast shantytowns on their margins. In the legal African "townships" such as Johannesburg's Soweto ("SOuth WEstern TOwnships") and in these squatter settlements the anti-apartheid movement burgeoned, and the strength of the African National Congress (ANC) movement grew. In February 1990 Nelson Mandela, imprisoned for 28 years on Robben Island, South Africa's Alcatraz, became a free man, and following the momentous first democratic election of 1994 he became president of an ANC-dominated government in Cape Town.

South Africa's Changed Map

Before this election could take place, however, South Africa's administrative geography had to be radically changed. The republic's political geography was a legacy of the Union period, when it was divided into four provinces:

the Cape, by far the largest and centered on the legislative capital of Cape Town; Natal, anchored by the port city of Durban; the Transvaal ("across the Vaal River"), focused on the administrative capital of Pretoria and including the great Johannesburg metropolis; and the Orange Free State, the Boer stronghold with Bloemfontein as its headquarters.

This structure was replaced by a new map of nine provinces (Fig. 6-12), creating a federal arrangement in which each province would have its own administration while being represented in the central government. The boundaries of Natal and the Orange Free State remained essentially unchanged, but Natal's name was changed to Kwazulu-Natal, recognizing the Zulu presence there, and the Orange Free State became simply the Free State. But the Cape Province was divided into four new provinces, one of them overlapping into the Transvaal, and the rest of the Transvaal was divided into three, one of which is Gauteng,

which includes the Johannesburg–Witwatersrand–Pretoria megalopolis.

A New Era Dawns

The 1994 election, based on this new layout, produced an ANC victory in seven of these nine provinces, excepting only Kwazulu-Natal, where the Zulu vote went to a local political movement called Inkatha, and the Western Cape, where a combination of white and Coloured voters outpolled the ANC. This was a very fortunate result, proving that the ANC was not all-powerful and giving minorities, including Coloureds and whites, reason to trust and participate in the new system.

The transition from Afrikaner domination to democratic government still was complicated by numerous circumstances. As it turned out, it was not white opposition that

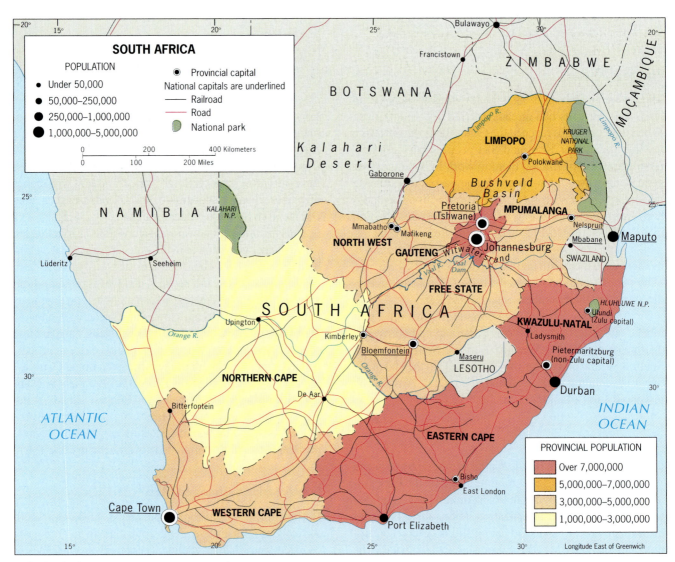

FIGURE 6-12 © H. J. de Blij, P. O. Muller, and John Wiley & Sons, Inc.

most threatened majority rule. The white extremist political parties collapsed relatively soon, and a brief terrorist episode was quickly halted. More consequential was the uncertain role of the largest single nation in South Africa, the Zulu, whose historic domain lies in Kwazulu-Natal, and whose prominent leader, Chief Buthelezi, at times hinted at secession if the new South Africa was not acceptable. In the runup to the election, violence killed thousands and threatened to precipitate a civil war. But that danger, too, was overcome, and the Zulu nation and its leaders remained in the fold.

In 1999, ANC leader Thabo Mbeki, who had served as President Mandela's deputy, became the country's second popularly elected president. In other African states, the succession from heroic founder-of-the-nation to political inheritor-of-the-presidency often has not gone well, but South Africa's new constitution proved its worth. The new president faced several disadvantages: inevitable comparisons to the incomparable Mandela; the rising tide of AIDS, which Mbeki controversially attributed to causes other than HIV; and the farm invasions in Zimbabwe (see p. 320) on which his leadership seemed to falter. Nevertheless, Mbeki's economic policies, social programs, and foreign initiatives (The Congo, Rwanda) have served his country well.

How the Economy Evolved: Diamonds and Gold

Undoubtedly the most serious immediate problems for the new South Africa are economic. Ever since diamonds were discovered at Kimberley in the 1860s, South Africa has been synonymous with minerals. The Kimberley finds, made in a remote corner of what was then the Orange Free State (the British soon annexed it to the Cape), set into motion a new economic geography. Rail lines were laid from the coast to the "diamond capital" even as fortune seekers, capitalists, and tens of thousands of African workers, many from as far afield as Lesotho, streamed to the site. One of the capitalists was Cecil Rhodes, of Rhodes Scholarship fame, who used his fortune to help Britain dominate Southern Africa.

Just 25 years after the diamond discoveries, prospectors found what was long to be the world's greatest goldfield on a ridge called the Witwatersrand (Fig. 6-12). This time the site lay in the so-called South African Republic (the Transvaal), and again the Boers were unable to hold the prize. Johannesburg became the gold capital of the world, and a new and even larger stream of foreigners arrived, along with a huge influx of African workers. Cheap labor enlarged the profits. Johannesburg grew explosively, satellite towns developed, and black townships mushroomed. The Boer War was only an interlude here on the mineral-rich Witwatersrand.

During the twentieth century, South Africa proved to be richer than had been foreseen. Additional goldfields were discovered in the Orange Free State. Coal and iron ore were found in abundance, which gave rise to a major iron and steel industry. Other metallic minerals, including chromium and platinum, yielded large revenues on world markets. Asbestos, manganese, copper, nickel, antimony, and tin were mined and sold; a thriving metallurgical industry developed in South Africa itself. Capital flowed into the country, white immigration grew, farms and ranches were laid out, and markets multiplied.

FROM THE FIELD NOTES

"Looking down on this enormous railroad complex, we were reminded of the fact that almost an entire continent was turned into a wellspring of raw materials carried from interior to coast and shipped to Europe and other parts of the world. This complex lies near Witbank in the eastern Rand, a huge inventory of freight trains ready to transport ores from the plateau to Durban and Maputo. But at least South Africa acquired a true transport network in the process, ensuring regional interconnections; in most African countries, railroads serve almost entirely to link resources to coastal outlets." © H. J. de Blij

www.conceptcaching.com

AMONG THE REALM'S GREAT CITIES . . . Johannesburg

SUBSAHARAN AFRICA DISPLAYS only one incipient conurbation, and Johannesburg, South Africa lies at the heart of it. Little more than a century ago, Johannesburg was a small (though rapidly growing) mining town based on the newly discovered gold reserves of the Witwatersrand.

Today Johannesburg forms the focus of a conurbation of more than 6 million, extending from Pretoria in the north to Vereeniging in the south, and from Springs in the east to Krugersdorp in the west. The population of metropolitan Johannesburg itself passed the 3 million mark in 2003 (it now totals 3.5 million), thereby overtaking Cape Town to become South Africa's largest metropolis.

Johannesburg's skyline is the most impressive in all of Africa, a forest of skyscrapers reflecting the wealth generated here over the past hundred years. Look southward from a high vantage point, and you see the huge mounds of yellowish-white slag from the mines of the "Rand," the so-called mine dumps, partly overgrown today, interspersed with suburbs and townships. In a general way, Johannesburg developed as a white city in the north and a black city in the south. Soweto, the black township, lies to the southwest; Houghton and other spacious, upper-class suburbs, formerly exclusively white residential areas, lie to the north.

Johannesburg has neither the scenery of Cape Town nor the climate and beaches of Durban. The city lies 1.5 kilometers above sea level, and its thin air often is polluted from smog created by automobiles, factories, mine-dump dust, and countless cooking fires in the townships and shantytowns that ring the metropolis.

Over the past century, the Johannesburg area produced nearly one-half of all the world's gold by value. But today,

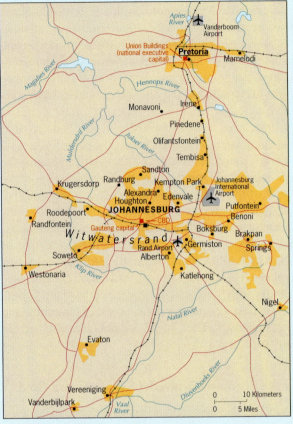

© H. J. de Blij, P. O. Muller, and John Wiley & Sons, Inc.

Johannesburg lies at the heart of an industrial, commercial, and financial complex whose name on the new map of South Africa is *Gauteng*.

Infrastructural Gains

South Africa's cities grew apace. Johannesburg was no longer just a mining town: it became an industrial complex and a financial center as well. The old Boer capital, Pretoria, just 50 kilometers (30 mi) north of the Witwatersrand, became the country's administrative center during apartheid's days. In the Orange Free State, major industrial growth matched the expansion of mining. While the core area developed megalopolitan characteristics, coastal cities expanded as well. Durban's port served not only the Witwatersrand but a wider regional hinterland as well. Cape Town was becoming South Africa's largest central city; its port, industries, and productive agricultural hinterland gave it primacy over a wide area.

The labor force for all this development, from mines to railroads and from farms to highways, came from the African peoples of South Africa (and, indeed, from beyond its borders as well). Draconian apartheid policies notwithstanding, workers from Moçambique, Zimbabwe, Botswana, and other neighboring countries sought jobs in South Africa even as millions of South African villagers also left their homes for the towns and cities. In the process they built the best infrastructure of any African country (see photo p. 317). But apartheid ruined South Africa's prospects. The cost of the separate development program was astronomical. Social unrest during the decade preceding the end of apartheid created a vast educational gap among youngsters. International sanctions against the race-obsessed regime damaged the economy.

The Sputtering Engine of Africa

During the upheavals of the last decade of the twentieth century—think of the Soviet Union, Yugoslavia, the Middle East—was any country more fortunate than South Africa? Surely a massive revolution would sweep apartheid away and bring an all-African regime to power? Had that happened, South Africa today would still be repairing the damage. But instead, largely through the unmatched statesmanship of Nelson Mandela, South Africa is governed by a multiracial party, the erstwhile architects of apartheid are in parliament as a legitimate opposition, and its economy is the largest and healthiest in the realm. Check Table G-1 (p. 42) and you will see that, on the African mainland, South Africa's per-capita GNI is in a class by itself. By some calculations, South Africa produces 45 percent of all of Subsaharan Africa's GDP.

In many respects South Africa is the most important country in Subsaharan Africa, and the entire realm's fortunes are bound up with it. No African country attracts more foreign investment or (still) foreign workers. None has the universities, hospitals, and research facilities. No other has the military forces capable of intervening in African trouble spots. Few have the free press, effective trade unions, independent courts, or financial institutions to match South Africa's. And with a population approaching 50 million (79 percent black, just under 10 percent white, 9 percent Coloured, 2.5 percent Asian), South Africa has a large, multiracial, and growing middle class.

But South Africa also faces challenges. When the African National Congress (ANC) won the 1994 elections (and subsequent democratic elections as well, another beacon for Africa), the country's African peoples harbored expectations not only of post-apartheid freedom but of better living conditions as well. And indeed the government made major progress: it began an old-age pension scheme, built nearly two million brick homes for the poor, provided free water to nearly 30 million people in remote areas—all this without destabilizing the economy. And through a series of laws, the ANC has made it mandatory for firms to hire more blacks and women. The growing African middle class strongly supports the ANC.

This progress notwithstanding, South Africa confronts long-range problems. The economy continues to depend far too much on the export of minerals and metals (diamonds, platinum, gold, iron, and steel) at a time when this entails risks at home and abroad. At home, those union-friendly labor laws, including rising wages, are making mining less profitable. Abroad, commodity prices are unreliable. In the apartheid year of 1970, South Africa produced nearly 70 percent of the world's gold; today it produces less than 15 percent. And the manufacturing sector of the economy remains too weak. Even while the black middle class is expanding, unemployment among blacks may be as high as 50 percent—and the gap between rich and poor is growing. Land reform, an urgent matter in a country where land alienation reached huge proportions, is too slow in the view of many. Add to this the scourge of AIDS and the government's failure to address this crisis effectively, and South Africa is shadowed by serious problems.

Core-Periphery Contrasts

On maps of development indices, South Africa is portrayed as a middle-income economy, but averages mean little in this country of strong internal core-periphery contrasts. In its great cities, industrial complexes, mechanized farms, and huge ranches, South Africa resembles a high-income economy, much like Australia or Canada. But outside the primary core area (centered on Johannesburg) and beyond the secondary cores and their linking corridors lies a different South Africa, where conditions are more like those of rural Zambia or Zimbabwe. In terms of such indices as life expectancy, infant and child mortality, overall health, nutrition, education, and many others, a wide range marks the country's population sectors. South Africa has been described as a microcosm of the world, exhibiting in a single state not only a diversity of cultures but also a wide array of human conditions. If South Africa can keep on course, this one-time pariah of apartheid will become a guidepost to a better world.

The Middle Tier

Between South Africa's northern border and the region's northern limit lie two groups of states: those with borders with South Africa and those beyond.

Five countries constitute the middle tier, neighboring South Africa: Zimbabwe, Namibia, Botswana, and the ministates of Lesotho and Swaziland (Fig. 6-11). As the map shows, four of these five countries are landlocked. Diamond-exporting (and middle-income) Botswana occupies the heart of the Kalahari Desert and surrounding steppe; despite its lucrative diamonds, most of its 1.8 million inhabitants are subsistence farmers. In 2006, Botswana was the most severely AIDS-afflicted country in all of tropical Africa. Lesotho and Swaziland, both traditional kingdoms, depend heavily on remittances from their workers in South African mines, fields, and factories.

The most important state in the middle tier undoubtedly is Zimbabwe (13.3 million), landlocked but well endowed with mineral and agricultural resources. Zimbabwe (the country is named after stone ruins in its interior) is mostly an elevated plateau between the Zambezi and Limpopo rivers, with the desert to the west and the Great Escarpment to the east. Its core area is defined by the mineral-rich Great Dyke and its environs, extending southwest from the vicinity of the capital, Harare, to the country's second city, Bulawayo. Copper, asbestos, and chromium (of which Zimbabwe is one of the world's leading sources) are among its major mineral exports, but Zimbabwe is not just an ore-exporting country. Its farms produce tobacco, tea, sugar, cotton, and other crops (corn is the staple).

Two nations form most of Zimbabwe's population: the Shona (74 percent) and the Ndebele (15 percent). A tiny minority of whites owned the best farmland and organ-

African cities are a jumble of "formal" and "informal" sectors and neighborhoods ranging from modern downtowns and leafy suburbs to shantytowns and slums. Many families that have moved from the countryside to the city have no choice but to rent (or build) a ramshackle dwelling lacking basic amenities. These poorest urban areas nevertheless have a vibrant economy as their occupants try to make a living as local vendors, repairers, builders, traders, or artisans. Often they find jobs in service or construction industries nearby without earning enough to allow them to move out of their abode. Most governments tolerate these neighborhoods, but not the president of Zimbabwe, who in 2005 ordered his capital city "cleaned up." Here, a group of residents of a shantytown in Harare watch the destruction, having hurriedly piled their belongings on the dirt road outside their dwellings—including a framed photograph of President Robert Mugabe, once a hero of Africa's liberation struggle and inheritor of a thriving economy, now the symbol of administrative failure and social repression.   © AP/World Wide Photos.

ized the agricultural economy. Mounting environmental and economic problems beginning in the 1980s led to rising social and political tensions. President Mugabe and his dominant party allowed white farms to be invaded by squatters who in some instances killed the owners. Later, Mugabe turned on the informal sector of the economy, prevalent throughout Africa, and ordered his henchmen to destroy the squatter homes and shops of some 700,000 urban residents, leaving them without shelter or livelihood (see photo on this page). An estimated 2 million Zimbabweans left the country during the rule of the "Saddam of Africa," but few other African leaders criticized Mugabe publicly. Meanwhile, Mugabe found a new friend in China, which in 2005 signed an agreement for economic and technical cooperation and financed a Confucius Institute at the University of Zimbabwe to offer Chinese language and culture courses.

Southern Africa's youngest independent state, Namibia (2.2 million), is a former German colony with a territorial peculiarity: the so-called Caprivi Strip linking it to the right bank of the Zambezi River (Fig. 6-11), another product of colonial partitioning. Administered by South Africa from 1919 to 1990, Namibia is named after one of the world's driest deserts. This state is about as large as Texas and Oklahoma, but only its far north receives enough moisture to permit subsistence farming, which is why most of the people live near the Angolan border. Mining in the Tsumeb area and ranching in the vast steppe country of the south form the main commercial activities. The capital, Windhoek, is centrally situated opposite Walvis Bay, the main port. German influence still lingers in what used to be called South West Africa, as does an Afrikaner presence from the apartheid period.

Orderly land reform has been the government's major objective, and since independence about 1 million hectares (2.5 million acres) of commercial farms have been bought and transferred, although funds to help the new landowners have been inadequate. Namibia will need international assistance to achieve its goal of purchasing 15 million hectares (37.5 million acres), to accommodate the 250,000 people awaiting their turn, and to significantly reduce the number of people living in poverty, still about 40 percent of the population in 2007.

The Northern Tier

In the four countries that extend across the northern tier of the region—Angola, Zambia, Malawi, and Moçambique—problems abound. Angola (16.7 million), formerly a Portuguese dependency, with its exclave of Cabinda (Fig. 6-11) had a thriving economy based on a wide range of mineral and agricultural exports at the time of independence in 1975. But then Angola fell victim to the Cold War, with northern peoples choosing a communist course and southerners falling under the sway of a rebel movement backed by South Africa and the United States. The results included a devastated infrastructure, idle farms, looting of diamonds, hundreds of thousands of casualties, and millions of landmines that continue to kill and maim. But Angola's oil wealth yields about U.S. $3 billion per year, and stability, if sustained, may attract investors to begin rebuilding this ruined country. Already, in 2007, Angola's capital, once-picturesque and then-ruined Luanda, was experiencing a building boom and road expansion, and high world oil prices, along with diamond sales, propelled the Angolan economy to a growth rate of more than 15 percent in 2006—the highest in all of Subsaharan Africa. But it will take much more than that for Angola's subsistence farmers to believe that better times have arrived.

On the opposite coast, the other major former Portuguese colony, Moçambique (20.7 million), fared poorly in a different way. Without Angola's mineral base and with limited commercial agriculture, Moçambique's chief asset was its relative location. Its two major ports, Maputo and Beira, handled large volumes of exports and imports for South Africa, Zimbabwe, and Zambia. But upon independence Moçambique, too, chose a Marxist course with dire economic and dreadful political consequences. Another rebel movement supported by South Africa caused civil conflict, created famines, and generated a stream of more than a million refugees toward Malawi. Rail and port facilities lay idle, and Moçambique at one time was ranked by the United Nations as the world's poorest country. In recent years, the port traffic has been somewhat revived, and Moçambique and South Africa are working on a joint Johannesburg-Maputo Development Corridor (Fig. 6-11), but it will take generations for Moçambique to recover.

Landlocked Zambia (12.3 million) is an African success story: it remained stable even during economically difficult times and avoided the turbulence engulfing its neighbors, and is now reaping some rewards as a result. Its boundaries a product of British imperialism, Zambia shares the minerals of the Copperbelt with The Congo's Katanga Province (Fig. 6-11). In recent years, high copper prices have brought substantial external income; refugee farmers from Zimbabwe boosted the tobacco industry; and stability and relatively open government were rewarded when donors forgave nearly U.S. $4 billion in foreign debt. Zambia still remains a very poor country; it is too dependent on a few commodities, its access to world markets remains difficult, and its farming sector needs irrigation systems. But Zambia exemplifies what is possible when an African state is run responsibly. Neighboring Malawi

(13.5 million), by comparison, has suffered from political instability and infighting ever since the death of its founding autocratic ruler, Hastings Banda. With a diversified but declining commercial agricultural economy and prevalent subsistence farming, Malawi faced an acute food crisis in 2005, but when foreign donors suggested that the politicians defer their scheming and concentrate on this emergency, they were told to stop meddling in the country's internal affairs.

Before leaving Southern Africa we should take note of Madagascar, the island-state that lies off the southeastern African coast. Despite its location, this country is neither Southern nor East African. Madagascar has historic ties to Southeast Asia and cultural links to both Asia and Africa (see box titled "Distinctive Madagascar" on the following page).

● EAST AFRICA

East Africa lies to the east of the row of African Great Lakes that extends from Lake Albert in the north to Lake Malawi in the south (Fig. 6-14). Here the land rises from the low-lying Congo Basin to the west to the East African Plateau which, at various levels of elevation and in varying degrees of dissection, stretches from the spectacular canyons of Ethiopia to the rolling uplands of southern Tanzania. There, the valley of the Rovuma River marks the limit of the region. Hills and valleys, fertile soils, and copious rains mark the transition in Rwanda and Burundi. Eastward the rainforest disappears and open savanna cloaks the countryside. Great volcanoes rise above a highland dissected by a rift valley. At the heart of the region lies Lake Victoria. In the north the surface rises above 3300 meters (10,000 ft), and so deep are the trenches cut by faults and rivers there that the land was called, appropriately, Abyssinia (now Ethiopia).

Five countries, in addition to the highland part of Ethiopia, form this East African region: Tanzania, Kenya, Uganda, Rwanda, and Burundi. Here the Bantu peoples that make up most of the population met Nilotic peoples from the north, including the Maasai. In the hills of Rwanda and Burundi a stratified society developed in which the minority Tutsi, in their cattle-owning kingdoms, dominated the Hutu peasantry. The Indian Ocean coast was the scene of many historic events: the arrival of Islam from Arabia, the visit of Ming Dynasty Chinese fleets in the 1400s, the quest for power by the Turks, the Arab slave trade, the European colonial competition. Here, too, developed the East African *lingua franca*, Swahili.

Distinctive Madagascar

MAPS SHOWING THE geographic regions of the Subsaharan Africa realm often omit Madagascar. And for good reason: Madagascar differs strongly from Southern Africa, the region to which it is nearest—just 400 kilometers (250 mi) away. It also differs from East Africa, from which it has received some of its cultural infusions. Madagascar is the world's fourth largest island, a huge block of Africa that separated from the main landmass 160 million years ago. About 2000 years ago, the first settlers arrived—not from Africa (although perhaps via Africa) but from Southeast Asia. Malay communities flourished in the interior highlands of the island, which resembles Africa in having a prominent eastern escarpment and a central plateau (Fig. 6-13). Here was formed a powerful kingdom, the empire of the Merina. Its language, Malagasy, of Malay-Polynesian origin, became the indigenous tongue of the entire island (Fig. 6-9).

The Malay and Indonesian immigrants brought Africans to the island as wives and slaves, and from this forced immigration evolved the African component in Madagascar's population of 18.8 million. In all, nearly 20 discrete ethnic groups coexist in Madagascar, among which the Merina (about 5 million) and Betsimisaraka (about 3 million) are the most numerous. Like mainland Africa, Madagascar experienced colonial invasion and competition. Portuguese, British, and French colonists appeared after 1500, but the Merina were well organized and resisted colonial conquest. Eventually Madagascar became part of France's empire, and French became the *lingua franca* of the educated elite.

Because of its Southeast Asian imprint, Madagascar's staple food is rice, not corn. It has some minerals, including chromite, iron ore, and bauxite, but the economy is weak, damaged by long-term political turmoil and burdened by rapid population growth. The infrastructure has crumbled; the "main road" from the capital to the nearest port (Fig. 6-13) is now a potholed 250 kilometers (150 mi) that takes 10 hours for a truck to navigate.

Meanwhile, Madagascar's unique flora and fauna retreated before the human onslaught. Madagascar's long-term isolation kept evolution here so distinct that the island is a discrete zoogeographic realm. Primates living on the island are found nowhere else; 33 varieties of lemurs are unique to Madagascar. Many species of birds, amphibians, and reptiles are also exclusive to this island. Their home, the rainforest, covered 168,000 square kilometers (65,000 sq mi) in 1950, but today only about one-third of it is left. Logging, introduced by the colonists, damaged it; slash-and-burn agriculture is destroying it; and severe droughts since 1980 have intensified the impact. Obviously, Madagascar should be a global conservation priority, but funds are limited and the needs are enormous. Malnutrition and poverty are powerful forces when survival is at stake for villages and families.

Madagascar's cultural landscape retains its Southeast Asian imprints, in the towns as well as the paddies. The capital, Antananarivo, is the country's primate city, its architecture and atmosphere combining traces of Asia and Africa. Poverty dominates the townscape here, too, and there is little to attract in-migrants (Madagascar is only 30 percent urbanized). But perhaps the most ominous statistic is the high natural increase rate (2.7 percent) of Madagascar's population.

FIGURE 6-13   © H. J. de Blij, P. O. Muller, and John Wiley & Sons, Inc.

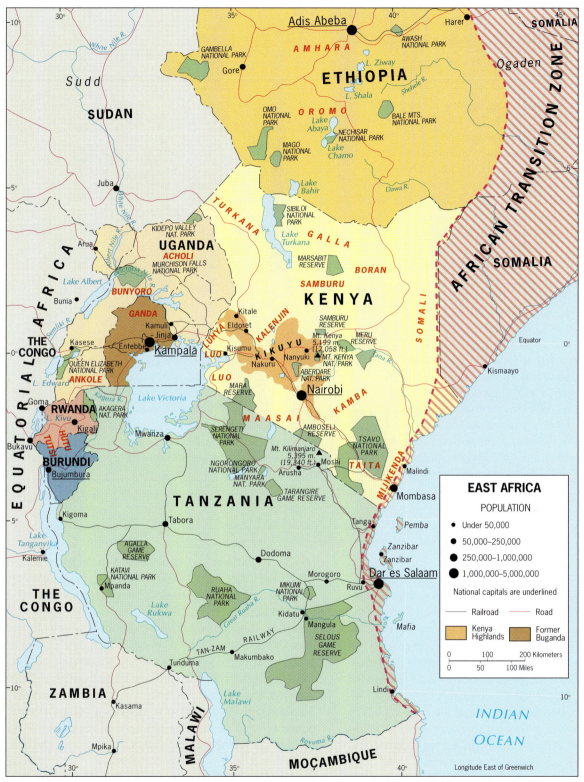

FIGURE 6-14   © H. J. de Blij, P. O. Muller, and John Wiley & Sons, Inc.

AMONG THE REALM'S GREAT CITIES . . . Nairobi

NAIROBI IS THE quintessential colonial legacy: there was no African settlement on this site when, in 1899, the railroad the British were building from the port of Mombasa to the shores of Lake Victoria reached it. However, it had something even more important: water. The fresh stream that crossed the railway line was known to the Maasai cattle herders as Enkare Nairobi (Cold Water).

The railroad was extended farther into the interior, but Nairobi grew. Indian traders set up shop. The British established their administrative headquarters here. When Kenya became independent in 1963, Nairobi naturally was the national capital.

Nairobi owes its primacy to its governmental functions, which ensured its priority through the colonial and independence periods, and to its favorable situation. To the north and northwest lie the Kenya Highlands, the country's leading agricultural area and the historic base of the dominant nation in Kenya, the Kikuyu. Beyond the rift valley to the west lie the productive lands of the Luo in the Lake Victoria Basin. To the east, elevations drop rapidly from Nairobi's 1600 meters (5000 ft), so that highland environs make a swift transition to tropical savanna that, in turn, yields to semiarid steppe.

A moderate climate, a modern city center, several major visitor attractions (including Nairobi National Park, on the city's doorstep), and a state-of-the-art airport have

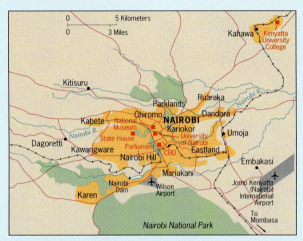

© H. J. de Blij, P. O. Muller, and John Wiley & Sons, Inc.

boosted Nairobi's fortunes as a major tourist destination, though wildlife destruction, security concerns, and political conditions have damaged the industry in recent years.

Nairobi is Kenya's principal commercial, industrial, and educational center. But its growth (to 3.2 million today) has come at a price: its modern central business district stands in stark contrast to the squalor in the shantytowns that house the countless migrants its apparent opportunities attract.

Kenya

Kenya is neither the largest nor the most populous country in East Africa, but over the past half-century it has been the dominant state in the region. Its skyscrapered capital at the heart of its core area, Nairobi, is the region's largest city; its port, Mombasa, is the region's busiest. During the 1950s, the Kikuyu nation led a vicious rebellion that hastened the departure of the British from the region.

After independence, Kenya chose a capitalist path of development, aligning itself with Western interests. Without major known mineral deposits, Kenya depended on coffee and tea exports and on a tourist industry based on its magnificent national parks (Fig. 6-14). Tourism became its largest single earner of foreign exchange, and Kenya prospered, apparently proving the wisdom of its capitalist course.

But serious problems arose. During the 1980s, Kenya had the highest rate of population growth in the world, and

population pressure on farmlands and on the fringes of the wildlife reserves mounted. Poaching became worrisome, and tourism declined. During the late 1990s, violent weather buffeted Kenya, causing landslides and washing away large segments of the crucial Nairobi-Mombasa Highway. This disaster was followed by a severe drought lasting several years, bringing famine to the interior. Meanwhile, government corruption siphoned off funds that should have been invested. Democratic principles were violated, and relationships with Western allies were strained. The AIDS pandemic brought another setback to a country that, in the early 1970s, had appeared headed for an economic takeoff.

Today, Kenya's prospects are uncertain. Geography, history, and politics have placed the Kikuyu (22 percent of the population of 36.5 million) in a position of power. But there are other major peoples (see Fig. 6-14) and several smaller ones. The Luhya, Luo, Kalenjin, and Kamba together constitute about 50 percent of the population, and on the territorial margins of the country there are peoples such

FROM THE FIELD NOTES

"Visiting the village in the Kenya Highlands where a graduate student was doing fieldwork on land reform, I took the long way and drove along the top of the eastern wall of the Eastern Rift Valley. Often the valley wall is not sheer but terraced, and soils on those terraces are quite fertile; also, the west-facing slopes tend to be well-watered. Here African farmers built villages and laid out communal plots, farming these lands in a well-organized way long before the European intrusion."   © H. J. de Blij

Concept Caching

www.conceptcaching.com

as the Maasai, Turkana, Boran, and Galla. Along the coast and in the east generally, there is a long-term Arab presence and a string of Muslim communities that are not well integrated into Kenya's social or political fabric. Mombasa, Malindi, and other points along the coast are tourist magnets, but tourist behavior and Muslim strictures sometimes clash. Core and periphery in Kenya thus are often at odds, and forging a political system that ensures democracy while representing the interests of these disparate sectors is Kenya's yet-unmet challenge.

Tanzania

Tanzania (a name derived from Tanganyika plus Zanzibar) is the biggest and most populous East African country (39.8 million). Its total area exceeds that of the region's other countries combined. Tanzania has been described as a country without a core because its clusters of population and zones of productive capacity lie dispersed—mostly on its margins on the east coast (where the capital, Dar es Salaam, is located), near the shores of Lake Victoria in the northwest, near Lake Tanganyika in the far west, and near Lake Malawi in the interior south. This is in sharp contrast to Kenya, which has a well-defined core area in the Kenya Highlands (centered on Nairobi in the heart of the country). Moreover, Tanzania is a country of many peoples, none numerous enough to dominate the state. About 100 ethnic groups coexist; one-third of the population, mainly those on the coast, are Muslims. (Tanzania has the distinction of having had both a Roman Catholic and a Muslim president during its period of independence.)

In contrast to Kenya, following independence Tanzania embarked on a socialist course toward development, including a massive farm collectivization program that was imposed without adequate planning. Communist China helped Tanzania construct a railroad, the Tan-Zam Railway, from Dar es Salaam to Zambia, but the project failed, as did an effort to move the capital from colonial Dar es Salaam to Dodoma in the interior (Fig. 6-14). The country's limited tourist infrastructure was allowed to degenerate, giving Kenya virtually the entire tourist market.

But Tanzania did achieve what several other East African countries could not: political stability and a degree of democracy. Thus Tanzania changed economic direction in the late 1980s without social turmoil and embarked on a market-oriented "recovery" program. The government, aided by foreign donors and investors, encouraged commercial agriculture, but although Tanzanian coffee, cotton, tobacco, and other farm products sold increasingly well on foreign markets, the great majority of Tanzanians remained subsistence farmers and the country stayed wretchedly poor. However, its transparent government, relatively low corruption level, and effective leadership (President Mkapa and his successor, President Kikwete, set the highest standards in the region) were rewarded with a growing and dependable stream of subsidies and investments. Today the economy is growing robustly, and the government is able to plan an expansion of commercial farming through investments in highways, irrigation, fertilizers, market mechanisms, and other needs. Meanwhile, Tanzania's tourist industry is back on track, its mining industry is expanding (gold is a major income earner, and

Tanzania may possess some of the world's largest nickel reserves), and this one-time basket-case of East Africa may yet rise to lead the region.

Uganda

Uganda (2008 population: 29.4 million) contained the most important African political entity in this region when the British arrived in the 1890s. This was the kingdom of Buganda (shown in dark brown in Fig. 6-14), which faced the north shore of Lake Victoria, had an impressive capital at Kampala, and was stable—as well as ideally suited for indirect rule over a large hinterland. The British established their headquarters at nearby Entebbe on the lake (thus adding to the status of the kingdom) and proceeded to organize their Uganda protectorate in accordance with the principles of indirect rule. The Baganda (the people of Buganda) became the dominant people in Uganda, and when the British departed, they bequeathed a complicated federal system to perpetuate Baganda supremacy.

16 **Landlocked states** tend to face particular problems, and Uganda is no exception. Dependent on Kenya for an outlet to the ocean, Uganda at independence in 1962 had better economic prospects than many other African countries. It was the largest producer of coffee in the British Commonwealth. It also exported cotton, tea, sugar, and other farm products. Copper was mined in the southwest, and an Asian immigrant population of about 75,000 dominated the country's commerce. Nevertheless, political disaster struck. Resentment over Baganda overlordship fueled revolutionary change, and a brutal dictator, Idi Amin, took control in 1971. He ousted the Asians, exterminated his opponents, and destroyed the economy. Eventually, in 1979, an invasion supported by neighboring Tanzania drove Amin from power, but by then Uganda lay in ruins. Recovery was slow, complicated by the AIDS pandemic, which struck Uganda with particular severity before enlightened government intervention reversed the tide.

In the 1990s, Uganda's interior location again afflicted its future: during a period of civil war in its northern neighbor, Sudan, Uganda was accused of supporting African rebels against their Muslim oppressors, and a costly conflict ensued. This destabilized the country's north, home of the Acholi people, and led to a second nightmare—the rise of a nominally Christian movement called the Lord's Resistance Army (LRA), which proclaimed that it wanted to rule Uganda according to the Ten Commandments. Known for unmatched brutality even in this zone of violent conflict, the LRA captured, conscripted, and enslaved children, drove some 2 million Ugandans, most of them Acholi, into squalid camps during the government's counterinsurgency campaign, and dashed Uganda's hopes for a return to post-Amin normalcy.

As if this were not enough, Uganda also found itself embroiled in the collapse of still another landlocked neighbor, Rwanda (Fig. 6-14), and in a civil war in The Congo. In both neighbors, Uganda took the side of Tutsi minorities under threat from powerful enemies, putting further strain on its limited financial resources. Situated at one of Africa's most volatile crossroads, Uganda faces a difficult future.

FROM THE FIELD NOTES

"It was 108 in the shade, but the narrow alleys of Zanzibar's Stone City felt even hotter than that. I spent some time here when I was working on my monograph on Dar es Salaam in the 1960s and had not been back. In those days, African socialism and *uhuru* were the watchwords; since then Tanzania has not done well economically. But here were signs of a new era: a People's Bank in a former government building, and on the old fort's tower a poster saying 'Think Digital Go Tritel.' Nearby, on the sandy beach where I relaxed 35 years ago, was evidence that Zanzibar had not escaped the ravages of AIDS. Now the sand served as a refuge for the sick, who were resting there. 'It's better, bwana, than the corridor of the clinic,' said a young man who could walk only a few steps at a time with the help of a cane, and who breathed with difficulty as he spoke. Here as everywhere in Subsaharan Africa, AIDS has severely strained already-limited facilities." © H. J. de Blij

www.conceptcaching.com

Rwanda and Burundi

Rwanda and Burundi would seem to occupy Tanzania's northwest corner (Fig. 6-14), and indeed they were part of the German colonial domain conquered before World War I. But during that war Belgian forces attacked the Germans from their Congo bases and were awarded these territories when the conflict ended in 1918. The Belgians used them as labor sources for their Katanga mines.

Rwanda (9.6 million) and Burundi (8.2 million) are physiographically part of East Africa, but their cultural geography is linked to the north and west. Here, Tutsi pastoralists from the north subjugated Hutu farmers (who had themselves made serfs of the local Twa [pygmy] population), setting up a conflict that was originally ethnic but became cultural. Certain Hutu were able to advance in the Tutsi-dominated society, becoming to some extent converted to Tutsi ways, leaving subsistence farming behind, and rising in the social hierarchy. These so-called moderate Hutu were—and are—often targeted by other Hutus, who resent their position in society. This longstanding discord, worsened by colonial policies, had repeatedly devastated both countries and, in the 1990s, spilled over into The Congo, generating the first interregional war in Subsaharan Africa.

An estimated 3.5 million people have perished as Hutu, Tutsi, Ugandan, and Congolese rebel forces have fought for control over areas of the eastern Congo, unleashing longstanding local animosities (such as those between the Hema and Lendu around Bunia) that worsened the death toll. Only massive international intervention could stabilize the situation, but the world turned a blind eye to the region's woes—again.

Highland Ethiopia

As Figure 6-14 shows, the East African region also encompasses the highland zone of Ethiopia, including the capital, Adis Abeba; the source of the Blue Nile, Lake Tana; and the Amharic core area that was the base of the empire that lost its independence only from 1935 to 1941. Ethiopia, mountain fortress of the Coptic Christians who held their own here, eventually became a colonizer itself. Its forces came down the slopes of the highlands and conquered much of the Islamic part of Africa's Horn, including present-day Eritrea and the Ogaden area, a Somali territory. (Geographically, these are parts of what we have mapped as the African Transition Zone [Fig. 6-10], which will be discussed in the final section of this chapter.)

Physiographically and culturally, highland Ethiopia is part of the East African region. But because Ethiopia was not colonized and because its natural outlets are to the Red Sea, not southward to Mombasa, effective interconnections between former British East Africa and highland Ethiopia never developed. But the Amhara and Oromo peoples of Ethiopia are Africans, not Arabs; nor have they been Arabized or Islamized as in northern Sudan and Somalia. The independence and secession of Eritrea in 1993 effectively landlocked Ethiopia, but for a few years there was cooperation and Ethiopia used Eritrean Red Sea ports (see Fig. 6-19). In 1998, however, a boundary dispute led to a bitter and costly war, and Ethiopia was forced to turn to Djibouti for a maritime outlet. This border conflict cost some 70,000 lives and did not end until UN mediation in 2000, when an agreement signed in Algiers gave Eritrea—in the view of Ethiopians—more than it deserved. By mid-decade, the arbitration still had not been completely accepted in Adis Abeba, but now a new problem loomed and Ethiopian troops were on (and across) the border with southern Somalia, where an Islamist militia was in the process of taking over the government. No matter how these issues turn out, Highland Ethiopia has adversaries on three sides and is likely to turn increasingly toward East Africa, however tenuous its surface links to the south may be today.

● EQUATORIAL AFRICA

The term *equatorial* is not just locational but also environmental. The equator bisects Africa, but only the western part of central Africa features the conditions associated with the low-elevation tropics: intense heat, high rainfall and extreme humidity, little seasonal variation, rainforest and monsoon-forest vegetation, and enormous biodiversity. To the east, beyond the Western Rift Valley, elevations rise, and cooler, more seasonal climatic regimes prevail. As a result, we recognize two regions in these lowest latitudes: Equatorial Africa to the west and (just discussed) East Africa to the east.

Equatorial Africa is physiographically dominated by the gigantic Congo Basin. The Adamawa Highlands separate this region from West Africa; rising elevations and climatic change mark its southern limits (see the *Cwa* boundary in Fig. G-8). Its political geography consists of eight states, of which The Congo (formerly Zaïre) is by far the largest in both territory (49 percent) and population (65 percent) (Fig. 6-15).

Five of the other seven states—Gabon, Cameroon, São Tomé and Príncipe, Congo, and Equatorial Guinea—all have coastlines on the Atlantic Ocean. The Central African Republic and Chad, the south of which is part of this region, are landlocked. In certain respects, the physical and human characteristics of Equatorial Africa extend even into southern

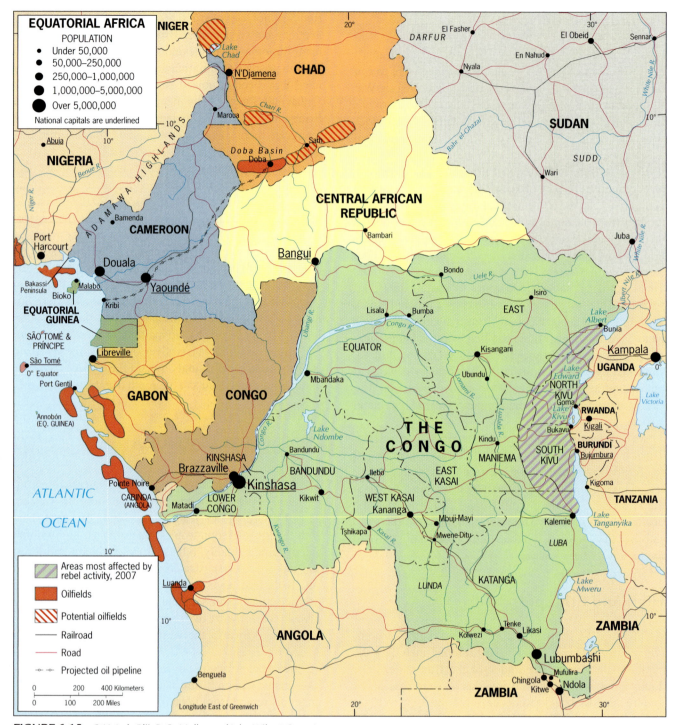

FIGURE 6-15 © H. J. de Blij, P. O. Muller, and John Wiley & Sons, Inc.

Sudan. This vast and complex region is in many ways the most troubled in the entire Subsaharan Africa realm.

The Congo

As the map shows, The Congo has but a tiny window (37 km/23 mi) on the Atlantic Ocean, just enough to accommodate the mouth of the Congo River. Oceangoing ships can reach the port of Matadi, inland from which falls and rapids make it necessary to move goods by road or rail to the capital, Kinshasa. This is not the only place where the Congo River fails as a transport route. Follow it upstream in Figure 6-15, and you note that other transshipments are necessary between Kisangani and Ubundu, and at Kindu. Follow the railroad south from Kindu, and you reach another narrow corridor of territory at the city of Lubumbashi. That vital part of Katanga Province contains most of The

Congo's major mineral resources, including copper and cobalt.

With a territory not much smaller than the United States east of the Mississippi, a population of 66.6 million, a rich and varied mineral base, and much good agricultural land, The Congo would seem to have all the ingredients needed to lead this region and, indeed, Africa. But strong centrifugal forces, arising from its physiography and cultural geography, pull The Congo apart. The immense forested heart of the basin-shaped country creates communication barriers between east and west, north and south. Many of The Congo's productive areas lie along its periphery, separated by enormous distances. These areas tend to look across the border, to one or more of The Congo's nine neighbors, for outlets, markets, and often ethnic kinship as well.

Crisis in the Interior

The Congo's civil wars of the 1990s started in one such neighbor, Rwanda, and spilled over into what was then still known as Zaïre. Rwanda, Africa's most densely populated country, has for centuries been the scene of conflict between sedentary Hutu farmers and invading Tutsi pastoralists. Colonial borders and practices worsened the situation (see the Issue Box titled "The Impact of Colonialism on Subsaharan Africa"), and after independence a series of terrible crises followed. In the mid-1990s, the latest of these crises generated one of the largest refugee streams ever seen in the world, and the conflict engulfed eastern (and later northern and western) Congo. The death toll will never be known, but estimates range from 3 to 4 million, a calamity that was not enough to propel the "international community" into concerted peacemaking action.

By 2005, a combination of power transfer in the capital, Kinshasa, negotiation among the rebel groups and the African states involved in various ways in the conflict, UN assistance, and exhaustion had produced a semblance of stability in all but some eastern areas of The Congo. In 2006, a democratic election in The Congo, a remarkable achievement under the circumstances, led to victory for Joseph Kabila, son of the (assassinated) ruler whose forces had swept westward into the capital of Kinshasa during the civil war, ushering in a new era in this troubled state. But in 2007, several warlords and militiamen continued to control parts of the eastern Congo (see Fig. 6-15), where dense rainforests enable them to operate and where illicit mining of valuable minerals supports their actions.

Across the River

To the west and north of the Congo and Ubangi rivers lie Equatorial Africa's other seven countries (Fig. 6-15). Two of them are landlocked. Chad, straddling the African Transition Zone as well as the regional boundary with West Africa, is one of Africa's most remote countries, although recent oil discoveries in the south are today changing this. Not only oil, but also the crisis in neighboring Darfur in Sudan is affecting Chad, with a major refugee influx and a rebel movement destabilizing this vulnerable state. The Central African Republic, chronically unstable and poverty-stricken, never was able to convert its agricultural potential and mineral resources (diamonds, uranium) into real progress. And one country consists of two small, densely forested volcanic islands: São Tomé and Príncipe, a ministate with a population of less than 250,000 whose economy is about to be transformed by recent oil discoveries.

The four coastal states present a different picture. All four possess oil reserves and share the Congo Basin's equatorial forests; oil and timber, therefore, rank prominently among their exports. In Gabon, this combination has produced Equatorial Africa's only upper-middle-income economy. Of the four coastal states, Gabon also has the largest proven mineral resources, including manganese, uranium, and iron ore. Its capital, Libreville (the only coastal capital in the region), reflects all this in its high-rise downtown, bustling port, and fast-growing squatter settlements.

Cameroon, less well endowed with oil or other raw materials, has the region's strongest agricultural sector by virtue of its higher-latitude location and high-relief topography. Western Cameroon is one of the more developed parts of Equatorial Africa and includes the capital, Yaoundé, and the port of Douala.

With five neighbors, Congo could be a major transit hub for this region, especially for The Congo if it recovers from civil war. Its capital, Brazzaville, lies across the Congo River from Kinshasa and is linked to the port of Pointe Noire by road and rail. But devastating power struggles have negated Congo's geographic advantages.

As Figure 6-15 shows, Equatorial Guinea consists of a rectangle of mainland territory and the island of Bioko, where the capital of Malabo is located. A former Spanish colony that remained one of Africa's least-developed territories, Equatorial Guinea, too, has been affected by the oil business in this area. Petroleum products now dominate its exports, but, as in so many other oil-rich countries, this bounty has not significantly raised incomes for most of the people.

One other territory would seem to be a part of Equatorial Africa: *Cabinda*, wedged between the two Congos just to the north of the Congo River's mouth. But Cabinda is one of those colonial legacies on the African map—it belonged to the Portuguese and was administered as part of Angola. Today it is an **exclave** of independent Angola, and a valuable one: it contains major oil reserves.

The Impact of Colonialism on Subsaharan Africa

COLONIALISM IS THE CULPRIT!

"If you want to see and feel the way colonialism ruined Africa, you don't have to go beyond the borders of my own country, the so-called Democratic Republic of the Congo, for the evidence. I'm a school teacher here in Kikwit, east of Kinshasa, and my whole life I've lived under the dictatorship left us by the Belgians and in the chaos before and after it. All you have to do is look at the map. Who can say that the political map of Africa isn't a terrible burden? Imagine some outside power, say the Chinese, coming into Europe and throwing the Germans and the French together in one country. There would never be peace! Well, that's what they did here. All over Africa they put old enemies together and parceled up our great nations. And now they tell us to get over our animosities and 'tribalism' and live in democratic peace.

"And let me tell you something about tribalism. Before the conquest, we had ethnic groups and clans like all other peoples have, from the Scots to the Siamese. And we certainly fought with each other. But mostly we got along, and there was lots of gray area between and among us. You weren't immediately labeled a Hutu or a Luba or a Bemba. The colonial powers changed all that. Suddenly we were tribalized. Whether you liked it or not, whether you filled the 'profile' or not, you were designated a Tutsi or a Hutu, a Ganda or a Toro. This enabled the European rulers to use the strong to dominate the weak, to use the rich to tax the poor, to police migrants and assemble labor. The tribalization of Africa was a colonial invention. It worsened our divisions and destroyed our common ground. And now we just have to get over it?

"Of course the colonialists exploited our mineral resources— I tell my pupils how The Congo enriched Europeans and other 'civilized' peoples outside of Africa. But they also grabbed our best land, turning food-producing areas into commercial plantations. Are you surprised that in Zimbabwe Mugabe wants to chase those white farmers off 'their' land? I'm not, and I'm with him all the way. If the colonialists hadn't ruined traditional agriculture, there wouldn't be any food shortages today.

"Look at my country. The war between the Hutu and the Tutsi that started during colonial times has now spread into The Congo. We've got so used to taking sides in 'tribal' conflicts that much of the east is out of control. And who are the losers? The poorest of the poor, who get robbed and killed by the Mai, the Interahamwe, and the Tutsis and their friends. What are the ex-colonial powers doing now to help us with the mess they left behind? Nothing! So don't complain to me about poor African leadership and our failure to carry out economic reforms dictated by some foreign bank. The poisonous legacy of colonialism will hold Africa back for generations to come."

COLONIALISM IS A SCAPEGOAT!

"We Africans have many factors to blame for our troubles, but colonialism isn't one of them. Many African countries have been independent for nearly two generations; most Africans were born long after the colonial era ended. Are we better off today than we were at the end of the colonial period? Are those military dictatorships that ran so many countries into the ground any better than the colonizers who preceded them? Are the ghosts of colonialists past rigging our elections, stifling the media? Is our endemic corruption the fault of lingering colonialism? Did the colonial governors set up Swiss bank accounts to which to divert development funds? Can extinct colonial regimes be blamed for the environmental problems we face? Did the bandits of interior Sierra Leone or eastern areas of The Congo learn their mutilation practices from colonialism? As a nurse in a Uganda hospital for AIDS victims, I say that it's time to stop blaming ancient history for our current failures.

"It's not that we Africans aren't at a disadvantage in this unforgiving world. Our postcolonial population explosion was followed by an HIV-induced implosion. Apartheid South Africa spread its malign influence far beyond its borders. The Cold War pitted regimes against rebels and, to use a colonial term, tribe against tribe. Droughts and other plagues of biblical proportions ravaged our continent when we were more vulnerable than ever. Prices for our raw materials on international markets fell when we most needed a boost. Foreign governments protected their farmers against competition from African producers. Those are legitimate complaints. But colonialism?

"Forget about blaming boundaries, white farmers, tribalism, religion, and other residues of colonialism for our present troubles. Look at what happened here in Uganda. A decade ago we were at the heart of the 'AIDS Belt' in tropical Africa, with one of the worst infection rates on the continent. Then came President Museveni, whose government launched a vigorous self-help campaign, educating people to the risks, urging restraint, distributing condoms, encouraging women to resist unwanted sex. Now Uganda's AIDS incidence is dropping and we are held up as an example for all of Africa to follow. While South Africa's President Mbeki reveals his confusion about the causes of AIDS, Zimbabwe's Mugabe is too busy chasing white farmers, and Kenya's Moi is trying to rig his succession, we in Uganda have leadership in an area where we need it desperately. That certainly beats blaming colonialism for everything.

"We have an African Union. If those colonial boundaries are so bad, let's change them, or do what the European Union is doing and begin to erase them. But first, let's isolate tyrants, nurture open democracy, and combat corruption— and stop blaming colonialism."

Regional ISSUE

Vote your opinion at www.wiley.com/college/deblij

WEST AFRICA

West Africa occupies most of Africa's Bulge, extending south from the margins of the Sahara to the Gulf of Guinea coast and from Lake Chad west to Senegal (Fig. 6-16). Politically, the broadest definition of this region includes all those states that lie to the south of Western Sahara, Algeria, and Libya and to the west of Chad (itself sometimes included) and Cameroon. Within West Africa, a rough division is sometimes made between the large, mostly steppe and desert states that extend across the southern Sahara (Chad could also be included here) and the smaller, better-watered coastal states.

Apart from once-Portuguese Guinea-Bissau and long-independent Liberia, West Africa comprises four former British and nine former French dependencies. The British-influenced countries (Nigeria, Ghana, Sierra Leone, and Gambia) lie separated from one another, whereas Francophone West Africa is contiguous. As Figure 6-16 shows, political boundaries extend from the coast into the interior, so that from Mauritania to Nigeria, the West African habitat is parceled out among parallel, coast-oriented states. Across these boundaries, especially across those between former British and former French territories, there is only limited interaction. For example, in terms of value, Nigeria's trade with Britain is about 100 times as great as its trade with nearby Ghana. The countries of West Africa are not interdependent economically, and their incomes are largely derived from the sale of their products on the non-African international market.

Given these cross-currents of subdivision within West Africa, why are we justified in speaking of a single West African region? First, this part of the realm has remarkable

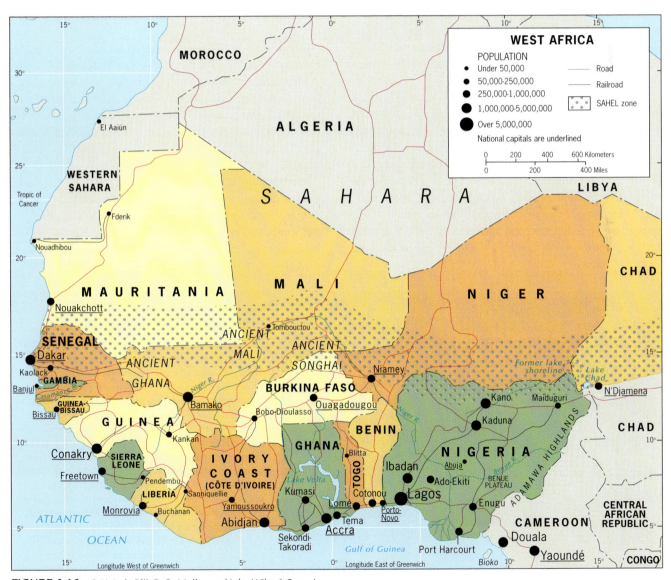

FIGURE 6-16   © H. J. de Blij, P. O. Muller, and John Wiley & Sons, Inc.

cultural and historical momentum. The colonial interlude failed to extinguish West African vitality, expressed not only by the old states and empires of the savanna and the cities of the forest, but also by the vigor and entrepreneurship, the achievements in sculpture, music, and dance, of peoples from Senegal to Nigeria's southeastern Iboland. Second, West Africa contains a set of parallel east-west ecological belts, clearly reflected in Figures G-7 and G-8, whose role in the development of the region is pervasive. As the transport-route pattern on the map of West Africa indicates, overland connections within each of these belts, from country to country, are poor; no coastal or interior railroad ever connected this tier of countries. Yet spatial interaction is stronger across these belts, and some north-south economic exchange does take place, notably in the coastal consumption of meat from cattle raised in the northern savannas. And third, West Africa received an early and crucial imprint from European colonialism, which—with its maritime commerce and slave trade—transformed the region from one end to the other. This impact reached into the heart of the Sahara and set the stage for the reorientation of the whole area, from which emerged the present patchwork of states.

Despite the effects of the slave trade, West Africa today is Subsaharan Africa's most populous region (Fig. G-9). In these terms, Nigeria (whose census results are in doubt, but with an estimated population of 141 million) is Africa's largest state; Ghana (23.7 million) and Ivory Coast (20.7 million) rank prominently as well. The southern half of the region, understandably, is home to most of the people. Mauritania, Mali, and Niger include too much of the unproductive Sahel's steppe and the arid Sahara to sustain populations comparable to those of Nigeria, Ghana, or Ivory Coast.

The peoples along the coast reflect the modern era that the colonial powers introduced: they prospered in their newfound roles as middlemen in the coastward trade. Later, they experienced the changes of the colonial period; in education, religion, urbanization, agriculture, politics, health, and many other endeavors, they adopted new ways. In contrast, the peoples of the interior retained their ties with a different era in African history. Distant and aloof from the main theater of European colonial activity and often drawn into the Islamic orbit, they experienced a significantly different kind of change. But the map reminds us that Africa's boundaries were not drawn to accommodate such contrasts. Both Nigeria and Ghana possess population clusters representing the interior as well as the coastal peoples, and in both countries the wide cultural gap between north and south has produced political problems.

Nigeria: West Africa's Cornerstone

Nigeria, the region's cornerstone, is home to more than 140 million people, by far the largest population total of any African country. When Nigeria achieved full independence from Britain in 1960, its new government was faced with the daunting task of administering a European political creation containing three major nations and nearly 250 other peoples ranging from several million to a few thousand in number.

For reasons obvious from the map, Britain's colonial imprint always was stronger in the two southern regions than in the north. Christianity became the dominant faith in the south, and southerners, especially the Yoruba, took a lead role in the transition from colony to independent state. The choice of Lagos, the port of the Yoruba-dominated

Ghana's coastal town of Elmina may be too small to show up on our regional map, but this was the site of the first European settlement in West Africa. It was a key node in Portuguese maritime trade and later became notorious during the slave trade era. Today it is a fishing port of about 20,000, but the industry is not doing well. The local fishers must compete with the larger foreign trawlers that harvest the waters off the coast of West Africa, and cold-storage facilities remain minimal, so that almost all of the local catch is smoked and sold on the local market. This keeps prices low, so that Elmina's economy has changed very little even after the government in 1983 introduced economy-boosting policies. Talk to the locals (fishing and related work are the livelihoods of over half the population), and you realize that it was the capital of Accra that benefited most from the new economic rules, not places like this.   © Richard J. Grant.

AMONG THE REALM'S GREAT CITIES . . . **Lagos**

IN A REALM that is only 34 percent urbanized, Lagos, former capital of federal Nigeria, is the exception: a teeming metropolis of 12.7 million, sometimes called the Calcutta of Africa.

Lagos evolved over the past three centuries from a Yoruba fishing village, Portuguese slaving center, and British colonial headquarters into Nigeria's largest city, major port, leading industrial center, and first capital. Situated on the country's southwestern coast, it consists of a group of low-lying islands and sand spits between the swampy shoreline and Lagos Lagoon. The center of the city still lies on Lagos Island, where the high-rises adjoining the Marina overlook Lagos Harbor and, across the water, Apapa Wharf and the Apapa industrial area. The city expanded southeastward onto Ikoyi Island and Victoria Island, but after the 1970s most urban sprawl took place to the north, on the western side of Lagos Lagoon.

Lagos's cityscape is a mixture of modern high-rises, dilapidated residential areas, and squalid slums. From the top of a high-rise one sees a seemingly endless vista of rusting corrugated roofs, the houses built of cement or mud in irregular blocks separated by narrow alleys. On the outskirts lie the shantytowns of the less fortunate, where shelters are made of plywood and cardboard and lack even the most basic facilities.

By world standards, Lagos ranks among the most severely polluted, congested, and disorderly cities. Mismanagement and official corruption are endemic. Laws, rules,

© H. J. de Blij, P. O. Muller, and John Wiley & Sons, Inc.

and regulations, from zoning to traffic, are flouted. The international airport is notorious for its inadequate security and for extortion by immigration and customs officers. In many ways, Lagos is a city out of control.

---

southwest, as the capital of a federal Nigeria (and not one of the cities in the more populous north) reflected British desires for the country's future. A three-region federation, two of which lay in the south, would ensure the primacy of the non-Islamic part of the state. But this framework did not last long. In 1967, the Ibo-dominated Eastern Region declared its independence as the Republic of Biafra, leading to a three-year civil war at a cost of 1 million lives. Since then, Nigeria's federal system has been modified repeatedly; today there are 36 States, and the capital has been moved from Lagos to centrally located Abuja (Fig. 6-17).

Fateful Oil

Large oilfields were discovered beneath the Niger Delta during the 1950s, when Nigeria's agricultural sector produced most of its exports (peanuts, palm oil, cocoa, cotton) and farming still had priority in national and State development plans. Soon, revenues from oil production dwarfed all other sources, bringing the country a brief period of pros-

perity and promise. But before long Nigeria's oil wealth brought more bust than boom. Misguided development plans now focused on grand, ill-founded industrial schemes and costly luxuries such as a national airline; the continuing mainstay of the vast majority of Nigerians, agriculture, fell into neglect. Worse, poor management, corruption, outright theft of oil revenues during military misrule, and excessive borrowing against future oil income led to economic disaster. The country's infrastructure collapsed. In the cities, basic services broke down. In the rural areas, clinics, schools, water supplies, and roads to markets crumbled. In the Niger Delta area, local people beneath whose land the oil was being exploited demanded a share of the revenues and reparations for ecological damage; the military regime under General Abacha responded by arresting and executing nine of their leaders. By the time the ruthless Abacha was succeeded by democratically elected President Olusegun Obasanjo, the situation in the Niger Delta had generated an armed resistance group named the Movement for the Emancipation of the Niger Delta (MEND), that carried

FIGURE 6-17 © H. J. de Blij, P. O. Muller, and John Wiley & Sons, Inc.

out attacks on pipelines, kidnapped oil-company workers, and committed other acts of sabotage. The delta has become one of the most dangerous oil-producing areas in the world for locals and workers alike.

On global indices of national well-being, Nigeria during the 1990s sank to the lowest rungs even as its oil production ranked it tenth in the world in that category, with the United States its chief customer. After 2000 some improvement occurred, but mismanagement and corruption slowed all progress and so did rapid population growth (which had eased up by 2007, but Nigeria remains a fast-growing country). Take a careful look at the data in Table G-1 and

you will see that this African as well as regional cornerstone state has a long way to go.

Islam Ascendant

Nigeria faces problems beyond the economic and demographic spheres. The free election of a president in 1999 raised the people's hopes, but trouble arose in another context: its historic cultural schism. Soon after the return to democracy, federal Nigeria's northern States, beginning with Zamfara, decided to proclaim **Sharia** (strict Islamic) law. When Kaduna State followed suit, riots between Chris-

huge territory on the Sahara's margins but containing small populations, are landlocked: Mali, Burkina Faso, Niger, and Chad. Figure G-8 shows clearly how steppe and desert conditions dominate the natural environments of these four interior states. Figure G-9 reveals the concentration of population in the steppe zone and along the ribbon of water provided by the Niger River. Scattered oases form the remaining settlements and anchor regional trade.

But even the coastal states do not escape the dominance of the desert over West Africa. Mauritania's environment is almost entirely desert. Senegal, as Figure 6-16 shows, is mostly a Sahel country; and not only northern Nigeria but also northern Benin, Togo, and Ghana have interior steppe zones. The loss of pastures to **desertification** (human-induced desert expansion) is a constant worry for the livestock herders there.

West Africa's states share the effects of the environmental zonation depicted in Figures G-7 and G-8, but they also have distinct regional geographies. Benin, Nigeria's neighbor, has a growing cultural and economic link with the Brazilian State of Bahía, where many of its people were taken in bondage and where elements of West African culture have survived. Ghana, once known as the Gold Coast, was the first West African state to achieve independence (1957), with a sound economy based on cocoa exports. Two grandiose postindependence schemes

Nigeria lies astride the Islamic Front, and strife between Muslims and Christians occurs intermittently, especially in the north, where Sharia law was adopted by 12 of the country's 36 States. When the Kaduna State government was debating the adoption of Sharia law, local Christians mounted a demonstration in opposition to the plan, resulting in a pitched battle and a major fire in the commercial core of the city. In this photo, Muslims are seen leaving a morning prayer service amid the rubble; this incident alone cost 200 lives. © AP/Wide World Photos

tians and Muslims devastated the old capital city of Kaduna (photo above). There, and in 11 other northern States (Fig. 6-17), the imposition of Sharia law led to the departure of thousands of Christians, intensifying the cultural fault line that threatens the cohesion of the country. Although Kaduna State temporarily repealed its decision, the Islamic revivalism now exhibited in the north raises the prospect that Nigeria, West Africa's cornerstone and one of Africa's most important states, may succumb to devolutionary forces arising from its location in the African Transition Zone. For Africa, that would be a calamity.

Coast and Interior

Nigeria is one of 17 states (counting Chad and offshore Cape Verde [not shown in Fig. 6-16]) that constitute the region of West Africa. Four of these countries, comprising a

Even in the poorer countries of the world, you see something that has become a phenomenon of globalization: gated communities. Widening wealth differences, security concerns, and real estate markets in societies formerly characterized by traditional forms of land ownership combined to produce this new element in the cultural landscape. This is the entrance to Golden Gate, the first private, gated community development in Accra, Ghana, started in 1993 as a joint venture between a Texas-based construction company and a Ghanaian industrial partner. © Richard J. Grant.

can be seen on its map: the port of Tema, which was to serve a vast West African hinterland, and Lake Volta, which resulted from the region's largest dam project. When neither fulfilled expectations, Ghana's economy collapsed. But in the 1990s, Ghana's military regime was replaced by democratic and stable government, and over the past decade Ghana has become a model for Africa. Even ethnic strife in the northern region in 2002 did not disrupt the country's progress, and its agriculture-dominated export economy, based on cocoa, received an unexpected boost when chaos descended on its chief West African competitor, neighboring Ivory Coast. In 2006, Ghana authorized the world's largest mining company to begin exploiting gold reserves northwest of Kumasi, the historic Ashanti capital, under strict compensatory and environmental conditions that set an example to the rest of the continent and, indeed, the world.

Ivory Coast (which, as we have stated, officially still goes by its French name, *Côte d'Ivoire*) translated three decades of autocratic but stable rule into economic progress that gave it lower-middle-income status based mainly on cocoa and coffee sales. Continued French involvement in the country's affairs contributed to this prosperity; the capital, Abidjan, reflected its comparative well-being. But in a familiar pattern, Ivory Coast's president-for-life first engineered the transfer of the capital to his home village, Yamoussoukro, and then spent tens of millions of dollars building a Roman Catholic basilica there to rival that of St. Peter's in Rome. It was dedicated just as the country's economy was slowing and its social conditions worsened.

By the late 1990s, Ivory Coast's economy had reverted to the low-income category (Fig. G-12). Worse, regional strife intensified during the run-up to a presidential election in 2000, when southern politicians objected to the ethnic origins and alleged Muslim sympathies of a northern candidate. In September 2002, a series of coordinated attacks by northern rebels failed to take the largest city and former capital, Abidjan, but succeeded in capturing major northern centers. The attacks cost hundreds of lives and confirmed the worst fears of citizens in this long-stable and comparatively prosperous country, where immigrants have long contributed importantly to the economy and society. At the turn of this century, the population included about 2 million Burkinabes (immigrants from Burkina Faso to the north), as many as 2 million Nigerians, about 1 million Malians, 500,000 Senegalese, and smaller groups of Ghanaians, Guineans, and Liberians. This cultural mix had been kept stable by strong central government and a growing economy, but when both of these mainstays failed, disorder followed. In 2004 French troops, sent to Ivory Coast to stabilize the situation and to facilitate negotiations as well as protect the French expatriate commu-

nity, suffered a deadly attack by the Ivoirian Air Force. This resulted in retaliatory action and led to assaults on French citizens in Abidjan and elsewhere. Most of the expatriate community departed, leaving Ivory Coast, its cocoa-based economy devastated, in dire straits, a West African success story with a tragic turn.

All of West Africa has been affected by what has happened in Liberia and Sierra Leone. Liberia, a country founded in 1822 by freed slaves who returned to Africa with the help of American colonization societies, was ruled by their descendants for more than six generations. Rubber plantations and iron mines made life comfortable for the "Americo-Liberians," but among the local peoples, resentment simmered. A military coup in 1980, in which the president was killed, was followed in 1989 by full-scale civil war that pitted ethnic groups against each other and drove hundreds of thousands of refugees into neighboring countries, including Ivory Coast, Guinea, and Sierra Leone. Monrovia, the capital (named after U.S. president James Monroe) was devastated; an estimated 230,000 people, almost 10 percent of the population, perished. In 1997, one of the rebel leaders, Charles Taylor, became president, but he was indicted by a UN war-crimes tribunal in 2003, just as he was negotiating a truce between his regime and other rebel groups. Nigerian forces and UN peacekeeping troops entered Monrovia in 2005 and finally succeeded in bringing stability back to this blighted country.

One of the countries affected by these events was Sierra Leone, Liberia's coastal neighbor, also founded as a haven for freed slaves, in this case by the British in 1787. Independent since 1961, Sierra Leone went the all-too-familiar route from self-governing Commonwealth member to republic to one-party state to military dictatorship. But in the 1990s a civil war brought untold horror to this small country even as refugees from Liberia's war were arriving. Rebels enabled by diamond sales fought supporters of the legitimate government in a struggle that devastated town and countryside alike and killed and mutilated tens of thousands. Eventually a combination of West African, United Nations, and British forces intervened in the conflict and resurrected a semblance of representative government, but not before Sierra Leone had sunk to dead last on the world's list of national well-being.

As Figure 6-16 shows, Guinea borders all three of the troubled countries just identified, and it has involved itself in the affairs of all three—receiving in return a stream of refugees in its border areas. In 2007, it continued to occupy a slice of Sierra Leone called the Yenga Zone. Guinea, with a population of 10.4 million, has been under dictatorial rule ever since its founding president, Sekou Touré, turned from father of the nation to usurper of the people. As Figure G-12 shows, Guinea is one of Africa's poorer states, but representative government and less corrupt administration could have made it a far better place. Guinea has significant gold deposits and may possess as much as one-third of the world's bauxite (aluminum-ore) reserves, and can produce far more coffee and cotton than it does; offshore lie productive fishing grounds. As the map shows, this is no ministate like neighboring Guinea-Bissau, the former Portuguese colony. Guinea's capital, Conakry, has a substantial hinterland reaching well into West Africa's interior, and, as a result, the country has a wide range of environments.

Senegal had the advantage of lying separated from Sierra Leone by Guinea, another Francophone country along the West African coast. Its Sahelian environmental problems notwithstanding, Senegal managed to convert its colonial advantage (its capital, Dakar, was the headquarters for France's West African empire) into lasting progress. By no means a rich country, Senegal depends for foreign income on farm exports (chiefly peanuts), fishing (the leading source), phosphate sales, and iron ore production. But its most valuable asset has been its democratic tradition, which has now lasted more than four decades. This has enabled Senegal to weather some storms, including a failed union with its English-speaking enclave, Gambia, and a secessionist movement in the southwestern Casamance District (Fig. 6-16).

Senegal is over 90 percent Muslim, but its population of 12.6 million is dominated by the Wolof, the largest ethnic group constituting about 43 percent of the total, the Fulani (23 percent), and the Serer (15 percent). The Wolof are concentrated in and around the capital and wield most of the power. Senegal's leaders have maintained close relationships with France, which is still Senegal's leading trading partner and financial supporter in times of need. Without oil, diamonds, or other lucrative income sources and with an overwhelmingly subsistence-farming population, Senegal nevertheless managed to achieve GNI levels that exceeded the region's average (see Table G-1). Here is proof that reasonably representative government and stability are greater assets than gold, liquid or otherwise.

Populous, multicultural, divided West Africa remains a region of farmers and herders (the herders range over the savannas and steppes of the north) along a tier of fast-changing environments between ocean and desert. Local village markets drive the traditional economy, and **periodic markets**—not open every day but operating every three, four, or more days—ensure that all villages participate in the exchange network. Traditions of this kind endure here, even as the cities beckon the farmers and burst at the seams. The region's great challenges are economic survival and nation-building, constrained by a boundary framework that is as burdensome as any in Subsaharan Africa.

THE AFRICAN TRANSITION ZONE

From our discussion of East, Equatorial, and West Africa, it is obvious that the northern margin of the realm we define as Subsaharan Africa is in turmoil. Along a zone from Ethiopia in the east to Guinea in the west, cultural and ethnic tensions tend to erupt into conflict, and often the conflict spreads to engulf large areas across international borders. It is also clear that much of this turmoil has to do with the religious transition from Islamic and Arabized Africa to the Africa where Christianity and traditional beliefs hold sway. This zone, called the *African Transition Zone*, epitomizes the way one geographic realm yields to another. But in terms of its instability, variability, and volatility, this particular transition between realms is unique.

The domino effect is at work again in Africa, with regime-abetted Muslim extremism, African resistance, and oil in the mix. The Arabized Muslim regime in Khartoum, Sudan, empowered by the oil recently discovered under its desert, settled its dispute with the Sudanese south but then turned, in western Darfur Province, on African farmers. Refugees flooded into neighboring Chad, pursued by Khartoum-supported *janjaweed* militias. Soon the political situation in Chad, itself recently endowed with oil revenues, was destabilized as opponents of the president saw their opportunity and marched on the capital, N'Djamena. By mid-2007, the rebels had been prevented from overthrowing the government, but Chad was in turmoil, another victim of the Islamic Front.   © Landov

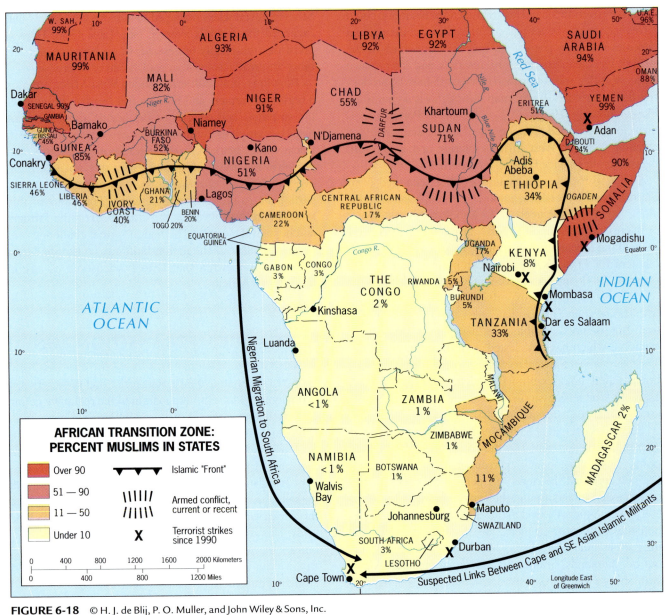

FIGURE 6-18   © H. J. de Blij, P. O. Muller, and John Wiley & Sons, Inc.

It is useful to look again at Figure 6-10, which shows that some entire countries lie within the African Transition Zone, such as Senegal, Burkina Faso, Eritrea, Djibouti, and Somalia. Others are bisected by it, including Ivory Coast, Nigeria, Chad, Sudan, and Ethiopia. Africa's colonial boundary framework was laid out even as Islam was on the march and paid little heed to the consequences. But it is Islam that defines the southern periphery of the African Transition Zone along a religious frontier sometimes referred to as the **Islamic Front** (Fig. 6-18).

As we noted earlier in this chapter, Islam arrived across the Sahara by caravan and up the Nile by boat, thus reaching and converting the peoples of the interior steppes and savannas on the southern side of the great desert. (This wide stretch of semiarid Africa is known today as the

Sahel, an Arabic word meaning "border" or "margin.") Then came Europe's colonial powers, and the modern politico-geographical map of Subsaharan Africa took shape. Muslims and non-Muslims were thrown together in countries not of their making. Vigorous Christian proselytism slowed the march of Islam, resulting in the distribution shown in Figure 7-2. The north-to-south transition is especially clear in the heart of the continent: Libya is 92 percent Muslim, Chad 55, the Central African Republic 17, and The Congo 1.6.

Conflict is the hallmark of much of the Islamic Front across Africa. In Sudan, a 30-year war between the Arab, Islamic north and the African, Christian-animist south, costing hundreds of thousands of lives and displacing millions, was finally settled in late 2004. Even so, a new and deathly conflict had arisen in Sudan's far-western Darfur Province (see p. 369) even as the southern war ended. In Ivory Coast, livestock owned by Muslim cattle herders from Burkina Faso had been trampling African farms for years before the north-south schism broke into open conflict in 2002. In Nigeria, Islamic revivalism in the North is clouding the entire country's prospects. But perhaps the most conflict-prone part of the African Transition Zone

lies in Africa's Horn, involving the historic Christian state of Ethiopia and its neighbors (Fig. 6-19).

Earlier we noted that highland Ethiopia where the country's core area, capital (Adis Abeba), and Christian heartland lie, is virtually encircled by dominantly Muslim societies. One poor-quality road, along which bandits rob bus passengers and law enforcement is virtually nonexistent, forms the only direct surface link between Adis Abeba

FIGURE 6-19   © H. J. de Blij, P. O. Muller, and John Wiley & Sons, Inc.

and Nairobi, the capital of Kenya to the south (Fig. 6-19). But, as Figure 6-19 reminds us, Ethiopia's eastern Ogaden is traditional Somali country and almost entirely Islamic, a vestige of a time when the rulers in Adis Abeba extended their power from their highland fortress over the plains below. Today, 34 percent of Ethiopia's population of 78.4 million is Muslim, and the Islamic Front is quite sharply defined (Fig. 6-18).

In 1998, Ethiopia and neighboring Eritrea fought a senseless border war over a few villages and a small strip of mainly desert territory, an aftermath of Eritrea's secession from Ethiopia. Once part of what may be called the Ethiopian Empire, Eritrea was under British administration during World War II and was made part of an Ethiopian Federation in 1952. After Ethiopia dissolved this federal arrangement in 1962 and absorbed Eritrea, Muslim as well as Christian Eritreans fought a war of secession. In 1993 their efforts were rewarded by international recognition of Eritrea's independence. This had the effect of landlocking Ethiopia, which had made Assab its key Red Sea port (Fig. 6-19). With the Muslim component of Eritrea's population of 4.9 million growing faster than the Christian sector, relationships between Ethiopia and Eritrea were tense, and the 1998 border war in fact was more than that. After spending U.S. $3 billion on the conflict and accepting foreign mediation in 2000, Ethiopia and Eritrea needed international food assistance to ward off starvation as a combined 16 million people suffered from hunger.

In 2002, a ministate in this part of the African Transition Zone, Djibouti, came to international attention as a result of the so-called War on Terror declared by the United States. Formerly a French colony and still host to a French military base, coastal Djibouti, ethnically divided between Afars and Issas but politically stable, proved to have a favorable relative location between troubled Somalia, vulnerable Eritrea, and crucial Ethiopia—and directly across the narrow Bab el Mandeb Strait (which separates the Red Sea and the Gulf of Adan) from Yemen, the ancestral home of Usama bin Laden.

But the key component in the eastern sector of the African Transition Zone is Somalia, where 9.4 million people, virtually all Muslim, live at the mercy of a desert-dominated climate that enforces cross-border migration into Ethiopia's Ogaden area in pursuit of seasonal pastures. As many as three to five million Somalis live permanently on the Ethiopian side of the border, but this is not the only division the Somali "nation" faces. In reality, the Somali people is an assemblage of five major ethnic groups fragmented into hundreds of clans engaged in an endless contest for power as well as survival.

Early during this decade, Somalia's condition as a *failed state* led to the country's fragmentation into three parts (Fig. 6-19). In the north, the sector called **Somaliland**, which had proclaimed independence in the 1990s and which remains by far the most stable of the three, functioned essentially as an African state although the "international community" will not recognize it as such. In the east, a conclave of local chiefs declared their territory to be separate from the rest of Somalia, called it **Puntland**, and asserted an unspecified degree of autonomy. In the south, where the official capital, Mogadishu, is located on the Indian Ocean coast, local secular warlords and Islamic militias continued their struggle for supremacy, the warlords supported by U.S. funding as part of the War on Terror. In June 2006, Islamic militias stormed into the capital and took control, ousting the warlords and proclaiming their determination to create an Islamic state. The combination of a failed state, Islamic fundamentalists in control, and al-Qaeda in the mix makes this part of the African Transition Zone a focus of international concern.

As this chapter has underscored, Africa is a continent of infinite social diversity, and Subsaharan Africa is a realm of matchless cultural history. Comparatively recent environmental change widened the Sahara and separated north from south; the arrival of Islam turned the north to Mecca and left the south to Christian proselytism. Millions of Africans were carried away in bondage; the cruelties of colonialism subjugated those who were left behind in a political straitjacket forged in Berlin. Africa's equatorial heart is an incubator for debilitating diseases second to none on Earth, but Africa's health problems never were a top priority in the medical world. When finally the richer countries abandoned their African empires, African peoples were left a task of reconstruction in which they not only got little help: they found their products uncompetitive on markets where rich-country producers basked in subsidies. We said at the beginning of this chapter that Africa's time and turn will come again, but it will take far more than is being done today to rectify half a millennium's malefactions.

What You Can Do

RECOMMENDATION: Write an op-ed piece! This is not merely a matter of writing a longer letter to the editor. An op-ed (opinion-editorial) essay is more challenging because you cannot assume the reader's familiarity with the topic. Therefore, you will want to write an introductory paragraph setting the stage and explaining what issue you plan to address, leading into a concisely argued position on the matter citing the necessary evidence, followed by a logical and judicious conclusion. Give yourself 1000 to 1200 words, and make every word count! It's great practice, and if the newspaper or magazine editor accepts your article, it will have much more impact than a letter to the editor. And, unlike a letter, you are able to title your op-ed piece. For example, you might try one under the headline Give African Farmers a Break! or another asking What Has Globalization Done for Africa? And here's a switch: readers are likely to write letters to the editor responding to your argument. They may even ask the editor to invite you to write again on another topic.

GEOGRAPHIC CONNECTIONS

1 Tourism is one of Africa's important industries, in some countries contributing substantially to the government's income and creating jobs where unemployment is high. But if you were a member of an organized tour group, you would soon discover that visitors to Subsaharan Africa tend to be more interested in the realm's remaining wildlife than in the people and their cultures. How would you go about adding some human interest to the itinerary? What aspects of the cultural geography of Subsaharan Africa would you like to see and why? In what ways are villages and small towns usually more evocative of regional cultures than large cities?

2 You have just been appointed to a key U.S.-government commission charged with the following responsibility: sufficient funds are available to give substantial assistance to just four countries in Subsaharan Africa, and the government hopes that the beneficial effects of this assistance will radiate out into neighboring states. The first task of the commission of which you are a member is to identify the four countries. If you had this decision to make, which four African states would you nominate, and what criteria, including geographic factors, would you employ to arrive at that selection?

Geographic Literature on Subsaharan Africa: The key introductory works on this realm were authored by Aryeety-Attoh, Bohannon & Curtin, Cole & de Blij, de Blij, Grove, and Stock. These works, together with a comprehensive listing of noteworthy books and recent articles on the geography of Africa south of the Sahara, can be found in the *References and Further Readings* section of this book's website at **www.wiley.com/college/deblij.**

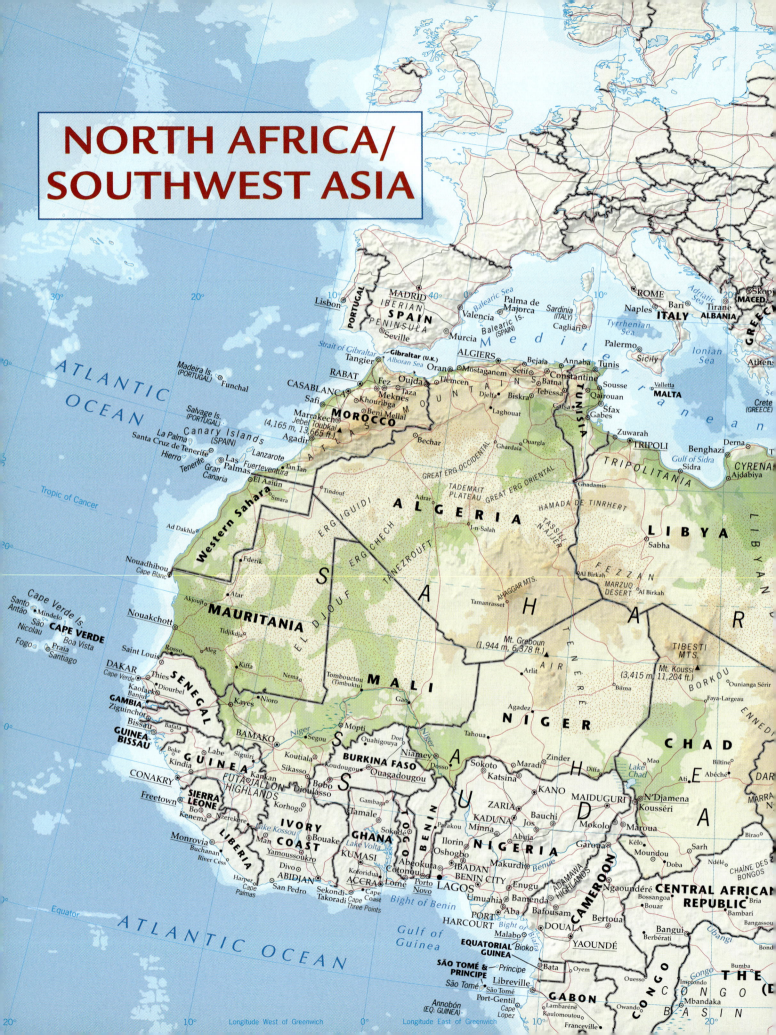

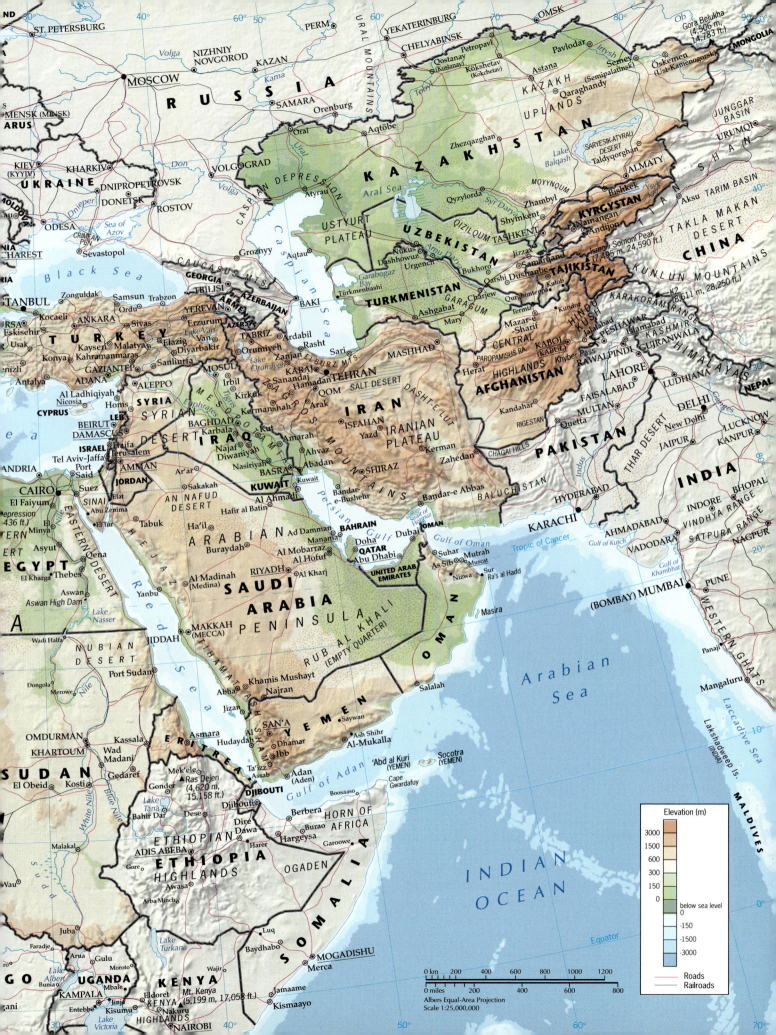

7

Where were these photos taken? Find out at www.wiley.com/college/deblij

Photos: © H. J. de Blij

FIGURE 7-1 *Map:* © H. J. de Blij, P. O. Muller, and John Wiley & Sons, Inc.

In This Chapter

● How Islam transformed a realm
● Where the oil is—and is not
● Foreign intervention, Muslim reaction
● Political theory and practical reality in Iraq
● The predicament of Israel
● Afghanistan and the War on Terror

FROM MOROCCO ON the shores of the Atlantic to the mountains of Afghanistan, and from the Horn of Africa to the steppes of inner Asia, lies a vast geographic realm of enormous cultural complexity. It stands at the crossroads where Europe, Asia, and Africa meet, and it is part of all three (Fig. 7-1). Throughout history, its influences have radiated to these continents and to practically every other part of the world as well. This is one of humankind's primary source areas. On the Mesopotamian Plain between the Tigris and Euphrates rivers (in modern-day Iraq) and on the banks of the Egyptian Nile arose several of the world's earliest civilizations. In its soils, plants were domesticated that are now grown from the Americas to Australia. Along its paths walked prophets whose religious teachings are still followed by billions. And at the opening of the twenty-first century, the heart of this realm is beset by some of the most bitter and dangerous conflicts on Earth.

Defining the Realm

Even the geographic name we assign to this realm is cumbersome, signifying that it encompasses large parts of two continents. Some regional geographers use the shorthand NASWA for it, a term that has found its way into the literature. Look again at Figure G-2, and you will see that this realm has an unmatched property: *centrality*. It lies at the core of the inhabited world, close to the heart of the Land Hemisphere. This geographic realm has more neighbors than any other. And, as we will see, its impact on the rest of the world continues in new ways.

Given its sprawling and varied geography, it is not surprising that this realm is sometimes characterized in a few words that purport to stress one or more of its dominant features. None of these is satisfactory, but all hold a kernel of truth. Here are four of them.

A "Dry World"?

Compare Figures 7-1 and G-8, and it appears that NASWA's expanse matches almost exactly that of the largest concentration of *BW* and *BS* climates, containing as it does the Sahara as well as the Arabian Desert. But add Figure G-9 to the picture, and it is clear that most of the realm's nearly

---

MAJOR GEOGRAPHIC QUALITIES OF

North Africa/Southwest Asia

1. North Africa and Southwest Asia were the scene of several of the world's great ancient civilizations, based in its river valleys and basins.

2. From this realm's culture hearths diffused ideas, innovations, and technologies that changed the world.

3. The North Africa/Southwest Asia realm is the source of three world religions: Judaism, Christianity, and Islam.

4. Islam, the last of the major religions to arise in this realm, transformed, unified, and energized a vast domain extending from Europe to Southeast Asia and from Russia to East Africa.

5. Drought and unreliable precipitation dominate natural environments in this realm. Population clusters exist where water supply is adequate to marginal.

6. Certain countries of this realm have enormous reserves of oil and natural gas, creating great wealth for some but doing little to raise the living standards of the majority.

7. The boundaries of the North Africa/Southwest Asia realm consist of volatile transition zones in several places in Africa and Asia.

8. Conflict over water sources and supplies is a constant threat in this realm, where population growth rates are high by world standards.

9. The Middle East, as a region, lies at the heart of this realm; and Israel lies at the center of the Middle East conflict.

10. Religious, ethnic, and cultural discord frequently cause instability and strife in this realm.

600 million people live where there is surface water—in the Nile Valley and Delta, along northwestern Africa's hilly Mediterranean coast (the *tell*, meaning "mound" in Arabic), along the eastern and northeastern shores of the Mediterranean Sea, in the basin of the Tigris and Euphrates rivers, in the uplands of western Iran and eastern Turkestan, and in far-flung oases throughout the realm. It is true that, in this realm, water is almost always at a premium (it is, for example, a major issue between the Israelis, who use far more of it per person, and their neighbors the Palestinians), and that millions of peasants struggle to make soil and moisture yield a small harvest. In one of humanity's ancient adaptations, nomadic peoples and their animals still circulate across dust-blown flatlands, stopping at oases that form islands of farming and trade in the vast drylands, but they are now a tiny—and dwindling—minority. Ancient as well as modern technology have enabled nearly 10 percent of the planet's population to defy these environmental limitations. Dams, canals, and wells bring water from rivers, snowy mountaintops, and aquifers (natural underground reservoirs) to hundreds of millions of people on the land as well as in the great cities. And in some countries desalinization plants paid for with oil money are making life in the "dry world" possible in ways unimagined just decades ago.

The "Middle East"?

This realm is also frequently called the Middle East. That must sound odd to someone in, say, India, who might think of a Middle West rather than a Middle East! The name, of course, reflects the biases of its source: the "Western" world, which saw a "Near" East in Turkey, a "Middle" East in Egypt, Arabia, and Iraq, and a "Far" East in China and Japan. Still, the term has taken hold, and it can be seen and heard in everyday usage by scholars, journalists, and members of the United Nations. Even so, it should be applied to only one of the regions of this vast realm, not to the realm as a whole. It is likely that the entire realm's centrality, referred to earlier, contributes to this misuse: look again at Figure G-13 and you will note that not only does NASWA lie at the heart of the inhabited world, but the *Middle East* region lies at the core of NASWA.

An "Arab World"?

North Africa/Southwest Asia is often referred to by another appellation: the Arab World. This term implies a uniformity that does not actually exist. First, the name *Arab* is applied loosely to the peoples of this area who speak Arabic and related languages, but ethnologists normally restrict it to certain occupants of the Arabian Peninsula—the Arab "source." In any case, the Turks are not Arabs, and neither are most Iranians or Israelis. Moreover, although the Arabic language prevails from Mauritania in the west across all of North Africa to the Arabian Peninsula, Syria, and Iraq in the east, it is not spoken in other parts of this realm (see Fig. G-11). In Turkey, for example, Turkish is the major language, and it has Ural-Altaic rather than Arabic's Semitic or Hamitic roots. The Iranian language belongs to the Indo-European linguistic family. Other "Arab World" languages that have separate ethnological identities are spoken by the Jews of Israel, the Tuareg people of the Sahara, the Berbers of northwestern Africa, and the peoples of the transition zone between North Africa and Subsaharan Africa to the south.

An "Islamic World"?

Finally, and perhaps most significantly, this realm is routinely referred to as the Islamic World. Indeed, the prophet Muhammad was born on the Arabian Peninsula in AD 571, and Islam, the religion to which he gave rise, spread across this realm and beyond after his death in 632. This great saga of Arab conquest saw Islamic armies, caravans, and fleets carry Muhammad's teachings across deserts, mountains, and oceans, penetrating Europe, converting the kings of West African states, threatening the Christian strongholds of Ethiopia, invading central Asia, conquering most of what is now India, and even reaching the peninsulas and islands of Southeast Asia. Today, the world's largest Muslim state is Indonesia, and the Islamic faith extends far outside the realm under discussion (Fig. 7-2). In this context, the term *Islamic World* is also misleading in that it suggests that there is no Islam beyond.

In any case, this World of Islam is not entirely Muslim either. Christian minorities continue to survive in all the regions and many of the countries of the NASWA realm. Judaism has its base in the Middle East region; and smaller religious communities, such as Lebanon's Druze and Iran's Bahais, many of them under pressure from Islamic regimes, diversify the religious mosaic.

Nevertheless, Islam's impact on this realm's cultural geography is pervasive. It ranges from dominant in such countries as Iran, Saudi Arabia, and Sudan to laminate (that is, more of a veneer) in several of the countries of Turkestan, where the former Soviet rulers discouraged the faith and where its comeback is still in progress—one of the criteria that allow us to identify Turkestan as a discrete region within this realm. The rise and impact of Islam lie at the heart of the story of this chapter.

States and Nations

Although Islam and its cultural expressions dominate this geographic realm, its political and social geographies are divided and often fractious. We will encounter states with strong internal divisions (Lebanon, Iraq), nations as yet without their own states (Palestinians, Kurds, Berbers), and territories still in the process of integration (Western Sahara, Palestine). No single state dominates this realm as Mexico eclipses the rest of Middle America or Brazil looms over South America. Of its nearly 30 states and territories, the three largest and in many ways the most important are Egypt in North Africa, Turkey on the threshold of Europe, and Iran at the margins of Turkestan. All three have populations between 70 and 80 million, modest by world standards. But at the other end of the continuum there are ministates such as Bahrain, Qatar, and Djibouti, each with fewer than one million inhabitants. It is important to study Figure 7-1 attentively, because the boundary framework—much of it inherited from the colonial era—lies at the root of many of the realm's current problems.

HEARTHS OF CULTURE

This geographic realm occupies a pivotal part of the world: here Eurasia, crucible of human cultures, meets Africa, source of humanity itself. Two million years ago, the ancestors of our species walked from East Africa into North Africa and Arabia and spread from one end of Asia to the other. Less than one hundred thousand years ago, *Homo sapiens* crossed these lands on the way to Europe, Australia, and, eventually, the Americas. Ten thousand years ago, human communities in what we now call the Middle East began to domesticate plants and animals, learned to irrigate their fields, enlarged their settlements into towns, and formed the earliest states. One thousand years ago, the heart of the realm was stirred and mobilized by the teachings of Muhammad and the Quran (Koran), and Islam was on the march from North Africa to India. Today this realm is a cauldron of religious and political activity, weakened by conflict but empowered by oil, plagued by poverty but fired by a wave of religious fundamentalism (Muslims prefer the term *religious revivalism*).

Dimensions of Culture

In the Introduction (pp. 18–23) we discussed the concept of culture and its regional expression in the cultural landscape. **1** **Cultural geography**, we noted, is a wide-ranging

and comprehensive field that studies spatial aspects of human cultures, focusing not only on cultural landscapes but also on **culture hearths**—the crucibles of civilization, the sources of ideas, innovations, and ideologies that changed regions and realms. Those ideas and innovations spread far and wide through a set of processes that we study under the rubric of **cultural diffusion**. Because we understand these processes better today, we can reconstruct ancient routes by which the knowledge and achievements of culture hearths spread (that is, *diffused*) to other areas. Another topic of cultural geography, also relevant in the context of the North Africa/Southwest Asia realm, is the **cultural environment** that a dominant culture creates. Human cultures exist in long-term accommodation with (and adaptation to) their natural environments, exploiting opportunities that these environments present and coping with the extremes they can impose. The study of the relationship between human societies and natural environments has become a separate branch of cultural geography called **cultural ecology**. As we will see, the North Africa/Southwest Asia realm presents many opportunities to investigate cultural geography in regional settings.

Rivers and Communities

In the basins of the major rivers of this realm (the Tigris and Euphrates of modern-day Turkey, Syria, and Iraq, and the Nile of Egypt) lay two of the world's earliest culture hearths (Fig. 7-3). *Mesopotamia*, "land amidst the rivers," had fertile alluvial soils, abundant sunshine, ample water, and animals and plants that could be domesticated. Here, in the Tigris-Euphrates lowland between the Persian Gulf and the uplands of present-day Turkey, arose one of humanity's first culture hearths, a cluster of communities that grew into larger societies and, eventually, into the world's first states. (Early state development probably was going on simultaneously in East Asia's river basins as well.) Mesopotamians were innovative farmers who knew when to sow and harvest crops, water their fields, and store their surplus. Their knowledge diffused to villages near and far, and a *Fertile Crescent* evolved extending from Mesopotamia across southern Turkey into Syria and the Mediterranean coast beyond (Fig. 7-3).

Mesopotamia

Irrigation was the key to prosperity and power in Mesopotamia, and urbanization was its reward. Among many settlements in the Fertile Crescent, some thrived, grew, enlarged their hinterlands, and diversified socially and occupationally; others failed. What determined success?

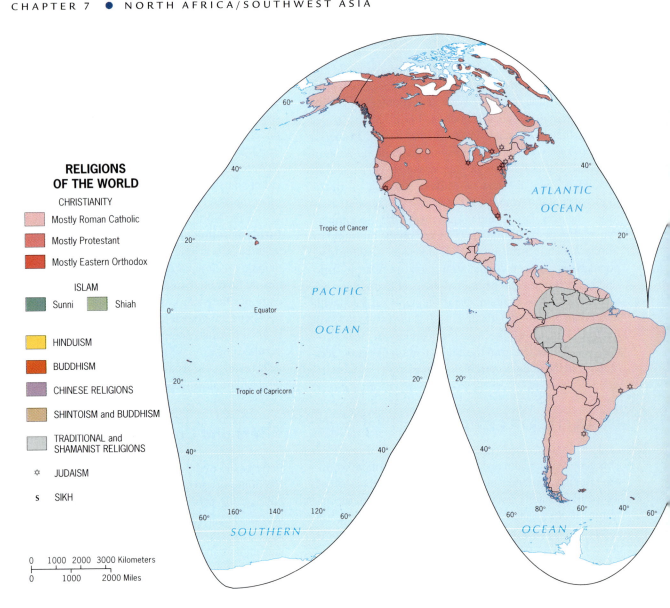

RELIGIONS OF THE WORLD

CHRISTIANITY
- Mostly Roman Catholic
- Mostly Protestant
- Mostly Eastern Orthodox

ISLAM
- Sunni
- Shiah

- HINDUISM
- BUDDHISM
- CHINESE RELIGIONS
- SHINTOISM and BUDDHISM
- TRADITIONAL and SHAMANIST RELIGIONS
- ✡ JUDAISM
- s SIKH

0 1000 2000 3000 Kilometers

0 1000 2000 Miles

FIGURE 7-2   © H. J. de Blij, P. O. Muller, and John Wiley & Sons, Inc.

6 One theory, the **hydraulic civilization theory**, holds that cities that could control irrigated farming over large hinterlands held power over others, used food as a weapon, and thrived. One such city, Babylon on the Euphrates River, endured for nearly 4000 years (from 4100 BC). A busy port, its walled and fortified center endowed with temples, towers, and palaces, Babylon for a time was the world's largest city.

Egypt and the Nile

Egypt's cultural evolution may have started even earlier than Mesopotamia's, and its focus lay upstream from (south of) the Nile Delta and downstream from (north of) the first of the Nile's series of rapids, or cataracts (Fig. 7-3). This part of the Nile Valley is surrounded by inhospitable desert, and unlike Mesopotamia (which lay open to all comers), the Nile provided a natural fortress here. The ancient Egyptians converted their security into progress. The Nile was their highway of trade and interaction; it also supported agriculture through irrigation. The Nile's cyclical ebb and flow was much more predictable than that of the Tigris-Euphrates river system. By the time Egypt finally fell victim to outside invaders (about 1700 BC), a full-scale urban civilization had emerged. Ancient Egypt's artist-engineers left a magnificent legacy of massive stone monuments, some of them containing treasure-filled crypts of

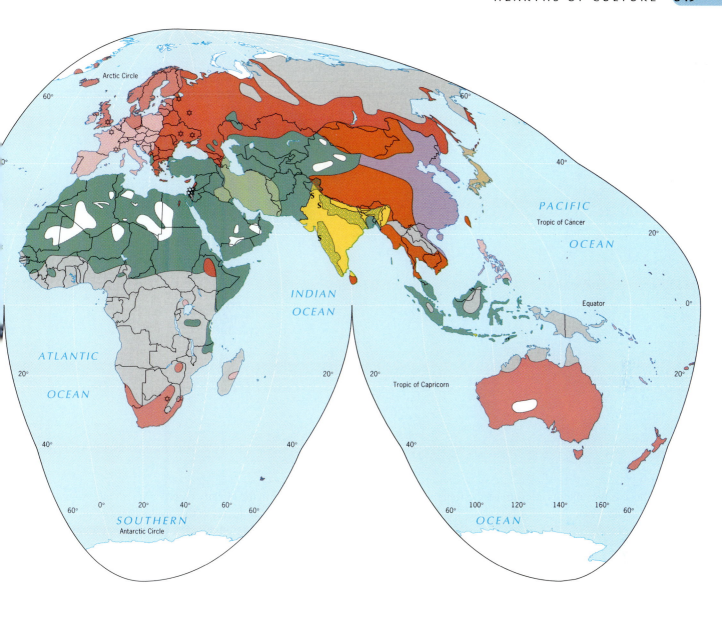

god-kings called pharaohs. These tombs have enabled archeologists to reconstruct the ancient history of this culture hearth.

The Indus Valley

To the east, separated from Mesopotamia by more than 1900 kilometers (1200 mi) of mountain and desert, lies the Indus Valley (Fig. 7-3). By modern criteria, this eastern hearth lies outside the realm under discussion here; but in ancient times it had cultural and commercial ties with the Tigris-Euphrates region. Mesopotamian innovations reached the Indus region early, and eventually the

cities of the Indus became power centers of a civilization that extended far into present-day northern India.

Today, the world continues to benefit from the accomplishments of the ancient Mesopotamians and Egyptians. They domesticated cereals (wheat, rye, barley), vegetables (peas, beans), fruits (grapes, apples, peaches), and many animals (horses, pigs, sheep). They also advanced the study of the calendar, mathematics, astronomy, government, engineering, metallurgy, and a host of other skills and technologies. In time, many of their innovations were adopted and then modified by other cultures in the Old World and eventually in the New World as well. Europe was the greatest beneficiary of these legacies of Mesopotamia and ancient

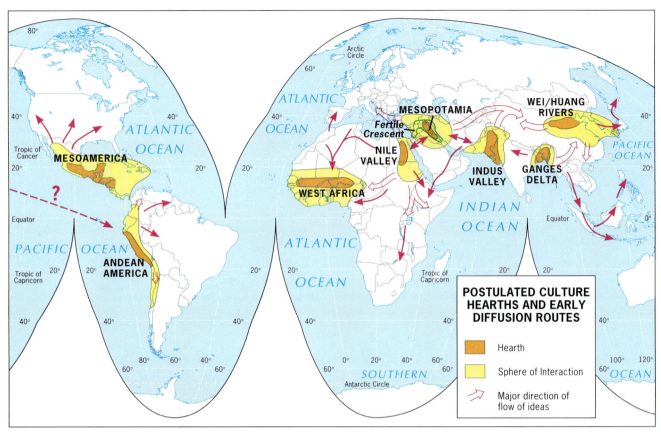

FIGURE 7-3 © H. J. de Blij, P. O. Muller, and John Wiley & Sons, Inc.

Egypt, whose achievements constituted the foundations of "Western" civilization.

Decline and Decay

As Figure G-8 reminds us, many of the early cities of this culture realm lay in what is today desert territory. Assuming that there were no good reasons to build large settlements in the middle of deserts, we may hypothesize that **climate change** sweeping over this region, not a monopoly over irrigation techniques, gave certain cities in the ancient Fertile Crescent an advantage over others. Climate change, associated with shifting environmental zones after the final Pleistocene glacial retreat, may have destroyed the last of the old civilizations. Perhaps overpopulation and human destruction of the natural vegetation contributed to the process. Indeed, some cultural geographers suggest that the momentous innovations in agricultural planning and irrigation technology were not "taught" by the seasonal flooding of the rivers but were forced on the inhabitants as they tried to survive changing environmental conditions.

The scenario is not difficult to imagine. As outlying areas began to fall dry and farmlands were destroyed, peo-ple congregated in the already crowded river valleys—and made every effort to increase the productivity of the land that could still be watered. Eventually overpopulation, destruction of the watershed, and perhaps reduced rainfall in the rivers' headwater areas dealt the final blow. Towns were abandoned to the encroaching desert; irrigation canals filled with drifting sand; croplands dried up. Those who could migrated to areas that were still reputed to be productive. Others stayed, their numbers dwindling, increasingly reduced to subsistence.

As old societies disintegrated, power emerged elsewhere. First the Persians, then the Greeks, and later the Romans imposed their imperial designs on the tenuous lands and disconnected peoples of North Africa/Southwest Asia. Roman technicians converted North Africa's farmlands into irrigated plantations whose products went by the boatload to Roman Mediterranean shores. Thousands of people were carried off as slaves to the cities of the new conquerors. Egypt was quickly colonized, as was the area we now call the Middle East. One region that lay distant, and therefore remote from these invasions, was the Arabian Peninsula, where no major culture hearth or large cities had emerged and where the turmoil had not affected Arab settlements and nomadic routes.

STAGE FOR ISLAM

In a remote place on the Arabian Peninsula, where the foreign invasions of the Middle East had had little effect on the Arab communities, an event occurred early in the seventh century that was to change history and affect the destinies of people in many parts of the world. In a town called Mecca (Makkah), about 70 kilometers (45 mi) from the Red Sea coast in the Jabal Mountains, a man named Muhammad in the year AD 611 began to receive revelations from Allah (God). Muhammad (571–632) was then in his early forties and had barely 20 years to live. Convinced after some initial self-doubt that he was indeed chosen to be a prophet, Muhammad committed his life to fulfilling the divine commands he believed he had received. Arab society was in social and cultural disarray, but Muhammad forcefully taught Allah's lessons and began to transform his culture. His personal power soon attracted enemies, and in 622 he fled from Mecca to the safer haven of Medina (Al Madinah), where he continued his work. This moment, the *hejira* ("migration"), marks the starting date of the Muslim era, Year 1 on Islam's calendar. Mecca, of course, later became Islam's holiest place.

The Faith

The precepts of Islam in many ways constituted a revision and embellishment of Judaic and Christian beliefs and traditions. All of these faiths have but one god, who occasionally communicates with humankind through prophets; Islam acknowledges that Moses and Jesus were such prophets but considers Muhammad to be the final and greatest prophet. What is Earthly and worldly is profane; only Allah is pure. Allah's will is absolute; Allah is omnipotent and omniscient. All humans live in a world that Allah created for their use but only to await a final judgment day.

Islam brought to the Arab World not only the unifying religious faith it had lacked but also a new set of values, a new way of life, a new individual and collective dignity. Islam dictated observance of the Five Pillars: (1) repeated expressions of the basic creed, (2) the daily prayer, (3) a month each year of daytime fasting (Ramadan), (4) the giving of alms, and (5) at least one pilgrimage in each Muslim's lifetime to Mecca (the *hajj*). And Islam prescribed and proscribed in other spheres of life as well. It forbade alcohol, smoking, and gambling. It tolerated polygamy, although it acknowledged the virtues of monogamy. Mosques appeared in Arab settlements, not only for the (Friday) sabbath prayer, but also as social gathering places to knit communities closer together. Mecca became the spiritual center for a divided, widely dispersed people for whom a collective focus was something new.

The Arab-Islamic Empire

Muhammad provided such a powerful stimulus that Arab society was mobilized almost overnight. The prophet died in 632, but his faith and fame spread like wildfire. Arab armies carrying the banner of Islam formed, invaded, conquered, and converted wherever they went. As Figure 7-4 shows, by AD 700 Islam had reached far into North Africa, into Transcaucasia, and into most of Southwest Asia. In the centuries that followed, it penetrated Southern and Eastern Europe, Central Asia's Turkestan, West Africa, East Africa, and South and Southeast Asia, even reaching China by AD 1000.

Routes of Diffusion

The spread of Islam provides a good illustration of a series of processes called **spatial diffusion**, focusing on the way ideas, inventions, and cultural practices propagate through a population in space and time. In 1952, Torsten Hägerstrand, a Swedish geographer, published a fundamental study of spatial diffusion entitled *The Propagation of Diffusion Waves*. He reported that diffusion takes place in two forms: **expansion diffusion**, when propagation waves originate in a strong and durable source area and spread outward, affecting an ever larger region and population; and **relocation diffusion**, in which an innovation, idea, or (for example) a virus is carried by migrants from the source to distant locations and diffuses from there. The recent global spread of AIDS is a case of relocation diffusion.

Islam on the March

Both expansion and relocation diffusion include several types of processes. The spread of Islam, as Figure 7-4 shows, initially proceeded by a form of expansion diffusion called **contagious diffusion** as the faith moved from village to village across the Arabian Peninsula and the Middle East. But Islam got powerful boosts when kings, chiefs, and other high officials were converted, who in turn propagated the faith downward through their bureaucracies to far-flung subjects. This form of expansion diffusion, called **hierarchical diffusion**, served Islam well.

But the map leaves no doubt that Islam later spread by relocation diffusion as well, notably to the Ganges Delta in South Asia, to present-day Indonesia, and to East Africa. That process continues to this day, but the heart of Islam remains in Southwest Asia. There, Islam became

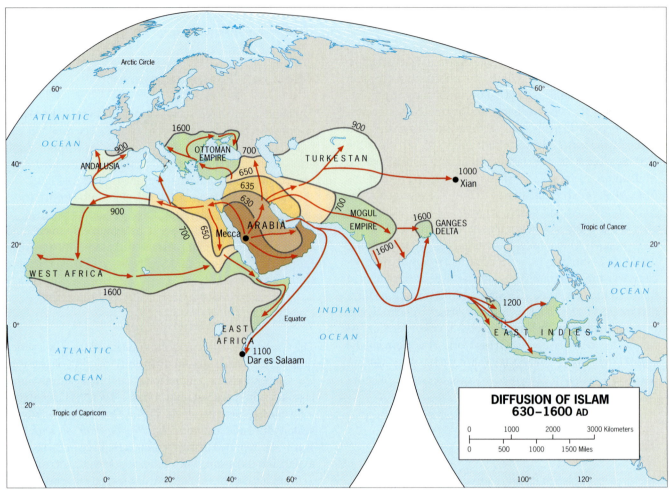

FIGURE 7-4   © H. J. de Blij, P. O. Muller, and John Wiley & Sons, Inc.

the cornerstone of an Arab Empire with Medina as its first capital. As the empire grew by expansion diffusion, its headquarters were relocated from Medina to Damascus (in present-day Syria) and later to Baghdad on the Tigris River (Iraq). And it prospered. In architecture, mathematics, and science, the Arabs overshadowed their European contemporaries. The Arabs established institutions of higher learning in many cities including Baghdad, Cairo, and Toledo (Spain), and their distinctive cultural landscapes united their vast domain (see box titled "The Flowering of Islamic Culture"). Non-Arab societies in the path of the Muslim drive were not only Islamized, but also Arabized, adopting other Arab traditions as well. Islam had spawned a culture; it still lies at the heart of that culture today.

As we noted, Islam's expansion eventually was checked in Europe, Russia, and elsewhere. But a map showing the total area under Muslim sway in Eurasia and Africa reveals the enormous dimensions of the domain affected by **13** **Islamization** at one time or another (Fig. 7-5). Islam continues to expand, now mainly by relocation diffusion. There are Islamic communities in cities as widely scattered

as Vienna, Singapore, and Cape Town, South Africa; Islam is also growing rapidly in the United States. With more than 1.4 billion adherents today, Islam is a vigorous and burgeoning cultural force around the world.

ISLAM DIVIDED

For all its vigor and success, Islam still fragmented into sects. The earliest and most consequential division arose after Muhammad's death. Who should be his legitimate successor? Some believed that only a blood relative should follow the prophet as leader of Islam. Others, a majority, felt that any devout follower of Muhammad was qualified. The first chosen successor was the father of Muhammad's wife (and thus not a blood relative). But this did not satisfy those who wanted to see a man named Ali, a cousin of Muhammad, made *caliph* (successor). When Ali's turn came, his followers, the *Shi'ites*, proclaimed that Muhammad finally had a legitimate successor. This offended the *Sunnis*, those

The Flowering of Islamic Culture

THE CONVERSION OF a vast realm to Islam was not only a religious conquest: it was also accompanied by a glorious explosion of Arab culture. In science, the arts, architecture, and other fields, Arab society far outshone European society. While the Western European remnants of the Roman Empire languished, Arab energies soared.

When the wave of Islamic diffusion reached the Maghreb (western North Africa), the Arabs saw, on the other side of the narrow Strait of Gibraltar, an Iberia ripe for conquest and ready for renewal. An Arab-Berber alliance, called the Moors, invaded Spain in 711 and controlled all but northern Castile and Catalonia before the end of the eighth century.

It took seven centuries for Catholic armies to recapture all of the Arabs' Iberian holdings, but by then the Muslims had made an indelible imprint on the Spanish and Por-

tuguese cultural landscape. The Arabs brought unity and imposed the rule of Baghdad, and their works soon overshadowed what the Romans had wrought.

Al-Andalus, as this westernmost outpost of Islam was called, was endowed with thousands of magnificent castles, mosques, schools, gardens, and public buildings. The ultimately victorious Christians destroyed most of the less durable art (pottery, textiles, furniture, sculpture) and burned the contents of Islamic libraries, but the great Islamic structures survived, including the Alhambra in Granada, the Giralda in Seville, and the Great Mosque of Córdoba, three of the world's greatest architectural achievements. While Spanish and Portuguese culture became Hispanic-Islamic culture, the Muslims were transforming their cities from Turkestan to the Maghreb in the image of Baghdad. The lost greatness of a past era still graces those townscapes today.

FROM THE FIELD NOTES

"Seville was a Muslim capital of Iberia, and walking the streets of the old city center is an adventure in cultural and historical geography. Seville fell to the Muslim invaders in 711, and the Muslims built a large mosque in the heart of town. Later the Christians, who ousted the Muslims in 1248, destroyed most of the mosque and built a huge cathedral on the site, but they saved the ornate minaret, made it the bell tower, and called it the Giralda (left). This minaret had been built by the Muslims to match the 67-meter (220-ft) minaret of the Koutoubia Mosque in Marrakech (photo p. 371). But the finest Islamic legacy in Seville surely is the Alcazar Palace (above), begun in the late twelfth century and embellished and finished by the Christians after their victory. Its arched façades and intricately carved walls remain a monument to the Muslim architects and artists who designed and created them." © H. J. de Blij

www.conceptcaching.com

Concept Caching

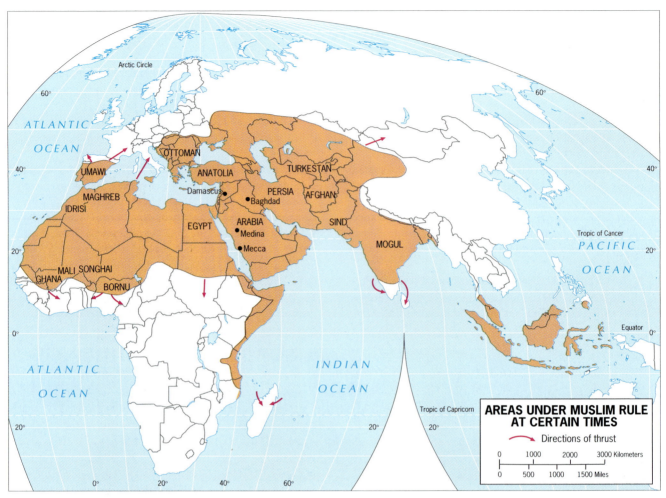

FIGURE 7-5 © H. J. de Blij, P. O. Muller, and John Wiley & Sons, Inc.

who did not see a blood relationship as necessary for the succession. From the beginning of this disagreement, the numbers of Muslims who took the Sunni side far exceeded those who regarded themselves as Shia (followers) of Ali. The great expansion of Islam was largely propelled by Sunnis; the Shi'ites survived as small minorities scattered throughout the realm. Today, about 85 percent of all Muslims are Sunnis.

The Strength of Shi'ism

But the Shi'ites vigorously promoted their version of the faith. In the early sixteenth century their work paid off: the royal house of Persia (modern-day Iran) made Shi'ism the only legal religion throughout its vast empire. That domain extended from Persia into lower Mesopotamia (modern Iraq), into Azerbaijan, and into western Afghanistan and Pakistan. As the map of religions (Fig. 7-2) shows, this created for Shi'ism a large **culture region** and gave the faith unprecedented strength. Iran remains the bastion of Shi'ism

in the realm today, and the appeal of Shi'ism continues to radiate into neighboring countries and even farther afield.

During the late twentieth century, Shi'ism gained unprecedented influence in the realm. In its heartland, Iran, a **shah** (king) tried to secularize the country and to limit the power of the **imams** (mosque officials); he provoked a revolution that cost him the throne and made Iran a Shi'ite Islamic republic. Before long, Iran was at war with neighboring, Sunni-ruled Iraq, and Shi'ite parties and communities elsewhere were invigorated by the new-found power of Shi'ism. From Arabia to Africa's northwestern corner, Sunni-ruled countries warily watched their Shi'ite minorities, newly imbued with religious fervor. Mecca, the holy place for both Sunnis and Shi'ites, became a battleground during the week of the annual pilgrimage, and for a time the (Sunni) Saudi Arabian government denied entry to Shi'ite pilgrims. Recently, that schism has healed somewhat, but intra-Islamic sectarian differences run deep.

Smaller Islamic sects further diversify this realm's religious landscape. Some of these sects play a disproportionately large role in the societies and countries of which they are a part, as we will see in our regional discussion.

Religious Revivalism in the Realm

Another cause of intra-Islamic conflict lies in the resurgence of religious fundamentalism, or as Muslims refer to **15** it, **religious revivalism**. In the 1970s, the imams in Shi'ite Iran wanted to reverse the shah's moves toward liberalization and secularization: they wanted to (and did) recast society in traditional, revivalist Islamic molds. An *ayatollah* (leader under Allah) replaced the shah in 1979; Islamic rules and punishments were instituted. Urban women, many of whom had been considerably liberated and educated during the shah's regime, resumed more traditional Islamic roles. Vestiges of Westernization, encouraged by the shah, disappeared. Under these new conditions, even the war against Iraq (1980–1990), which began as a conflict over territory, became a holy war that cost more than a million lives.

Islamic revivalist fundamentalism did not rise in Iran alone, nor was it confined to Shi'ite communities. Many Muslims—Sunnis as well as Shi'ites—in all parts of the realm disapproved of the erosion of traditional Islamic values, the corruption of society by European colonialists and later by Western modernizers, and the declining power of the faith in the secular state. As long as economic times were good, such dissatisfaction remained submerged. But when jobs were lost and incomes declined, a return to fundamental Islamic ways became more appealing.

Muslim Against Muslim

This set Muslim against Muslim in all the regions of the realm (see the Issue Box entitled "Islam in the Twenty-First Century: Revive or Reform?"). Revivalists fired the faith with a new militancy, challenging the status quo from Afghanistan to Algeria. The militants forced their governments to ban "blasphemous" books, to resegregate the sexes in schools, to enforce traditional dress codes, to legitimize religious-political parties, and to heed the wishes of the *mullahs* (teachers of Islamic ways). Militant Muslims proclaimed that democracy inherited from colonialists and adopted by Arab nationalists was incompatible with the rules of the Quran (Koran).

The rift between moderate Muslims and militant revivalists began to spill over into other parts of the world decades ago, from Pakistan to the African Transition Zone and from the Caucasus to the Philippines. Even while Islamic countries such as Iran and Algeria struggled to overcome the instability and damage arising from their internal conflicts, militant activists planned and carried out attacks against the allies of the moderates—oil-guzzling Western countries guilty of military intervention, propping up unrepresentative regimes, mistreatment of immigrant Muslims, and cultural corruption. Before the worst of these attacks (through mid-2007) destroyed the World Trade Center in New York and severely damaged the Pentagon outside Washington, D.C., the French had foiled a similar assault on the Eiffel Tower in Paris, the Russians had suffered hundreds of casualties in Moscow and Transcaucasia, and U.S. targets were hit in the Middle East, the Arabian Peninsula, East Africa, and elsewhere.

Back to Basics

Although it has been argued that these attacks do not represent Islam as a religion in conflict with the West, they do have a religious context. During the 1990s, following the Cold War defeat of Soviet forces in Afghanistan and the subsequent abandonment of that country by the United States and its allies, an organization named al-Qaeda emerged, dedicated to punishing the perceived enemies of Islamic peoples by the most effective means at its disposal: terrorist attacks. However, the leader and chief financier of this movement, Usama bin Laden, had a bigger agenda. His ultimate goal was to overthrow the regime of rich princes ruling his home country, Saudi Arabia, accusing them of collusion with the infidel United States, of apostasy (faithlessness), and worse. Bin Laden wanted to make Saudi Arabia, birthplace of Muhammad and guarantor of holy Mecca, a true Islamic state.

Bin Laden was not the first Saudi to wish to return his country to its Islamic roots. During the eighteenth century an Islamic theologian named Muhammad ibn Abd al-Wahhab (1703–1792) founded a movement to bring Islamic society on the peninsula back to the most fundamental, traditional, and puritanical form of the faith. So strict were his teachings that he was expelled from his home town in 1744—a fateful exile because he moved to a place that happened to be the headquarters of Ibn Saud, ruler over a sizeable swath of the Arabian Peninsula. Ibn Saud formed an alliance with al-Wahhab, who had inherited a fortune when his wife died; between them, they set out on a campaign of conquest that made the Saudi dynasty the rulers of the region and **Wahhabism** the dominant form of Islam. **16** Even the relatively liberal Ottoman sultans could not suppress either the Saudis or the Wahhabis, and in 1932, when the modern Kingdom of Saudi Arabia was established, Wahhabism became its official and dominant form of Islam.

Despite its reference to the name of the founder, Wahhabism today is a term used mostly by non-Muslims and outsiders. Followers of Wahhabi doctrines call themselves *Muwahhidun* or "Unitarians" to signify the strict and fundamental nature of their beliefs. These are the doctrines

Islam in the Twenty-First Century: Revive or Reform?

ISLAMIC REVIVAL IS THE ONLY WAY

"We Muslims must return to, and protect, the fundamental tenets of our faith. Too many of us have strayed, lured and seduced by the vulgarities of Western and other foreign ways. Consider our history. The Prophet set us on course, the Quran was and is the source of all truth, and the message took root from Mecca to Morocco to Malaya. Don't forget: Allah's revelations came many centuries after other religions—Hinduism, Buddhism, Christianity—had captured the minds of millions. But many saw that Islam is the only way, and today practicing Muslims (which means every believer) outnumber Christians, millions of whom never see the inside of a church and do not live pious lives.

"As a devout Muslim living in Egypt and having suffered the consequences of my devotion, let me say this: we Muslims have tried to live harmoniously with other faiths. We not only brought science and enlightenment to the world we converted, but we lived in peace with non-Arabs. In Islamic Andalusia, Jewish communities were safe and secure. It was the Christian Catholics who invented the horrors of the Inquisition. From West Africa to Central Asia, we proved our willingness to live in peace with infidels. So what did we get in return? The 'three c's'—the crusades, colonialism, and commercial exploitation! From my American friends I understand that there's an expression, 'three strikes and you're out.' Well, you non-Muslims are out. And so are those Muslims who still cooperate with foreigners, including the president of my own country and the entire Saudi royal family. I agree with Sheikh Hamoud: whoever supports the infidel against Muslims is himself an infidel. When you sell your soul for oil dollars or foreign aid, you're subject to a fatwa.

"Yes, I'm a revivalist, or what you people call a fundamentalist. Our salvation lies in a return to the strictest rules of Islam. Every Muslim must adhere to every tenet of the faith. Youngsters should learn the Quran by heart: it is the source of all truth beyond which nothing else matters. Women should know and keep their place as it is accorded to them by the Quran. Sharia law should be imposed by all Islamic countries, so that the punishment of transgressors will serve as an example to others. We should dress conservatively, resist the degenerate temptations of other cultures, and follow the teachings of our religious leaders. Let me ask you this: are we better off today than when Sheikh al-Wahhab and Ibn Saud were forging the Islamic state of Arabia? I would say not. We have to go back to our roots to undo the damage that has been done to us.

"Ayatollah Khomeini's Islamic revolution failed in Iran, and our Algerian brothers were prevented from making Algeria an Islamic state, but these were just harbingers of things to come. Our goal is Saudi Arabia, where it all started fourteen centuries ago—and where we will make the new start that will transform the world."

ISLAM NEEDS A REFORMATION

"How is it that we Muslims, with our rich cultural and scientific heritage, wealth of oil, history of interaction with other societies, and central location in the world are held back today by autocratic, corrupt, and incompetent governments, interfering and medieval clerics, stultifying traditions, and internal discord? As a practicing Muslim who is also a realist, I can tell you from my perspective as a Tunisian working in the tourist industry that we need Islamic reform, not the revivalism about which I keep hearing.

"Let's be objective about this. Look at the world's social and economic statistics, and you'll see that countries where Islam is the dominant religion tend to rank near the bottom—not always, but far too often for this to be a coincidence. The world has been swept by a wave of democratization, but not here. The globe is in the grip of a revolution in information and communication technology, but not here. The planet has a mosaic of religions, each professing to be the real path to salvation, and each bedeviled by a fundamentalist fringe, but not here. Here the fundamentalists are a constituency, not a fringe. And they are using tactics ranging from intimidation to terrorism to maintain and increase their clout.

"It all boils down to this: freedom. We need the freedom to debate the issues that confront us without fear of fatwas or imprisonment. In no other religious society can you be sentenced to death for expressing an opinion about God or scriptures. It shouldn't happen here. We need freedom of the press and other media, which are more restricted here than anywhere else on Earth. We need freedom of association without fear of persecution. But most of all, we need the right to debate the role of religion in our societies. There is nothing incompatible between Islamic societies and others around the world, but you cannot designate every non-Muslim as an infidel and then expect to have cordial and trusting relations.

"Islam needs a reformation, a reinterpretation of its inspired writings in light of modern times. And the mullahs need to get out of the business of controlling government, which is their constant aim. The separation of mosque and state is indispensable to our progress.

"Our religion-suffused educational system produces graduates who are unable to use what they learned to enter the globalizing job market. I can visualize those kids in the Taliban's madrassas doing only one thing, day after day—'studying' to recite the Quran by heart. Even where schools don't go to that extreme, they tend to produce unemployable youngsters whose joblessness and frustration cause them to turn to the mullahs—who are all too happy to direct their anger toward the West.

"And our reformation will have to address the plight of women. As a woman, it infuriates me that half of all women in the Arab world can neither read nor write, that they have unequal citizenship, and that they play virtually no role in government. Nowhere in the world is the domination of men more absolute, and I refuse to believe that the Prophet would approve of this. We need a reformation, not a so-called revival."

Regional ISSUE

Vote your opinion at www.wiley.com/college/deblij

Usama bin Laden accuses the Saudi princes of having violated, and these are the harsh teachings so often cited in the Western media as emanating from the pulpits of Saudi Arabia's revivalist mosques.

Islam and Other Religions

Two additional major faiths had their sources in the Levant (the area extending from Greece eastward around the Mediterranean coast to northern Egypt), and both were older than Islam. Islam's rise submerged many smaller Jewish communities, but the Christians, not the Jews, waged centuries of holy war against the Muslims, seeking, through the Crusades, not only to drive Islam back but to reestablish Christian communities where they had dominated before the Islamic expansion. The aftermath of that campaign still marks the realm's cultural landscape today. A substantial Christian minority (about one-fourth of the population) remains in Lebanon, and Christian minorities also survive in Israel, Syria, Egypt, and Jordan. Strained relations between the long-dominant Christian minority in Lebanon and the Muslim majority (itself divided into five sects) contributed to the disastrous armed struggle that engulfed this country in the 1970s and 1980s.

But the most intense conflict in modern times has pitted the Jewish state, Israel, against its Islamic neighbors near and far. Israel's United Nations-sponsored creation in 1948 precipitated more than a half-century of intermittent strife and attempts at mediation; it also caused friction among Islamic states in the region. Jerusalem—holy city for Judaism, Christianity, and Islam—lies in the crucible of this confrontation.

The Ottoman Aftermath

It is a geographic twist of fate that Islam's last great advance into Europe resulted in the European occupation of Islam's very heartland. The Ottomans (named after their leader, Osman I), based in what is today Turkey, conquered Constantinople (now Istanbul) in 1453 and pushed into Eastern Europe. Soon Ottoman forces were on the doorstep of Vienna; they also invaded Persia, Mesopotamia, and North Africa (Fig. 7-6). The Ottoman Empire under Suleyman the Magnificent, who ruled from 1522 to 1560, was the most

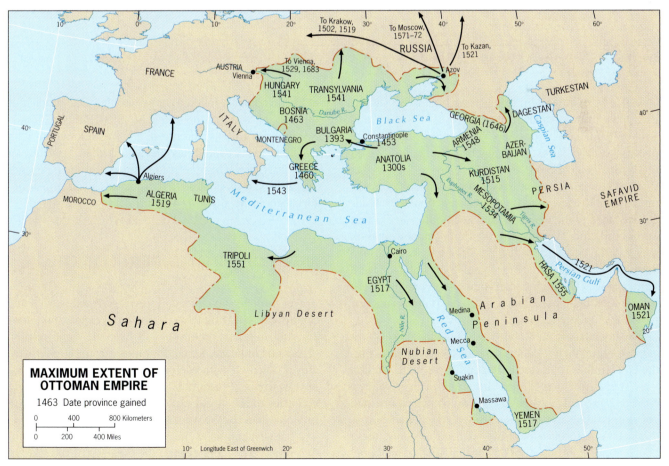

FIGURE 7-6 © H. J. de Blij, P. O. Muller, and John Wiley & Sons, Inc.

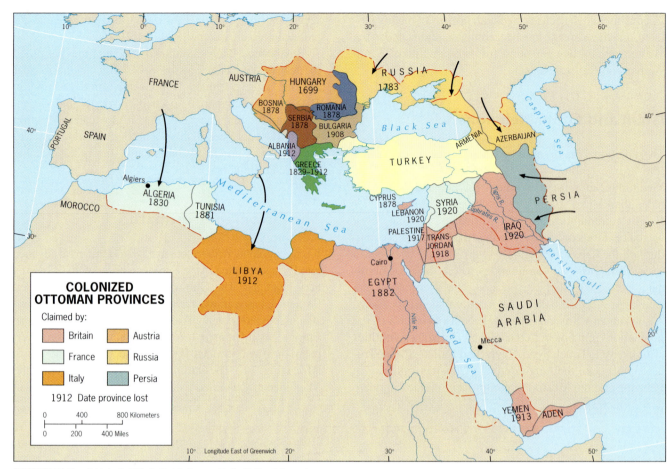

FIGURE 7-7 © H. J. de Blij, P. O. Muller, and John Wiley & Sons, Inc.

powerful state in western Eurasia. As Figure 7-6 shows, its armies even advanced toward Moscow, Kazan, and Krakow from their base at Azov (near present-day Rostov in Russia). The Turks also launched marine attacks on Sicily, Spain, and France.

The Ottoman Empire survived for more than four centuries, but it lost territory as time went on, first to the Hungarians, then to the Russians, and later to the Greeks and Serbs until, after World War I, the European powers took over its provinces and made them colonies—colonies we now know by the names of Syria, Iraq, Lebanon, and Yemen (Fig. 7-7). As the map shows, the French and the British took large possessions; even the Italians annexed part of the Ottoman domain.

The boundary framework that the colonial powers created to delimit their holdings was not satisfactory. As Figure G-9 reminds us, this realm's population of 581 million is clustered, fragmented, and strung out in river valleys, coastal zones, and crowded oases. The colonial powers laid out long stretches of boundary as ruler-straight lines across uninhabited territory; they saw no need to adjust these boundaries to cultural or physical features in the landscape. Other boundaries, even some in desert zones, were poorly

defined and never marked on the ground. One product of this process was Iraq, delimited by (then) British Colonial Secretary Winston Churchill in 1921 as a kingdom with Baghdad as its capital; another was Jordan, centered on Amman. Later, when the colonies had become independent states, such boundaries led to quarrels, even armed conflicts, among neighboring Muslim states.

THE POWER AND PERIL OF OIL

Travel through the cities and towns of North Africa and Southwest Asia, talk to students, shopkeepers, taxi drivers, and migrants, and you will hear the same refrain: "Leave us in peace, let us do things our traditional way. Our problems—with each other, with the world—seem always to result from outside interference. The stronger countries of the world exploit our weaknesses and magnify our quarrels. We want to be left alone."

That wish might be closer to fulfillment were it not for two relatively recent events: the creation of the state of Israel and the discovery of some of the world's largest oil reserves. We will discuss the founding of Israel in the regional section of this chapter. Here we focus on the realm's most valuable export product: oil.

Location and Dimensions of Known Reserves

About 30 of the world's countries have significant oil reserves. But five of these countries, all located in the North Africa/Southwest Asia realm, in combination possess larger reserves (ca. 77 percent of the world total) than the rest combined. The estimated reserves of this Big Five in 2005, averaged from several sources including the governments of the countries themselves and reported in billions of barrels, were as follows: (1) Saudi Arabia, 267; (2) Iraq, 114; (3) United Arab Emirates, 101; (4) Kuwait, 97; and (5) Iran, 91. The next-ranking country, Venezuela, had known reserves below 80 billion barrels; Russia below 60; the United States below 40; and Libya, Mexico, Nigeria, China, and Ecuador below 25. No other country's known reserves approached 20 billion barrels as of 2005. It should also be noted that Canada may possess the largest oil reserves of all—but most of this supply is contained in Alberta's *tar sands* (see p. 199), which requires complex and expensive recovery and refining processes. But with oil prices at unprecedentedly high levels, Canada's production is likely to grow.

In general terms, oil (and associated natural gas) exists in this realm in three discontinuous zones (Fig. 7-8). The most productive of these zones extends from the southern and southeastern part of the Arabian Peninsula northwestward around the rim of the Persian Gulf, reaching into Iran and continuing northward into Iraq, Syria, and southeastern Turkey, where it peters out. The second zone lies across North Africa and extends from north-central Algeria eastward across northern Libya to Egypt's Sinai Peninsula, where it ends. The third zone begins on the margins of the realm in eastern Azerbaijan, continues eastward under the Caspian Sea into Turkmenistan and Kazakhstan, and also reaches into Uzbekistan, Tajikistan, Kyrgyzstan, and Afghanistan.

Producers and Consumers

Saudi Arabia is the world's largest oil exporter, but in recent years Russia has risen to second place. As we noted in Chapter 2, oil and natural gas are by far Russia's most valuable commodities, and the Russian state direly needs the revenues they produce. Thus Russia, with modest (though expanding) known reserves, is vigorously exporting to international markets and is now a leading factor in the global energy picture. In combination, however, production by the countries of the North Africa/Southwest Asia realm far exceeds that from all other sources. The United States, not surprisingly, is the world's leading importer even while it consumes almost all of its own production.

As Figure G-12 indicates, the production and export of oil and gas has elevated several of this realm's countries into the higher-income categories. But petroleum wealth also has enmeshed these Islamic societies and their governments in global strategic affairs. When regional conflicts create instability in producing countries that have the potential to disrupt supply lines, powerful consumers are tempted to intervene—and have done so.

Colonial Legacy

When the colonial powers laid down the boundaries that partitioned this realm among themselves, no one knew about the riches that lay beneath the ground. A few wells had been drilled, and production in Iran had begun as early as 1908 and in Egypt's Sinai Peninsula in 1913. But the major discoveries came later, in some cases after the colonial powers had already withdrawn. Some of the newly independent countries, such as Libya, Iraq, and Kuwait, found themselves with wealth undreamed of when the Turkish Ottoman Empire collapsed. As Figure 7-8 shows, however, others were less fortunate. A few countries had (and still have) potential. The smaller, weaker emirates and sheikdoms on the Arabian Peninsula always feared that powerful neighbors would try to annex them (Kuwait faced this prospect in 1990 when Iraq invaded it). The unevenly distributed oil wealth, therefore, created another source of division and distrust among Islamic neighbors.

A Foreign Invasion

The oil-rich countries of the realm found themselves with a coveted energy source but without the skills, capital, or equipment to exploit it. These had to come from the Western world and entailed what many tradition-bound Muslims feared most: a strong foreign presence on Islamic soil, foreign intervention in political as well as economic affairs, and penetration of Islamic societies by the vulgarities of Western ways.

Many Muslims' worst fears came true in Iran in the 1950s, when the government of Iran tried to control British oil exploitation centered on Abadan and the Persian Gulf. The imperious ways of the British, the luxurious life of the European expatriates compared to the abject poverty of Iranian workers, and the unfair terms of the concession under which the oil was exported all contributed to the rise of nationalist sentiment in Iran, whose leaders appealed to the United States for help in negotiating a better deal with

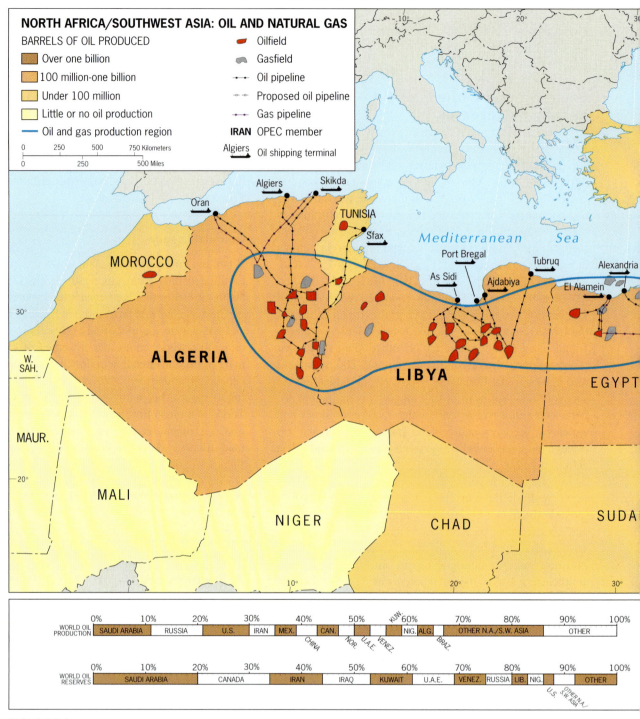

FIGURE 7-8 © H. J. de Blij, P. O. Muller, and John Wiley & Sons, Inc.

the British. American President Harry Truman had some sympathy for the Iranians, but his successor Dwight Eisenhower worried about Iran turning toward communism. Capitalizing on political disputes within Iran, CIA operatives engineered the overthrow of Premier Mohammad Mosaddeq, restoring to power the young shah and setting in motion a sequence of events that would lead to the demise of Iran's monarchy and the proclamation of an Islamic Republic in 1979.

Even when foreign intervention was more subtle, it intensified clashes between traditional and modern. Oil revenues created cultural landscapes in which gleaming skyscrapers towered over ornate mosques and historic neighborhoods. The social chasms between rich, well-connected elites and less fortunate citizens bred resentment. Involvement of Western armies in regional conflicts produced the unthinkable: foreign armies on Arabia's hallowed soil. To many Muslims, all this violated the most

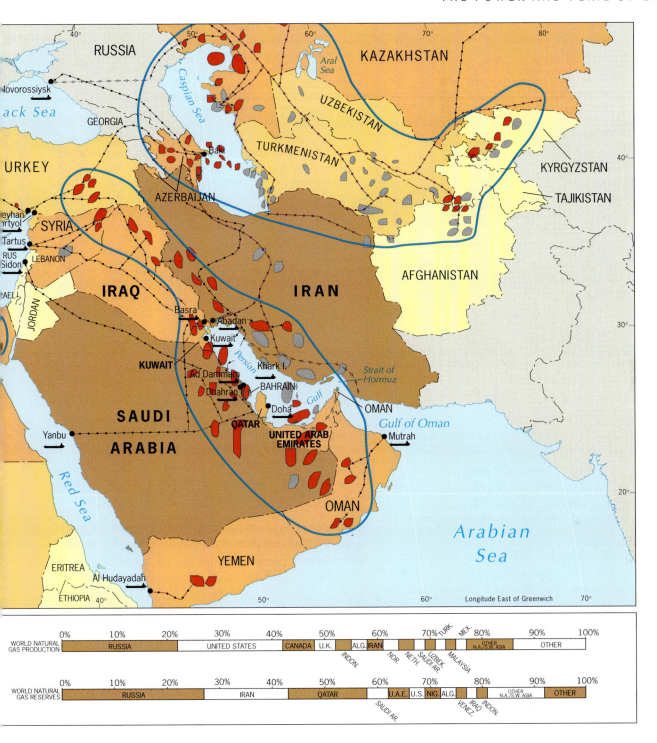

basic tenets of their faith, and some retaliated with violence of their own. That violence, in the form we define as terrorism, has become one of the challenges of our time.

Oil's Geographic Impact

In geographic terms, the impact of oil, its location, production, and sale, has produced massive change in this realm. Here are several key consequences:

1. **High Incomes.** When oil prices on international markets are high, several countries in this realm rank as the highest-income societies in the world. Even when oil prices decline, virtually all the petroleum-exporting states remain at least in the upper-middle-income category (Fig. G-12).

2. **Modernization.** Notably in Saudi Arabia but also in Kuwait and UAE (see photo p. 362), huge oil revenues have been used to modernize infrastructure. Futuristic

When it comes to Dubai's explosive (mostly vertical) growth, you have to be careful about estimates, but it is undoubtedly spectacular. According to press reports, Dubai's construction site is so large that 70 percent of the world's construction cranes are being used here (China may demand a recount). When Dubai's grand design reaches completion, United Arab Emirates sources say, it will "dwarf Manhattan." This already-dated photo is of the foundations of what will become the world's tallest building by far, the Burj Dubai, nearly four-fifths of a kilometer (a half-mile) high and twice as tall as the World Trade Center was. Encircling Dubai's high-rise center are ultramodern entertainment complexes, race tracks, golf courses, tennis pavilions—even an air-conditioned ski slope. Not quite the image the world has of the Arabian Peninsula region! © Matt Shonefeld/Redux Pictures.

city skyscrapers are only one manifestation of this; modern ports, airports, and superhighways are others.

3. *Industrialization.* Farsighted governments among those with oil wealth, realizing that petroleum reserves will not last forever, have invested some of their income in industrial plants that will outlast the era of oil. Petrochemical industries, plastics fabrication, and desalinization plants are among these facilities.

4. *Intra-Realm Migration.* The oil wealth has attracted millions of workers from less favored parts of the realm to work in the oilfields, in the ports, and in many other, mainly menial capacities. This has brought many Shi'ites to the countries of eastern Arabia; hundreds of thousands of Palestinians also work there as laborers. Saudi Arabia, with a population of 25.4 million, has more than 5 million foreign workers.

5. *Inter-Realm Migration.* The willingness of workers from such countries as Pakistan, India, and Sri Lanka to work for wages even lower than those the oil industry pays has attracted a substantial flow of temporary immigrants from outside the realm. These workers serve mostly as domestics, gardeners, refuse collectors, and the like.

6. *Regional Disparities.* Oil wealth and its manifestations in the cultural landscape create strong contrasts with areas not directly affected. The ultramodern east coast of Saudi Arabia is a world apart from large areas of its interior, where it becomes a land of desert, oasis, and camel, of vast distances, slow change, and isolated settlements. This phenomenon affects all oil-rich countries.

7. *Foreign Investment.* To some degree, governments and Arab businesspeople have invested oil-generated wealth in foreign countries. These investments have created a network of international involvement that links many of this realm's countries not only to the economies of foreign states, but also to growing Arab (and thus Islamic) communities in those states.

Oil brought this realm into contact with the outside world in ways unforeseen just a century ago. Oil has strengthened and empowered some of its peoples; it has dislocated and imperiled others. It has truly been a double-edged sword.

POINTS TO PONDER

● The world's tallest building by far, four-fifths of a kilometer (one half-mile) high, is rising over the Emirate of Dubai on the Arabian Peninsula (photo above).

● By the middle of 2007, as many as 3 million Iraqis had fled to neighboring countries and more than 2 million had become internal refugees as a result of the war started by the United States in 2003.

● Afghanistan still produces almost 90 percent of the world's annual harvest of opium, with revenues estimated to exceed half the country's GNI.

● The Aral Sea is coming back, in part as a result of a World Bank-financed 13-kilometer (8-mi)-long dike designed to initially raise the water level in its northern sector.

Regions of the Realm

Identifying and delimiting regions in this vast geographic realm is quite a challenge. Population clusters are widely scattered in some areas, highly concentrated in others. Cultural transitions and landscapes—internal as well as peripheral—make it difficult to discern a regional framework. This, furthermore, is a highly changeable realm and always has been. Several centuries ago it extended into Eastern Europe; now it reaches into Central Asia, where an Islamic **cultural revival** (the regeneration of a long-dormant culture through internal renewal and external infusion) is well under way.

17

The following are the regional components of this far-flung realm today (Fig. 7-9):

1. *Egypt and the Lower Nile Basin.* This region in many ways constitutes the heart of the realm as a whole. Egypt (together with Iran and Turkey) is one of the realm's three most populous countries. It is the historic focus of this part of the world and a major political and cultural force. It shares with its southern neighbor, Sudan, the waters of the lower Nile River.

2. *The Maghreb and Its Neighbors.* Western North Africa (the Maghreb) and the areas that border it also form a region, consisting of Algeria, Tunisia, and Morocco at the center, and Libya, Chad, Niger, Mali, and Mauritania along the broad periphery. The last four of these countries also lie astride or adjacent to the broad transition zone where the Arab-Islamic realm of northern Africa merges into Subsaharan Africa.

3. *The Middle East.* This region includes Israel, Jordan, Lebanon, Syria, and Iraq. In effect, it is the crescent-like zone of countries that extends from the eastern Mediterranean coast to the head of the Persian Gulf.

4. *The Arabian Peninsula.* Dominated by the large territory of Saudi Arabia, the Arabian Peninsula also includes the United Arab Emirates (UAE), Kuwait, Bahrain, Qatar, Oman, and Yemen. Here lies the source and focus of Islam, the holy city of Mecca; here, too, lie many of the world's greatest oil deposits.

5. *The Empire States.* We refer to this region as the Empire States because two of the realm's giants, both the centers of historic empires, dominate its geography. Turkey, heart of the Ottoman Empire, now is the realm's most secular state and aspires to joining the European Union. Shi'ite Iran, the core of the former Persian Empire, has become an Islamic republic. To the north, Azerbaijan, once a part of the Persian Empire, lies in the turbulent, Muslim-infused Transcaucasian Transition Zone. To the southwest, the island of Cyprus is divided between Turkish and Greek spheres, the Turkish sector being another remnant of imperial times.

6. *Turkestan.* Turkish influence ranged far and wide in southwestern Asia, and following the Soviet collapse that influence proved to be durable and strong. In the five former Soviet Central Asian republics the strength and potency of Islam vary, and the new governments deal warily (and sometimes forcefully) with Islamic revivalists. Russian influence continues, democratic institutions remain weak, and Western (chiefly American) involvement has increased since the War on Terror began and Afghanistan—which also forms part of this region—became a major target of that campaign.

MAJOR CITIES OF THE REALM

City	Population* (in millions)
Algiers, Algeria	3.5
Almaty, Kazakhstan	1.2
Baghdad, Iraq	6.3
Beirut, Lebanon	1.9
Cairo, Egypt	11.8
Casablanca, Morocco	3.2
Damascus, Syria	2.5
Istanbul, Turkey	10.2
Jerusalem, Israel	0.7
Khartoum, Sudan	4.9
Riyadh, Saudi Arabia	4.7
Tehran, Iran	7.6
Tel Aviv-Jaffa, Israel	3.2
Tashkent, Uzbekistan	2.3
Tunis, Tunisia	2.2

*Based on 2008 estimates.

● EGYPT AND THE LOWER NILE BASIN

Egypt occupies a pivotal location in the heart of a realm that extends over 9600 kilometers (6000 mi) longitudinally and some 6400 kilometers (4000 mi) latitudinally. At the northern end of the Nile and of the Red Sea, at the eastern end of the Mediterranean Sea, in the northeastern corner of Africa across from Turkey to the north and Saudi Arabia to the east, adjacent to Israel, to Islamic Sudan, and to militant Libya, Egypt lies in the crucible of this realm. Because it owns the Sinai Peninsula, Egypt, alone among states on the African continent, has a foothold in Asia, a foothold that gives it a coast overlooking the strategic Gulf of Aqaba (the

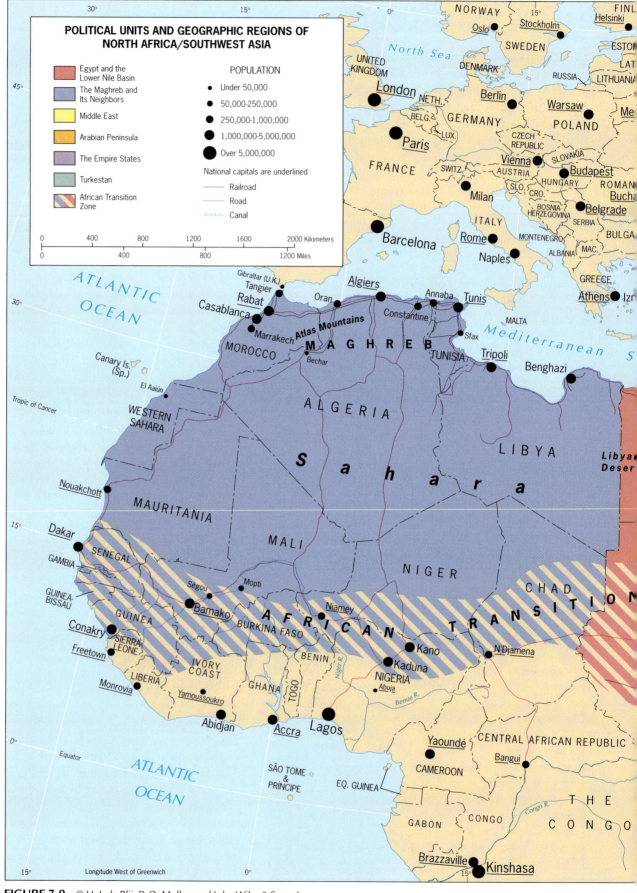

FIGURE 7-9   © H. J. de Blij, P. O. Muller, and John Wiley & Sons, Inc.

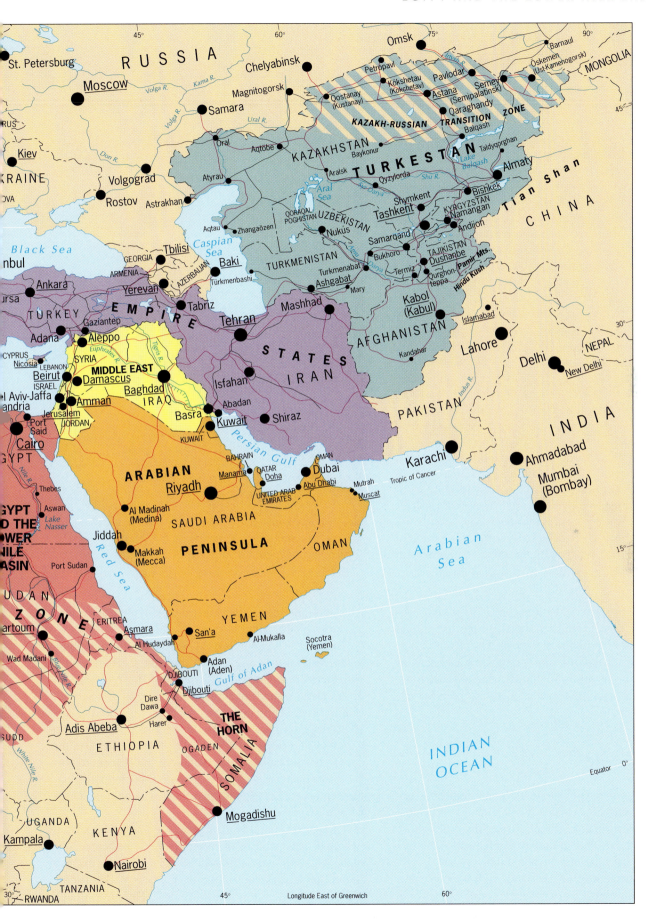

AMONG THE REALM'S GREAT CITIES . . .   CAIRO

STAND ON THE roof of one of the high-rise hotels in the heart of Cairo, and in the distance you can see the great pyramids, monumental proof of the longevity of human settlement in this area. But the present city was not founded until Muslim Arabs chose the site as the center of their new empire in AD 969. Cairo became and remains Egypt's primate city, situated where the Nile River opens into its delta, home today to more than one-seventh of the entire country's population.

Cairo at 11.8 million ranks among the world's 20 largest urban agglomerations, and it shares with other cities of the poorer world the staggering problems of crowding, inadequate sanitation, crumbling infrastructure, and substandard housing. But even among such cities, Cairo is noteworthy for its stunning social contrasts. Along the Nile waterfront, elegant skyscrapers rise above carefully manicured, Parisian-looking surroundings. But look eastward, and the urban landscape extends gray, dusty, almost featureless as far as the eye can see. Not far away, more than a million squatters live in the sprawling cemetery known as the City of the Dead (photo p. 368). On the outskirts, millions more survive in overcrowded shantytowns of mud huts and hovels.

And yet Cairo is the dominant city not only of Egypt but for a wider sphere; it is the cultural capital of the Arab World, with centers of higher learning, splendid museums, world-class theater and music, and magnificent mosques and Islamic learning centers. Although Cairo always has been primarily a center of government, administration, and religion, it also is a river port and an industrial com-

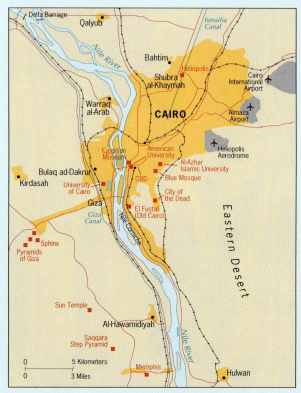

© H. J. de Blij, P. O. Muller, and John Wiley & Sons, Inc.

plex, a commercial center, and, as it sometimes seems, one giant bazaar. Cairo is the heart of the Arab World, a creation of its geography, and a repository of its history.

---

northeasternmost arm of the Red Sea). Egypt also controls the Suez Canal, vital link between the Indian and Atlantic oceans and lifeline of Europe. The capital, Cairo (*Al Qahira*), is the realm's largest city and a leading center of Islamic civilization. We hardly need to further justify Egypt's designation (together with northern Sudan) as a discrete region.

Gift of the Nile

Egypt's Nile is the aggregate of two great branches upstream: the White Nile, which originates in the streams that feed Lake Victoria in East Africa, and the Blue Nile, whose source lies in Lake Tana in Ethiopia's highlands. The two Niles converge at Khartoum in modern-day Sudan.

About 95 percent of Egypt's 78.6 million people live within 20 kilometers (12 mi) of the great river's banks or in its delta (Fig. 7-10). It has always been this way: the Nile

rises and falls seasonally, watering and replenishing soils and crops on its banks.

The ancient Egyptians used **basin irrigation**, building fields with earthen ridges and trapping the floodwaters with their fertile silt, to grow their crops. That practice continued for thousands of years until, during the nineteenth century, the construction of permanent dams made it possible to irrigate Egypt's farmlands year round. These dams, with locks for navigation, controlled floods, expanded the country's cultivable area, and allowed the farmers to harvest more than one crop per year on the same field. In a single century, all of Egypt's farmland was brought under **perennial irrigation**.

The greatest of all Nile projects, the Aswan High Dam (which began operating in 1968), creates Lake Nasser, one of the world's largest artificial lakes (Fig. 7-10). As the map shows, the lake extends into Sudan, where 50,000 people had to be resettled to make way for it. The Aswan High Dam increased Egypt's irrigable land by nearly 50 percent

thousands of people living nearby. By blocking most of the natural fertilizers in the Nile's annual floodwaters, the dam necessitated the widespread use of artificial fertilizers, proving very costly to small farmers and damaging to the natural environment. And the now fertilizer- and pesticide-laden Nile no longer supports the fish fauna offshore, reducing the catch and depriving coastal populations of badly needed proteins.

Valley and Delta

Egypt's elongated oasis along the Nile, just 5 to 25 kilometers (3 to 15 mi) wide, broadens north of Cairo across a delta anchored in the west by the great city of Alexandria and in the east by Port Said, gateway to the Suez Canal. The delta contains extensive farmlands, but it is a troubled area today. The ever more intensive use of the Nile's water and silt upstream is depriving the delta of much needed replenishment. And the low-lying delta is geologically subsiding, raising fears of salt-water intrusion from the Mediterranean Sea that would damage soils here.

Egypt's millions of subsistence farmers, the *fellaheen*, still struggle to make their living off the land, as did the peasants of the Egypt of five millennia past. Rural landscapes seem barely to have changed; ancient tools are still used, and dwellings remain rudimentary. Poverty, disease, high infant mortality rates, and low incomes prevail. The Egyptian government is embarked on grandiose plans to expand its irrigated acreage, but without modernization and greater efficiency in the existing farmlands, the future is dim.

Subregions of Egypt

Egypt has six subregions, mapped in Figure 7-10. Most Egyptians live and work in Lower (i.e., northern) and Middle Egypt (regions ① and ②), the country's core area anchored by Cairo and flanked by the leading port and second industrial center, Alexandria. The economy has benefited from further oil discoveries in the Sinai (region ⑥) and in the Western Desert (region ④), so that Egypt now is self-sufficient and even exports some petroleum. Cotton and textiles are the other major source of external income, but the important tourist industry has repeatedly been hurt by Islamic extremists. As the population mushrooms, the gap between food supply and demand widens, and Egypt must import grain. Since the

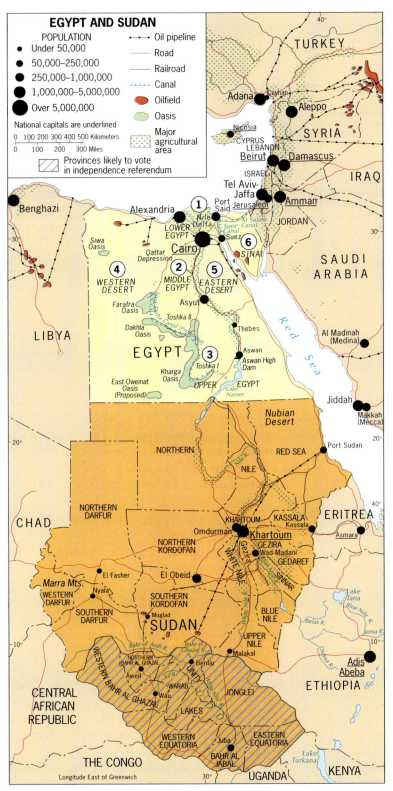

FIGURE 7-10 © H. J. de Blij, P. O. Muller, and John Wiley & Sons, Inc.

and today provides the country with about 40 percent of its electricity. But, as is so often the case with megaprojects of this kind, the dam also produced serious problems. Snail-carried schistosomiasis and mosquito-transmitted malaria thrived in the dam's standing water, afflicting hundreds of

FROM THE FIELD NOTES

"On the eastern edge of central Cairo we saw what looked like a combination of miniature mosques and elaborate memorials. Here lie buried the rich and the prominent of times past in what locals call the 'City of the Dead.' But we found it to be anything but a dead part of the city. Many of the tombs here are so large and spacious that squatters have occupied them. Thus the City of the Dead is now an inhabited graveyard, home to at least one million people. The exact numbers are impossible to determine; indeed, whereas metropolitan Cairo in 2008 has an official population of just under 12 million, many knowledgeable observers believe that 17 million is closer to the mark."   © H. J. de Blij

Concept Caching

www.conceptcaching.com

late 1970s, Egypt has been a major recipient of U.S. foreign aid.

Egypt today is at a crossroads in more ways than one. Its planners know that reducing the high birth rate would improve the demographic situation, but revivalist Muslims object to any programs that promote family planning. Its accommodation with Israel helps ensure foreign aid but divides the people. Its government faces a fundamentalist challenge. Egypt's future, in this crucial corner of the realm, is uncertain.

Divided Sudan

As Figure 7-10 shows, Egypt is flanked by two countries that have posed challenges to its leadership: Sudan to the south and Libya to the west. Sudan, more than twice as large as Egypt and with 43.5 million people, lies centered on the confluence of the White Nile (from Uganda) and the Blue Nile (from Ethiopia). Here the twin capital, Khartoum-Omdurman, anchors a large agricultural area where cotton was planted during colonial times. The British administration combined northern Sudan, which was Arabized and Islamized, with a large area to the south, which was African and where many villagers had been Christianized. After the British left, the regime in Khartoum wanted to impose its Islamic rule on this portion of the African Transition Zone that formed southern Sudan, and a bitter civil war ensued.

The cost in human lives and dislocation of this 30-year war is incalculable. The United Nations estimates that as many as 2 million people died, and as recently as 2003 as many as 4 million people were refugees in their own country. But international mediation finally brought relief in 2005 in the form of a peace agreement whereby the leader of the rebellious South would become vice president of a

federal Sudan and the people of the South would, six years hence, be able to vote on independence.

Sudan has a 500-kilometer (300-mi) coastline on the Red Sea, where Port Sudan lies almost directly across from Jiddah and Makkah (Mecca) in Saudi Arabia. The country's economy for many years was typical of the energy-poor periphery, exchanging sheep, cotton, and sugar for oil, with Saudi Arabia the main trading partner. The pursuit of war in the African Transition Zone impoverished the Islamic regime in Khartoum, and per-capita income in Sudan was one of the world's lowest. All that changed in the 1990s.

Oil in the Desert

During the 1980s it became clear that significant oil reserves lie in Sudan, and during the 1990s discoveries were made in Kordofan Province and elsewhere, including major deposits in the embattled South (Fig. 7-10). This complicated the political as well as the economic situation. In short order Sudan became an exporter rather than an importer of oil. Foreign companies from Canada to China arrived to explore and exploit. The government saw its income rise dramatically, allowing it to buy more weaponry. Local peoples beneath whose lands the oil was found were ruthlessly dispossessed and pushed away. Ultramodern high-rises in the Khartoum-Omdurman area reflected a new era of wealth and privilege.

Leaders in the South viewed the oil bonanza as a ticket to self-sufficiency (the peace agreement stipulated that the South would receive half of local oil revenues, but the records are kept by the North, and trust has always been in short supply). But Southerners were divided over the independence issue. Would it be better to be a part of a prospering Sudan or to build an independent but landlocked state for themselves? In 2007, some Southern leaders were talking about oil pipelines to Kenya's Indian Ocean coast to avoid any dependence on Sudan. Sudan's oil bonanza will have major political-geographical consequences either way.

Tragedy in Darfur

One consequence is already evident: the terrible conflict in Darfur. This western province on the border with Chad and the Central African Republic (Fig. 7-10; see also Fig. 6-18) was engulfed by terror almost as soon as the Southern war was quieted by the 2005 agreement. As the map shows, Darfur (Arabic for "House of the Fur") is divided into three segments. The south and part of the west are inhabited by the Islamized but ethnically African Fur people. The north is predominantly Arab. Apparently, the centuries-old Islamization of the African Fur was not enough for the regime in Khartoum, which blamed these sedentary farmers for sympathizing with anti-Khartoum rebels. Soon the province's camel- and cattle-herding Arabs were invading the Africans' lands, supported by the army and by an informal militia named the *janjaweed* that burned villages, destroyed farms, and killed at will (see photo below).

By mid-2007, more than 2.5 million people had been driven from their lands, over 300,000 had been killed, the United Nations was trying to feed hundreds of thousands at risk of starvation, the conflict had spread into neighboring Chad and the Central African Republic, and the Khartoum regime, flush with money and influence with governments interested in its oil, could ignore the world's appeals. The tragedy of Sudan forms a reminder of the capacity of a rapacious, oil-empowered regime to destroy the lives and hopes of millions under its territorial sway.

● THE MAGHREB AND ITS NEIGHBORS

The countries of northwestern Africa are collectively called the **Maghreb**, but the Arab name for them is more elaborate than that: *Jezira-al-Maghreb*, or "Isle of the West," in recognition of the great Atlas Mountain range rising like a huge island from the Mediterranean Sea to the north and the sandy flatlands of the immense Sahara to the south.

The countries of the Maghreb (which is sometimes spelled *Maghrib*) are Morocco, last of the North African kingdoms; Algeria, a secular republic beset by the religious-political problems we noted earlier; and Tunisia, smallest and most Westernized of the three (Fig. 7-11). Neighboring Libya, facing the Mediterranean between the Maghreb and Egypt, is unlike any other North African country: an oil-rich desert state whose population is almost entirely clustered in settlements along the coast.

Atlas Mountains

Whereas Egypt is the gift of the Nile, the Atlas Mountains facilitate settlement of the Maghreb. These high ranges wrest from the rising air enough orographic rainfall to sustain life in the intervening valleys, where good soils support productive farming. From the vicinity of Algiers eastward along the coast into Tunisia, annual rainfall averages more than 75 centimeters (30 in), a total more than three times as high as that recorded for Alexandria in Egypt's

The "international community" has been slow to respond to the crisis in Darfur, arguing over the question of whether the assault by Arabized Muslim militias on sedentary African farmers in this western province of Sudan technically constitutes genocide, and relying on an undermanned, poorly equipped, and inadequately mandated African Union force to stop the violence. The result can be seen in this photograph of the northern Darfur village of Bandago, where virtually every dwelling has been burned by *janjaweed* militias empowered by Khartoum's Islamic regime. By the middle of 2007, more than 2.5 million people had been driven from their homes, leaving behind charred remains like this; more than 300,000 had died; and this part of the African Transition Zone faced famine and dislocation for many years to come, whatever the result of the power struggles afflicting them.   © AP/Wide World Photos

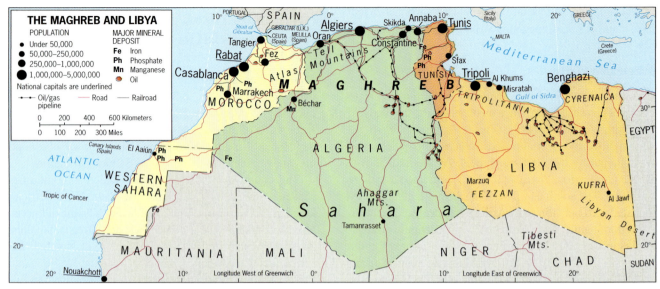

FIGURE 7-11 © H. J. de Blij, P. O. Muller, and John Wiley & Sons, Inc.

delta. Even 250 kilometers (150 mi) inland, the slopes of the Atlas still receive over 25 centimeters (10 in) of rainfall. The effect of the topography can be read on the world map of precipitation (Fig. G-7): where the highlands of the Atlas terminate, desert conditions immediately begin.

The Atlas Mountains trend southwest-northeast and begin in Morocco as the High Atlas, with elevations close to 4000 meters (13,000 ft). Eastward, two major ranges dominate the landscapes of Algeria proper: the Tell Atlas to the north, facing the Mediterranean, and the Saharan Atlas to the south, overlooking the great desert. Between these two mountain chains, each consisting of several parallel ranges and foothills, lies a series of intermontane basins (analogous to South America's Andean *altiplanos* but at lower elevations), markedly drier than the northward-facing slopes of the Tell Atlas. In these valleys, the *rain shadow* effect of the Tell Atlas is reflected not only in the steppe-like natural vegetation but also in land-use patterns: pastoralism replaces cultivation, and stands of short grass and bushes blanket the countryside.

Colonial Impact

During the colonial era, which began in Algeria in 1830 and lasted until the early 1960s, well over a million Europeans came to settle in North Africa—most of them French, and a large majority bound for Algeria—and these immigrants soon dominated commercial life. They stimulated the renewed growth of the region's towns; Casablanca, Algiers, Oran, and Tunis became the urban foci of the colonized territories. Although the Europeans dominated trade and commerce and integrated the North African countries with France and the European Mediterranean world, they did

not confine themselves to the cities and towns. They recognized the agricultural possibilities of the favored parts of the tell (the lower Tell Atlas slopes and narrow coastal plains that face the Mediterranean) and established thriving farms. Not surprisingly, agriculture here is Mediterranean. Algeria soon became known for its vineyards and wines, citrus groves, and dates; Tunisia has long been one of the world's leading exporters of olive oil; and Moroccan oranges went to many European markets.

The Maghreb Countries

Between desert and sea, the Maghreb states display considerable geographic diversity. Morocco, a conservative kingdom in a revolutionary region, is tradition-bound and weak economically. Its core area lies in the north, anchored by four major cities; but the Moroccans' attention is focused on the south, where the government is seeking to absorb its neighbor, **Western Sahara** (a former Spanish dependency with about 400,000 inhabitants, many of them immigrants from Morocco). Even if this campaign is successful, it will do little to improve the lives of most of Morocco's 32.7 million people. Hundreds of thousands have emigrated to Europe, many of them via Spain's two tiny exclaves on the Moroccan coast, Ceuta and Melilla (Fig. 7-11).

Algeria, rich in oil and natural gas, fought a bitter war of liberation against the French colonizers and has been in turmoil virtually ever since, with devastating economic and social consequences. Algiers, Algeria's primate city, is centrally situated along its Mediterranean coast and is home to 3.5 million of the country's 34.6 million citizens—but there are more Algerians in France than in the capital today. Conflict between those wanting to make Algeria an Islamic

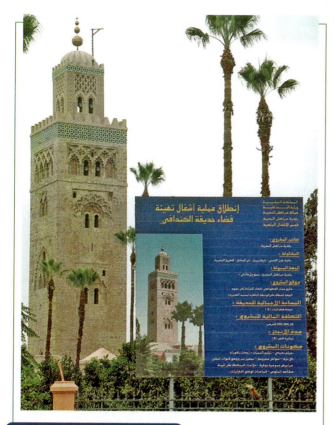

FROM THE FIELD NOTES

"Having seen the *Giralda* in Seville, Spain (photo p. 353), I was interested to view the tower of the Koutoubia Mosque in Marrakech, Morocco, also built in the twelfth century and closely resembling its Andalusian counterpart. Some 67 meters (220 ft) tall, the Koutoubia minaret was built by Spanish slaves and is a monument to the heyday of Islamic culture of the time. I soon discovered that it is not as easy as it looks to get a clear view of the tower, and a boy offered his help, guiding me around the block on which the mosque is situated. He was well informed on the building's history and on the varying fortunes of his home town. It was European mispronunciation of the name 'Marrakech' that led to Morocco being called Morocco, he reported. And Marrakech was Morocco's capital for a very long time, and still should be. 'But back to the mosque,' I said. 'What does the sign say?' He smiled. 'You'll just have to learn some Arabic, just as we know French,' he said. Obviously a teacher in the making, I thought."
© H. J. de Blij

www.conceptcaching.com

republic and the military-backed regime cost an estimated 100,000 lives; in 2005, the government sought to put this episode to rest by holding a referendum on the question of a general amnesty (it was approved). In recent years, new strife has threatened in the Kabylia region east of Algiers involving Algeria's Berber minority.

The smallest of the Maghreb states, Tunisia, lies at the eastern end of the region. As Table G-1 shows, Tunisia in many ways outranks surrounding countries. Stability, national cohesion, and the maximization of limited natural resources (no oil reserves here comparable to Algeria or Libya) have yielded a higher GNI, higher social indicators, and a lower population growth rate (among its total of 10.3 million) than elsewhere in the Maghreb. Most of the country's productive capacity lies in the north, in the hinterland of its large, historic capital, Tunis. As the authoritarian national government slowly loosens its grip, Tunisia's relations with Europe—already the strongest in all of North Africa—continue to improve.

Almost rectangular in shape, Libya (6.2 million) lies on the Mediterranean Sea between the Maghreb states and Egypt (Fig. 7-11). What limited agricultural possibilities exist lie in the district known as Tripolitania in the northwest, centered on the capital, Tripoli, and in the northeast in Cyrenaica, where Benghazi is the urban focus. But it is oil, not farming, that drives the economy of this highly urbanized country. The oilfields lie well inland from the Gulf of Sidra, linked by pipelines to coastal terminals. Libya's two interior corners, the desert Fezzan district in the southwest and the Kufra oasis in the southeast, are connected to the coast by two-lane roads subject to sandstorms.

Adjoining Saharan Africa

As Figure 7-11 shows, the Maghreb countries and Libya adjoin several desert-dominated states in the African Transition Zone. These are African countries with strong Islamic imprints, but as Figure 6-18 shows, they lie north of the Islamic Front with only one exception, Chad. Vast and sparsely populated Mauritania is the most Islamized of these countries, with about half its population of 3.4 million concentrated in and near the capital, Nouakchott, base of a small fishing fleet. Neighboring Mali depends on the waters of the upper Niger River, on which its capital of Bamako lies. A multiparty, democratic state, Mali is also multicultural, with a sizeable minority adhering to traditional beliefs and a small Christian minority. To the east of Mali lies Niger, which shares boundaries with Algeria and Libya; like Mali, this is one of the world's least urbanized countries (21 percent). Unlike Mali, however, Niger has only a short stretch of the middle course of the Niger River, where its capital, Niamey, is situated. And Chad, directly south of Libya, has the lowest percentage of Muslim inhabitants but the strongest division between Islamized north and Christian/animist south. As Figure 6-18 reminds us, the capital, N'Djamena, lies on the southern edge of the African Transition Zone, directly on the Islamic Front. Beginning in 2004, even as Chad was experiencing an oil boom in the south, a cultural conflict in Sudan's neighboring Darfur Province has sent more than 250,000 refugees across its eastern border.

THE MIDDLE EAST: CRUCIBLE OF CONFLICT

The regional term *Middle East*, we noted earlier, is not satisfactory, but it is so common and generally used that avoiding it creates more problems than it solves. It originated when Europe was the world's dominant realm and when places were "near," "middle," and "far" from Europe: hence a Near East (Turkey), a Far East (China, Japan, Korea, and other countries of East Asia), and a Middle East (Egypt, Arabia, Iraq). If you check definitions used in the past, you will see that the terms were applied inconsistently: Syria, Lebanon, Palestine, even Jordan sometimes were included in the "Near" East, and Persia and Afghanistan in the "Middle" East.

Today, the geographic designation *Middle East* has a more specific meaning. And at least half of it has merit: this region, more than any other, lies at the middle of the vast Islamic realm (Fig. 7-9). To the north and east of it, respectively, lie Turkey and Iran, with Muslim Turkestan beyond the latter. To the south lies the Arabian Peninsula. And to the west lie the Mediterranean Sea and Egypt, and the rest of North Africa. This, then, is the pivotal region of the realm, its very heart.

Five countries form the Middle East (Fig. 7-12): Iraq, largest in population and territorial size, facing the Persian Gulf; Syria, next in both categories and fronting the

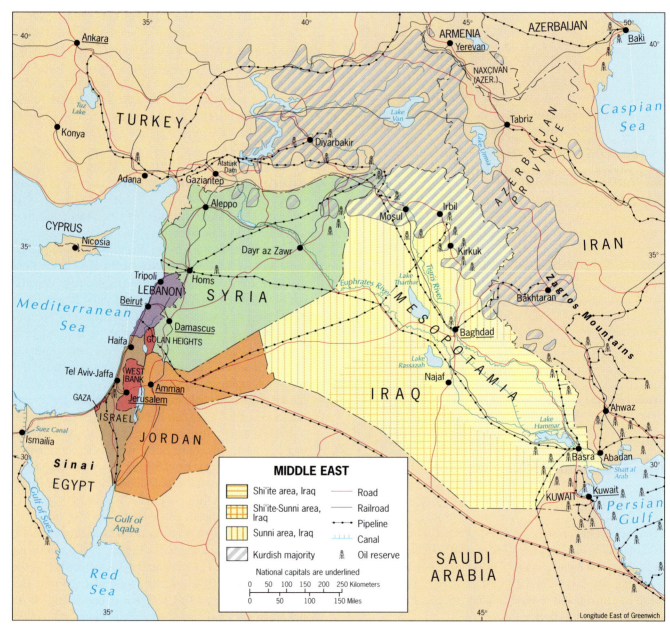

FIGURE 7-12   © H. J. de Blij, P. O. Muller, and John Wiley & Sons, Inc.

Mediterranean; Jordan, linked by the narrow Gulf of Aqaba to the Red Sea; Lebanon, whose survival as a unified state has come into question; and Israel, Jewish nation in the crucible of the Muslim world. Because of the extraordinary importance of this region in world affairs, we focus in some detail on issues of cultural, economic, and political geography in the discussion that follows.

Iraq: War and Its Consequences

In the immediate aftermath of the 9/11 terrorist attacks in the United States in 2001, American forces overthrew the ruling revivalist Taliban regime in Afghanistan that had given refuge to Usama bin Laden, who plotted the assault. Soon, however, Afghanistan lost priority among U.S. concerns as Iraq was designated a security risk based on intelligence reports of its terrorist links, its alleged arsenal of weapons of mass destruction (WMD), its murderous regime, and the regional threat it posed. In March 2003 the United States attacked and invaded Iraq and following a brief struggle ended the rule of the Sunni-Muslim clique led by Saddam Hussein. While the search for WMD continued, the American government now faced the task of stabilizing Iraq, reconstructing its damaged infrastructure, and reconstituting its government.

It is therefore important to understand key aspects of the regional geography of California-sized Iraq and its official population of 31.2 million (which may in fact be three million less because of unregistered emigration). The country constitutes nearly 60 percent of the total area of the Middle East as we define it and has more than 40 percent of its population. With its world-class oilfields, major gas reserves, and large areas of irrigable farmland, Iraq also is best endowed of all the region's countries with natural resources. Iraq is heir to the early Mesopotamian states and empires that emerged in the Tigris-Euphrates Basin, and the country is studded with matchless archeological sites and museum collections, to which disastrous damage was done during and after the 2003 invasion from combat and looting.

The War's Wider Regional Impact

Please take a careful look at Figure 7-12, showing the location of Iraq in the region we have defined as the Middle East, which reveals that Iraq has six immediate neighbors. Perhaps the most obvious feature of this map is the lavender-striped area that not only covers northeastern Iraq but also large parts of Turkey and Iran as well as smaller parts of adjacent countries. This is the area where the majority of the inhabitants are Kurds, who have been discriminated against by all the countries in which they live. About 14 million Kurds are in Turkey, some 8 million in Iran, and about 4 million in Iraq, and together they form one of the world's largest **stateless nations**, divided **18** and often exploited by those who rule over them (see box titled "A Future Kurdistan?").

A Future Kurdistan?

POLITICAL MAPS OF the region do not show it, but where Turkey, Iraq, and Iran meet, the cultural landscape is not Turkish, Iraqi, or Iranian. Here live the Kurds, a fractious and fragmented nation of at least 25 million (their numbers are uncertain). More Kurds live in Turkey than in any other country (perhaps as many as 14 million); possibly as many as 8 million in Iran; about half that number in Iraq; and smaller numbers in Syria, Armenia, and even Azerbaijan (see Fig. 7-12).

The Kurds have occupied this isolated, mountainous, frontier zone for over 3000 years. They are a nation, but they have no state; nor do they enjoy the international attention that peoples of other *stateless nations* (such as the Palestinians) receive. Turkish and Iraqi repression of the Kurds, and Iranian betrayal of their aspirations, briefly make the news but are soon forgotten. Relative location has much to do with this: spatial remoteness and the obstacles created by the ruling regimes inhibit access to their landlocked domain.

Many Kurds dream of a day when their fractured homeland will become a nation-state. Most would agree that the city of Diyarbakir, now in southeastern Turkey, would become the capital. It is the closest any Kurdish town comes to a primate city, although the largest urban concentration of Kurds today is in the shantytowns of Istanbul, where more than 3 million have migrated. In their heartland, meanwhile, the Kurds, their shared goals notwithstanding, are a divided people whose intense disunity has thwarted their objectives; and, as in the case of Europe's Basques, a small minority of extremists has tended to use violence in pursuit of their aims.

The Kurds—without oil reserves, without a seacoast, without a powerful patron, without a global public relations machine—are victims of a conspiracy between history and geography. One of the world's largest stateless nations keeps hoping for a deliverance that is unlikely to come.

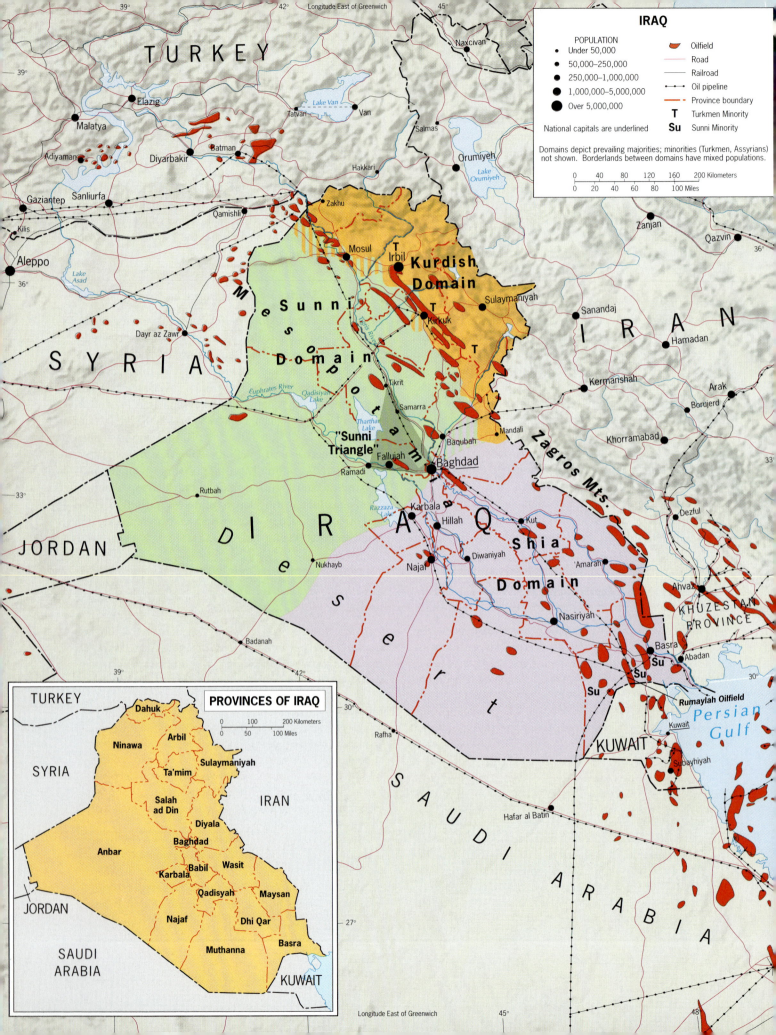

The Kurdish population not only in Iraq but also in several of Iraq's neighbors is only one reason why it is important to study Figure 7-12 closely. Turkey is just one of the countries bordering war-torn Iraq that are directly concerned over the outcome of the struggle that started in 2003. Also anxious is Iran to the east, which fought a bitter war with Iraq in the 1980s but which shares Shia Islam with a majority of Iraqis. To the west, Syria was Iraq's ally during Saddam's despotic rule, and like Iraq under Saddam, Syria itself still is ruled by a strongman and his minority clique. To the south, Iraq is bordered by Kuwait, the country it invaded in 1990, triggering the first Gulf War, and by Saudi Arabia, separated from populated Iraq by a wide swath of desert but troubled by the impact any resolution of Iraq's problems may have at home. And to the southwest lies Jordan, which gave Iraq some help during the first Gulf War but later turned its back not only on Saddam but also on the Sunni insurgents who waged war on American and allied forces following Saddam's ouster.

With so many land neighbors, it is no surprise that Iraq is very nearly landlocked, and Figure 7-13 at left shows how short and congested its single outlet to the Persian Gulf is (that outlet was a factor in Saddam's 1990 decision to try to conquer and annex Kuwait). As a result, a network of pipelines across Iraq's neighbors must carry much of Iraq's oil export volume to coastal terminals in other countries—another reason these countries have an interest in what happens to Iraq in the years ahead.

Cultural and Political Geography in Discord

Let us now carefully examine the regional map of Iraq (Fig. 7-13). Note that Iraq's two vital rivers, the Tigris and Euphrates, rise in Turkey. The map shows that Iraq is divided into three cultural domains: that of the Shia majority in the south, the Sunni minority in the northwest, and the Kurdish minority in the hilly northeast. Not far from the geographic center of Iraq lies the capital, Baghdad, astride the Tigris River at the heart of the country's core area. The Baghdad urbanized area has a population today of between 6 and 7 million, a vast and often violent multicultural agglomeration (see photo, p. 376) of about one quarter of the entire country's population (Fig. 7-14). Only the Shia south, which extends from the eastern suburbs of Baghdad all the way south to Basra and the shores of the Persian Gulf, has more people than Baghdad. The Sunni northwest, including the "Sunni Triangle" where Saddam's clan had its base and the turbulent province of Anbar (see inset map in Fig. 7-13) has between 5 and 6 million, and the Kurdish sector about 4 million. Of these three cultural domains, the Kurdish sector, the most remote of the three from the worst of Iraq's violent postwar

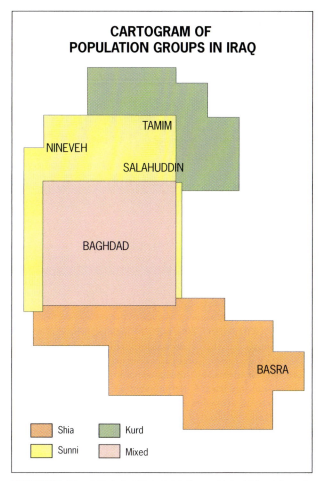

CARTOGRAM OF POPULATION GROUPS IN IRAQ

TAMIM

NINEVEH

SALAHUDDIN

BAGHDAD

BASRA

Shia   Kurd

Sunni   Mixed

FIGURE 7-14   © H. J. de Blij, P. O. Muller, and John Wiley & Sons, Inc.

transition, shows what might have been in the rest of the country: comparative stability and security have produced progress and prosperity.

Figure 7-13 also shows the location of Iraq's major oilfields, most of which lie in the Shia domain and in the Kurdish sector and its borderland, creating an obstacle to any plan to create a three-part federation out of Iraq's political geography. The Sunni sector would be the loser; some acceptable way to allocate oil income to the Sunni provinces would have to be designed.

Before the U.S.-led invasion of 2003, Iraq was held together by the brutal regime headed by Saddam Hussein, which ruled through terror and mass murder (chemical weapons killed thousands of Kurds; mass graves contained tens of thousands of Shi'ites). The intervention and subsequent termination of that regime unleashed forces ranging from sectarian strife and ethnic cleansing to insurgency and terrorism, even as Iraq's population repeatedly voted in elections that, by mid-2006, had produced a first representative government. Whether Iraq can achieve a transition to stability and democracy remains an open question.

FIGURE 7-13   © H. J. de Blij, P. O. Muller, and John Wiley & Sons, Inc.

The tragedy of Baghdad is reflected in its maimed urban landscape. Even during the terrible regime of the dictator Saddam Hussein and his two rapacious sons, when order was imposed through murderous enforcement, the city functioned. Schools and colleges were open, Sunnis and Shi'ites mixed and in some cases even married despite an informal segregation that made some neighborhoods synonymous with culture, like "Saddam City," the hated name for a Shi'ite enclave now named "Sadr City" after the son of a revered Shi'ite cleric. But now, some five years into the U.S.-led intervention, Baghdad is a city of corrosive and violent borders. Families of mixed marriage are being thrown out of their houses (or worse). Streets never crossed have become sectarian boundaries—as shown in this photo of central Baghdad's Sybaa Street (temporarily deserted between incessant street battles), which now divides the Sunni neighborhood of Fadhil on the left from the Shi'ite-dominated area of Sadriya on the right. Sunnis and Shi'ites in many professions now interact almost exclusively with colleagues of the same sect. Shi'ite laborers no longer cross the Tigris River to work in Sunni neighborhoods. Sunnis hesitate to go to hospitals because Shi'ites loyal to Moktada al-Sadr run the Health Ministry. From garbage collection to street repair, Baghdad has become a mosaic of dysfunction, a casualty of poorly planned external intervention and pent-up internal division.   © Robert Nickelsberg/Getty Images.

Syria

Syria, too, is ruled by a minority. Although Syria's population of 20.5 million is about 75 percent Sunni Muslim, the ruling elite comes from a smaller Shi'ite sect based there, the Alawites. Leaders of this powerful minority have retained control over the country for decades, at times by ruthless suppression of dissent. In 2000, president-for-life Hafez al-Assad died and was succeeded by his son Bashar, signaling a continuation of the political status quo. For 25 years, part of this status quo was the occupation and control of neighboring Lebanon, but in 2005 this came to an end.

Like Lebanon and Israel, Syria has a Mediterranean coastline where crops can be raised without irrigation. Behind this densely populated coastal zone, Syria has a much larger interior than its neighbors, but its areas of productive capacity are widely dispersed. Damascus, in the southwest corner of the country, was built on an oasis and is considered to be the world's oldest continuously inhabited city. It is now the capital of Syria, with a population of 2.5 million.

The far northwest is anchored by Aleppo, the focus of cotton- and wheat-growing areas in the shadow of the Turkish border. Here the Orontes River is the chief source of irrigation water, but in the eastern part of the country the Euphrates Valley is the crucial lifeline. It is in Syria's interest to develop its eastern provinces, and recent discoveries of oil there will contribute to this. But the war in neighboring Iraq has caused problems along Syria's eastern border, which will delay development.

Jordan

Jordan, Syria's southern neighbor, is a classic case (and victim) of changing relative location. This desert kingdom was a product of the Ottoman collapse, but when Israel was created it lost its window on the Mediterranean Sea, the (now Israeli) port of Haifa, previously in the British-administered Mandate of Palestine. Following its independence in 1946 with about 400,000 inhabitants, newly created Israel bequeathed Jordan some 500,000 West Bank Palestinians and, later, a huge inflow of refugees. Today Palestinians outnumber original residents by more than two to one in the population of 5.9 million. It may be said that the 47-year rule by King Hussein, ending in 1999, was the key centripetal force that held the country together.

With an impoverished capital, Amman, without oil reserves, and possessing only a small and remote outlet to the Gulf of Aqaba, Jordan has survived with U.S., British, and other aid. It lost its West Bank territory in the 1967 war with Israel, including its sector of Jerusalem (then the kingdom's second-largest city). No third country has a greater stake in a settlement between Israel and the Palestinians than Jordan.

Lebanon

The Middle East map suggests that Lebanon has significant geographic advantages in this region: a lengthy coastline on the Mediterranean Sea; a world-class capital, Beirut, on its shoreline; oil terminals on its seafront; and a major capital (Syria's Damascus) in its hinterland. The map at the scale of Figure 7-12 cannot reveal still another asset: the fertile, agriculturally productive Bekaa Valley in the eastern interior.

French colonialism in the post-Ottoman era created a territory that in 1930 was about equally Muslim and Christian. Beirut became known as the "Paris of the Middle East." After independence in 1946, Lebanon functioned as a democracy and did well economically as the region's leading banking and commercial center. But the Muslim sector of the population grew much faster than the Christian one, and in the late 1950s the Muslims launched their first rebellion against the established, Christian-dominated order. Further upheaval resulted from the arrival of several massive waves of Palestinian refugees, and in 1975 Lebanon fell apart in a civil war that wrecked Beirut, devastated the economy, and left the country at the mercy of Syrian forces that entered Lebanon to take control. In its wake came the emergence of the Iran-sponsored terrorist movement known as Hizbollah which, based in the dominantly Shi'ite south of Lebanon facing the border with Israel, also became a political force that gained representation in Lebanon's parliament.

For so small a country (4.0 million, including more than 400,000 naturalized as well as noncitizen Palestinians), Lebanon is riven with religious and ethnic factions. After the Syrian forces were ousted in 2005 under United Nations auspices, a Sunni-led government, supported by a Christian faction, began the task of reconstruction. But in mid-2006 a Hizbollah team kidnapped two Israeli soldiers and killed several others, provoking massive Israeli retaliation that destroyed much of Lebanon's infrastructure and left the south devastated, though not incapacitated either politically or logistically: Iranian money ensured that Hizbollah would continue as a force in the state. Too many external powers are involved in Lebanon's affairs, it seems, for the country to have a chance at achieving stability and progress.

Israel and the Palestinian Territories

Israel, the Jewish state, lies at the heart of the Arab World (Fig. 7-9). Since 1948, when Israel was created as a homeland for the Jewish people on the recommendation of a United Nations commission, the Arab-Israeli conflict has overshadowed all else in the region.

Figure 7-15 helps us understand the complex issues involved here. In 1946 the British, who had administered this area in post-Ottoman times, granted independence to what was then called "Transjordan," the kingdom east of the Jordan River. In 1948, the orange-colored area became the U.N.-sponsored state of Israel—including, of course, land that had long belonged to Arabs in this territory called Palestine.

As soon as Israel proclaimed its independence, neighboring Arab states attacked it. Israel, however, not only held its own but pushed the Arab forces back beyond its borders, gaining the green areas shown in Figure 7-15. Meanwhile, Transjordanian armies crossed the Jordan River and annexed the yellow-colored area named the West Bank, including part of the city of Jerusalem. The king called his newly enlarged country Jordan.

More conflict followed. In 1967 a week-long war produced a major Israeli victory: Israel took the Golan Heights from Syria, the West Bank from Jordan, and the Gaza Strip from Egypt, and conquered the whole Sinai Peninsula all the way to the Suez Canal. In later peace agreements, Israel returned the Sinai but not the Gaza Strip. In 2005, however, the Israeli government decided to remove all Jewish settlements from Gaza, yielding this territory to the Palestinian Authority (Fig. 7-15, left inset map).

All this strife produced a huge outflow of Palestinian Arab refugees and displaced persons. The Palestinian Arabs constitute another of the region's stateless nations; about 1.2 million continue to live as Israeli citizens within the borders of Israel, but more than 2.5 million are in the West Bank and another 1.5 million in the Gaza Strip (the Golan Heights population is comparatively insignificant). Others live in neighboring and nearby countries including Jordan (ca. 3 million), Syria (460,000), Lebanon (430,000), and Saudi Arabia (320,000); another quarter-million reside in Iraq, Egypt, Kuwait, and Libya. Many have been assimilated into the local societies, but tens of thousands continue to live in refugee camps. In 2007 the scattered Palestinian population was estimated to total more than 9.5 million.

Israel is about the size of Massachusetts and has a population of 7.4 million (including its 1.2 million Arab citizens), but because of its location amid adversaries and its strong international links, these data do not reflect Israel's importance. Israel has built a powerful military even as its regional Arab neighbors have grown stronger; its policies arouse Arab passions; and Palestinian aspirations for a

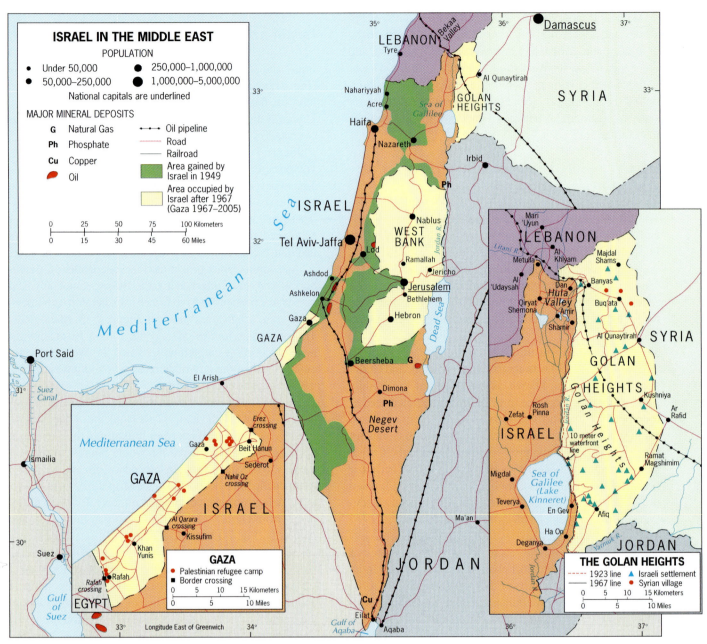

FIGURE 7-15 © H. J. de Blij, P. O. Muller, and John Wiley & Sons, Inc.

territorial state have yet to bear fruit. As a democracy with strong Western ties, Israel has been a recipient of massive U.S. financial aid, and U.S. foreign policy has been to seek a permanent accommodation between the Jews and Palestinians, recently in the form of a "two-state solution" that would create a Palestinian state alongside Israel. In wider context, the United States seeks a general accommodation between Israel and its Arab neighbors as well as between Jews and Arabs generally.

The geographic obstacles to such an accommodation include the following:

1. ***The West Bank.*** Even after its capture by Israel in 1967, the West Bank might have become a Palestinian homeland (and possibly a state), but Jewish immigration to the area made such a future difficult. In 1977 only 5000 Jews lived in the West Bank; by 2007 there were over 400,000, making up more than 15 percent of the population and creating a seemingly inextricable jigsaw of Jewish and Arab settlements (Fig. 7-16). This, in the view of many analysts, has created a situation that may render a two-state solution unattainable, with troubling implications for the future.

FIGURE 7-16 © H. J. de Blij, P. O. Muller, and John Wiley & Sons, Inc.

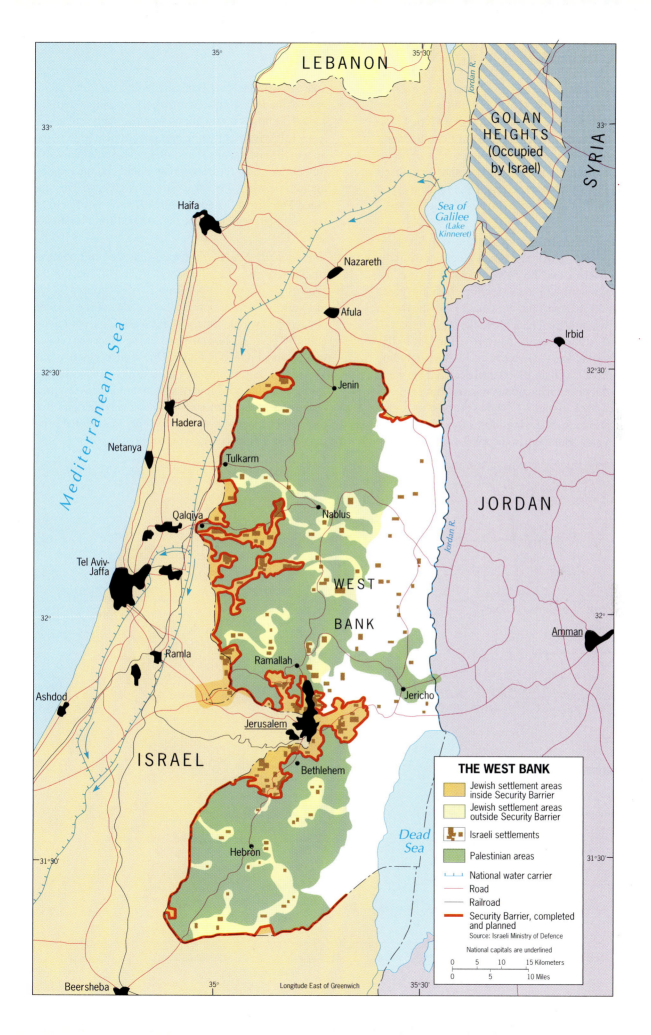

THE WEST BANK

Jewish settlement areas inside Security Barrier

Jewish settlement areas outside Security Barrier

Israeli settlements

Palestinian areas

National water carrier

Road

Railroad

Security Barrier, completed and planned

Source: Israeli Ministry of Defence

National capitals are underlined

0   5   10   15 Kilometers

0   5   10 Miles

379

2. *The Golan Heights.* The right inset map in Figure 7-15 suggests how difficult the Golan Heights issue is. The "heights" overlook a large area of northern Israel, and they flank the Jordan River and crucial Lake Kinneret (the Sea of Galilee), the water reservoir for Israel. Relations with Syria are not likely to become normalized until the Golan Heights are returned, but in democratic Israel the political climate may make ceding this territory impossible.

3. *Jerusalem.* The United Nations intended Tel Aviv to be Israel's capital, and Jerusalem an international city. But the Arab attack and the 1948–1949 war allowed Israel to drive toward Jerusalem (see Fig. 7-15, the green wedge into the center of the West Bank). By the time a cease-fire was arranged, Israel held the western part of the city, and Arab forces the eastern sector. But in this eastern sector lay major Jewish historic sites, including the Western Wall. Still, in 1950 Israel declared the western sector of Jerusalem its capital, making this, in effect, a *forward capital*. Figure 7-17 shows the position of the (black) armistice line, leaving most of the Old City in Jordanian hands. But then, in the 1967 war, Israel con-quered all of the West Bank, including East Jerusalem; in 1980 the Jewish state reaffirmed Jerusalem's status as capital, calling on all nations to move their embassies from Tel Aviv. Meanwhile, the government redrew the map of the ancient city, building Jewish settlements in a ring around East Jerusalem that would end the old distinction between Jewish west and Arab east. This enraged Palestinian leaders, who still view Jerusalem as the eventual headquarters of a hoped-for Palestinian state.

4. *The Security Fence.* In response to the infiltration of suicide terrorists, Israel has now walled off most of the West Bank along the "Security Barrier" border shown in Figure 7-16. The new wall does not conform to the *de facto* postwar boundary of the West Bank, cutting off about 10 percent of its territory and in effect annexing it (and its inhabitants) to Israel. This reinforcement of the West Bank border imposes a hardship on Palestinians (photo at right), for some of whom it runs between their homes and their farmlands. Palestinians demand that the wall be taken down; Israelis reply that the Palestinian Authority must control the terrorists.

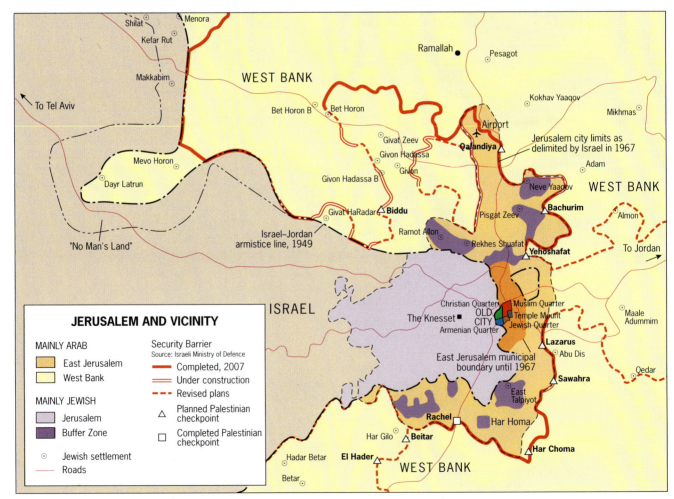

FIGURE 7-17 © H. J. de Blij, P. O. Muller, and John Wiley & Sons, Inc.

The Israeli Security Fence (officially known as the Security Barrier) against Palestinian suicide bombers forms an ugly reminder of the depth of division between the two peoples. Here the concrete wall marks the border between the Palestinian West Bank town of Abu Dis and Jewish Jerusalem. As the map on page 380 shows, the Security Barrier does not follow the West Bank boundary but puts sizeable parts of the West Bank on the Israeli side. Palestinians found themselves cut off from schools and farmlands, and even the Israeli Supreme Court ruled that parts of it were needlessly disruptive to Palestinians' lives, and ordered changes—including the segment shown here. Pedro Ugarte/AFP/Getty Images News and Sport Services

Israel lies at the center of a fast-moving geopolitical storm; the four issues raised above are only part of the overall problem (others include water rights, compensation for land expropriation, and the Palestinians' "right to return" to pre-refugee abodes). Now, in the age of nuclear, chemical, and biological weapons and longer-range missiles, Israel's search for an accommodation with its Arab neighbors is a race against time.

THE ARABIAN PENINSULA

The Arabian Peninsula is one of the world's most clearly defined regions. Flanked on three sides by water and on the fourth by only two countries (Jordan and Iraq), this expanse of deserts and mountains contains the birthplace of Islam, the largest known concentration of oil reserves in the world, and the country whose name is synonymous with oil and revivalist Islam, Saudi Arabia. Six other countries share the peninsula with the Saudi state. Clockwise from

the north, these are Kuwait, Bahrain, Qatar, the United Arab Emirates (UAE), Oman, and Yemen. All of these names should be familiar to Americans. The United States led a war to liberate Kuwait from Iraq's invaders in the 1991 Gulf War. Bahrain and Qatar provided indispensable assistance following the U.S.-led invasion of Iraq in 2003. The United Arab Emirates is home to the "Hong Kong of Arabia," Dubai. Oman overlooks one of the world's narrowest and most dangerous strategic straits at the entrance to the Persian Gulf. Yemen, where the *U.S.S. Cole* was incapacitated by suicide bombers, has a key location alongside another narrow strait (at the mouth of the Red Sea) just across from Africa's violent Transition Zone.

Saudi Arabia

Saudi Arabia itself has barely more than 25 million inhabitants (including about 5 million expatriate workers) in its vast territory, but we can see the kingdom's importance in Figure 7-8: the Arabian Peninsula contains the Earth's

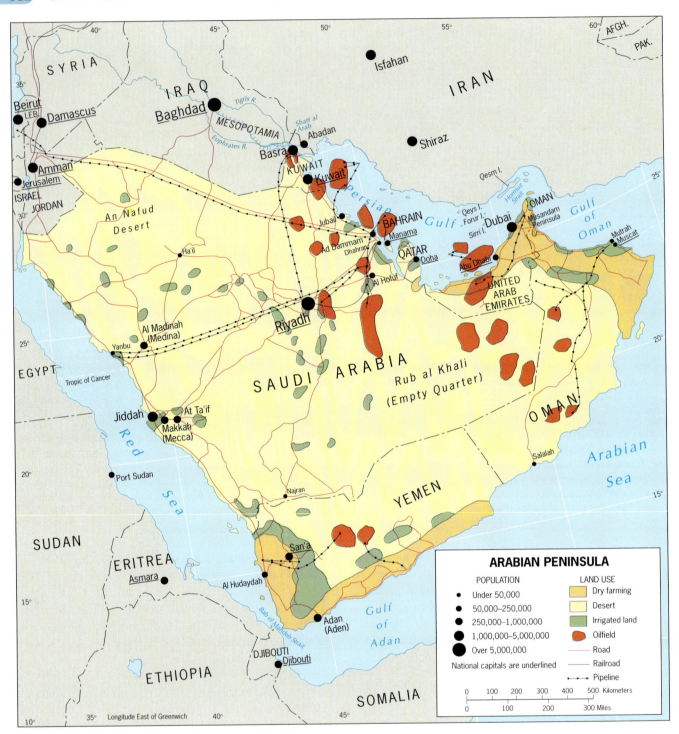

FIGURE 7-18 © H. J. de Blij, P. O. Muller, and John Wiley & Sons, Inc.

largest concentration of known petroleum reserves. Saudi Arabia occupies most of this area and by some estimates may possess as much as one-quarter of the world's oil deposits. As Figure 7-18 shows, these reserves lie in the eastern part of the country, particularly along the Persian Gulf coast and in the Rub al Khali (Empty Quarter) to the south.

As a region, the Arabian Peninsula is environmentally dominated by a desert habitat and politically dominated by the kingdom of Saudi Arabia (Fig. 7-18). With 2,150,000 square kilometers (830,000 sq mi), Saudi Arabia is the realm's fourth biggest state; only Kazakhstan, Algeria, and Sudan are larger territorially. But in terms of population,

Saudi Arabia ranks far down the list; indeed, the six small neighbors on its periphery combined have almost 10 million more inhabitants (35 to 25.4 million) than Saudi Arabia. And within its small southern corner, Yemen alone has nearly three million more citizens than Saudi Arabia excluding the latter's guest workers.

Figure 7-18 reveals that most economic activities in Saudi Arabia are concentrated in a wide belt across the "waist" of the peninsula, from the oil boomtown of Dhahran on the Persian Gulf through the national capital of Riyadh in the interior to the Mecca-Medina area near the Red Sea. A modern transportation and communications network has recently been completed. But in the more remote zones of the interior, Bedouin nomads still ply their ancient caravan routes across the vast deserts. For decades, Saudi Arabia's royal families were virtually the sole beneficiaries of their country's wealth, and there was hardly any impact on the lives of villagers and nomads. When the oil boom arrived in the 1950s, foreign laborers were brought in (today there are about 5 million, many of them Shi'ite) to work in the oilfields, ports, factories, and as servants. The east boomed, but the rest of the country lagged behind.

Disparities and Uncertainties

Efforts to reduce Saudi Arabia's regional economic disparities have been impeded by the enormous cost of bringing water from deep-seated sources to the desert surface to stimulate agriculture, by the sheer size of the country, and by the high rate of population growth (2.7 percent). Nonetheless, housing, health care, and education have seen major improvement, and industrialization also has been stimulated in such places as Jubail on the Persian Gulf as well as the Red Sea ports of Yanbu, near Medina, and Jiddah, near Makkah (Mecca).

Saudi Arabia's conservative monarchism, official friendship with the West, and social contrasts resulting from its economic growth have raised political opposition, for which no adequate channels exist. Relations with the United States were affected by the role of Saudi suicide terrorists in 9/11, Saudi financial support for Wahhabist extremists, anti-American sermons by revivalist clerics, discord over the war in Iraq, and other issues arising in the wake of the events of 2001 and 2003. Saudi Arabia's oil bonanza, like that of other countries, carries its risks as well as rewards. The world price of oil tends to fluctuate, and heavy dependence on high prices leads to public expectations of social benefits that cannot be met when the price falls. Saudi Arabia (and other energy exporters) experienced such a downturn as the twenty-first century opened, long forgotten in the more recent resurgence that has empowered rulers from Khartoum to Caracas. Meanwhile, with the Iraq War on one border,

Shi'ite Iran's nuclear ambitions looming over the Gulf, and pressures for political change at home, long-term stability is in doubt in this, the birthplace of the Prophet Muhammad.

On the Periphery

Five of Saudi Arabia's six neighbors on the Arabian Peninsula face the Persian and Oman gulfs (Fig. 7-18) and are monarchies in the Islamic tradition. All five also derive substantial revenues from oil. Their populations range from 0.7 to 5.0 million in addition to hundreds of thousands of foreign workers; these are not strong or (OPEC aside) influential states. They do, however, display considerable geographic diversity. Kuwait at the head of the Persian Gulf, almost cuts Iraq off from the open sea, an issue the Gulf War did not settle. This oil-rich emirate's population, including its guest workers, has a Sunni majority but a substantial (30 percent) Shia minority. Kuwait's transitional location is further reflected by its social geography: even while Islamic militants sought to gain political power, women were allowed to register and vote for the first time in the 2007 parliamentary elections. Bahrain, a tiny island-state with dwindling oil reserves, has a population of approximately 700,000 with a substantial Shia majority but a Sunni-dominated regime; economic problems and social unrest arising from resentment against the large foreign-born labor force have been mitigated by military and trade agreements with the United States. Neighboring Qatar consists of a small peninsula jutting out into the Persian Gulf, a featureless, sandy wasteland made habitable by oil (now declining in importance) and natural gas (rising simultaneously, with growing exports to Europe and North America). With a strong Sunni majority, this emirate is trying to establish itself as an Arab World focus for international economic, social, and democratic reform.

The United Arab Emirates (UAE), a federation of seven emirates, faces the Persian Gulf between Qatar and Oman. In each emirate the reigning sheik is an absolute monarch, and the seven sheiks together form the Supreme Council of Rulers with a rotating presidency. Abu Dhabi has most of the oil in this federation, but all eyes are on *Dubai* (Dubayy) and its huge port, which is increasingly called the Hong Kong of the Arabian Peninsula. A burgeoning financial center, high-technology hub, shipping base, ultramodern city, residential magnet, and tourist haven, Dubai has become the synonym for modernization and comparative liberalism in this region (see cover photo). World-class sports events and other activities of international significance draw attention to an emirate that promises stability and prosperity.

FROM THE FIELD NOTES

"The port of Mutrah, Oman, like the capital of Muscat nearby, lies wedged between water and rock, the former encroaching by erosion, the latter crumbling as a result of tectonic plate movement. From across the bay one can see how limited Mutrah's living space is, and one of the dangers here is the frequent falling and downhill sliding of large pieces of rock. It took about five hours to walk from Mutrah to Muscat; it was extremely hot under the desert sun but the cultural landscape was fascinating. Oil also drives Oman's economy, but here you do not find the total transformation seen in Kuwait or Dubai. Townscapes (as in Mutrah) retain their Arab-Islamic qualities; modern highways, hotels, and residential areas have been built, but not at the cost of the older and the traditional. Oman's authoritarian government is slowly opening the country to the outside world after long-term isolation." © H. J. de Blij.

Concept Caching

www.conceptcaching.com

The eastern corner of the Arabian Peninsula is occupied by the Sultanate of Oman, another absolute monarchy, centered on the capital, Muscat. Figure 7-18 shows that Oman consists of two parts: the large eastern corner of the peninsula and a small but critical cape to the north, the Mu-sandam Peninsula, that protrudes into the Persian Gulf to form a narrow *choke point* (see box titled "Choke Points: Danger on the Sea Lanes"). Here the narrow Hormuz Strait separates the Arabian Peninsula from Iran. Tankers that leave the other Gulf states must negotiate this narrow chan-

Choke Points: Danger on the Sea Lanes

THE HORMUZ STRAIT between the Arabian Peninsula and Iran is only one of dozens of maritime choke points that have taken on added significance during this time of war on terrorism. A **choke point** is defined as the narrowing of an international waterway causing marine traffic congestion, requiring reduced speeds and/or sharp turns, and increasing the risk of collision as well as vulnerability to attack.

For many years, Southeast Asia's Strait of Malacca between the Malay Peninsula and the Indonesian island of Sumatera (Sumatra) has been a notorious choke point because reduced speeds in the Strait give pirates the opportunity to board vessels and plunder them or worse, kill their crews, and take them over. Since the 1980s, hundreds of acts of piracy have made this the world's least safe waterway, especially for smaller ships.

One of the world's busiest choke points is the Strait of Gibraltar between Spain and Morocco, which was in the news in late 2002 when an al-Qaeda document was found that referred to plans for an attack on Western ships slowing down through this 60-kilometer (35-mi)-long funnel that narrows to 13 kilometers (8 mi) in width and is subject to high winds and fast currents. Another risky choke point is the Bab el Mandeb Strait between Africa and the Arabian Peninsula, or, more specifically, between Djibouti and Yemen, at the southern entrance to the Red Sea. Barim (Perim) Island partially blocks this 30-kilometer (18-mi)-wide strait. Oil tankers, one of which was attacked by al-Qaeda in 2002, and freighters as well as cruise ships approach this strait via all-too-predictable routes.

The English Channel between Britain and France, Turkey's Bosporus between the Black Sea and the Mediterranean Sea, Indonesia's Sunda Strait between the Java Sea and the Indian Ocean, the Strait of Magellan for ships rounding the southern end of South America, and the Hainan Strait between southernmost China's Leizhou Peninsula and Hainan Island are just some of the many other choke points on the world map.

And some choke points are artificial. The Suez Canal between the Red Sea and the Mediterranean Sea and the Panama Canal between Atlantic (Caribbean) and Pacific waters are the busiest and most prominent, but many other route-shortening canals are vulnerable choke points at a time when the risk of hostile action looms larger than the risk of collision.

nel at slow speed, and during politically tense times warships have had to protect them. Iran's claim to several small islands near the Strait that are owned by the UAE is a potential source of dispute. Oman, with only limited oil resources that are likely to run out in two decades, has used its income frugally to improve infrastructure (a good road system and clean water supplies for all outlying areas) and to prepare for an oil-less future. A giant container port and free-trade zone have been built at Salalah, and major investments have been made in luxury tourist facilities, capitalizing on the country's sunshine and beaches. Oman also has gone further than most of the region's countries in attracting foreign investment and privatizing facilities such as its international airport.

At the southern tip of the Arabian Peninsula lies what is in many ways the key state on the region's periphery: Yemen. As Figure 7-18 shows, Yemen's boundary with Saudi Arabia has only recently been delimited in a zone where there are known to be oil reserves. Yemen today is the product of a merger between two neighbors, North Yemen and South Yemen, which agreed to join in 1990. San'a, the interior capital in the north, became the new country's official seat of government; Adan (Aden), the port on the southern coast, continues to anchor the south. Given Yemen's complex cultural geography (the population of 23 million, almost the size of Saudi Arabia's, is 60 percent Sunni, 40 percent Shi'ite), the country has made remarkable progress toward representative government; the minister for human rights is a woman, Wahiba Fare. In terms of political geography, Yemen lies across the narrow Bab el Mandeb Strait from the Horn of Africa, an extremely strategic location at a choke point on the route between the Indian Ocean and the Red Sea that leads to the Suez Canal. It is also the ancestral home of the terrorist leader Usama bin Laden, whose family moved to Saudi Arabia before he was born; in October 2000, the port of Adan was the scene of a successful terrorist attack on a U.S. warship, the *U.S.S. Cole.*

The Arabian Peninsula is a region in difficult and dangerous transition. In the south, monarchical, oil-rich, polarizing Saudi Arabia stands in stark contrast to democratic, agricultural, poorer Yemen. Historic conflicts, cultural disharmony, and latent territorial disputes, now complicated by the threat of terrorism, make this a particularly troubled region.

● THE EMPIRE STATES

Two major states, both with imperial histories, dominate the region that lies immediately to the north of the Middle East and Persian Gulf: Turkey and Iran (Fig. 7-9). Al-

though they share a short border and are both Islamic countries, they display significant differences as well. Even their versions of Islam are different: Turkey is a dominantly Sunni state, whereas Iran is the heartland of Shi'ism. Turkey's leaders have established satisfactory relations with Israel; Iran's president promised to "wipe Israel off the map." In recent years, Turkey has been trying to negotiate entry into the European Union. Iran's goals have been quite different as the country has defied international efforts to constrain its nuclear ambitions.

Two smaller countries are inextricably bound up with these two regional powers: island *Cyprus* to the southwest, divided today between Turks and Greeks in a way that threatens Turkey's European ambitions (see p. 95); and oil-rich, Caspian Sea-bordering *Azerbaijan* to the north, with ethnic and religious ties to Iran but economic links elsewhere (see p. 148).

Turkey

Earlier in this chapter we chronicled the historical geography of the Ottoman Empire, its expansion, cultural domination, and collapse. By the beginning of the twentieth century, the country we now know as Turkey lay at the center of this decaying and corrupt state, ripe for revolution and renewal. This occurred in the 1920s and thrust into prominence a leader who became known as the father of modern Turkey: Mustafa Kemal, known after 1933 as Atatürk, meaning "Father of the Turks."

Capitals Old and New

The ancient capital of Turkey was Constantinople (now Istanbul), located on the Bosporus, part of the strategic straits connecting the Black and Mediterranean seas. But the struggle for Turkey's survival had been waged from the heart of the country, the Anatolian Plateau, and it was here that Atatürk decided to place his seat of government. Ankara, the new capital, possessed certain advantages: it would remind the Turks that they were (as Atatürk always said) Anatolians; it lay nearer the center of the country than Istanbul; and it could therefore act as a stronger unifier (Fig. 7-19). Istanbul lies on the threshold of Europe, with the minarets and mosques of this largest and most varied Turkish city rising above a townscape that resembles Eastern Europe.

Although Atatürk moved the capital eastward and inward, his orientation was westward and outward. To implement his plans for Turkey's modernization, he initiated reforms in almost every sphere of life within the country. Islam, formerly the state religion, lost its official status. The state took over most of the religious schools that had controlled education. The Roman alphabet replaced the

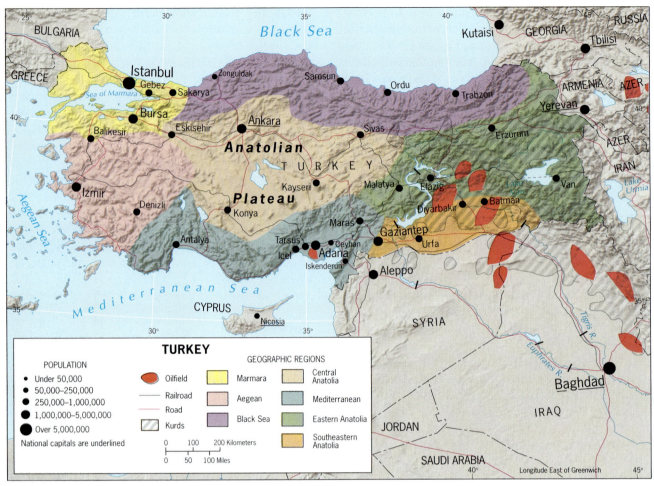

FIGURE 7-19   © H. J. de Blij, P. O. Muller, and John Wiley & Sons, Inc.

Arabic. A modified Western code supplemented Islamic law. Symbols of old—growing beards, wearing the fez—were prohibited. Monogamy was made law, and the emancipation of women was begun. The new government emphasized Turkey's separateness from the Arab World, and it has remained aloof from the affairs that engage other Islamic states.

As Figure 7-19 indicates, Turkey is a mountainous country of generally moderate relief; it also exhibits considerable environmental diversity ranging from steppe to highland (Fig. G-8). On the dry Anatolian Plateau villages are small, and subsistence farmers grow cereals and raise livestock. Coastal plains are not large, but they are productive and densely populated. Textiles (from home-grown cotton) and farm products dominate the export economy, but Turkey also has substantial mineral reserves, some oil in the southeast, massive dam-building projects on the Tigris and Euphrates rivers, and a small steel industry based on domestic raw materials. Normally self-sufficient in staples, Turkey imports oil and gas, chemical products, automobiles, and machinery—a pattern that is changing as

its economy records rapid growth in most (recent) years. Germany, Turkey's leading trade partner and home to as many as 3 million Turkish (many of them Kurdish) immigrants, has not been among Europe's most enthusiastic advocates of Turkish admission to the EU, a political complication that reflects current tensions in Germany over immigration issues. Meanwhile, Turkey's tourist industry is thriving as the country's incomparable archeological, architectural, and scenic attractions are enhanced by improving infrastructure. Turkey is an international, globalizing state in many ways far ahead of any and all other countries in this realm.

Armenians and Kurds

But old issues still cloud Turkey's horizons. From before Atatürk's time, the Turks had a history of mistreating minorities. Soon after the outbreak of World War I, the (pre-Atatürk) regime decided to expel all the Armenians, concentrated at the time in Turkey's northeast. Nearly two million Turkish Armenians were uprooted and bru-

AMONG THE REALM'S GREAT CITIES . . . ISTANBUL

FROM BOTH SHORES of the Sea of Marmara, northward along the narrow Bosporus toward the Black Sea, sprawls the fabulous city of Istanbul, known for centuries as Constantinople, headquarters of the Byzantine Empire, capital of the Ottoman Empire, and, until Atatürk moved the seat of government to Ankara in 1923, capital of the modern Republic of Turkey as well.

Istanbul's site and situation are incomparable. Situated where Europe meets Asia and where the Black Sea joins the Mediterranean, the city was built on the requisite seven hills (as Rome's successor), rising over a deep waterway that enters the Bosporus from the west, the famous Golden Horn. A sequence of empires and religions endowed the city with a host of architectural marvels that give it, when approached from the water, an almost surreal appearance—and what is seen today is a mere remnant of history's accretion, survivals of earthquakes, fires, and combat.

Turkey's political capital may have moved to Ankara, but Istanbul remains its cultural and commercial headquarters. It also is the country's leading urban magnet, luring millions from the poverty-stricken countryside. In the heart of the city known as Stamboul and in the "foreign" area north of the Golden Horn called Beyoglu, modern buildings vie for space and harbor views, blocking the vistas that long made Istanbul's cultural landscape unique. On the outskirts,

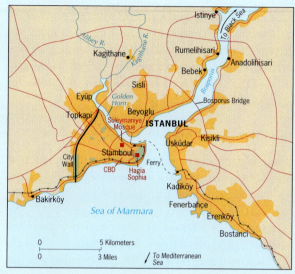

© H. J. de Blij, P. O. Muller, and John Wiley & Sons, Inc.

shantytowns emerge so rapidly that Istanbul's population is doubling every 15 years at current rates, having reached just over 10 million in 2007. Istanbul's infrastructure is crumbling under this unprecedented influx, its growth is virtually without planning controls, and its legacy of two millennia of cultural landscapes is threatened.

---

tally forced out. An estimated 600,000 died in a campaign that still arouses anti-Turkish emotions among Armenians today.

In more recent times, Turkey has been criticized for its treatment of its large and regionally concentrated Kurdish population (a nation discussed in the box on p. 373). About one-fifth of Turkey's population of 75.6 million is Kurdish, and successive Turkish governments have mishandled relationships with this minority nation, even prohibiting the use of Kurdish speech and music in public places during one especially repressive period. The historic Kurdish homeland lies in southeastern Turkey, centered on Diyarbakir, but millions of Kurds have moved to the shantytowns around Istanbul—and to jobs in the countries of the European Union. The Kurdish issue has been an obstacle in Turkey's aspirations to join the EU, whose regulations relating to human rights and minority protections are strict. The Turks have made substantial progress: the death penalty has been abolished, and Kurds have been given freedom to express their culture in ways long forbidden. But some Kurds still resort to violence to redress old (or new) grievances, and Turkish forces still pursue some

of them across international borders. Reports of mistreatment in jails and abrogations of due process continue to surface. In a related human-rights arena, protections for women subject to Islamic family traditions have been strengthened, although these are more difficult to ensure in remote rural areas than in urban communities.

Another cultural issue in modern Turkey involves one of Islam's many sects, in this case the Alevis. Most of Turkey's Muslims are Sunnis, and some are orthodox and strict. The Alevis, who number about 20 million (more than 25 percent of the population), are Shi'ites—but they practice a rather nonchalant version of the faith. While orthodox Sunnis worship in mosques, live by Islam's daily rules, and fast during the daytime during Ramadan, Alevis do not gather in mosques and do not follow the rules of the Ramadan fast. This has led to serious cultural conflict, including murder and arson. The historic center of Turkey's Alevi sect lies in Sivas, in the east-central part of the country northwest of the Kurdish zone. Like the Kurds, many Alevis have left their cultural base, where they are under pressure from orthodox Sunnis, for the more cosmopolitan cities.

In Turkey's general election in 2002, a significant event occurred: a party with Islamist roots led by a politician with a record of revivalist statements led the voting and became the governing majority. The leader who would be prime minister, Recep Tayyip Erdogan, immediately proclaimed his and his party's intention to retain Turkey's secular status. In previous elections, Islamist parties were either prevented from running or were barred by the army from taking office. The 2002 election was thus a milestone for Turkey and a boost for its EU prospects.

EU Prospects and Problems

Of continuing concern in and for Turkey is the situation on the island of *Cyprus*. In 2003, in advance of the EU expansion of the following year that would include Cyprus, the Turkish and Greek sides were offered a UN plan that would have created a confederation of so-called component states, allowing the island to join the EU during its 2004 expansion (see inset map on p. 89). The anticipation was that the Turkish Cypriots, who constitute about 20 percent of the population but stood to lose territory in the deal, would vote against it and that the Greeks would support it. But when the vote was held, it was the Turkish side, encouraged by Ankara, that supported it while the Greek side, to the dismay of Europeans, voted against it. The EU agreed to incorporate Cyprus nonetheless, but it was not a stellar day for Greek Cypriots' political reputation.

These days almost every development in Turkey is in one way or another overshadowed by the EU accession issue. As European opposition (especially Germany's) is reported in the Turkish media, the level of public support for EU membership drops—from three-quarters of those polled a few years ago to just one-third in mid-2007. As in the EU itself, it is the leadership that favors membership; ordinary people have growing doubts. The fate of Turkey's initiative has far-reaching ramifications. If an Islamic country as sophisticated and modernizing as Turkey cannot be accommodated by a Europe that was willing to accept Christian Romania and Bulgaria with all their shortcomings, many Turks argue, then it is Islam, not human rights or minority protection or any other criterion the Europeans cite, that is making the key difference. So, in 2007, it appeared that Turkey might turn its back on Europe before Europe shut the door to Turkey. Either way, the consequences of such a diplomatic failure would be enormous.

Iran

Iran, Turkey's neighbor to the east, has its own history of conquest and empire, collapse and revival. Oil-rich and vulnerable, Iran already was just a remnant of a once-vast regional empire that waxed and waned, from the Danube to the Indus, over more than two millennia. Long known to the outside world as *Persia* (although the people had called themselves Iranians for centuries), the state was renamed *Iran* by the reigning shah (king) in 1935. His intention was to stress his country's Indo-European heritage as opposed to its Arab and Mongol infusions.

As we noted earlier, democracy was stirring in oil-rich Iran during the 1950s when the United States undermined a popular prime minister and helped reinstate the degenerating monarchy. In 1971, the then-reigning shah and his family celebrated the 2500th anniversary of Persia's royal rule with unmatched splendor. By 1979, revolution had engulfed Iran, and Shi'ite fundamentalists, with widespread public support, drove the shah from power. The monarchy was replaced by an Islamic republic ruled by an ayatollah, and a frightful wave of retribution followed. Today, despite a veneer of democracy, Islamists continue to control the state.

Crucial Location, Dangerous Terrain

As Figure 7-9 shows, Iran, again like Turkey, occupies a critical area in this turbulent realm. It controls the entire corridor between the Caspian Sea and the Persian Gulf. To the west it adjoins Turkey and Iraq, both historic enemies. To the north (west of the Caspian Sea) Iran borders Azerbaijan and Armenia, where once again Muslim confronts Christian. To the east Iran meets Pakistan and Afghanistan, and east of the Caspian Sea lies volatile Turkmenistan. Iran, as Figure 7-20 demonstrates, is a country of mountains and deserts. The heart of the country is an upland, the Iranian Plateau, that lies surrounded by even higher mountains, including the Zagros in the west, the Elburz in the north along the Caspian Sea coast, and the mountains of Khurasan to the northeast. The Iranian Plateau therefore is actually a huge highland basin marked by salt flats and wide expanses of sand and rock. The highlands wrest some moisture from the air, but elsewhere only oases break the arid monotony—oases that for countless centuries have been stops on the area's caravan routes.

In eastern Iran, neighboring Pakistan, and Afghanistan, people still move with their camels, goats, and other livestock along routes that are almost as old as the human history of this realm. Usually they follow a seasonal and an annual cycle, visiting the same pastures year after year, pitching their tents near the same stream. It is a lifestyle especially associated with this realm: **nomadism**. In Iran **20** as elsewhere, nomads are not aimless wanderers across boundless plains. They know their terrain intimately, and they carefully judge how long to linger and when to depart based on many years of experience along the route.

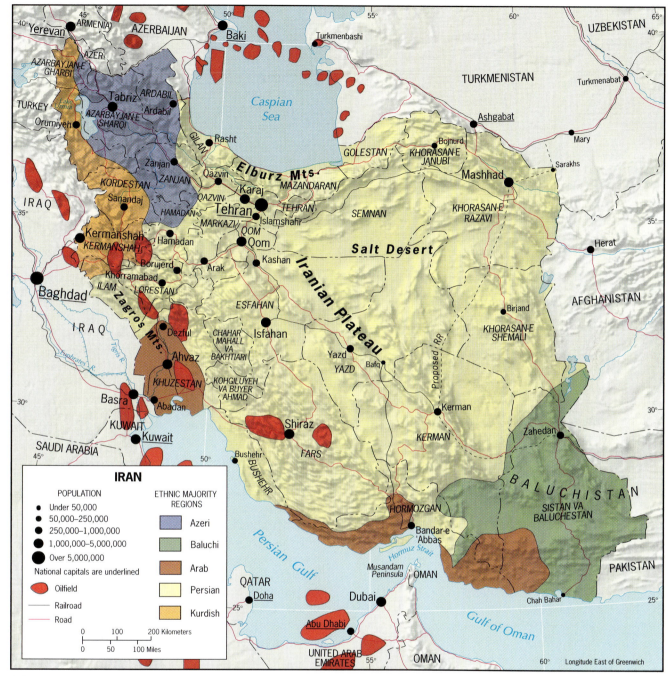

FIGURE 7-20   © H. J. de Blij, P. O. Muller, and John Wiley & Sons, Inc.

City and Countryside

In ancient times, Persepolis in southern Iran (located near the modern city of Shiraz) was the focus of a powerful Persian kingdom, a city dependent on *qanats*, underground tunnels carrying water from moist mountain slopes to dry flatland sites many kilometers away. Today, Iran's population of 72.0 million is 67 percent urban, and the capital, Tehran, lies far to the north, on the southern slopes of the Elburz Mountains. This mushrooming metropolis of 7.6 million, lying at the heart of modern Iran's core area, still depends in part on the same kinds of qanats that sustained Persepolis more than 2000 years ago. As such, Tehran symbolizes the internal contradictions of Iran: a country in which modernization has taken hold in the cities, but little has changed in the vast countryside, where the mullahs led their peasant followers in a revolution that overthrew a monarchy and installed a theocracy.

The Ethnic-Cultural Map

Iranians dominate Iran's affairs, but Iran is an ethnically unified state no more than Turkey is. Figure 7-20 shows that three corners of the country are inhabited by non-Iranians, two in the west and one in the east. Iran's northwestern corner is part of greater Kurdistan, the multistate region in which Kurds are in the majority and of which southeastern Turkey also is a part (Fig. 7-12). Between the Kurdish area and the Caspian Sea the majority population is Azeri, who may number as many as 15 million and whose kinspeople form the majority in neighboring Azerbaijan to the north.

In the southwest, the minority is Arab and is concentrated in Khuzestan Province. During the war of the 1980s, Iraq's Saddam Hussein had his eye on this part of Iran, whose Arab minority often complained of mistreatment by the Iranian regime, and where a large part of Iran's huge oil reserve is located (note the situation of the export terminal of Abadan in Fig. 7-20). The Arab minority here is among Iran's poorest and most restive, in a province whose stability is key to the country's economy.

The third non-Iranian corner of Iran lies in the southeast, where a coastal Arab minority centered on the key port of Bandar-e Abbas on the Strait of Hormuz gives way eastward to a scattered and still-nomadic population of Baluchis who have ties to fellow ethnics in neighboring Pakistan (see Fig. 7-22). On the Iranian side of the border, this is a remote and presently rather inconsequential area, but in Pakistani Baluchistan the government faces a persistent rebellion by tribespeople who feel slighted by official neglect (see p. 423). Iran's ethnic-cultural map, therefore, is anything but simple.

Energy and Conflict

As Figure 7-8 shows, Iran has a large share of the oil riches in this part of the world. Petroleum and petroleum products provide about 90 percent of the country's income. The reserves lie in a zone along the southwestern periphery of Iran's territory, and Abadan became its "oil capital" near the head of the Persian Gulf. But Iran is a large and populous country, and the wealth oil generated could not transform it the way the last shah intended, a transformation that had it occurred might have staved off the revolution. Modernization remained but a veneer: in the villages away from Tehran's polluted air, the holy men continued to dominate the lives of ordinary Iranians. As elsewhere in the Muslim world, urbanites, villagers, and nomads remained enmeshed in a web of production and profiteering, serfdom, and indebtedness that has always characterized traditional society here. The revolution swept this system away, but it

did not improve the lot of Iran's millions. A devastating war with Iraq (1980–1990), into which Iran ruthlessly poured hundreds of thousands of its young men, sapped both the coffers and energies of the state. When it was over, Iran was left poorer, weaker, and aimless, its revolution spent on unproductive pursuits.

In the early years of the twenty-first century, evidence abounded that the people of Iran remain divided between conservatives determined to protect the power of the mullahs and reformers intent on modernizing and liberalizing Iranian society. In 2002, a conservative court sentenced a university professor to death for publishing a proposal for an Islamic "enlightenment"; thousands of students took to the streets in protest. That such a sentence could be handed down at all is indicative of the gulf between the dogmatic and the rational in this historic and civilized society. (In the end, the death sentence was not carried out.)

Iran has long been accused of supporting terrorist activities and organizations, and even of using its embassies in planning and executing terrorist attacks (Argentina has accused Tehran of such operations in the bombing of Jewish targets in Buenos Aires in 1992 and 1994). In 2002, President Bush designated Iran as part of his "axis of evil." Shortly thereafter, it became evident that Iran was developing nuclear-energy capacity with the potential to enrich uranium that could become part of a nuclear-weapons program. Iran's huge oil reserves would seem to make nuclear energy redundant, and when Iran also tested longer-range missiles in 2004, concern over its intentions grew in Europe as well as the United States (Russia, citing its commercial interests, provided technical assistance for the construction of the nuclear plant at Bushehr [photo p. 391]). This concern was heightened by the geography of Iran's nuclear development program: unlike Iraq, which in the 1980s had concentrated all of its nuclear operations in a single facility at Osirak (which was completely destroyed by one Israeli air strike in 1983), the Iranians dispersed their facilities throughout the country and buried much of their nuclear infrastructure deep underground. By early 2005, difficult negotiations led by European countries and involving visits to Iran by representatives of the International Atomic Energy Agency, produced only conditional concessions and unsatisfactory commitments.

Iran's rise as a potential nuclear power should not surprise its neighbors or negotiators. Although revolutionary Iran following the 1979 fall of the shah disavowed its Persian imperial memories and ambitions, this does not mean that its regime's interests now stop at its borders. Iran already has one nuclear neighbor, Pakistan, and it has another in near-neighbor Israel. Iran remains a regional force with cultural links to minorities in Pakistan, Afghanistan, Iraq, and Azerbaijan. It is flanked on two sides by countries

conquerors could not expunge them, as the names on the modern map prove (Fig. 7-21). The latest conquerors, the Russian czars and their communist successors, created Soviet Socialist Republics named after the majority peoples within their borders. Thus the Kazakhs, Turkmen, Kyrgyz, and other Turkic peoples retained some geographic identity in what was Soviet Central Asia.

Central Asia—*Turkestan*—is a still-changing region. In some areas, the cultural landscapes of neighboring realms extend into it, for example, in northern (Russian) Kazakhstan. In other areas, Turkestan extends into adjacent realms, as in Uyghur-influenced western China (Xinjiang). Some areas once penetrated by Turkic peoples are no longer dominated by them, for instance, Afghanistan. And certain peoples now living in Turkestan are not of Turkic ancestry, notably the Tajiks. This is a fractious region in sometimes turbulent transition.

The Ethnic Mosaic

The underlying cultural-geographic reason for this contentiousness—not only in Turkestan but in neighboring regions in Southwest and South Asia as well—is illustrated in Figure 7-22. This detailed map of ethnolinguistic groups actually is a generalization of an even more complex mosaic of peoples and cultures. Every cultural domain on the map also includes minorities that cannot be mapped at this scale, so that people of different faiths, languages, and ways of life rub shoulders everywhere. Often this results in friction, and at times such friction escalates into ethnic conflict.

Regional Framework

As we define it, Turkestan includes most or all of six states: (1) *Kazakhstan*, territorially larger than the other five combined but situated astride an ethnic transition zone; (2) *Turkmenistan*, with important frontage on the Caspian Sea and bordering Iran and Afghanistan; (3) *Uzbekistan*, the most populous state and situated at the heart of the region; (4) *Kyrgyzstan*, wedged between powerful neighbors and chronically unstable; (5) *Tajikistan*, regionally and culturally divided as well as strife-torn; and (6) *Afghanistan*, engulfed in war almost continuously since it was invaded by Soviet forces in 1979.

During their hegemony over Central Asia, the Soviets tried to suppress Islam and install secular regimes (this was their objective when they invaded Afghanistan as well), but today Islam's revival is one of the defining qualities of this region. From Almaty to Samarqand, mosques

During the first decade of the twenty-first century, Iraq and Iran lay at the center of world attention and concern—Iraq because of America's armed invasion, the overthrow of dictator Saddam Hussein, and the Sunni-based insurgency that followed; and Iran because of its determination to achieve nuclear-power capacity, the suspicion that its nuclear objectives went beyond electricity generation, and the election of President Mahmoud Ahmadinejad. Iran's nuclear infrastructure already is formidable, as shown by this satellite image of just one facility near Bushehr on the Persian Gulf; other installations are deep underground. The prospect of a nuclear-armed Iran at the heart of Eurasia troubles many members of the "international community," but it has not been enough to generate a unified and sufficiently firm response. © DIGITAL GLOBE/Reuters/Corbis

under the control of the "Great Satan," the United States. Tehran has economic interests in oil reserves and pipelines in the Caspian Basin. Iran's revolution ended a monarchy, but it did not extinguish all ties to its imperial past.

● TURKESTAN

For centuries Turkish (Turkic) peoples held sway over a vast central Asian domain that extended from Mongolia and Siberia to the Black Sea. Propelled by population growth and energized by Islam, they penetrated Iran, defeated the Byzantine Empire, and colonized much of Eastern Europe. Eventually, their power declined as Mongols, Chinese, and Russians invaded their strongholds. But these

FIGURE 7-21 © H. J. de Blij, P. O. Muller, and John Wiley & Sons, Inc.

are being repaired and revived, and Islamic dress again is part of the cultural landscape. National leaders have made high-profile visits to Mecca; most are sworn into office on the Quran. All of Central Asia's countries now observe Islamic holidays. In other ways, too, Turkestan reflects the norms of this realm: in its dry-world environments and the clustering of its population, its mountain-fed streams irrigating farms and fields, its sectarian conflicts, its oil-based economies. It also is a region where democratic government remains an elusive goal.

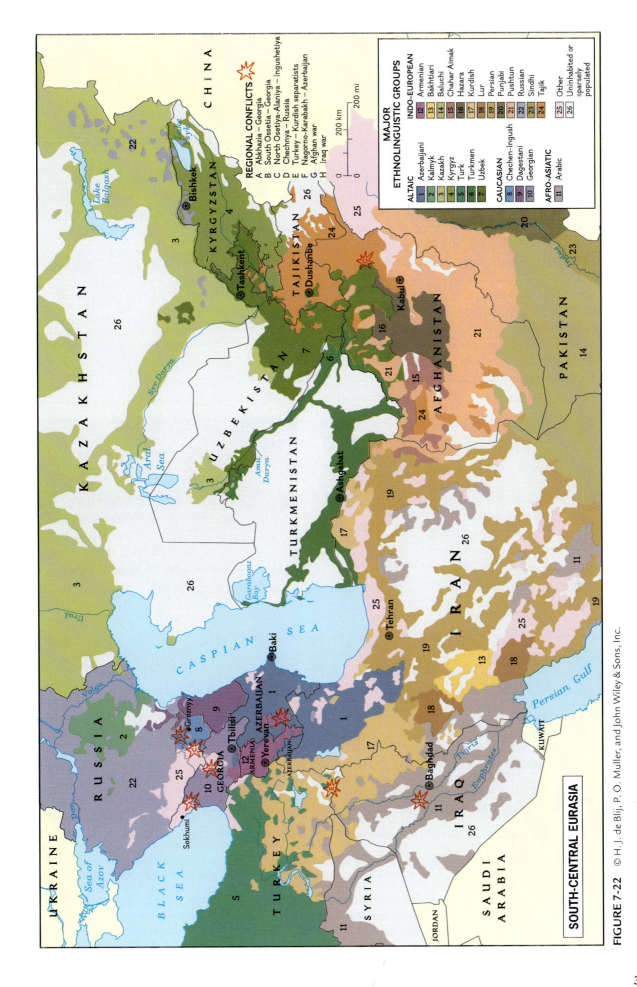

MAJOR ETHNOLINGUISTIC GROUPS

ALTAIC
- 1 Azerbaijani
- 2 Kalmyk
- 3 Kazakh
- 4 Kyrgyz
- 5 Turk
- 6 Turkmen
- 7 Uzbek

CAUCASIAN
- 8 Chechen-Ingush
- 9 Dagestani
- 10 Georgian

AFRO-ASIATIC
- 11 Arabic

INDO-EUROPEAN
- 12 Armenian
- 13 Bakhtiari
- 14 Baluchi
- 15 Chahar Aimak
- 16 Hazara
- 17 Kurdish
- 18 Lur
- 19 Persian
- 20 Punjabi
- 21 Pushtun
- 22 Russian
- 23 Sindhi
- 24 Tajik

- 25 Other
- 26 Uninhabited or sparsely populated

REGIONAL CONFLICTS
- A Abkhazia – Georgia
- B South Ossetia – Georgia
- C North Ossetia-Alaniya – Ingushetiya
- D Chechnya – Russia
- E Turkey – Kurdish separatists
- F Nagorno-Karabakh – Azerbaijan
- G Afghan war
- H Iraq war

SOUTH-CENTRAL EURASIA

FIGURE 7-22 ©H. J. de Blij, P. O. Muller, and John Wiley & Sons, Inc.

393

The States of Former Soviet Central Asia

Kazakhstan is the territorial giant of the region and borders two greater giants: Russia and China (Fig. 7-21). During the Soviet period, northern Kazakhstan was heavily Russified and became, in effect, part of Russia's Eastern Frontier (see Fig. 2-11). Rail and road links crossed the area mapped as the Kazakh-Russian Transition Zone, connecting the north to Russia. The Soviets made Almaty, in the heart of the Kazakh domain, the territory's capital. Today the Kazakhs are in control, and they in turn have moved the capital to Astana, right in the heart of the Kazakh-Russian Transition Zone, where more than 4.5 million Russians (who constitute 30 percent of the total national population of 15.5 million) still live. Clearly, Astana is another *forward capital*.

Figure 7-21 reveals Kazakhstan's situation as a corridor between the Caspian Basin's oil reserves and China. Oil and gas pipelines partially completed across Kazakhstan will eliminate, or greatly reduce, China's dependence on oil carried by tankers along distant sea lanes.

Uzbekistan, the most populous state in this region (27.0 million), occupies the heart of Turkestan and borders every other state in it. Uzbeks not only make up 80 percent of this population, but also form substantial minorities in several neighboring states. The capital, Tashkent, lies in the far eastern core area of the country, where most of the people live in towns and farm villages, and the crowded Fergana Valley is the focus. In the west lies the shrinking Aral Sea (photos below), whose feeder streams were diverted into cotton fields and croplands during the Soviet occupation; heavy use of pesticides contaminated the groundwater, and countless thousands of local people suffered severe medical problems as a result.

In the post-Soviet period, Islamic (Sunni) revivalism has become a problem for Uzbekistan. *Wahhabism*, the particularly virulent form of Sunni fundamentalism, took root in eastern Uzbekistan; government efforts to suppress

The Aral Sea is a saltwater lake on the boundary between Kazakhstan and Uzbekistan, both former Soviet Socialist Republics. During the communist period, the lake's two inflowing rivers were diverted for irrigation, resulting in the dramatic shrinkage illustrated by these two Landsat images. The image on the left was taken in the 1960s, when the Aral Sea covered an area of about 68,000 square kilometers (26,300 sq mi). Thirty years later, the lake had been cut in two and the total water area combined was less than 33,700 square kilometers (13,000 sq mi). The surface had dropped about 15 meters (50 ft) and a large island in what is now known as the "Greater Sea" (as opposed to the "Lesser Sea" to the north) threatened to become another land bridge. The fate of the Aral Sea is but one chapter in Soviet-era mismanagement of the environment in the headlong rush to increase agricultural productivity. The Kazakh and Uzbek governments, with assistance from international agencies, are now in the process of restoring the flow of the two feeder rivers to arrest the Aral Sea's decline, but it is unlikely that a significant rise in the water level can be achieved.   © Worldsat International.

it have only increased its strength in the countryside. In turn, the country's regime has taken on increasingly dictatorial habits. In 2005, a public demonstration in Andijon was put down with many casualties.

Turkmenistan, the autocratic desert republic that extends all the way from the Caspian Sea to the borders of Afghanistan, has a population of just 5.5 million, of which about three-quarters are Turkmen. During Soviet times, the communist planners began work on a massive project: the Garagum (Kara Kum) Canal designed to bring water from Turkestan's eastern mountains into the heart of the desert. Today the canal is 1100 kilometers (700 mi) long, and it has enabled the cultivation of some 1.2 million hectares (3 million acres) of cotton, vegetables, and fruits. The plan is to extend the canal all the way to the Caspian Sea, but meanwhile Turkestan has hopes of greatly expanding its oil and gas output from Caspian Basin reserves. But look again at Figure 7-21: Turkmenistan's relative location is not advantageous for export routes.

Kyrgyzstan's topography and political geography are reminiscent of the Caucasus. Shown in yellow in Figure 7-21, Kyrgyzstan lies intertwined with Uzbekistan and Tajikistan to the point of exclaves and enclaves along its borders. The Kyrgyz, for whom the Soviets established this "republic," constitute barely 50 percent of the population of 5.3 million. Uzbeks and other minorities form a complex cultural geography; mountains and valleys isolate communities and make nation-building difficult. The agricultural economy is weak, consisting of pastoralism in the mountains and farming in the valleys. About 75 percent of the people profess allegiance to Islam, and Wahhabism has gained a strong foothold here. The town of Osh is often referred to as the headquarters of the movement in Turkestan.

Tajikistan's mountainous scenery is even more spectacular than Kyrgyzstan's, and here, too, topography is a barrier to the integration of a multicultural society. The Tajiks, who constitute about 65 percent of the population of 7.3 million, are of Persian (Iranian), not Turkic, origin and speak a Persian (and thus Indo-European) language. Most Tajiks, despite their Persian affinities, are Sunni Muslims, not Shi'ites. Small though Tajikistan is, regionalism plagues the state: the government in Dushanbe often is at odds with the barely connected northern area (see Fig. 7-21), a hotbed not only of Islamic revivalism but also of anti-Tajik, Uzbek activism.

Fractious Afghanistan

Afghanistan, the southernmost country in this region, exists because the British and Russians, competing for hegemony in this area during the nineteenth century, agreed to tolerate it as a cushion, or **buffer state**, between them. This is how Afghanistan acquired the narrow extension leading from the main territory eastward to the Chinese border—the Vakhan Corridor (Fig. 7-23). As the colonialists delimited it, Afghanistan adjoined the domains of the Turkmen, Uzbeks, and Tajiks to the north, Persia (now Iran) to the west, and the western flank of British India (now Pakistan) to the east.

Landlocked and Fractured

Geography and history seem to have conspired to divide Afghanistan. As Figure 7-23 shows, the towering Hindu Kush range dominates the center of the country, creating three broad environmental zones: the relatively well-watered and fertile northern plains and basins; the rugged, earthquake-prone central highlands; and the desert-dominated southern plateaus. Kabol (Kabul), the capital, lies on the southeastern slope of the Hindu Kush, linked by narrow passes to the northern plains and by the Khyber Pass to Pakistan.

Across this variegated landscape moved countless peoples: Greeks, Turks, Arabs, Mongols, and others. Some settled here, their descendants today speaking Persian, Turkic, and other languages. Others left archeological remains or no trace at all. The present population of Afghanistan (32.7 million) has no ethnic majority. This is a country of minorities in which the Pushtuns (or Pathans) of the east are the most numerous but make up barely 40 percent of the total. The second-largest minority are the Tajiks, a world away across the Hindu Kush, concentrated in the zone near Afghanistan's border with Tajikistan. The Hazaras of the central highlands and the south, the Uzbeks and Turkmen in the northern border areas, the Baluchi of the southern deserts, and other, smaller groups scattered across this riven country create one of the world's most complex cultural mosaics (Fig. 7-22). Two major languages, Pushtun and Dari (the local variant of Persian), plus several others create a veritable Tower of Babel here.

Costs of Conflict

Episodes of conflict have marked the history of Afghanistan, but none was as costly as its involvement in the Cold War. Following the Soviet intervention of 1979, the United States supported the Muslim opposition, the *Mujahideen* ("strugglers"), with modern weapons and money, and the Soviets were forced to withdraw. Soon the factions that had been united during the anti-Soviet campaign were in conflict, delaying the return of some 4 million refugees who had fled to Pakistan and Iran. The situation resembled the pre-Soviet

past: a feudal country with a weak and ineffectual government in Kabol.

In 1994, what at first seemed to be just another warring faction appeared on the scene: the so-called **Taliban** ("students of religion") from religious schools in Pakistan. Their avowed aim was to end Afghanistan's chronic factionalism by instituting strict Islamic law. Popular support in the war-weary country, especially among the Pushtuns, led to a series of successes, and by 1996 the Taliban had taken Kabol.

The Taliban's imposition of Islamic law was so strict and severe that Islamic as well as non-Islamic countries objected. Restrictions on the activities of women ended their professional education, employment, and freedom of movement, and had a devastating impact on children as well. Public amputations and stonings enforced the Taliban's code. In the process, Afghanistan became a haven for groups of revolutionaries whose goals went far beyond those of the Taliban: they plotted attacks on Western interests throughout the realm and threatened Arab regimes they deemed compliant with Western priorities. Taliban-ruled, cave-riddled, remote and isolated Afghanistan was an ideal locale for these outlaws. Already in possession of arms and ammunition (Soviet as well as American) left over from the Cold War, they also benefited from Afghanistan's huge, illicit opium trade (photo at right). By 2005, according to United Nations estimates, Afghanistan produced about 87 percent of the world's opium. Much of the revenue had found its way into the coffers of the conspirators.

If you are a farmer in Afghanistan, the most profitable crop is the poppy, so named for the reason obvious from this photo. The poppy crop yields opium, from which heroin and morphine are derived. While morphine has medicinal applications, heroin is a popular narcotic sold on illicit markets around the world. Various regimes—colonial (British), interventionist (Soviet), domestic (yes, even the Taliban!), anti-terror (NATO), and now the Karzai government—have tried and failed to stem the tide of poppy cultivation in Afghanistan, which today accounts for about 87 percent of the world's opium production and an estimated 60 percent of Afghanistan's GNI. As Figure 7-23 shows, poppy cultivation is widespread throughout southern and northern Afghanistan. These women stand in a field about 500 kilometers (300 mi) north of Kabol, the capital. © Landov LLC.

Failed State, Terrorist Base

With such resources, the militants were able to organize and launch attacks against several Western targets in the realm and elsewhere, but events took a fateful turn in 1996 that would empower them as never before. During the conflict with the Soviets, the Mujahideen cause had been supported not only by the United States but also by a devout and revivalist Saudi Muslim named Usama bin ("son of") Laden. The child of a construction billionaire who had more than 50 children including 22 sons, Usama graduated from a university in Jiddah in 1979 and headed for Afghanistan with an inherited fortune estimated at about $300 million to help the anti-Soviet campaign. Well-connected in Saudi Arabia and now a pivotal figure in Afghanistan, bin Laden saw his fame as well as his war chest grow as the Soviet intervention collapsed. Following the withdrawal of the communist forces, he returned to Saudi Arabia, where he denounced his government for allowing U.S. troops on Saudi soil during the Gulf War. The Saudi regime responded by stripping him of his citizenship and expelling him, and bin Laden fled to Sudan.

There he set up several legitimate businesses to facilitate his now-global financial transactions, but he also established terrorist training camps. Under international pressure, the Khartoum regime ousted him in 1996, and bin Laden returned to a country he knew would welcome him again: Afghanistan.

Bin Laden's fateful return to Afghanistan coincided with the Taliban's conquest of Kabol, and now he helped its forces push northward into the fertile and productive northern plains. Meanwhile, a terrorist organization named **al-Qaeda** took root in the country, a global network that would further the aims of the revolutionaries once loosely allied. Afghanistan became al-Qaeda's headquarters, and bin Laden its director; its exploits were funded by numerous Muslim sources and ranged from terrorist-training in local camps to lethal attacks on American targets, including a warship in Yemen's port of Adan and two U.S. embassies in East Africa.

On February 26, 1993, terrorists exploded a massive car bomb in the basement garage of the World Trade

FIGURE 7-23 © H. J. de Blij, P. O. Muller, and John Wiley & Sons, Inc.

Center in New York, but their objective—to topple the 110-story tower—failed. Even as those responsible went on trial, al-Qaeda's leaders were planning the suicide attacks of September 11, 2001 that destroyed the buildings, killed thousands, and caused billions of dollars in damage. Several weeks later, United States and British forces, with the acquiescence of Pakistan, attacked both the Taliban regime and the al-Qaeda infrastructure in Afghanistan, the first stage of a global "War on Terror" that would leave no country unaffected. Proof of bin Laden's and al-Qaeda's complicity in the September 11 assault was found in videotape and documentary form.

The Taliban and al-Qaeda leaderships may not have counted on Pakistan's compliance with Western demands, but they surely knew what the consequences of 9/11 would be for Afghanistan and its people. Once again a foreign power would invade the country and set its political course, the associated upheaval enabling rapacious warlords to exploit the peoples of border provinces far from the capital. But as it turned out, Afghanistan made significant progress toward representative government; most of the warlords were defeated or coopted, and many of the refugees returned. In mid-2007, remnants of the Taliban continued to harass American and other foreign troops in Afghanistan, the leadership of al-Qaeda (probably hiding in western Pakistan) had not been apprehended, and the heroin trade continued to flourish. Still, Afghanistan was in far better shape than it was under the Soviet occupation or during Taliban rule. Sustained stability may finally permit recovery of this long-suffering buffer state.

What You Can Do

RECOMMENDATION: Seize a campus opportunity! If you are taking this course while studying at a college or university (rather than online), you may be living and working in the most multicultural social environment you will ever encounter. Every campus administration provides ways to meet and interact with foreign students, and using this chance to do so can be as valuable an educational experience as any you will have, and may lead to lasting friendships, perhaps even visits in years to come. Foreign students tend to be pleased when their hosts express an interest in their home countries, and their reactions to what they encounter here can be interesting and enlightening. Ask them if they've ever taken any geography courses back home!

GEOGRAPHIC CONNECTIONS

1 If you were to travel through the Muslim world, visiting universities and schools, you would frequently see a version of Figure 7-5 (Areas Under Muslim Rule at Certain Times) on office walls and in classrooms. Using your knowledge of spatial diffusion processes (covered on pp. 351–352), explain how this map came about. What areas on the map has Islam "lost"? Where does Islam appear to be gaining? Why do some of Islam's most significant gains not appear on this map?

2 The United States and its allies invaded Iraq in 2003 for several reasons, including the overthrow of the country's dictatorial regime. Some geographers argued that, among the contingency plans for the country following the military campaign, consideration should have been given to the "Bosnia Model"—the option to temporarily divide Iraq into three parts or perhaps four (the fourth being the capital, Baghdad) to allow for variable rates of adjustment to the occupation, followed eventually by reunification. With the support of the maps in Chapter 7, briefly characterize the cultural geography of Iraq and comment on its traditional regionalism in this context. What might be the responses of Iraq's neighbors to such a plan?

Geographic Literature on North Africa/Southwest Asia: The key introductory works on this realm were authored by Anderson, Beaumont et al., Drysdale & Blake, Fisher, Held, and Longrigg. These works, together with a comprehensive listing of noteworthy books and recent articles on the geography of North Africa and Southwest Asia, can be found in the *References and Further Readings* section of this book's website at **www.wiley.com/college/deblij.**

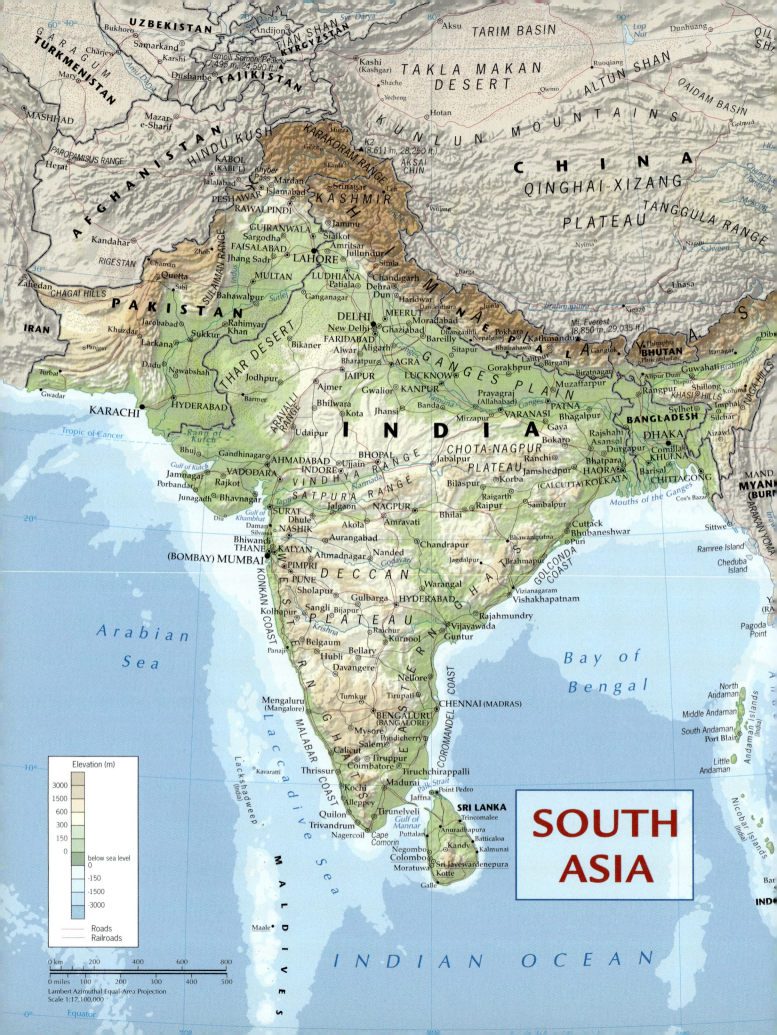

CONCEPTS, IDEAS, AND TERMS

1. Wet monsoon
2. Social stratification
3. Refugees
4. Population geography
5. Population distribution
6. Population density
7. Physiologic density
8. Rate of natural population change
9. Demographic transition
10. Population explosion
11. Forward capital
12. Marchland
13. Caste system
14. Intervening opportunity
15. Natural hazards
16. Failed state
17. Insurgent state

REGIONS

- PAKISTAN
- INDIA
- BANGLADESH
- MOUNTAINOUS NORTH
- SOUTHERN ISLANDS

Where were these photos taken? Find out at www.wiley.com/college/deblij

In This Chapter

- The monsoon still is the key to India's fortunes
- Two nuclear powers quarrel over Kashmir
- Hindu nationalism rising in multicultural India
- India's growing economic power: American jobs in peril?

FIGURE 8-1 *Map:* © H. J. de Blij, P. O. Muller, and John Wiley & Sons, Inc. *Photos:* © H. J. de Blij.

FOR MORE THAN two decades now, the eyes of the world have been on East Asia—on the dramatic developments that transformed China from an isolated backwater to an economic power and a force in global affairs. But significant changes have also come to South Asia, and especially to the gigantic state that dominates this multicultural realm: India. South Asia may be territorially small (it is less than half the size of East Asia), but before long this will become the world's most populous realm, and India the world's most populous country.

South Asia, therefore, merits our close attention. Our daily lives will be affected increasingly by what happens in this crowded, restive part of the world. There is more to South Asia than India: Pakistan, Bangladesh, Nepal, and Sri Lanka also form important parts of this realm (Fig. 8-1). Take a look at the population numbers in Table G-1: just the two countries of Bangladesh and Pakistan together have a larger population than the United States. From the mountainous kingdom of Bhutan to the tourist beaches of the Maldives, and from the terrorist-infested frontier with Afghanistan to the dark forest trails along the border with Myanmar (Burma), South Asia is a world unto itself.

Defining the Realm

South Asia lies wedged between the snowcapped peaks of the Himalayas to the north and the waters of the Indian Ocean to the south. The triangular peninsula of India points dagger-like into those waters and divides them into the dangerous Bay of Bengal to the east and the historic Arabian Sea to the west (Fig. 8-1). On land, the peninsula gives way westward to the deserts of Baluchistan and Iran as well as the mountains of Afghanistan. Eastward, it ends in the forested mountains separating India from Myanmar.

Neither formidable mountains nor forbidding deserts stopped foreign influences from penetrating this realm,

MAJOR GEOGRAPHIC QUALITIES OF

South Asia

1. South Asia is clearly defined physiographically, and much of the realm's boundary is marked by mountains, deserts, and the Indian Ocean.

2. South Asia's great rivers, especially the Ganges, have for tens of thousands of years supported huge population clusters.

3. South Asia covers just over 3 percent of the Earth's land area but contains nearly 23 percent of the world's human population.

4. South Asia's population continues to grow at annual rates higher than the world average. It will become the world's most populous geographic realm during the next decade.

5. Poverty remains endemic in South Asia, despite pockets of progress. Hundreds of millions of people are plagued by inadequate nutrition and poor overall health.

6. British commercial penetration and colonial domination left strong geographic imprints on this realm, ranging from its boundary framework to its cultural landscapes.

7. South Asia's annual monsoon continues to dominate life for hundreds of millions of subsistence and commercial farmers. Failure of the monsoon cycle spells economic crisis.

8. Despite its physiographic demarcation, invading armies and cultures, from ancient Greeks to modern British and from Mongols to Muslims, diversified this realm's cultural mosaic.

9. Several of the world's great religions and many minor faiths have strong bases in South Asia, including Hinduism, Islam, and Buddhism. Religion remains a powerful force in political and economic life.

10. India, the world's largest democracy and a functioning federation, is the dominant state in this realm but has contentious relations with several of its neighbors.

11. Kashmir, the mountainous territory in the northern frontier zone between India and Pakistan, is a dangerous source of friction between these two nuclear powers.

and across the ocean came religious proselytizers and European imperialists. As a result, South Asia today is a patchwork of religions, languages, traditions, and cultural landscapes. But unlike Africa, where the contest among seven colonial powers produced a hodgepodge of nearly 50 dependencies that later became independent countries, South Asia fell under the sway of just one such power: Britain (Portugal did hold on to a small colony in Goa). When independence came to South Asia in the 1940s, the British wanted to transfer their authority to the whole of the British Empire in South Asia. But local demands, especially from Muslims in the west and east, from a pair of kingdoms in the north, and from Buddhists on the southern island then called Ceylon (now Sri Lanka), led to negotiations that produced the map in Figure 8-2.

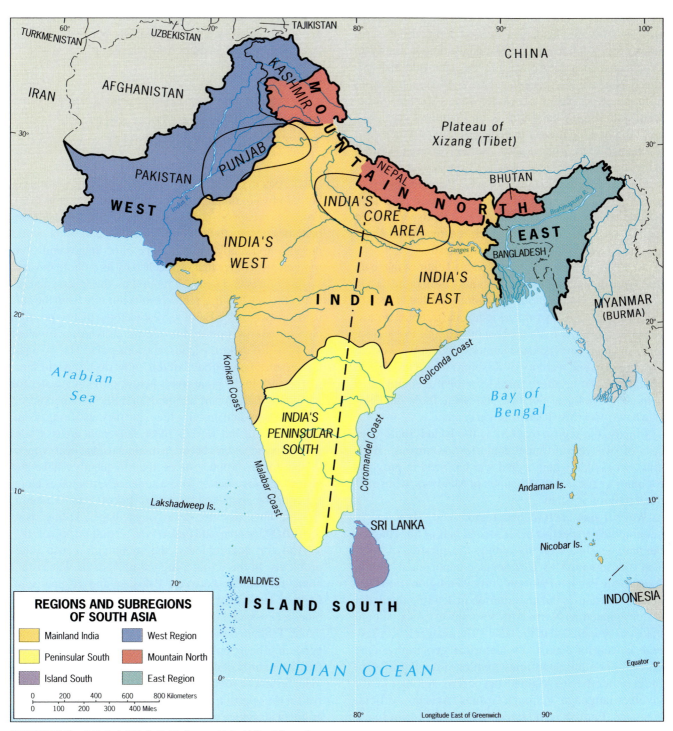

REGIONS AND SUBREGIONS OF SOUTH ASIA

Mainland India
Peninsular South
Island South
West Region
Mountain North
East Region

FIGURE 8-2   © H. J. de Blij, P. O. Muller, and John Wiley & Sons, Inc.

As the map shows, India is the cornerstone of South Asia, its core area anchored in the basin of the Ganges River, the historic heart of this realm. India is a region as well as a state, with subregions based on cultural and other criteria to be discussed later in this chapter. To the west lies Islamic Pakistan, whose lifeline, the Indus River and its tributaries, creates a core area in the Punjab. But note that the Punjab subregion extends across the border with India. What you see here is an aftermath of an enormous human tragedy. When independence approached, Muslim leaders in the west demanded the creation of an Islamic state. A boundary drawn hastily in 1947 between a future Muslim Pakistan and a dominantly Hindu India precipitated the sudden migration of millions of people in the border area—Muslims westward and Hindus eastward—with countless casualties and innumerable human tragedies. In South Asia's east, the British had earlier drawn a border between (Hindu) Bengal and (dominantly Muslim) East Bengal, and that border became the boundary of what became known as East Pakistan, the political connection based on shared Islamic values. In 1971, however, East Pakistan severed its link with West Pakistan (which renamed itself Pakistan) and became independent Bangladesh.

As Figure 8-2 shows, three separated entities make up the region defined as the Mountainous North. From east to west, these are the traditional kingdom of Bhutan, as isolated a country as the world has today; the troubled state of Nepal, where royal rule failed but the struggle to establish representative government continues; and, at the western end, the disputed territory of Kashmir, where the British boundary-making effort failed to resolve a complicated cultural and political situation, and where India and Pakistan have repeatedly been in armed conflict.

Finally, the region mapped as the Island South consists of two very different countries: dominantly Buddhist Sri Lanka, where a civil war has been in progress for many years, and the Republic of Maldives, whose official religion is Islam.

We will view each of these regions and subregions in geographic perspective in the second half of this chapter, but here we should briefly address the question of Pakistan's designation as a regional component of South Asia. Since Islam is Pakistan's official religion (India has none) and Islam is the key criterion in defining the realm we designated as North Africa/Southwest Asia, should Pakistan be included in the latter? The answer lies in several aspects of Pakistan's historical geography. One criterion is ethnic continuity, which links Pakistan to India rather than to Afghanistan or Iran. Another factor involves language: although Urdu is Pakistan's official language, English is the *lingua franca*, as it is in India. Still another, of course, is Pakistan's evolution as part of the British South Asian Empire. Furthermore, the boundary between Pakistan and India does not signify the eastern frontier of Islam in Asia. As we will see, nearly 165 million of India's more than 1 billion citizens are Muslims (which is just about as many as there are in all of Pakistan), and millions live very close to the Indian side of the border whose creation cost so many lives in 1947. And not only are Pakistan and India linked in the cultural-historical arena: they are locked in a deadly and dangerous embrace in embattled Kashmir. Pakistan today remains part of a realm that changes not in the Punjab, but at the Khyber Pass, the highland gateway to Afghanistan.

SOUTH ASIA'S PHYSIOGRAPHY: FROM FERTILE FARMLAND TO TERRORIST HIDEOUT

South Asia may be small as world geographic realms go, but its physiographic diversity is unmatched (Fig. 8-3). From snowcapped peaks to tropical forests and from bone-dry deserts to lush farmlands, this part of the world presents a virtually endless range of ecologies and environments—a range that is matched by its cultural mosaic.

As Figures G-4 and G-5 remind us, the spectacular relief in the north of this realm results initially from the collision of two of the Earth's great tectonic plates, pushing parts of the crust upward into elevations where permanent snow and ice make the landscape look polar. The swing of the seasons melts enough of this snow in spring and summer to sustain the great rivers below, providing water for farmlands that support hundreds of millions of people.

Thus the mountains give life—but they also bring death. This is one of the world's most dangerous areas, where people live in valleys between mountains that bear witness to the forces below, forces that frequently shake the ground. On October 8, 2005 a series of earthquakes along the collision zone struck northern Pakistan and Kashmir, opening huge fissures, destroying villages, killing more than 70,000 people, and making over 3 million homeless.

And for governments, the mountains cause other problems. In the remote, rugged uplands of Waziristan, along the border between Pakistan and Afghanistan, fugitive leaders of the terrorist al-Qaeda organization hide in caves, and members of the deposed extremist Islamic Taliban movement blend in with fiercely independent tribespeople who resist government scrutiny. In the distant hills of desert Baluchistan, a separatist movement festers beyond the reach of Pakistan's government. And at the opposite corner of South Asia, the Indian government struggles to control various rebel movements in the rugged mountain spurs of Nagaland.

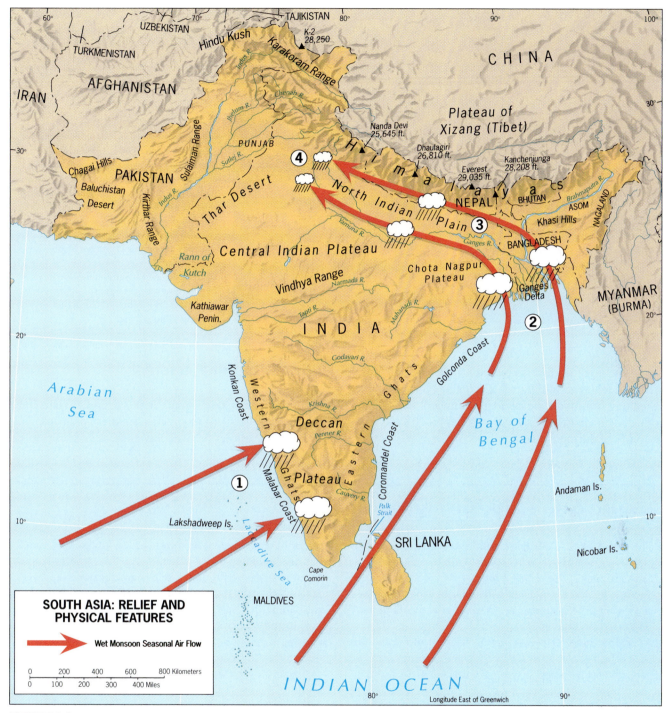

FIGURE 8-3 © H. J. de Blij, P. O. Muller, and John Wiley & Sons, Inc.

The Monsoon

Physical geography, therefore, is crucial here in South Asia—but not just on and below the ground. What happens in the atmosphere is crucial as well. The name "South Asia" is almost synonymous with the term *monsoon*, because the annual rains that come with its onset, usually in June, are indispensable to subsistence as well as commercial agriculture in the realm's key country, India (see box titled "South Asia's Life-Giving Monsoon"). As Figure 8-3 shows, the monsoon brings life-giving rains to the southwestern (Malabar) coast, and a second arm from the Bay of Bengal turns and sweeps across northern India toward Pakistan, losing power (and moisture) as it proceeds. This means that amply watered eastern India and Bangladesh grow rice as their staple

The arrival of the annual rains of the wet monsoon transforms the Indian country-side. By the end of May, the paddies lie parched and brown, dust chokes the air, and it seems that nothing will revive the land. Then the rains begin, and blankets of dust turn into layers of mud. Soon the first patches of green appear on the soil, and by the time the monsoon ends all is green. The photograph on the left, taken just before the onset of the wet monsoon in the State of Goa, shows the paddies before the rains begin; three months later the countryside looks as on the right.   © Steve McCurry/ Magnum Photos, Inc.

crop, but drier northwestern India and Pakistan raise wheat. Look again at Figure 8-2 and note the dashed line extending from India's core area southward. India's historic division between east and west has much to do with the paths of the monsoon.

Physiographic Regions

Figure 8-3 underscores South Asia's overall division into three zones: northern mountains, southern plateaus, and, in between, a tract of river lowlands. It is useful to be

South Asia's Life-Giving Monsoon

THE DYNAMICS OF daily summer sea breezes are well known. The hot sun heats up the land, the air over this surface rises, and cool air from over the water blows in across the beach, replacing it. Teeming beaches and high-cost seafront apartments attest to the efficacy of this natural air conditioning.

Under certain circumstances, this kind of circulation can affect whole regions. When an entire landmass heats up, a huge low-pressure system forms over it. This system can draw vast volumes of air from over the ocean onto the land. It is not just a local, daily phenomenon. It takes months to develop, but once it is in place, it also persists for months. When the inflow of moist oceanic air starts over South Asia, the **wet monsoon** has arrived. It may rain for 60 days or more. The countryside turns green, the paddies fill, and another dry season's dust and dirt are washed away. The region is reborn (see photo pair above).

Not all continents or coasts experience monsoons. A particular combination of topographic and atmospheric circumstances is needed. Figure 8-3 shows how it works in

South Asia. It is June. For months, a low-pressure system has been building over northern India. Now, this cell has become so powerful that it draws air from over a large area of the warm, tropical Indian Ocean toward the interior. Some of that moisture-laden air is forced upward against the Western Ghats ① , cooling as it rises and condensing large amounts of rainfall. Other streams of air flow across the Bay of Bengal and get caught up in the convection over northeastern India and Bangladesh ② . Seemingly endless rain now inundates a much larger area, including the whole North Indian Plain. The mountain wall of the Himalayas stops the air from spreading and the rain from dissipating ③ . The air moves westward, drying out as it flows toward Pakistan ④ .

After persisting for weeks, the system finally breaks down and the wet monsoon gives way to periodic rains and, eventually, another dry season. Then the anxious wait begins for the next year's monsoon, for without it India would face disaster. In recent years, the monsoon has shown signs of irregularity and sometimes partial failure. In India, life hangs by a meteorological thread.

aware of the traditional names of the four great stretches of coastline—Konkan, Malabar, Coromandel, and Golconda—each of which has distinctive relief. Approximately where the Konkan and Malabar coasts meet lies Goa, once the sole non-British colony on the peninsula and today a thriving tourist mecca.

Northern Mountains

The northern mountains extend from the Hindu Kush and Karakoram ranges in the northwest through the Himalayas in the center (Mount Everest, the world's tallest peak, lies in Nepal) to the ranges of Bhutan and the Indian State of Arunachal Pradesh in the east. Dry and barren in the west on the Afghanistan border, the ranges become green and tree-studded in Kashmir, forested in the lower sections of Nepal, and even more densely vegetated in Arunachal Pradesh. Transitional foothills, with many deeply eroded valleys cut by rushing meltwaters, lead to the river basins below.

River Lowlands

The belt of river lowlands extends eastward from Pakistan's lower Indus Valley (the area known as Sind) through the wide plain of the Ganges Valley of India and on across the great double delta of the Ganges and Brahmaputra in Bangladesh (Fig. 8-1). In the east, this physiographic region often is called the North Indian Plain. To the west lies the lowland of the Indus River, which rises in Tibet, crosses Kashmir, and then bends southward to receive its major tributaries from the Punjab ("Land of Five Rivers").

Southern Plateaus

Peninsular India is mostly plateau country, dominated by the massive **Deccan**, a tableland built of basalt that poured out when India separated from Africa during the breakup of Gondwana (see Fig. 6-3). The Deccan (meaning "South") tilts to the east, so that its highest areas are in the west and the major rivers flow into the Bay of Bengal. North of the Deccan lie two other plateaus, the Central Indian Plateau to the west and the Chota Nagpur Plateau to the east (Fig. 8-3). On the map, note the Eastern and Western Ghats ("hills") that descend from Deccan plateau elevations to the narrow coastal plains below. Onshore winds of the annual wet monsoon bring ample rain to the Western Ghats. As a result, here lies one of India's most productive farming areas and one of southern India's largest population concentrations.

LOCALS AND INVADERS

For tens of thousands of years, South Asia has seen peoples (and their cultural attributes and technologies) arrive, organize, compete, succeed, fail, move out, and change. Although this can be said of other world geographic realms, South Asia has a special place in the historical geography of humankind. Recent research indicates that the Ganges Basin was a crucial stop on the great out-of-Africa migration that carried our species from the shore of the Red Sea to the coast of Australia. Here, perhaps as long as 70,000 years ago, environmental conditions during the Wisconsinan Glaciation enabled groups of hunter-gatherers to stabilize and expand. From here, they went on to Southeast Asia and Australia and, scholars conclude, emigrated toward the Middle East and Europe as well. The Ganges Basin may have been a crucible of humanity.

Indus Valley Civilization

Whatever the distant past, we know that a complex and technologically advanced civilization had emerged in the Indus Valley by about 5000 years BC, encompassing a large area of the lower Indus River Basin and, apparently, succeeding a still-earlier culture (Fig. 8-4). The Indus Valley Civilization (IVC) was centered on two major cities, Harappa and Mohenjo-Daro, which may have been capitals during different periods of its history; in addition, there were more than 100 smaller towns and villages. All this indicates that here was an early state in which political centralization had been achieved; the culture was literate, there appear to have been commercial contacts with Mesopotamia, and technologies such as irrigation, flood control, and well construction left their imprints. Today, it is known that IVC influences extended far to the west and eastward as far as the Delhi area, confirming the long-held suspicion that this was a foundation for the later Iron Age civilization that arose in what is now India. The locals apparently called their state **Sindhu** in the language that developed late in the IVC period, and both **Indus** (for the river) and **India** (for the later state) may derive from this name.

A combination of environmental problems resulting from destructive floods and foreign invasion destroyed the IVC, apparently quite suddenly, about 3500 years ago. The invaders, **Aryans** (peoples speaking Indo-European languages based in what is today Iran) destroyed urban civilization in this northwestern corner of South Asia for centuries to come; but in the south, on and around the Kathiawar Peninsula you see in Figure 8-4, vestiges of the IVC survived. As the Iron Age dawned in India, acculturated

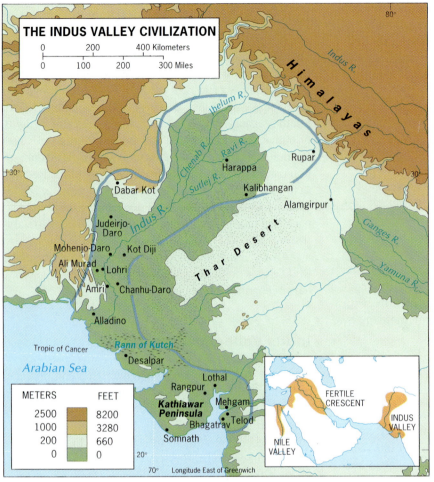

THE INDUS VALLEY CIVILIZATION

FIGURE 8-4   © H. J. de Blij, P. O. Muller, and John Wiley & Sons, Inc.

and others all had their place. Those in the lowest castes, deemed to be there because they deserved it, were worst off, without hope of advancement and at the mercy of those higher up on the social ladder. Meanwhile, India's political geography evolved into a growing number of small states ruled by kings in consort with Brahmans.

Hinduism's restrictive and punitive qualities led others to seek a better way, and in the sixth century BC (thus more than 2500 years ago) a prince born into one of India's northern kingdoms proclaimed that he had found it. Prince Siddhartha, better known as Buddha, gave up his royal rights to teach religious salvation through meditation, the rejection of Earthly desires, and a reverence for all forms of life. In truth, his teachings did not have a major impact on the Hindu-dominated society of his time, but he traveled widely, left followers in many parts of India, and was not forgotten. Centuries later, the ruler of a powerful Indian state decided to make **Buddhism** the state religion, and Buddhism spread far and wide.

Aryans began the process of welding the Ganges Basin's isolated tribes and villages into a new organized system, and urbanization made a comeback.

Incipient India

The Aryans brought their language (Sanskrit) and a new social order to the vast riverine flatlands of northern India, and they introduced their own religious belief system, **Vedism**. Out of the texts of Vedism and local creeds there arose a new belief system—**Hinduism**—and with it a new way of life. Starting about 3000 years ago, a combination of regional integration, the organization of villages into controlled networks, and the emergence of numerous small city-states produced a hierarchy of power among the people, a ranking from the very powerful (Brahmans—highest-order priests) to the weakest (peasants). Thus a **[2]** complex **social stratification** was part of Hinduism's **caste system**, in which soldiers, merchants, artists, toolmakers,

Earlier Invaders

The cultural geography of South Asia bears witness to the numerous times outsiders invaded and put their stamp on entire regions of this realm. Invasions, we noted, have taken place almost throughout human history; the Persia-based Aryans were preceded by others about whom little or nothing is known. The map of one aspect of regional culture, the distribution of languages, reflects this conundrum (Fig. 8-5). When you compare this map to the world map of language families (Fig. G-11) it is clear that the languages of South Asia (shown in yellow in Fig. 8-5) lie at the eastern end of a language family that extends from the British Isles to the shores of the Bay of Bengal. That is why Hindi, the major domestic language of India, belongs to the same language family as Persian, Italian, and English. But look at the south: in what is now southern India, the indigenous languages are not Indo-European. They belong to a family called the Dravidian, evidence of migrations that long preceded the heyday of the Indus Valley Civilization. But these are

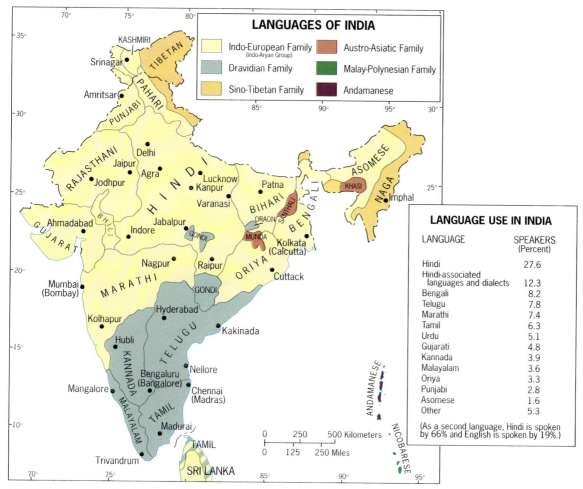

LANGUAGES OF INDIA

Indo-European Family (Indo-Aryan Group)	Austro-Asiatic Family
Dravidian Family	Malay-Polynesian Family
Sino-Tibetan Family	Andamanese

LANGUAGE USE IN INDIA

LANGUAGE	SPEAKERS (Percent)
Hindi	27.6
Hindi-associated languages and dialects	12.3
Bengali	8.2
Telugu	7.8
Marathi	7.4
Tamil	6.3
Urdu	5.1
Gujarati	4.8
Kannada	3.9
Malayalam	3.6
Oriya	3.3
Punjabi	2.8
Asomese	1.6
Other	5.3

(As a second language, Hindi is spoken by 66% and English is spoken by 19%.)

FIGURE 8-5 © H. J. de Blij, P. O. Muller, and John Wiley & Sons, Inc.

not fossil languages: they remain vibrant today and have long literary histories. Telugu, Tamil, Kanarese (Kannada), and Malayalam are spoken by some 250 million people, a basis for recognizing a southern subregion in peninsular India (Fig. 8-2).

Aśoka's Mauryan Empire

After the Aryans from Persia came the Greeks under Alexander the Great, who in the late fourth century BC made an incursion that would not change the realm but did demonstrate its vulnerability to invasion: part of his huge army came through the Khyber Pass, and other forces crossed the mountains and hills to the north. Alexander never made it to the Ganges Basin, however, and now, for the first time, a local state grew into a regional power. Starting in what is today the Indian State of Bihar, this expanding state came to control the whole of the North Indian Plain (see Fig. 8-3). When the emperor Aśoka took control of it, he headed what was to be

the first empire in South Asia—reaching as far west as the Indus Valley, thereby incorporating the Punjab, and as far east as the great double delta of the Ganges and the Brahmaputra at the head of the Bay of Bengal. And to the south, this Mauryan Empire extended to the latitude of the modern city of Bengaluru (Bangalore).

Aśoka was no ordinary leader. For 40 years during the middle of the third century BC, he guided and expanded his vast empire *not* by conquest and subjugation, but through persuasion and teaching. A devout Buddhist, he deployed a veritable army of missionaries throughout his domain, all preaching the virtues of peace and stability. In the process, he disseminated Indian culture far and wide, and he did not stop at the frontiers of the Mauryan state. Aśoka's emissaries spread the faith as far afield as Southwest and Southeast Asia, ensuring Buddhism's survival after his death. Today, Sri Lanka, the island state off the southern tip of India, is a dominantly Buddhist country, a legacy of its great benefactor—and Buddhism now has an estimated 350 million adherents around the globe (see Fig. 7-2).

Fracture and Fragmentation

As is so often the case when an inspiring leader has unified a state for a long time, the Mauryan Empire fell apart after Aśoka's death around 230 BC. In the more remote outposts, local princes took control and ceased to recognize the new regime, which itself was in dispute over matters of succession. Buddhism was in retreat, Hinduism in ascendance, and South Asia set on a course that was to mark it for many centuries to come: a course of disunity and strife. Only during the Gupta Empire (ca. AD 320–540), again centered in present-day Bihar, did a lengthy dynasty achieve regional unification and nurture a great flowering of Indian culture. Great Sanskrit epics, superb Hindu art, and major accomplishments in science and technology marked this formative period. But the Gupta period was the exception, not the rule, and thus South Asia was vulnerable to renewed invasions.

The Power of Islam

In the late tenth century, Islam came rolling like a giant tide across South Asia, spreading across Persia and Afghanistan, through the high-mountain passes, into the Indus Valley, across the Punjab, and into the Ganges Basin, converting virtually everyone in the Indus Valley and foreshadowing the emergence, many centuries later, of the Islamic Republic of Pakistan. By the early thirteenth century, the Muslims had established a long-surviving and powerful sultanate at Delhi, which expanded across much of the peninsula, absorbing Indian states in the process and ruling over an empire larger even than the Mauryan Empire. The Muslims also came by sea, arriving at the Ganges-Brahmaputra Delta and spreading the faith from the east as well as the west, in the process laying the foundation of today's dominantly Muslim state of Bangladesh.

Why did millions of Hindus convert to Islam? It was a combination of compulsion and attraction. The princes of India's local states faced the choice of cooperation or annihilation, and thus the elites of the native states found it prudent to convert, just like the leaders of West Africa's states did. And among the lower castes of Hindu society, Islam was a welcome alternative to the rigid socio-religious hierarchy in which they were trapped at the bottom. Thus Islam was the faith of the rulers and of the disadvantaged, a powerful force in the heartland of Hinduism.

The Muslims, like other conquerors, also fought among themselves. In the early 1500s a descendant of Genghis Khan named Babur put his forces in control of Kabol (Kabul) in Afghanistan, and from that base he penetrated the Punjab and challenged the Delhi Sultanate. In the 1520s, his Islamicized Mongol armies ousted the Delhi rulers and established the Mughal (Mongol) Empire.

By all accounts, Mughal rule was remarkably enlightened, especially under the leadership of Babur's grandson Akbar, who expanded the empire by force but adopted tolerant policies toward Hindus under his sway; Akbar's grandson, Shah Jahan, made his mark on India's cultural landscape through such magnificent architectural creations as the Taj Mahal and the Great Mosque of Delhi. But then religious bigotry took over, infighting intensified, and by the eighteenth century the Mughal Empire was in decline. Maratha, a Hindu state in the west, expanded not only into the peninsular south but also northward toward Delhi, capturing the allegiance of local rulers and weakening Islam's hold. Fractured India now lay open to still another foreign intrusion, this time from Europe.

Reflecting on more than seven centuries of Islamic rule in South Asia, it is remarkable that Islam never achieved proportional dominance over the realm as a whole. While Pakistan is more than 96 percent Muslim and Bangladesh 86 percent, India—where the Delhi Sultanate and the Mughal Empire were centered—remains less than 15 percent Muslim today; thus fewer than one-third of all South Asians are Muslims. By contrast, more than 50 percent of the realm's population is Hindu, and about 75 percent of all Indians are Hindu. Islam arrived like a giant tide, but Hinduism stayed afloat and outlasted the invasion.

The European Intrusion

Although the European invasion of South Asia gained momentum after 1800, European colonialists had been on South Asia's shores for a very long time—in fact, ever since the arrival of the Portuguese seafarer Vasco da Gama in 1498, even before the advent of the Delhi Sultanate. But the Portuguese, Dutch, French, and British were less interested in territorial acquisition than in control over trade, and eventually the British got the better of their competitors. Exploiting local rivalries and even defeating a combined Mughal-French force in 1757, the British East India Company (EIC) took control not only of the trade with Europe in spices, cotton, and silk goods, but also India's longstanding commerce with Southeast Asia, long in the hands of Indian, Arab, and Chinese merchants. In effect, the EIC became India's colonial administration.

In time, however, the EIC faced problems that made it increasingly difficult to combine commerce with admin-

ists saw this as competition, and soon India was exporting raw materials and importing manufactured goods—from Europe, of course. Local industries declined and Indian merchants lost their markets.

Unifying their realm was a tougher task for the British. In 1857 about 2 million square kilometers (750,000 sq mi) of South Asian territory still lay beyond British control, including hundreds of entities that had been guaranteed autonomy by the EIC during its administration. These "Native States," ranging in size from a few hectares to Hyderabad's 200,000 square kilometers (80,000 sq mi), were assigned British advisors, but in fact India was a near-chaotic amalgam of modern colonial and traditional feudal systems.

Colonialism did produce assets for India. The country was bequeathed one of the best transport networks of the colonial domain, especially the railroad system (although the network focused on interior-to-seaport linkages rather than fully interconnecting the various parts of the country). British engineers laid out irrigation canals through which millions of acres of land were brought into cultivation. Settlements that had been founded by Britain developed into major cities and bustling ports, led by Bombay (now Mumbai), Calcutta (now Kolkata), and Madras (now Chennai). These three cities are still three of India's largest urban centers, and their cityscapes bear the unmistakable imprint of colonialism (see photo at left). Modern industrialization, too, was brought to India by the British on a limited scale. In education, an effort was made to combine English and Indian traditions; the Westernization of India's elite was supported through the education of numerous Indians in Britain. Modern practices of medicine were also introduced. Moreover, the British administration tried to eliminate features of Indian culture that were deemed undesirable by any standards—such as the burning alive of widows on their husbands' funeral pyres, female infanticide, child marriage, and the caste system. Obviously, the task was far too great to be achieved in barely three generations of colonial rule, but independent India itself has continued these efforts where necessary.

FROM THE FIELD NOTES

"More than a half-century after the end of British rule, the centers of India's great cities continue to be dominated by the Victorian-Gothic buildings the colonizers constructed here. Here is evidence of a previous era of globalization, when European imprints transformed urban landscapes. Walking the streets of Mumbai (the British called it Bombay) you can turn a corner and be forgiven for mistaking the scene for London, double-deckered buses and all. One of the British planners' major achievements was the construction of a nationwide railroad system, and railway stations were given great prominence in the urban architecture. I had walked up Naoroji Road, having learned to dodge the wild traffic around the circles in the Fort area, and watched the throngs passing through Victoria Station. Inside, the facility is badly worn, but the trains continue to run, bulging with passengers hanging out of doors and windows."
© H. J. de Blij.

Concept Caching

www.conceptcaching.com

istration. Eventually, a mutiny among Indian troops in the service of the EIC led to the abolition of the company. The British government took over in 1857 and maintained its rule (*raj*) until 1947.

Colonial Transformation

When the British took power over South Asia, they controlled a realm with considerable industrial development (notably in metal goods and textiles) and an active trade with both Southwest and Southeast Asia. The colonial-

Partition

Even before the British government decided to yield to Indian demands for independence, it was clear that British India would not survive the coming of self-rule as a single political entity. As early as the 1930s, the idea of a separate state (that would become Pakistan) was being promoted by Muslim activists. They circulated pamphlets arguing that British India's Muslims were a nation distinct from the Hindus and that a separate state consisting of Sind, Punjab, Baluchistan, Kashmir, and a portion of Afghanistan should be created from the British South Asian Empire in this area. The first formal demand for such partitioning was made in 1940, and, as later elections proved, the idea had almost universal support among the realm's Muslims.

As the colony moved toward independence, a political crisis developed: India's majority Congress Party would not even consider partition, and the minority Muslims refused to participate in any future unitary government. But partition would not be a simple matter. True, Muslims were in the majority in the western and eastern sectors of British India, but Islamic clusters were scattered throughout the realm (Fig. 8-6A). Any new boundaries between Hindus and Muslims to create an Islamic Pakistan and a Hindu India would have to be drawn right through areas where both sides coexisted. People by the millions would be displaced.

The consequences of this migration for the social geography of India were far-reaching, especially in the northwest (Fig. 8-6B). Comparing the 1931 and 1951 distributions in Figure 8-6, you can see the dimensions of the losses in the Indian Punjab and in what is today the State of Rajasthan. (Because Kashmir was mapped as

three entities before the partition and as one afterward, the change there represents an administrative, not a major numerical, alteration.) Even in the east a Muslim exodus occurred, as reflected on the map by the State of West Bengal, adjacent to Bangladesh, where the Muslim component in the population declined substantially.

The world has seen many **refugee** migrations but none involving so many people in so short a time as the one resulting from British India's partition (which occurred on independence day, August 15, 1947). Scholars who study the refugee phenomenon differentiate between "forced" and "voluntary" migrations, but, as this case underscores, it is not always possible to separate the two. Many Muslims, believing they had no choice, feared for their future in India and joined the stampede. Others had the means and the ability to make a decision to stay or leave, but even these better-off migrants undoubtedly sensed a threat. In any event, massive as the refugee movement was, it left far more Muslims in India than departed. As Figure 8-6 shows, the percentage of Muslims in India's total population (then about 360 million) declined even below what it had been in 1931, but India's Muslim minority has always had a high rate of

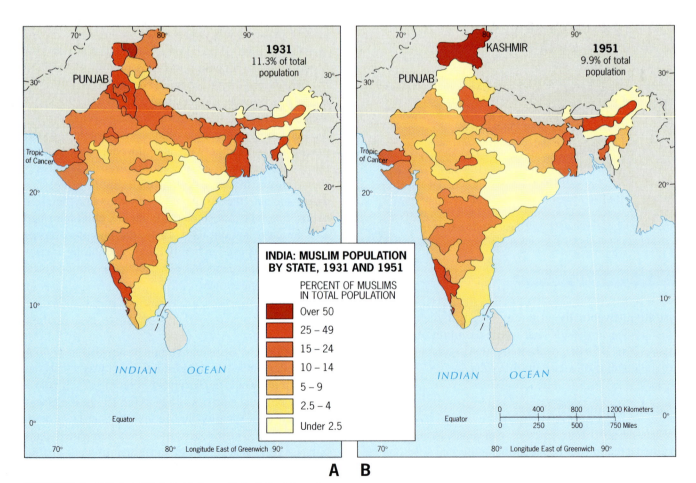

1931
11.3% of total population

1951
KASHMIR
9.9% of total population

PUNJAB

INDIA: MUSLIM POPULATION BY STATE, 1931 AND 1951

PERCENT OF MUSLIMS IN TOTAL POPULATION

- Over 50
- 25 – 49
- 15 – 24
- 10 – 14
- 5 – 9
- 2.5 – 4
- Under 2.5

INDIAN OCEAN

A   B

FIGURE 8-6 © H. J. de Blij, P. O. Muller, and John Wiley & Sons, Inc.

Flight was one response to the 1947 partition of what had been British India, resulting in one of the greatest mass population transfers in human history. Here, two trainloads of eastbound Hindu refugees fleeing (then) West Pakistan arrive at the station in Amritsar, the first city inside India. © Corbis-Bettmann.

population growth. By 2007 it surpassed 160 million, 14 percent of the total population—the largest cultural minority in the world and almost as large as Pakistan's entire population (174 million).

And what about Hindus who found themselves on the "wrong" side of the border at the time of partition? The great majority of Hindus preferred to live in a secular democracy than under Islamic rule, so they, too, crossed the border (photo above). The Hindu component of present-day Pakistan may have neared 20 percent in 1947 but is only about 1 percent today; in Bangladesh (which was East Pakistan at the time of partition) it declined from 30 percent to about 12 percent today. The end of British rule in South Asia changed the realm's cultural geography forever.

SOUTH ASIA'S POPULATION DILEMMA

South Asia, as we noted earlier, is on course to become the most populous geographic realm on Earth, and India the most populous country, perhaps as early as the end of the next decade (see Table G-1). That prospect has serious implications not only for this realm, but for the world as a whole. Over the past three decades, population growth in many parts of the world has been slowing, and some countries in Europe and Asia are even reporting that their populations are actually decreasing. The countries where this is happening tend to be the richer countries. This means that an ever-larger proportion of the new births in this century will be occurring in already poverty-afflicted regions. This has troubling ramifications, and nowhere is the trouble more serious than in South Asia.

All this is part of the field of **population geography**, in which research focuses on the dimensions, distribution, growth, and other aspects of human population in a country, region, or realm as this relates to soils, climates, land ownership, social conditions, economic development, and other factors. Population geography, therefore, is the *geography of demography*. As usual, the map is a powerful ally in this geographic research.

In the Introduction chapter we referred to the various ways maps can reveal **population distribution**, even at the generalized global scale, Figure G-9 (pp. 18–19) shows how crowded South Asia is—despite the existence of sparsely peopled deserts and highlands even in this comparatively small realm. You can actually discern the courses of the great Ganges, Indus, and other rivers because of the immense concentrations of population that inhabit their basins. But even more telling than distribution is **population density**, that is, the number of people per unit area (such as a square kilometer or square mile) in a country, province, or physiographic zone such as a river basin or mountain belt. Please take a careful look at the data displayed for South Asia on page 43, because you will see that population density appears in two columns. The first of these shows what we call the *arithmetic* population density for a country, which is simply the number of people per square kilometer (in this case) for each country. Truth be told, this figure is not very meaningful despite its appearance in gazetteers and atlases, for obvious reasons. Take the case of Pakistan: when tens of millions of people cluster in river basins and valleys and almost nobody lives in deserts and high mountains, what does 218 people per square kilometer really mean? Not much, which is why geographers prefer to use the **physiologic density** measure that reports the number of people in a country per unit of arable land area (suitable for agriculture). According to this measure, note that Bangladesh and Sri Lanka, not India, are the most densely populated countries of this realm (the very high Maldives number reflects the virtual absence of farmland on its tiny tropical islands).

Population Change

As noted above, geographers are especially interested in
8 the **rate of natural population change** in a country, region, or realm, that is, the number of births minus the number of deaths, usually reported as a percentage (as we do in Table G-1) or as an index per thousand of the population. These data can reveal much about the condition and prospects of a country. We have already seen how stagnant population growth produces problems of aging, how too-rapid population decline (such as Russia's) signals serious societal problems, and how a slower decline (as in South America) can be helpful in economic development.

Population geographers analyze how population growth rates change as a region's economy evolves. The higher-income economies have gone through the so-
9 called **demographic transition**, a four-stage sequence that took them from high birth rates and high death rates in preindustrial times to low death rates and even lower birth rates today (Fig. 8-7). Stages 2 and 3 in this model,
10 with high birth rates and low death rates, form the **population explosion** that caused such great alarm during the

twentieth century when death rates in industrializing and urbanizing countries dropped sharply, but birth rates took longer to follow suit. In 1900 the planet's human population was approximately 1.5 billion; by 2000 it had surpassed 6 billion. During the population explosion of the 1950s and 1960s, some scholars predicted a global calamity with as many as 10 to 12 billion people on Earth by the end of the century.

Prospects for South Asia

It did not come to that, but some areas of the world face population (and associated economic) problems that will afflict them for decades to come, no matter what the global averages predict. And South Asia may be the most severely burdened of all.

When the British ruled India during the nineteenth century, the country still was in the first stage, with high birth rates and high death rates; the high death rates were caused not only by a high incidence of infant and child mortality but also by famines and epidemics. As Figure 8-7 indicates, the population during Stage 1 does not grow or decline much, but it is not stable. Famines and disease outbreaks kept erasing the gains made during better times. But then India entered the second stage. Birth rates remained high, but death rates declined because medical services improved (soap came into widespread use), food distribution networks became more effective, farm production expanded, and urbanization developed. In the 1920s, India's population still was growing at a rate of only 1.04 percent, but by the 1970s, that rate had shot up to 2.22 percent per year (Fig. 8-8). Note that India gained 28 million people during the 1920s but a staggering 135 million during the 1970s.

Has India entered the third stage, when the death rate begins to level off and birth rates decline substantially, narrowing the gap and slowing the annual increase? The rate of increase suggests that it has: from 2.22 percent during the 1970s, it dropped to 2.11 percent in the 1980s and then to 1.88 percent during the 1990s (Fig. 8-8). But India has another problem. During its population explosion, its numbers grew so large that even a declining rate of natural change continues to add ever greater numbers to its total. In Figure 8-8, we see that while

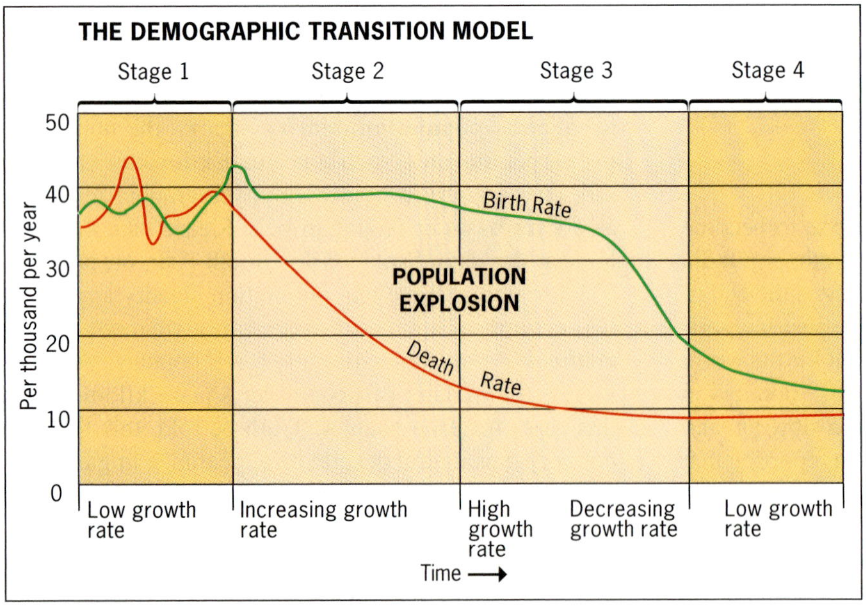

FIGURE 8-7 © H. J. de Blij, P. O. Muller, and John Wiley & Sons, Inc.

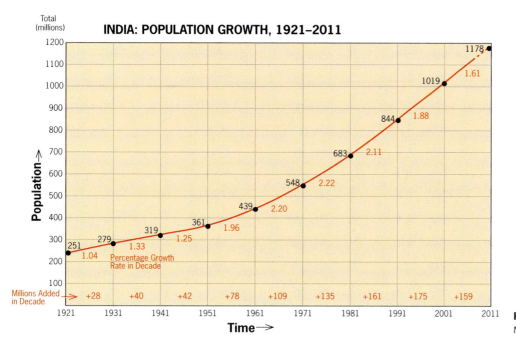

INDIA: POPULATION GROWTH, 1921–2011

FIGURE 8-8 © H. J. de Blij, P. O. Muller, and John Wiley & Sons, Inc.

the decadal rate of increase dropped from 2.22 to 1.88 percent between 1971 and 2001, the millions added grew from 135 in the 1970s to 161 in the 1980s to 175 during the 1990s—taking the total past 1 billion during 1999. If India has indeed entered the third stage of the demographic transition, it will not feel its effects for some time.

Some population geographers theorize that the populations of all countries will eventually stabilize at some level, just as Europe's did. Certain governments, notably China's, have instituted regulations to limit family size, but this policy is more easily implemented by dictatorships than democracies. And even if such stabilization were just one doubling of its current population away, India still would have an astronomical total of nearly two and a half billion residents.

India: Internal Geographic Variation

Statistics for a country as large as India tend to lose their usefulness unless they are put in geographic context. In its demographic as in so many of its other aspects, there is not just one India but several regionally different and distinct Indias. Figure 8-9 takes population growth down to the State level and provides a comparison between the census periods of 1981–1991 and 1991–2001 (the names of India's States are shown in Fig. 8-12, and their populations are given in Table 8-1).

During the period from 1981 to 1991, the highest growth rates were recorded in the States of the northeast, and only eight States had growth rates below 2 percent.

But between 1991 and 2001, no fewer than 11 States, including Goa, had growth rates below 2 percent. Comparing these two maps tells us that little has changed in India's heartland, where the populous States of West Bengal and Madhya Pradesh show slight decreases but Uttar Pradesh and Bihar display similarly slight increases. The most important reductions in rates of natural increase are recorded in parts of the northeast, the east, and the south. In the tip of the peninsula, Kerala and neighboring Tamil Nadu have growth rates lower than that of the region's leading country, Sri Lanka.

One reason for these contrasts lies in India's federal system: individual States pursued population-control policies to reduce population growth, including mass sterilizations. Another reason, as we note in more detail in the regional section, relates to spatial differences in India's development. As in the world at large, India's economically-better-off States have lower rates of population growth.

On the Periphery

India is a Hindu-dominated, officially secular country in which population policies can be debated and implemented. With 76 percent of South Asia's total population, its demographic circumstances obviously affect the entire realm. India's population growth rate remains far higher than the global rate (1.7 to 1.2 percent, respectively). But the growth rates of Pakistan and Bangladesh are substantially higher than India's (in the case of Pakistan at 2.4, double the world rate and one of the highest among the more populous countries on Earth). In both countries, the

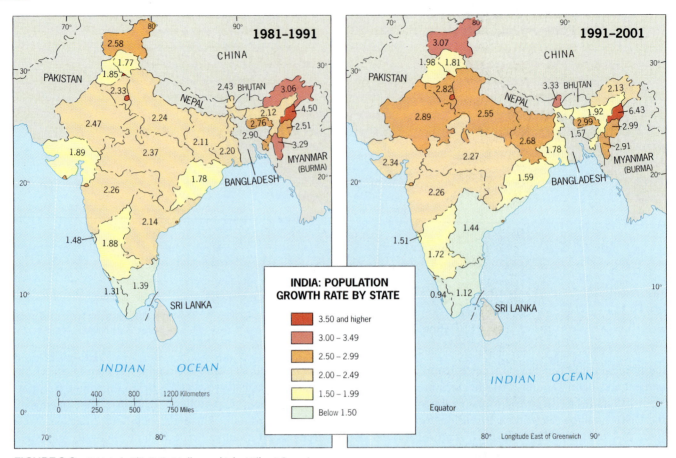

FIGURE 8-9   © H. J. de Blij, P. O. Muller, and John Wiley & Sons, Inc.

predominance of Islamic traditions, low levels of urbanization and modernization, and the inferior status of women (check the percentage of female literacy in Table G-1!) are among the factors contributing to this troubling situation.

When the world's geographic realms are viewed in demographic context, South Asia and Subsaharan Africa appear to face similar problems. But we should remember that the total population of all of Subsaharan Africa's countries is only half of South Asia's, in an area almost five times as large. South Asia's demographic condition is one of its key diagnostic properties.

SOUTH ASIA'S BURDEN OF POVERTY

In recent years, optimistic television news reports and articles in the popular press have been proclaiming a new era for South Asia, marked by rising growth rates for the realm's national economies, rewards from globalization and modernization, and increasing integration into the global economy. India, obviously the key to the realm,

has been described as "India Shining" during this wave of enthusiasm.

And indeed, a combination of circumstances ranging from America's involvement with Pakistan in the campaign against terrorism to the real estate and stock market booms in India suggest that a new era has arrived. But consider this: more than two-thirds of India's nearly 1.2 billion people continue to live in poverty-stricken rural areas, their villages and lives virtually untouched by what is happening in the cities (where, by the way, tens of millions of urban dwellers inhabit some of the world's poorest slums). Fully a third of Pakistan's population lives in abject poverty; female literacy is below 30 percent. Half of the people of Bangladesh, and nearly half those in the realm as a whole, live on the equivalent of one U.S. dollar per day or less. It is estimated that half the children in South Asia are malnourished and underweight, a majority of them girls—this at a time when the world is able to provide adequate calories for all its inhabitants, if not adequately balanced daily meals.

Why is poverty so severe in this realm, comparable in this respect only to Subsaharan Africa? We have already noted more than one reason: rapid population growth in a realm with already high physiologic densities; a continued and direct dependence, by hundreds of millions of subsistence farmers, on the vagaries of climate and weather; cultural traditions that put females at a perilous disadvantage. But there are other reasons, including land ownership (in Pakistan's province of Sind, for example, 7 percent of landowners hold more than 40 percent of the land; in India, hundreds of millions of peasants have no control over the land they cultivate) plus government inefficiency, corruption, and venality. Even as children go hungry, powerful farm lobbies pressure the government to buy and store their grain, raising prices and thus increasing the market cost for poor villagers unable to purchase enough of it. India's nationwide subsidized food distribution system, set up nearly a half-century ago to deal with food crises, suffers from disorganization, corruption, and theft. And in this era of globalization, the government is also under pressure from international organizations that lend India money to limit subsidies to consumers. In the pages that follow, the specter of poverty haunts even the most optimistic projections for this troubled realm.

THE LATEST INVASIONS

While it may be premature to forecast a dramatic economic rise of the sort that has thrust China onto the world stage, there can be no doubt that South Asia today commands the world's attention. Even before the events of 9/11, the end of the Cold War had altered the geopolitics of Eurasia. Before the 1990s India had tilted toward Moscow, but the new era saw renewed linkages between India and the United States propelled by business as well as strategic motivations. Following the 2001 terrorist attacks in New York and Washington, the United States secured Pakistan's support in the War on Terror in return for massive financial assistance and support for Pakistan's authoritarian ruler (who had come to power through a coup in 1999). To get an idea of the consequences, consider this: in 2001, Pakistan's economy grew by just 2 percent (substantially below its rate of population growth); in 2006, it grew by 7 percent. Meanwhile, India became a bastion for information technology and outsourcing industries, a factor in the fortunes of many large American companies. In 2005, during a presidential visit to India, the United States signed an agreement to grant India full civil nuclear energy cooperation, encompassing fuel supplies and technology transfers, and easing energy-poor India's way toward nuclear-energy development. Since both Pakistan and India possess nuclear weapons, the United States had withheld such cooperation because India never signed international non-proliferation agreements.

Thus South Asia today is witnessing still another of its historic invasions. But this time it is an invasion of investment, commerce, technology, custom, and connectivity—in short, globalization. As we noted, hundreds of millions of people are as yet virtually unaffected by this penetration, but from the textile sweatshops of Bangladesh to the telephone banks of Bengaluru (Bangalore) you can see the future. And in the streets, where shops carry ever more international goods, you see the increasing blend of the traditional and the external. It would be impractical to draw a map of all this, because the mosaic is too intricate. India's teeming Mumbai (Bombay) has become a global metropolis, part of the international network of world cities. On the other hand, Pakistan's Lahore remains traditional and locally focused. Chaotic, dangerous Karachi is somewhere in between. The ricefields of Bangladesh and the grinding rural poverty of India's adjoining West Bengal State are a world apart from the expansive wheatfields of the Punjab.

The latest—and current—invasion of South Asia is in the process of linking this realm to the wider world as never before, not even during colonial times. As we focus on the regions of this realm, we should remember that the fortunes of the twenty-first century world will be linked directly to the fate of these constituent parts.

POINTS TO PONDER

- More than 60 years after independence, India's changing of geographic names continues. Bangalore is now Bengaluru, Cochin is Kochi. Not everyone in India is happy about these name changes.

- In 2006 India and the United States signed a landmark agreement allowing India to import civilian nuclear technology in exchange for opening key nuclear facilities to international inspections. India already has a nuclear weapons arsenal.

- Vegetarian Indians are occasionally eating meat in ever-larger numbers, a consequence of globalization. Loosening observance of religious strictures, cheaper available meat (especially chicken), and increasing foreign travel appear to be driving the change.

- Populous India contains the world's largest minority, the 162-million-strong Muslim sector. But apart from a small, affluent elite, Muslims remain near the bottom of the social ladder.

Regions of the Realm

● PAKISTAN: ON SOUTH ASIA'S WESTERN FLANK

If India is the dominant entity in South Asia, why focus first on Pakistan? There are several reasons, both historic and geographic. Here in Pakistan lay South Asia's earliest urban civilizations, whose innovations radiated into the great peninsula. Here lies South Asia's Muslim frontier, contiguous to the great Islamic realm to the west and irrevocably linked to the enormous Muslim minority to its east. Pakistan's cultural landscapes bear witness to its transitional location. Teeming, disorderly Karachi is the typical South Asian city; as in India, the largest urban center lies on the coast. Historic, architecturally Islamic Lahore is reminiscent of the scholarly centers of Muslim Southwest Asia. In Pakistan's east, the partition boundary divides a Punjab that is otherwise continuous, a land of villages, wheatfields, and irrigation ditches. In the northwest, Pakistan resembles Afghanistan in its huge migrant populations and its mountainous frontier. And in the far north, Pakistan and India are locked in a deadly conflict over Jammu and Kashmir. The western flank is South Asia's most critical region.

This was perhaps never more true than it is today. Pakistan lies on the eastern periphery of the vast contiguous Islamic realm that extends from Casablanca to Kashmir, in the shadow of Afghanistan and in the fulcrum of the War on Terror. Islamic Pakistan's coherence and stability have become crucial at a time when the global struggle against terrorism is entangling its leaders with Western power and priorities. Pakistan's role in this struggle, dictated by its chief of state, does not have the approval of tens of millions of Pakistanis, who evince their distaste by voting for militant Islamic parties, voicing support for the vanquished Taliban movement, and giving refuge to fugitive al-Qaeda operatives in their homes. The future of Pakistan is once again in the balance, and with it the stability of South Asia.

Gift of the Indus

If, as is so often said, Egypt is the gift of the Nile, then Pakistan is the gift of the Indus. The Indus River and its principal tributary, the Sutlej, nourish the ribbons of life that form the heart of this populous country (Fig. 8-10). Territorially, Pakistan is not large by Asian standards; its area is about the same as that of Texas plus Louisiana. But Pakistan's population of 173.9 million makes it one of the world's ten most populous states. Among Muslim countries (officially, it is known as the Islamic Republic of Pakistan), only Southeast Asia's Indonesia is larger, but Indonesia's Islam has long been less militant than Pakistan's. Pakistan has been in the forefront of support for Islamic revivalism; it lies near the heart of central Asia; it has the nuclear bomb. Pakistan is a giant in the world of Islam.

Pakistan lies like a gigantic wedge between Iran and Afghanistan to the west and India to the east. This wedge extends from the Arabian Sea in the south, where it is widest, to the snowcapped mountains of the north. Here in the far north of Pakistan, the boundary framework is complex and jurisdictions are uncertain. India, China, Pakistan, and, until recently, the former Soviet Union have all vied for control over parts of this mountainous northern frontier. The Soviets, by virtue of their temporary rule over Afghanistan, controlled the Vakhan Corridor (the long narrow segment of land lying just south of Tajikistan in Fig. 8-10). The Chinese have claimed areas to the east. But the major territorial conflict has been between Pakistan and India over the areas of Jammu and Kashmir. This issue has plagued relations between these two countries for decades (see box titled "The Problem of Kashmir").

Postcolonial Reorientation

When Pakistan became an independent state following the partition of British India in 1947, its capital was Karachi on

MAJOR CITIES OF THE REALM

City	Population* (in millions)
Ahmadabad, India	5.5
Bengaluru (Bangalore), India	6.9
Chennai (Madras), India	7.3
Colombo, Sri Lanka	0.7
Delhi–New Delhi, India	16.4
Dhaka, Bangladesh	13.9
Hyderabad, India	6.5
Karachi, Pakistan	12.7
Kathmandu, Nepal	1.0
Kolkata (Calcutta), India	15.0
Lahore, Pakistan	6.8
Mumbai (Bombay), India	19.3
Varanasi, India	1.4

*Based on 2008 estimates.

FIGURE 8-10   © H. J. de Blij, P. O. Muller, and John Wiley & Sons, Inc.

the south coast, near the western end of the Indus Delta. As the map shows, however, the present capital is Islamabad, near the larger city of Rawalpindi in the north, not far from Kashmir. By moving the capital from the "safe" coast to the embattled interior and by placing it on the doorstep of contested territory, Pakistan announced its intent to stake a claim to its northern frontiers. And by naming the city Islamabad, Pakistan proclaimed its Muslim foundation, here in the face of the Hindu challenge. This politico-geographical use of a national capital can be assertive, and **11** Islamabad exemplifies the principle of the **forward capital**.

At independence, Pakistan had a bounded national territory, a capital, a cultural core, and a population—but it had few centripetal forces to bind state and nation. The disparate regions of Pakistan shared the Islamic faith and

an aversion for Hindu India, but little else. Karachi and the coastal south, the desert of Baluchistan, the city of Lahore and the Punjab, the rugged northwest along Afghanistan's border, and the mountainous far north are worlds apart, and a Pakistani nationalism to match that of India at independence did not exist. Successive Pakistani governments, civilian as well as military, turned to Islam to provide the common bond that history and geography had denied the nation. In the process, Pakistan became one of the world's most theocratic states; its common law, based on the English model, was gradually transformed into a Quranic (Koranic) system with Islamic Sharia courts and associated punishments.

But even Islam itself is not unified in restive Pakistan. About 77 percent of the people are Sunni Muslims, and

The Problem of Kashmir

KASHMIR IS A territory of high mountains surrounded by Pakistan, India, China, and, along more than 50 kilometers in the far north, Afghanistan (Fig. 8-11). Although known simply as Kashmir, the area actually consists of several political divisions, including the State properly referred to as Jammu and Kashmir (one of the 562 Indian States at the time of independence) and the administrative areas of Gilgit in the northwest and Ladakh (including Baltistan) in the east. The main conflict between India and Pakistan over the final disposition of this territory has focused on the southwest, where the State of Jammu and Kashmir is located.

When partition took place in 1947, the existing States of British India were asked to decide whether they wanted to be incorporated into India or Pakistan. In most of the States, the local ruler made this decision, but Kashmir was an unusual case. It had about 5 million inhabitants at the time, nearly half of them Muslims concentrated in the basin known as the Vale of Kashmir, where the capital, Srinagar, is located (Fig. 8-11). Another 45 percent of the people, dominantly Hindu, were concentrated in Jammu, which leads down the foothill slopes of the Himalayas to the edge of the Punjab. A small minority of Buddhists made the area called Ladakh their mountainous home.

But in the State of Jammu and Kashmir the ruler was Hindu, not Muslim, although the overall population was more than 75 percent Muslim. Thus the ruler had to make a difficult decision in 1947—to go with Pakistan and thereby exclude the State from Hindu India, or to go with India and thereby incur the wrath of most of the people. Hence the maharajah of Kashmir sought to remain outside both Pakistan and India and to retain autonomous status. This decision was followed, after the partitioning of India and Pakistan, by a Muslim uprising against Hindu rule in Kashmir. The maharajah asked for Indian help, and Pakistan's forces came to the aid of the Muslims. After

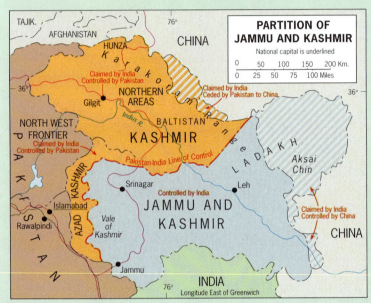

FIGURE 8-11   © H. J. de Blij, P. O. Muller, and John Wiley & Sons, Inc.

the Shiah minority numbers about 20 percent. Sunni fanatics intermittently attack Shi'ites, leading to retaliation and creating grounds for subsequent revenge.

Despite the Islamization of Pakistan's plural society, it remains a strongly regionalized country in which Urdu is the official language and English is still the *lingua franca* of the elite. Yet several other major languages prevail in diverse parts, and lifeways vary from nomadism in Baluchistan to irrigation farming in the Punjab to pastoralism in the northern highlands.

A Long Way to Go

To govern so diverse and fractious a country would test any system, and so far Pakistan has failed the test. Democratically elected governments have repeatedly squandered their opportunities, only to be overthrown by military coups; a military regime was in power when the United States launched its War on Terror in late 2001, and Pakistan lay in the crosshairs. Becoming an American ally boosted the economy, but Pakistan's fundamental problems may in fact have worsened. The economic boom has not filtered down to the poor; literacy rates are not rising; health conditions are not improving significantly; national institutions are weak (for example, there are only about 2 million registered taxpayers in a country of nearly 175 million); and one consequence of the War on Terror is that Pakistanis, who have always been overwhelmingly secular in their political choices, are joining Islamic parties in larger numbers.

Meanwhile, too little is being done to confront a growing water-supply crisis, an insurgency festers in Baluchistan, the army is incapable of establishing control over mountainous Waziristan (where al-Qaeda and Taliban organizations maintain hideouts), the issue of Kashmir costs Pakistan dearly, and relations with neigh-

more than a year's fighting and through the intervention of the United Nations, a cease-fire line left Srinagar, the Vale, and most of Jammu and Kashmir (including nearly four-fifths of the territory's population) in Indian hands (Fig. 8-11). Eventually, this line began to appear on maps as the final boundary settlement, and Indian governments have proposed that it be so recognized.

Why should two countries, whose interests would be served by peaceful cooperation, allow a distant mountainland to trouble their relationship to the point of war? There is no single answer to this question, but there are several areas of concern for both sides. In the first place, Pakistan is wary of any situation whereby India would control vital irrigation waters needed in Pakistan. As the map shows, the Indus River, the country's lifeline, crosses Kashmir. Moreover, other tributary streams of the Indus originate in Kashmir, and it was in the Punjab that Pakistan learned the lessons of dealing with India for water supplies. Second, the situation in Kashmir is analogous to the one that led to the partition of the subcontinent: Muslims are under Hindu domination. The majority of Kashmir's people are Muslims, so Pakistan argues that free choice would deliver Kashmir to the Islamic Republic. A free plebiscite is what Pakistanis have sought—and the Indians have thwarted. Furthermore, Kashmir's connections with Pakistan before partition were much stronger than those between Kashmir and India, although India has invested heavily in improving its links to Jammu and Kashmir since the military stalemate. In recent years, it did seem more likely that the cease-fire line (now called the Line of Control) would indeed become a stable boundary between India and Pakistan. (The incorporation of the State of Jammu and Kashmir into the Indian federal union was accomplished as far back as 1975, when India was able to reach an agreement with the State's chief minister and political leader.)

The drift toward stability was reversed in the late 1980s when a new crisis engulfed Kashmir. Extremist Muslim groups demanding independence escalated a long-running insurgency in 1988. That brought a swift crackdown by the Indian military, whose harsh tactics prompted the citizenry to support the separatists far more strongly. In the early 1990s, Pakistan charged the Indians with widespread human rights violations, and the Indians accused the Pakistanis of inciting secession and supplying arms. Between 1990 and 1995, nearly 10,000 people were killed in the sporadic fighting that continues to afflict this scenic area. Among the casualties was the Kashmiri Pandit (Hindu) community of about 350,000 living in the Vale, which suffered as many as 1000 deaths and left the area en masse, many ending up in refugee camps in a more secure part of the territory. In 1995, Muslim extremists began a campaign that involved capturing and killing foreign visitors, a move that was condemned by representatives of both sides. Nevertheless, the memory of three wars between India and Pakistan defied every attempt at compromise. But all this was overshadowed by the implications of the 1998 test explosion of nuclear bombs by both India and Pakistan. This development transformed Kashmir from a problem frontier to a potentially calamitous flashpoint.

boring India, which if they were satisfactory would bring enormous benefits, remain conflicted. And looming over all this are the nuclear capabilities of both Pakistan and India and their potential to trigger disaster in this contentious part of the world.

Subregions

Punjab

Pakistan's core area is the Punjab (Fig. 8-10), the Muslim heartland across which the postcolonial boundary between Pakistan and India was superimposed. (As a result, India also has a region called Punjab, sometimes spelled Panjab there.) Pakistan's Punjab is home to more than 55 percent of the country's population. In the triangle formed by the Indus River and its tributary, the Sutlej, live almost 100 million people. Punjabi is the language here, and intensive farming of wheat is the mainstay.

Three cities anchor this core area: Lahore, the outstanding center of Islamic culture in the realm, Faisalabad, and Multan. Lahore, now home to 6.8 million people, lies close to the India–Pakistan border. Founded about 2000 years ago, Lahore was situated favorably to become a great Muslim center during the Mughal period, when the Punjab was a corridor into India. As a center of royalty, Lahore was adorned with numerous magnificent buildings, including a great fort, multiple palaces, and several mosques displaying superb stonework and marble embellishments. The site of an old university and many majestic gardens, Lahore is the cultural focus of Islam. The city received hundreds of thousands of refugees and grew rapidly after partition. Lahore did lose its eastern hinterland, but its new role in independent Pakistan sustained its growth.

Punjab's relationship with Pakistan's other three provinces is one of the country's weak points. Both the governments and peoples of the other three provinces feel uneasy about the dominance of Punjab, the populous, powerful core of the country, from which most of the army is drawn.

Sind

The lower Indus River is the key to life in Sind (Fig. 8-10), but the Punjab controls the waters upstream, which is one of the issues dividing this country. When the Punjab-dominated regime proposes to build dams across the Indus and its tributaries, Sindhis (who make up about one-fourth of Pakistan's population) are reminded of their underrepresentation in government and talk of greater autonomy, even secession.

The ribbon of fertile, irrigated alluvial land along the Indus, where the British laid out irrigation systems, makes Sind a Pakistani breadbasket for wheat and rice. Commercially, cotton is king here, supplying textile factories in the cities and towns; textiles account for more than half of Pakistan's exports by value.

But the dominant presence in southern Sind is the chaotic, crime-ridden megacity of Karachi, with its stock market, dangerous streets, crowded beaches, and poverty-stricken shantytowns, a place of searing contrasts under a broiling sun. Karachi grew explosively during and after the partition of 1947, when refugee Muhajirs from across the new border with India streamed into this urban area, setting off riots and gang battles whose aftermath still simmers. With little effective law enforcement, Karachi later became a hotbed of terrorist activity, but somehow the city functions as Pakistan's (and Afghanistan's) major maritime outlet and seat of the provincial government.

North West Frontier

As Figure 8-10 shows, the North West Frontier Province lies wedged between the powerful Punjab to the east and troubled Afghanistan to the west, with the territory long known as the Tribal Areas intervening in the south (see box titled "Pakistan's Defiant Tribal Areas").

The province's mountainous physical geography reflects its remoteness and the isolation of many of its people. Mountain passes lead to Afghanistan; the Khyber Pass, already noted as the historic route of invaders, is legendary (see photo below). Coming from Afghanistan, the road takes you directly to the provincial capital, Peshawar, which lies in a broad, alluvium-filled and fertile valley where wheat and corn drape the countryside.

Here again Pakistan had to deal with a huge influx of immigrants: during the Soviet occupation of Afghanistan in the 1980s and later during the Taliban regime, several million Pushtun refugees streamed through the Khyber and other passes into refugee camps. There they found themselves subject to political pressure to demand autonomy within Pakistan, a case of ***irredentism*** that was relieved in part by United Nations assistance to the migrants. Following the ouster of the Taliban in late 2001, the great majority of Pushtuns went home. The aptly named North West Frontier Province remains a conservative, deeply religious, militant territory, where Islamic political parties and move-

FROM THE FIELD NOTES

"Driving through the Khyber Pass that links Afghanistan and Pakistan was a riveting experience. From the dry and dusty foothills on the Afghanistan side, with numerous ruins marking historic battles, I traversed the rugged Hindu Kush via multiple hairpin turns and tunnels into Pakistan. This is one of the most strategic passes in the world, but for invading armies it was no easy passage: defensive forces had the advantage in these treacherous valleys. But now the roads and tunnels facilitate the movement of refugees, drugs, and arms, and they are used by militant separatist forces from both sides of the border. Pakistan's North West Frontier Province, and especially the city of Peshawar, are hotbeds of these activities."   © Barbara A. Weightman.

Concept Caching

www.conceptcaching.com

Pakistan's Defiant Tribal Areas

WHY IS THERE a zone designated "Tribal Areas" (Fig. 8-10) in a country otherwise subdivided into provinces? It started during colonial times, when the British assigned a special status to the obstinate Pushtun villages and their chiefs living in the remote hills and mountains on the Afghan border. The "Tribal Agencies" were the responsibility of an official "agent" who exercised control over the local chiefs through ample reward and harsh punishment—and little or no accountability. There were seven of them, some very small, several large enough to appear on a relatively small-scale map such as Figure 8-10. North and South Waziristan were among the largest.

Today, the Tribal Areas (as they are now called) continue to resist authority. The Pakistan government left them pretty much alone; the army had never even entered the Tribal Areas until after 9/11, when the search for al-Qaeda and Taliban members hiding there required it. But neither Pakistan's military nor U.S. helicopters could root out the fugitive terrorists or the Taliban, and the Tribal Areas continue to be impenetrable and ungovernable.

ments are proportionately stronger than in any other part of the country and where obstruction of national policies (including the War on Terror) is a common goal.

Baluchistan

As Figure 8-10 indicates, Baluchistan is by far the largest of Pakistan's four provinces, accounting for (not including Kashmir) nearly half of the national territory but inhabited only by an estimated 8.5 million people or 5 percent of the country's population. For a sense of its terrain, take a look at Figure 8-1; much of this vast territory is desert, with mostly barren mountains that wrest some moisture from the air only in the northeast. Sheep raising is the leading livelihood here, and wool the primary export. In the far north, Baluchistan abuts the Tribal Areas and Afghanistan. The provincial capital, Quetta, lies in this zone. This province could easily be called the "South West Frontier Province."

However, Baluchistan is not unimportant. Beneath its parched surface lie possibly substantial reserves of oil as well as coal, and already the province produces most of Pakistan's natural gas. Yet 90 percent of local residents have no energy supply themselves. And there is groundwater in several areas, but 8 out of 10 Baluchis do not have access to clean water. For decades, short-lived local rebellions have signaled dissatisfaction with the government, but recently a more serious insurgency, under the banner of the Baluchistan Liberation Army (BLA), has taken a heavier toll and appears more durable (photo at right).

Part of the problem involves Pakistan's plan to build a major port and energy terminal at Gwadar on Baluchistan's southwest coast. Not only would Gwadar serve the region; it could potentially become a transshipment point for oil and gas from remote sources such as Iran and the Caspian Basin destined for markets in East Asia. China dispatched engineers to plan and build this port with a loan to Pakistan, but Baluchis complained that they were left out of the process. In 2004 three Chinese engineers were killed and a dozen wounded in a terrorist attack at Gwadar. The future of this province remains very much in doubt.

Both India and Pakistan tend to blame each other for problems ranging from territorial disputes to insurgent and terrorist actions. India and Pakistan confront each other over Kashmir, but Pakistan faces other, more domestic challenges, including a rebellion in the southwestern province of Baluchistan. This photograph suggests how protective this area's terrain can be for insurgents, in this case a group of rebels protecting their leader (center) in cave-riddled topography in the mountains around Dera Bugti. Baluchistan's Shi'ite minority objects to Islamabad's policies, but Pakistan's president has accused India of inciting them.   © AFP/Getty Images News and Sport Services.

Who Should Govern Kashmir?

KASHMIR SHOULD BE PART OF PAKISTAN!

"I don't know why we're even debating this. Kashmir should and would have been made part of Pakistan in 1947 if that colonial commission hadn't stopped mapping the Pakistan-India boundary before they got to the Chinese border. And the reason they stopped was clear to everybody then and there: instead of carrying on according to their own rules, separating Muslims from Hindus, they reverted to that old colonial habit of recognizing "traditional" States. And what was more traditional than some Hindu potentate and his minority clique ruling over a powerless majority of Muslims? It happened all over India, and when they saw it here in the mountains they couldn't bring themselves to do the right thing. So India gets Jammu and Kashmir and its several million Muslims, and Pakistan loses again. The whole boundary scheme was rigged in favor of the Hindus anyway, so what do you expect?

"Here's the key question the Indians won't answer. Why not have a referendum to test the will of all the people in Kashmir? India claims to be such a democratic example to the world. Doesn't that mean that the will of the majority prevails? But India has never allowed the will of the majority even to be expressed in Kashmir. We all know why. About two-thirds of the voters would favor union with Pakistan. Muslims want to live under an Islamic government. So people like me, a Muslim carpenter here in Srinagar, can vote for a Muslim collaborator in the Kashmir government, but we can't vote against the whole idea of Indian occupation.

"Life isn't easy here in Srinagar. It used to be a peaceful place with boats full of tourists floating on beautiful lakes. But now it's a violent place with shootings and bombings. Of course we Muslims get the blame, but what do you expect when the wishes of a religious majority are ignored? So don't be surprised at the support our cause gets from Pakistan across the border. The Indians call them terrorists and they accuse them of causing the 60,000 deaths this dispute has already cost, but here's a question: why does it take an Indian army of 600,000 to keep control of a territory in which they claim the people prefer Indian rule?

"Now this so-called War on Terror has made things even worse for us. Pakistan has been forced into the American camp, and of course you can't be against 'terrorism' in Afghanistan while supporting it in Kashmir. So our compatriots on the Pakistani side of the Line of Control have to stay quiet and bide their time. But don't underestimate the power of Islam. The people of Pakistan will free themselves of collaborators and infidels, and then they will be back to defend our cause in Kashmir."

KASHMIR BELONGS TO INDIA!

"Let's get something straight. This stuff about that British boundary commission giving up and yielding to a maharajah is nonsense. Kashmir (all of it, the Pakistani as well as the Indian side) had been governed by a maharajah for a century prior to partition. What the maharajah in 1947 wanted was to be ruled by neither India nor Pakistan. He wanted independence, and he might have gotten it if Pakistanis hadn't invaded and forced him to join India in return for military help. As a matter of fact, our Prime Minister Nehru prevailed on the United Nations to call on Pakistan to withdraw its forces, which of course it never did. As to a referendum, let me remind you that a Kashmir-wide referendum was (and still is) contingent on Pakistan's withdrawal from the area of Kashmir it grabbed. And as for Muslim 'collaborators,' in the 1950s the preeminent leader on the Indian side of Kashmir was Sheikh Muhammad Abdullah (get it?), a Muslim who disliked Pakistan's Muslim extremism even more than he disliked the maharajah's rule. What he wanted, and many on the Indian side still do, is autonomy for Kashmir, not incorporation.

"In any case, Muslim states do not do well by their minorities, and we in India generally do. As far as I am concerned, Pakistan is disqualified from ruling Kashmir by the failure of its democracy and the extremism of its Islamic ideology. Let me remind you that Indian Kashmir is not just a population of Hindus and Muslims. There are other minorities, for example, the Ladakh Buddhists, who are very satisfied with India's administration but who are terrified at the prospect of incorporation into Islamic Pakistan. You already know what Sunnis do to Shi'ites in Pakistan. You are aware of what happened to ancient Buddhist monuments in Taliban Afghanistan (and let's not forget where the Taliban came from). Can you imagine the takeover of multicultural Kashmir by Islamabad?

"To the Muslim citizens of Indian Kashmir, I, as a civil servant in the Srinagar government, say this: look around you, look at the country of which you are a subject. Muslims in India are more free, have more opportunities to participate in all spheres of life, are better educated, have more political power and influence than Muslims do in Islamic states. Traditional law in India accommodates Muslim needs. Women in Muslim-Indian society are far better off than they are in many Islamic states. Is it worth three wars, 60,000 lives, and a possible nuclear conflict to reject participation in one of the world's greatest democratic experiments?

"Kashmir belongs to India. All inhabitants of Kashmir benefit from Indian governance. What is good for all of India is good for Kashmir."

Regional **ISSUE**

Vote your opinion at www.wiley.com/college/deblij

Pakistan's Prospects

Table G-1 tells the story of Pakistan's social circumstances, but not the entire story. What began as a corner of South Asia that would be a haven for Muslims in a realm of Hindus has become one of the world's ten most populous countries, the largest Islamic state on the Eurasian mainland, a nuclear power, and, officially, a partner against terrorism. It is a country of enormous cultural contrasts, where the modern and medieval exist side by side, where you hear people of one province call those of another (but not themselves) "Pakistanis," and where a sense of nationhood is still elusive among many people whose loyalties to family, clan, and village are stronger. Take another look at Figure 8-1 to be reminded how much regional geography has to do with this: Pakistan, with neighbors such as Iran, Afghanistan, disputed and divided Kashmir, and India has been what geographers call a **marchland**, an area of uncertain boundaries entered or crossed by so many armies, refugees, and migrants that its identity has become blurred. From the Muhajirs of Karachi to the Pushtuns of the Tribal Areas, from rebellious Baluchis to disaffected Sindhis, Pakistan carries the scars of its relative location.

Location and Livelihoods

Nothing heals such scars as effectively as a thriving economy, representative government, dependable and transparent institutions, rising social indicators, and confidence in the future. But Pakistan has experienced so many cycles of progress and failure that confidence is in short supply. And yet there are areas of progress against all odds: a combination of expanded irrigation and Green Revolution farming techniques has allowed Pakistan in recent years to export some rice (although wheat imports continue), and the country's legitimate economy has shown substantial growth. Exports include not only cotton-based textiles but also carpets, tapestries, and leather goods. Domestic manufacturing remains limited, but Pakistan has built its own steel mill near Karachi. Meanwhile, the authorities struggle to control Pakistan's growing production and trade in opium and hashish; neighboring Afghanistan remains the world's largest source of this illegal commerce, which supported the Taliban regime in the recent past. The spillover effect continues, and Pakistan has many remote corners where poppy fields yield high returns and trade routes are well established. In this as in so many other spheres, Pakistan displays the contradictory symptoms of a state in transition.

Emerging Regional Power

A country's underdevelopment is not necessarily a barrier against its emergence as a power to be reckoned with in international affairs. When Pakistan became independent in 1947, the country was weak, disorganized, and divided. Just six decades later, Pakistan is a major military force and a nuclear power.

To say that Pakistan in 1947 was a divided country is no exaggeration. In fact, upon independence, present-day Pakistan was united with present-day Bangladesh, and the two countries were called West Pakistan and East Pakistan, respectively. As we have noted, the basis for this scheme was Islam: in Bangladesh, too, Islam is the state religion. Between the two Islamic wings of Pakistan lay Hindu India. But there was little else to unify the easterners and westerners, and their union lasted less than 25 years. In 1971, a costly war of secession, in which India supported East Pakistan, led to the collapse of this unusual arrangement. East Pakistan, upon independence in 1971, took the name **Bangladesh**, and since there was no longer any need for a "West" Pakistan, that qualifier was dropped and the name **Pakistan** remained on the map.

India's encouragement of independence for Bangladesh emphasized the continuing tension between Pakistan and India, which had already led to war in 1965, to further conflict in the 1970s over Jammu and Kashmir, and to flare-ups over other issues. (The continuing dispute over Kashmir is elaborated further in the Issues Box entitled "Who Should Govern Kashmir?") During the Cold War, India tilted toward Moscow, and Pakistan found favor in Washington because of its strategic location adjacent to Afghanistan. Still, armed conflict between the two South Asian countries seemed to be a regional matter—until the early 1990s, when their arms race took on ominous nuclear proportions. Both India and Pakistan had been developing nuclear weaponry, and in 1998 India exploded several nuclear bombs at its test site, a provocation that put intense pressure on Pakistan to follow suit. This took place only days later, and the notion of an "Islamic Bomb" had become reality. Since then, the specter of nuclear war has hung over the conflicts that continue to embroil Pakistan and India, a concern not just for the South Asian realm but for the world as a whole. No longer merely a decolonized, divided, and disadvantaged country trying to survive, Pakistan has taken a crucial place in the political geography of a geographic realm in turbulent transition.

● INDIA ASTIR

If you have been reading the press and watching television over the past few years, you have seen the growing attention being paid to India—not just in North America but around the world. The *New York Times* in late 2005 began a series under the title "India Accelerating." *Newsweek* in March 2006 published a special feature under the banner headline "India Rising." Three months later a leading British newsmagazine, *The Economist*, produced a Survey of India introduced by an editorial asking "Can India Fly?" State visits by an Indian prime minister to Washington, an American president to New Delhi, and a senior Chinese leader to India, all within a matter of months, reflected India's growing global importance. Meanwhile, CNN kept up a drumbeat of reportage about American jobs departing for India as U.S. corporations saved money by "outsourcing." India is on the move—but will India become the next economic superpower?

Giant of the Realm

Certainly India has the dimensions to make the world pay attention. Not only does it occupy three-quarters of the great land triangle of South Asia: India also is poised to overtake China to become the world's most populous country. Already, India is the planet's largest democracy, a federation of 28 States and several additional Territories with a 2008 population of 1.158 billion.

That India has endured as a unified country is a politico-geographical miracle. India is a cultural mosaic of immense ethnic, religious, linguistic, and economic diversity and contrast; it is a state of many nations. The period of British colonialism gave India the underpinnings of unity: a single capital, an interregional transport network, a *lingua franca*, a civil service. Upon independence in 1947, India adopted a federal system of government, giving regions and peoples some autonomy and identity, and allowing others to aspire to such status. Unlike Africa, where federal systems failed and where military dictatorships replaced them, India remained essentially democratic and retained a federal framework in which States have considerable local authority.

This political, democratic success has been achieved despite the presence of powerful centrifugal forces in this vast, culturally diverse country. Relations between the Hindu majority and the enormous Muslim minority, better in some States than in others, have at times threatened to destabilize the entire federation. Local rebellions, demands by some minorities for their own States, frontier wars, even involvement in a foreign but nearby civil war (in Sri Lanka) have buffeted the system—which has bent but not collapsed. India has succeeded where others in the postcolonial world have failed.

This success has not been matched in the field of economics, however. After more than 60 years of independence, India remains a very poor country, and not all of this can be blamed on the colonial period or on population growth, although overpopulation remains a strong impediment to improvement of living standards. Much of it re-

sults from poor and inconsistent economic planning, too much state ownership of inefficient industries, excessive government control over economic activities, bureaucratic suppression of initiative, corruption, and restraints against foreign investment. As we shall presently see, a few bright spots in some of India's States contrast sharply to the overwhelming poverty of hundreds of millions.

States and Peoples

The map of India's political geography shows a federation of 28 States, 6 Union Territories (UTs), and 1 National Capital Territory (NCT) (Fig. 8-12). The federal government retains direct authority over the UTs, all of which are small in both territory and population. The NCT, however, includes Delhi and the capital, New Delhi, and has almost 17 million inhabitants.

Postcolonial Restructuring

The political spatial organization shown in Figure 8-12 is mainly the product of India's restructuring following independence from Britain. Its State boundaries reflect the broad outlines of the country's cultural mosaic: as far as possible the system recognizes languages, religions, and cultural traditions. Indians speak 14 major and numerous minor languages, and while Hindi is the official language (and English is the *lingua franca*), it is by no means universal. The map is the product of endless compromise—endless because demands for modifications of it continue to this day. As recently as late 2000, the federal government authorized the creation of three new States. In the northeast lie very small States established to protect the local traditions of small populations; minority groups in the larger States ask why they should not receive similar recognition.

Staggering Numbers

With only 28 States for a national population of 1.158 billion, several of India's States contain more people than many countries of the world (Table 8-1). As Figure 8-12 shows, the (territorially) largest States lie in the heart of the country and on the great southward-pointing peninsula. Uttar Pradesh (190 million, according to the 2008 population estimate) and Bihar (about 94 million) constitute much of the Ganges River Basin and are the core area of modern India (see box titled "Solace and Sickness from the Holy Ganges"). Maharashtra (almost 108 million), anchored by the great coastal city of Bombay (renamed Mumbai in 1996), also has a population larger than that of most countries. West Bengal, the State that adjoins Bangladesh, contains 87 million residents,

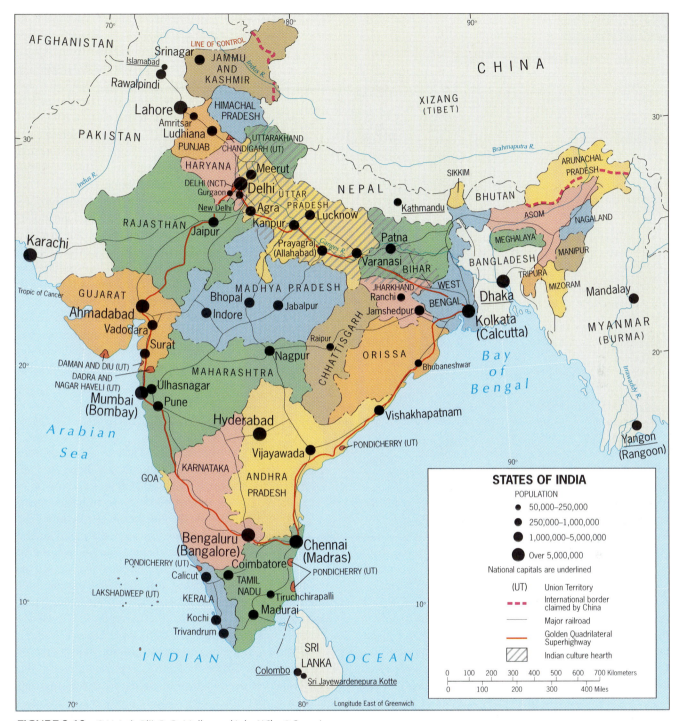

FIGURE 8-12   © H. J. de Blij, P. O. Muller, and John Wiley & Sons, Inc.

15 million of whom live in its urban focus, Calcutta (re-named Kolkata in 2000).

These are staggering numbers, and they do not decline much toward the south. Southern India consists of four States linked by a discrete history and by their distinct Dravidian languages. Facing the Bay of Bengal are Andhra Pradesh (more than 82 million) and Tamil Nadu (66 million), both part of the hinterland of the city of Madras (renamed Chennai in 1997) and located on the coast near their joint border. Facing the Arabian Sea are Karnataka (almost 58 million) and Kerala (34 million). Kerala, often at odds with the federal government in New Delhi, has long had the highest literacy rate in India and one of the lowest rates of population growth owing to strong local government and strictly enforced policies. "It's a matter of geography," explained a teacher in the Kerala city of Kochi. "We are here about as far away as you can get from the capital, and we make our own rules."

TABLE 8-1

India: Population by State

State	1991 Census	2001 Census	Estimated 2008
Andhra Pradesh	66,508,000	75,728,000	82,375,000
Arunachal Pradesh	865,000	1,091,000	1,198,000
Asom (Assam)	22,414,000	26,638,000	29,435,000
Bihar	86,374,000	82,879,000	93,633,000
Chhattisgarh*		20,796,000	23,269,000
Goa	1,170,000	1,344,000	1,596,000
Gujarat	41,310,000	50,597,000	56,626,000
Haryana	16,464,000	21,083,000	24,171,000
Himachal Pradesh	5,171,000	6,077,000	6,595,000
Jammu and Kashmir	7,719,000	10,070,000	11,257,000
Jharkhand*		26,909,000	30,181,000
Karnataka	44,977,000	52,734,000	57,550,000
Kerala	29,099,000	31,839,000	33,802,000
Madhya Pradesh	66,181,000	60,385,000	68,737,000
Maharashtra	78,937,000	96,752,000	107,972,000
Manipur	1,837,000	2,167,000	2,364,000
Meghalaya	1,775,000	2,306,000	2,530,000
Mizoram	690,000	891,000	970,000
Nagaland	1,210,000	1,989,000	2,171,000
Orissa	31,660,000	36,707,000	39,655,000
Punjab	20,282,000	24,289,000	26,722,000
Rajasthan	44,006,000	56,473,000	64,534,000
Sikkim	406,000	541,000	591,000
Tamil Nadu	55,859,000	62,111,000	66,106,000
Tripura	2,757,000	3,191,000	3,491,000
Uttar Pradesh	139,112,000	166,053,000	190,254,000
Uttarakhand*		8,480,000	9,511,000
West Bengal	68,078,000	80,221,000	86,995,000
National Capital Territory**		13,783,000	16,955,000
Union Territories	1,998,000	2,670,000	3,491,000

*Established 2000
**Established 1993

The Northern Peripheries

As Figure 8-12 shows, India's smaller States lie mainly in the northeast, on the far side of Bangladesh, and in the northwest, toward Jammu and Kashmir. North of Delhi, India is flanked by China and Pakistan, and physical as well as cultural landscapes change from the flatlands of the Ganges to the hills and mountains of spurs of the Himalayas. In the State of Himachal Pradesh, forests cover the hillslopes and relief reduces living space; less than 7 million people live here, many in small, comparatively isolated clusters. Before independence and political consolidation, the colonial government called this area the "Hill States."

The map becomes even more complex in the distant northeast, beyond the narrow corridor between Bhutan and Bangladesh. The dominant State here is Asom (Assam), home to 30 million, famed for its tea plantations and important because its oil and gas production amounts to more than 40 percent of India's total.

In the Brahmaputra Valley, Asom resembles the India of the Ganges. But in almost all directions from Asom, things change. To the north, in sparsely populated Arunachal Pradesh (1.2 million), we are in the Himalayan offshoots again. To the east, in Nagaland (2.2 million), Manipur (2.4 million), and Mizoram (1 million), lie the forested and terraced hillslopes that separate India from Myanmar (Burma). This is an area of numerous ethnic groups (more than a dozen in Nagaland alone) and of frequent rebellion against Delhi's government. And to the south, the States of Meghalaya (2.5 million) and Tripura (3.5 million), hilly and still wooded, border the teeming floodplains of Bangladesh. Here in the country's northeast, where peoples are always restive and where population growth is still soaring, India faces one of its strongest regional challenges.

China's Latent Claims

Here, too, India faces a challenge from its powerful neighbor China. As Figure 8-12 shows, China's version of its southern border with India lies deep inside the Northeast, even coinciding with the Asom boundary part of the way. This dispute has its origins in the Simla Conference of 1913–1914, when the British negotiated a treaty with Tibet that defined their boundary in accordance with the terms laid down on the map by their chief negotiator, Sir Henry McMahon. Essentially, this **McMahon Line**, as it came to be known, ran along the most prominent crestline of the Himalayas. However, China's representatives at the conference refused to sign the agreement, arguing that Tibet was part of China and had no power to enter into treaties with foreign governments. In 1962, Chinese armed forces crossed the border and occupied Indian territory, but then withdrew. Today the matter remains unresolved.

India's Changing Map

As we noted earlier, 90 years of British rule in India—the *raj*—created an intricate mosaic of direct and indirect rule. Geographically, British authority prevailed in coastal

Solace and Sickness from the Holy Ganges

STAND ON THE banks of the Ganges River in Varanasi, Hinduism's holiest city, and you will see people bathing in holy water, drinking it, and praying as they stand in it—while the city's sewage flows into it nearby, and the partially cremated corpses of people and animals float past. It is one of the world's most compelling—and disturbing—sights.

The Ganges (*Ganga*, as the Indians call it) is Hinduism's sacred river. Its ceaseless flow and spiritual healing power are Earthly manifestations of the Almighty. Therefore, tradition has it, the river's water is immaculate, and no amount of human (or other) waste can pollute it. On the contrary: just touching the water can wash away a believer's sins.

At Varanasi, Prayagraj (Allahabad), and other cities and towns along the Ganges, the river banks are lined with Hindu temples, decaying ornate palaces, and dozens of wide stone staircases called *ghats*. These stepped platforms lead down to the water, enabling thousands of bathers to enter the river. They come from the city and from afar, many of them pilgrims in need of the healing and spiritual powers of the water. It is estimated that more than a million people enter the river somewhere along its 2600-kilometer (1600-mile) course every day. During religious festivals, the number may be ten times as large.

By any standards, the Ganges is one of the world's most severely polluted streams, and thousands among those who enter it become ill with diarrhea or other diseases; many die. In 1986, then-Prime Minister Rajiv Gandhi launched a major scheme to reduce the level of pollution in the river, a decade-long construction program of sewage treatment plants and other facilities. In the mid-1980s, Gandhi was told, nearly 400 million gallons of sewage and other wastes were being disgorged into the Ganges every day. The plan called for the construction of nearly 40 sewage treatment plants in riverfront cities and towns.

Many Hindus, however, opposed this costly program to clean up the Ganges. To them, the holy river's spiritual purity is all that matters. Getting physically ill is merely incidental to the spiritual healing power contained in a drop of Ganga's water.

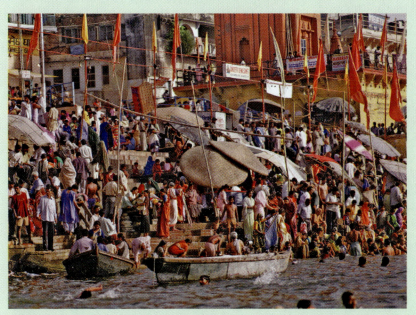

The stone steps leading into the Ganges' waters in Varanasi, India. Varanasi is India's holiest city, and millions descend these ghats every year.   © David Zimmerman/ Masterfile.

---

areas, in the major river valleys where large populations clustered, and in strategic border zones where outside threats required direct rule. Most of the historic Princely States in the interior were controlled by British "advisors" to local princes whose positions depended on their acceptance of indirect rule. The British began experimenting with representative government at an early stage, allowing the formation of an Indian National Congress in 1885, transferring some political power to elected provincial officials in 1919, and approving elected provincial legislatures in 1935. Indian leaders, notably Jawaharlal Nehru, wanted to achieve independence through a united front, but this was thwarted by the Muslim League, which demanded the es-

tablishment of a separate Islamic state. When sovereignty came in 1947, there was chaos and conflict—and India's administrative map needed immediate restructuring.

After independence, the government started this process by phasing out the still-privileged Princely States. Next, the government reorganized the country on the basis of its major regional languages (see Fig. 8-5). Hindi, then spoken by more than one-third of the population, was designated the country's official language, but the Indian constitution gave national status to 13 other major languages, including the four Dravidian languages of the south. English, it was anticipated, would become India's common language, its **lingua franca** at government, administrative, and business levels. Indeed, English not only remained the language of national administration but also became the chief medium of commerce in growing urban India. English was the key to better jobs, financial success, and personal advancement, and the language constituted a common ground in higher education.

Devolutionary Pressures

The newly devised framework based on the major regional languages, however, proved to be unsatisfactory to many communities in India. In the first place, many more languages are in use than the 14 that had been officially recognized. Demands for additional States soon arose. As early as 1960, the State of Bombay was divided into two language-based States, Gujarat and Maharashtra.

This devolutionary pressure has continued throughout India's existence as an independent country. In 2000, the three newest States were recognized: Jharkhand, carved from southern Bihar State on behalf of 18 poverty-stricken districts there; Chhattisgarh, where tribal peoples had been agitating since the 1930s for separation from the State of Madhya Pradesh; and Uttarakhand (originally named Uttaranchal), which split from India's most populous, Ganges Basin, core-area State of Uttar Pradesh on the basis of its highland character and lifeways (Fig. 8-12).

For many years India has faced quite a different set of cultural-geographic problems in its northeast, where numerous ethnic groups occupy their own niches in a varied, forest-clad topography. The Naga, a cluster of peoples whose domain had been incorporated into Asom (Assam) State, rebelled soon after India's independence. A protracted war brought federal troops into the area; after a truce and lengthy negotiations, Nagaland was proclaimed a State in 1961. That led the way for other politico-geographic changes in India's problematic northeastern wing.

The Sikhs

A further dilemma involved India's Sikh population. The Sikhs (the word means "disciples") adhere to a religion that was created about five centuries ago to unite warring Hindus and Muslims into a single faith. This faith's principles rejected the negative aspects of Hinduism and Islam, and it gained millions of followers in the Punjab and adjacent areas. During the colonial period, many Sikhs supported British administration of India, and by doing so they won the respect and trust of the British, who employed tens of thousands of Sikhs as soldiers and policemen. By 1947, there was a large Sikh middle class in the Punjab. When independence came, many left their rural homes and moved to the cities to enter urban professions. Today, they still exert a strong influence over Indian affairs, far in excess of the less than 2 percent of the population (about 20 million) they constitute.

For a time after India's independence, the Sikhs created India's most serious separatist problem, demanding the formation of an autonomous State they wanted to call **Khalistan**. This campaign entailed violence that not only rocked India but also reached into Sikh communities as far away as Canada. There, in the worst terrorist act in that country's history, Sikhs in 1985 placed bombs aboard an Air India jumbo jet, killing 329 as the plane exploded over the North Atlantic on its way back to India. But once again India's capacity to accommodate divisive forces eventually overcame the Sikhs' devolutionary drive. The government's decision to divide the original State of Panjab (Punjab) into a Sikh-dominated northwest—which retained the name Punjab—and a Hindu-dominated southeast (Haryana) created the Sikh stronghold this minority had wanted and that turned out to be crucial in defusing the separatist movement.

The Muslims

When India became a sovereign state and the great population shifts across its borders had ended, the country was left with a Muslim minority of about 35 million widely dispersed throughout the country. By 1991, that minority had grown to nearly 99 million, representing 11.7 percent of the total population (Fig. 8-13A). The 2001 Census of India reported a Muslim population of more than 138 million, constituting 13.4 percent and growing at a rate of nearly 4 million per year. This indicated a rate of increase in the Muslim sector of about 2.8 percent compared to India's overall rate of 1.7 percent, that is, two-thirds higher than the national average. As Table G-1 indicates, this is one of the world's fastest rates of natural increase; only a few countries are growing as rapidly, and none in South Asia ranks higher than 2.4 percent. The latest estimates for 2007 give a Muslim population total of 162 million (14 percent of the national total).

The map displaying the 2001 distribution of Muslims in India (Fig. 8-13B) reveals actual numbers as well as percentages and indicates that the core-area State of Uttar Pradesh and the Bangladesh-bordering State of West Bengal each contain more than 20 million Muslim inhabitants, while another two States, core-area Bihar and west-central Maharashtra, each contain more than 10 million. In terms of percentage of the total State population, however, Jammu and Kashmir leads with 67 percent, followed by Asom in northeastern India with 31 percent. Also noteworthy is Kerala in the far south, with nearly 25 percent of its population Muslim. India's great geographic advantage

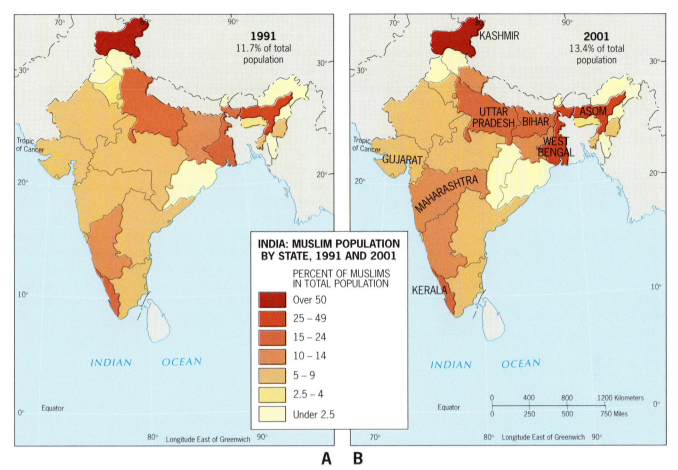

FIGURE 8-13   © H. J. de Blij, P. O. Muller, and John Wiley & Sons, Inc.

is that its Muslim minority is not regionally concentrated, precluding the kind of secession movement that often develops when communities with different cultures are spatially divided (as we shall see, Sri Lanka in this realm suffers from exactly that kind of problem).

Relations between the Hindu majority and the Muslim minority are complex. What makes the news is conflict—for example, in 2002 when Muslims attacked a train carrying Hindus to a contested holy site in Ayodhya, killing dozens, which was followed by retaliation in several towns (but not others) in Gujarat. What does not make the news is that a Muslim population of more than 160 million lives and participates in the kind of democracy that is all but unknown in the Muslim world itself. In the first decade of the twenty-first century, a Hindu nationalist party was the leading force in the federal government, but a Muslim scientist, the man who developed India's first nuclear weapons, was the country's president. On the other hand, many Muslims have been angered by the Hindu-inspired name changes appearing on India's map, most recently the substitution of Prayagraj for the venerable Muslim name of Allahabad in the State of Uttar Pradesh.

Concern over links between local Muslim subversives and foreign terrorists is changing Hindu attitudes, as is reflected by growing cooperation between Israeli and Indian intelligence operations and official statements linking a recent terrorist attack on the Indian Parliament not only to Pakistan (the usual suspect) but also to domestic collaborators. Pakistan's prominent role in the War on Terror is seen as a contradiction in terms by Indians facing continuous terrorism in their sector of Kashmir. Achieving a lasting accommodation of the Muslim minority will be a key challenge for India in the decades ahead.

Centrifugal Forces: From India to Hindustan?

In Chapter 1 we introduced the concept of centrifugal and centripetal forces affecting the fabric of the state. No country in the world exhibits greater cultural diversity than India, and variety in India comes on a scale unmatched anywhere else on Earth. Such diversity spells strong centrifugal forces, although, as we will see, India also has powerful consolidating bonds.

Class and Caste

Among the centrifugal forces, Hinduism's stratification of society into castes remains pervasive. Under Hindu dogma, *castes* are fixed layers in society whose ranks are based on ancestries, family ties, and occupations. The caste system may have its origins in the early social divisions into priests and warriors, merchants and farmers, craftspeople and servants; it may also have a racial basis, for the Sanskrit term for caste is color. Over the centuries, its complexity grew until India had thousands of castes, some with a few hundred members, others containing millions. Thus, in city as well as in village, communities were segregated according to caste, ranging from the highest (priests, princes) to the lowest (the untouchables). The term *untouchable* has such negative connotations that some scholars object to its use. Alternatives include *dalits* (oppressed), the common term in Maharashtra State but coming into general use; *harijans* (children of God), which was Gandhi's designation, still widely used in the State of Bihar; and *Scheduled Castes*, the official government label.

A person was born into a caste based on his or her actions in a previous existence. Hence, it would not be appropriate to counter such ordained caste assignments by permitting movement (or even contact) from a lower caste to a higher one. Persons of a particular caste could perform only certain jobs, wear only certain clothes, worship only in prescribed ways at particular places. They or their children could not eat, play, or even walk with people of a higher social status. The untouchables occupying the lowest tier were the most debased, wretched members of this rigidly structured social system. Although the British ended the worst excesses of the caste system, and postcolonial Indian leaders—including Mohandas (Mahatma) Gandhi (the great spiritual leader who sparked the independence movement) and Jawaharlal Nehru (the first prime minister)—worked to modify it, a few decades cannot erase centuries of class consciousness. In traditional India, caste provided stability and continuity; in modernizing India, it constitutes an often painful and difficult legacy.

Today we can discern a geography of caste—a degree of spatial variation in its severity. Cultural geographers estimate that about 15 percent of all Indians are of lower caste, about 40 percent of backward caste (one important rank above the lower caste), and some 18 percent of upper caste, at the top of which are the Brahmans, men in the priesthood. (The caste system does not extend to the Muslims or Sikhs, although converted Christians are affected, which is why these percentages do not add up to 100.) The colonial government and successive Indian governments have tried to help the lowest castes. This effort has had more effect in urban than in rural areas of

India. In the isolated villages of the countryside, the untouchables often are made to sit on the floor of their classroom (if they go to school at all); they are not allowed to draw water from the village well because they might pollute it; and they must take off their shoes, if they wear any, when they pass higher-caste houses. But in the cities, untouchables have reserved for them places in the schools, a fixed percentage of State and federal government jobs, and a quota of seats in national and State legislatures. Gandhi, who took a special interest in the fate of the untouchables in Indian society, accomplished much of this reform.

The caste system remains a powerful centrifugal force, not only because it fragments society but also because efforts to weaken it often result in further division. Gandhi himself was killed, only a few months after independence, by a Hindu fanatic who opposed his work for the least fortunate in Indian society. But progress is being made, and while efforts to help the poorest are not always popular among the better-off, the future of India depends on it.

Hindutva

Another growing centrifugal force in India has to do with a concept known as *hindutva* or Hinduness—a desire to remake India as a society in which Hindu principles prevail. This concept has become the guiding agenda for a political party that became a powerful component of the federal government, and it is variously expressed as Hindu nationalism, Hindu patriotism, and Hindu heritage. This naturally worries Muslims and other minorities, but it also concerns those who understand that India's secularism, its separation of religion and state, is indispensable to the survival of its democracy. *Hindutva* enthusiasts want to impose a Hindu curriculum on schools, change the flexible family law in ways that would make it unacceptable to Muslims, inhibit the activities of non-Hindu religious proselytizers, and forge an India in which non-Hindus are essentially outsiders. Moderate Hindus and non-Hindus in India oppose such notions, which are as divisive as any India has faced. They nonetheless acknowledged the appeal of the Bharatiya Janata Party (BJP), which swept to power on a Hindu-nationalist platform, and realized that only the constraints of a coalition government kept the party from pursuing its more extreme goals. Recent State and national elections have seen a decline in the fortunes of the BJP, reflecting an internal party struggle between moderates and *hindutva* hardliners—and also confirming the continuing strength and vitality of Indian democracy.

The radicalization of Hinduism and the infusion of Hindu nationalism into federal politics might be seen by outsiders as threats to India's unity, but Indian voters have not rushed to embrace these initiatives. During the most recent election campaign, the opposition Congress Party accused the BJP of polarizing India, citing the recent spate of name changes on India's map as one example of this divisive effect. The BJP narrowly lost its majority in the 2004 elections, and its prime minister resigned, yielding to a successor from the opposition Congress Party. But the Congress Party does not have a stellar record of governance, so the pendulum could swing back again. Increasingly, however, the key issues are likely to revolve around the management of India's diverse and growing economy, and the *hindutva* campaign may well take a back seat.

Communist Insurgency

Over the past several years there have been troubling signs of still another divisive force in India: communist- (avowedly Maoist-) inspired rebellions that appear to be growing into a coordinated revolutionary campaign (see photo below). It is known in India as the *Naxalite* movement, named after a town in the State of West Bengal where it was founded in the 1960s to oppose, by force of arms, the still-new government in New Delhi. In the decades that followed, India's police managed to control and at times virtually wipe out the Naxalites, but in recent years they have been making a comeback. Based in some of India's most remote, poorest, and disaffected areas of States such as Bihar, Jharkhand, Chhattisgarh, and even Andhra Pradesh, the Naxalites have had notable successes against India's law enforcement, including liberating prisoners from jails, blowing up railroad tracks, laying land mines, and fighting pitched battles against police.

A look at Figure 8-12 shows that one of these States where the Naxalites have had growing impact lies adjacent to Nepal, where a Maoist insurgency took control of a sizeable area in the west, ruined the economy, and destabilized the national government at a cost of many thousands of lives. A potential linkup between the Naxalites of India and the Maoists of Nepal would raise great concern in India. Already, the Naxalites are reported to have a presence in 13 of India's 28 States, and the Indian government has acknowledged that nearly 80 of the country's 602 administrative districts (the next level below that of a State) are severely affected. Independent observers report that the actual number is twice as large.

The Naxalites appeal to the lower-caste poorest of the poor, and one of their objectives is to disrupt the burgeoning Indian economy. By damaging infrastructure, targeting mines and factories, and threatening contractors, the movement's impact on India's economic prospects has the potential to create serious difficulties.

Islamic Extremism

Comparatively successful as the integration of India's Muslim communities into the fabric of the Indian state has been, the risk of Islamic violence, directed against Indian society in general, is rising. For many years India has suffered from Islamic terrorism, attributable to the longstanding quarrel with Pakistan over Kashmir, to discord with Islamic Pakistan in general, and to occasional incidents of sectarian violence of the kind that followed the razing of a mosque in the town of Ayodhya to make way for a Hindu temple. At times such attacks have brought the two neighbors close to war, for example, in 2001 when Islamic terrorists with roots in Pakistan assaulted the Indian Parliament. In July 2006, a series of simultaneous, coordinated attacks struck several trains

The ghost of Mao Zedong may be dissipating in China, but it lives on in India. Shown in this remarkable 2006 photo is a brigade of Mao-inspired rebels marching during what they claimed was their ninth "convention," deep in the forest of the State of Chhattisgarh 1500 kilometers (930 mi) southeast of New Delhi. These Naxalites, as they are called in India, have a presence in nearly half of India's 28 States, and they prove capable of damaging infrastructure and defying law enforcement. On March 15, 2007, they attacked a police station in Orissa State, killing more than 50 officers and destroying the post. They claim to support the causes of India's dalits and demand land and jobs for the poor in India's rural east. ©AP/World Wide Photos.

along a commuter line in Mumbai (Bombay), killing about 200 people. While Indian investigators found the usual Pakistani links (some of the perpetrators had attended Pakistani terrorist-training camps), others were local Mumbai residents. Among those arrested were a doctor and a software engineer, middle-class Muslim residents of a multicultural city who had been drawn into the circle of Islamic terrorism.

This development has serious implications for India, where the overwhelming majority of Muslim citizens has resisted involvement in extremist causes, even after provocations such as the deathly anti-Muslim riots of 2002 in the State of Gujarat. But now it seems that a few educated Muslims are joining local terrorist cells with links to Pakistan and perhaps other Islamic countries. Their terrorist acts spur investigations that offend ordinary and peaceful Indian Muslims, radicalizing a number of them and creating a growing market for Islamic militancy. It is as yet too early to gauge the potential impact of this development on a country that has long and justly prided itself on its multicultural democracy and multisectarian stability, but the portents for India's political, social, and economic geography are obviously serious.

Centripetal Forces

In the face of all these divisive forces, what bonds have kept India unified for so long? Without question, the dominant binding force in India is the cultural strength of Hinduism, its sacred writings, holy rivers, and influence over Indian life. For most Indians, Hinduism is a way of life as much as it is a faith, and its diffusion over virtually the entire country (regardless of the Muslim, Sikh, and Christian minorities) brings with it a national coherence that constitutes a powerful antidote to regional divisiveness. Over the long term, however, the key ingredients of this Hinduism have been its gentility and introspection, radical outbursts notwithstanding.

Democracy

Another centripetal force lies in India's democratic institutions. In a country as culturally diverse and as populous as India, reliance on democratic institutions has been a birthright ever since independence, and democracy's survival—raucous, often corrupt, always free—has been a crucial unifier.

Furthermore, communications are better in much of India than in many other countries in the global periphery, and the continuous circulation of people, ideas, and goods helps bind the disparate state together. Before independence, opposition to British rule was a shared philosophy, a strong centripetal force. After independence,

the preservation of the union was a common objective, and national planning made this possible.

Accommodation

India's capacity for accommodating major changes and its flexibility in the face of regional and local demands are also a centripetal force. Boundaries have been shifted; internal political entities have been created, relocated, or otherwise modified; and secessionist demands have been handled with a mixture of federal power and cooperative negotiation. Indians in South Asia have accomplished what Europeans in former Yugoslavia could not, and India's history of success is itself a centripetal force.

Education

Still another centripetal force in India is education. The country takes great pride in its high literacy rates, which in urban India exceed 96 percent for both males and females; these numbers are substantially above those of neighboring mainland countries and reflect Indians' determination to avail themselves of every educational opportunity. Private institutes teaching English abound in the cities, and when opportunities for the outsourcing of service jobs arose on the global economic scene, India had the educated workforce to seize them.

Leadership

Finally, no discussion of India's binding forces would be complete without mentioning the country's strong leadership. Gandhi, Nehru, and their successors did much to unify India by the strength of their compelling personalities. For many years, leadership was a family affair: Nehru's daughter, Indira Gandhi, twice took decisive control (in 1966 and 1980) after weak governments, and her son, Rajiv Gandhi (who, in 1991, like his mother seven years earlier, also was assassinated), was prime minister in the late 1980s. Since then, political leadership of India has been less dynastic—and also less cohesive.

Urbanization

India is famous for its great and teeming cities, but India is not yet an urbanized society. Only 29 percent of the population lived in cities and towns in 2008—but in terms of sheer numbers, that 29 percent amounts to 336 million people, more than the entire population of the United States.

India's rate of urbanization is on the upswing. People by the hundreds of thousands are arriving in the already overcrowded cities, swelling urban India by about 5 per-

FROM THE FIELD NOTES

"Searing social contrasts abound in India's overcrowded cities. Even in Mumbai (Bombay), India's most prosperous large city, hundreds of thousands of people live like this, in the shadow of modern apartment buildings. Within seconds we were surrounded by a crowd of people asking for help of any kind, their ages ranging from the very young to the very old. Somehow this scene was more troubling here in well-off Mumbai than in Kolkata (Calcutta) or Chennai (Madras), but it typified India's urban problems everywhere."
© H. J. de Blij.

www.conceptcaching.com

cent annually, three times as fast as the overall population growth. Not only do the cities attract as they do everywhere: many villagers are driven off the land by the desperate conditions in the countryside. As villagers manage to establish themselves in Mumbai or Kolkata or Chennai, they help their relatives and friends to join them in squatter settlements that often are populated by newcomers from the same area, bringing their language and customs with them and cushioning the stress of the move.

As a result, India's cities display staggering social contrasts. Squatter shacks without any amenities at all crowd against the walls of modern high-rise apartments and condominiums (see photo above). Hundreds of thousands of homeless roam the streets and sleep in parks, under bridges, on sidewalks. As crowding intensifies, social stresses multiply. Disorder never seems far from the surface; sporadic rioting, often attributable to rootless urban youths unable to find employment, has become commonplace in India's cities.

Urban Tradition and Evolution

India's modern urbanization has its roots in the colonial period, when the British selected Calcutta (Kolkata), Bombay (Mumbai), and Madras (Chennai) as regional trading centers and fortified ports. Madras was fortified as early as 1640; Bombay (1664) had the situational advantage of being the closest of all Indian ports to Britain; and Calcutta (1690) lay on the margin of India's largest population cluster and had the most productive hinterland, to which the Ganges Delta's countless channels connected it. This natural transport network made Calcutta an ideal colonial headquarters, but the population of Bengal was often rebellious. In 1912 the British moved their colonial government from Calcutta to the safer interior city of New Delhi, built adjacent to the old Mughal capital of Delhi.

Figure 8-12 displays the distribution of major urban centers in India. Except for Delhi–New Delhi, the largest

India is urbanizing rapidly, but the road network linking its cities remains inadequate. Now India is building its first major interstate highway, a four-lane loop that will connect the key cities of Delhi, Mumbai, Chennai, and Kolkata, and, in the process, more than a dozen of its States. Shown here is a section of this "Golden Quadrilateral" in the State of Rajasthan. However, it will take more than a superhighway with a grandiose name to overcome the barriers to efficient linkages among India's cities. Lorry (truck) drivers are stopped for long waits dozens of times as they complete the circuit for paperwork at State borders, "inspections," and other formalities. By the way, why do Indian motorists drive on the left? ©Tyler Hicks/*The New York Times*/Redux.

cities have coastal locations: Kolkata dominates the east, Mumbai the west, and Chennai the south. But urbanization also has expanded in the interior, notably in the core area. By 2008, India had nearly 40 cities with populations exceeding 1 million, but road links among these (and other) cities remain grossly inadequate. Now, in the first decade of the twenty-first century, India is constructing a nationwide four-lane expressway that will link the four anchors of this urban system (Delhi, Mumbai, Chennai, and Kolkata) and in the process connect 15 other major cities along this national route called the *Golden Quadrilateral* (Fig. 8-12). The impacts of this project are multiple: it is expanding urban hinterlands, commuters are using it to travel farther to work than ever before, once-remote villages now have a link to markets, and it is accelerating the rural-to-urban migration flow that will transform India in this century (see photo above). In this edition of our book we are still able to say that India is a land of 600,000 villages in a country whose population is less than one-third urbanized. Future editions will chronicle the emergence of a new India with fewer villages, ever-larger metropolitan complexes, and an expanding number of megacities.

Economic Geography

Improving India's infrastructure, though, is not enough to overcome all serious impediments to the country's economic advancement. Take another look at Figure 8-12

and note that the Golden Quadrilateral crosses numerous State boundaries. In the United States, we are used to seeing thousands of trucks on the interstate highways, crossing from one State into another without slowing down; there are truck-weighing stations along these expressways, but in the newest ones all the truck has to do is slow down enough for electronic surveillance to record its passing.

In India, truck drivers face a very different experience. It can take as long as nine days, including more than 30 hours waiting at State-border checkpoints and tollbooths, for a loaded truck to travel from Kolkata to Mumbai via Chennai, or less than the distance from Los Angeles to New Orleans on I-10. Drivers are subject to daunting piles of paperwork and repeated demands for bribes, and when mechanical problems occur it may take days to get the truck back on the road. It will take more than expanding India's highway network to improve the circulation the country so desperately needs.

Globalization

When you arrive in any Indian city, you are struck by the number of small shops everywhere—tiny businesses wedged into every available space in virtually every non-public building along every street. Even the upper, walkup floors are occupied by shops, their advertising signs suspended from windows and balconies. According to a study by the Indian government's Department of Consumer Affairs, India has the highest density of retail outlets of any country in the world with 15 million shops (compared to well below 1 million in the United States, where the marketplace is 13 times richer). After farming, the retail sector is India's largest provider of jobs.

What keeps all these small stores in business? Most of them earn very little and can afford to stay open only because they are part of what economic geographers call the *informal sector*: they are essentially unregistered, pay no rent and probably no taxes, use family labor, have been handed down through generations, and survive because India's economic geography, bound by longstanding protective government regulation, has been slow to change. When you are in India you may wonder how so many shops can stay in business, but the answer is in the throngs of people on the sidewalks (spilling over into the clogged traffic in the streets). Not many of these people are wealthy, but all of them need basic goods and some can afford small luxuries, and so the shops tend to be busy all day.

Now imagine what would happen if India suddenly opened its doors to large international retailing companies like Wal-Mart. An invasion of foreign superstores would force millions of India's small shops to close, throw countless workers into unemployment, and de-

AMONG THE REALM'S GREAT CITIES . . .   Mumbai (Bombay)

ANOTHER HISTORIC NAME is disappearing from the map: Bombay. In precolonial times, fishing folk living on the seven small islands at the entrance to this harbor named the place after their local Hindu goddess, Mumbai. The Portuguese, first to colonize it, called it Bom Bahia, "Beautiful Bay." The British, who came next, corrupted both to Bombay, and so it remained for more than three centuries. Now local politics has taken its turn. Leaders of a Hindu nationalist party that control the government of the State of which Bombay is the capital, Maharashtra, passed legislation to change its name back to Mumbai. In 1996, India's federal government approved this change.

Mumbai's 19.3 million people make this India's largest city. Maharashtra is India's economic powerhouse, the State that leads the country in virtually every respect. Locals dream of an Indian Ocean Rim of which Mumbai would be the anchor.

As such, Mumbai is a microcosm of India, a burgeoning, crowded, chaotic, fast-moving agglomeration of humanity (see photo, p. 435). The Victorian-Gothic architecture of the city center is a legacy of the British colonial period (see photo, p. 411). Shrines, mosques, temples, and churches evince the pervasive power of religion in this multicultural society. Street signs come in a bewildering variety of scripts and alphabets. Creaking double-decker buses compete with oxwagons and handcarts on the congested roadways. The throngs on the sidewalks—businesspeople, holy men, sari-clad women, beggars, clerks, homeless wanderers—spill over into the streets.

Mumbai is an urban agglomeration of city-sized neighborhoods, each with its own cultural landscape. The seven islands have been connected by bridges and causeways, and the resulting Fort District still is the center of the city, with many of its monuments and architectural landmarks. Marine Drive leads to the wealthy Malabar Hill area, across Back Bay. Northward lie the large Mus-

© H. J. de Blij, P. O. Muller, and John Wiley & Sons, Inc.

lim districts, the Sikh neighborhoods, and other ethnic and cultural enclaves. Beyond are some of the world's largest and poorest squatter settlements. It is situation, not site, that gave Bombay primacy in South Asia. The opening of the Suez Canal in 1869 made Bombay the nearest Indian port to Europe. Today, Mumbai is the focus of India's fastest-growing economic zone—not yet a tiger but on the move.

stroy ways of life that have put most urban residents within walking distance of their daily needs. And yet this is what India faces. This is the era of globalization, and large-scale organized retailing, long held back by a combination of restraints (such as the nationwide distribution problems discussed above, but also legal and cultural obstacles), is emerging within India itself. This means that competition from foreign companies will somehow infiltrate the market, and in this respect the Indian government faces only one key question: how to keep control of the process through gradual deregulation. That issue is complicated, however, by India's chaotic democracy.

Even if New Delhi approves appropriate legislation, individual States may counter it by imposing their own regulations and restrictions.

Still, the statistics show that the process is already well underway. India's middle class, now estimated to include some 300 million people (which equals the entire U.S. population), is growing rapidly, and something very new is beginning to appear in the cities: shopping malls. As recently as 2000, this vast country containing one-sixth of humankind did not have a single shopping center! By 2005, there were more than 100, and by the end of 2007 more than 350 were in operation. You will

FROM THE FIELD NOTES

"The streets of India's cities often seem to be one continuous market, with people doing business in open storefronts, against building walls, or simply on the sidewalk. Here the formal and informal sectors of India's economy intermingle. I walked this way in Delhi every morning, and the store selling mattresses and pillows was always open. But the women in the foreground, selling handkerchiefs and other small items from a portable iron rack, were sometimes here, and sometimes not; one time I ran into them about a half-mile down the road. As I learned one tumultuous day, every time the government tries to exercise some control over the street hawkers and sidewalk sellers, massive opposition results and chaos can ensue. A few days later, everything is as it was. Change comes very slowly here."   © H. J. de Blij.

www.conceptcaching.com

see American fast-food restaurants among the establishments in these malls as well as the brand names of numerous other foreign companies, proving that globalization has already breached India's walls. The question remains: can India's economic transformation be achieved without severe social dislocation?

FROM THE FIELD NOTES

"Travel into rural India, and you soon grasp the realities of Indian agriculture. Human and animal labor predominate; farming methods and equipment are antiquated. At this village I watched the women feed scrawny sticks of sugarcane into a rotating press turned slowly by a pair of bullocks prodded by a boy. In this family enterprise, the grandfather is responsible for boiling the liquid sugar in a huge pan over the fire pit (upper right). The dehydrated sugar is then molded into large round cakes to be sold at the local market. As I observed this slow and inefficient process, I understood better why India is among the world's leaders in terms of sugarcane acreage—and among the world's last in terms of yields."
© Barbara A. Weightman.

www.conceptcaching.com

Farming's Enduring Importance

Agriculture, we noted earlier, still provides more jobs in India than any other economic sector, and India's fortunes (and misfortunes) remain strongly tied to farming. For all the emphasis on India's cities and middle-class growth, about 70 percent of the people still live on (and from) the land, traditional farming methods persist, yields per hectare remain among the world's lowest, and hunger and malnutrition still afflict millions even as grain surpluses accrue. The relatively few areas of modernization, as in the wheatlands of the Punjab, are islands in a sea of stagnation. Thus the agricultural sector and indeed the entire country is directly vulnerable to environmental variations, with all the risks that entails. Furthermore, land reform has essentially failed; roughly one-quarter of India's entire cultivated area, including much of the best land, is still owned by less than 5 percent of the country's landholders, who have much political influence and obstruct redistribution. Perhaps half of all rural families own less than one hectare (2.5 acres) or no land at all: an estimated 175 million live and work as tenants, always uncertain of their fate.

Getting produce to markets—if there is any to sell—is a struggle for millions of farmers. In 2007, almost half of India's 600,000 villages could not be accessed by truck or

AMONG THE REALM'S GREAT CITIES . . . **Kolkata (Calcutta)**

CALCUTTA IS SYNONYMOUS with all that can go wrong in large cities: poverty, dislocation, disease, pollution, crime, corruption. To call a city the Calcutta of its region is to summarize urban catastrophe. West Bengal's leaders had this in mind when they proposed that the city's local name, Kolkata, be restored. The city got its dreadful reputation (as well as its Anglicized name) during colonial times, when plague, malaria, and other diseases claimed countless thousands in legendary epidemics.

The British chose the site, 130 kilometers (80 mi) up the Hooghly River from the Bay of Bengal and less than 9 meters (30 ft) above sea level, not far from some unhealthy marshes but well placed for commerce and defense. When the British East India Company was granted freedom of trade in the populous hinterland, Calcutta's heyday began; when (in 1772) the British made it their colonial capital, the city prospered. The British sector of the city was drained and raised, and so much wealth accumulated here that Calcutta became known as the "city of palaces." Outside the British town, rich Indian merchants built magnificent mansions. Beyond lay neighborhoods that were often based on occupational caste, whose names are still on the map today (such as Kumartuli, the potters' district). Almost everywhere, on both banks of the Hooghly, lay the huts and hovels of the poorest of the poor. Searing social contrasts characterized Calcutta.

The twentieth century was not kind to Calcutta. In 1912 the British moved their colonial capital to New Delhi. The 1947 partition that created Pakistan also created then East Pakistan (now Bangladesh), cutting off a large part of Calcutta's hinterland and burdening the city with a flood of refugees. The Indian part of the city had arisen virtually without any urban planning, and the influx

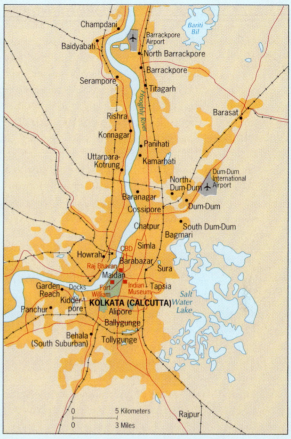

© H. J. de Blij, P. O. Muller, and John Wiley & Sons, Inc.

created almost unimaginable conditions. Today, Kolkata counts 15 million residents, including as many as 500,000 homeless. Beyond the façade of the downtown lies what may be the sickest city of all.

---

car, and in this era of modern transportation animal-drawn carts still far outnumber motor vehicles nationwide.

Figure 8-14 maps India's agriculture and reflects the climatic distribution shown in Figure G-8, the rainfall pattern in Figure G-7, and the monsoonal cycle depicted in Figure 8-3. Rice dominates along the Arabian Sea-facing southwestern coast and in the monsoon-drenched peninsular northeast; where drier conditions develop, wheat and other grains prevail. Although, as Table G-1 reminds us, physiologic densities in India are lower than in neighboring Bangladesh to the east, the comparison is deceptive because Indian farming, especially in the rice-growing zones and despite the effects of the Green Revolution, is so inefficient. For India's economic transformation to have

real impact in the rural areas, nothing less than a technological revolution is required.

The Energy Problem

If you have a friend in (or from) India, you know someone who is familiar with power outages. They are a way of life in India, where electricity demand routinely exceeds available supply, where governments cannot bring themselves to require customers to pay for the actual cost of the power they consume, and where power grids, generating equipment, and other related infrastructure are in bad shape. All this while hundreds of millions of villagers still have no electricity supply at all.

Delhi New and Old

FLY DIRECTLY OVER the Delhi–New Delhi conurbation into the regional airport, and you may not see the place at all. A combination of smog and dust creates an atmospheric soup that can limit visibility to a few hundred feet for weeks on end. Relief comes when the rains arrive, but Delhi's climate is mostly dry. The tail-end of the wet monsoon reaches here during late June or July, but altogether the city only gets about 60 centimeters (25 in) of rain a year. When the British colonial government decided to leave Calcutta and build a new capital city adjacent to Delhi, conditions were different. South of the old city lay a hill about 15 meters (50 ft) above the surrounding countryside, on the right bank of the southward-flowing Yamuna tributary of the Ganges. Compared to Calcutta's hot, swampy environment, Delhi's was agreeable. In 1912 it was not yet a megacity. Skies were mostly clear. Raisina Hill became the site of a New Delhi.

This was not the first time rulers chose Delhi as the seat of empire. Ruins of numerous palaces mark the passing of powerful kingdoms. But none brought to the Delhi area the transformation the British did. In 1947 the Indian government decided to keep its headquarters here. In 1970 the metropolitan population exceeded 4 million. By 2008 it was 16.4 million.

Delhi is popular as a seat of government for the same reason as its present expansion: the city has a fortuitous relative location. The regional topography creates a narrow corridor through which all land routes from north-

© H. J. de Blij, P. O. Muller, and John Wiley & Sons, Inc.

western India to the North Indian Plain must pass, and Delhi lies in this gateway. Thus the twin cities not only contain the government functions; they also anchor the core area of this populous country.

Old Delhi was once a small, traditional, homogeneous town. Today Old and New Delhi form a multicultural, multifunctional urban giant.

And yet electrical power is key to India's modernization. Already, foreign companies doing business in India sometimes import their own generators, but others are discouraged from investing in factories and other facilities because the power supply is so unreliable. Most of India's electricity is generated in thermal plants burning coal, oil, or natural gas; about 25 percent comes from hydroelectric sources, and about 3 percent from nuclear plants. The problems are many: India has substantial coal deposits, but the railroads cannot handle the transport to power plants. So India must import coal, but the ports do not have the required capacity. And India possesses only limited oil and natural gas reserves. Add to this an increasingly inadequate national power-supply grid in a country with a still-exploding population and even faster-growing demand, and you have trouble.

A key remedy, of course, lies in increased oil and gas imports, but here India runs into geopolitical problems. While the American government was eager to give India leeway in

the nuclear arena, the United States made it clear that it does not like India's plan to buy Iranian natural gas via a pipeline across Pakistan. India's other options lie in interior Asia, but those sources are more distant and pipeline construction would involve further diplomatic complications. Once again, the other alternative—importing oil and gas via tankers and ports—is constrained by inadequate infrastructure.

Limitations on Manufacturing Growth

Given these problems, the geography of India's manufacturing (Fig. 8-15) is changing too slowly for the country's needs. The map is a legacy of colonial times, with coastal Mumbai, Kolkata, and Chennai anchoring major industrial zones and textile industries—the entry-level industry of disadvantaged countries—dominating the scene. India's information technology (IT) industries, centered in and around Bengaluru (Bangalore), Mumbai, and Delhi, draw much international attention, and software and IT services account for more than one-quarter

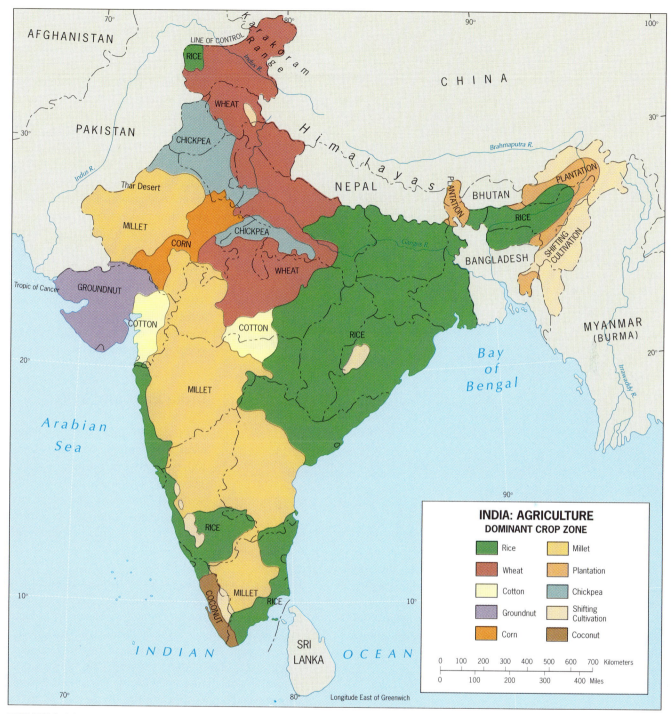

FIGURE 8-14   © H. J. de Blij, P. O. Muller, and John Wiley & Sons, Inc.

of merchandise exports by value, but what India needs far more are manufacturing industries competitively selling goods on world markets, putting tens of millions to work, and transforming the economy (as has happened in China, which we discuss in Chapter 9). It is almost unbelievable that as many people worked in manufacturing in 1991 as in 2001 (the two most recent census years) while China's industrial workforce more than quadrupled during the same period.

Could India follow in China's footsteps? Some economic geographers suggest that India might leapfrog China and move quickly from an "underdeveloped" to a "postindustrial" services-led economy. Certainly India has the requisite intellectual clout: Indian corporations have set up hundreds of IT schools in China where tens of thousands of Chinese students are learning the business. But to reach India's hundreds of millions of potential wage-earners, India needs a vigorous expansion of its

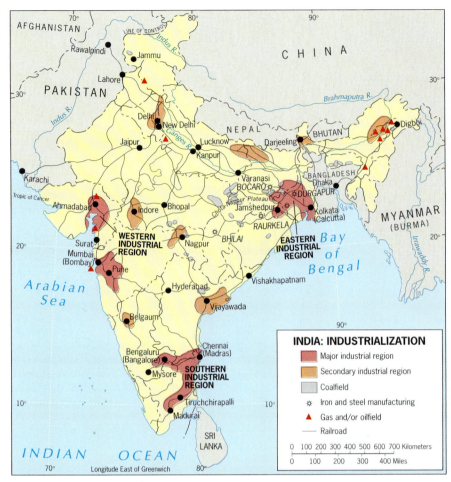

FIGURE 8-15   © H. J. de Blij, P. O. Muller, and John Wiley & Sons, Inc.

it is likely to rank third. Over the past several years, the economy has grown by an average of 7 percent—no match for China, but far ahead of the growth rates in most other regions of the world. As we will see in the next chapter, China's dramatic growth resulted from decisions at the top, a transformation planned and implemented in controlled detail. In India, the economy is growing from the bottom up, with all the traditional chaos that makes India a country like no other. As with China's provinces, some of India's States will advance ahead of others, and India's incredible socioeconomic contrasts will intensify. But over time, India could achieve what China has—hitherto—not: an economic and cultural geography of consensus.

India East and West

The most commonly cited, and most clearly evident, regional division of India is between north and south. The north is India's heartland, the south its Dravidian appendage; the north speaks Hindi as its *lingua franca*, the south prefers English over Hindi; the north is bustling and testy, the south seems slower and less agitated.

But there is another, as yet less obvious, but potentially more significant divide across India. In Figure 8-15, draw a line from Lucknow, on the Ganges River, south to Madurai, near the southern tip of the peninsula (also review Figure 8-2 on page 403). To the west of this line, India is showing signs of economic progress, the kind of economic activity that has brought Pacific Rim countries such as Thailand and Indonesia a new life. To the east, India has more in common with less promising countries also facing the Bay of Bengal: Bangladesh and Myanmar (Burma).

As with other regional divides, there are exceptions to our east-west delineation. Indeed, our map seems to suggest that much of India's industrial strength lies in the east. But what the map cannot reveal is the profitability of those industries. True, the east is rich in iron and coal, but the heavy industries built by the state in the 1950s are now outdated, uncompetitive, and in decline. The hinterland of Kolkata contains India's Rustbelt. The government keeps many industries going but at a high cost. Old industries, such as carpetmaking and cottonweaving,

secondary industries, those that make goods (beyond textiles) and sell them at home and abroad. Here's how it may happen: when an economy churns along the way China's has over the past three decades, labor and production costs tend to rise. That causes manufacturers to look for places where cheaper labor will reduce such costs. India, with its long history of local manufacturing, huge domestic market, and vast reservoir of capable labor, would then take its turn on the world stage.

Improving Prospects

Already, India is on the move, if not yet in the dramatic terms of some of those popular media cited earlier. A middle class of more than 300 million people demands goods ranging from mobile phones (in mid-2006, 2.5 million new subscribers were signing up each month) to motor bikes (10,000 per day were being sold in mid-decade). Chennai is becoming the Indian automobile center: in 2003 a South Korean manufacturer started selling cars from its local factory not only in India but also exporting them to Europe, and BMW in 2007 was building an assembly plant nearby. India's economy today is the world's sixth-largest; by 2020

continue to use child labor to remain viable. The State of Bihar represents the stagnation that afflicts much of India east of our line: by several measures it ranks among the poorest of the 28 States.

Compare this to western India. The State of Maharashtra, the hinterland of Mumbai, leads India in many categories, and Mumbai leads Maharashtra. Many smaller, private industries have emerged here, manufacturing goods ranging from umbrellas to satellite dishes and from toys to textiles. Across the Arabian Sea lie the oil-rich economies of the Arabian Peninsula. Hundreds of thousands of workers from western India have found jobs there, sending money back to families from Punjab to Kerala. More importantly, many have used their foreign incomes to establish service industries back home. Outward-looking western India, in contrast to the inward-looking east, has begun to establish other ties to the outside world. Satellite links have enabled Bengaluru (Bangalore) to become the center of a growing software-producing complex reaching world markets. The beaches of Goa, the small coastal State immediately to the south of Maharashtra, appeal to the tourist markets of Europe. This is, in fact, a classic case of **intervening opportunity** because resorts have sprung up along Goa's shore, and European tourists who once went to the more distant Maldives and Seychelles are coming to Goa. Maharashtra's economic success also has spilled over into Gujarat to the north, and even landlocked Rajasthan (the next State to the north) is experiencing the beginnings of what, by Indian standards, is a boom.

The boom has created political problems, however. Not only is Maharashtra State a rising economic power; it also is the base of a strong Hindu nationalist political movement whose leaders object to foreign intrusions and have blocked major development projects and other enterprises. They halted a huge industrial scheme about halfway through and closed a fast-food operation that they deemed incompatible with local culture. Such clashes between foreign interests and domestic traditions continue even as India's economy forges ahead.

Nevertheless, India's east-west divide shows a growing contrast that puts the west far ahead. The hope is that Maharashtra's success will spread northward and southward along the Arabian Sea coast and will ultimately diffuse eastward as well. But for this to happen, India will have to bring its population spiral under control.

● BANGLADESH: CHALLENGES OLD AND NEW

On the map of South Asia, Bangladesh looks like another State of India: the country occupies the area of the double delta of India's great Ganges and Brahmaputra rivers,

India's leading high-tech center is Bengaluru (formerly Bangalore: another of India's ongoing place-name changes), and the Indian company Infosys Technologies, based here, is India's leading software exporter. This Silicon Valley-like campus in suburban Electronics City is the center of the company's booming outsourcing business (Bank of America and Citigroup are among its American customers). Bengaluru, as Figure 8-12 shows, lies near the intersection of the states of Karnataka, Andhra Pradesh, and Tamil Nadu. While some skilled workers have emigrated to Europe, Australia, and America, where their skills earn them more than at home, many more choose to stay here and benefit from the higher local incomes the IT industries pay. Thus Bengaluru attracts India's best and brightest, whose children attend the city's superior schools. And this is a pleasant place to live: at 900 meters (3000 ft) above sea level, even the summer weather is tolerable. Bengaluru, however, is growing so rapidly that its infrastructure cannot keep up, so that congestion, traffic jams, and commuter times are all growing as well. Sound familiar?   ©Reuters/Corbis.

and India almost completely surrounds it on its landward side (Fig. 8-16). But Bangladesh is an independent country, born in 1971 after its brief war for independence against Pakistan, with a territory about the size of Wisconsin. Today it remains one of the poorest and least developed countries on Earth, with a population of 152.2 million that is growing at an annual rate of 1.9 percent.

A Vulnerable Territory

Not only is Bangladesh a poor country; it also is highly susceptible to damage from **natural hazards**. During the twentieth century, eight of the ten deadliest natural disasters in the entire world struck this single country. Most recently, a 1991 cyclone (as hurricanes are called in this part of the world) killed over 150,000 people.

The reasons for Bangladesh's vulnerability can be deduced from Figures 8-16 and 8-1. Southern Bangladesh is the deltaic plain of the Ganges–Brahmaputra river system, combining fertile alluvial soils that attract farmers with the low elevations that endanger them when water rises. The shape of the Bay of Bengal forms a funnel that

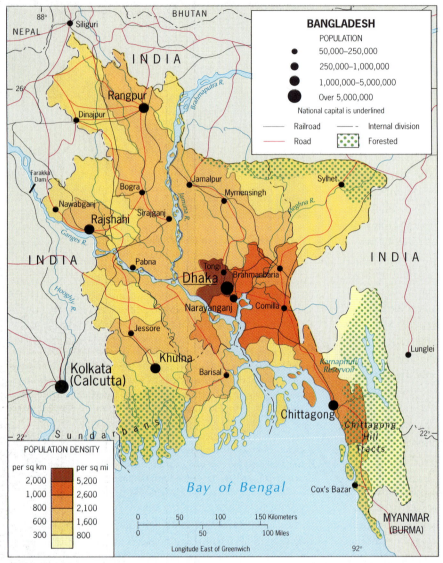

FIGURE 8-16 © H. J. de Blij, P. O. Muller, and John Wiley & Sons, Inc.

the world (1678 people per square kilometer/4346 per square mile), and only higher-yielding varieties of rice and the introduction of wheat in the crop rotation (where climate allows) have improved diets and food security. But diets remain poorly balanced, and, overall, the people's nutrition is unsatisfactory. The textile industry provides most of Bangladesh's foreign revenues, but the once-thriving jute industry continues its decline. The discovery of a natural gas reserve is now the subject of a national debate: home consumption or money-making export?

Bangladesh is a dominantly Muslim society, but not one dominated by revivalists. For example, 30 seats in the national legislature are reserved for women. Its relations with neighboring India have at times been strained over water resources (India's control over the Ganges that is Bangladesh's lifeline), cross-border migration (one-sixth of the population is Hindu), and transit between parts of India across Bangladesh's north (refer to Figure 8-15 to see the reason). All the disadvantages of the global periphery afflict this populous, powerless country where survival is the leading industry and all else is luxury.

sends cyclones and their storm surges of wind-whipped water barreling into the delta coast. Without money to build seawalls, floodgates, elevated shelters in sufficient numbers, or adequate escape routes, hundreds of thousands of people are at continuous risk, with deadly consequences. And as if this is not enough, millions of people have been found to be exposed to excessive (natural) arsenic in the drinking water from their wells.

Limits to Opportunity

Bangladesh remains a nation of subsistence farmers. Urbanization is at only 23 percent; Dhaka, the megacity capital, and the cities of Chittagong, Rangpur, Khulna, and Rajshahi are the only urban centers of consequence. Moreover, Bangladesh has one of the highest physiologic densities in

● THE MOUNTAINOUS NORTH

As Figures 8-1 and 8-2 show, a tier of landlocked countries and territories lies across the mountainous zone that walls India off from China. One of them, Kashmir, is in a state of near-war. Another, Sikkim, was absorbed by India in 1975 and made into one of its federal States. But Nepal and Bhutan retain their independence.

Nepal, northeast of India's Hindu core, has a population of 27.2 million and is the size of Illinois. It has three geographic zones (Fig. 8-17): a southern, subtropical, fertile lowland called the Terai; a central belt of Himalayan foothills with swiftly flowing streams and deep valleys; and the spectacular high Himalayas themselves (topped by Mount Everest) in the north. The capital, Kathmandu, lies in the east-central part of the country in an open valley of the central hill zone.

Nepal is materially poor but culturally rich. The Nepalese are a people of many sources, including India, Tibet, and interior Asia. About 85 percent are Hindu, and Hinduism is the country's official religion; but Nepal's Hinduism is a unique blend of Hindu and Buddhist ideals. Thousands of temples and pagodas ranging from the simple to the ornate grace the cultural landscape, especially in the valley of Kathmandu, the country's core area. Although over a dozen languages are spoken, 90 percent of the people also speak Nepali, a language related to Indian Hindi.

As the data in Table G-1 suggest, Nepal is a troubled country suffering from severe underdevelopment, with the lowest GNI in the realm. It also faces strong centrifugal social and political forces. Environmental degradation, crowded farmlands and soil erosion, and deforestation scar the countryside (see photo next page). The Himalayan peaks form a world-renowned tourist attraction, but tourist spending in Nepal, always relatively modest, has been cut back because of the current revival of Maoist-communist terrorism, not only in the western hills where it has long been active but also in the capital itself.

Nepal's political geography has long been troubled. The end of absolute monarchy in 1991 and the advent of democracy did not end the country's regional divisions: the southern Terai with its tropical lowlands is a world apart from the hills of central Nepal, and the peoples of the west have origins and traditions different from those in the east. Moreover, in 2001 the assassination of Nepal's king threatened the disintegration of the state, narrowly averted by the leadership of the Nepal Congress Party. Then in 2002 the country's stability was imperiled again as the Maoist insurgency spread. By mid-decade, Nepal was in effect a **failed state**, its government **16** unable to control the insurgency, its economy collapsed, and its future hanging in the balance.

Mountainous Bhutan, wedged between India and China's Tibet (Fig. 8-17), is the only other buffer between Asia's giants. In landlocked, fortress-like Bhutan, time seems to have stood still. Bhutan is officially a constitutional monarchy, but its king rules the country with virtually absolute power; economic subsistence and political allegiance are the norms of life for most of the population of just over 900,000. Thimphu, the capital, has about 60,000 inhabitants. The symbols of Buddhism, the state religion, dominate its cultural landscape. Social tensions arise from the large but diminishing Nepalese minority, most of whom are Hindus and some of whom have been persecuted by the dominant Bhutia.

Forestry, hydroelectric power, and tourism all have potential here, and Bhutan has considerable mineral resources. But isolation and inaccessibility preserve traditional ways of life in this mountainous *buffer state*.

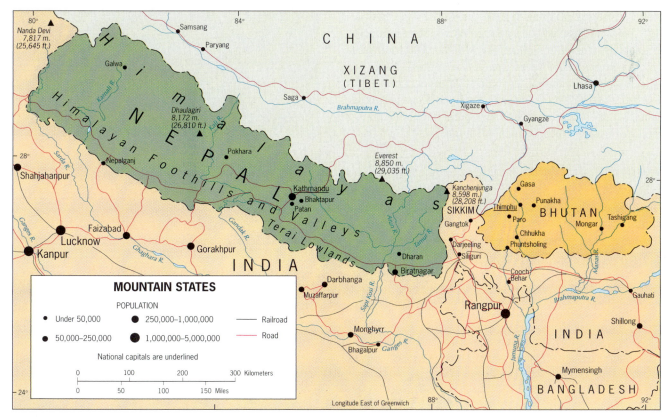

FIGURE 8-17 © H. J. de Blij, P. O. Muller, and John Wiley & Sons, Inc.

Human-induced environmental degradation is vividly displayed on Nepal's highest inhabitable slopes. Against the spectacular backdrop of the main Himalaya range north of Kathmandu, population pressures in the teeming valleys have forced farming and fuel-wood extraction to ever higher elevations. The inevitable result is deforestation and when the dense vegetation and binding roots are removed, copious summer-monsoon rainwater is free to cascade down the steep, denuded mountainsides. This massive runoff soon produces myriad gullies that further destabilize slopes by triggering mass movements of soil and rock. The lower half of this photo exhibits the evidence of this erosional process, which affects the more level terrain in the foreground as well. Ominously, we can also discern human dwellings creeping upslope, which portends a final onslaught on the single remaining forested ridge. © Galen Rowell/Mountain Light Photography, Inc.

● THE SOUTHERN ISLANDS

As Figure 8-1 shows, South Asia's continental landmass is flanked by several sets of islands: Sri Lanka off the southern tip of India, the Maldives in the Indian Ocean to the southwest, and the Andaman Islands (belonging to India) marking the eastern edge of the Bay of Bengal.

The Maldives consists of more than a thousand tiny islands whose combined area is just 300 square kilometers (115 sq mi) and whose highest elevation is less than 2 meters (6 ft) above sea level. Its population of just over 300,000 from Dravidian and Sri Lankan sources is now 100 percent Muslim, one-quarter of which is concentrated on the capital island named Maale. The Maldives might be unremarkable, except that, as Table G-1 shows, this country has the realm's highest GNI per capita. The locals have translated their palm-studded, beach-fringed islands into a tourist mecca that attracts tens of thousands of mainly European visitors annually.

FROM THE FIELD NOTES

"Some countries have to take warnings of global warming more seriously than others. As we approached the Maldives, the islands lay like lilypads on the surface of a pond. No part of this country's natural surface lies more than 6 feet (less than 2 meters) above sea level. The upper floors of the buildings in the capital, Maale, form the Maldives' highest points. Almost any rise in sea level would threaten this Indian Ocean outpost of South Asia." © H. J. de Blij.

www.conceptcaching.com

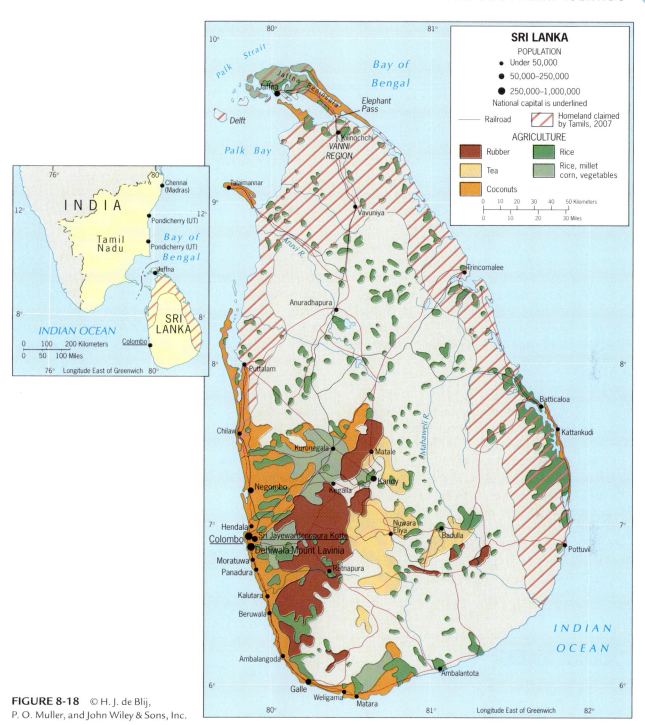

FIGURE 8-18 © H. J. de Blij,
P. O. Muller, and John Wiley & Sons, Inc.

The Maldives' low elevation has been mentioned repeatedly in assessments of the future impact of global warming, which would produce rising sea levels. Those risks became sudden reality on December 26, 2004 when the Indian Ocean tsunami generated off Indonesia swept over the islands, killing more than 100 residents as well as tourists and destroying resort facilities along the shore as well as inland. In mid-2007 it was too early to gauge the overall impact on the Maldives' realm-leading economy, but the tourist industry was badly damaged.

Sri Lanka: South Asian Tragedy

Sri Lanka (known as Ceylon before 1972), the compact, pear-shaped island located just 35 kilometers (22 mi) across the Palk Strait from India, became independent from Britain in 1948 (Fig. 8-18). There were good reasons to create a separate sovereignty for Sri Lanka. This is neither a Hindu nor a Muslim country: the majority of its population of just over 20 million people—more than 70 percent—are Buddhists. Furthermore, unlike India or

Pakistan, Sri Lanka is a plantation country, and commercial farming still is the mainstay of the agricultural economy.

The great majority of Sri Lanka's people are descended from migrants who came to this island from northwest India beginning about 2500 years ago. Those migrants introduced the advanced culture of their source area, building towns and irrigation systems and bringing Buddhism. Today, their descendants, known as the Sinhalese, speak a language (Sinhala) that belongs to the Indo-European language family of northern India.

The Dravidians who lived on the mainland, just across the Palk Strait, came later and in far smaller numbers—until the British colonialists intervened. During the nineteenth century the British brought hundreds of thousands of Tamils to work on their tea plantations, and soon a small minority became a substantial segment of Ceylonese society. The Tamils brought their Dravidian tongue to the island and introduced their Hindu faith. At the time of independence, they constituted more than 15 percent of the population; today they total about 18 percent.

Hope and Disaster

When Ceylon became independent, it was one of the great hopes of the postcolonial world. The country had a sound economy and a democratic government, and it was renowned for its tropical beauty. Its reputation soared when a massive campaign succeeded in eradicating malaria and when family-planning campaigns reduced population growth while the rest of the realm was experiencing a population explosion. Rivers from the cool, forested interior highlands fed the paddies that provided ample rice; crops from the moist southwest paid the bills, and the capital, Colombo, grew to reflect the optimism that prevailed.

In the midst of this glowing scenario, the seeds of disaster were already being sown. Sri Lanka's Tamil minority soon began proclaiming its sense of exclusion, demanding better treatment from the Sinhalese majority. Although the government recognized Tamil as a "national language" in 1978, sporadic violence marked the Tamil campaign, and in 1983 full-scale civil war began. Now many in the Tamil community demanded a separate Tamil state to encompass the north and east of the country (see Fig. 8-18), and a rebel army called the Tamil Tigers confronted Sri Lanka's national forces.

The sequence of events fits the model of the evolution of the **insurgent state** discussed in Chapter 5 (pp. 259–260). **17** In Sri Lanka, the *equilibrium* stage was reached in the early 1990s, when the Tamil Tigers claimed the Jaffna Peninsula and set up their headquarters there. The *counteroffensive* stage is now in progress, but the Tamil forces strike at will, and a negotiated settlement, now being pursued with the help of Norwegian mediators, appears inevitable. Even if the Tamils do not succeed in securing the independent state they call Eelam, they will likely force the government to make some territorial concessions. And even the devastation from the 2004 Indian Ocean tsunami, which struck Sri Lanka especially hard and killed an estimated 35,000 people, most of them along the exposed southern and eastern coast, could not dampen the hostility between the two sides. Accusations of theft of relief supplies and intimidation

FROM THE FIELD NOTES

"Sri Lanka is a dominantly Buddhist country, the religion of the majority Sinhalese. Most areas of the capital, Colombo, have numerous reminders of this in the form of architecture and statuary: shrines to the Buddha, large and small, abound. But walk into the Tamil parts of town, and the cultural landscape changes drastically. This might as well be a street in Chennai or Madurai: elaborate Hindu shrines vie for space with storefronts and Buddhist symbols are absent. The people here seemed to be less than enthused about the Tamil Tigers' campaign for an independent state. 'This would never be a part of it anyway,' said the fellow walking toward me as I took this photograph. 'We're here for better or worse, and for us the situation up north makes it worse.' But, he added, Sri Lankan governments of the past had helped create the situation by discriminating against Tamils." © H. J. de Blij.

www.conceptcaching.com

The costly struggle between the national government and the Tamil Tigers over the latter's demand for an independent state has turned Sri Lanka into an armed camp with soldiers on guard at public buildings, major intersections, airports, railroad stations, and other potential targets even as others engage in actual combat elsewhere. The four soldiers in this photo protect a major road in Vavuniya, a town and district in contention on the highway linking the south to the rebel strongholds of the north. It is a familiar sight throughout this strife-ridden, scenically beautiful but tragically divided country.   © Landov LLC.

of survivors occurred even as both parties faced a common enemy.

Once-promising Sri Lanka is paying dearly for its politicians' failures. The cost of the civil war is enormous.

The economic consequences are incalculable, with the tourist industry having been reduced to a fraction of what it could be. And South Asia has lost a beacon of opportunity and progress.

What You Can Do

RECOMMENDATION: Seek an internship! If you are willing to trade your time for some experience and perhaps even education and training, consider looking for an internship where you can use your knowledge of geography while learning more about it. Should you decide to become a geography major or take a minor, you may be eligible for one of the National Geographic Society's internships in Washington, D.C. The Society places interns in many of its operations, from the Geography Bee to National Geographic Maps. Inquire about this and other internship opportunities at the office of the Geography Department at your college or university.

GEOGRAPHIC CONNECTIONS

1 One of the current issues in the United States involves the so-called outsourcing of jobs: the transfer of salaried employment from American to foreign locations. Many economists say that this is a logical outgrowth of globalization, that it involves comparatively few jobs, and that it helps U.S. companies survive competition to which they might otherwise succumb—costing more jobs than the outsourcing did in the first place. India has become a major player in the outsourcing of certain work. What are these jobs? Where in India do they cluster and why? Do language and/or time-zone location have anything to do with India's success, and if so, how?

2 Six decades after India and Pakistan ceased to be part of the British colonial empire, the contrasts between India and Pakistan—demographic, economic, social, political—could hardly be greater despite the fact that they are neighbors. What, in your view, are the key differences between the two South Asian states in these contexts? Can you identify crucial geographic factors that would account for such enormous dissimilarities? Given their shared histories, do you see any common ground between Pakistan and India today that might form a basis for their future convergence and cooperation?

Geographic Literature on South Asia: The key introductory works on this realm were authored by Farmer, Spate & Learmonth, Spencer & Thomas, Weightman, and Wolpert. These works, together with a comprehensive listing of noteworthy books and recent articles on the geography of South Asia, can be found in the References and Further Readings section of this book's website at **www.wiley.com/college/deblij.**

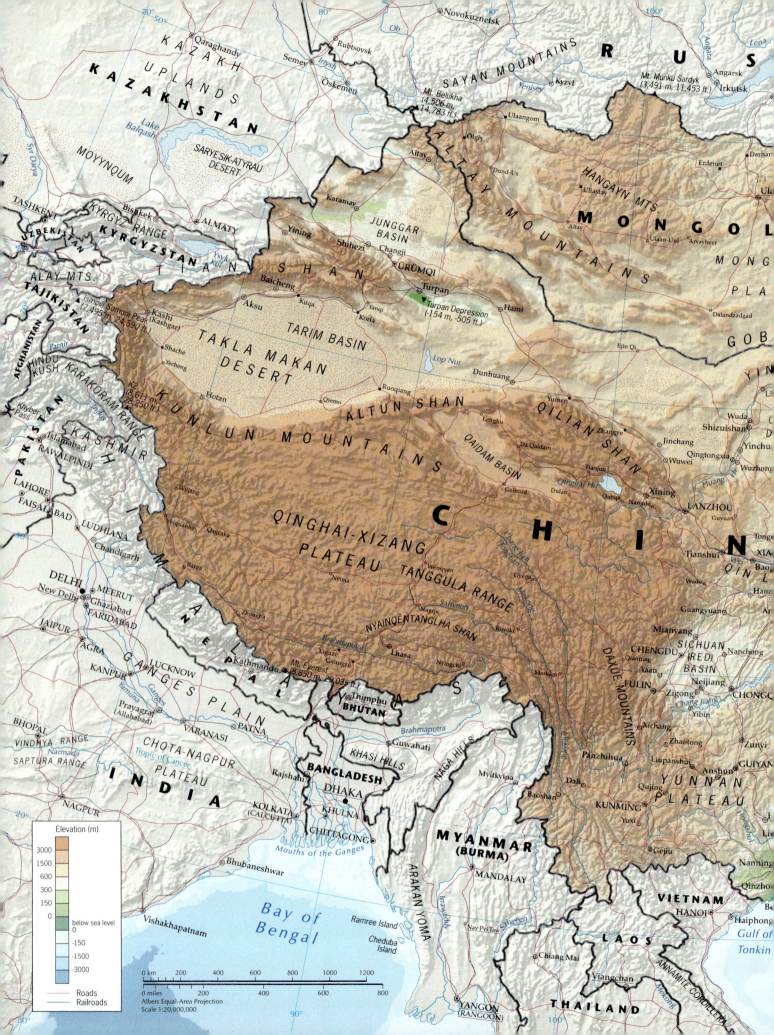

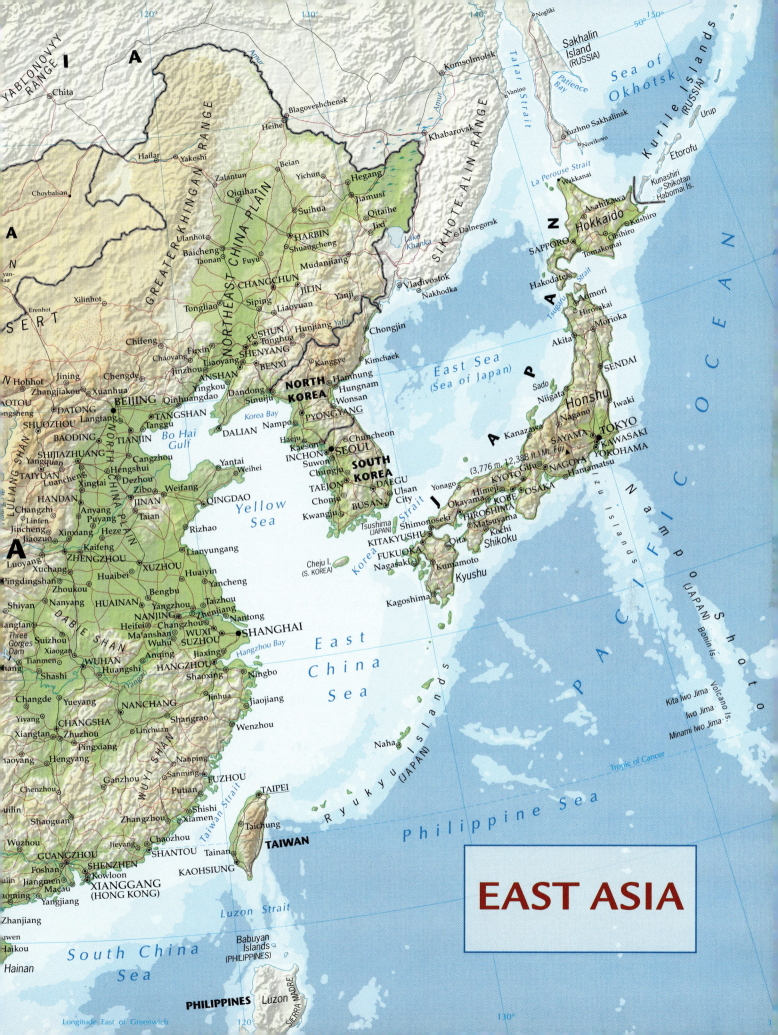

EAST ASIA

9

Where were these photos taken? Find out at www.wiley.com/college/deblij

In This Chapter

- From Xia to Xiamen: China's ancient roots, modern transformation
- China's global economic challenge
- Angry nationalism rising in China as well as Japan
- Rebounding Japan: Still the region's powerhouse
- How geography could help reunite the Korean Peninsula
- Signs of progress over the issue of Taiwan

Photos: © H. J. de Blij

FIGURE 9-1   *Map:* © H. J. de Blij, P. O. Muller, and John Wiley & Sons, Inc.

EAST ASIA IS A geographic realm like no other. At its heart lies the world's most populous country. On its periphery lies one of the globe's most powerful national economies. Along its coastline, on its peninsulas, and on its islands an economic boom has transformed cities and countrysides. Its interior contains the world's highest mountains and vast deserts. It is a storehouse of raw materials. The basins of its great rivers produce food that can sustain more than a billion people.

Defining the Realm

The East Asian geographic realm consists of six political entities: China, Japan, South Korea, North Korea, Mongolia, and Taiwan. Note that we refer here to political entities rather than states. In changing East Asia, the distinction is significant. The island of Taiwan, whose government calls this the Republic of China, functions as a democratic state but is regarded by mainland China (the communist People's Republic of China) as an out-of-control province that must one day be reunited with the motherland. Some foreign government officials do not always get this right. When the president of the People's Republic visited the White House in 2006 and the Chinese national anthem was about to be played during the welcoming ceremony, the announcer introduced it as the anthem of the Republic of China. The mistake was embarrassing, the guests were insulted, and the lesson—once again—was that geographic literacy is important. Fortunately, the band had the right music.

As we define it, East Asia lies between the sparsely peopled Siberian expanses of Russia to the north and the populous countries that make up South and Southeast Asia to the south. The realm extends from lofty snowcapped peaks and vast empty deserts in central Asia to earthquake-prone, crowded Pacific islands offshore (Fig. 9-1). Environmental diversity is one of its hallmarks.

Currently, East Asia is experiencing a massive geographic transformation that we will study from several

MAJOR GEOGRAPHIC QUALITIES OF

East Asia

1. East Asia is encircled by snowcapped mountains, vast deserts, cold climates, and Pacific waters.

2. East Asia was one of the world's earliest culture hearths, and China is one of the world's oldest continuous civilizations.

3. East Asia is the world's most populous geographic realm, but its population remains strongly concentrated in its eastern regions.

4. China, the world's largest nation-state demographically, is the current rendition of an empire that has expanded and contracted, fragmented and unified many times during its long existence.

5. China today remains a strongly rural society, and its vast eastern river basins feed hundreds of millions in a historic pattern that continues today.

6. China's sparsely peopled western regions are strategically important to the state, but they lie exposed to minority pressures and Islamic influences.

7. Along China's Pacific frontage an economic transformation is taking place, affecting all the coastal provinces and creating an emerging Pacific Rim region.

8. Increasing regional disparities and fast-changing cultural landscapes are straining East Asian societies.

9. Japan, the economic giant of the East Asian realm, has a history of colonial expansion and wartime conduct that still affects international relations here.

10. East Asia may witness the rise of the world's next superpower as China's economic and military strength and influence grow—and if China avoids the devolutionary forces that fractured the Soviet Union.

11. The political geography of East Asia contains a number of flashpoints that can generate conflict, including Taiwan, North Korea, and several island groups in the realm's seas.

Places and Names

As in other parts of the world, European colonists in China wrote down the names of places and people as they heard them—and often got them wrong. The Wade-Giles system put such place names as Peking, Canton, and Tientsin on the map, but that's not how the Chinese knew them. In 1958, the communist regime replaced the foreign version with the *pinyin* system, based on the pronunciation of Chinese characters in Northern Mandarin, which was to become the standard form of the Chinese language throughout China. The world now became familiar with these same three cities written as Beijing, Guangzhou, and Tianjin. Among prominent Chinese geographic names, only one has not gained universal recognition: Tibet, called Xizang by China but referred to by its old name on many maps for reasons that will become obvious later in this chapter.

Pinyin usage may be standard in China today, but China remains a country of many languages and dialects. Not only do minorities speak many different languages, but Mandarin-speakers use numerous dialects—so that when you sit down with some Chinese friends who have just met, it may take just one sentence for one to say to the other "ah, you're from Shanghai!"

Migration is changing this, of course. China's population is on the move, and many residents of Shanghai today were not born in the city. But China's language map is anything but simple. And not just place names, but personal names, too, were revised under pinyin rules. The communist leader who used to be called Mao Tse-tung in the Western media is now Mao Zedong. Also remember that the Chinese write their last name first: President Hu Jintao is Jintao to friends and Mr. Hu to others. To the Chinese, it is Bush George and Cheney Dick, not the other way around.

---

viewpoints because it affects our daily lives in numerous ways. Take a look at a geography textbook written 35 years ago, and you will find that neither text nor index will refer to the regional name **Pacific Rim**. Today, that term signifies the fastest and most dramatic economic and social transfiguration in the history of humankind, and its continuing development is headline news. Travel through China's coastal provinces and cities from Dalian in the north to Hong Kong (Xianggang) in the south, and you will see what economic geographers describe as the greatest public works project on Earth (for Chinese spellings, be sure to see the box above titled "Places and Names"). Whether in Beijing, Shanghai, Guangzhou, or dozens of other cities on or near the Pacific coast, forests of futuristic skyscrapers rise where traditional neighborhoods, villages, and farmers resisted change for countless centuries. China's economic juggernaut set in motion the greatest human migration in the country's history as millions moved from countryside to city in search of a better life.

Before China's Pacific Rim became the geographic subregion it is today, Japan built a gigantic economy with global connections and was the only highly developed country in East Asia. Japanese capital fueled the emergence of other economies in East Asia (such as Hong Kong) and Southeast Asia (including Singapore), but Japanese investment in China's Pacific Rim played a key role in the process we will chronicle in this chapter. China and Japan have some serious issues to resolve, ranging from lingering arguments over what happened during World War II to growing disputes over Pacific islands, but there can be no doubt

about their common interest in the economic miracle that has transformed East Asia in little more than one generation.

No matter how powerful Japan's influence was in the evolution of East Asia's Pacific Rim, China is the realm's colossus, with more than ten times Japan's population and 25 times its area. One of the world's oldest (perhaps the oldest) continuous civilizations, China already was a major culture hearth when Japan was still an isolated frontier inhabited by the Ainu. Migrations that originated in China advanced through the Korean Peninsula and reached the islands. Philosophies, religions, and cultural traditions (including, for example, urban design and architecture) diffused from China to Korea and Japan, and to other locales on the margins of the Chinese hearth. Over the past millennium, China repeatedly expanded to acquire empires, only to collapse in chaos when its center failed to maintain control. European colonial powers took such opportunities to partition China among themselves. The Russians and the Japanese pushed their empires deep into China. But always China recovered, and today China may again be poised to expand its sphere of influence and, indeed, to take its place as a world power in the twenty-first century.

NATURAL ENVIRONMENTS

Figure 9-1 dramatically illustrates the complex physical geography of the East Asian realm. In the southwest lie ice-covered mountains and plateaus, the Earth's crust in this region crumpled up like the folds of an accordion. A gigantic collision of tectonic plates is creating this landscape as the Indian Plate pushes northward into the underbelly of the Eurasian Plate (Fig. G-4). The result is some of the world's most spectacular scenery, but snow, ice, and cold are not the only dangers to human life here. Earthquakes and tremors occur almost continuously, causing landslides and avalanches. As the map shows, the high mountains and plateaus widen from a relatively narrow belt in the Karakoram to form Xizang's (Tibet's) vast plateau, flanked by the Himalayas to the south. Then, east of Tibet, the mountain ranges converge again and bend southward into Southeast Asia, where they lose their high relief.

Physiography and Population

As Figure G-9 shows, this Asian interior is one of the world's most sparsely populated areas, but it is nevertheless critical to the lives of hundreds of millions of people. In these high mountains, fed by the melting ice and snow, rise the great rivers that flow eastward across China and southward across Southeast and South Asia. Throughout the Holocene, these rivers have been eroding the uplands and depositing their sediments in the lowlands, in effect creating the alluvium-filled basins that now sustain huge populations. Fertile alluvial soils and adequate growing seasons, combined with ample water and millions of hands to sow the wheat and plant the rice, have allowed the emergence of one of the great population concentrations on Earth.

Figure G-8 records this high-elevation zone of East Asia as category *H*, highland climate, which is by far the largest area of its kind in the world. Here one criterion overpowers the rules of climatic classification: altitude. The mountains wrest virtually all moisture from air masses moving northward and block their path into interior Asia. As Figure G-8 shows, desert conditions prevail in East Asia's northern interior. Figure 9-2 names two of the more famous ones: the Takla Makan of far western China and the Gobi of Mongolia.

Physiography, therefore, has much to do with East Asia's population distribution, but even the more habitable

EAST ASIA: PHYSIOGRAPHY

0 200 400 600 800 1000 1200 Kilometers

0   200   400   600 Miles

FIGURE 9-2   © H. J. de Blij, P. O. Muller, and John Wiley & Sons, Inc.

and agriculturally productive east has its limitations. The northeast suffers from severe continentality, with long and bitterly cold winters. High relief encircles the river basins north of the Yellow Sea and dominates much of the northern part of the Korean Peninsula, creating strong local environmental contrasts. South Korea, as Figure G-8 shows, experiences relatively moderate conditions comparable to those of the U.S. Southeast; North Korea has a harsh continental climate like that of North Dakota.

In general, coastal, peninsular, and insular East Asia possess more moderate climates than the interior. Like South Korea, southern Japan and southeastern China have humid temperate climates, and southernmost Taiwan and Hainan Island even have areas of tropical (*A*) conditions. Proximity to the ocean tends to moderate climatic environments, and coastal East Asia proves the point.

Three Great Rivers

Before we focus on East Asia's human geography, it is useful to look at a map of this realm's complex physical stage (Fig. 9-2). From the high-relief interior come three major river systems that have played crucial roles in the human drama. In the north, the **Huang He** (Yellow River) arises deep in the high mountains, crosses the Ordos Desert and the Loess Plateau, and deposits its fertile sediments in the vast North China Plain, where East Asia's earliest states emerged. In the center, the **Chang Jiang** (Long River), called the *Yangzi* downstream, crosses the Sichuan Basin and the Three Gorges, where a huge dam project is under way, and waters extensive ricefields in the Lower Chang Basin. And in the south, the **Xi Jiang** (West River) originates on the Yunnan Plateau and becomes the *Pearl River* in its lowest course. Its estuary, flanked by several of China's largest urban-industrial complexes, has become one of the hubs of the evolving Pacific Rim.

Further scrutiny of Figure 9-2 indicates that a fourth river system plays a role in China: the *Liao River* in the northeast and its basin, the Northeast China Plain. As the map suggests, however, the Liao is not comparable to the great rivers to its south, its course being shorter and its basin, in this higher-latitude area, much smaller. As we will see, China's Northeast was for some time the country's industrial, not agricultural, heartland.

Interior Environments

Looking again toward the interior, note the Loess Plateau south of the Ordos Desert, where the Huang He makes its giant loop. **Loess** is a fertile, windblown deposit composed of rock pulverized by glaciers (see photo at right). Add water (in this case the middle Huang and its tributaries) and an adequate growing season, and a sizeable population will

arise. To the south, deep in the interior, lies the Sichuan (Red) Basin, crossed by the Chang Jiang. This basin has supported human communities for a long time, and you can actually make out its current population cluster on the world population map (Fig. G-9). The Sichuan Basin, encircled as it is by mountains, is one of the world's most clearly defined physiographic regions, and the concentration of its approximately 120 million inhabitants reflects that definition.

Still farther to the south lies the Yunnan Plateau, source of the tributaries that feed the Xi River. Much of southeastern China has comparatively high relief; it is hilly and in

FROM THE FIELD NOTES

"From the train, traveling from Beijing to Xian, we had a memorable view of the Loess Plateau. Loess is a fine-grained dust formed from rocks pulverized by glacial action, blown away by persistent winds, and deposited in sometimes well-defined locales. Loess covers much of the North China Plain, which is what makes it so fertile, but there it is not as thick as in the Loess Plateau in the middle basin of the Huang He (Yellow River). Here the loess averages 75 meters (250 ft) in thickness and in places reaches as much as 180 meters (600 ft). Because loess has some very distinctive physical properties, it tends to create unusual landscapes. Through a complicated physical process following deposition, loess develops the capacity to stand upright in walls and columns, and resists collapse when it is excavated. As a result the landscape looks terraced: streams cut deep and steep-sided valleys. The Loess Plateau is a physiographic region, but it is a cultural region too. Hundreds of thousands of people have literally dug their homes into vertical faces of the kind shown here, creating cave-like, multi-room dwellings with wooden exterior doors." © H. J. de Blij.

www.conceptcaching.com

places mountainous. This high relief has helped limit contacts between China and Southeast Asia.

Along the Coast

East Asia's Pacific margin is a jumble of peninsulas and islands. The Korean Peninsula looks like a near-bridge from Asia to Japan, and indeed it has served as such in the past. The Liaodong and Shandong peninsulas protrude into the Yellow Sea, which continues to silt up from the sediments of the Huang and Liao rivers. Off the mainland lie the islands that have played such a crucial role in the modern human geography of Asia and, indeed, the world: Japan, Taiwan, and Hainan. Japan's environmental range is expressed by cold northern Hokkaido and warm southern Kyushu, but Japan's core area lies on its main island, Honshu. As Figure 9-1 shows, myriad smaller islands flank the mainland and dot the East and South China seas. As we will discover, some of these smaller islands have major significance in the human geography of this realm.

HISTORICAL GEOGRAPHY

Consider this: there is no evidence that *Homo sapiens*, modern humans, reached the Americas any earlier than 40,000 years ago, and many paleoanthropologists argue that the only available evidence indicates that humans crossed from Eurasia to Alaska a mere 14,000 to 13,000 years ago. On the opposite side of the Pacific, however, the story of hominid and human settlement spans hundreds of thousands of years—perhaps 1 million.

What took so long? If hominids like *Homo erectus* left their ancestral African homelands and migrated as far away as eastern Asia more than half a million years ago, and if modern humans followed them and reached Australia some 50,000 years ago, what kept the hominids from the Americas altogether, and what delayed the humans?

No satisfactory answers have yet been found to questions like these. The width of the Pacific Ocean may have doomed maritime migrants, although trans-Pacific migration may eventually have occurred. The frigid temperatures along the northern route from Siberia to Alaska may have stopped hominids and humans alike. But the Pleistocene was punctuated by warm interglaciations; why did it take until the early Holocene for humans to make the Bering Strait crossing?

All this implies that for hundreds of thousands of years East Asia was a cul-de-sac, a dead-end for migrants out of Africa. Many archeological sites in the realm have yielded evidence of *Homo erectus*, including what is perhaps the most famous of all: Peking Man. In a cave near the Chinese capital, Beijing, an archeologist in 1927 found a single tooth he recognized as representing a hominid. Later excavations proved him right as parts of more than three dozen skeletons were unearthed. They proved that Peking Man and his associates made stone and bone tools, had a communal culture, controlled the use of fire, hunted wildlife, and cooked their meat. Later arrivals undoubtedly challenged older communities, and the frontier of settlement must have expanded across the river basins of the east. But when *Homo sapiens*, modern humans, arrived in East Asia, the hominids could not compete. Current anthropological theory holds that the better-equipped humans eliminated the hominids, perhaps between 60,000 and 40,000 years ago, after which the human migration into the Americas could begin.

In the context of East Asia we should mention a minority view. Some anthropologists state that the evidence supports the notion that modern humanity developed *not* from one stock in Africa, but from four stocks in widely separated parts of the world: Africa, western Eurasia (the Caucasoids), Australia (the Australoids), and eastern Eurasia (the Mongoloids). According to this idea, today's East Asians trace their ancestry to Peking Man and beyond.

Early Cultural Geography

Whatever the outcome of the debate over human origins in East Asia, it is clear that humans have inhabited the plains and river basins, foothills, and islands of this realm for a very long time. Hunting sustained both the hominids and the early human communities; fishing drew them to the coasts and onto the islands. The first crossing into Japan may have occurred as long as 10,000 to 12,000 years ago, possibly much longer, when the Jomon people, a Caucasoid population of uncertain geographic origins, entered the islands; their modern descendants, the Ainu, spread throughout the Japanese archipelago. Today, only about 20,000 persons living in northernmost Hokkaido trace their ancestry to Ainu sources.

About 2300 years ago the Yayoi people, rice farmers who had settled in Korea, appear to have crossed by boat to Kyushu, Japan's southernmost island, from where they advanced northward. The Ainu, who subsisted by fishing, trapping, and hunting, were driven back, but gene-pool studies show that much mixing of the groups took place; they also show that the Yayoi invasion was followed by other incursions from the Asian mainland. By then, powerful dynastic states had already arisen in what is today China, and early Chinese culture traits thus found their way into Japan through the process we know as relocation diffusion.

Mainland Cultures

On the Asian mainland, plant and animal domestication had begun as early as anywhere on Earth. We in the Western world take it for granted that these momentous processes began in what we now call the Middle East and diffused from the Fertile Crescent to other parts of Eurasia and the rest of the world. But the taming of animals and the selective farming of plants may have begun as early, or earlier, here in East Asia. As in Southwest Asia, the fertile alluvial soils of the great river basins and the ebb and flow of stream water created an environment of opportunity, and millet and rice were being harvested between 7000 and 8000 years ago.

Even during this Neolithic period of increasingly sophisticated stone tools, East Asia was a mosaic of regional cultures. Their differences are revealed by the tools they made and the decorations on their bowls, pots, and other utensils. An especially important discovery of two 8000-year-old pots in the form of a silkworm cocoon, from China's Hebei Province, suggests a very ancient origin for one of the region's leading historic industries.

As noted earlier, plant and animal domestication produced surpluses and food storage, enabling population growth and requiring wider regional organization. Here as elsewhere during the Neolithic, settlements expanded, human communities grew more complex, and power became concentrated in a small group, an *elite*.

Formation of States and Dynasties

This process of **state formation** is known to have occurred in only a half-dozen regions of the world, and China was one of these. But evidence about China's earliest states has long been scarce. Today, however, archeologists are focusing on the lower Yi-Luo River Valley in the western part of Henan Province, where the first documented Chinese dynasty, the Xia Dynasty (2200–1770 BC), existed. The capital of this ancient state, Erlitou, has been found, and archeologists now refer to the Xia Dynasty as the Erlitou culture. Secondary centers are being discovered, and Erlitou tools and implements in a wider area prove that the Xia Dynasty represents a substantial state.

All early states were ruled by elites, but China's political history is chronicled in **dynasties** because here the succession of rulers came from the same line of descent, sometimes enduring for centuries. In the transfer of power, family ties counted for more than anything else. Dynasties were overthrown, but the victors did not change this system. Dynastic rule lasted into the twentieth century.

The Xia Dynasty may have been the earliest Chinese state, but it lay in the area where, later, more powerful dynastic states arose: the North China Plain. Here the tenets of what was to become Chinese society were implanted

early and proved to be extremely durable. From this culture hearth ideas, innovations, and practices diffused far and wide. From agriculture to architecture, poetry to porcelain-making, influences radiated southward into Southeast Asia, westward into interior Asia, and eastward into Korea and Japan. In the North China Plain lay the origins of what was to become the Middle Kingdom, which its citizens considered the center of the world.

Table 9-1 summarizes the sequence of dynasties that may have begun with Xia more than four millennia ago and ended in 1911 when the last emperor of the Qing (Manchu) Dynasty, a boy 6 years old, was forced to abdicate the Chinese throne. As the table shows, every dynasty contributed importantly to the development of Chinese society, sometimes progressively through enlightened policies and efficient administration, at other times regressively when capricious cruelty, authoritarian excesses, and xenophobia prevailed. During those 4000 years, China evolved into the world's most populous nation, always rebounding after disastrous famines and floods, its territory covering more than 80 percent of the East Asian realm (Fig. 9-3).

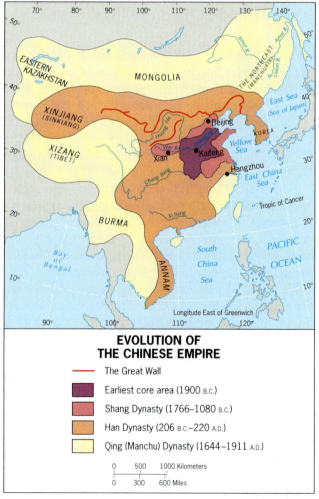

EVOLUTION OF THE CHINESE EMPIRE

— The Great Wall

Earliest core area (1900 B.C.)

Shang Dynasty (1766–1080 B.C.)

Han Dynasty (206 B.C.–220 A.D.)

Qing (Manchu) Dynasty (1644–1911 A.D.)

| 0 | 500 | 1000 Kilometers |
| 0 | 300 | 600 Miles |

FIGURE 9-3 © H. J. de Blij, P. O. Muller, and John Wiley & Sons, Inc.

POINTS TO PONDER

- China has the world's deadliest mining industry. Accidents kill more than 5000 Chinese coal miners annually. Coal-fired power plants produce the bulk of China's electricity.

- Shanghai has built a vast underground bunker complex capable of sheltering 200,000 people who could survive a nuclear attack or other crisis for two weeks.

- Russia's President Putin has promised Japan that it will build a 4000-kilometer (2600-mile) oil pipeline from Siberia to a terminal on the East Sea (Sea of Japan) directly across from Honshu that will deliver 1.6 million barrels of oil a day.

- China's high-elevation railroad across mountains and permafrost linking Beijing to Lhasa in Xizang (Tibet) is an engineering marvel, but it represents just one part of a larger effort to bring its far-west regions more firmly into the national fold.

Dominant as China is in its regional sphere, there is more to East Asia than the Chinese giant. To the north, the landlocked state of Mongolia is what remains of the time when Mongol armies conquered much of Eurasia (as Table 9-1 shows, even China fell under the Mongol sway). To the east, the Korean Peninsula escaped incorporation into China. Offshore, Japan first surpassed China as a military power, colonizing Manchuria (as it was then called) and penetrating deep into China's heartland. Later, Japan outclassed China as an economic power, achieving world status while China languished under Maoist communism. And as the twenty-first century dawned, Taiwan remained a separate political entity, its detachment from China resulting from the struggle that brought the communists to power in Beijing.

Regions of the Realm

East Asia presents us with an opportunity to illustrate the changeable nature of regional geography. Our regional delimitation is based on current circumstances, and it predicts ways the framework may change. It is anything but static. In the first decade of the twenty-first century, we can identify five geographic regions in the East Asian realm (Fig. 9-4). These are:

1. *China Proper.* Almost any map of China's human geography—population distribution, urban centers, surface communications, agriculture, or industry—emphasizes the strong concentration of Chinese activity in the country's eastern sector. This is the "real" China, where its great cities, populous farmlands, and historic sources are located. Long ago, scholars called this *China Proper*, and it is a good regional designation. But China is a large and complex country, and a number of subregions are nested within China Proper. Some of these, such as the North China Plain and the Sichuan Basin, are old and well-established geographic units. One in particular is new: China's Pacific Rim, still growing, yet poorly defined, and shown as "formative" in Figure 9-4.

2. *Xizang (Tibet).* The high mountains and plateaus of Xizang, ruled by China but still widely known by its older name of Tibet, form a stark contrast to teeming China Proper. Here, next to one of the world's largest and most populous regions, lies one of the emptiest and, in terms of inhabited space, smallest regions.

MAJOR CITIES OF THE REALM

City	Population* (in millions)
Beijing (Shi), China	11.3
Busan, South Korea	3.5
Chongqing (Shi), China	6.5
Guangzhou, China	9.1
Hong Kong (Xianggang) SAR, China	7.3
Macau SAR, China	0.5
Nanjing, China	3.8
Osaka, Japan	11.3
Seoul, South Korea	9.6
Shanghai (Shi), China	15.3
Shenyang, China	4.9
Shenzhen, China	7.8
Taipei, Taiwan	2.5
Tianjin (Shi), China	7.4
Tokyo, Japan	26.7
Wuhan, China	7.3
Xian, China	4.0

*Based on 2008 estimates.

TABLE 9-1

The Chinese Dynasties, 2200 BC to AD 1911

Dynasty Name(s)	Date(s)	Areas Governed	Major Features	Geographical Impact
Xia	c2200–c1770 BC	Small part of Huang River Basin, centered in Yi-Luo tributary.	First dynastic state (?); Neolithic technology, sophisticated stone tools.	Beginnings of stream diversion, irrigation.
Shang (latter part Yin)	c1766–c1080 BC	Huang-Wei River confluence across Shandong and Henan Provinces. Anyang capital.	Neolithic to Bronze Age transition. Timber houses with thatched roofs, beginnings of Chinese writing, moon-cycle-based calendar. Superb bronze vases and other implements.	Unifies large region in lower Huang River Basin.
Zhou	c1027–221 BC Two periods: Spring and Autumn c1027–481 BC; Warring States 475–221 BC	A people centered in Shaanxi Province to the west of the Shang area overthrow the Shang rulers and expand their domain eastward. Warring States is a period of feudalism.	Formative period for China. Irrigation systems, iron smelting, horses, expanded farm production. Major towns develop. Writing system is established. Chopsticks come into use. Daoism and Confucianism arise; Daoism centers on mysticism, individualism, the personal "way," and Confucianism on duties, community standards, respect for government.	Taoism diffuses widely throughout East Asia, reaches South Asia, and influences Buddhism. Confucianism becomes China's guiding philosophy for more than 2000 years. Building of *Great Wall* begins.
Qin (Ch'in)	221–206 BC	Most of North China Plain, south into Lower Chang Basin; first consolidation of large Chinese state.	Time of bureaucratic dictatorship after ferment of Zhou: Confucian Classics burned, Taoism combated. Much of Great Wall is built at huge cost in lives. Cruel despotism, excesses. Emperor's tomb discovered at Xian in 1976 with 6000 terracotta life-sized men and horses.	Ruler Ch'in (Qin)'s name immortalized as *China*.
Han	206 BC–AD 220	Major territorial expansion adds Xinjiang in interior Asia, Vietnam in Southeast Asia.	Second formative dynasty for China; restoration of Confucian principles and a flowering of culture. First use of eunuchs by authoritarian government, planting a long-term weakness in administration. Xian becomes one of the greatest cities of the ancient world. Chinese refer to themselves as the *People of Han*.	About the same time, same size as the Roman Empire. *Silk Road* carries goods from China across inner Asia to Syria and on to Rome.

Period from AD 220–580 witnessed disunion and division. Important developments included the arrival of Buddhism in East Asia and the diffusion of Chinese influence and ideas into Korea and across Korea into Japan.

Sui	AD 581–618	North and South China reunified, Xian rebuilt and greatly expanded. Conflict with Turks in Central Asia and with Koreans in east.	Brief but important dynasty with revival of Confucian rituals and practices in education. Brutal but modernizing regime conducts a census, establishes a penal code, and engages in massive public works.	Construction of key segments of Grand Canal links Huang and Yangzi Rivers and thus connects northern and southern economies.
Tang	AD 618–907	Defeat of Turks in Central Asia. Campaign to put down rebellions in the south succeeds, though northern frontier remains unstable.	A golden age for China and a third formative dynasty. Enormous and efficient bureaucracy dictates virtually all aspects of life. Xian now the cultural capital and largest city in the world. *Buddhism* thrives, pagodas arise everywhere. The arts, Chinese and foreign, flourish.	Arab and Persian seafarers visit Chinese ports, the Silk Route is loaded with trade, and China seems set for an international era. It is not to be: *Islam* arrives in Central Asia and Tang armies are defeated by Arabs in 751. The Silk Route breaks down. In the east, however, Chinese influences permeate Korea and penetrate Japan.

TABLE 9-1

The Chinese Dynasties, 2200 BC to AD 1911 (*Continued*)

Dynasty Name(s)	Date(s)	Areas Governed	Major Features	Geographical Impact
The 907–960 period is known as the Five Dynasties because five leaders tried to establish dynasties in northern China in quick succession, each failing in turn. In southern China it is known as the time of the Ten Kingdoms, for ten regimes ruled various parts of the region. Cultural continuity prevails despite political instability.				
Song (Northern (Southern	AD 960–1279 AD 960–1127) AD 1127–1279)	Rulers consolidate the Five (northern) Dynasties, then begin capturing the Ten (southern) Kingdoms. They are unable to control the Liao state north of the Great Wall. Mongols invade, drive Song rulers southward, and end dynasty in 1279.	Preoccupied with organization and administration, Song rulers create a competent, efficient *civil service*. A defensive rather than an expansionist period. An era of rich cultural and intellectual achievement in mathematics, astronomy, mapmaking. Paper and movable type are invented, as is *gunpowder*. Large ships are built. The arts, from porcelain-making to poetry, continue to thrive. Practice of footbinding begins. Improved rice varieties feed estimated 100 million inhabitants.	Capital is moved from northern Kaifeng on Huang River to Hangzhou in the south as Song rulers lose control over the north. Mongol invasion is followed by Kublai Khan's choice of Beijing as his capital (1272). Mongols use Chinese authoritarian and bureaucratic traditions to forge their own dynastic rule. Several cities have populations exceeding 1 million.
Yuan (Mongol)	AD 1264–1368	Controls eastern Asia from Siberia in the north to the Vietnam border in the south, including present-day Russian Far East and most of Korea.	Dynasty begins in north before Song ends in south. Repressive regime, social chasms between Mongol rulers and Chinese ruled. Deep and enduring hostility between Chinese and Mongols. Cultural *isolationism* results; Mongols are acculturated to Chinese norms. Marco Polo visits East Asia.	Most of present-day Mongolia is part of the Chinese sphere for the first time. Mongol effort to enter Japan fails. Mongol power over much of Eurasia stimulates trade and flow of information. Beijing grows into major city.
Ming	AD 1368–1644	Rules over all eastern China from Amur River in north to Red River (Vietnam) in south, as far west as Inner Mongolia and Yunnan. At various times incorporates North Korea, most of Mongolia, even Myanmar (Burma).	Chinese rule again after Mongol Yuan dynasty. Stable but autocratic and inward-looking regime. Major advances in science and technology. In first half of fifteenth century, Chinese oceangoing vessels, larger than any in Europe, explore Pacific and Indian Ocean waters and reach East Africa. Farming expands, silk and cotton industries improve, printing plants multiply. Defensive walls are built in the north and around cities, including the Forbidden City in Beijing, the Ming capital. But mismanagement, infighting, eunuch influence in the palaces, and anticommercialism weaken Ming rule.	Early Ming fleets herald an era of exploration and potential colonialism, but the impact of the Little Ice Age creates environmental crisis forcing massive grain redistribution from central to northern China. The great ships are burned, barges are built, and Grand Canal is expanded to cope with food crisis. Political system erodes and central authority falters.
Qing (Manchu)	AD 1644–1911	*Largest* China-centered empire ever incorporates Mongolia, much of Turkestan, Xizang (Tibet), Myanmar, Indochina, Korea, Taiwan.	Foreign rule again: Manchu, a people with Tatar links living in present-day Northeast China, seize an opportunity and take control of Beijing. Improbably, a group numbering about 1 million rule a nation of several hundred million. It is done by keeping Ming systems of administration and retaining (and rewarding) Ming officials. Expansion creates an empire. Population pressure and the concentration of land ownership in fewer hands create problems worsened by floods and famines. European powers and Japan force concessions on weakening Qing rulers; defeats in war lead to revolution and collapse.	Manchu (Qing) rulers create the map that today forms the justification for China's claims to Xizang (Tibet), Taiwan, and the central Eurasian interior; at the height of its power, the Qing empire forces the Russians to recognize Manchu authority over China's Northeast as far north as the Argun River. Latent claims to Russia's Far East are based on Manchu predominance there.

FIGURE 9-4 © H. J. de Blij, P. O. Muller, and John Wiley & Sons, Inc.

3. *Xinjiang.* The vast desert basins and encircling mountains of Xinjiang form a third East Asian region. Again, physical as well as human geographic criteria come into play: here China meets Islamic Central Asia.

4. *Mongolia.* The desert state of Mongolia forms East Asia's fourth region. Like Tibet, landlocked Mongolia, vast but sparsely peopled, stands in stark contrast to populous China Proper.

5. *Jakota Triangle.* East Asia's fifth region is defined by its economic geography. *Ja*pan, South *Ko*rea, and *Tai*wan (the acronym "Jakota" derives from the first two letters of each) were transformed by Pacific Rim economic developments during the second half of the twentieth century. Its recent emergence foreshadows further changes in the decades ahead as Korean unification becomes a possibility and contrasts between the Jakota Triangle and Pacific Rim China diminish.

CHINA PROPER

When we in the Western world chronicle the rise of civilization, we tend to focus on the historical geography of Southwest Asia, the Mediterranean, and Western Europe. Ancient Greece and Rome were the crucibles of culture; Mediterranean and Atlantic waters were the avenues of its diffusion. China lay remote, so we believe, barely connected to this Western realm of achievement and progress. When an Italian adventurer named Marco Polo visited China during the thirteenth century and described the marvels he saw there, his work did little to change European minds. Europe was and would always be the center of civilization.

The Chinese, naturally, take a different view. Events on the western edge of the great Eurasian landmass were deemed irrelevant to theirs, the most advanced and refined culture on Earth. Roman emperors were rumored to be powerful, and Rome was a great city, but nothing could match the omnipotence of China's rulers. Certainly the Chinese city of Xian (which had a different name at the time) far eclipsed Rome as a center of sophistication. Chinese civilization existed long before ancient Greece and Rome emerged, and it was still there long after they collapsed. China, the Chinese teach themselves, is eternal. It was, and always will be, the center of the civilized world.

We should remember this notion when we study China's regional geography because 4000 years of Chinese culture and perception will not change overnight— not even in a generation. Time and again, China overcame the invasions and depredations of foreign intruders, and afterward the Chinese would close off their vast country against the outside world. Just 35 years ago, in the early 1970s, there were just a few *dozen* foreigners in the entire country with its (then) nearly 1 billion inhabitants. The institutionalization of communism required this insularity, and even the Soviet advisors had been thrown out. But by the early 1970s, China's rulers decided that an opening to the Western world would be advantageous, and so U.S. President Richard Nixon was invited to visit Beijing. That historic occasion, in 1972, ended this latest period of isolation—as always, on China's terms. Since then, China has been open to tourists and businesspeople, teachers, and investors. Tens of thousands of Chinese students have been sent to study at American and other Western institutions. Long-suppressed ideas flowed into China, and a pro-democracy movement arose and climaxed in 1989. China's rulers knew that their violent repression of this movement would anger the world, but that did not matter because they deemed foreign condemnation irrelevant. Foreigners in China had done much worse. Moreover, Westerners had no business interfering in China's domestic affairs.

RELATIVE LOCATION

Throughout their nation's history, the Chinese have at times decided to close their country to any and all foreign influences; the most recent episode of exclusion occurred just a few decades ago. Exclusion is one of China's recurrent traditions, made possible by China's relative location and Asia's physiography. In other words, China's "splendid isolation" was made possible by geography.

Earlier we noted the role of relief and desert in encircling the culture hearth of East Asia, but equally telling is the factor of distance. Until recently, China lay far from the modern source areas of innovation and change. True,

China—as the Chinese emphasize—was itself such a hearth, but China's contributions to the outside world remained limited, essentially, to finely made arts and crafts. China did interact with Korea, Japan, Taiwan, and parts of Southeast Asia, and eventually millions of Chinese emigrated to neighboring countries. But compare these regional links to those of the Arabs, who ranged worldwide and who brought their knowledge, religion, and political influence to areas from Mediterranean Europe to Bangladesh and from West Africa to Indonesia. Later, when Europe became the center of intellectual and material innovation, China found itself farther removed, by land or sea, than almost any other part of the world.

Today, modern communications notwithstanding, China still is distant from almost anywhere else on Earth. Going by rail from Beijing to Moscow, the capital of China's Eurasian neighbor, involves a tedious journey that takes the better part of a week. Direct surface connections with India are, for all practical purposes, nonexistent. Overland linkages with Southeast Asian countries, though improving, remain tenuous.

But for the first time in its history, China now lies near a world-class hearth of technological innovation and financial power: Japan. This proximity to an industrial and financial giant is critical for the momentous economic developments taking place in China's coastal provinces. Japanese investments and business partnerships have transformed the economic landscape of Pacific-coast China. American and European trade links also are important, but Japan's role has been crucial. Japan's economic success set the Pacific Rim engine in motion, and Japan's best financial years happened to coincide with China's reopening to foreigners in the 1970s. Geographic and economic circumstances combined to transform the map and made "Pacific Rim" a household word around the world.

EXTENT AND ENVIRONMENT

China's total area is slightly smaller than that of the United States including Alaska: each country has about 9.6 million square kilometers (3.7 million sq mi). As Figure 9-5 reveals, the longitudinal extent of China and the 48 contiguous U.S. States also is similar. Latitudinally, however, China is considerably wider. Miami, near the southern limit of the United States, lies halfway between Shanghai and Guangzhou. Thus China's lower-latitude southern region takes on characteristics of tropical Asia. In the Northeast, too, China incorporates much of what in North America would be Quebec and Ontario. Westward, China's land area becomes narrower and physiographic similarities increase. But, of course, China has no west coast.

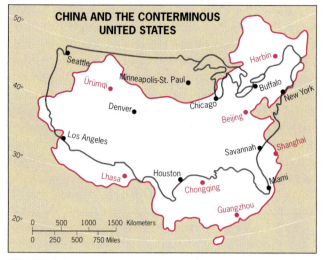

CHINA AND THE CONTERMINOUS UNITED STATES

FIGURE 9-5 © H. J. de Blij, P. O. Muller, and John Wiley & Sons, Inc.

Now compare the climate maps of China and the United States in Figure 9-6 (which are enlargements of the appropriate portions of the world climate map in Fig. G-8). Note that both have a large southeastern climatic region marked *Cfa* (that is, humid, temperate, warm-summer), flanked in China by a zone of *Cwa* (where winters become drier). Westward in both countries, the *C* climates yield to colder, drier climes. In the United States, moderate *C* climates develop again along the Pacific coast. China, however, stays dry and cold as well as high in elevation at equivalent longitudes.

Note especially the comparative location of the U.S. and Chinese *Cfa* areas in Figure 9-6. China's lies much farther to the south. In the United States, the *Cfa* climate extends beyond 40° North latitude, but in China, cold and generally winter-dry *D* climates take over at the latitude of Virginia. Beijing has a warm summer but a bitterly cold and long winter. Northeast China, in the general latitudinal range of Canada's lower Quebec and Newfoundland, is much more severe than its North American equivalent. Harsh environments prevail over vast regions of China, but, as we will see, nature compensates in spectacular fashion. From the climatic zone marked *H* (for highlands) in the west come the great life-giving rivers whose wide basins contain enormous expanses of fertile soils. Without this water supply, China would not have a population more than four times that of the United States.

EVOLVING CHINA

Even if China is not the world's longest continuous civilization (Egypt may claim this distinction), no other state on Earth can trace its cultural heritage as far back as China

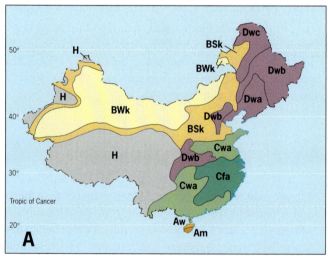

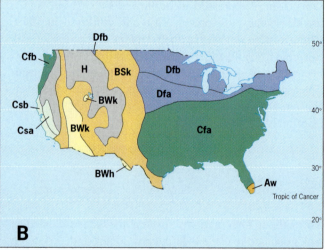

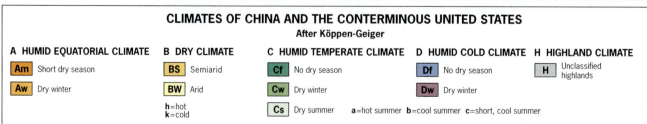

CLIMATES OF CHINA AND THE CONTERMINOUS UNITED STATES
After Köppen-Geiger

A HUMID EQUATORIAL CLIMATE	B DRY CLIMATE	C HUMID TEMPERATE CLIMATE	D HUMID COLD CLIMATE	H HIGHLAND CLIMATE
Am Short dry season	**BS** Semiarid	**Cf** No dry season	**Df** No dry season	**H** Unclassified highlands
Aw Dry winter	**BW** Arid	**Cw** Dry winter	**Dw** Dry winter	
	h=hot **k**=cold	**Cs** Dry summer **a**=hot summer **b**=cool summer **c**=short, cool summer		

FIGURE 9-6 © H. J. de Blij, P. O. Muller, and John Wiley & Sons, Inc.

Kongfuzi (Confucius)

2 Confucius (*Kongfuzi* or *Kongzi* in pinyin) was China's most influential philosopher and teacher. His ideas dominated Chinese life and thought for over 20 centuries. Kongfuzi was born in 551 BC and died in 479 BC. Appalled at the suffering of ordinary people during the Zhou Dynasty, he urged the poor to assert themselves and demand explanations for their harsh treatment by the feudal lords. He tutored the indigent as well as the privileged, giving the poor an education that had hitherto been denied them and ending the aristocracy's exclusive access to the knowledge that constituted power.

Kongfuzi's revolutionary ideas extended to the rulers as well as the ruled. He abhorred supernatural mysticism and cast doubt on the divine ancestries of China's aristocratic rulers. Human virtues, not godly connections, should determine a person's place in society, he taught. Accordingly, he proposed that the dynastic rulers turn over the reins of state to ministers chosen for their competence and merit. This was another Kongzi heresy, but in time this idea came to be accepted and practiced.

His Earthly philosophies notwithstanding, Kongfuzi took on the mantle of a spiritual leader after his death. His thoughts, distilled from the mass of philosophical writing (including Daoism) that poured forth during his lifetime, became the guiding principles of the formative Han Dynasty. The state, he said, should not exist just for the power and pleasure of the elite; it should be a cooperative system for the well-being and happiness of the people.

With time, a mass of writings evolved, much of which Kongfuzi never wrote. At the heart of this body of literature lay the *Confucian Classics*, 13 texts that became the basis for education in China for 2000 years. From government to morality and from law to religion, the *Classics* were Chinese civilization's guide. The entire national system of education (including the state examinations through which everyone, poor or privileged, could enter the civil service and achieve political power) was based on the *Classics*. Kongfuzi championed the family as the foundation of Chinese culture, and the Classics prescribe a respect for the aged that was a hallmark of Chinese society.

But Kongfuzi's philosophies also were conservative and rigid, and when the colonial powers penetrated China, Kongzi's *Classics* came face to face with practical Western education. For the first time, some Chinese leaders began to call for reform and modernization, especially of teaching. Kongzi principles, they said, could guide an isolated China, but not China in the new age of competition. But the Manchu rulers resisted this call, and the Nationalists tried to combine Kongzi and Western knowledge into a neo-Kongzi philosophy, which was ultimately an unworkable plan.

The communists who took power in 1949 attacked Kongzi thought on all fronts. The *Classics* were abandoned, indoctrination pervaded education, and, for a time, even the family was viewed as an institution of the past. Here the communists miscalculated. It proved impossible to eradicate two millennia of cultural conditioning in a few decades. When China entered its post-Mao period, public interest in Kongfuzi surged, and the shelves of bookstores again sag under the weight of the *Analects* and other Confucian and post-Kongfuzi writings. The spirit of Kongfuzi will pervade physical and mental landscapes in China for generations to come.

---

can. China's fortunes rose and fell, but over more than 40 centuries its people created a society with strong traditions, values, and philosophies. Kongfuzi (Confucius), whose teachings and writing still influence China, lived during the Zhou Dynasty, 2500 years ago (see box titled "Kongfuzi [Confucius]"). The Chinese still refer to themselves as the "People of Han," the dynasty that marks the breakdown of the old feudal order, the rise of military power, the unification of a large empire, the institution of property rights, the flourishing of architecture, the arts, and the sciences, and the development of trade along the Silk Route (Table 9-1). The Han Dynasty reigned 2000 years ago. The walled city of Xian, then called Ch'angan, was the early Han capital and one of the greatest cities of the ancient world.

As we try to gain a better understanding of China's present political, economic, and social geography, we should not lose sight of the China that endured for thousands of years under a sequence of imperial dynasties as a single entity. Larger and more populous than Europe, China had its divisive feudal periods, but always it came together again, ruled—mostly dictatorially—from a strong center as a unitary state. China was, and remains, a predominantly rural society, a powerless and subservient population controlled by an often ruthless bureaucracy. Retributions, famines, epidemics, and occasional uprisings took heavy tolls in the countryside. But in the cities, Chinese culture flourished, subsidized by the taxes and tribute extracted from the hinterlands. As China grew, it incorporated minorities ranging from Koreans and Mongols to Uyghurs and Tibetans; in the far south, it now includes several peoples with Southeast Asian affinities. Like the Chinese citizenry itself, these minorities have experienced both benevolent government and brutal subjugation. But **sinicization** has always been China's wish: to endow them with the elements of Chinese culture.

A Century of Convulsion

When the European colonialists appeared in East Asia, China long withstood them with a self-assured superiority based on the strength of its culture and the reassuring continuity of the state. There was no market for the British East India Company's rough textiles in a country long used to finely fabricated silks and cottons. There was little interest in the toys and trinkets the Europeans produced in the hope of barter for Chinese tea and porcelain. Even key European inventions, such as the mechanical clock, though considered amusing and entertaining, were ignored and even deprecated as irrelevant to Chinese culture.

The self-confident Ming emperors even beat the Europeans at their own game. In the fourteenth century, great oceangoing vessels sailed the South China Sea; Chinese fleets carrying as many as 20,000 men reached Southeast and South Asia and even East Africa. Several times as large and far more sophisticated than anything built in Europe, China's ships, together with their sailors and the goods they carried, demonstrated the potential of the Middle Kingdom. But suddenly China faced a crisis at home: the impact of the onset of the Little Ice Age that also affected Europe. Cold and drought decimated the wheat harvest on the North China Plain, and boats were needed by the thousands to carry rice from the Yangzi/Chang Basin to the hungry north. That spelled the end of China's long-range maritime explorations; it may have changed the historical geography of the world.

Even when Europe's sailing ships made way for steam-driven vessels and newer and better European products (including weapons) were offered in trade for China's tea and silk, China continued to reject European imports and resisted commerce in general. (When the Manchus took control of Beijing in 1644, their bows and arrows proved superior to Chinese-manufactured muskets, which were so heavy and hard to load that they were almost useless.) The Chinese kept the Europeans confined to small peninsular outposts, such as Macau, and minimized interaction with them. Long after India had succumbed to mercantilism and economic imperialism, China maintained its established order. This was no surprise to the Chinese. After all, they had held a position of undisputed superiority in their Celestial Kingdom as long as could be remembered, and they had dealt with foreign invaders before.

A (Lost) War on Drugs and Its Aftermath

All this confidence was shattered during the Manchu (Qing) Dynasty, China's last imperial regime. The Manchus, who had taken control in Beijing as a small minority and who had grafted their culture onto China's, had the misfortune of reigning when the standoff with Europeans ended and the balance of power shifted in favor of the colonialists.

FROM THE FIELD NOTES

"To pass through the gate from Tiananmen Square and enter the Manchu Emperors' Forbidden City was a riveting experience. Look back now toward the gate, and you see throngs of Chinese visiting what less than a century ago was the exclusive domain of the Qing Dynasty's absolute rulers—so absolute that unauthorized entry was punished by instant execution. The vast, walled complex at the heart of Beijing not only served as the imperial palace; it was a repository of an immense collection of Chinese technological and artistic achievements. Some of this is still here, but most of all you are awed by the lingering atmosphere, the sense of what happened here in these exquisitely constructed buildings that epitomized what high Chinese culture could accomplish. Inevitably you think of the misery of millions whose taxes and tribute paid for what stands here." © H. J. de Blij.

AMONG THE REALM'S GREAT CITIES . . . **Xian**

THE CITY KNOWN today as Xian is the site of one of the world's oldest urban centers. It may have been a settlement during the Shang-Yin Dynasty more than 3000 years ago; it was a town during the Zhou Dynasty, and the Qin emperor was buried here along with 6000 life-sized terracotta soldiers and horses, reflecting the city's importance. During the Han Dynasty the city, then called Ch'angan, was one of the greatest centers of the ancient world, the Rome of ancient China. Ch'angan formed the eastern terminus of the Silk Route, a storehouse of enormous wealth. Its architecture was unrivalled, from its ornamental defensive wall with elaborately sculpted gates to the magnificent public buildings and gardens at its center.

Situated on the fertile loess plain of the upper Wei River, Ch'angan was the focus of ancient China during crucial formative periods. After two centuries of Han rule, political strife led to a period of decline, but the Sui emperors rebuilt and expanded Ch'angan when they made it their capital. During the Tang Dynasty, Ch'angan again became a magnificent city with three districts: the ornate Palace City; the impressive Imperial City, which housed the national administration; and the busy Outer City containing the homes and markets of artisans and merchants.

After its Tang heyday the city again declined, although it remained a bustling trade center. During the Ming Dynasty it was endowed with some of its architectural landmarks, including the Great Mosque marking the arrival of Islam; the older Big Wild Goose Pagoda dates from the influx of Buddhism. After the Ming period, Ch'angan's name was changed to Xian (meaning "Western Peace"), then to Siking, and in 1943 back to Xian again.

Having been a gateway for Buddhism and Islam, Xian in the 1920s became a center of Soviet communist ideol-

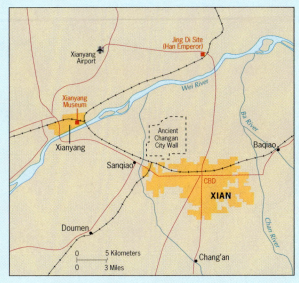

© H. J. de Blij, P. O. Muller, and John Wiley & Sons, Inc.

ogy. The Nationalists, during the struggle against the Japanese, moved industries from the vulnerable east to Xian, and when the communists took power, they enlarged Xian's industrial base still further. The present city (population: 4.0 million) lies southwest of the famed tombs, its cultural landscape now dominated by a large industrial complex that includes a steel mill, textile factories, chemical plants, and machine-making facilities. Little remains (other than some prominent historic landmarks) of the splendor of times past, but Xian's location on the railroad to the vast western frontier of China sustains its long-term role as one of the country's key gateways.

On two fronts in particular, the economic and the political, the European powers destroyed China's invincibility. Economically, they succeeded in lowering the cost and improving the quality of manufactured goods, especially textiles, and the handicraft industries of China began to collapse in the face of unbeatable competition. Politically, the demands of the British merchants and the growing English presence in China led to conflicts. In the early part of the nineteenth century, the central issue was the importation into China from British India of opium, a dangerous and addictive intoxicant. Opium was destroying the very fabric of Chinese culture, weakening the society, and rendering China easy prey for colonial profiteers. As the Manchu government moved to stamp out the opium trade in 1839, armed hostilities broke out, and soon the Chinese found

themselves losing a war on their own territory. The First Opium War (1839–1842) ended in disaster: China's rulers were forced to yield to British demands, and the breakdown of Chinese sovereignty was under way.

Drugs and Defeat

British forces penetrated up the Chang Jiang and controlled several areas south of it (Fig. 9-7); Beijing hurriedly sought a peace treaty by which it granted leases and concessions to foreign merchants. In addition, China ceded Hong Kong Island to the British and opened five ports, including Guangzhou (Canton) and Shanghai, to foreign commerce. No longer did the British have to accept a status that was inferior to the Chinese in order to do business;

FIGURE 9-7 © H. J. de Blij, P. O. Muller, and John Wiley & Sons, Inc.

henceforth, negotiations would be pursued on equal terms. Opium now flooded into China, and its impact on Chinese society became even more devastating. Fifteen years after the First Opium War, the Chinese again tried to stem the disastrous narcotic tide, and the foreigners who had attached themselves to their country again defeated them. Now the government legalized cultivation of the opium poppy in China itself. Chinese society was disintegrating;

the scourge of this drug abuse was not defeated until after the revival of Chinese power in the twentieth century.

Colonial Indignities

But before China could reassert itself, much of what remained of its independence steadily eroded (see box titled "Extraterritoriality"). In 1898 the Germans obtained

Extraterritoriality

4

DURING THE NINETEENTH century, as China weakened and European colonial invaders entered China's coastal cities and sailed up its rivers, the Europeans forced China to accept a European doctrine of international law—**extraterritoriality**. Under this doctrine, foreign states and their representatives are immune from the jurisdiction of the country in which they are based. Today, this applies to embassies and diplomatic personnel. But in Qing (Manchu) China, it went far beyond that.

The European, Russian, and Japanese invaders established as many as 90 treaty ports—extraterritorial enclaves in China's cities under unequal treaties enforced by gunboat diplomacy. In their "concessions," diplomats and traders were exempt from Chinese law. Not only port areas but also the best residential suburbs of large cities were declared to be "extraterritorial" and made inaccessible to Chinese citizens. In the city of Guangzhou (Canton in colonial times), Sha Mian Island in the Pearl River was a favorite extraterritorial enclave. A sign at the only bridge to the island stated, in English and Cantonese, "No Dogs or Chinese."

Christian missionaries fanned out into China, their residences and churches fortified with extraterritorial security. In many places, Chinese found themselves unable to enter parks and buildings without permission from foreigners. This involved a loss of face that contributed to bitter opposition to the presence of foreigners—a resentment that exploded in the Boxer Rebellion of 1900.

After the collapse of the Qing Dynasty in 1911, the Chinese Nationalists negotiated an end to all commercial extraterritoriality in China Proper; the Russians, however, would not yield in then-Manchuria. Only Hong Kong and Macau retained their status as colonies.

When China's government in 1980 embarked on a new economic policy that gave major privileges and exemptions to foreign firms in certain coastal areas and cities, opponents argued that this policy revived the practice of extraterritoriality in a new guise. This issue remains a sensitive one in a China that has not forgotten the indignities of the colonial era.

---

a lease on Qingdao on the Shandong Peninsula, and the French acquired a sphere of influence in the far south at Zhanjiang (Fig. 9-7). The Portuguese took Macau; the Russians obtained a lease on Liaodong in the Northeast as well as railway concessions there; even Japan got into the act by annexing the Ryukyu Islands and, more importantly, Formosa (Taiwan) in 1895.

After four millennia of recurrent cultural cohesion, economic security, and political continuity, the Chinese world lay open to the aggressions of foreigners whose innovative capacities China had negated to the end. Now the ships flying European flags lay in the ports of China's coasts and rivers, but China had not learned to manufacture the cannons to blast them out of the water. The smokestacks of foreign factories rose over the townscapes of its great cities, and no Chinese force could dislodge them. The Japanese, seeing China's weakness, invaded Korea. The Russians entered Manchuria, the home base of the Manchus. The foreign invaders even took to fighting among themselves, as Japan and Russia did in Manchuria in 1904. (Today this part of China is called the ***Northeast***, the name Manchuria having been—understandably—rejected.)

A New China Emerges

In the meantime, organized opposition to the foreign presence in China was gathering strength, and the twentieth century opened with a large-scale revolt against all outside elements. Bands of revolutionaries roamed both cities and countryside, attacking not only the hated foreigners but also Chinese who had adopted Western cultural traits. Known as the Boxer Rebellion (after a loose translation of the Chinese name for these revolutionary groups), this 1900 uprising was put down with much bloodshed by an international force consisting of British, Russian, French, Italian, German, Japanese, and American soldiers. Simultaneously, another revolutionary movement (led by the nationalist, Sun Yat-sen) was gaining support, aimed against the Manchu leadership itself. In 1911, the emperor's garrisons were attacked all over China, and in a few months the 267-year-old Qing Dynasty was overthrown. Indirectly, it too was a casualty of the foreign intrusion, and it left China divided and disorganized.

The fall of the Manchus and the proclamation of a republican government in China did little to improve the country's overall position. The Japanese captured Germany's holdings on the Shandong Peninsula, including the city of Qingdao, during World War I. When the victorious European powers met at Versailles in 1919 to divide the territorial spoils, they affirmed Japan's rights in the area. This led to yet another Chinese effort to counter the foreign scourge. Nationwide protests and boycotts of Japanese goods were organized in what became known as the May Fourth Movement. One participant in these demonstrations was a charismatic young man named Mao Zedong.

Nationalists and Communists

During the chaotic 1920s, the Nationalists and the Communist Party at first cooperated, with the remaining foreign presence their joint target. After Sun Yat-sen's death in 1925, Chiang Kai-shek became the Nationalists' leader, and by 1927 the foreigners were on the run, escaping by boat and train or falling victim to rampaging Nationalist forces. But soon the Nationalists began purging communists even as they pursued foreigners, and in 1928, when Chiang established his Nationalist capital in the city of Nanjing, it appeared that the Nationalists would emerge victorious from their campaigns. They had driven the communists ever deeper into the interior, and by 1933 the Nationalist armies were on the verge of encircling the last communist stronghold in the area of Ruijin in Jiangxi Province.

This led to a momentous event in Chinese history: the **Long March**. Nearly 100,000 people—soldiers, peasants, leaders—marched westward from Ruijin in 1934, a communist column that included Mao Zedong and Zhou Enlai. The Nationalist forces rained attack after attack on the marchers, and of the original 100,000, about three-quarters were killed. But new sympathizers joined along the way (see the route marked on Fig. 9-7), and the 20,000 survivors found a refuge in the mountainous interior of Shaanxi Province, 3200 kilometers (2000 miles) away. There, they prepared for a renewed campaign that would bring them to power.

Japan in China

Although many foreigners fled China during the 1920s and 1930s, others seized the opportunity presented by the contest between the Nationalists and communists. The Japanese took control over the Northeast, and when the Nationalists proved unable to dislodge them, they set up a puppet state there, appointed a Manchu ruler to represent them, and called their possession Manchukuo.

The inevitable full-scale war between the Chinese and the Japanese broke out in 1937, with the Nationalists bearing the brunt of it (which gave the communists an opportunity to regroup). The gray boundary on Figure 9-7 shows how much of China the Japanese conquered. The Nationalists moved their capital to Chongqing, and the communists controlled the area centered on Yanan. China had been broken into three pieces.

The Japanese committed unspeakable atrocities in their campaign in China. Millions of Chinese citizens were shot, burned, drowned, subjected to gruesome chemical and biological experiments, and otherwise wantonly victimized. The "Rape of Nanking," an orgy of murder, rape, torture, and pillage following the fall of Nanjing to the Japanese Army in late 1937, was perhaps the most heinous of Japan's war crimes during this period. Years later, when China's economic reforms of the 1980s and 1990s led to a renewed Japanese presence in China, the Chinese public and its leaders called for Japan to acknowledge and apologize for these wartime abuses. In Japan, this pitted apologists against strident nationalists, a dispute that continues today. The Chinese remain dissatisfied with Japan's failure to apologize unconditionally for its actions, and the book is not yet closed on this sensitive issue.

Communist China Arises

After the U.S.-led Western powers defeated Japan in 1945, the civil war in China quickly resumed. The United States, hoping for a stable and friendly government in China, sought to mediate the conflict but at the same time recognized the Nationalists as the legitimate government. The United States also aided the Nationalists militarily, destroying any chance of genuine and impartial mediation. By 1948, it was clear that Mao Zedong's well-organized militias would defeat Chiang Kai-shek. Chiang kept moving his capital—back to Guangzhou, seat of Sun Yat-sen's first Nationalist government, then back to Chongqing. Late in 1949, after a series of disastrous defeats in which hundreds of thousands of Nationalist forces were killed, the remnants of Chiang's faction gathered Chinese treasures and valuables and fled to the island of Taiwan. There, they took control of the government and proclaimed their own Republic of China.

Meanwhile, on October 1, 1949, standing in front of the assembled masses at the Gate of Heavenly Peace on Beijing's Tiananmen Square, Mao Zedong proclaimed the birth of the People's Republic of China.

CHINA'S HUMAN GEOGRAPHY

After more than a half-century of communist rule, China is a society transformed. It has been said that the year 1949 actually marked the beginning of a new dynasty not so different from the old, an autocratic system that dictated from the top. In that view, Mao Zedong simply bore the mantle of his dynastic predecessors. Only the family lineage had fallen away; now communist "comrades" would succeed each other.

And certainly some of China's old traditions continued during the communist era, but in many other ways Chinese society was totally overhauled. Benevolent or otherwise, the dynastic rulers of old China headed a country in which—for all its splendor, strength, and cultural richness—the fate of landless people and of serfs often was undescribably miserable; in which floods, famines, and diseases could decimate the populations of entire regions without any help from the state; in which local lords could (and often did) repress the people with impunity; in which children were sold and brides were bought. The European intrusion made things even worse, bringing slums, starvation, and deprivation to millions who had moved to the cities.

The communist regime, dictatorial though it was, attacked China's weaknesses on many fronts, mobilizing virtually every able-bodied citizen in the process. Land was taken from the wealthy; farms were collectivized; dams and levees were built with the hands of thousands; the threat of hunger for millions receded; health conditions improved; child labor was reduced. But China's communist planners also made terrible mistakes. The *Great Leap Forward*, requiring the reorganization of the peasantry into communal brigade teams to speed industrialization and make farming more productive, had the opposite effect and so disrupted agriculture that between 20 and 30 million people died of starvation between 1958, when the program was implemented, and 1962, when it was abandoned.

Mao ruled China from 1949 to 1976, long enough to leave lasting marks on the state. Another of his communist dictums had to do with population. Like the Soviets (and influenced by a horde of Soviet advisors and planners), Mao refused to impose or even recommend any population policy, arguing that such a policy would represent a capitalist plot to constrain China's human resources. As a result, China's population grew explosively during his rule.

Yet another costly episode of Mao's rule was the so-called *Great Proletarian Cultural Revolution*, launched by Mao Zedong during his last decade in power (1966–1976). Fearful that Maoist communism was being contaminated by Soviet "deviationism" and worried about his own stature as its revolutionary architect, Mao unleashed a campaign against what he viewed as emerging elitism in society. He mobilized young people living in cities and towns into cadres known as Red Guards and ordered them to attack "bourgeois" elements throughout China, criticize Communist Party officials, and root out "opponents" of the system. He shut down all of China's schools, persecuted untrustworthy intellectuals, and encouraged the Red Guards to engage in what he called a renewed "revolutionary experience." The results were disastrous: Red Guard factions took to fighting among themselves, and anarchy, terror, and economic paralysis followed. Thousands of China's leading intellectuals died, moderate leaders were purged, and teachers, elderly citizens, and older revolutionaries were tortured

to make them confess to "crimes" they did not commit. As the economy suffered, food and industrial production declined. Violence and famine killed as many as 30 million people as the Cultural Revolution spun out of control. One of those who survived was a Communist Party leader who had himself been purged and then been reinstated—Deng Xiaoping. Deng was destined to lead the country in the post-Mao period of economic transformation.

Political and Administrative Subdivisions

Before we investigate the emerging human geography of contemporary China, we should acquaint ourselves with the country's political and administrative framework (Fig. 9-8). For administrative purposes, China is divided into the following units:

  4 Central-Government-Controlled Municipalities (*Shi's*)
  5 Autonomous Regions (A.R.'s)
 22 Provinces
  2 Special Administrative Regions (SAR's)

The four central-government-controlled *Municipalities* are the capital, Beijing; its nearby port city, Tianjin; China's largest metropolis, Shanghai; and the Chang River port of Chongqing, in the interior. These *Shi's* form the cores of China's most populous and important subregions, and direct control over them from the capital entrenches the central government's power.

We should note that the administrative map of China continues to change, posing problems for geographers. The city of Chongqing was made a *Shi* in 1996, and its "municipal" area was enlarged to incorporate not only the central urban area but a huge hinterland covering all of eastern Sichuan Province. As a result, the "urban" population of Chongqing is officially 30 million, making this the world's largest metropolis—but in truth, as shown in the cities population table on page 459, the central urban area contains only 6.5 million inhabitants. And because Chongqing's population is officially *not* part of the province that borders it to the west (Fig. 9-8), the official population of Sichuan declined by 30 million when the *Chongqing Shi* was created. So despite what Table 9-2 suggests, the population within Sichuan Province's outer borders remains the largest of any of China's administrative divisions.

The five *Autonomous Regions* were established to recognize the non-Han minorities living there. Some laws that apply to Han Chinese do not apply to certain minorities. As we saw in the case of the former Soviet Union, however,

FIGURE 9-8   © H. J. de Blij, P. O. Muller, and John Wiley & Sons, Inc.

demographic changes and population movements affect such regions, and the policies of the 1940s may not work in the twenty-first century. Han Chinese immigrants now outnumber several minorities in their own regions. The five Autonomous Regions (A.R.'s) are: (1) Nei Mongol A.R. (Inner Mongolia); (2) Ningxia Hui A.R. (adjacent to Inner Mongolia); (3) Xinjiang Uyghur A.R. (China's northwest corner); (4) Guangxi Zhuang A.R. (far south, bordering Vietnam); and (5) Xizang A.R. (Tibet).

China's 22 *Provinces*, like U.S. States, tend to be smallest in the east and largest toward the west. The territorially smallest are the three easternmost provinces on China's coastal bulge: Zhejiang, Jiangsu, and Fujian. The two largest are Qinghai, flanked by Tibet, and Sichuan, China's Midwest.

As with all large countries, some provinces are more important than others. The Province of Hebei nearly surrounds Beijing and occupies much of the core of the country. The Province of Shaanxi is centered on the great an-

cient city of Xian. In the southeast, momentous economic developments are occurring in the Province of Guangdong, whose urban focus is Guangzhou. When, in the pages that follow, we refer to a particular province or region, Figure 9-8 is a useful locational guide.

In 1997, the British dependency of Hong Kong (Xianggang) was taken over by China and became the country's first **Special Administrative Region (SAR)**. In 1999, Portugal similarly transferred Macau, opposite Hong Kong on the Pearl River Estuary, to Chinese control, creating the second SAR under Beijing's administration.

Population Issues

Table 9-2 underscores the enormity of China's population: many Chinese provinces have more inhabitants than do most of the world's countries. With a 2008 population of 1.325 billion, China remains the largest nation on Earth.

TABLE 9-2

China: Population by Major Administrative Divisions* (in millions)

Provinces	1990 Population	2000 Population	2005 Population**
Anhui	56.2	59.9	61.2
Fujian	30.0	34.7	35.3
Gansu	22.4	25.6	25.9
Guangdong	62.8	86.4	91.9
Guizhou	32.4	35.3	37.3
Hainan	6.6	7.9	8.3
Hebei	61.1	67.4	68.4
Heilongjiang	35.2	36.9	38.2
Henan	85.5	92.6	93.7
Hubei	54.0	56.3	57.1
Hunan	60.7	62.4	63.2
Jiangsu	67.1	74.4	74.7
Jiangxi	37.7	41.4	43.1
Jilin	24.7	26.9	27.2
Liaoning	39.5	41.9	42.2
Qinghai	4.5	5.2	5.4
Shaanxi	32.9	36.1	37.2
Shandong	84.4	90.8	92.4
Shanxi	28.8	33.0	33.5
Sichuan	107.2	83.3	82.1
Yunnan	37.0	42.9	44.4
Zhejiang	41.5	46.8	48.9

Autonomous Region

Guangxi Zhuang	42.2	44.9	46.6
Nei Mongol	21.5	23.2	23.7
Ningxia Hui	4.7	5.6	6.0
Xinjiang Uyghur	15.2	19.3	20.1
Xizang	2.2	2.6	2.8

*Population data for the four centrally administered municipalities and the Hong Kong (Xianggang) and Macau Special Administrative Regions are shown in the table on page 459.
**Source: 1% Population Sample Survey, National Bureau of Statistics of China, November 1, 2005.

We should view these huge numbers in the context of China's population growth rate. High rates of population growth dim national development prospects in many countries in the periphery, but China has a special problem. Even a modest growth rate—such as today's 0.6 percent annually—still adds nearly 8 million to the total every year.

Mao Zedong, as we have noted, refused to institute policies that would reduce the rate of population increase because he believed (encouraged by Soviet advisors) that this would play into the hands of his capitalist enemies. Fast-growing populations, he said, constituted the only asset many poor, communist, or socialist countries had. If the capitalist world wanted to reduce growth rates, it must be an anticommunist plot.

The One-Child Policy and Its Impact

But other Chinese leaders, realizing that rapid population growth would stymie their country's progress, tried to stem the tide through local propaganda and education. Some of these leaders lost their positions as "revisionists," but their point was not lost on China's future leaders. After Mao's death, China embarked on a vigorous population-control program. In the early 1970s, the annual rate of natural increase was about 3 percent; by the mid-1980s, it was down to 1.2 percent. Families were ordered to have one child only, and those who violated the policy were penalized by losing tax advantages, educational opportunities, and even housing privileges. As we noted, China's census bureau now officially reports a moderate growth rate of 0.6 percent.

The one-child policy was vigorously policed throughout China by Communist Party members, leading to serious excesses. The number of abortions, sometimes forced, rose sharply; female infanticide increased. During the 1980s, when reports of these byproducts of the policy reached the outside world, the regime began to institute certain exceptions (minorities were not subject to the one-child limit). Couples living in rural areas were allowed to have a second child if their firstborn was female. Certain farming and fishing families could apply for permission to raise a second child in view of future labor needs. Since 2001, urban couples could have two children if both the father and mother came from one-child families.

Chinese demographers report that these **restrictive population policies** have produced 300 million "non-births" over the past 30 years; that is, China's population would be more than 1.6 billion today if these policies had not been instituted. They estimate that the fertility rate declined from 2.3 children per woman in 1960 to 1.6 in 2006, which would mean that China's population will level off at mid-century and then begin a steady decline. This will entail a rapid increase in the proportion of the population over 65 years old, with consequences similar to those in Europe—a swift reduction in the number of workers to support these old-age pensioners. And this is not some remote, distant worry: China is already having problems meeting its pension obligations.

China's one-child policy may have prevented a population explosion, but because families preferred male heirs it also led to an excess of males over females in the population pyramid. The official ratio at birth over the past 30 years is 118 boys for every 100 girls; this is a troubling statistic because it implies selective abortion, female infanticide, and child neglect. Another concern arising from this ratio imperils the future: millions of young men will be unable to find wives, and geographers who study this phenomenon, and its precedents elsewhere, report that such situations tend to induce governments to expand their armed forces and, sometimes, to find reasons to give those forces a purpose.

As a result, China in mid-decade was engaged in an internal debate over the one-child policy, and Chinese newspapers and magazines carried articles proposing a substitute two-child policy, with no exceptions, even in urban areas. Such a change would raise China's fertility rate and keep the population growing beyond 2050, but the advantages would include a larger number of younger workers to support the aged and a reduction in the skewed male-female ratio that threatens to undermine this vast country's future social fabric.

The Minorities

We asserted earlier that China remains an empire; its government controls territories and peoples that are in effect colonized. The ethnolinguistic map (Fig. 9-9) apparently confirms this proposition. This map should be seen in context, however. It does not show local-area majority populations but instead reveals where minorities are concentrated. For example, the Mongolian population in China is shown to be clustered along the southeastern border of Mongolia in the Autonomous Region called Nei Mongol (see Fig. 9-8). But even in that A.R., the Han Chinese, not the Mongols, are now in the majority.

Nonetheless, the map gives definition to the term *China Proper* as the home of the People of Han, the ethnic (Mandarin-speaking) Chinese depicted in tan and light orange in Figure 9-9. From the upper Northeast to the border with Vietnam and from the Pacific coast to the margins of Xinjiang, this Chinese majority dominates. When you compare this map to that of population distribution (to be discussed shortly), it will be clear that the minorities constitute only a small percentage of the country's total. The Han Chinese form the largest and densest clusters.

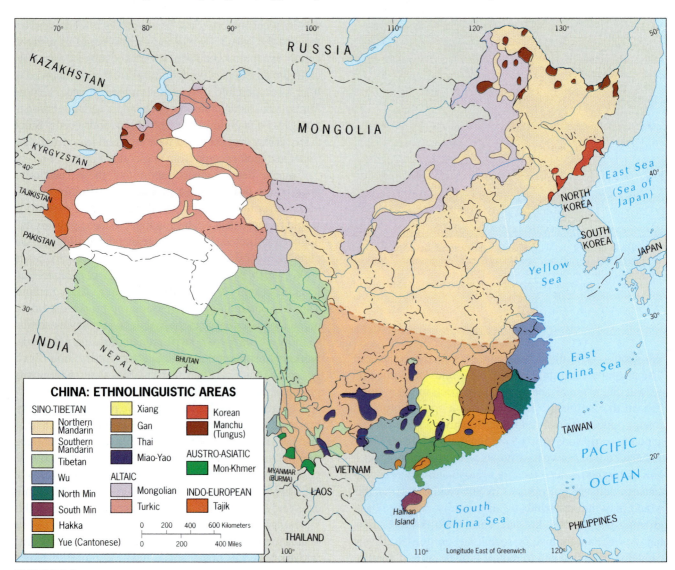

FIGURE 9-9 © H. J. de Blij, P. O. Muller, and John Wiley & Sons, Inc.

In any case, China controls non-Chinese areas that are vast, if not populous. The Tibetan group numbers under 3 million, but it extends over all of settled Xizang. Turkic peoples inhabit large areas of Xinjiang. Thai, Vietnamese, and Korean minorities also occupy areas on the margins of Han China. As we will see later, the Southeast Asian minorities in China have participated strongly in the Pacific Rim developments on their doorsteps. Hundreds of thousands have migrated from their Autonomous Regions to the economic opportunities along the coast.

Numerically, Chinese dominate in China to a far greater degree than Russians dominated their Soviet Empire. But territorially, China's minorities extend over a proportionately larger area. The Ming and Manchu rulers bequeathed the People of Han an empire.

PEOPLE AND PLACES OF CHINA PROPER

A map of China's population distribution (Fig. 9-10) reveals the continuing relationship between the physical environment and its human occupants. In technologically advanced countries, we have noted, people shake off their dependence on what the land can provide; they cluster in cities and in other areas of economic opportunity. This depopulates rural areas that may once have been densely inhabited. In China, that stage has not yet been reached. Although China has large cities, as in India a sizeable majority of the people (63 percent) still live on—and from—the land. Thus the map of population distribution reflects the livability and productivity of China's basins, lowlands, and plains. Compare Figures 9-6A and 9-10, and China's continuing dependence on soil, water, and warmth will be evident.

The population map also suggests that in certain areas environmental limitations are being overcome. Industrialization in the Northeast, irrigation in the Inner Mongolia Autonomous Region, and oil-well drilling in Xinjiang enabled millions of Chinese to migrate from China Proper into these frontier zones, where they now outnumber the indigenous minorities.

Nonetheless, physiography and demography remain closely related in China. To grasp this relationship, it is useful to compare Figures 9-2 (physiography) and 9-10 (population) as our discussion proceeds. On the population map, the darker the color, the denser the population: in places we can follow the courses of major rivers through this scheme. Look, for example, at China's Northeast. A ribbon of population follows the Liao and the Songhua rivers. Also note the huge, nearly circular population concentration on the western edge of the red zone;

this is the Sichuan Basin, in the upper course of the Chang Jiang. Four major river basins contain more than three-quarters of China's 1.3 billion people:

1. The Liao-Songhua Basin or the Northeast China Plain
2. The Lower Huang He (Yellow River) Basin, known as the North China Plain
3. The Upper and Lower Basins of the Chang Jiang (Long River; lower course known as the Yangzi)
4. The Basins of the Xi (West) and Pearl River

Not all the people living in these river plains are farmers, of course. China's great cities also have developed in these populous areas, from the industrial centers of the Northeast to the Pacific Rim upstarts of the South. Both Beijing and Shanghai lie in major river basins. And, as the map shows, the hilly areas between the river basins are not exactly sparsely populated either. Note that the hill country south of the Chang Basin (opposite Taiwan) still has a density of 50 to 100 people per square kilometer (130 to 260 per sq mi).

Northeast China Plain

The Northeast China Plain is the heartland of China's Northeast, ancestral home of the Manchus who founded China's last dynasty, battleground between Japanese and Russian invaders, Japanese colony, heartland of communist industrial development, and now losing ground to the industries of the Pacific Rim. This region used to be called Manchuria. As the administrative map (Fig. 9-8) shows, there are three provinces here: Liaoning in the south, facing the Yellow Sea; Jilin in the center; and Heilongjiang, by far the largest, in the north.

When you are familiar with the vast farmlands of the North China Plain, the southernmost part of the Northeast China Plain looks so similar that it seems to be a continuation. Near the coast, farms look as they do near the mouth of the great Huang He; to the east, the Liaodong Peninsula looks like the Shandong Peninsula across the water. That impression soon disappears, however. Travel northward, and the Northeast China Plain reveals the effects of cold climate and thin soils. Farmlands are patchy; smokestacks seem to rise everywhere. This is industrial, not primarily farming, country. But the landscape looks like a Rustbelt: equipment often is old and outdated, transport systems are inadequate, buildings are in disrepair. Stream, land, and air pollution is dreadful. The Northeast in some ways resembles industrial zones in former communist Eastern Europe.

The Northeast has experienced many ups and downs. Although Japanese colonialism was ruthless and exploitive, Japan did build railroads, roads, bridges, factories, and other components of the regional infrastructure. (At one

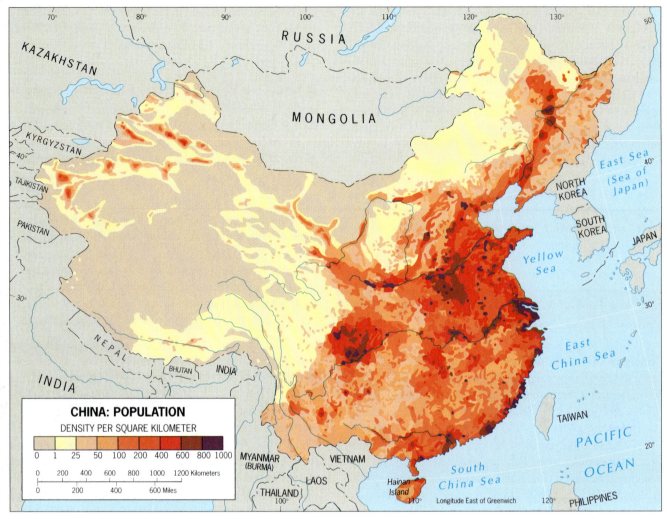

FIGURE 9-10 © H. J. de Blij, P. O. Muller, and John Wiley & Sons, Inc.

time, half of the entire railroad mileage in China was in the Northeast.) After the Japanese were ousted, the Soviets looted the area of machinery, equipment, and other goods. During the late 1940s, the Northeast was a ravaged frontier. But then the communists took power and made the industrial development of the Northeast a priority. From the 1950s until the 1970s, the Northeast led the nation in manufacturing growth. Its population, just a few million in the 1940s, mushroomed to more than 100 million. Towns and cities grew exponentially.

Minerals and Manufacturing

All this growth was based on the Northeast's considerable mineral wealth (Fig. 9-11). Iron-ore deposits and coalfields lie concentrated in the Liao Basin; aluminum ore, ferroalloys, lead, zinc, and other metals are plentiful, too. As the map shows, the region also is well endowed with oil reserves; the largest lie in the Daqing Reserve be-

tween Harbin and Qiqihar. The communist planners made Changchun capital of Jilin Province, the Northeast's automobile manufacturing center. Shenyang, capital of Liaoning Province, became the leading steelmaking complex. Fushun and Anshan were assigned other functions, and the communists encouraged Han Chinese to migrate to the Northeast to find employment there. To attract them, the government built apartments, schools, hospitals, recreational facilities, even old-age homes—all to speed the region's industrial development.

For a time, it worked. During the 1970s, the Northeast was responsible for fully one-quarter of the entire country's industrial output. Stimulated by the needs of the growing cities and towns, farms expanded and crops diversified. The Songhua River Basin in the hinterland of Harbin, capital of Heilongjiang Province, became a productive agricultural zone despite the environmental limitations here. The Northeast was a shining example of what communist planning could accomplish.

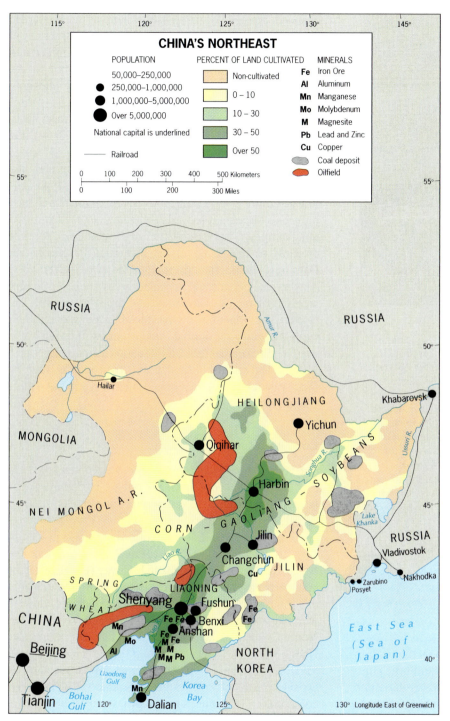

FIGURE 9-11   © H. J. de Blij, P. O. Muller, and John Wiley & Sons, Inc.

Rustbelt in Search of Rebound

7 Then came the **economic restructuring** of the post-Mao era. Under the new, market-driven economic order, the state companies that had led China's industrial resurgence were unable to adapt. Suddenly they were aging, inefficient, obsolete. Official policy encouraged restructuring and privatization, but the huge state factories could

not (and much of their labor force would not) comply. Today, the Northeast's contribution to China's industrial output has fallen below 10 percent as the production of factories in the Southeast, from Shanghai to Guangdong, rises annually.

Economic geographers, however, will not write off the prospects of this northeastern corner of China; this area, as we noted, has had its downturns before. The Northeast still contains a storehouse of resources, including extensive stands of oak and other hardwood forests on the mountain slopes that encircle the Northeast China Plain. The locational advantages of the major centers are undiminished. Harbin, for example, lies at the convergence of five railroads and is situated at the head of navigation on the Songhua River, providing a link to the northeast corner of the region and, beyond the Amur River, to Russia's Khabarovsk. Moreover, when the Koreas reunite, producing what may become the western Pacific Rim's second Japan, and when the Russian Far East takes off (Vladivostok and Nakhodka are the natural ports for the upper Northeast), this northern frontier of China may yet reverse its decline and enter still another era of economic success.

North China Plain

Take the new superhighway between the port of Tianjin and the capital, Beijing, or the train from Beijing southward to Anyang, and you see a physical landscape almost entirely devoid of relief and a cultural landscape that repeats itself endlessly. Villages, many showing evidence of the collectivization program of the early communist era, pass at such regular intervals that some kind of geometry seems to control the cultural landscape. Carefully diked canals stretch in all directions. Rows of trees mark fields and farms. Depending on the season, the air is clear, as it is during the moist summer when the wheat crop covers the countryside like a carpet, or choked with dust, which can happen almost any

FROM THE FIELD NOTES

"We flew over the North China Plain for an hour, and one could not miss the remarkably regular spacing of the villages on this seemingly table-flat surface. Many of these villages became communes during the communist reorganization of China's social order. The elongated buildings (many added since the 1950s) were each designed to contain one or more extended families. After the demise of Mao and his clique, the collectivization effort was reversed and the emphasis returned to the small nuclear family. In this village and others we overflew, the long sides of the buildings almost always lay east-west, providing maximum sun exposure to windows and doors, and enhancing summer ventilation from longitudinal breezes. The more distant view looks hazy, because the North China Plain during the spring is wafted by loess-carrying breezes from the northwest. On the ground, a yellow-gray hue shrouds the atmosphere until a passing weather front temporarily clears the air." © H. J. de Blij.

www.conceptcaching.com

other time of the year. The soil here is a fertile mixture of alluvium and loess (deposits of fine, windborne silt or dust of glacial origin), and the slightest breeze stirs it up. In the early spring the dust hangs like a fog over the landscape, covering everything with a fine powder. Add this to urban pollution and the cities become as unhealthy as any in the world.

Agricultural Colossus

For all we see, there are geographic explanations. The geometry of the settlement pattern results from the flatness of the uniformly fertile surface and the distance villagers could walk to neighboring villages and back during daylight. The canals form part of a vast system of irrigation

channels designed to control the waters of the Huang He (Yellow River). The rows of trees serve as windbreaks. And while wheat dominates the crops of the North China Plain, you will also see millet, sorghum, soybeans, cotton for the textile industry, tobacco (the Chinese are heavy smokers), and fruits and vegetables for the urban markets.

The North China Plain is one of the world's most heavily populated agricultural areas. Figure 9-10 shows that most of it has a density of more than 400 people per square kilometer (1000 people per sq mi), and in some parts the density is twice as high. Here, the ultimate hope of the Beijing government lay less in land redistribution than in raising yields through improved fertilization, expanded irrigation facilities, and the more intensive use of labor. A series of dams on the Huang River, including the Xiaolangdi Dam upstream in Henan Province, now reduce the flood danger, but outside the irrigated areas the ever-present problem of rainfall variability and drought persists. The North China Plain has not produced any substantial food surplus even under normal circumstances; thus when the weather turns unfavorable, the situation soon becomes precarious. The specter of famine may have receded, but the food situation is still uncertain in this very critical part of China Proper.

Core Area Cities

The North China Plain is an excellent example of the concept of a national **core area**. Not only is it a densely populated, highly productive agricultural zone, but it also is the site of the capital and other major cities, a substantial industrial complex, and several ports, among which Tianjin ranks as one of China's largest (Fig. 9-12). Tianjin, on the Bohai Gulf, is linked by rail and highway (less than a two-hour drive) to Beijing. Like many of China's harbors, that of Tianjin's river port is not particularly good; but Tianjin is well situated to serve not only the northern sector of the North China Plain and the capital, but also the Upper Huang Basin and Inner Mongolia beyond (see Fig. 9-1). Tianjin, again like several other Chinese ports, had its modern start as a treaty port, but the city's major growth awaited communist rule. For decades it was a center for light industry and a flood-prone harbor, but after 1949 the communists constructed a new artificial port and flood canals. They also chose Tianjin as a site for major industrial development and made large investments in the chemical industry (in which Tianjin still leads China), iron and steel production, heavy machine manufacturing, and textiles. Today, with a population of 7.4 million, Tianjin is China's fifth-largest metropolis and the center of one of its leading industrial complexes.

Northeast of Tianjin lies the smaller city of Tangshan. In 1976, a devastating earthquake, with its epicenter between Tangshan and Tianjin, killed an estimated 750,000 people,

FIGURE 9-12 © H. J. de Blij, P. O. Muller, and John Wiley & Sons, Inc.

the deadliest natural disaster of the twentieth century. So inept was the government's relief effort in nearby Beijing that the widespread resentment it provoked may have helped end the Maoist period of communist rule in China.

Beijing, unlike Tianjin, is China's political, cultural, and educational center. Its industrial development has not matched Tianjin's. The communist administration did, however, greatly expand the municipal area of Beijing, which (as we noted) is not controlled by the Province of Hebei but is directly under the central government's authority. In one direction, Beijing was enlarged all the way to the Great Wall—50 kilometers (30 mi) to the north—so that the "urban" area includes hundreds of thousands of farmers. Not surprisingly, this has circumscribed an enormous total population, enough to rank Beijing, with its 11.3 million inhabitants, among the world's megacities.

AMONG THE REALM'S GREAT CITIES . . . **Beijing**

BEIJING (11.3 MILLION), capital of China, lies at the northern apex of the roughly triangular North China Plain, just over 160 kilometers (100 mi) from its port, Tianjin, on the Bohai Gulf. Urban sprawl has reached the hills and mountains that bound the Plain to the north, a defensive barrier fortified by the builders of the Great Wall. You can reach the Great Wall from central Beijing by road in about an hour.

Although settlement on the site of Beijing began thousands of years ago, the city's rise to national prominence began during the Mongol (Yuan) Dynasty, more than seven centuries ago. The Mongols, preferring a capital close to their historic heartland, endowed the city with walls and palaces. Following the Mongols, later rulers at times moved their capital southward, but the government always returned to Beijing. From the third Ming emperor onward, Beijing was China's focus; it was ideally situated for the Manchus when they took over in 1644. During the twentieth century, China's Nationalists again chose a southern capital, but when the communists prevailed in 1949 they reestablished Beijing as their (and China's) headquarters.

Ruthless destruction of historic monuments carried out from the time of the Mongols to the communists (and now during China's modernization) has diminished, but not destroyed, Beijing's heritage. From the Forbidden City of the Manchu emperors (photo on p. 466) to the fifteenth-century Temple of Heaven, Beijing remains an open-air museum and the cultural focus of China. Suc-

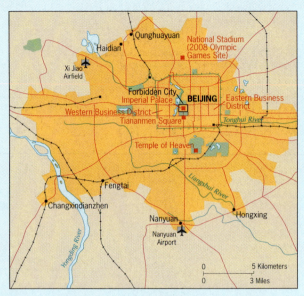

© H. J. de Blij, P. O. Muller, and John Wiley & Sons, Inc.

cessive emperors and aristocrats, moreover, bequeathed the city numerous parks and additional recreational spaces that other cities lack. Today Beijing is being transformed as a result of China's new economic policies (photo below). A forest of high-rises towers above the retreating traditional cityscape, avenues have been widened, and a beltway has been built. A new era dawns in this historic capital.

After Beijing won the right to stage the 2008 Olympic Games, the city—already the scene of a vast construction project—redoubled its determination to show the world that China has the capacity to meet all challenges. From building new facilities to cleaning up its notoriously polluted air, Beijing went through still another transformation. This photo shows the Olympic Stadium under construction in 2005, less than three years before the Olympic Torch, en route via Mount Everest, was due to enter the grounds. © Redux Pictures.

Inner Mongolia

Northwest of the core area as defined by the North China Plain, along the border with the state of Mongolia, lies Inner Mongolia, administratively defined as the Nei Mongol Autonomous Region (see Fig. 9-8). Originally established to protect the rights of the approximately 5 million Mongols who live outside the Mongolian state, Inner Mongolia has been the scene of massive immigration by Han Chinese. Today Chinese outnumber Mongols here by nearly four to one. Near the Mongolian border, Mongols still traverse the steppes with their tents and herds, but elsewhere irrigation and industry have created an essentially Chinese landscape. The A.R.'s capital, Hohhot, has been eclipsed by Baotou on the Huang He, which supports a corridor of farm settlements as it crosses the dry land here on the margins of the Gobi and Ordos deserts. With about 24 million inhabitants, Inner Mongolia cannot compare to the huge numbers that crowd the North China Plain, but recent mineral discoveries have boosted industry in Baotou and livestock herding is expanding. Nei Mongol may retain its special administrative status, but it functions as part of Han China in all but name.

Basins of the Chang/Yangzi

In contrast to the contiguous, flat agricultural-urban-industrial North China Plain defined by the sediment-laden Huang River, the basins and valleys of the Chang Jiang (Long River) display variation in elevation and relief. The Chang River, whose lower-course name becomes the Yangzi, is an artery like no other in China. Near its mouth lies the country's largest city, Shanghai. Part of its middle course is now being transformed by a gigantic engineering project. Farther upstream, the Chang crosses the populous, productive Sichuan Basin. And unlike the Huang, the Chang Jiang is navigable to oceangoing ships for over 1000 kilometers (600 mi) from the coast all the way to Wuhan (Fig. 9-12). Smaller ships can reach Chongqing, even after dam construction. Several of the Chang River's tributaries also are navigable, so that 30,000 kilometers (18,500 mi) of water transport routes serve its drainage basin. Thus the Chang Jiang constitutes one of China's leading transit corridors. With its tributaries it handles the trade of a vast area, including nearly all of middle China and sizeable parts of the north and south. The North China Plain may be the core area of China, but in many ways the Lower Chang Basin is its heart.

Early in Chinese history, when the Chang's basin was being opened up and rice and wheat cultivation began, a canal was built to link this granary to the northern core of old China. Over 1600 kilometers (1000 mi) long, this was the longest artificial waterway in the world, but during the nineteenth century it fell into disrepair. Known as the Grand Canal, it was dredged and rebuilt when the Nationalists controlled eastern China. After 1949, the communist regime continued this restoration effort, and much of the

FROM THE FIELD NOTES

"China's new consumer culture is perhaps nowhere more evident than in burgeoning, modernizing, globalizing Shanghai. I walked today down Nanjing Road, exactly 20 years after I first saw it in 1982, but only four years after my last excursion here. The rate of transformation is increasing. Thousands of locals have taken up the fast-food habit and are eating at Kentucky Fried Chicken (the colonel's face is everywhere) or McDonald's (I counted five of the latter along my route). A section of Nanjing Road has been made into a pedestrian mall, and here too is evidence of a new habit: private automobile transportation. Car sales lined the street, and potential buyers thronged the displays. Given the existing congestion on local roads, where will these cars fit?" © H. J. de Blij.

canal is now again open to barge traffic, supplementing the huge fleet of vessels that hauls domestic interregional trade along the east coast (Fig. 9-12).

Pivotal Corridor

As Figure 9-13 shows, the Lower Chang Basin is an area of both rice and wheat farming, offering further proof of its pivotal situation between south and north in the heart of China Proper. Shanghai lies at the coastal gateway to this productive region on a small tributary of the Chang (now Yangzi) River, the Huangpu. The city has an immediate hinterland of some 50,000 square kilometers (20,000 sq mi)—smaller than West Virginia—containing more than 50 million people. About two-thirds of this population are farmers who produce food, silk filaments, and cotton for the city's industries.

Travel upriver along the Yangzi/Chang Jiang, and you meet an unending stream of vessels large and small, including numerous "barge trains"—as many as six or more barges pulled by a single tug in an effort to save fuel (they

are slow, but time is not the primary concern). The traffic between Shanghai and Wuhan—Wuhan is short for Wuchang, Hanyang, and Hankou, a coalescing conurbation of three cities—makes the Yangzi one of the world's busiest waterways. Smoke-belching boat engines create a more or less permanent plume of pollution here, worsening the regional smog created by the factories of Nanjing and others perched on the waterfront. Here China's Industrial Revolution is in full gear, with all its environmental consequences.

Three Gorges Dam

Above Wuhan the river traffic dwindles because the depth of the Chang reduces the size of vessels that can reach Yichang. River boats carry coal, rice, building materials, barrels of fuel, and many other items of trade. But the middle course of the Chang River is becoming more than a trade route. In the northward bend of the great river between Yichang and Chongqing, where the Chang has cut deep troughs on its way to the coast, one of the world's largest engineering projects is in progress (photo p. 484). It has sev-

FIGURE 9-13 © H. J. de Blij, P. O. Muller, and John Wiley & Sons, Inc.

AMONG THE REALM'S GREAT CITIES... Shanghai

SAIL INTO THE mouth of the great Yangzi River, and you see little to prepare you for your encounter with China's largest city (population: 15.3 million). For that, you turn left into the narrow Huangpu River, and for the next several hours you will be spellbound. To starboard lies a fleet of Chinese warships. On the port side you pass oil refineries, factories, and neighborhoods. Next you see an ultramodern container facility, white buildings, and rust-free cranes, built by Singapore. Soon rusty tankers and freighters line both sides of the stream. High-rise tenements tower behind the cluttered, chaotic waterfront where cranes, sheds, boatyards, piles of rusting scrap iron, mounds of coal, and stacks of cargo vie for space. In the river, your boat competes with ferries, barge trains, and cargo ships. Large vessels, anchored in midstream, are being offloaded by dozens of lighters tied up to them in rows. The air is acrid with pollution. The noise—bells, horns, whistles—is deafening.

What strikes you is the vastness and sameness of Shanghai's cityscape, until you pass beneath the first of two gigantic suspension bridges. Suddenly, everything changes. To the left, or east, lies Pudong, an ultramodern district with the space-age Oriental Pearl Television Tower rising like a rocket on its launchpad (photo, p. 497) above a forest of modern, glass-and-chrome skyscrapers that make the Huangpu look like Hong Kong's Victoria Harbor. To the right, along the Bund, stand the remnants of Victorian buildings, monuments to the British colonialists who made Shanghai a treaty port and started the city on its way to greatness. Everywhere, construction cranes rise above the cityscape. Shanghai looks like one vast construction zone, complete with a 460-meter (1510-ft) office tower—called the World Financial Center—that is one of the world's tallest buildings.

The Chinese spent heavily to improve infrastructure in the Shanghai area, including brand-new Pudong International Airport, connected to the city by the world's first mag-lev (magnetic levitation) train—the fastest anywhere at 267 mph (430 kph). Pudong has become a magnet for

© H. J. de Blij, P. O. Muller, and John Wiley & Sons, Inc.

foreign investment, attracting as much as 10 percent of the annual total for the entire country. Shanghai's income is rising much faster than China's, and the Yangzi River Delta is becoming a counterweight to the massive Pearl River Delta development in China's South. Among other things, Shanghai is becoming China's "motor city," complete with a Formula One race track seating 200,000 at the center of an automobile complex where all levels of the industry, from manufacturing to sales, are being concentrated.

In 2010 Shanghai will host World Expo on the banks of the Huangpu River (the mag-lev track will be extended to the site), and the city's planners, who have already transformed the place from just another Open City to a symbol of the New China, will have an opportunity to show the world what has been accomplished here in just one generation.

---

eral names: the Sanxia Project, the Chang Jiang Water Transfer Project, the Three Gorges Dam, and the New China Dam. It is most commonly called the Three Gorges Dam because it is here that the great river flows through a 240-kilometer (150-mi)-long series of steep-walled valleys less than 100 meters (330 ft) wide. Near the lower end of this natural trough, a dam now rises to over 180 meters (600 ft) above the valley floor to a width of 2 kilometers (1.3 mi), creating a reservoir that inundates the historic Three Gorges

and extends more than 600 kilometers (over 400 mi) upstream. China's project engineers say that the dam will end the river's rampaging flood cycle, enhance navigation, stimulate development along the new lake perimeter, and provide at least one-tenth, and perhaps as much as one-eighth, of the country's electrical power supply. As such it will transform the heart of China, which is why many Chinese like to compare this gigantic project to the Great Wall and the Grand Canal, and call it the New China Dam.

The massive Three Gorges Dam on the Chang River above Yichang in Hubei Province under construction in mid-2006, when the dam was authorized to impound water to the 156-meter (512-ft)-high level. This is the world's largest-ever hydroelectric project and began providing power supplies in 2003. Capacity since then has steadily expanded, and the facility is on target to reach full production in 2009. © Chine Nouvelle/Sipa Press.

The Sichuan Basin

As Figure 9-1 shows, all this development is taking place downstream from Chongqing, the city near the eastern margin of the great and populous Sichuan Basin that was made, along with its hinterland, a *Shi* in the hierarchy of China's administrative units. Obviously this city is strongly affected by the dam's construction, which disrupted its river trade (boats up to 1000 tons could reach Chongqing previously). A diversion channel around the Three Gorges Dam will restore and enhance Chongqing's port and transit functions, and the city figures prominently in an important economic design for China: the expansion of its coastal economic success into the deep interior. We return to this topic later, but consider Chongqing's relative location on the western margin of the Yangzi's great economic corridor and on the eastern edge of the vast and productive Sichuan Basin, now upstream from the country's greatest engineering project and about midway between Kunming to the south and Xian to the north. This city is likely to become the most important growth pole in China's interior.

Sichuan, its basin crossed by the Chang (called the Yangzi downstream) after the river emerges from the mountains of the west, frequently was a problem province for both China's dynastic rulers and the communist regime. Its thriving capital, Chengdu, was one of China's most active centers of dissent during the 1989 pro-democracy movement. With about 120 million inhabitants (including Chongqing) today, Sichuan Province would be one of the world's ten largest states, but it does not even contain 10 percent of China's population. Sichuan's fertile, well-watered soils yield a huge variety of crops: grains (rice, wheat, corn), soybeans, tea, sugarcane, and many fruits and vegetables.

In many ways Sichuan is a country unto itself, isolated by topography and distance from the government in Beijing. The people here are Han Chinese, but cultural landscapes reveal the area's proximity to Southeast Asia. Along the great riverine axis of the Chang/Yangzi, China changes character time and again.

Basins of the Xi (West) and Pearl River

As Figure 9-12 demonstrates, the Xi River and its basins are no match for the Chang Jiang or Huang He, not even for the Liao. This southernmost river even seems to have the wrong name: Xi means West! It reaches the coast in a complex area where it forms a delta immediately adjacent to the estuary of the Pearl River. In this subtropical part of China, local relief is higher than in the lowlands and basins of the center and north, so that farmlands are more confined. Especially in the interior areas, water supply is a recurrent problem. On the other hand, its warm climate permits the double-cropping of rice. Regional food production, however, has never approached that of the North China Plain or the Lower Chang Basin. And, as Figure 9-10 shows, the population in this southern part of China is smaller than that in the more northerly heartlands.

Again unlike the Huang and Chang rivers, which rise in snow-covered interior mountains, the Xi is a shorter stream whose source is on the Yunnan Plateau (see Fig. 9-2). Just up the coast from its delta, however, lies one of the most important areas of modern China. As we will note later, the mouth of the Pearl River is flanked by some of China's fastest-growing economic complexes, including Guangzhou, capital of Guangdong Province, Hong Kong, the former British dependency, and adjoining Shenzhen, one of the world's fastest-growing cities.

This is all the more remarkable because, as the maps emphasize, South China is not especially well endowed with geographic advantages, locational or otherwise. The Pearl River is navigable for larger ships only to Guangzhou, and the Xi for small river boats no farther than Wuzhou. Guangdong Province, the key province of southern China, lies far from China's heartlands. Nor is South China blessed with the resources that propelled the industrial development of the Northeast and other parts of China. Indeed, one of the principal objectives of the Three Gorges Dam project is to provide South China with ample electricity. As Figure 9-14 shows, northern, central, and western China have most of the coal, oil, and natural gas resources, not the South.

China Proper, with its populous and powerful subregions, is in many ways the dominant region in the East Asian geographic realm. But China Proper is only one part of China. To the west lie two vast regions with relatively sparse populations: Xizang (Tibet) and Xinjiang. And here, in this remote western periphery adjoining Turkestan, China meets Muslim Central Asia (Fig. 9-15).

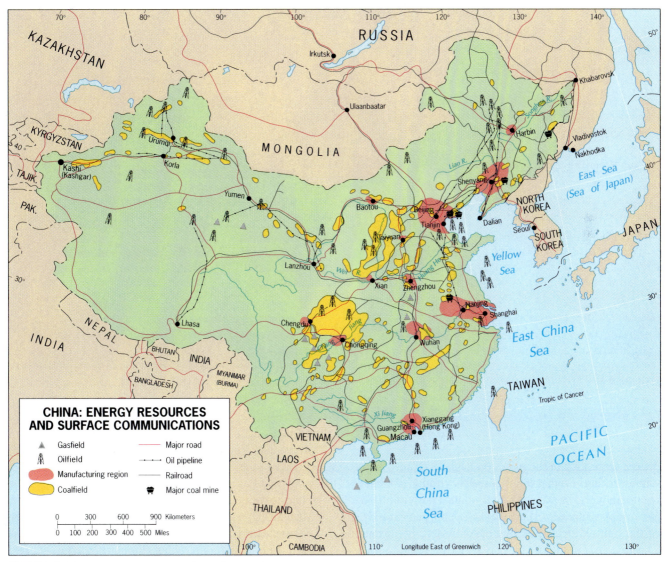

FIGURE 9-14   © H. J. de Blij, P. O. Muller, and John Wiley & Sons, Inc.

FIGURE 9-15 © H. J. de Blij, P. O. Muller, and John Wiley & Sons, Inc.

● XIZANG (TIBET)

Take a close look at the photo on page 487, and you are witness to an event that signifies the penetration of China's most remote corner: the historic Buddhist outpost of Tibet (*Xizang* in Chinese parlance). The opening in mid-2006 of the railroad that links Beijing with Xizang's capital, Lhasa, across some of the Earth's most forbidding terrain, symbolizes China's determination to bring Tibet into Beijing's orbit.

Remote, isolated, encircled by snowcapped mountains, and cut by deep valleys, Tibet for centuries was the site of fortress-like Buddhist monasteries whose monks paid allegiance to their supreme leader, the Dalai Lama. Tibetan society was tradition-bound and changeless, but came under nominal Chinese control in 1720, during the Qing (Manchu) Dynasty. When the Qing order collapsed in the late nineteenth century, Tibet found itself free of Chinese rule and remained autonomous (though powerless) during the chaos that enveloped China in the first half of the twentieth century.

A symbolic event took place in China during July 2006: the inaugural journey of a train linking China's capital, Beijing, with Lhasa, the capital of Xizang (Tibet). Laid by 100,000 construction workers across icy mountains, the final 1150 kilometers (700 mi) of the track constitutes an engineering marvel, climbing as high as 5072 meters (16,640 ft) across permafrost-ridden terrain. No other railway in the world reaches such altitudes; the carriages of the "Sky Train," with a total capacity of 900 passengers, are sealed and pressurized like the fuselage of an airplane. But the train has its detractors. This $4.2 billion project, they say, will facilitate Chinese penetration of Xizang and enhance Beijing's economic and military control over the remote territory while Tibet's infrastructure needs, such as roadbuilding and electrification, are neglected. It is evidence that China will spend whatever it takes to bring in larger numbers of Han Chinese to overwhelm the local Buddhist population, suppress Tibetan separatism, and promote "modernization." As such, the train to Xizang symbolizes the future as seen in Beijing, not Lhasa.   © AP/World Wide Photo/Color China Photo.

A new Chinese invasion of Tibet followed shortly after the communist victory: in 1950 Chinese forces took control of Xizang, and in 1959 they crushed a Tibetan uprising, ousted the Dalai Lama, and emptied the monasteries, in the process destroying much of Tibet's cultural heritage. Formal annexation followed in 1965, and Xizang was given the status of Autonomous Region. After Mao's death in 1976, the Chinese made some amends, returning looted religious treasures, permitting reconstruction of monasteries, and allowing Buddhist religious life to resume. But villagers continued to rebel, the Dalai Lama campaigned in exile for his people and faith, and for all intents and purposes Xizang remained what it had been since annexation—a colony of the Chinese Empire, an occupied society. (It is also worth noting that present-day Tibet is only part of the Buddhist culture area's earlier domain, which included what is now western China's Qinghai Province [see Fig. 9-8].)

As in other remote parts of China, Beijing's objective has been to integrate its minority peoples into the Chinese state, to establish a strong Chinese presence, and to display the virtues and assets of Chinese culture. In the case of Xizang, this was especially difficult because of the region's remoteness. Less than three million Tibetan people live scattered across a vast, frigid, mountainous and high-plateau area consisting not only of Tibet itself but also the neighboring Qaidam Basin in Qinghai Province (refer again to Fig. 9-9, and also note Fig. 9-15). Difficult roads through mountain passes and steep-sided valleys, seasonally impassable, slowed sinicization here more than anywhere else in the empire. But a mountain-defying railway would facilitate Chinese penetration, display China's technological prowess, and confirm the permanence of China's presence in Xizang.

China has good reasons to establish itself firmly in Xizang and the Qaidam Basin. Hydroelectric power potential, mineral deposits only partially known, and, in the Qaidam Basin, oil reserves and coal deposits make this more than just a cultural mission.

XINJIANG

China's northwestern corner consists of the giant Xinjiang-Uyghur Autonomous Region, constituting over one-sixth of the country's land area and situated in a strategic part of central Asia (Fig. 9-15). Linked to China Proper via the vital Hexi Corridor in Gansu Province to the east, Xinjiang is China's largest single administrative area.

Even today, however, after more than a half-century of vigorous sinicization, Chinese remain a minority here. About 40 percent of the population of just over 20 million is Chinese (up from 5 percent in 1949), and the Chinese control virtually all aspects of life. The majority are Muslim Uyghurs, who make up approximately half the total, and Kazakhs, Kyrgyz, and others with cultural affinities across the western border to all five republics of former Soviet Central Asia.

The physical geography of Xinjiang is dominated by high mountain ranges and vast basins (Fig. 9-2). The southern margin is defined by the Kunlun Shan (Mountains) beyond which lies Xizang (Tibet). Across the north-center of the region lies the mountain wall of the Tian Shan. Between the Kunlun and Tian Shan extends the vast, dry Tarim Basin within which lies the Takla Makan Desert, and north of the Tian Shan lies another depression, the Junggar Basin. Rivers rising on the mountain slopes disappear beneath the sediments of the basins, sustaining a ring of oases where their waters come near the surface or where wells and *qanats* make irrigation possible (Fig. 9-16).

Energy and Technology

This would not seem to be the most favorable geography, but Xinjiang has other assets that China exploits. The Qing (Manchu) armies that fought the nomadic Muslim warlords here in the 1870s saw the agricultural possibilities and planted the first crops. The communist rulers after 1949 exiled political opponents as well as criminals to notorious prison camps in this remote frontier. Then Xinjiang turned out to have extensive reserves of oil and natural gas (Figs. 9-15, 9-16). Here, too, China could experiment with space technology, rocketry, and nuclear weaponry far from densely populated areas (and shielded from foreign spies). And now, as Figure 9-15 reminds us, China's westernmost borders lie relatively close to the great energy reservoirs of

the Caspian Sea Basin—and only one neighbor lies in the way. Indeed, pipeline construction across Kazakhstan to Xinjiang is in progress.

Xinjiang's evolving human geography is revealed by Figure 9-16. Most of the Han Chinese are concentrated in the north-central cluster of cities and towns between the capital, Ürümqi, and the oil center of Karamay. Of special interest here is the city of Shihezi, founded by the Chinese army and now China's model for the region. Over 90 percent of its population of 400,000 are ethnic Chinese; Shihezi's comparative modernity and orderly life stand in sharp contrast to the poorer, often squalid towns and settlements outside the A.R.'s Chinese heartland. Also centered on the Ürümqi-Karamay area is Xinjiang's largest agricultural zone, where cotton and fruits yield export revenues. As the map shows, a ring of oases and clusters of irrigated agriculture encircles both the Tarim and Junggar basins.

Beijing has made extensive efforts to integrate Xinjiang into the Chinese state. Both a railroad and highway link Ürümqi and Lanzhou via the Hexi Corridor. Westward from Ürümqi, a railroad crosses the border with Kazakhstan and connects to the Öskemen-Almaty line; another railroad, opened in 2000, connects the capital to Kashi (Kashgar) in the distant southwest (Fig. 9-16). But vast spaces in Xinjiang remain distant and inaccessible. As the map shows, just one road, recently upgraded, loops around the vast southern and eastern flank of the Tarim Basin from Kashi via Hotan to Korla.

China's western frontier region, therefore, is a place of challenges as well as opportunities. Here a regime that from time to time displays intolerance of "foreign" belief systems meets a majority Muslim population that resists incorporation and whose leaders sometimes call for secession. Here China is having its brush with the War on Terror, accusing some of these Muslims of al-Qaeda sympathies, if not activism. Here the gap between comparatively well-off Chinese and poorer minorities is large—and spatially vivid. Here, too, China has unresolved boundary disputes with neighbors, including Tajikistan and India. On the other hand, Xinjiang's energy resources (also including some coal), agricultural and industrial potential, and relative location make this a promising frontier, a symbol for the rest of central Asia of what Chinese administration can accomplish.

CHINA'S PACIFIC RIM: EMERGING REGION

From Dalian on the doorstep of the Northeast to Hainan Island off the country's southernmost point, China is booming. In the cities, old neighborhoods are being bulldozed to

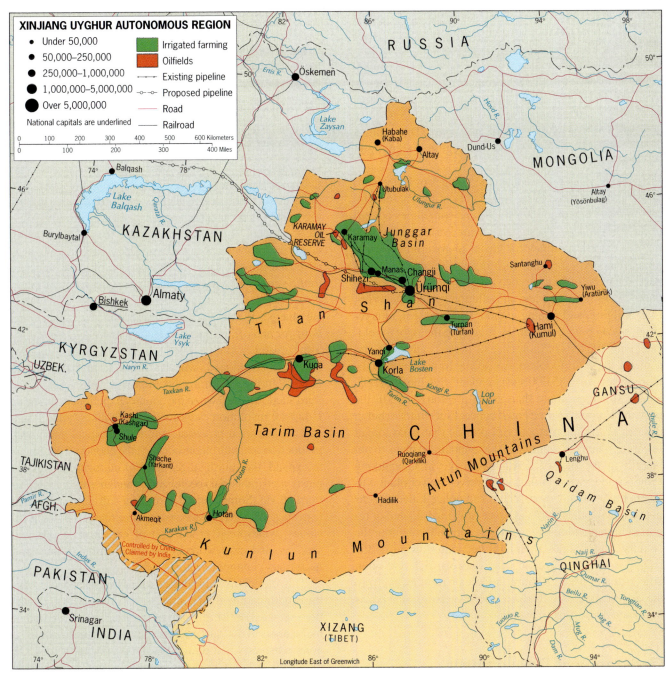

FIGURE 9-16   © H. J. de Blij, P. O. Muller, and John Wiley & Sons, Inc.

make way for modern skyscrapers. In the countryside of the coastal provinces, thousands of factories employing millions of workers who used to farm the land are changing life in the towns and villages. New roads, airports, power plants, and dams are being built. It is an uneven, sometimes chaotic process that is affecting certain parts of China's coast more than others. It is penetrating parts of China's interior, but meanwhile social gaps are widening: between advantaged and disadvantaged, well-off and poor, urbanite and villager. It is creating regional economic disparities

that China has never seen before. It is also generating a discrete geographic region facing China's Pacific coast that functionally has more in common with Japan, South Korea, and Taiwan than with Guizhou or Gansu.

In the Introduction, we discussed issues relating to the geography of development, emphasizing that, in this complex, globalizing world, the concept of developed versus underdeveloped countries (DCs and UDCs, respectively) is no longer tenable. By almost any measure, China remains what used to be called a UDC. As Figure G-12 shows, it barely ranks as a lower-middle-income economy. But China now has its own core-periphery duality: in its Pacific coast provinces, per-capita incomes today are considerably above the national average GNI of U.S. $6600 while hundreds of millions of villagers in China's remote interior earn far less. In its burgeoning Pacific Rim, China resembles a modernizing economy with "developed" characteristics. In much of its interior, it does not. This contrast is one reason a discrete Pacific Rim region is forming in China Proper.

In recent years the people in the villages of China's interior, and even those in still-rural areas of the Pacific Rim region, have begun to give voice (and sometimes action) to their grievances. These range from corrupt communist mayors extorting money for favors to government-sanctioned land expropriation enabling well-connected businessmen to build factories on villagers' fields (see bottom photo, p. 497). Stream and air pollution by such factories, road building, and mining without local consent, plus restrictions on personal liberties, including the right to relocate, generate tens of thousands of protests annually, and some of these turn violent. According to police officials, there were nearly 75,000 such incidents in China in 2005, about 20 percent more than in the previous year. In the past, the Chinese regime could conceal such unrest, but in this era of the Internet and cell phones that is no longer feasible. When the regime in 2006 announced plans to put landfill in the bay of a fishing village in Guangdong Province in order to build a wind-driven power plant, villagers mounted a demonstration and police killed dozens in just one incident that was never fully investigated or reported in the media. But word of it spread throughout the country, and China's rulers are becoming aware of the risks posed by the growing gap between rich and poor, powerful and weak. Reform in China's rural areas, from the governance of villages to the ownership of land, is long (and dangerously) overdue.

Geography of Development

Many yardsticks are used to gauge a country's development, including, as we noted, GNI per capita; percentage of workers in farms, factories, or other kinds of employment; amount of energy consumed; efficiency of transport and communications; use of manufactured metals like aluminum and steel in the economy; productivity of the labor force; and social measures such as literacy, nutrition, medical services, and savings rates. All these data are calculated per person, yielding an overall measure that ranks the country among all those providing statistics. These data do not, however, tell us *why* some countries, or parts of countries, exhibit the level of development they do.

The **geography of development** leads us into the study of raw-material distributions, environmental conditions, cultural traditions, historic factors (such as the lingering effects of colonialism), and forces of location. In the emergence of China's Pacific Rim, relative location played a major role, as we will see.

Multistage Development Model

Geographers tend to view the development process spatially, whereas other scholars focus on structural aspects. In the 1960s, economist Walt Rostow formulated a global model of the development process that is still being discussed today. This model suggested that all developing countries follow an essentially similar path through five interrelated growth stages. In the earliest of these stages, a *traditional society* engages mainly in subsistence farming, is locked in a rigid social structure, and resists technological change. When it reaches the second stage, *preconditions for takeoff*, progressive leaders move the country toward greater flexibility, openness, and diversification. Old ways are abandoned, workers move from farming into manufacturing, and transport improves. This leads to the third stage, *takeoff*, when the country experiences a type of industrial revolution. Sustained growth takes hold, industrial urbanization proceeds, and technological and mass-production breakthroughs occur. If the economy continues to expand, the fourth stage, *drive to maturity*, brings sophisticated industrial specialization and increasing international trade. Some countries reach a still more advanced fifth stage of *high mass consumption* marked by high incomes, widespread production of consumer goods and services, and most workers employed in the tertiary and quaternary economic sectors.

This model takes little account of core-periphery contrasts within individual countries. In China, takeoff conditions (stage 3) exist in much of the Pacific Rim, but other major areas of this vast country remain in stage 1—an economic disparity that dominates the provincial-level GDP map (Fig. 9-17). China has passed through its transition from tradition-bound to progressive leadership, but the effect of that leadership's reformist policies has not been the same in all parts of the country. The most fertile ground lies along the Pacific, where China has a history of contact with foreign enterprises, where the country is most open to the world, and from where many Chinese departed for Southeast Asian (and other) countries. These **Overseas Chinese** were positioned to play a crucial role in China's Pacific Rim development: many had left kin and community there, and those links facilitated the investment of hundreds of millions of dollars when the opportunity arose. Many of the Pacific Rim's factories were built by means of this repatriated money.

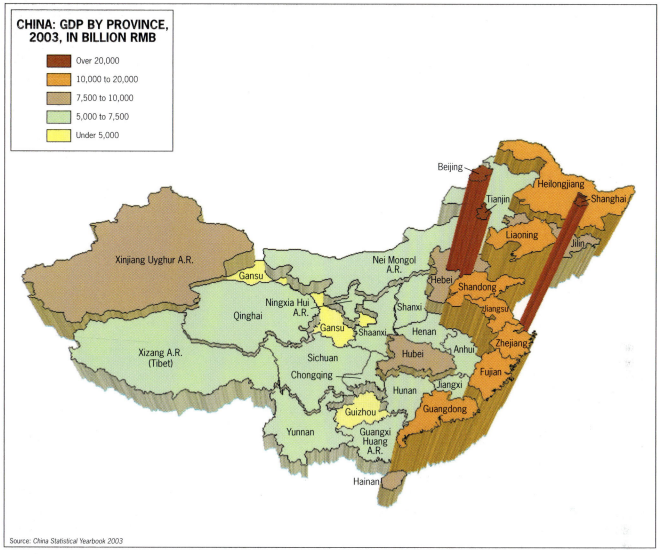

FIGURE 9-17   © H. J. de Blij, P. O. Muller, and John Wiley & Sons, Inc.

China's New Economic Geography

China's communist era has witnessed the metamorphosis of the world's most populous nation. Despite the dislocations of collectivization, communization, the Cultural Revolution, power struggles, policy reversals, and civil conflicts, China as a communist state in the aftermath of Mao managed to stave off famine and hunger, control its population growth rate, strengthen its military capacity, improve its infrastructure, and enhance its world position. It took over Hong Kong from the British and Macau from the Portuguese, closing the chapter on colonialism. When economic turbulence struck its Pacific Rim and Southeast Asian neighbors, China withstood the onslaught. Its trade with neighbors near and far is growing. Foreign investment in China has mushroomed. And still its government espouses communist dogma.

How could the government manipulate the coexistence of communist politics and market economics? That was the key question confronting Deng Xiaoping and his comrades when they took power in 1979. In terms of ideology, this objective would seem unattainable; an economic "open-door" policy would surely lead to rising pressures for political democracy.

Economic Zones of the Pacific Rim

But Deng thought otherwise. If China's economic experiments could be spatially separated from the bulk of the country, the political impact would be kept at bay. At the outset, the new economic policies would apply mainly to China's bridgehead on the Pacific Rim, leaving most of the vast country comparatively unaffected. Accordingly, the government introduced a complicated system of

FIGURE 9-18 © H. J. de Blij, P. O. Muller, and John Wiley & Sons, Inc.

11 **Special Economic Zones (SEZs)**, so-called *Open Cities*, and *Open Coastal Areas*, which would attract technologies and investments from abroad and transform the economic geography of eastern China (Fig. 9-18).

In these economic zones, investors are offered many incentives. Taxes are low. Import and export regulations are eased. Land leases are simplified. The hiring of labor under contract is allowed. Products made in the economic zones may be sold on foreign markets and, under some restrictions, in China as well. Even Taiwanese enterprises may operate here. And profits made may be sent back to the investors' home countries.

When Deng's government made the decisions that would reorient China's economic geography, location was a prime consideration. Beijing wanted China to participate in the global market economy, but it also wanted to cause as little impact on interior China as possible—at least in the first stages. The obvious answer was to position the Special Economic Zones along the coast. Initially, the government established four SEZs, all with particular locational properties (Fig. 9-18):

1. *Shenzhen*, adjacent to then-booming British Hong Kong on the Pearl River Estuary in Guangdong Province

2. *Zhuhai*, across from then still-Portuguese Macau, also on the Pearl River Estuary in Guangdong Province

3. *Shantou*, opposite southern Taiwan, a colonial treaty port, also in Guangdong Province, source of many Chinese now living in Thailand

4. *Xiamen*, on the Taiwan Strait, also a colonial treaty port (then known as Amoy in the local dialect), in Fujian Province, source of many Chinese now based in Singapore, Indonesia, and Malaysia

In 1988 and 1990, respectively, two additional SEZs were proclaimed:

5. *Hainan Island*, declared an SEZ in its entirety, its potential success linked to its location near Southeast Asia

6. *Pudong*, across the river from Shanghai, China's largest city, different from other SEZs because it was a giant state-financed project designed to attract large multinational companies

And the process continues. in 2006, still another SEZ was authorized:

7. *Binhai New Area*, the coastal zone of the northern port city of Tianjin, a long-established Open City with considerable foreign investment, now elevated to SEZ status, projected to outperform even Shanghai–Pudong and Shenzhen itself (see photo below).

Foreign investors doing business in China quickly learn that rules and regulations vary by geographic designation—that is, the SEZs provide the greatest advantages, the Open Cities somewhat less so, and the Open Coastal Areas least (but even this ranking is inconsistent). The SEZs enjoy strong official support in Beijing; the Open Coastal Cities were seen as catalysts for development and chosen for their size, location, overseas trading history, links to emigrated Chinese, infrastructure, and pool of local talent and labor. If history is a guide, the Binhai New Area will become China's next *growth pole*, because the country's leaders in Beijing favor it. This should remind us that while China may be seeking ways to spread the wealth of the Pacific Rim into its distant interior, they are not finished ex-

China's latest "new engine of growth," the Binhai New Area centered on China's fifth-largest city, Tianjin, stretches for 150 kilometers (90 mi) along the western Bohai Gulf waterfront. This development zone will be larger than either Shanghai's Pudong or Shenzhen, adjacent to Hong Kong. Special tax incentives and cheap land for factory building will, according to its planners, make this a manufacturing powerhouse for products ranging from automobiles to aircraft and from chemicals to microchips. The Binhai New Area is more than just another growth pole along China's Pacific Rim. It reflects a longstanding rivalry between China's communist-imbued, dogmatic, and bureaucratic north and its freewheeling competitors to the south in Shanghai and the Pearl River Delta. After local rivalries and policy disagreements were overcome, Beijing and Tianjin finally were able to agree on the launch of a development project that will change the economic geography of the region they share.   © Eye Press.

ploiting the opportunities still beckoning on the Pacific coast itself.

CHINA'S PACIFIC RIM TODAY

When China's planners laid out the SEZs, they held the highest hopes for Shenzhen, immediately adjacent to the thriving British dependency of Hong Kong. They knew that the kinds of attractions the SEZs offered to foreign investors would entice many factory owners paying Hong Kong wages to move across the border to Shenzhen. Successful Chinese living in Southeast Asia could use Hong Kong as a base for investment in Shenzhen; Taiwanese, even during the political standoff between Beijing and Taipei, could channel investments into this favored SEZ.

Those hopes were not disappointed. In the late 1970s, Shenzhen was just a sleepy fishing and duck-farming village of about 20,000. By 2008, its population neared 8 million—the fastest growth of any urban center on Earth in human history. Thousands of factories, large and small, moved into the SEZ from Hong Kong. Workers by the hundreds of thousands came from Guangdong Province, Guangxi Zhuang A.R., and beyond to look for jobs. High-rise buildings housing banks, corporate offices, hotels, and other service facilities made Shenzhen look like a new Hong Kong (photo p. 495).

Shenzhen's economic impact was far-reaching. Persuading Hong Kong industries to cross the border precipitated a sharp modification of Hong Kong's own economy; attracting investments from Overseas Chinese, mainly in Southeast Asia, opened financial coffers long closed to China; and enabling Taiwanese investors to participate in China's Pacific Rim boom created links with the "wayward province" that have economic as well as political benefit. By its very success (in its 1990s heyday, it produced 15 percent of all of China's exports by value), Shenzhen signaled the way for the other SEZs and Open Cities.

Guangdong Province and the Pearl River Estuary

As Figure 9-18 shows, eight provinces (including Hainan Island) and one Autonomous Region face the Pacific and form the basis for recognizing an emerging Pacific Rim region in eastern China. But among these provinces, one stands out by making the largest contribution by far to the national economy: Guangdong Province. Although Hong Kong and Macau, both SARs, are not administratively a part of Guangdong Province, they are functionally crucial parts of the vast economic complex that fringes the estuary of the Pearl River (Fig. 9-19). Indeed, without Hong Kong, there would not have been a Shenzhen of the kind described above.

The economic core of Guangdong Province, the ***Pearl River Delta (PRD)***, consists of Guangzhou, the province's capital; the Hong Kong SAR; the Shenzhen SEZ; the Zhuhai SEZ; the Macau SAR; and the latest juggernaut, fast-growing Zhongshan (Fig. 9-19). The PRD now has three cities with more than 5 million inhabitants; by 2010, it will have at least a dozen more containing between 500,000 and 5 million. Per-capita incomes here are estimated to be ten times the national average.

But this Pearl River Delta hub suffers from growing problems of inadequate infrastructure, worsening pollution, power outages, labor shortages (yes, labor shortages in China!), and resulting increases in labor costs. This is making the PRD less attractive to investors, who are looking at the Shanghai-centered Yangzi River Delta as an alternative. Paradoxically, this may stimulate a process China needs: the expansion of its coastal economic "miracle" into the deep interior. Leaders in Guangdong Province propose the extension of the PRD's model of economic cooperation into nine provinces of China's South, all the

way west to Yunnan and Sichuan. A network of road and railroad links among these provinces would allow raw materials and labor to circulate far more efficiently (there are no labor shortages in the interior) and move finished products not only to ports but also to the huge domestic market. A recent report in *The Economist* suggests that such an integrated region would encompass about 460 million people (one-third of the national population) on one-fifth of China's land producing nearly 40 percent of China's economic output, including one-third of all its exports.

The PRD as an economic heartland will need Beijing's help in making its regional aspirations come true, and all signs indicate that such support will be forthcoming. Already, the government has begun to grant SEZ-like privileges to parts of Yunnan Province around the capital, Kunming. Similar plans also exist for areas in Sichuan: the rapid expansion of Shanghai and the Yangzi River Delta will feed into the burgeoning economic corridor that leads through Wuhan toward Chongqing, providing further impetus to the westward spread of Pacific Rim productivity.

The Pearl River Delta as well as the Yangzi River Delta fit the model of the **regional state** as defined by the Japan- **12**

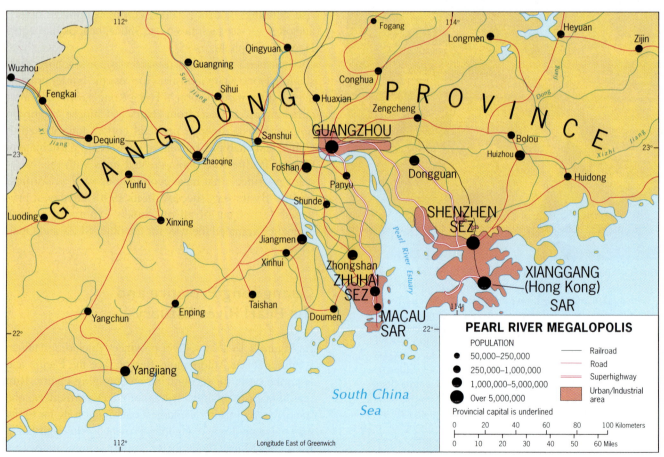

FIGURE 9-19 © H. J. de Blij, P. O. Muller, and John Wiley & Sons, Inc.

ese economist Kenichi Ohmae. These "natural economic zones," as he called them, defy old political borders and are shaped by the global economy of which they are parts; their representatives deal directly with foreign partners and negotiate the best terms they can with the national governments under which they operate (Europe and North America, we noted, have several examples of this kind). Both the Pearl River Delta and the Yangzi River Delta have now moved so far ahead of China as a whole that they have become political as well as economic forces in the country.

Hong Kong's Role

The process we have been discussing started with the economic success of Hong Kong while it was still a British dependency, empowered by a unique set of geographic circumstances. Here, on the left bank of the Pearl River's estuary, more than 7 million people (97 percent of Chinese descent) crowded onto 1000 square kilometers (400 sq mi) of fragmented, hilly territory under the tropical sun and created an economy larger than that of a hundred countries.

Hong Kong (the name means "Fragrant Harbor") was ceded by the Chinese to Britain during the colonial period in three parts: the islands in 1841, the Kowloon Peninsula in 1860, and the New Territories (a lease) in 1898 (Fig. 9-20). Until the 1950s, Hong Kong was just another trading colony, but then the Korean War and the United Nations embargo on communist China cut Hong Kong off from its hinterland. With a huge labor force and no do-

mestic resources, the colony needed an alternative—and it came in the form of a textile industry based on imported raw materials. Soon Hong Kong textiles were selling on worldwide markets; capital was invested in light industries; and the industrial base diversified into electrical equipment, appliances, and countless other consumer goods. Meanwhile, Hong Kong quietly served as a back door to China, which rewarded the colony by providing fresh water and staple foods. The colony had become an **economic tiger**.

Hong Kong also became one of the world's leading financial centers. When China changed course and embarked on its "Open Door" policy in the 1980s, Hong Kong was ready. Hong Kong factory owners moved their plants to Shenzhen, where wages were even lower; Hong Kong banks financed Shenzhen industries; and Hong Kong's economy adjusted once again to a new era. Hong Kong's port, however, continued to transfer Shenzhen's products. During the 1990s, about 25 percent of all of China's trade passed through Hong Kong harbor.

When the British lease on Hong Kong expired in 1997 and the British agreed to yield authority to China, the Chinese promised to allow Hong Kong's freewheeling economy to continue, in effect creating "one country, two systems." At midnight on June 30, Hong Kong became the Xianggang Special Administrative Region (SAR). True to the letter (if not always the spirit) of this commitment, the name "Hong Kong" has continued in general use. And on December 20, 1999 China took control over the last remaining European colony—Portuguese *Macau*—and made it a

Just 30 years ago, Shenzhen was a town of 20,000 with pig pens and fish ponds across the ironclad border with Hong Kong. Today it is an SEZ metropolis of 7.8 million with numerous links to the adjacent SAR—the fastest-growing urban agglomeration in human history. And Shenzhen is only one growth pole on China's burgeoning Pacific Rim. To be sure, the skyscrapers seen here include hotels, offices, and banks—only a fortunate minority among the city's workers can afford to live in apartments such as those in the left foreground. But this scene symbolizes China's headlong economic rush. © Gareth Brown/Corbis Images.

FIGURE 9-20 © H. J. de Blij, P. O. Muller, and John Wiley & Sons, Inc.

two decades ago were clogged with bicycles are now jammed with automobiles. Carmakers from all over the world are in China to capitalize on this mushrooming market, and before long China's domestic industry will undoubtedly challenge Japanese, American, and European brands on their home turf. Meanwhile, people by the millions are streaming toward the cities in one of the greatest rural-to-urban migrations in the history of humanity, which in mid-2007 required the *monthly* addition of a metropolis the size of Orlando (2 million) just to keep pace! Not surprisingly, many of urban China's traditional neighborhoods are being destroyed in the process as a new cultural landscape evolves with astonishing speed.

Coast and Interior

Inevitably, the concentration of industrial and commercial expansion in the coastal

SAR as well, guaranteeing (as in Hong Kong) the continuation of its existing social and economic system for 50 years.

The Burgeoning Pacific Rim

Not only did China grant economic advantages to corporations doing business in the coastal SEZs: China also made enormous infrastructure investments to enhance the efficiency of these growth poles. Although the old SEZ system still prevails officially, the entire coastal zone is being transformed at a staggering pace. New airports, bridges, superhighways, and port facilities are springing up; cities are being transformed by office towers, residential skyscrapers, ultramodern hotels, and gated communities. Beijing's metamorphosis befits its role as capital; Shanghai's **Pudong** district, across the Huangpu River from the preserved colonial riverfront, rivals Shenzhen in the speed of its construction and makes Shanghai look like the most futuristic city in the world (photo p. 497). Roads that just

provinces intensifies the regional disparities that have always marked this vast country (Fig. 9-17). China's planners are now addressing this potentially dangerous problem by channeling economic development inland, thereby to reverse the growth of the gap between coast and interior. The key initiative involves the Yangzi-Chang corridor and includes the Three Gorges Dam, the city of Chongqing, the Sichuan Basin, and Yunnan Province, where SEZ-type privileges are being extended to corporations as they are along the coast. As Figures 9-8 and 9-18 show, however, this project involves mainly central and southern China, but in many ways the greatest disparities lie in the interior of the north, in Shanxi, Shaanxi, and Gansu provinces. Channeling development westward in the north will prove far more difficult. Nonetheless, the need to improve the living conditions of farmers and others in the interior provinces is growing. China must restrain the influx of migrants from countryside to city, but this makes things worse—and can only be a temporary remedy. Closing the prosperity gap will be the lasting solution.

Old and new in Shanghai. The colonial buildings of the Bund, Shanghai's old frontage on the Huangpu River, stand in sharp contrast to the ultramodern district of **Pudong**, symbol of China's Pacific Rim development. The futuristic Oriental Pearl Television Tower looks like a rocket set to take off; its viewing platform affords one of China's most dramatic vistas. Two huge globes flanking the "launchpad" leave no doubt as to China's role in the world. To the right, the top of the 88-story Mao Observation Tower, containing a hotel with the world's highest atrium, will soon be outranked by an even taller structure. There was a time when the skylines of American cities symbolized the New World's power and modernity, but now it is Asia's turn to erect skyscrapers showing the world that a new era is in the making. © Magnum Photos, Inc.

CHINA: GLOBAL SUPERPOWER?

China in the early twenty-first century appears likely to become more than an economic force of world proportions: China also appears on course to achieve global superpower stature. As long as it retains its autocratic form of government (which allowed it to impose draconian population policies and comprehensive economic experiments without having to consult the electorate), China will be able to practice the kind of state capitalism that—in another guise—made South Korea an economic power. Unlike Japan or South Korea, however, China has no constraints on its military power. Its People's Liberation Army served as security forces during the pro-democracy turbulence of 1989; China's military is the largest standing army in the world, with about 3 million soldiers and some 1.2 million reserves. Its weaponry is being updated, and the government has made a major investment in its military by raising its armed forces budget significantly since 2000. Already, China is a nuclear power and has a growing arsenal of medium-range and intercontinental ballistic missiles.

During the twentieth century, the United States and the Soviet Union were locked in a 45-year Cold War that repeatedly risked nuclear conflict. That fatal exchange never happened, in part because it was a struggle between superpowers who understood each other comparatively well. While the politicians and military strategists were plotting, the cultural doors never closed: American audiences listened to Prokofiev and Shostakovich, watched Russian ballet, and read Tolstoy and Pasternak even as the Soviets applauded Van Cliburn, read Hemingway, and lionized American dissidents. In short, it was an *intracultural* Cold War, which reduced the threat of mutual destruction.

Here is a dramatic image of the "urban front," as geographers refer to the encroachment of cities and their suburbs on farmland nearby. This farm family rests outside their home near Lishui, about 130 kilometers (80 mi) from the coastal city of Wenzhou in Zhejiang Province, as truckloads of earth are dumped on their crops while builders convert their land into a construction site. Farmers are not always compensated fairly for the land they lose, but they have little power to slow the relentless advance of China's urbanization. Sometimes they remain on the village land until the day their homes are bulldozed away. © NG Image Collection.

The twenty-first century may witness a far more dangerous geopolitical struggle in which the adversaries may well be the United States and China. U.S. power and influence still prevail in the western Pacific, but it is easy to discern areas where Chinese and American interests will diverge (Taiwan is only one example). American bases in Japan, thousands of American troops in South Korea, and American warships in the East and South China seas are potential grounds for dispute. All this might generate the world's first *intercultural* Cold War, in which the risk of fatal misunderstanding is incalculably greater than it was during the last.

How can such a Cold War be averted? Trade, scientific and educational links, and cultural exchanges are obvious remedies: the stronger our interconnections, the less likely is a deepening conflict. We Americans should learn as much about China as we can, to understand it better, to appreciate its cultural characteristics, and to recognize the historico-geographical factors underlying China's views of the West.

For more than 40 centuries, China has known authoritarianism of both the brutal and the benevolent kind, has been fractured by regionalism only to unify again, and has depended on communalism to survive environmental and despotic depredations. For nearly 25 centuries Confucianism has guided it. Time and again, China has been opened to, and ruled by, foreigners, and time and again it has retreated into isolationism when things went wrong. Such a reversal may no longer be possible, given what we have seen in this chapter.

But now a renewed force is rising in modernizing China: nationalism. A combination of factors drives this process, ranging from broad issues such as Taiwan to incidents like the mid-air collision in 2001 between what the Chinese regarded as a U.S. spy plane and one of China's jet fighters near the island of Hainan. One of the broad issues has geographic dimensions: China disputes Japanese ownership of islands in the East China Sea (and the surrounding waters beneath which oil reserves may be found). China's attitude toward Japan is further inflamed by Japanese refusals to acknowledge atrocities committed in China during World War II, and by a recent Japanese prime minister's repeated visits to a war memorial containing the remains of Japanese war criminals. In 2005, angry anti-Japanese demonstrations throughout China followed the appearance of a Japanese history textbook containing a whitewashed version of Japanese wartime actions. Meanwhile, American scorn for Chinese human-rights failures inflames nationalist sentiments still more.

China today is on the world stage, globally engaged and internationally involved. Constraining the forces that tend to lead to competition and conflict between China and the United States must be a priority in the decades ahead, whether the issue is Taiwan, North Korea, or Japan. At stake are regional stability and world peace.

● MONGOLIA

Between China's Inner Mongolia and Xinjiang to the south and Russia's Eastern Frontier region to the north lies a vast, landlocked, isolated country called Mongolia that constitutes a discrete region of East Asia (Fig. 9-4). With only 2.7 million inhabitants in an area larger than Alaska, Mongolia is a steppe- and desert-dominated vacuum between two of the world's most powerful countries.

From the late 1600s until the revolutionary days of 1911, Mongolia was part of the Chinese Empire. With Soviet support the Mongols held off Chinese attempts to regain control, and in the 1920s the country became a People's Republic on the Soviet model. Free elections in 1990 ushered in a new political era, but Mongolia's economy, still largely based on animal products (chiefly cashmere wool), is troubled by mismanagement and environmental problems.

The map reflects Mongolia's difficulties. Despite its historic associations, ethnic affinities, and cultural involvement with China, Mongolia's incipient core area, including the capital of Ulaanbaatar, adjoins Russia's Eastern Frontier. The country's 800,000 herders and their millions of sheep follow nomadic tracks along the fringes of the vast Gobi Desert. When the cold of a Siberian winter descends on the steppes, as happened twice earlier in this decade, human and livestock losses are severe and the economy breaks down.

Over the past five years, China's involvement in Mongolia has intensified as Russia's has declined. Copper, iron, zinc, and coal reserves attract Chinese companies; road, bridge, and railroad building funded by China are in the works; and today more than 1000 Chinese businesses are active in Mongolia. Public opinion polls reflect concern on the part of Mongolians over China's role as the leading foreign investor and trade partner, but Mongolia has accepted what many regard as the inevitable. A recent *New York Times* article reported another side of the Chinese coin: a government official is quoted as stating that Mongolia had 4 million deer before 1990, but now very few are left "because Chinese, who like deer antlers, shot them one by one."

Mongolia's Soviet period has yielded to a time of Chinese involvement, and Mongolia's relatively high rate of economic growth is undoubtedly the result of its proximity to China's markets. But Mongolia is a weak country

14 in a vulnerable part of inner Asia, a **buffer state** wedged between populous former and future superpowers. Will it escape the fate of other Asian buffers?

materials from all over the world, voluminous exports, and global financial linkages. It also is a region of social problems, political uncertainties, and economic vulnerabilities.

● THE JAKOTA TRIANGLE REGION

We turn now to a region that exemplifies the future of East Asia. Along the Pacific Rim, from Japan (through South Korea, Taiwan, China's Guangdong) to Singapore, rapid economic development has transformed community and society. In China, as we saw, a Pacific Rim region is in the making, still discontinuous today but likely to extend all along the coast in the future. In Japan, South Korea, and Taiwan we can observe that future—today. That is why we rec-

15 ognize an East Asian region we call the **Jakota Triangle**, consisting of *Ja*pan, *Ko*rea (South Korea at present but probably a reunited Korea later), and *Ta*iwan (Fig. 9-21). This is a region of great cities, huge consumption of raw

JAPAN

When we assess China's prospects of becoming a superpower, we should remember what happened in Japan in the nineteenth century. In 1868, a group of reform-minded modernizers seized power from an old guard, and by the end of the century Japan was a military and economic force. From the factories in and around Tokyo and from urban-industrial complexes elsewhere poured forth a stream of weapons and equipment the Japanese used to embark on colonial expansion. By the mid-1930s, Japan lay at the center of an empire that included all of the Korean Peninsula, the whole of China's Northeast (which the Japanese called Manchukuo), the Ryukyu Islands, and Taiwan, as well as the southern half of Sakhalin Island (called Karafuto). Not even a disastrous earthquake, which destroyed much of Tokyo in 1923 and killed 143,000 people, could slow the Japanese drive.

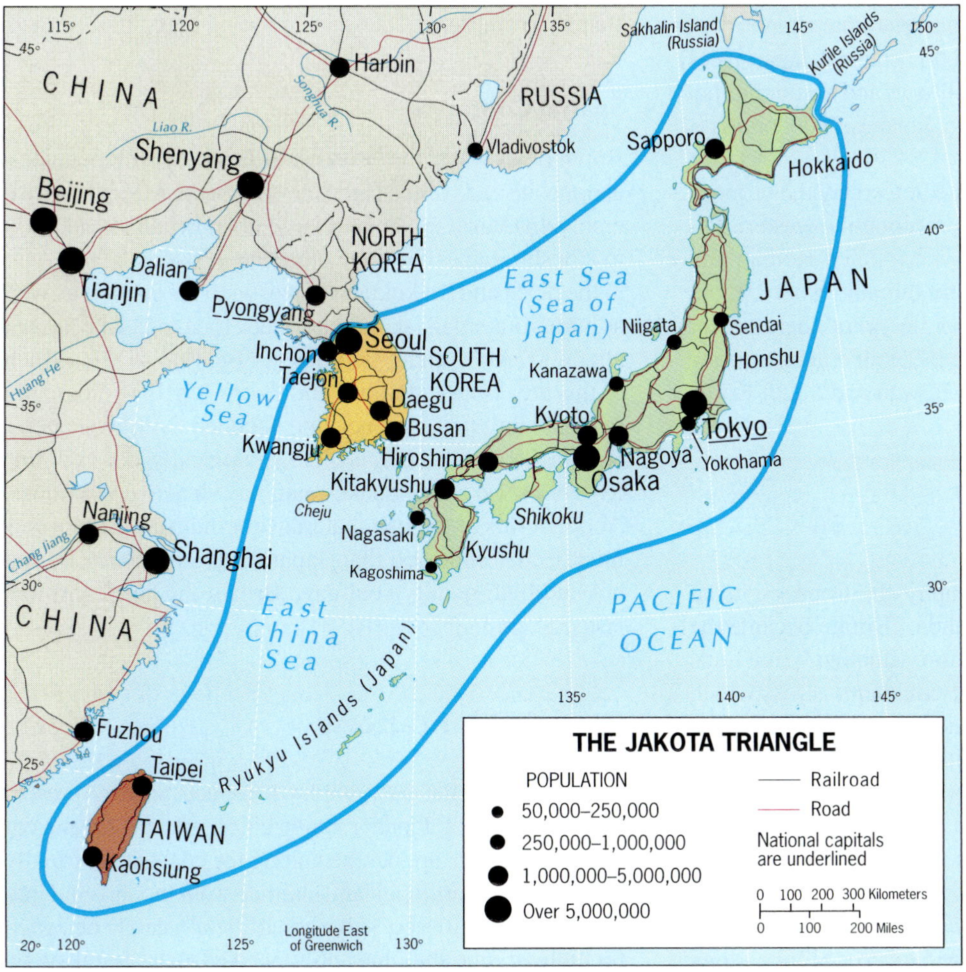

FIGURE 9-21 © H. J. de Blij, P. O. Muller, and John Wiley & Sons, Inc.

Colonial Wars and Recovery

World War II saw Japan expand its domain farther than the architects of the 1868 "modernization" could have anticipated. By early December 1941, Japan had conquered large parts of China Proper, all of French Indochina to the south, and most of the small islands in the western Pacific. Then, on December 7, 1941, Japanese-built aircraft carriers moved Tokyo's warplanes within striking range of Hawai'i, and the surprise attack on Pearl Harbor underscored Japan's confidence in its war machine. Soon the Japanese overran the Philippines, the (then) Netherlands

East Indies, Thailand, and British Burma and Malaya, and drove a wide corridor through the heart of China to the border with Vietnam.

A few years later, Japan's expansionist era was over. Its armies had been driven from virtually all its possessions, and when American nuclear bombs devastated two Japanese cities in 1945, the country lay in ruins. But once again, Japan, aided this time by an enlightened U.S. postwar administration, surmounted disaster.

Japan's economic recovery and its rise to the status of world economic superpower was the success story of the second half of the twentieth century. Japan lost the war and its empire, but it scored many economic victories in a new global arena. Japan became an industrial giant, a technological pacesetter, a fully urbanized society, a political power, and an affluent nation. No city in the world today is without Japanese cars in its streets; few photography stores lack Japanese cameras and film; laboratories the world over use Japanese optical equipment. From microwave ovens to DVD players, from oceangoing ships to camcorders, Japanese goods flood the world's markets. Even today, while experiencing a severe downturn, the Japanese economy remains the second largest national economy in the world. As we will see later in this chapter, Japan faces several challenges, economic, political, and demographic. But even while in recession, the Japanese economy remains formidable.

Japan's brief colonial adventure helped lay the groundwork for other economic successes along the western Pacific Rim. The Japanese ruthlessly exploited Korean and Formosan (Taiwanese) natural and human resources, but they also installed a new economic order there. After World War II, this infrastructure facilitated an economic transition—and soon made both Taiwan and South Korea competitors on world markets.

British Tutelage

In discussing the economic geography of Europe, we noted how the comparatively small island of Britain became the crucible of the Industrial Revolution, which gave it a huge head start and advantage over the continent, across which the Industrial Revolution spread decades later. Britain was not large, but its insular location, its local raw materials, the skills of its engineers, its large labor force, and its social organization combined to endow it not only with industrial strength but also with a vast overseas empire.

After the modernizers took control of Japan in 1868—an event known as the *Meiji Restoration* (the return of "enlightened rule" centered on the Emperor Meiji)—they turned to Britain for guidance in reforming their nation and its economy. In the decades that followed, the British advised the Japanese on the layout of cities and the construction of a railroad network, on the location of industrial plants, and on the organization of education. The British influence is still visible in the Japanese cultural landscape today: the Japanese, like the British, drive on the left side of the road. Consider how this affects the effort to open the Japanese market to U.S. automobiles!

The Japanese reformers of the late nineteenth century undoubtedly saw many geographic similarities between Britain and Japan. At that time, most of what mattered in Japan was concentrated on the country's largest island, Honshu (literally, "mainland"). The ancient capital, Kyoto, lay in the interior, but the modernizers wanted a coastal, outward-looking headquarters. So they chose the town of Edo, on a large bay where Honshu's eastern coastline turns sharply (Fig. 9-22). They renamed the place *Tokyo* ("eastern capital"), and little more than a century later it was the largest urban agglomeration in the world. Honshu's coasts were near mainland Asia, where raw materials and potential markets for Japanese products lay. The notion of a greater Japanese empire followed naturally from the British example.

Spatial Contrasts

But in other ways, the British and Japanese archipelagoes, at opposite ends of the Eurasian landmass, differed considerably. In total area, Japan is larger than Britain. In addition to Honshu, Japan has three other large islands—Hokkaido to the north and Shikoku and Kyushu to the south—as well as numerous small islands and islets, for a total land area of about 377,000 square kilometers (146,000 sq mi). Much of this area is mountainous and steep-sloped, geologically young, earthquake-prone, and studded with volcanoes. Britain has lower relief, is older geologically, does not suffer from severe earthquakes, and has no active volcanoes. And in terms of the raw materials for industry, Britain was much better endowed than Japan. Self-sufficiency in iron ore and high-quality coal gave Britain the head start that lasted a century.

Japan's Limitations

Japan's high-relief topography has been an ever-present challenge. All of Japan's major cities, except the ancient capital of Kyoto, are perched along the coast, and virtually all lie partly on artificial land claimed from the sea. Sail into Kobe harbor, and you will pass artificial islands designed for high-volume shipping and connected to the mainland by automatic space-age trains. Enter Tokyo Bay, and the re-

FIGURE 9-22 © H. J. de Blij, P. O. Muller, and John Wiley & Sons, Inc.

fineries and factories to your east and west stand on huge expanses of landfill that have pushed the bay's shoreline outward. With 128.1 million people, the vast majority (79 percent) living in towns and cities, Japan uses its habitable living space intensively—and expands it wherever possible.

As Figure 9-23 shows, farmland in Japan is both limited and regionally fragmented. Urban sprawl has invaded much cultivable land. In the hinterland of Tokyo lies the Kanto Plain; around Nagoya, the Nobi Plain; and surrounding Osaka, the Kansai District—each a major farming zone under relentless urban pressure. All three of these plains lie within Japan's fragmented but well-defined core area (delimited by the red line on the map), the heart of Japan's prodigious manufacturing complex.

FIGURE 9-23 © H. J. de Blij, P. O. Muller, and John Wiley & Sons, Inc.

From Realm to Region

For nearly a half-century, Japan's seemingly insurmountable economic lead, its almost unique (for so large a population) ethnic homogeneity, its insular situation, and its cultural uniformity justified its designation as a discrete geographic realm, not a region of a greater East Asia. But rapid change on the rim of the western Pacific—both within Japan and beyond it—has eroded the bases for this distinction. In the mid-1990s, Japan's economic engine faltered while those of its Pacific Rim neighbors accelerated; its social fabric was torn by unprecedented acts of violence; and its foreign relations were soured by disputes over wartime atrocities in China and Korea. Japan stood alone in the vanguard of Pacific Rim developments, but it is no longer discrete.

Early Directions

On the map, the Korean Peninsula seems to extend like an unfinished bridge from the Asian mainland toward southern Japan. Two Japanese-owned islands, Tsushima and Iki, in the Korea Strait almost complete the connection (Fig. 9-22). In fact, Korea was linked to Japan in prehistoric times, when sea levels were lower. Peoples and cultural infusions reached Japan from Asia time and again, even after rising waters inundated the land bridge between them.

The Japanese islands were inhabited even before the ancestors of the modern Japanese arrived from Asia. As we noted earlier, the Ainu, a people of Caucasian ancestry, had established themselves on all four of the major islands thousands of years before. But the Ainu could not hold their ground against the better organized and armed invaders, and they were steadily driven northward until they retained a mere foothold on Hokkaido. Today, the last vestiges of Ainu ancestry and culture are disappearing.

By the sixteenth century, Japan had evolved a highly individualistic culture, characterized by social practices and personal motivations based on the Shinto belief system. People worshiped many gods, venerated ancestors, and glorified their emperor as a divine figure, infallible and omnipotent. The emperor's military ruler, or *shogun*, led the campaign against the Ainu and governed the nation. The capital, Kyoto, became an assemblage of hundreds of magnificent temples, shrines, and gardens (photo below). In their local architecture, modes of dress, festivals, theater, music, and other cultural expressions, the Japanese forged a unique society. Handicraft industries, based on local raw materials, abounded.

Nonetheless, in the mid-nineteenth century, Japan hardly seemed destined to become Asia's leading power. For 300 years the country had been closed to outside influences, and Japanese society was stagnant and tradition-bound. When the European colonizers first appeared on Asia's shores, Japan tolerated, even welcomed, their merchants and missionaries. But as the European presence in East Asia grew stronger, the Japanese began to shut their doors. Toward the end of the sixteenth century, the emperor decreed that all foreign traders and missionaries should be expelled. Christianity (especially Roman Catholicism) had gained a considerable foothold in Japan, but now this was feared as being a prelude to colonial conquest. Early in the seventeenth century, the shogun launched a massive and bloody campaign that practically stamped out Christianity.

Determined not to suffer the same fate as the Philippines, which by then had fallen to Spain, the Japanese henceforth allowed only minimal contact with Europeans. A few Dutch traders, confined to a little island near the city of Nagasaki, were for many decades the sole representatives of Europe in Japan. Thus Japan entered a long period of isolation, lasting past the middle of the nineteenth century.

Japan could maintain its aloofness, while other areas were being enveloped by the colonial tide, because of its strong central government and well-organized military, the natural protection its islands provided, and its remoteness along East Asia's difficult northern coast. Also,

FROM THE FIELD NOTES

"The city of Kyoto, chronologically Japan's second capital (after Nara; before Tokyo), is the country's principal center of culture and religion, education and the arts. Tree-lined streets lead past hundreds of Buddhist temples; tranquil gardens provide solace from the bustle of the city. I rode the bullet train from Tokyo and spent my first day following a walking route recommended by a colleague, but got only part of the way because I felt compelled to enter so many of the temple grounds and gardens. And not only Buddhism, but also Shinto makes its mark on the cultural landscape. I passed under a *torii*—a gateway usually formed by two wooden posts topped by two horizontal beams turned up at their ends—which signals that you have left the secular and entered the sacred, and found this beautiful Shinto shrine with its orange trim and olive-green glazed tiles." © H. J. de Blij.

www.conceptcaching.com

Japan's isolation was far less splendid than that of China, whose exquisite silks, prized teas, and skillfully made wares attracted traders and usurpers alike.

When Japan finally came face to face with the new weaponry of its old adversaries, it had no answers. In the 1850s, the steel-hulled "black ships" of the American fleet sailed into Japanese harbors, and the Americans extracted one-sided trade agreements. Soon the British, French, and Dutch were also on the scene, seeking similar treaties. When there was local resistance to their show of strength, the Americans quickly demonstrated their superiority by shelling parts of the Japanese coast. By the late 1860s, even as Japan's modernizers were about to overturn the old order, no doubt remained that Japan's protracted isolation had come to an end.

A Japanese Colonial Empire

When the architects of the Meiji Restoration confronted the challenge to build a Japan capable of competing against powerful adversaries in a changing world, they took stock of the country's assets and liabilities. Material assets, they found, were limited. To achieve industrialization, coal and iron ore were needed. In terms of coal, there was enough to support initial industrialization. Coalfields in Hokkaido and Kyushu were located near the coast; since the new industries also were on the coast, cheap water transportation was possible. As a result, the shores of the (Seto) Inland Sea (Fig. 9-23) became the sites of many factories. But there was little iron ore, certainly nothing like what was available domestically in Britain, and not enough to sustain the massive industrialization the reformers had in mind. This commodity would have to be purchased overseas and imported.

On the positive side, manufacturing—light manufacturing of the handicraft type—already was widespread in Japan. In cottage industries and in community workshops, the Japanese produced textiles, porcelain, wood products, and metal goods. The small ore deposits in the country were enough to supply these local industries. Power came from human arms and legs and from wheels driven by water; the chief source of fuel was charcoal. Importantly, Japan did have an industrial tradition and an experienced labor force that possessed appropriate manufacturing skills. All this was not enough to lead directly to industrial modernization, but it did hold promise for capital formation. The community and home workshops were integrated into larger units, hydroelectric plants were built, and some thermal (coal-fired) power stations were constructed in critical areas.

For the first time, Japanese goods (albeit of the light manufactured variety) began to compete with Western products on international markets. The Japanese planners resisted any infusion of Western capital, however. Although Western support would have accelerated industrialization, it would have cost Japan its economic autonomy. Instead, farmers were more heavily taxed, and the money thus earned was poured into the industrialization effort.

Another advantage for Japan's planners lay in the country's military tradition. Although the shogun's forces had been unable to repel the invasions of the 1850s, this was the result of outdated equipment, not lack of manpower or discipline. So while its economic transformation gathered momentum, the military forces, too, were modernized. Barely more than a decade after the Meiji Restoration, Japan laid claim to its first Pacific prize: the Ryukyu Islands (1879). A Japanese colonial empire was in the making.

When Japan gained control over its first major East Asian colonies, Taiwan and Korea, its domestic raw-material problems were essentially solved, and a huge labor pool fell under its sway as well. High-grade coal, high-quality iron ore, and other resources were shipped to the factories of Japan; and in Taiwan as well as in Korea, the Japanese built additional manufacturing plants to augment production. This, in turn, provided the equipment to sustain the subsequent drive into China and Southeast Asia.

Modernization

The reformers who set Japan on a new course in 1868 probably did not anticipate that three generations later their country would lie at the heart of a major empire sustained by massive military might. They set into motion a process of **modernization**, but they managed to **16** build on, not replace, Japanese cultural traditions. We in the Western world tend to equate modernization with Westernization: urbanization, the spread of transport and communications facilities, the establishment of a market (money) economy, the breakdown of local traditional communities, the proliferation of formal schooling, the acceptance and adoption of foreign innovations. In the non-Western world, the process often is viewed differently. There, "modernization" is seen as an outgrowth of colonialism, the perpetuation of a system of wealth accumulation introduced by foreigners driven by greed. In this view, the local elites who replaced the colonizers in the newly independent states only continue the disruption of traditional societies, not their true modernization. Traditional societies, they argue, can be modernized without being Westernized.

In this context, Japan's modernization is unique. Having long resisted foreign intrusion, the Japanese did not achieve the transformation of their society by importing a Trojan horse; it was done by Japanese planners, building on the existing Japanese infrastructure, to fulfill Japanese objectives. Certainly Japan imported foreign technologies and adopted innovations from the British and others, but the Japan that was built, a unique combination of modern and traditional elements, was basically an indigenous achievement.

Relative Location

Japan's changing fortunes over the past century reveal the **17** influence of **relative location** in the country's development. When the Meiji Restoration took place, Britain, on the other side of the Eurasian landmass, lay at the center of a global empire. The colonization and Europeanization of the world were in full swing. The United States was still a developing country, and the Pacific Ocean was an avenue for European imperial competition. Japan, even while it was conquering and consolidating its first East Asian colonies (the Ryukyus, Taiwan, Korea), lay remote from the mainstream of global change.

Then Japan became embroiled in World War II and dealt severe blows to the European colonial armies in Asia. The Europeans never recovered: the French lost Indochina, and the Dutch were forced to abandon their East Indies (now Indonesia). When the war ended, Japan was defeated and devastated, but at the same time the Japanese had done much to diminish the European presence in the Pacific Basin. Moreover, the global situation had changed dramatically. The United States, Japan's trans-Pacific neighbor, had become the world's most powerful and wealthiest country, whereas Britain and its global empire were fading. Suddenly Japan was no longer remote from the mainstream of global action: now the Pacific was becoming the avenue to the world's richest markets. Japan's relative location—its situation relative to the economic and political foci of the world—had changed. Therein lay much of the opportunity the Japanese seized after the postwar rebuilding of their country.

Japan's Spatial Organization

Imagine this: 128 million people crowded into a territory the size of Montana (population: 975,000), most of it mountainous, subject to frequent earthquakes and volcanism, with no domestic oilfields, little coal, few raw materials for industry, and not much level land for farming. If Japan today were an underdeveloped country in need of food relief and foreign aid, explanations would abound: overpopulation, inefficient farming, energy shortages.

True, only an estimated 18 percent of Japan's national territory is designated as habitable. And Japan's large population is crowded into some very big cities. Moreover,

FROM THE FIELD NOTES

"Visiting the Peace Memorial Park in Hiroshima is a difficult experience. Over this site on August 6, 1945 began the era of nuclear weapons use, and the horror arising from that moment in history, displayed searingly in the museum, is an object lesson in this time of nuclear proliferation and enhanced risk. In the museum is a model of the city immediately after the explosion (the red ball marks where the detonation occurred), showing the total annihilation of the entire area at an immediate loss of more than 80,000 people and the death from radiation of many more subsequently. In the park outside, the Atomic Bomb Memorial Dome, the only building to partially survive the blast, has become the symbol of Hiroshima's devastation and of the dread of nuclear war." © H. J. de Blij.

www.conceptcaching.com

Concept Caching

Japan's agriculture is not especially efficient. But Japan defeated the odds by calling on old Japanese virtues: organizational efficacy, massive productivity, dedication to quality, and adherence to common goals. Even before the Meiji Restoration, Japan was a tightly organized country of some 30 million citizens. Historians suggest that Edo, even before it became the capital and was renamed Tokyo, may have been the world's largest urban center. The Japanese were no strangers to urban life, they knew manufacturing, and they prized social and economic order.

Areal Functional Organization

All this proved invaluable to the modernizers when they set Japan on its new course. The new industrial growth of the country could be based on the urban and manufacturing development that was already taking place. As we noted, Japan does not possess major domestic raw-material sources, so no substantial internal reorganization was necessary. However, some cities were better sited and enjoyed better situations relative to those limited local resources and, more importantly, to external sources of raw materials than others. As Japan's regional organization took shape, a hierarchy of cities developed; Tokyo took and kept the lead, but other cities grew rapidly into industrial centers.

This process was governed by a geographic principle **18** that Allen Philbrick called **areal functional organization**, a set of five interrelated tenets that help explain the evolution of regional organization, not only in Japan but throughout the world. Human activity, Philbrick reasoned, has spatial focus. It is concentrated in some locale, whether a farm or factory or store. Every one of these establishments occupies a particular location; no two of them can occupy exactly the same spot on the Earth's surface (even in high-rises, there is a vertical form of absolute location). Nor can any human activity proceed in total isolation, so that interconnections develop among these various establishments. This system of interconnections grows more complex as human capacities and demands expand. Each system (for example, farmers sending crops to market and buying equipment at service centers) forms a unit of areal functional organization. In the Introduction, we referred to functional regions as systems of spatial organization; we can map Philbrick's units of areal functional organization as regions. These regions evolve because of what he called "creative imagination" as people apply their total cultural experience and their technological know-how when they organize and rearrange their living space. Finally, Philbrick suggests, we can recognize levels of development in areal functional organization, a ranking of places and regions based on the type, extent, and intensity of exchange.

In the broadest sense, we can divide regions of human organization into three categories: subsistence, transitional, and exchange. Japan's areal organization reflects the exchange category. Within each of these categories, we can also rank individual places on the basis of the number and kinds of activities they generate. Even in regions where subsistence activities dominate, some villages have more interconnections than others. The map of Japan (Fig. 9-23) showing its resources, urban settlements, and surface communications can tell us much about Japan's economy. It looks just like maps of other parts of the world where an exchange type of areal organization has developed—a hierarchy of urban centers ranging from the largest cities to the tiniest hamlets, a dense network of railways and roads connecting these places, and productive agricultural areas near and between the urban centers.

For Japan, however, the map shows us something else: Japan's external orientation, its dependence on foreign trade. All primary and secondary regions lie on the coast. Of all the cities in the million-size class, only Kyoto lies in the interior (Fig. 9-22). If we deduced that Kyoto does not match Tokyo–Yokohama, Osaka–Kobe, or Nagoya in terms of industrial development, that would be correct— the old capital remains a center of small-scale light manufacturing. Actually, Kyoto's ancient character has been deliberately preserved, and large-scale industries have been discouraged. With its old temples and shrines, its magnificent gardens, and its many workshop and cottage industries, Kyoto remains a link with Japan's premodern past.

Leading Economic Regions

As Figure 9-23 shows, Japan's dominant region of urbanization and industry (along with productive agriculture) is the **Kanto Plain**, which contains about one-third of the Japanese population and is focused on the Tokyo–Yokohama–Kawasaki metropolitan area. This gigantic cluster of cities and suburbs (the world's second largest urban agglomeration), interspersed with intensively cultivated farmlands, forms the eastern anchor of the country's elongated and fragmented core area. Besides its flatness, the Kanto Plain possesses other advantages: its fine natural harbor at Yokohama, its relatively mild and moist climate, and its central location with respect to the country as a whole. (The region's only disadvantage is its vulnerability to earthquakes—see box titled "When the Big One Strikes.") It has also benefited enormously from Tokyo's designation as the modern capital, which coincided with Japan's embarkation on its planned course of economic development. Many industries and businesses chose Tokyo as their headquarters in view of

AMONG THE REALM'S GREAT CITIES . . . Tokyo

MANY URBAN AGGLOMERATIONS are named after the city that lies at their heart, and so it is with the second largest of all: Tokyo (26.7 million). Even its longer name—Tokyo-Yokohama-Kawasaki—does not begin to describe the congregation of cities and towns that form this massive, crowded metropolis that encircles the head of Tokyo Bay and continues to grow, outward and upward.

Near the waterfront, some of Tokyo's neighborhoods are laid out in a grid pattern. But the urban area has sprawled over hills and valleys, and much of it is a maze of narrow, winding streets and alleys. Circulation is slow, and traffic jams are legendary. The train and subway systems, however, are models of efficiency—although during rush hours you must get used to the *shirioshi* pushing you into the cars to get the doors closed.

At the heart of Tokyo lies the Imperial Palace with its moats and private parks. Across the street, buildings retain a respectful low profile, but farther away Tokyo's skyscrapers seem to ignore the peril of earthquakes. Nearby lies one of the world's most famous avenues, the Ginza, lined by department stores and luxury shops. In the distance you can see an edifice that looks like the Eiffel Tower, only taller: this is the Tokyo Tower (see photo, p. 509), a multipurpose structure designed to test lighter Japanese steel, transmit television and radio signals, detect Earth tremors, monitor air pollution, and attract tourists.

Tokyo is the epitome of modernization, but Buddhist temples, Shinto shrines, historic bridges, and serene gar-

© H. J. de Blij, P. O. Muller, and John Wiley & Sons, Inc.

dens still grace this burgeoning urban complex, a cultural landscape that reflects Japan's successful marriage of the modern and the traditional.

the advantages of proximity to the government's decision makers.

The Tokyo–Yokohama–Kawasaki conurbation has become Japan's leading manufacturing complex, producing more than 20 percent of its annual output. The raw materials for all this industry, however, come from far away. For example, the Tokyo area is among the chief steel producers in Japan, using iron ores from the Philippines, Malaysia, Australia, India, and even Africa; most of the coal is imported from Australia and North America, and the petroleum comes from Southwest Asia and Indonesia. The Kanto Plain cannot produce nearly enough food for its massive resident population. Imports must come from Canada, the United States, and Australia as well as other areas in Japan. Thus Tokyo depends completely on its external trading ties for food, raw materials, and markets for its products, which run the gamut from children's toys to high-precision optical equipment to the world's largest oceangoing ships.

The second-ranking economic region in Japan's core area is the Osaka–Kobe–Kyoto triangle—also known as the *Kansai District*—located at the eastern end of the Inland Sea (Fig. 9-23). The Osaka–Kobe conurbation had an advantageous position with respect to Manchuria (Manchukuo) during the height of Japan's colonial empire. Situated at the head of the Inland Sea, Osaka was the major Japanese base for the China trade and for the exploitation of Manchuria, but it suffered when the empire was destroyed and it lost its trade connections with China after World War II. Kobe (like Yokohama, which is Japan's chief shipbuilding center) has remained one of the country's busiest ports, handling both Inland Sea traffic and extensive overseas linkages. Kyoto, as we noted, remains much as it was before Japan's great leap forward, a city of small workshop industries. The Kansai District is also an important farming area. Rice, of course, is the most intensively grown crop in the warm,

When the Big One Strikes

THE TOKYO–YOKOHAMA–KAWASAKI urban area is the second biggest metropolis on Earth (only Mexico City is larger). But Tokyo is more than a large city: it constitutes the most densely concentrated financial and industrial complex in the world. Two-thirds of Japan's businesses worth more than U.S. $50 million are clustered here, many with vast overseas holdings. More than half of Japan's huge industrial profits are generated in the factories of this gigantic metropolitan agglomeration.

But Tokyo has a worrisome environmental history because three active tectonic plates are converging here (Fig. 9-24). All Japanese know about the "70-year" rule: over the past three-and-a-half centuries, the Tokyo area has been struck by major earthquakes roughly every 70 years—in 1633, 1703, 1782, 1853, and 1923. The Great Kanto Earthquake of 1923 set off a firestorm that swept over the city and killed an estimated 143,000 people. Tokyo Bay virtually emptied of water; then a tsunami (seismic sea wave) roared back in, sweeping everything before it. That Japan could overcome this disaster was evidence of the strength of its economy.

Today, Tokyo is more than a national capital. It is a global financial and manufacturing center in which so much of the world's wealth and productive capacity are concentrated that an earthquake comparable to the one of 1923 would have a calamitous effect worldwide. Ominously, Tokyo at the outset of our new century is a much more vulnerable place than the Tokyo of the 1920s. True, building regulations are stricter, and civilian preparedness is better. But whole expanses of industries have been built on landfill that will liquefy; the city is honeycombed by

underground gas lines that will rupture and stoke countless fires; congestion in the area's maze of narrow streets will hamper rescue operations; and many older high-rise buildings do not have the structural integrity that has lately emboldened builders to erect skyscrapers of 50 stories and more. Add to this the burgeoning population of the Kanto Plain—which surpasses 40 million on what may well be the most dangerous 4 percent of Japan's territory—and we realize that the next big earthquake in the Tokyo area will not be a remote, local news story.

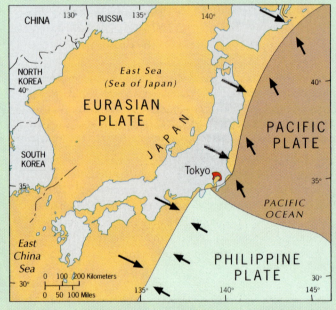

FIGURE 9-24   © H. J. de Blij, P. O. Muller, and John Wiley & Sons, Inc.

---

moist lowlands, but this agricultural zone is smaller than the one on the Kanto Plain. Here is another huge concentration of people that must import food.

The Kanto Plain and Kansai District, as Figure 9-23 indicates, are the two leading primary regions within Japan's core. Between them lies the **Nobi Plain** (also called the *Chubu District*), focused on the industrial metropolis of Nagoya, Japan's leading textile producer. The map indicates some of the Nagoya area's advantages and liabilities. The Nobi Plain is larger than the lowlands of the Kansai District; thus its agricultural productivity is greater, though not as great as that of the Kanto Plain. But Nagoya has neither Tokyo's centrality nor Osaka's position on the Inland Sea: its connections to Tokyo are via the tenuous Sen-en Coastal Strip. Its westward connections are better, and the Nagoya area may be coalescing with the Osaka–Kobe conurbation. Still, the quality of Nagoya's port is not nearly

as good as Tokyo's Yokohama or Osaka's Kobe, and it has had silting problems.

Westward from the three regions just discussed—which together constitute what is often called the *Tokaido* megalopolis—extends the Seto Inland Sea, along whose shores the remainder of Japan's core area is continuing to develop. The most impressive growth has occurred around the western entry to the Inland Sea (the Strait of Shimonoseki), where **Kitakyushu**—a conurbation of five cities on northern Kyushu—constitutes the fourth Japanese manufacturing complex and primary economic region. Road and railway tunnels connect Honshu and Kyushu, but the northern Kyushu area does not have an urban-industrial equivalent on the Honshu side of the strait. The Kitakyushu conurbation includes Yawata, site of northwest Kyushu's (rapidly declining) coal mines. The first steel plant in Japan was built there on the basis of this coal; for many years it was the country's largest. The advantages of transportation here at the western end of the Inland Sea are obvious: no place in Japan is better situated to do business with Korea and China. As relations with mainland Asia expand, this area will reap many benefits. Elsewhere on the Inland Sea coast, the Hiroshima–Kure urban area has a manufacturing base that includes heavy industry. And on the coast of the Korea Strait, Fukuoka and Nagasaki are the principal centers—respectively, an industrial city and a center of large shipyards.

Only one major Japanese manufacturing complex lies outside the belt extending from Tokyo in the east to Kitakyushu in the west—the secondary region centered on Toyama on the Sea of Japan (East Sea). The advantage here is cheap power from nearby hydroelectric stations, and the cluster of industries reflects it: paper manufacturing, chemical factories, and textile plants. Figure 9-23 also gives an inadequate picture of the variety and range of industries that exist throughout Japan, many of them oriented to local (and not insignificant) markets. Thousands of manufacturing plants operate in cities and towns other than those shown on this map, even on the cold northern island of Hokkaido, which is connected to Honshu by the Seikan rail tunnel, the world's longest, beneath the treacherous Tsugaru Strait.

The map of Japan's areal organization shows that four primary regions dominate its core area. Each of them is primary because each duplicates to some degree the contents of the others. Each contains iron and steel plants, is served by one major port, and lies in or near a large, productive farming area. What the map does not show is that each also has its own external connections for the overseas acquisition of raw materials and sale of finished products. These linkages may even be stronger than those among the four internal regions of the core area. Only for Kyushu and its coal have domestic raw materials affected the nature and location of heavy manufacturing, and these resources are almost depleted. In the structuring of the country's areal functional organization, therefore, more than just the contents of Japan itself is involved. In this respect Japan is not unique: all countries that have exchange-type organization must adjust their spatial forms and functions to the external interconnections required for progress. But for few countries is this truer than Japan.

East and West, North and South

The previous discussion shows that Japan has its own Pacific Rim: almost all of Honshu's economic expansion is occurring on the east coast. Japanese geographers often call this eastern zone **Omote Nippon** ("Front Japan"), the Pacific-facing, megalopolitan, high-tech, dynamic sector that embodies the modern economic superpower—as opposed to **Ura Nippon** ("Back Japan"), the comparatively underdeveloped periphery of unkempt farms and poor villages bordering the Sea of Japan (East Sea) to the west. In a country that needs food as much as Japan does, the derelict state of farmlands in *Ura Nippon* is of much concern to the country's planners.

Tokyo, at the center of one of the largest metropolises in the world, continues to change. Land-filling and bridge-building in the bay continue; skyscrapers sprout amid low-rise neighborhoods in this earthquake-prone area; traffic congestion worsens. The red-painted Tokyo Tower, a beacon in this part of the city, was modeled on the Eiffel Tower in Paris but, as a billboard at its base announces, is an improvement over the original: lighter steel, greater strength, less weight. Tokyo Bay, part of which can be seen from this vantage point, was the scene of one of history's most costly environmental disasters during the earthquake of 1923. A giant tsunami (seismic sea wave) swept up the bay from the epicenter even as landfills liquefied and buildings sank into the mud. The death toll exceeded 140,000; a much larger population than lived in Tokyo in 1923 is now at risk of a repeat. © Yann Arthus-Bertrand/Photo Researchers.

The congestion on Honshu is another concern. Efforts to link Hokkaido and Inland-Sea-facing Shikoku to the "mainland" have produced world-record tunnels and bridges such as the Seikan Tunnel and the Akashi Kaikyo Bridge (Fig. 9-23), designed to bring remote areas into the national sphere and to diffuse Honshu's economic primacy beyond the main island. This effort has had limited success, especially in Hokkaido, which constitutes nearly one-quarter of Japan's total area, but contains only about 5 percent of the national population. Farming remains the main pursuit on Hokkaido, producing wheat, dairy products, and meat; fishing and tourism also do well. But Hokkaido has not translated its improved linkages into competitive industries; locals will tell you that there is a "subsidy mentality" here, that Tokyo's investments have led to complacency rather than initiative. Only the Sapporo area, with its automobile parts and electronics factories, shares in the Honshu-led export boom.

Food and Population

Japan's economic modernization so occupies center stage that we can easily forget the country's considerable achievements in agriculture. Japan's planners, who are as interested in closing the food gap as in expanding industries, have created extensive networks of experiment stations to promote mechanization, optimal seed selection and fertilizer use, and information services to distribute knowledge about enhancing crop yields to farmers as rapidly as possible. Although this program has succeeded, Japan faces the unalterable reality of its stubborn topography: it simply lacks sufficient land to farm. Populous Japan has one of the highest physiologic population densities—2607 per square kilometer (6752 per square mile)—on Earth. (We discuss this concept on pp. 413–414.)

The Japanese go to extraordinary lengths to maximize their limited farming opportunities, making huge investments in research on higher-yielding rice varieties, mechanization, irrigation, terracing, fertilization, and other practices. More than 90 percent of farmland is assigned to food crops. Hilly terrain is given over to cash crops, such as tea and grapes (Japan has a thriving wine industry). Vegetable gardens ring all the cities.

But all this costs money, and Japan's food prices are high. Millions of Japanese have moved from the farms to the cities, depopulating the countryside, which needs farmers. Japan's political system gives underpopulated rural areas disproportionate influence, and produce prices are kept artificially high, all to induce farmers to stay on the land. It has not worked. Between 1920 and 2000, full-time farmers declined from 50 percent to under 4 percent of the labor force. Yet the policies persist. In the early 1990s,

American farmers proved that they could provide rice to Japanese consumers at one-sixth the price the Japanese were paying for home-grown rice. But the government resisted pressure to open its markets to imports.

With their national diet so rich in starchy foods like rice, wheat, barley, and potatoes, the Japanese need protein to balance it. Fortunately, they can secure enough of it, not by buying it abroad but by harvesting it from rich fishing grounds near the Japanese islands. With their customary thoroughness, the Japanese have developed a fishing industry that is larger than that of the United States or any of the long-time fishing nations of northwestern Europe, and it now supplies the domestic market with a second staple after rice. Although mention of the Japanese fishing industry brings to mind a fleet of ships scouring the oceans and seas far from Japan, most of this huge catch (about one-seventh of the world's annual total) comes from waters within a few dozen kilometers of Japan itself. Where the warm Kuroshio and Tsushima currents meet colder water off Japan's coasts, a rich fishing ground yields sardines, herring, tuna, and mackerel in the warmer waters and cod, halibut, and salmon in the seas to the north. Japan's coasts have about 4000 fishing villages, and tens of thousands of small boats ply the waters offshore to bring home catches that are distributed to local and city markets. The Japanese also practice *aquaculture*—the "farming" of freshwater fish in artificial ponds and flooded paddy (rice) fields, of seaweeds in aquariums, and of oysters, prawns, and shrimp in shallow bays. They are even experimenting with cultivating algae for their food potential.

When you walk the streets of Japan's cities, you will notice something that Japanese vital statistics are reporting: the Japanese are getting taller and heavier. Coupled with this change is evidence that heart disease and cancer are rising as causes of mortality. The reason seems to lie in the changing diets of many Japanese, especially the younger people. When rice and fish were the staples for virtually everyone, and the calories available were limited but adequate, the Japanese as a nation were among the world's healthiest and longest-lived. But as fast-food establishments diffused throughout Japan, and Western (especially American) tastes for red meat and fried food spread among children and young adults, their combined impact on the population was soon evident. Coupled with the heavy cigarette smoking that prevails in all the Asian countries of the Pacific Rim, this is changing the region's medical geography.

Japan's Pacific Rim Prospects

As the twenty-first century unfolds, Japan remains the economic giant on the western Pacific Rim—but a giant with an uncertain future. Japanese products still dominate world markets, Japanese investments span the globe, but in the late 1990s Japan's economy faltered. Its growth rate declined. Many companies scaled back their operations; some went bankrupt. The banking system proved to be vulnerable. The first Pacific Rim economic tiger was ailing.

Japan's problems stemmed from a combination of circumstances at home and abroad. Overseas, the collapse of Southeast Asian currencies and economies hurt Japanese investments there. From automobile plants to toy factories, long-profitable operations turned into liabilities. In Japan itself, financial mismanagement, long tolerable because of the booming economy, now took its toll. In addition, competition from still-healthy economies, notably from Taiwan but also from struggling South Korea, undercut Japanese goods on world markets.

In the long run, all this may prove to have been a temporary setback, but Japan is unlikely to regain its former economic dominance on the Pacific Rim. Japan's dependence on foreign oil is a potential weakness that could further depress its economy should international supply lines be disrupted. Western competition for Japan's markets is growing. The countless investments Japan made around the world during its time of plenty, from skyscrapers and movie studios to golf courses and hotels, have lost much of their value and must be sold at huge losses. While Japan reorganizes, American and European businesses buy its assets and take advantage of its problems.

Were you to make a field trip to Japan to study its economic geography, none of this might be immediately evident. The cities still bustle with activity, raw materials continue to arrive at the ports, container traffic still stirs the docks. Even in a downturn, this mighty economy outproduces all others on the western Pacific Rim. And, indeed, that observation is relevant to the future. Japan's infrastructure has the capacity to propel an economic rebound. But it needs new opportunities.

Politico-Geographical Dilemma

One of these opportunities lies in the Russian Far East and beyond, in Siberia. As we noted in Chapter 2, the Russian Far East lies directly across from northern Honshu and Hokkaido and is a storehouse of raw materials awaiting exploitation. For this, the Japanese are ideally located—but political relations with the former Soviet Union, and now with Russia, have stymied all initiatives (see box titled "The Wages of War"). Japan should be one of the major beneficiaries of any substantial oil reserves near Sakhalin (see map, p. 144), but again the necessary connections may not be possible. Japan also has a major potential opportunity in Korea, especially if a unification process takes hold there. But relations with Korea continue to be eroded by Korean memories of Japanese behavior during World War II and by Japan's refusal to acknowledge its misdeeds.

The Wages of War

JAPAN AND THE former Soviet Union never signed a peace treaty to end their World War II conflict. Why? There are four reasons, and all of them are on the map just to the northeast of Japan's northernmost large island, Hokkaido (Fig. 9-22). Their names are Habomai, Shikotan, Kunashiri, and Etorofu. The Japanese call these rocky specks of the Kurile Island chain their "Northern Territories." The Soviets occupied them late in the war and never gave them back to Japan. Now they are part of Russia, and the Russians have not given them back either.

The islands themselves are no great prize. During World War II, the Japanese brought 40,000 forced laborers, most of them Koreans, to mine the minerals there. When the Red Army overran them in 1945, the Japanese were ordered out, and most of the Koreans fled. Today the population of about 50,000 is mostly Russian, many of them members of the military based on the islands and their families. At their closest point the islands are only 5 kilometers (3 mi) from Japanese soil, a constant and visible reminder of Japan's defeat and loss of land. Moreover, territorial waters bring Russia even closer, so that the islands' geostrategic importance far exceeds their economic potential.

Attempts to settle the issue have failed. In 1956, Moscow offered to return the tiniest two, Shikotan and Habomai, but the Japanese declined, demanding all four islands back. In 1989, then-Soviet President Mikhail Gorbachev visited Tokyo in the hope of securing an agreement. The Japanese, it was widely reported, offered an aid-and-development package worth U.S. $26 billion to develop Russia's eastern zone—its Pacific Rim and the vast resources of the eastern Siberian interior. This would have begun the transformation of Russia's Far East, stimulated the ports of Nakhodka and Vladivostok, and made Russia a participant in the spectacular growth of the western Pacific Rim.

But it was not to be. Subsequently, Russian presidents Yeltsin and Putin also were unable to come to terms with Japan on this issue, facing opposition from the islands' inhabitants and from their own governments in Moscow. And so World War II, more than 60 years after its conclusion, continues to cast a shadow over this northernmost segment of the western Pacific Rim.

Population Change

All this must be seen against the backdrop of Japan's changing society. The population of 128 million is aging rapidly, and population geographers expect it to rise only slightly before the end of this decade, when it will begin to decline. This decline will increase quite rapidly after 2025, to about 100 million in 2050 and only 67 million in 2100. Already, for some time, Japan has faced a labor shortage (although the recent economic downturn produced rising unemployment, a novelty in the postwar period). Over the long term, Japan will need millions of immigrants to keep its economy going, but for culturally homogeneous, ethnically conscious Nippon, which has historically resisted immigration, this will be a wrenching reality. Still, Japanese leaders will have no choice. The country's "graying" will strain welfare systems and tax burdens.

Modernization, meanwhile, is further stressing Japan's society. The Japanese have a strong tradition of family cohesion and veneration of their oldest relatives, but this tradition is breaking down. Younger people are less willing than their parents and grandparents to accept overcrowded housing, limited comforts, and little privacy. In Japan, for all its material wealth, family homes tend to be small, cramped, often flimsily built, and sometimes lacking the basic amenities other high-income countries consider indispensable. With their disposable income, the Japanese during the boom economy came to rank among the world's most-traveled tourists, and, having seen how Americans and Europeans of similar income levels live, they returned home dissatisfied.

As we just noted, Japan remains one of the most uniform, culturally homogeneous societies in today's interactive world. The Chinese closed their doors to foreigners intermittently; the Japanese have done so far more effectively. Other than the dwindling Ainu, there are no indigenous minorities. Koreans were brought to Japan during World War II to serve as forced laborers, but otherwise Japanese residency and citizenship imply an almost total ethnic sameness. During periods of labor shortage, the notion to invite foreign workers would not be entertained. When the United States, after the Vietnam War, sought new homes for hundreds of thousands of Vietnamese refugees, Japan accepted just a few hundred after much negotiation. Such jealously guarded ethnic-cultural uniformity is unlikely to be viable in the globalizing world of the twenty-first century.

Crossroads

Japan's future in the geopolitical framework of the western Pacific is also uncertain. More than six decades after the end of World War II, tens of thousands of American forces still are based on Japanese soil. China and South Korea accept this arrangement, which constrains Japan's rearming. But in Japan, where nationalism is rising, domestic military preparedness is a growing political issue. Over the past ten years, however, Japan's sense of security was shaken by developments in communist North Korea, which in 1998 fired a missile across Honshu into the Pacific Ocean and was suspected of developing nuclear weapons capacity; by admissions from North Korea that it had kidnapped Japanese citizens from streets and beaches and forced them to teach Japanese to North Korean spies; and by the appearance of armed North Korean vessels in Japanese territorial waters. In 2003, the Japanese parliament approved legislation providing for the participation of Japanese troops in the U.S.-led intervention in Iraq, a move that contradicted public opinion as reflected by surveys but strengthened American-Japanese government ties. Japan also joined the United States and three other countries (China, Russia, and South Korea) to negotiate an end to North Korea's nuclear threat.

Japan's economic success and political stability may have been facilitated by enlightened U.S. postwar policy, but the future will be different. In 2002, China overtook the United States as the leading exporter to the Japanese market; today, Japanese companies compete vigorously against American corporations in China's Pacific Rim. Japan continues to have image problems among Chinese and Koreans (a Japanese soccer victory in Beijing in 2004 led to a near-riot), but at the highest levels cooperation is growing. On the other hand, Japan and China will become fierce competitors over critical energy supplies. In 2005, the Japanese tried to outbid the Chinese over the routing of oil and natural gas pipelines from Russia: the Chinese wanted to position these pipelines to link to terminals in its Northeast, whereas Japan wanted to route them to terminals on the Sea of Japan (East Sea) at Nakhodka, directly across from northern Honshu. Japan today is the world's second-largest oil consumer, but China's consumption is rising rapidly and heightened competition for Russian exports is inevitable. This will obviously have an effect on Japanese-Chinese relations, political as well as economic, in the decades ahead.

Japan, therefore, is a pivotal nation on the western Pacific Rim, an economic superpower that stands at social and political crossroads whose future will profoundly influence not only East Asia and the Pacific, but the world as a whole.

KOREA

On the Asian mainland, directly across the Sea of Japan (East Sea), lies the peninsula of Korea (Fig. 9-25), a territory about the size of the State of Idaho, much of it moun-

FIGURE 9-25   © H. J. de Blij, P. O. Muller, and John Wiley & Sons, Inc.

of the 38th parallel) to the forces of the Soviet Union and South Korea to those of the United States. In effect, Korea traded one master for two new ones. The country was not reunited for the rest of the century because North Korea immediately fell under the communist ideological sphere and became a dictatorship in the familiar (but in this case extreme) pattern. South Korea, with massive American aid, became part of East Asia's capitalist perimeter. Once again, it was the will of external powers that prevailed over the desires of the Korean people.

War and Aftermath

In 1950, North Korea sought to reunite the country by force and invaded South Korea across the 38th parallel. This attack drew a United Nations military response led by the United States. Thus began the devastating Korean War (1950–1953) in which North Korea's forces pushed far to the south only to be driven back across its own half of Korea almost to the Chinese border. Then China's Red Army entered the war and drove the UN troops southward again. A cease-fire was arranged in 1953, but not before the people and the land had been ravaged in a way that was unprecedented even in Korea's violent past. The cease-fire line shown in Figure 9-25 became a heavily fortified *de facto* boundary. For five decades, until very recently, virtually no contact of any kind occurred across this border, which continues to divide families and economies alike.

As a result, the Jakota Triangle, as a region of East Asia, must include South Korea but exclude North Korea (see box titled "High-Risk North Korea"). Times may change, as they did in formerly divided Germany, but by the turn of this century South Korea, with 45 percent of Korea's land area but with two-thirds of the Korean population, had emerged as one of the economic tigers on the Pacific Rim. If it were an undivided country, Korea might have achieved even more because North and South Korea exhibit what geographers call **regional complementarity**. This con- **19** dition arises when two adjacent regions complement each

tainous and rugged, and containing a population of 73 million. Unlike Japan, however, Korea has long been a divided country, and the Koreans a divided nation. For uncounted centuries Korea has been a pawn in the struggles of more powerful neighbors. It has been a dependency of China and a colony of Japan. When it was freed from Japan's oppressive rule at the end of World War II (1945), the victorious Allied powers divided Korea for administrative purposes. That division gave North Korea (north

North Korea's communist regime has impoverished and starved its citizens and stifled freedoms, but in recent years its policies in the nuclear and missile arenas have become matters of international concern. By its own admission, North Korea has a (still-small) arsenal of nuclear weapons; the country also has tested missiles, one of which overflew Japan and plunged into the Pacific beyond. Diplomatic efforts to curb Pyongyang's nuclear ambitions, involving six nations most directly concerned, had not fully succeeded by mid-2007, even as satellite images showed the expansion of North Korea's nuclear-facility infrastructure. This is one of the country's older installations, the Yongbyon Nuclear Center north of the capital. As the map on page 513 shows, North Korea's nuclear and missile network extends across much of the country. © AP/Wide World Photos

cient; the country was stagnant. But huge infusions of aid, first from the United States and then Japan, coupled with the reorganization of farming and stimulation of industries (even if they lost money), produced a dramatic turnaround. Large feudal estates were parceled out to farm families in 3-hectare (7.5-acre) plots, and a program of massive fertilizer importation was begun. Production rose to meet domestic needs and in some years yielded surpluses.

The industrialization program was based on modest local raw materials, plentiful and capable labor, ready overseas (especially American) markets, and continuing foreign assistance. Despite corrupt dictatorial rule, political instability, and social unrest, South Korean regimes managed to sustain a rapid rate of economic growth that placed the country, by the late 1980s, among the world's top ten trading powers. To get the job done, the government borrowed heavily from overseas, and it controlled banks and large industries. South Korea's growth resulted from **state capitalism** rather than free-enterprise capitalism, but it had impressive results. South Korea became the world's leading shipbuilding nation; its automobile industry grew rapidly; and its iron and steel and chemical industries began to thrive. It has placed less emphasis, however, on smaller, high-technology industries, which are the strengths of other Pacific Rim economic tigers. So the future remains uncertain.

Three Industrial Areas

South Korea (population: 48.9 million) prospers today, and we can see its economic prowess on the map (Fig. 9-25). The capital, Seoul, with about 10 million inhabitants, now ranks among the world's megacities and anchors a huge industrial complex facing the Yellow Sea at the waist of the Korean Peninsula. Hundreds of thousands of farm families migrated to the Seoul area after the end of the Korean War (today only 18 percent of South Koreans remain on the land). They also moved to Busan, the nucleus of the country's second largest manufacturing zone, located on the Korea Strait opposite the western tip of Honshu. And the government-supported, urban-industrial drive here continues. Just 30 years ago, Ulsan City, 65 kilometers (40 mi) north of Busan along the coast, was a fishing center with perhaps 50,000 inhabitants; today its population exceeds a million, nearly half of them the families of workers in the Hyundai automobile factories and the local shipyards and docks.

The third industrial area shown in Figure 9-25, anchored by the city of Kwangju, has advantages of relative location that will spur its development, but it faces domestic problems. Kwangju lies at the center of the *Cholla* region, the southwest corner of South Korea (the southeast is called *Kyongsang*). For more than three decades beginning in the

other in economic-geographic terms. Here, North Korea has raw materials that the industries of South Korea need; South Korea produces food the North needs; North Korea produces chemical fertilizers farms in the South need.

After the mid-twentieth century, however, the two Koreas were cut off from each other, and they developed in opposite directions. North Korea, whose large coal and iron ore deposits attracted the Japanese and whose hydroelectric plants produce electricity that the South could use, carried on what limited trade it generated with the Chinese and the Soviets. South Korea's external trade links were with the United States, Japan, and Western Europe.

South Korea

In the early postwar years, there were few indications that South Korea would emerge as a major economic force on the Pacific Rim and, indeed, on the world stage. Over 70 percent of all workers were farmers; agriculture was ineffi-

High-Risk North Korea

THE KOREAN PENINSULA is physiographically well-defined, and its people—the Korean nation—have occupied it for thousands of years. But internal political division, foreign cultural intrusion, and colonial subjugation have created a turbulent history. Now Koreans are divided again, this time by ideology. Even as the capitalist South became one of the Pacific Rim's economic tigers, North Korea remained an isolated, stagnant state on a Stalinist-communist model abandoned long ago even in the former Soviet Union itself. For over a half-century, North Korea has been ruled by two men: Kim Il Sung, the "Great Leader," and his son Kim Jong Il, the "Dear Leader." Total allegiance to the dictators is ensured by an omnipotent police and military. Self-imposed isolation is North Korea's hallmark. The state controls all aspects of life in North Korea.

Economic policies under the communist regime have been disastrous. Per-capita incomes in the North (population: 23.5 million) are less than one-tenth of those in the South; poverty is rampant. Infant mortality is more that 8 times higher than in South Korea. Industrial equipment is 25 years out of date. Exports are negligible. When a combination of failed agricultural policies and environmental extremes created famine in the late 1990s and early 2000s, the regime resisted relief efforts, even from South Korea. As during the Maoist period in China, the world will never know the number of casualties.

North Korea did progress on one front: nuclear capability and associated weaponry. In 1999, it fired a test missile across northern Japan, heightening international concern. The North Koreans used their nuclear threat to negotiate a multibillion-dollar technical aid program to build reactors for peaceful purposes, with Japan and South Korea among the major contributors. But in late 2002, a North Korean source admitted that his country had indeed pursued a nuclear weapons program and was now in possession of "at least one" nuclear bomb. The North Koreans refused to allow visitors to inspect its nuclear facilities at Yongbyon, 100 kilometers (60 mi) due north of the capital, Pyongyang (photo p. 514). Other known nuclear installations are mapped in Figure 9-25.

President George W. Bush, in his 2002 State of the Union address, called North Korea part of the "axis of evil" to be dealt with as part of the War on Terror. But while Iraq was forced to admit UN weapons inspectors, North Korea, with 37,000 U.S. troops along its border with the South, remained exempt. Meanwhile, North Korea acknowledged having kidnapped Japanese youngsters from public places in Japan and having forced them to teach Japanese language and culture to North Korean spies. In 2002 this admission led to authorized visits to Japan by some of the survivors and a furor among the Japanese people.

Even as the "Dear Leader" was able to reject U.S. and Japanese demands that nuclear operations at Yongbyon be reviewed, he approved some small gestures toward his southern neighbor. Beginning in 2000, a few hundred members of families separated by the 1953 cease-fire line were allowed a few hours to meet each other. He also allowed the preliminary design of a railroad that would link North and South across the narrow demilitarized zone (DMZ). After authorizing the creation of a barbed-wire-enclosed "free-trade zone" in the far northeast at Rajin-Sonbong, where South Korean cruise ships could dock, Kim Jong Il approved another one in the northwest, at Sinuiji, in 2002. But when the Chinese billionaire who was to take charge of this free-trade zone was arrested for corruption in China, that plan fell apart.

Desperate North Koreans flee their country across the Chinese border and on the seas; in foreign countries they invade embassy compounds and plead for asylum. In 2007 the reunification of the two Koreas seemed far off; but so did German reunification in 1985, and then it happened a few years later. But the reunification of the two Koreas will entail far greater difficulties than that of the Germanies. North Korea is far worse off even than former communist East Germany; the cost will be staggering. South Koreans will face far greater sacrifices than West Germans did, with far fewer resources. Still, the boundary superimposed on the Korean nation is a leftover of the Cold War, and its elimination and the reintegration of North Korea into the international community is a laudable goal.

1960s, the army generals who ruled South Korea all came from Kyongsang, and through state control of banks and other monopolies they bestowed favors on Kyongsang companies while denying access to potential competitors in Cholla. The regime in Seoul also thwarted infrastructure development in Cholla, even to the point of impeding plans for improved railroad links between Kwangju and the capi-

tal. This regional friction led to protests in Cholla and violent repression by the military rulers, the aftermath of which is fading only slowly in this part of the now-democratic country. These historic discriminations seem to be ending as South Korea's democracy matures and regional differences subside. One reason for this has to do with infrastructural improvements: a new high-speed train system,

AMONG THE REALM'S GREAT CITIES . . . Seoul

© H. J. de Blij, P. O. Muller, and John Wiley & Sons, Inc.

SEOUL (9.6 MILLION), located on the Han River, is ideally situated to be the capital of all of Korea, North and South (its name means capital in the Korean language). Indeed, it served as such from the late fourteenth century until the early twentieth, but events in that century changed its role. Today, the city lies in the northwest corner of South Korea, for which it serves as capital; not far to the north lies the tense, narrow demilitarized zone (DMZ) that contains the cease-fire line with North Korea. That line cuts across the mouth of the Han, depriving Seoul of its river traffic. Its ocean port, Incheon, has emerged as a result (Fig. 9-25).

Seoul's undisciplined growth, attended by a series of recent accidents including the failure of a major bridge over the Han and the collapse of a six-story department store, reflects the unbridled expansion of the South Korean economy as a whole (photo p. 517), as well as the political struggles that carried the country from autocracy to democracy. Central Seoul lies in a basin surrounded by hills to an elevation of about 300 meters (1000 ft), and the city has sprawled outward in all directions, even toward the DMZ (see photo, p. 3). An urban plan designed in the early 1960s was overwhelmed by the immigrant inflow.

During the period of Japanese colonial control, Seoul's surface links to other parts of the Korean Peninsula were improved, and this infrastructure played a role in the city's later success. Seoul is not only the capital but also the leading industrial center of South Korea, exporting sizeable quantities of textiles, clothing, footwear, and (increasingly) electronic goods. South Korea thrives as an economic tiger on the Pacific Rim, and Seoul is its heart.

the Korea Train Express (KTX), now links Seoul with both Mokpo in the Cholla region and Busan in the Kyongsang region (Fig. 9-25), and there are plans to link Kwangju with Daegu, thereby establishing an interregional link in the south. South Koreans know what the bullet-train network did for Japan, and they expect the KTX to accelerate the integration of their country as well.

Pacific Rim Opportunities

Another sign of South Korea's political maturation is the government's proposal to move the capital from congested Seoul to an area about 150 kilometers (90 mi) to the south, near Taejon in Chungchong Province (Fig. 9-25). Leaders argue that moving the government to a more centralized location will be costly but will enhance efficiency, but as the plan gains momentum look for politicians and others to raise obstacles. This will ultimately have to be a majority decision; the relocation project is slated to begin by 2008, and time is growing short.

Meanwhile, South Korea has moved far beyond the days when giant corporate conglomerates (known as *chaebol*) conspired to suppress competition, controlled the banks, and dictated economic policy to the government. It was state capitalism at its most extreme, and it proved to have weaknesses similar to planned communist economies. After the system collapsed and South Korea proved its capacity to thrive in the competitive world of the Pacific Rim, the country became one of the region's economic tigers—not without economic problems, but with boundless potential. Seoul's metropolitan economy has become a high-tech hub; the adjacent port city of Incheon is now spawning Songdo, a special economic zone designed to rival Shanghai's Pudong. China has surpassed the United States as South Korea's leading trade partner, and Songdo lies only about 300 kilometers (200 mi) across the Yellow Sea from the Chinese mainland—symbolizing the new century as well as the new Pacific Rim era that is transforming all countries in this realm.

FROM THE FIELD NOTES

"South Korea was one of the four 'economic tigers' marking the upsurge of the western Pacific Rim in the 1980s and 1990s, its products ranging from computers to cars selling on world markets and its GNI rising rapidly. Walking through the prosperous central business district of globalizing Seoul you see every luxury name brand on display amid plentiful evidence of wealth and well-being. But you also see evidence of the survival of traditions long lost elsewhere, in the huge number of bookstores (South Koreans still read voraciously), in the stores selling traditional as well as modern musical instruments, in the school uniforms. And in other ways: when I got on a bus with all seats taken, a boy got up and offered me his place." © H. J. de Blij

Concept Caching

www.conceptcaching.com

TAIWAN

The third component of the region we have called the Jakota Triangle is another, and more durable, success story on the Pacific Rim: Taiwan. Along with the Penghu (Makung) Islands in the Taiwan Strait and the islands of Jinmen Dao (Quemoy) near Xiamen and Matsu Tao (Matsu) near Fuzhou at the coast of mainland China, Taiwan is regarded by Beijing's communist regime as a Chinese province temporarily "wayward," that is, disobedient. After the reunification of Portuguese Macau with China in 1999, Taiwan remained the only major unresolved territorial issue in China's postcolonial, postwar sphere.

Peoples and Rulers

Beijing's claim to Taiwan has strengths as well as weaknesses. Han people settled on Taiwan, but the island had been settled thousands of years earlier by Malay-Polynesian groups who had split into mountain (east) and plains (west) inhabitants. When the Chinese emigrated from famine-stricken Fujian Province to Taiwan in the seventeenth century, they soon displaced and assimilated the plains people, but the mountain aborigines held out until the Japanese colonial invasion. Formal Chinese control began in 1683 when the expansionist Qing (Manchu) Dynasty made Taiwan part of Fujian Province. Rice and sugar became important exports to the mainland, and by the 1840s Taiwan's population was an estimated 2.5 million. In 1886 the island was declared a province of China, but by then the British had established two treaty ports there, the French had blockaded it, and the Japanese had sent expeditions to it. Nine years later, in 1895, China ceded Taiwan to Japan in the treaty settlement of the Sino-Japanese War, and it became a Japanese colony.

When China's aging Manchu Dynasty was overthrown following the rebellion of 1911, a Nationalist government in 1912 proclaimed the Republic of China (ROC), led by Sun Yat-sen. Naturally, this government wanted to oust all colonialists, not only Europeans but also Japanese, and not just from the mainland but from Taiwan as well. On Taiwan, however, the Japanese held on, using the island as a source of food and raw materials as well as a market for Japanese products. To ensure all this, the Japanese launched a prodigious development program involving road and railroad construction, irrigation projects, hydroelectric schemes, mines (mainly for coal), and factories. Farmlands were expanded and farming methods improved. While mainland China was engulfed in conflict, Taiwan remained under Japanese control. Not until 1945, when Japan was defeated in World War II, did Taiwan become an administrative part of China again. But only briefly. As we noted earlier, the Nationalists, under the leadership of Chiang Kai-shek, now faced the rising tide of communism on the mainland, and the ROC government fought a losing war. In 1949, the flag of the Republic of China was taken to a last stronghold: offshore Taiwan. There, Chiang and his followers, with their military might, weapons, and wealth taken from the mainland, established what they (and the world, led by the United States) proclaimed as the legitimate government of all China. On the island, the Nationalists, helped by the United States, began a massive reconstruction of the war-damaged infrastructure. In the international arena, the ROC represented the country; in the still-young United Nations, the Nationalists of Taiwan occupied China's seat. On the huge mainland, however, it was the communists in Beijing who ruled and called their country the People's Republic of China (PRC). Thus was born the two-China dilemma which still confronts the international community. When the ROC was ousted from the United Nations in 1971 to make way for the PRC, legitimacy was also transferred.

Thriving Economic Tiger

All this might have remained an internal dispute but for two major developments: Taiwan's rise as an economic tiger on the Pacific Rim, coupled with remarkable strides toward democracy made in the ROC; and the emergence on the world political and economic stage of communist-ruled China following decades of isolation. Today, the PRC's power is growing rapidly, and Taiwan's heirs to the ROC face an uncertain future; but it is the ROC, not the PRC, that has managed to combine economic success with democratization.

Taiwan, as Figure 9-26 shows, is not a large island. It is smaller than Switzerland but has a population much larger (22.9 million), most of it concentrated in an arc lining the western and northern coasts. The Chungyang Mountains, an area of high elevations (some over 3000 meters [10,000 ft]), steep slopes, and dense forests, dominate the eastern half of the island. Westward, these mountains yield to a zone of hilly topography and, facing the Taiwan Strait, a substantial coastal plain. Streams from the mountains irrigate the paddyfields, and farm production has more than doubled since 1950 even as hundreds of thousands of farmers left the fields for work in Taiwan's expanding industries.

Today, the lowland urban-industrial corridor of western Taiwan is anchored by the capital, Taipei (Taibei), at the island's northern end and rapidly growing Kaohsiung (Gaoxiong) in the far south.* The Japanese developed Chilung (Jilong), Taipei's outport, to export nearby coal, but now the raw materials flow the other way. Taiwan imports raw cotton for its textile industry, bauxite (for aluminum) from Indonesia, oil from Brunei, and iron ore from Africa. Taiwan has a developing iron and steel industry, nuclear power plants, shipyards, a large chemical industry, and modern transport networks. Increasingly, however, Taiwan is exporting products of its high-technology industries: personal computers, telecommunications equipment, and precision electronic instruments. Taiwan has enormous brainpower, and many foreign firms join in the research and development carried on in such places as Hsinchu (Xinzhu) in the north, where the government has helped establish a technopole centered on the microelectronics and personal computer industries. In the south, the "science city" of Tainan specializes in microsystems and information technology.

The establishment of China's Special Economic Zones has gone a long way toward solving the economic problems arising from the political discord between China and Taiwan. In the almost-anything-goes environment of the SEZs, Taiwanese business owners are permitted to build or buy factories just as "real" foreigners are, and as we noted earlier, the Xiamen SEZ was situated directly across from Taiwan for just this purpose. As a result, thousands of Taiwanese companies now operate in China even as political tensions continue.

In some ways, Taiwan's emergence as an economic tiger on the Pacific Rim is even more spectacular than Japan's. True, Taiwan received much Western assistance, but it got out of debt faster than anyone expected. Today, annual per-capita income exceeds U.S. $16,000, which is higher than that in many European countries. Taiwan's trade surplus has yielded tens of billions of dollars in reserves, which is being used to further its own development and to invest overseas. Taiwan, for example, is the largest foreign investor in the economy of Vietnam. But although Taiwan weathered the regional economic crisis of the late 1990s rather well, the new century has been problematic.

FIGURE 9-26 © H. J. de Blij, P. O. Muller, and John Wiley & Sons, Inc

---

*The Taiwanese have retained the old Wade-Giles spelling of place names. China now uses the *pinyin* system (see box on p. 454). Names in parentheses are written according to the pinyin system.

Chinese Rule or Independence for Taiwan?

SHOULD CHINA RULE TAIWAN?

"I am a graduate student in political geography here at East China Normal University, and I've been listening to your lecture about Chinese-American relations and the disagreement we have over Taiwan. You talk about this as though you have a right to stand between two parts of China, and with the power you have right now, I guess you can. You said that you hoped that mainland Chinese would become more 'reasonable' in their attitude toward the island province, but let me (as my professor would say) erect a hypothetical.

"Imagine that America wakes up one morning and discovers that a band of thugs have flown several 747s, loaded with weapons of all kinds, including those of mass destruction ready for use, into the San Juan, Puerto Rico airport. After taking the airport amid many casualties, they move into the city and enter administration buildings, threatening mass annihilation unless the local government resigns. By evening, Puerto Rico is under their control, American and Puerto Rican flags have been lowered and replaced by the banner of an 'independent' Puerto Rico.

"In Washington, your Congress meets and, realizing that the entire population of the island is at risk, makes the only move it can under the circumstances: it orders your armed forces to assemble a fleet to head for Puerto Rico, to blockade the island so as to create a basis for negotiation and a platform for possible military action. The fleet assembles near Miami and heads eastward through the channel between the Bahamas and Cuba.

"A day before reaching Puerto Rican waters, the path of the hastily assembled American fleet is blocked by a larger fleet from China, which has sailed from its base at Colón near the northern end of the Panama Canal. The commander of the American fleet communicates with his Chinese counterpart. 'We cannot let you pass,' states the Chinese admiral. 'We have not yet determined whether the invaders of Puerto Rico were invited by the people there to liberate them. We may have common ground with them. Please avoid an unnecessary conflict by going back to your bases.'

"How, sir, do you suppose the American people would feel toward China if a version of this scenario took place? They would be outraged at China's interference in a domestic matter. They would order your navy to attack. Well, we in China have had something like this happen with Taiwan. Thugs took it over. American aircraft carriers appeared in the Taiwan Strait. We've 'avoided conflict' for many decades, but that won't last forever. One way or another, Taiwan will submit to Beijing's rule. The Chinese people demand it."

Regional **ISSUE**

TAIWAN SHOULD BE INDEPENDENT

"Instead of harping constantly on the 'reunification' issue, why don't Beijing's rulers get their own house in order first? Their political system is straight out of the discarded past, they're apparently allergic to democracy, and they can't seem to accept or accommodate any disobedience or dissent. I represent a Taiwanese high-tech corporation and I fly frequently between Taipei and Hong Kong and visit the so-called People's Republic via that route—and I can tell you, the Chinese could learn a lot about democracy, openness, freedom, transparency, and honesty from the Taiwanese. Politically, that 'One Country, Two Systems' promise in Hong Kong is eroding as we speak. Economically, the Chinese ought to acknowledge that a lot of the 'Chinese miracle' on the Pacific Rim is actually a 'Taiwanese miracle.' Without our money, skills, discipline, and rigor there wouldn't be much of a miracle, especially if you also subtract what Southeast Asia's Chinese have contributed.

"Of course, Taiwan is a great rallying point for Beijing's autocrats. Gives them a chance to rail against the Western world, against America, against social excesses (we even have freedom of religion in Taiwan, imagine that!), against international conspiracies. But the truth is that the voters of Taiwan long ago ousted the 'thugs' who took us over, forged a thriving economy and a vibrant democracy, and made this island an example of what might have been across the Taiwan Strait.

"Could I say a few words about the 'historic' justification for China's claim to Taiwan? First of all, this island was peopled by Southeast Asian groups long before any Chinese arrived here, so if it's a matter of who got here first, we're last on the list. Beijing didn't integrate Taiwan politically until the 1680s, after groups of hungry citizens started to cross the Strait. That's pretty late in China's vaunted 5000-year history. Taiwan didn't become a province until 1886, which lasted exactly nine years. From 1895 to 1945 Taiwan belonged to Japan as part of the war settlement, and then Beijing took it back only to lose it again four years later to the 'thugs' under Chiang Kai-shek. That makes 13 years as part of the mainland system and 103 years of something else.

"For a country that has expanded and contracted for millennia, has gained and lost territory from Tibet to Taiwan, you would think that China could live with an ethnically compatible, economically successful, politically stable, democratic and independent neighbor. Those stiff bureaucrats in Beijing should have better things to do than threaten force to 'reincorporate' their exemplary neighbor."

Vote your opinion at www.wiley.com/college/deblij

The U.S. and Japanese markets for high-tech exports, particularly computer chips, shrank markedly. Thousands of firms long profitable in Taiwan had to move their operations to China's SEZs, where labor costs were much lower, to stay in business. This reduced economic growth from 6.3 percent in 2000 to 1.1 percent in 2001 and subsequently into negative territory, while unemployment, basically unheard of in Taiwan, shot up to nearly 5 percent. All this happened while mainland China's economy continued to grow at an annual rate that exceeded 5 percent. It was a shock to Taiwan's confidence as an East Asian economic tiger.

Geopolitical Concerns

Taiwan's political future remains clouded. The Nationalists who made the island their base in 1949 established a Republic of China that was authoritarian, but the ROC has evolved into a democratic state. This was not an easy process, but when Taiwan's own democracy movement culminated in 1991, it produced not a massacre, but free elections for a new parliament. Taiwan's leaders now proclaim that the direct and popular election of their president in 1996 was the first such vote not just in Taiwan or the ROC, but in the 4000-year history of the Chinese state. Taiwan thus presents itself as an alternative model to the communist-ruled PRC, proving to the world that democracy is compatible with Chinese culture and tradition. Nevertheless, the international community does not recognize Taiwan as an independent state, and a substantial segment of the Taiwanese electorate itself opposes any move toward outright sovereignty. Beijing dictates that Taiwan is off limits to U.S. cabinet members. In 1995, when the United States permitted the ROC's president to visit the American university of which he is a graduate, so that he might attend a class reunion, relations between Washington and Beijing

plummeted. In 2000, when Taiwan's voters elected a president who, during the campaign, had proclaimed Taiwan's "equality" with mainland China, Beijing placed 600 missiles aimed at Taiwan on its shores opposite the island. The president then took to calling his country "Taiwan" rather than the ROC, and urged his fellow Taiwanese to do so as well; he also took a vote on the question of Taiwan's right to hold a referendum on independence.

The political geography of Taiwan, therefore, is at odds with its economic geography. The ROC stood firm against communism in the 1960s and 1970s; but its long-term ally, the United States, was required to publicly commit itself to a one-China policy, the one being the PRC, the communist giant. Even as goods and capital flow between Taiwan and the mainland, Beijing engages in threatening acts (such as nearby missile tests) when democracy in the ROC takes another step forward.

Today, Taiwan's political options are as limited as its economic horizons are endless. The communist Chinese have made it clear that a move toward independence, based on the likely outcome of any referendum, would lead to intervention. Though militarily well prepared, Taiwan could not long hold off the PRC's enormous armed forces. In any case, independence for Taiwan has philosophical disadvantages for the leaders of the ROC, as it would diminish their claim to legitimacy as the alternative model for a reunified China. In practical terms, conflict on the Pacific Rim would be disastrous for both sides, whatever course it were to take.

And so Taiwan's future hangs in the balance, a wayward province with national aspirations, a quasi-state with global connections, a test of Chinese capacity for compromise and accommodation (see the Issue Box on p. 519 entitled "Chinese Rule or Independence for Taiwan?"). In this era of devolutionary forces, the fate of Taiwan will be a harbinger of the world order of the twenty-first century.

What You Can Do

RECOMMENDATION: Join a Research Team! Many of the faculty members of your college or university, the professors you meet in class, have another life: they perform research and publish their findings in their discipline's scholarly literature. Some faculty may be working every summer at a base in some remote locale in a distant land; others will be using the library and the Internet. Often, when you read their publications, you will note that these are joint ventures (some articles in academic journals may have three, four, or even more authors). Such long-term research is likely to require lots of labor. If a professor's course or field is of special interest to you, ask her or him whether she or he is doing research in this particular area, and if so whether you can be of any assistance. Such help may start out in the library, but soon you may be working on interview teams or, who knows, joining a team of scientists overseas. Make your interest known, and an opportunity is likely to arise. You will learn things you would never learn in a classroom!

GEOGRAPHIC CONNECTIONS

1 East Asia is a realm of great cities, including many of the largest in the world. Names like Tokyo, Beijing, and Shanghai represent the megacity phenomenon like few others. But look at East Asia in Table G-1, and you will find that the countries of this realm display a range of urbanization levels from high (Japan and South Korea are in the lead) to low (China, for all its burgeoning cities, remains the least urbanized state in the realm). What geographic factors help explain these variable levels of urbanization among East Asia's countries? Why does China have more cities in the one-million-plus population category than any other country in the realm and yet the lowest percentage of urbanization? Do you expect the urban geography of this realm to change, and if so, where and how?

2 Consider the core-periphery and globalization concepts (discussed in the Introduction) in the context of the East Asian realm. If you were asked to define what region or regions of East Asia presently form part of the global "core," what criteria would you use and where would this core component appear on the map? In those areas that form part of the periphery, what has been the impact of globalization? Has globalization drawn these areas closer to the core, benefiting them, or has the economic gap widened between core and periphery in East Asia?

Geographic Literature on East Asia: The key introductory works on this realm were authored by Burks, Chow, Hodder, Hsieh & Lu, Karan, Pannell & Ma, Pannell & Veeck, Smith, Veeck et al., and Weightman. These works, together with a comprehensive listing of noteworthy books and recent articles on the geography of China, Japan, and other components of East Asia, can be found in the *References and Further Readings* section of this book's website at **www.wiley.com/college/deblij.**

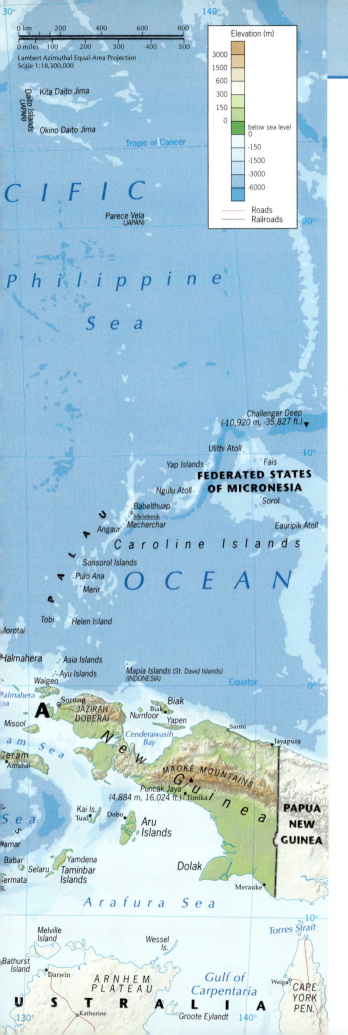

FIGURE 10-1 © H. J. de Blij, P. O. Muller, and John Wiley & Sons, Inc.

Where were these photos taken? Find out at **www.wiley.com/college/deblij**

In This Chapter

Photos: © H. J. de Blij.

SOUTHEAST ASIA . . . the very name roils American emotions. Here the United States owned its only major colony. Here American forces triumphed over Japanese enemies. Here the United States fought the first war it ever lost. Here Islamic terrorists killed hundreds, drawing American troops and intelligence agents into the fray.

Here, too, American companies invested heavily when the Pacific Rim's economic opportunities beckoned and dormant economies turned into potential economic tigers. Southeast Asia, once remote and stagnant, has taken center stage in our globalizing world.

Southeast Asia's human drama plays itself out on a volatile, unsteady physical stage. Repeatedly in human history, gigantic volcanic eruptions here have had global consequences; one of them, the Toba eruption about 72,000 years ago, plunged the planet into darkness and, scientists conclude, came close to extinguishing humankind. In 1815, people around the world woke up one April morning to skies that darkened as the day went on, the result of ash and soot propelled into the stratosphere by the eruption of the Indonesian volcano named Tambora. A year later, that continuing eruption had shrouded the Earth in a blanket so thick that there literally was no summer. Crops would not ripen, and hunger and starvation afflicted millions. The human population at the time was just 1 billion; consider what would happen today under similar circumstances, with the human population nearly seven times as large.

Those globe-girdling Indonesian eruptions killed many hundreds of thousands of Indonesians living in the shadows of the realm's great volcanoes; at least another 30,000 perished during Krakatau's relatively minor eruption in 1883. But volcanic activity is only one of the hazards of this realm's physical geography. Here churns some of the

MAJOR GEOGRAPHIC QUALITIES OF

Southeast Asia

1. Southeast Asia extends from the peninsular mainland to the archipelagos offshore. Because Indonesia controls part of New Guinea, its functional region reaches into the neighboring Pacific geographic realm.

2. Southeast Asia, like Eastern Europe, has been a shatter belt between powerful adversaries and has a fractured cultural and political geography shaped by foreign intervention.

3. Southeast Asia's physiography is dominated by high relief, crustal instability marked by volcanic activity and earthquakes, and tropical climates.

4. A majority of Southeast Asia's more than half-billion people live on the islands of just two countries: Indonesia, with the world's fourth-largest population, and the Philippines. The rate of population increase in the Insular region of Southeast Asia exceeds that of the Mainland region.

5. Although the overwhelming majority of Southeast Asians have the same ancestry, cultural divisions and local traditions abound, which the realm's divisive physiography sustains.

6. The legacies of powerful foreign influences, Asian as well as non-Asian, continue to affect the cultural landscapes of Southeast Asia.

7. Southeast Asia's political geography exhibits a variety of boundary types and several categories of state territorial morphology.

8. The Mekong River, Southeast Asia's Danube, has its source in China and borders or crosses five Southeast Asian countries, sustaining tens of millions of farmers, fishing people, and boat owners.

9. The realm's giant in terms of territory as well as population, Indonesia, has not asserted itself as the dominant state because of mismanagement and corruption; but Indonesia has enormous potential.

Earth's most mobile crust, its colliding and subducting tectonic plates creating a steady roar of quakes and tremors (refer back to Figs. G-4 and G-5). Earthquakes with epicenters on land claim many victims, but even those occurring beneath the sea pose terrible risks. On December 26, 2004 one of those submarine earthquakes, in a subduction zone west of the Indonesian island of Sumatera,* generated a **tsunami** that thrust walls of water onto Indian Ocean shores near and far, sweeping away entire towns and villages as well as their populations and killing as many as 300,000 people from nearby Sumatera to distant Somalia in Africa's Horn. Southeast Asia's fragmented map is the product of geologic forces and human risk-taking.

Defining the Realm

Southeast Asia is a realm of peninsulas and islands, a corner of Asia bounded by India on the northwest and China on the northeast (Fig. 10-1). Its western coasts are washed by the Indian Ocean, and to the east stretches the vast Pacific. From all these directions, Southeast Asia has been penetrated by outside forces. From India came traders; from China, settlers; from across the Indian Ocean, Arabs to engage in commerce and Europeans to build empires; and from across the Pacific, the Americans. Southeast Asia has been the scene of countless contests for power and primacy—the competitors have come from near and far.

Southeast Asia's geography in some ways resembles that of Eastern Europe. It is a mosaic of smaller countries on the periphery of one of the world's largest states. It has been a **buffer zone** between powerful adversaries. It is a **shatter belt** in which stresses and pressures from without and within have fractured the political geography. Like Eastern Europe, Southeast Asia exhibits great cultural diversity. It is a realm of hundreds of cultures, numerous languages and dialects, and several major religions.

*As in Africa, names and spellings have changed with independence. In this chapter we will use the contemporary spellings, except when we refer to the colonial period. Thus Indonesia's four major islands are Jawa, Sumatera, Kalimantan (the Indonesian part of Borneo), and Sulawesi. The Dutch called them Java, Sumatra, Dutch Borneo, and Celebes, respectively.

REGIONAL FRAMEWORK

The Southeast Asian realm borders South Asia and East Asia on land and the Pacific and Austral realms at sea (Fig. 10-2). Note that one political entity that forms part of it, Indonesia, extends beyond the realm's borders. The easternmost province of Indonesia is the western half of the Pacific island of New Guinea, where indigenous cultures are Papuan, not Indonesian. Today Indonesia rules what is in effect a Pacific island colony called Papua. We discuss this territory as part of our assessment of Indonesia, but we will also consider all of New Guinea as a component of the Pacific Realm in Chapter 12.

Because the politico-geographical map (Fig. 10-2) is so complicated, it should be studied attentively. One good way to strengthen your mental map of this realm is to follow the mainland coastline from west to east. The westernmost state in the realm is Myanmar (called Burma before 1989 and still referred to by that name), the only country in Southeast Asia that borders both India and China. Myanmar shares the "neck" of the Malay Peninsula with Thailand, heart of the *Mainland* region. The south of the peninsula is part of Malaysia—except for Singapore, at the very tip of it. Facing the Gulf of Thailand is Cambodia. Still moving generally eastward, we reach Vietnam, a strip of land that extends all the way to the Chinese border. And surrounded by its neighbors is landlocked Laos, remote and isolated. This

FIGURE 10-2   © H. J. de Blij, P. O. Muller, and John Wiley & Sons, Inc.

leaves the islands that constitute *Insular* Southeast Asia: the Philippines in the north and Indonesia in the south, and between them the offshore portion of Malaysia, situated on the largely Indonesian island of Borneo. Also on Borneo lies the ministate of Brunei, small but, as we will see, important in the regional picture. And finally, a new state has just appeared on the map in the realm's southeastern corner: East Timor, a former colony of Portugal annexed by Indonesia in 1976 and released to United Nations supervision in 1999. After a transition period, East Timor became an independent state in 2002 (its government wants the country to be known as *Timor-Leste*, the Portuguese way of writing it).

No Dominant State—Yet

No dominant state has arisen among the 11 countries of Southeast Asia. The largest country in terms of population as well as territory, Indonesia, has as-yet unrealized potential to overshadow the rest but not in the foreseeable future. Neither did any single, dominant core of indigenous culture develop here as it did in East Asia. In the river basins and on the plains of the mainland, as well as on the islands offshore, a flowering of cultures produced a diversity of societies whose languages, religions, arts, music, foods, and other achievements formed an almost infinitely varied mosaic—but none of those cultures rose to imperial power. The European colonizers forged empires here, often by playing one state off against another; the Europeans divided and ruled.

Out of this foreign intervention came the modern map of Southeast Asia; only Thailand (formerly Siam) survived the colonial era as an independent entity. Thailand was useful to two competing powers, the French to the east and the British to the west. It was a convenient buffer, and while the colonists carved pieces off Thailand's domain, the kingdom endured.

Indeed, the Europeans accomplished what local powers could not: the formation of comparatively large, multicultural states that encompassed diverse peoples and societies and welded them together. Were it not for the colonial intervention, it is unlikely that the 17,000 islands of far-flung Indonesia would today comprise the world's fourth-largest country in terms of population. Nor would the nine sultanates of Malaysia have been united, let alone with the peoples of northern Borneo across the South China Sea. For good or ill, the colonial intrusion consolidated a realm of few culture cores and numerous ministates into less than a dozen countries.

PHYSICAL GEOGRAPHY

As Figure 10-3 shows, Southeast Asia is a realm in which high relief dominates the physiography. From the Arakan Mountains in western Myanmar (Burma) to the glaciers (yes, glaciers!) of the Indonesian part of New Guinea, elevations rise above 3300 meters (10,000 ft) in many places.

When you take the road from Semarang on central Jawa's north coast toward Yogyakarta near the south coast, you have a constant companion to your left: towering Mount Merapi, its near-perfect volcanic cone rising above the densely-peopled, fertile rural area below. Merapi is an active volcano, and the people in its shadow are at continuing risk—not just from volcanic eruptions, but also from associated earthquakes in this especially unstable part of the Pacific Ring of Fire. In the spring of 2006, Mount Merapi came to life and thousands were evacuated from its most hazardous slopes. Then, in May, a severe earthquake, probably related to the volcanic activity, killed thousands.
© Palani Mohan/Getty Images

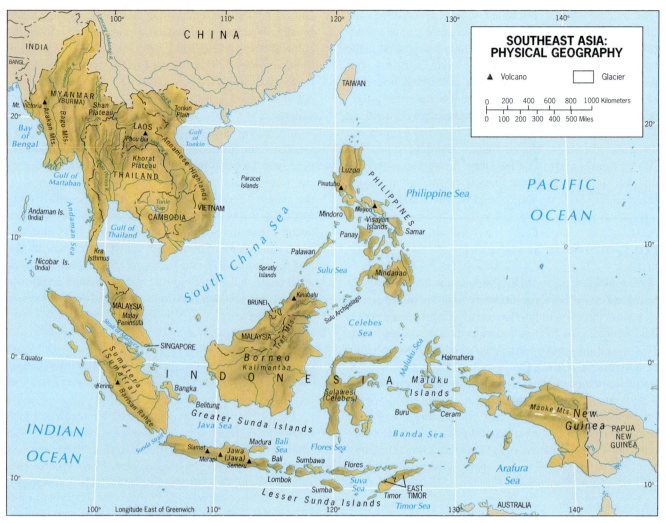

FIGURE 10-3 © H. J. de Blij, P. O. Muller, and John Wiley & Sons, Inc.

In Myanmar, Mount Victoria is such a peak; in northern Laos, Phou Bia comes close; on the Indonesian island of Sumatera (Sumatra under the old spelling) the tallest volcano is Kerinci; on Jawa (Java) Mounts Slamet and Semeru tower over the countryside; and on Borneo Mount Kinabalu, not a volcano but an erosional remnant, is higher than any other mountain in this realm. In the Philippines, Mount Pinatubo at 1477 meters (4874 ft) is not especially high, but this volcano's eruptions have had major impact on the country and, in 1991, on global climate.

The relief map is a further reminder of the overlap here between the Pacific Rim and the Pacific Ring—the Pacific Rim of economic and political development and the Pacific Ring of Fire with its earthquakes and volcanic eruptions (see photo, p. 527). If you examine Figure G-5 closely, you will note that all the islands of this realm are affected by the latter except one, the largely Indonesian island of Borneo. Borneo is a slab of ancient crust, a mini-continent pushed high above sea level by tectonic forces and eroded into its present mountainous topography.

Four Mainland Rivers

As Figures 10-1 and 10-3 underscore, rivers rise in the highland backbones of the islands and peninsulas, and deposit their sediments as they wind their way toward the coast; the physiography of Sumatera demonstrates this unmistakably. The volcanic hills, plateaus, and better-drained lowlands are fertile and, in the warmth of tropical climates, can yield multiple crops of rice.

On the peninsular mainland we see a pattern that is already familiar: rivers rising in the Asian interior that create alluvial plains and deltas. The Mekong River is the Chang/Yangzi of Southeast Asia: you can trace it all the way from China via Laos, Thailand, and Cambodia into southern Vietnam, where it forms a massive and populous delta. In the west, Myanmar's key river is the Irrawaddy; Thailand's is the Chao Phraya. In the north, the Red River Basin is the breadbasket of northern Vietnam.

No survey of the physical geography of Southeast Asia would be complete without reference to the realm's seas, gulfs, straits, and bays. Irregular and indented coastlines such as these, with thousands of islands near and far, create difficult problems when it comes to drawing maritime boundaries (the islands also form havens for rebels and criminals). As we note later, Southeast Asia has one of the most complex maritime boundary frameworks in the world, and it is rife with potential for conflict.

POPULATION GEOGRAPHY

Compared to the huge population numbers and densities in the habitable regions of South Asia and China, demographic totals for the countries of Southeast Asia, with the exception of Indonesia's 232 million, seem modest. Again, comparisons with Europe come to mind. Three countries—Thailand, the Philippines, and Vietnam—have populations between 65 and 90 million. Laos, quite a large country territorially (comparable to the United Kingdom), had just 6.4 million inhabitants in 2008. Cambodia, substantially larger than Greece, had 14.7 million. It is noteworthy that of Southeast Asia's 582 million inhabitants, more than 55 percent live on the islands of Indonesia and the Philippines, leaving the realm's mainland countries with less than 40 percent of the population.

The Ethnic Mosaic

Southeast Asia's peoples come from a common stock just as (Caucasian) Europeans do, but this has not prevented the emergence of regionally or locally discrete ethnic or cultural groups. Figure 10-4 displays the broad distribution of ethnolinguistic groups in the realm, but be aware that this is a generalization. At the scale of this map, numerous small groups cannot be depicted.

The map shows the rough spatial coincidence, on the mainland, between major ethnic group and modern political state. The Burman dominate in the country once known as Burma (today Myanmar); the Thai occupy the state once known as Siam (now Thailand); the Khmer form the nation of Cambodia and extend northward into Laos; and the Vietnamese inhabit the long strip of territory facing the South China Sea.

Territorially, by far the largest population is classified in Figure 10-4 as Indonesian, the inhabitants of the great archipelago that extends from Sumatera west of the Malay Peninsula to the Malukus (Moluccas) in the east and from the Lesser Sunda Islands in the south to the Philippines in the north. Collectively, all these peoples—the Filipinos, Malays, and Indonesians—shown in Figure 10-4 are known as Indonesians, but they have been divided by history and

politics. Note, on the map, that the Indonesians in Indonesia itself include Javanese, Madurese, Sundanese, Balinese, and other large groups; hundreds of smaller ones are not shown. In the Philippines, too, island isolation and contrasting ways of life are reflected in the cultural mosaic. Also part of this Indonesian ethnic-cultural complex are the Malays, whose heartland lies on the Malay Peninsula but who form minorities in other areas as well. Like most Indonesians, the Malays are Muslims, but Islam is a more powerful force in Malay society than, in general, in Indonesian culture.

In the northern part of the mainland region, numerous minorities inhabit remote parts of the countries in which the Burman (Burmese), Thai, and Vietnamese dominate. Those minorities, as a comparison of Figures 10-2 and 10-4 indicates, tend to occupy areas on the peripheries of their countries, away from the core areas, where the terrain is mountainous and the forest is dense, and where the governments of the national states do not have complete control. This remoteness and sense of detachment give rise to notions of secession, or at least resistance to governmental efforts to establish authority, often resulting in bitter ethnic conflict.

Immigrants

Figure 10-4 also reminds us that, again like Eastern Europe, Southeast Asia is home to major ethnic minorities from outside the realm. On the Malay Peninsula, note the South Asian (Hindustani) cluster. Hindu communities with Indian ancestries exist in many parts of the peninsula, but in the southwest they form the majority in a small area. In Singapore, too, South Asians form a significant minority. These communities arose during the colonial period, but South Asians had arrived in this realm many centuries earlier, propagating Buddhism and leaving architectural and cultural imprints on places as far away as Jawa and Bali.

The Chinese

By far the largest immigrant minority in Southeast Asia, however, is Chinese. The Chinese began arriving here during the Ming and early Qing (Manchu) dynasties, but the largest exodus occurred during the late colonial period (1870–1940), when as many as 20 million immigrated. The European powers at first encouraged this influx, using the Chinese in administration and trade. But soon these **Overseas Chinese** began to move into the major cities, where they established "Chinatowns" and gained control over much of the commerce. By the time the Europeans tried to reduce Chinese immigration, World War II was about to start and the colonial era would soon end.

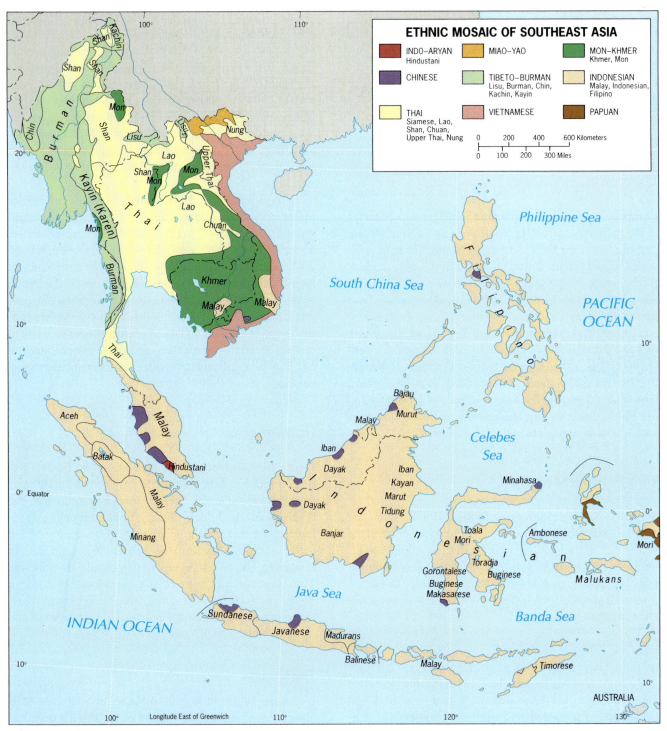

FIGURE 10-4 © H. J. de Blij, P. O. Muller, and John Wiley & Sons, Inc.

Today, Southeast Asia is home to as many as 30 million Overseas Chinese, more than half the world total. Their lives have often been difficult. The Japanese relentlessly persecuted those Chinese who lived in Malaya during World War II; during the 1960s, Chinese in Indonesia were accused of communist sympathies, and hundreds of thousands were killed. In the late 1990s, Indonesian mobs again attacked Chinese and their property, this time because of their relative wealth and because many Chinese became Christians during the colonial period, now targeted by Islamic throngs.

FROM THE FIELD NOTES

"It was a relatively cool day in Hanoi, which was just as well because the drive from Haiphong had taken nearly four hours, much of the time waiting behind oxcarts, and the countryside was hot and humid. On my way to an office along Hang Bai Street I ran into this group of youngsters and their teacher on their way to the Ho Chi Minh memorial. He right away realized the level of my broken French, and explained patiently that he taught history, and took his class on a field trip several times a year. 'They get all dressed up and it is a big moment for them,' he said, 'and there is a lot of history to be visited in Hanoi.' The kids had seen me take this picture and now they crowded around. I told them that I noticed that rouge seemed to be a favorite color in their outfits. 'But that should not surprise you, monsieur,' said the teacher, smiling. 'This is Vietnam, communist Vietnam, and red is the color that rallies us!' In the city at the opposite end of the country named after Ho Chi Minh (founder of his country's Communist Party, president from 1945 to 1969, victor over the French colonizers, and leader in the war against the United States), my experiences with teachers and students had been quite different, to the point that most called their city Saigon. But here in Hanoi the ideological flame still burned brightly." © H. J. de Blij.

www.conceptcaching.com

Figure 10-5 shows the migration routes and current concentrations of Chinese in Southeast Asia. Most originated in China's Fujian and Guangdong provinces, and many invested much of their wealth in China when it opened up to foreign businesses. The Overseas Chinese of Southeast Asia played a major role in the economic miracle of the Pacific Rim.

HOW THE POLITICAL MAP EVOLVED

The leading colonial competitors in Southeast Asia were the Dutch, French, British, and Spanish (with the last replaced by the Americans in their stronghold, the Philippines). The Japanese had colonial objectives here as well, but these came and went during the course of World War II.

The Dutch acquired the greatest prize: control over the vast archipelago now called Indonesia (formerly the Netherlands East Indies). France established itself on the eastern flank of the mainland, controlling all territory east of Thailand and south of China. The British conquered the Malay Peninsula, gained power over the northern part of the island of Borneo, and established themselves in Burma as well. Other colonial powers also gained footholds but not for long. The exception was Portugal, which held on to its eastern half of the island of Timor (Indonesia) until after the Dutch had been ousted from their East Indies.

Figure 10-6 shows the colonial framework in the late nineteenth century, before the United States assumed control over the Philippines. Note that while Thailand survived as an independent state, it lost territory to the British in Malaya and Burma and to the French in Cambodia and Laos.

The Colonial Imprint

French Indochina

The colonial powers divided their possessions into administrative units as they did in Africa and elsewhere. Some of these political entities became independent states when the colonial powers withdrew. France, one of the mainland's leading colonial powers, divided its Southeast Asian empire into five units. Three of these units lay along the east coast: Tonkin in the north next to China, centered on the basin of the Red River; Cochin China in the south, with the Mekong Delta as its focus; and between these two, Annam. The other two French territories were Cambodia, facing the Gulf of Thailand, and Laos, landlocked in the interior. Out of these five French dependencies there emerged the three states of *Indochina*. The three east-coast territories ultimately became one state, Vietnam; the other two (Cambodia and Laos) each achieved separate independence.

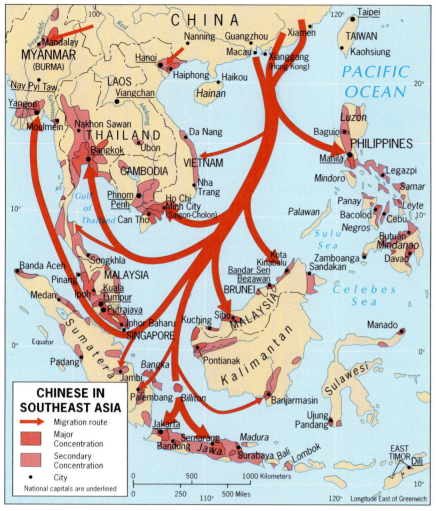

FIGURE 10-5 © H. J. de Blij, P. O. Muller, and John Wiley & Sons, Inc.

British Imperialism

The British ruled two major entities in Southeast Asia (Burma and Malaya) in addition to a large part of northern Borneo and many small islands in the South China Sea. Burma was attached to Britain's Indian Empire; from 1886 until 1937 it was governed from distant New Delhi. But when British India became independent in 1947 and split into several countries, Burma was not part of the grand design that created West and East Pakistan (the latter now Bangladesh), Ceylon (now Sri Lanka), and India. Instead, in 1948 Burma (now Myanmar) was given the status of a sovereign republic.

In Malaya, the British developed a complicated system of colonies and protectorates that eventually gave rise to the equally complex, far-flung Malaysian Federation. Included were the former Straits Settlements (Singapore was one of these colonies), the nine protectorates on the Malay Peninsula (former sultanates of the Muslim era), the British dependencies of Sarawak and Sabah on the is-

land of Borneo, and numerous islands in the Strait of Malacca and the South China Sea. The original Federation of Malaysia was created in 1963 by the political unification of recently independent mainland Malaya, Singapore, and the former British dependencies on the largely Indonesian island of Borneo. Singapore, however, left the Federation in 1965 to become a sovereign city-state, and the remaining units were later restructured into peninsular Malaysia and, on Borneo, Sarawak and Sabah. Thus the term *Malaya* properly refers to the geographic area of the Malay Peninsula, including Singapore and other nearby islands; the term *Malaysia* identifies the politico-geographical entity of which Kuala Lumpur is the capital city.

Netherlands "East Indies"

The Hollanders took control of the "spice islands" through their Dutch East India Company, and the wealth that they extracted from what is today Indonesia brought the Netherlands its Golden Age. From the mid-seventeenth to the late-eighteenth century, the Dutch could develop their East Indies sphere of influence almost without challenge, for the British and French were preoccupied with the Indian subcontinent. By playing the princes of Indonesia's states against one another in the search for economic concessions and political influence, by placing the Chinese in positions of responsibility, and by imposing systems of forced labor in areas directly under its control, the Company had a ruinous effect on the Indonesian societies it subjugated. Java (Jawa), the most populous and productive island, became the focus of Dutch administration; from its capital at Batavia (now Jakarta), the Company extended its sphere of influence into Sumatra (Sumatera), Dutch Borneo (Kalimantan), Celebes (Sulawesi), and the smaller islands of the East Indies. This was not accomplished overnight, and the struggle for territorial control was carried on long after the Dutch East India Company had yielded its administration to the Netherlands government. Dutch colonialism thus threw a girdle around Indonesia's more than 17,000 islands, paving the way for the creation of the realm's largest and most populous state (232 million today).

FIGURE 10-6   © H. J. de Blij, P. O. Muller, and John Wiley & Sons, Inc.

From Spain to the United States

In the colonial tutelage of Southeast Asia, the Philippines, long under Spanish domination, had a unique experience. As early as 1571, the islands north of Indonesia were under Spain's control (they were named for Spain's King Philip II). Spanish rule began when Islam was reaching the southern Philippines via northern Borneo. The Spaniards spread their Roman Catholic faith with great zeal, and between them the soldiers and priests consolidated Hispanic dominance over the mostly Malay population. Manila, founded in 1571, became a profitable waystation on the route between southern China and western Mexico (Acapulco usually was the trans-Pacific destination for the galleons leaving Manila's port). There was much profit to be made, but the indigenous people shared little in it. Great landholdings were awarded to loyal Spanish civil servants and to men of the church. Oppression eventually yielded revolution, and Spain was confronted with a major uprising when the Spanish-American War broke out elsewhere in 1898.

As part of the settlement of that war, the United States replaced Spain in Manila. That was not the end of the revolution, however. The Filipinos now took up arms against their new foreign ruler, and not until 1905, after terrible losses of life, did American forces manage to "pacify" their new dominion. Subsequently, U.S. administration in the Philippines was more progressive than Spain's had been. In 1934, Congress passed the Philippine Independence Law, providing for a ten-year transition to sovereignty. But before independence could be arranged, World War II intervened. In 1941, Japan conquered the islands, temporarily ousting the Americans; U.S. forces returned in 1944 and, with strong Filipino support, defeated the Japanese in 1945. The agenda for independence was resumed, and in 1946 the sovereign Republic of the Philippines was proclaimed.

Today, all of Southeast Asia's states are independent, but centuries of colonial rule have left strong cultural imprints. In their urban landscapes, their education systems, their civil service, and countless other ways, this realm still carries the marks of its colonial past.

Cultural-Geographic Legacies

The French, who ruled and exploited a crucial quadrant of Southeast Asia, had a name for their empire: *Indochina*. That name would be appropriate for the rest of the realm as well because it suggests its two leading Asian influences for the past 2000 years. Periodic expansions of the Chinese Empire, notably along the eastern periphery, as well as the arrival of large numbers of Overseas Chinese settlers (discussed on pp. 529–531; map p. 532), infused the region with cultural norms from the north. The Indians came from the west by way of the sea, as their trading ships plied the coasts, and settlers from India founded colonies on Southeast Asian shores in the Malay Peninsula, on the lower Mekong Plain, on Jawa and Bali, and on Borneo.

Religious Infusions from South Asia

With the migrants from the Indian subcontinent came their faiths: first Hinduism and Buddhism, later Islam. The Muslim religion, promoted by the growing number of Arab traders who appeared on the scene, became the dominant religion in Indonesia (where nearly 90 percent of the population adheres to Islam today). But in Myanmar, Thailand, and Cambodia, Buddhism remained supreme, and in all three countries the overwhelming majority of the people are now adherents. In culturally diverse Malaysia, the Malays

are Muslims (to be a Malay *is* to be a Muslim), and almost all Chinese are Buddhists; but most Malaysians of Indian ancestry remain Hindus. Although Southeast Asia has generated its own local cultural expressions, most of what remains in tangible form has resulted from the infusion of foreign elements. For instance, the main temple at Angkor Wat, constructed in Cambodia during the twelfth century, remains a monument to the Indian architecture of that time.

The *Indo* part of Indochina, then, refers to the cultural imprints from South Asia: the Hindu presence, the importance of Buddhism (which came to Southeast Asia via Sri Lanka [Ceylon] and its seafaring merchants), the influences of Indian architecture and art (especially sculpture), writing and literature, and social structures and patterns.

Chinese Imprints and Influence

The *China* in the name Indochina signifies the role of the Chinese here. Chinese emperors coveted Southeast Asian lands, and China's power reached deep into the realm. Social and political upheavals in China sent millions of sinicized people southward. Chinese traders, pilgrims, seafarers, fishermen, and others sailed from southeastern China to the coasts of Southeast Asia and established settlements there. Over time, those settlements attracted more Chinese emigrants, and Chinese influence in the realm grew (see map p. 532). Not surprisingly, relations between the Chinese settlers and the earlier inhabitants of Southeast Asia have at times been strained, even violent. The Chinese presence in Southeast Asia is long-term, but the invasion has continued into modern times. The economic power of Chinese minorities and their role in the political life of the realm have led to conflicts.

The Chinese initially profited from the arrival of the Europeans, who stimulated the growth of agriculture, trade, and industries; here the Chinese found opportunities they lacked at home. They harvested rubber, found jobs on the docks and in the mines, cleared the bush, and transported goods in their sampans. They brought useful skills with them, and as tailors, shoemakers, blacksmiths, and fishermen, they prospered. The Chinese were also astute in business; soon, they not only dominated the region's retail trade but also held prominent positions in banking, industry, and shipping. Thus they have always been far more important than their modest numbers in Southeast Asia would suggest. The Europeans used them for their own designs but at times found the Chinese to be stubborn competitors—so much so that eventually they tried to impose restrictions on Chinese immigration. When the United States took control of the Philippines, it also sought to stop the influx of Chinese into those islands.

When the European colonial powers withdrew and Southeast Asia's independent states emerged, Chinese pop-

FROM THE FIELD NOTES

"Like most major Southeast Asian cities, Bangkok's urban area includes a large and prosperous Chinese sector. No less than 14 percent of Thailand's population of 66 million is of Chinese ancestry, and the great majority of Chinese live in the cities. In Thailand, this large non-Thai population is well integrated into local society, and intermarriage is common. Still, Bangkok's 'Chinatown' is a distinct and discrete part of the great city. There is no mistaking Chinatown's limits: Thai commercial signs change to Chinese, goods offered for sale also change (Chinatown contains a large cluster of shops selling gold, for example), and the urban atmosphere, from street markets to bookshops, is dominantly Chinese. This is a boisterous, noisy, energetic part of multicultural Bangkok, a vivid reminder of the Chinese commercial success in Southeast Asia."
© H. J. de Blij.

Concept Caching

www.conceptcaching.com

ulation sectors ranged from nearly 50 percent of the total in Malaysia (in 1963) to barely over 1 percent in Myanmar. In Singapore today, Chinese constitute 77 percent of the population of 4.6 million; when Singapore seceded from Malaysia in 1965, the Chinese component in Malaysia was substantially reduced (it is about 25 percent today). In Indonesia, the percentage of Chinese in the total population is not high (no more than 3 percent), but the Indonesian population is so large that even this small percentage indicates a Chinese sector of almost 7 million. In 1998, this tiny minority reportedly controlled 70 percent of Indonesia's economy. In Thailand, on the other hand, many Chinese have married Thais, and the Chinese minority of about 14 percent has become a cornerstone of Thai society, dominant in trade and commerce (photo above).

In general, Southeast Asia's Chinese communities remained aloof and formed their own separate societies in the cities and towns. They kept their culture and language alive through social clubs, schools, and even residential suburbs that, in practice if not by law, were Chinese in character. At one time they were caught in the middle between the Europeans and Southeast Asians when the local people were hostile toward both white people and the Chinese. Since the withdrawal of the Europeans, however, the Chinese have become the main target of this antagonism, which remained strong because of the Chinese involvement in money lending, banking, and trade monopolies (see the Issue Box entitled "The Chinese Presence in Southeast Asia"). Moreover, there is the specter of an imagined or real Chinese political imperialism along Southeast Asia's northern flanks.

The *China* in Indochina, therefore, represents a diversity of penetrations. Southern China was the source of most of the old invasions, and Chinese territorial consolidation provided the impetus for successive immigrants. Although Indian cultural influences remained strong, people throughout Southeast Asia adopted Chinese modes of dress, plastic arts, types of houses and boats, and other cultural attributes. During the past century, and especially during the last half-century, renewed Chinese immigration brought skills and energies that propelled these minorities to comparative wealth and influence.

SOUTHEAST ASIA'S POLITICAL GEOGRAPHY

Southeast Asia's diverse and fractured natural landscapes, as well as varied cultural influences, have combined to create a complex political geography. As we noted earlier, *political geography* focuses on the spatial expressions of political behavior; we were introduced to several aspects of it when we examined Western Europe's supranationalism, Eastern Europe's balkanization, Pakistan's irredentism toward Kashmiri Muslims, and the geopolitics of eastern Eurasia. Many political geographers have studied the fundamental causes of the cyclical rise and decline of states; their answers have ranged from environmental changes to biological forces. Friedrich Ratzel (1844–1904) conceptualized the state as a biological organism whose life, from

The Chinese Presence in Southeast Asia

THE CHINESE ARE TOO INFLUENTIAL!

"It's hard to imagine that there was a time when we didn't have Chinese minorities in our midst. I think I understand how Latin Americans [sic] feel about their 'giant to the north.' We've got one too, but the difference is that there are a lot more Chinese in Southeast Asia than there are Americans in Latin America. The Chinese even run one country because they are the majority there, Singapore. And if you want to go to a place where you can see what the Chinese would have in mind for this whole region, go there. They're knocking down all the old Malay and Hindu quarters, and they've got more rules and laws than we here in Indonesia could even think of. I'm a doctor here in Bandung and I admire their modernity, but I don't like their philosophy.

"We've had our problems with the Chinese here. The Dutch colonists brought them in to work for them, they got privileges that made them rich, they joined the Christian churches the Europeans built. I'm not sure that a single one of them ever joined a mosque. And then shortly after Sukarno led us to independence they even tried to collaborate with Mao's communists to take over the country. They failed and many of them were killed, but look at our towns and villages today. The richest merchants and the money lenders tend to be Chinese. And they stay aloof from the rest of us.

"We're not the only ones who had trouble with the Chinese. Ask them about it in Malaysia. There the Chinese started a full-scale revolution in the 1950s that took the combined efforts of the British and the Malays to put down. Or Vietnam, where the Chinese weren't any help against the enemy when the war happened there. Meanwhile they get richer and richer, but you'll see that they never forget where they came from. The Chinese in China boast about their coastal economy, but it's coastal because that's where our Chinese came from and where they sent their money when the opportunity arose.

"As a matter of fact, I don't think that China itself cares much for or about Southeast Asia. Have you heard what's happening in Yunnan Province? They're building a series of dams on the Mekong River, our major river, just across the border from Laos and Burma. They talk about the benefits and they offer to destroy the gorges downstream to facilitate navigation, but they won't join the Mekong River Commission. When the annual floods cease, what will happen to Tonle Sap, to fish migration, to seasonal mudflat farming? The Chinese do what they want—they're the ones with the power and the money."

Regional ISSUE

THE CHINESE ARE INDISPENSABLE!

"Minorities, especially successful minorities, have a hard time these days. The media portray them as exploiters who take advantage of the less fortunate in society, and blame them when things go wrong, even when it's clear that the government representing the majority is at fault. We Chinese arrived here long before the Europeans did and well before some other 'indigenous' groups showed up. True, our ancestors seized the opportunity when the European colonizers introduced their commercial economy, but we weren't the only ones to whom that opportunity was available. We banded together, helped each other, saved and shared our money, and established stable and productive communities. Which, by the way, employed millions of locals. I know: my family has owned this shop in Bangkok for five generations. It started as a shed. Now it's a six-floor department store. My family came from Fujian, and if you added it all up we've probably sent more than a million U.S. dollars back home. We're still in touch with our extended family and we've invested in the Xiamen SEZ.

"Here in Thailand we Chinese have done very well. It is sometimes said that we Chinese remain aloof from local society, but that depends on the nature of that society. Thailand has a distinguished history and a rich culture, and we see the Thais as equals. Forgive me, but you can't expect the same relationship in East Malaysia. Or, for that matter, in Indonesia, where we are resented because the 3 percent Chinese run about 60 percent of commerce and trade. But we're always prepared to accommodate and adjust. Look at Malaysia, where some of our misguided ancestors started a rebellion but where we're now appreciated as developers of high-tech industries, professionals, and ordinary workers. This in a Muslim country where Chinese are Christians or followers of their traditional religions, and where Chinese regularly vote for the Malay-dominated majority party. So when it comes to aloofness, it's not us.

"The truth is that the Chinese have made great contributions here, and that without the Chinese this place would resemble South or Southwest Asia. Certainly there would be no Singapore, the richest and most stable of all Southeast Asian countries, where minorities live in peace and security and where incomes are higher than anywhere else. In fact, mainland Chinese officials come to Singapore to learn how so much was achieved there. Imagine a Southeast Asia without Singapore! It's no coincidence that the most Chinese country in Southeast Asia also has the highest standard of living."

Vote your opinion at www.wiley.com/college/deblij

birth through expansion and maturation to eventual senility and collapse, mirrors that of any living thing. Ratzel's **5** **organic theory** of state development held that nations, being aggregates of human beings, would over the long term live and die as their citizens did.

Other political geographers sought to measure the strength of the forces that bind states (centripetal forces) and that divide them (centrifugal forces), and thus to assess a state's chances to survive separatism of the kind Canada and Sri Lanka are experiencing today, to name just two examples. Even in today's era of modern warfare, a stretch of water still affects the course of events. Taiwan would not be what it is without the 150 kilometers (90 mi) of water that separates it from mainland China. Singapore's secession from (then) Malaya was facilitated because Singapore is an island; had it been attached to the Malay Peninsula, the centrifugal forces of separatism might not have prevailed. The role of physical geography in political events remains powerful.

In this section we pay particular attention to the boundaries and territories of the component states of Southeast Asia because this realm's political geography displays virtually every possible attribute of both. We focus first on the borders and then on the states' territorial shapes.

The Boundaries

Boundaries are sensitive parts of a state's anatomy: just as people are territorial about their individual properties, so nations and states are sensitive about their territories and borders. The saying that "good fences make good neighbors" certainly applies to states, but, as we know, the boundaries between states are not always "good fences."

Boundaries, in effect, are contracts between states. **6** That contract takes the form of a treaty that contains the **definition** of the boundary in the form of elaborate description. Next, cartographers perform the **delimitation** of the treaty language, drawing the boundary on official, large-scale maps. Throughout human history, states have used those maps to build fences, walls, or other barriers in a process called **demarcation**.

Classifying Boundaries

Once established, we can classify boundaries geographically. Some are sinuous, conforming to rivers (see photo below) or mountain crests (**physiographic**) or coinciding with breaks or transitions in the cultural landscape (**anthropogeographic**). As any world political map shows, many boundaries are simply straight lines, delimited without reference to physical or cultural features. These **geometric** boundaries can produce problems when the cultural landscape changes where they exist.

In general, the boundaries of Southeast Asia were better defined than those of several other postcolonial areas of the world, notably Africa, the Arabian Peninsula, and Turkestan. The colonial powers that established the original treaties tried to define boundaries to lie in remote and/or sparsely peopled areas, for example, across interior Borneo. Nevertheless, certain Southeast Asian boundaries have produced problems, among them the geometric boundary between

FROM THE FIELD NOTES

"I stood on the Laotian side of the great Mekong River which, during the dry season, did not look so great! On the opposite side was Thailand, and it was rather easy for people to cross here at this time of the year. But, the locals told me, it is quite another story in the wet season. Then the river inundates the rocks and banks you see here, it rushes past, and makes crossing difficult and even dangerous. The buildings where the canoes are docked are built on floats, and rise and fall with the seasons. The physiographic-political boundary between Thailand and Laos lies in the middle of the valley we see here."
© Barbara A. Weightman.

www.conceptcaching.com

Papua, the portion of New Guinea ruled by Indonesia, and Papua New Guinea, the eastern component of the island.

Even on a small-scale map of the kind we use in this chapter, we can categorize the boundaries of this realm. A comparison between Figures 10-2 and 10-4 reveals that the boundary between Thailand and Myanmar over long segments is anthropogeographic, notably where the name *Kayin (Karen)*, the Myanmar minority, appears in Figure 10-4. Figure 10-1 shows that a large segment of the Vietnam–Laos boundary is physiographic-political, coinciding with the Annamite Cordillera (Highlands).

Genetic Classification System

Boundaries also can be classified genetically, that is, as their evolution relates to the cultural landscapes they traverse. A leading political geographer, Richard Hartshorne (1899–1992), proposed a four-level *genetic boundary classification*. All four of these boundary types can be observed in Southeast Asia.

Certain boundaries, Hartshorne reasoned, were defined and delimited before the present-day human landscape developed. In Figure 10-7 (upper-left map), the boundary

between Malaysia and Indonesia on the island of Borneo is an example of the first boundary type, the **antecedent boundary**. Most of this border passes through sparsely inhabited tropical rainforest, and the break in settlement can even be detected on the small-scale world population map (Fig. G-9).

A second category of boundaries evolved as the cultural landscape of an area took shape, part of the ongoing process of accommodation. These **subsequent boundaries** are represented in Southeast Asia by the map in the upper right of Figure 10-7, which shows in some detail the border between Vietnam and China. This border is the result of a long process of adjustment and modification, the end of which may not yet have come.

The third category involves boundaries drawn forcibly across a unified or at least homogeneous cultural landscape. The colonial powers did this when they divided the island of New Guinea by delimiting a boundary in a nearly straight line (curved in only one place to accommodate a bend in the Fly River), as shown in the lower-left map of Figure 10-7. The **superimposed boundary** they delimited gave the Netherlands the western half of New Guinea. When Indonesia became independent in 1949, the Dutch

GENETIC POLITICAL BOUNDARY TYPES

FIGURE 10-7 © H. J. de Blij, P. O. Muller, and John Wiley & Sons, Inc.

did not yield their part of New Guinea, which is peopled mostly by ethnic Papuans, not Indonesians. In 1962 the Indonesians invaded the territory by force of arms, and in 1969 the United Nations recognized its authority there. This made the colonial, superimposed boundary the eastern border of Indonesia and had the effect of extending Indonesia from Southeast Asia into the Pacific Realm. Geographically, all of New Guinea forms part of the Pacific Realm.

10   The fourth genetic boundary type is the so-called **relict boundary**—a border that has ceased to function but whose imprints (and sometimes influence) are still evident in the cultural landscape. The boundary between the former North and South Vietnam (Fig. 10-7, lower-right map) is a classic example: once demarcated militarily, it has had relict status since 1976 following the reunification of Vietnam in the aftermath of the Indochina War (1964–1975).

Southeast Asia's boundaries have colonial origins, but they have continued to influence the course of events in postcolonial times. Take one instance: the physiographic boundary that separates the main island of Singapore from the rest of the Malay Peninsula, the Johor Strait (see Fig. 10-13). That physiographic-political boundary facilitated, perhaps crucially, Singapore's secession from the state of Malaysia. Without it, Malaysia might have been persuaded to stop the separation process; at the very least, territorial issues would have arisen to slow the sequence of events. As it was, no land boundary needed to be defined. The Johor Strait demarcated Singapore and left no question as to its limits.

State Territorial Morphology

Boundaries define and delimit states; they also create the mosaic of often interlocking territories that give individual countries their shape, also known as their *morphology*.
11   The **territorial morphology**

of a state affects its condition, even its survival. Vietnam's extreme elongation has influenced its existence since time immemorial. As we will see, Indonesia has tried to redress its fragmentation into thousands of islands by promoting unity through the "transmigration" of Jawanese from the most populous island to many of the others.

Political geographers identify five dominant state territorial configurations, all of which we have encountered in our world regional survey but which we have not categorized until now. All but one of these shapes are represented in Southeast Asia, and Figure 10-8 provides the terminology and examples:

• **Compact states** have territories shaped somewhere **12** between round and rectangular, without major indentations. This encloses a maximum amount of territory within a minimum length of boundary. Southeast Asian example: Cambodia.

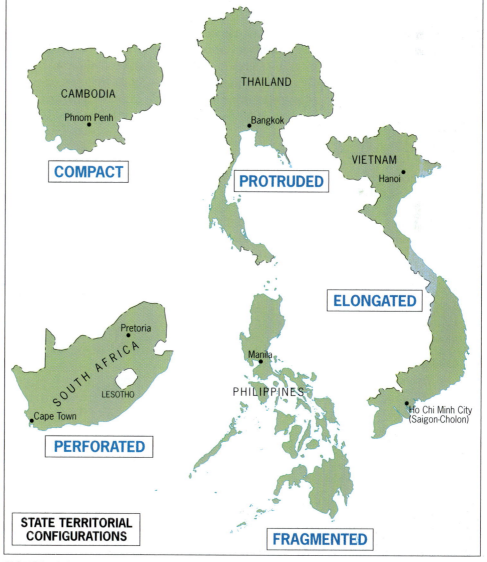

FIGURE 10-8   © H. J. de Blij, P. O. Muller, and John Wiley & Sons, Inc.

13 • **Protruded states** (sometimes called *extended*) have a substantial, usually compact territory from which extends a peninsular corridor that may be landlocked or coastal. Southeast Asian examples: Thailand, Myanmar.

14 • **Elongated states** (also called *attenuated*) have territorial dimensions in which the length is at least six times the average width, creating a state that lies astride environmental and/or cultural transitions. Southeast Asian example: Vietnam.

15 • **Fragmented states** consist of two or more territorial units separated by foreign territory or by water. Subtypes are mainland-mainland, mainland-island, and island-island. Southeast Asian examples: Malaysia, Indonesia, Philippines.

16 • **Perforated states** completely surround the territory of other states, so that they have a "hole" in them. No Southeast Asian example; the most illustrative current case is South Africa, perforated by Belgium-sized Lesotho (also see map on p. 316).

In the discussion that follows, we will have frequent occasion to refer to this geographic property of Southeast Asia's states. For so comparatively small a realm with so few states, Southeast Asia displays a considerable variety of state morphologies. When we link these features to other geographic aspects (such as relative location), we obtain useful insights into the regional framework.

POINTS TO PONDER

● Myanmar is becoming increasingly important to China. An oil pipeline across Myanmar from a west-coast terminal will carry Middle East oil to Yunnan Province. China's only military base on the Indian Ocean lies in Myanmar.

● *Pesantren*, the hate-dispensing Islamic schools that are the *madrassas* of Indonesia, are radicalizing a significant minority of the country's youngsters.

● The nation-building project in East Timor has essentially failed. The country is independent, but its people are mired in poverty and its institutions are failing.

● Thriving Singapore wants to lease one or two of Indonesia's more than 17,000 islands in order to expand its commercial, research, financial, and other operations.

One point of caution: states' territorial morphologies do not determine their viability, cohesion, unity, or lack thereof; they can, however, influence these qualities. Cambodia's compactness has not ameliorated its divisive political geography, for example. But as we will find in the pages that follow, shape plays a key role in the still-unfolding political and economic geography of Southeast Asia.

Regions of the Realm

Southeast Asia's first-order regionalization must be based on its mainland-island fragmentation. But as we have noted there are physiographic, historical, and cultural reasons to include the Malaysian (southern) part of the Malay Peninsula in the Insular region, as shown in Figure 10-2. Using the political framework as our grid, we see that the regions of Southeast Asia are constituted as follows:

Mainland region Vietnam, Cambodia, Laos, Thailand, Myanmar (Burma)

Insular region Malaysia, Singapore, Indonesia, East Timor, Brunei, Philippines

● MAINLAND SOUTHEAST ASIA

Five countries form the Mainland region of Southeast Asia (Fig. 10-2): two of them protruded, sharing a narrow peninsula; one elongated along the Pacific coast; one compact and facing the Gulf of Thailand; and one landlocked. One religion, Buddhism, dominates cultural landscapes, but this is a multicultural, multiethnic region. Although still one of the less urbanized regions of the world, the Mainland region contains several major cities and the pace of urbanization is increasing. And as Figure 10-2 shows, two countries—Vietnam and Myanmar—possess more than one core area each. We approach the region from the east.

MAJOR CITIES OF THE REALM

City	Population* (in millions)
Bangkok, Thailand	6.8
Hanoi, Vietnam	4.5
Ho Chi Minh City (Saigon), Vietnam	7.5
Jakarta, Indonesia	14.5
Kuala Lumpur, Malaysia	1.5
Manila, Philippines	11.3
Phnom Penh, Cambodia	1.6
Singapore, Singapore	4.6
Viangchan, Laos	0.7
Yangon, Myanmar (Burma)	4.4

*Based on 2008 estimates.

ELONGATED VIETNAM

Vietnam's economic transformation from postwar stagnation to present expansion is one of the realm's—and the world's—great geographic stories. In the aftermath of the war that ended in American (and South Vietnamese) defeat in 1975, devastated Vietnam could only watch as China's Pacific Rim burgeoned, its products flooding world markets and foreign investment rushing in. The United States placed an embargo on the country, isolating it and stifling its economy. The Vietnamese government experimented with communism even as an estimated two million citizens left by any means they could (hundreds of thousands of "boat people" perished in the process), and for a few years this onetime breadbasket of the region faced famine. Then, in the mid-1980s, Vietnamese leaders decided to try what they called *doi-moi* economic reforms—a more gradual opening of their country to globalization's forces than China had adopted. By 1995, Vietnam's exports, from coffee to shrimps, from rice to footwear, and including electronic equipment, boats, and even some crude oil, were growing faster than China's. In that significant year, Washington during the Clinton Administration moved to normalize relations with Hanoi, opening an embassy and encouraging economic ties. In 2005, Vietnam's economy grew by more than 8 percent (population growth was leveling off at around 1.3 percent), and Vietnamese economists now anticipate that growth during this entire first decade of the century will average 8 percent or more.

Vietnam, as Table G-1 indicates, has a long way to go, but the Vietnamese government is determined not to allow uncontrolled capitalism to create the kinds of wealth disparities among its 87 million citizens that are now seen in China. And indeed, the severest form of poverty—people living on the equivalent of one dollar a day or less—has dropped dramatically in Vietnam from more than 50 percent in 1990 to under 10 percent in the middle of this decade even as high-school attendance has more than doubled to exceed 70 percent.

Elongation and Integration

Travel up the crowded road from the northern port of Haiphong to the capital, Hanoi, or sail up the river to the southern metropolis officially known as Ho Chi Minh City—but called **Saigon** by almost everyone there—and you are quickly reminded of the cultural effects of Vietnam's elongation (Fig. 10-9). The French colonizers delimited Vietnam as a 2000-kilometer (1200-mi) strip of land extending from the Chinese border in the north to the tip of the Mekong Delta in the south. Substantially smaller than California, this coastal belt was the domain

of the Vietnamese (Fig. 10-4). The French recognized that Vietnam, whose average width is under 250 kilometers (150 mi), was not a homogeneous colony, so they divided it into three units: (1) Tonkin, land of the Red River Delta and centered on Hanoi in the north; (2) Cochin China, the Mekong Delta region and centered on Saigon in the south; and (3) Annam, focused on the ancient city of Hué, in the middle (Fig. 10-6). Today, the Vietnamese prefer to use **Bac Bo**, **Nam Bo**, and **Trung Bo**, respectively, to designate these areas.

The Vietnamese (or Annamese, also Annamites, after their cultural heartland) speak the same language, although the northerners can easily be distinguished from southerners by their accent. As elsewhere in their colonial empire, the French made their language the **lingua franca** of Indochina, but their tenure was cut short by the Japanese, who invaded Vietnam in 1940. During the Japanese occupation, Vietnamese nationalism became a powerful force, and after the Japanese defeat in 1945, the French could not regain control. In 1954, the French suffered a disastrous final trouncing on the battlefield at Dien Bien Phu in the far northwest and were ousted from the country.

Even after its forces routed the colonizers, however, Vietnam did not become a unified state. Separate regimes took control: a communist one in Hanoi and a noncommunist counterpart in Saigon. Vietnam's pronounced elongation had made things difficult for the French; now it played its role during the postcolonial period. Note, in Figure 10-9, that Vietnam is widest in the north and south, with a narrow "waist" in its middle zone. North and South Vietnam were worlds apart.

The Indochina War

Many Americans still remember how the United States became involved in the inevitable conflict between communists and noncommunists in Vietnam during the 1960s and 1970s. At first, American military advisors were sent to Saigon to help the shaky regime there cope with communist insurgents. When the tide turned against the South, the United States committed military forces and equipment to halt the further spread of communism (see box titled "Domino Theory"); at one time, more than 500,000 U.S. soldiers were in Vietnam.

The conduct of the war, its mounting casualties, and its apparent futility created severe social tensions in the United States. If you had been a student during that period, your college experience would have been radically different from what it is today. Protest rallies, "teach-ins," anti-draft demonstrations, "sit-ins," marches, strikes, and even hostage-taking disrupted campus life. The Indochina War threatened the stability of American society. It drove an American president (Lyndon Johnson) from office in 1968

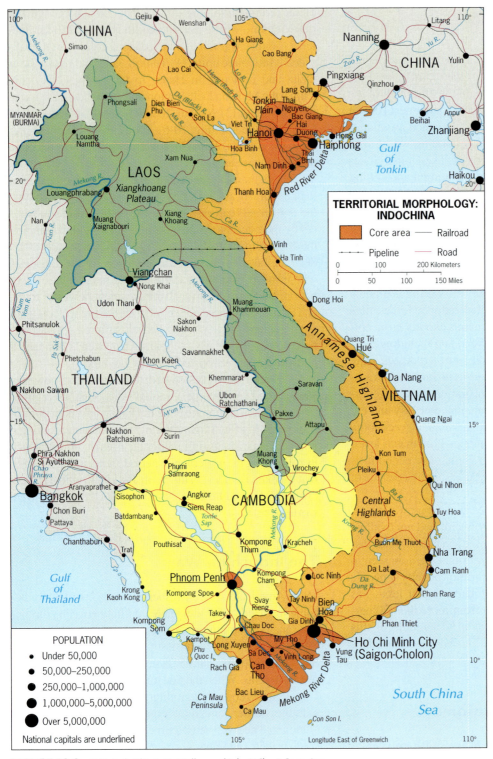

TERRITORIAL MORPHOLOGY: INDOCHINA

- ▨ Core area
- ─── Railroad
- ╂─╂ Pipeline
- ─── Road

0   100   200 Kilometers
0   50   100   150 Miles

POPULATION
- ● Under 50,000
- ● 50,000–250,000
- ● 250,000–1,000,000
- ● 1,000,000–5,000,000
- ● Over 5,000,000

National capitals are underlined

FIGURE 10-9   © H. J. de Blij, P. O. Muller, and John Wiley & Sons, Inc.

copter left from the roof of the U.S. embassy in Saigon amid scenes of desperation and desertion, it marked the end of a sequence of events that closely conformed to the *insurgent state model* we introduced in the context of contemporary Colombia in Chapter 5 (p. 260).

Vietnam in Transition

During the difficult years following the end of the war, Vietnam's prospects looked bleak not only because of the huge death toll (an estimated 3 million Vietnamese died during the conflict), devastated infrastructure, and chaotic emigration, but also because Hanoi found itself in a territorial conflict with China and faced still more strife. Not only Vietnamese, but tens of thousands of Chinese citizens of Vietnam fled the country. It was difficult for distant Hanoi to establish effective control over Saigon (renamed Ho Chi Minh City in honor of the North's wartime leader), and soon the two cities and their respective core areas began to grow apart. Indeed, most residents of Ho Chi Minh City continued to call their city Saigon; with about 8 percent of the country's population and a far larger share of its best-educated people, metropolitan Saigon recovered faster than Hanoi, and its old entrepreneurial spirit quickly revived. In the late 1980s,

and destroyed the electoral chances of his vice president (Hubert Humphrey), who would not disavow the U.S. role in the war.

In 1975, the Saigon government fell, and the United States was ousted—just a little over two decades after the French defeat at Dien Bien Phu. When the last heli-

studying Saigon and Hanoi in turn was like visiting two different Southeast Asian countries. Saigon bustled with bicycle and scooter traffic, businesses of all kinds lined streets and alleys, modern buildings were going up, its river port a beehive of activity. A Special Economic Zone on the Chinese model, and a New Saigon business and residential

Domino Theory

DURING THE INDOCHINA War (1964–1975), it was U.S. policy to contain communist expansion by supporting the efforts of the government of South Vietnam to defeat communist insurgents. Soon, the war engulfed North Vietnam as U.S. bombers attacked targets north of the border between North and South. In the later phases of the war, conflict spilled over into Laos and Cambodia. U.S. warplanes also used bases in Thailand. Like dominoes, one country after another fell to the ravages of the war or was threatened by it.

Some scholars warned that this domino effect could eventually affect not only Thailand but also Malaysia, Indonesia, and Burma (today Myanmar): the whole Southeast Asian realm, they predicted, could be destabilized. But, as we know, that did not happen. The war remained confined to Indochina. And the domino "theory" seemed invalid.

But is the theory totally without merit? Unfortunately, some political geographers to this day mistakenly define this idea in terms of communist activity. Communist insurgency, however, is only one way a country may be destabilized (Peru came close to collapse from it). But right-wing rebellion (Nicaragua's Contras), ethnic conflict (Bosnia), religious extremism (Algeria), and even economic and environmental causes can create havoc in a country. Properly defined, the **domino theory** holds that destabilization from any cause in one country can result in the collapse of order in a neighboring country, triggering a chain of events that can affect a series of contiguous states in turn.

A recent instance of the domino effect comes from Equatorial Africa, where the collapse of order in Rwanda spilled over into The Congo (then Zaïre) and also affected Burundi, Congo, and Angola.

17

FROM THE FIELD NOTES

"Although modernization has changed the skylines of Vietnam's major cities (Saigon more so than Hanoi), most of the country's urban areas look and function much as they have for generations. Main arteries throb with commerce and traffic (still mostly bicycles and mopeds); side streets are quieter and more residential. No high-rises here, but the problems of dense urban populations are nevertheless evident. Every time I turned into a side street, the need for street cleaning and refuse collection was obvious, and drains were clogged more often than not. 'We haven't allowed our older neighborhoods to be destroyed like they have in China,' said my colleague from Hanoi University, 'but we also don't have an economy growing so fast that we can afford to provide the services these people need.' Certainly the Pacific Rim contrast between burgeoning coastal China and slower-growing Vietnam—a matter of communist-government policy—is evident in their cities. 'As you see, nothing much has changed here,' she said. 'Not much evidence of globalization. But we're basically self-sufficient, nobody goes hungry, and the gap between the richest and the poorest here in Vietnam is a fraction of what it is now in China. We decided to keep control.' You would find more agreement on these points here in Hanoi than in Saigon, but the Vietnamese state is stable and progressing."   © H. J. de Blij.

Concept Caching

AMONG THE REALM'S GREAT CITIES . . . Saigon

OFFICIALLY, IT IS known as Ho Chi Minh City, renamed in 1975 after the ouster of American forces, but only bureaucrats and atlases call it that. To the people here, it is still **Saigon**, city on the Saigon River, served by, it seems, a thousand companies using the name. Saigon Taxi, Saigon Hotel, Saigon Pharmacy—there is hardly a city block without the name Saigon on some billboard.

The phenomenon is significant. Hanoi's northerners ordained the change, but this still is South Vietnam, and many southerners see the North as peopled by slow, dogmatic communists obstructing the hopes and plans for their vibrant city. In truth, Saigon may be nearly twice as large as Hanoi (as well as Mainland Southeast Asia's largest city), but like the capital it suffers from decades of deterioration, a broken-down infrastructure, and undependable public services. When, in the early 1990s, businesspeople could find no modern accommodations or working telephones, an Australian firm built an eight-story floating hotel with its own power plant, towed it up the Saigon River, and parked it on the CBD waterfront.

Therein lies Saigon's great advantage: large oceangoing vessels can reach the city in a three-hour sail up the meandering river. The Chinese developed its potential: in the 1960s, Chinese merchants controlled more than half of the South's exports and nearly all of its textile factories and foreign exchange. Saigon-Cholon (Cholon means "great market") was a Chinese success story, but after the war the Hanoi regime nationalized all Chinese-owned

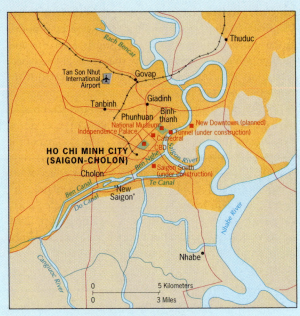

© H. J. de Blij, P. O. Muller, and John Wiley & Sons, Inc.

firms. Tens of thousands of middle-class Chinese left the city, its economy devastated.

Today, Saigon (population: 7.5 million) is steadily recovering. A modern skyline is rising above the colonial one. A large Special Economic Zone has been created south of the city. Factories are being built. The Pacific Rim era is under way. Will the South rise again?

---

district, signified the new era. Hanoi, by comparison, was quiet and staid (photos p. 543), its port a four-hour crawl away along a congested, two-lane road.

Today, two decades later, those differences and disparities are fading. Hanoi's answer to New Saigon is Tu Liem, a modern business-and-residential complex on the city's southwestern periphery complete with four-lane roads and modern amenities. Major hotels and a world-class convention center are changing the capital's skyline. The road to the port of Haiphong now has four lanes, and travel time is down to 90 minutes on a good day. But Saigon and environs still lead, in 2006 contributing more than 25 percent of Vietnam's industrial production and paying about one-third of its taxes. A new, vast development named Saigon South is under construction, and a tunnel is being dug beneath the Saigon River. The planners in Saigon do not have just Vietnam in mind when they talk about expansion and growth: they want nothing less than the transformation of Ho Chi Minh City into a regional hub for all of Southeast Asia.

During colonial times, central Saigon grew into a gracious and elegant city on the right bank of the Saigon River. Because a bridge was never built, the opposite side of the river remained essentially undeveloped (see map above). With the need for a new downtown, and in anticipation of 50-percent population growth over the next decade, Ho Chi Minh City's leaders hope to build a world-class city center across the river, taking the pressure off what remains of the historic city—much as Shanghai's Bund was saved by the development of Pudong (see box on p. 483). More than 30 years after the end of its disastrous war, Vietnam is reaching its potential on its own terms.

COMPACT CAMBODIA

Former French Indochina contained two additional entities—Cambodia and Laos. In Cambodia, the French possessed one of the greatest treasures of Hinduism: the city of Angkor, capital of the ancient Khmer Empire, and the temple complex known as Angkor Wat. The Khmer Empire prevailed here from the ninth to the fifteenth centuries, and King Suryavarman II built Angkor Wat to symbolize the universe in accordance with the precepts of Hindu cosmology. When the French took control of this area during the 1860s, Angkor lay in ruins, sacked long before by Vietnamese and Thai armies. The French created a protectorate, began the restoration of the shrines, established Cambodia's permanent boundaries, and restored the monarchy under their supervision.

Geographically, Cambodia enjoys several advantages, most notably its compact shape. Compact states enclose a maximum of territory within a minimum of boundary and are without peninsulas, islands, or other remote extensions of the national spatial framework. Cambodia had the further advantage of strong ethnic and cultural homogeneity: 90 percent of its 14.7 million inhabitants are Khmers, with the remainder divided between Vietnamese and Chinese. As Figure 10-9 shows, Southeast Asia's greatest river, the Mekong, enters Cambodia from Laos and crosses it from north to south, creating a great bend before flowing into southern Vietnam (see box titled "The Mighty Mekong"). Phnom Penh, the country's capital, lies on the river. The ancient capital of Angkor lies in the northwest, not far from the Tonle Sap ("Great Lake"), a major lake linked to, and filled by, the waters of the Mekong. Cambodia has a coastline on the Gulf of Thailand, but its core area lies in the interior; the Mekong is more important than the Gulf. Kompong Som, the port at the end of the railroad from Phnom Penh, has only limited facilities.

Postwar Problems

Cambodia's present social geography continues to suffer from the after-effects of the Indochina War. Neither its compactness nor its isolation could overcome the disadvantage of relative location near a conflict that was bound to spill across its borders (Fig. 10-9). Internal disharmony fueled by external interference led in 1970 to the ouster of the last king by the military; in 1975, that regime was itself overthrown by communist revolutionaries, the so-called Khmer Rouge. These new rulers embarked on a course of terror and destruction in order to reconstruct Kampuchea (as they called Cambodia) as a rural society. They drove townspeople into the countryside where they had no place to live or work, emptied hospitals and sent the sick and dying into the streets, outlawed religion and family, and in the process killed as many as 2 million Cambodians (out of

The Mighty Mekong

FROM ITS SOURCE among the snowy peaks of China's Qinghai and Xizang, the Mekong River rushes and flows some 4200 kilometers (2600 mi) to its delta in southernmost Vietnam. This "Danube of Southeast Asia" crosses or borders five of the realm's countries (see photo, p. 537), supporting rice farmers and fishing people, forming a transport route where roads are few, and providing electricity from dams upstream. Tens of millions of people depend on the waters of the Mekong, from subsistence farmers in Laos to apartment dwellers in China. The Mekong Delta in southern Vietnam is one of the realm's most densely populated areas and produces enormous harvests of rice.

But problems loom. China is building a series of dams across the Lancang (as the Mekong is called there) to supply Yunnan Province with electricity. Although such hydroelectric dams should not interfere with water flow, countries downstream worry that a severe dry spell in the interior would impel the Chinese to slow the river's flow to keep the reservoirs full. Cambodia is concerned over the future of the Tonle Sap, a large natural lake filled by the Mekong (Fig. 10-9). In Vietnam, farmers worry about salt water invading the paddies should the Mekong's level

drop. And the Chinese may not be the only dam builders in the future: Thailand has expressed an interest in building a dam on the Thai-Laos border where it is defined by the Mekong.

In such situations, the upstream states have an advantage over those downstream. Several international organizations have been formed to coordinate development in the Mekong Basin, including the Mekong River Commission (MRC) founded half a century ago. China has offered to sell electricity from its dams to Thailand, Laos, and Myanmar. Coordinated efforts to reduce logging in the Mekong's drainage basin have had some effect. After consultations with the MRC, Australia built a bridge linking Laos and Thailand. There is even a plan to make the Mekong navigable from Yunnan to the coast, creating an alternative outlet for interior China.

Sail the Mekong today, however, and you are struck by the slowness of development along this artery. Wooden boats, thatch-roofed villages, and teeming paddies mark a river still crossed by antiquated ferries and flanked by few towns. Of modern infrastructure, one sees little. And yet the Mekong and its basin form the lifeline of mainland Southeast Asia's dominantly rural societies.

a population of 8 million). In the late 1970s, Vietnam, having won its own war, invaded Cambodia to drive the Khmer Rouge away. But this action led to new terror, and a stream of refugees crossed the border into Thailand. Eventually, remnants of the Khmer Rouge managed to establish a base in the northwest of Cambodia, where their murderous leader, Pol Pot, who was never brought to account for his actions, committed suicide in 1998.

Cambodia's postwar trauma continues today. Once self-sufficient and able to feed others, it now must import food. Rice and beans are the subsistence crops, but the dislocation in the farmlands set production back severely, and continuing strife in the countryside has disrupted supply routes for years.

Cambodians hope that tourism, focused chiefly on the great temples at Angkor, will become a mainstay of the economy. But political stability is a prerequisite for this industry, and it has not yet returned to this ancient kingdom.

LANDLOCKED LAOS

North of Cambodia lies Southeast Asia's only landlocked country, Laos (Fig. 10-9). Interior and isolated, Laos changed little during 60 years of French colonial administration (1893–1953). Then, along with other French-ruled areas, it became an independent state. Soon this well-entrenched, traditional kingdom fell victim to rivalries between traditionalists and the victorious communists, and the old order collapsed.

Laos has no fewer than five neighbors, one of which is the East Asian giant, China. The Mekong River forms a long stretch of its western boundary, and the important sensitive border with Vietnam to the east lies in mountainous terrain. With 6.4 million people (about 55 percent of them ethnic Lao, related to the Thai of Thailand), Laos lies surrounded by comparatively powerful states. The country has no railroads, just a few miles of paved roads, and very little industry; it is only 21 percent urbanized (the capital, Viangchan, lies on the Mekong and has an oil pipeline to Vietnam's coast). Laos remains the region's poorest and most vulnerable entity.

PROTRUDED THAILAND

In virtually every way, Thailand is the leading state of the Mainland region. In contrast to its neighbors, Thailand has been a strong participant in the Pacific Rim's economic development. Its capital, Bangkok, is the second-largest urban center in the Mainland region and one of the world's most prominent primate cities. The country's population, 66.1 million in 2008, is growing at the second slowest rate in this geographic realm (only fully urbanized Singapore grows more slowly). Over the past few decades, only political instability and uncertainty have inhibited economic progress. Thailand is a constitutional monarchy; its moves toward stable democracy have been halting. Although Thailand is a constitutional monarchy with an elected parliament, its progress toward stable democracy has a history of setbacks. The latest of these occurred in 2006, when the armed forces ousted a controversial prime minister and took control amid the usual promises of a return to representative government.

Thailand is the textbook example of a protruded state. From a relatively compact heartland, in which lie the core area, capital, and major areas of productive capacity, a 1000-kilometer (600-mi) corridor of land, in places less than 30 kilometers (20 mi) wide, extends southward to the border with Malaysia (Fig. 10-10). The boundary that defines this protrusion runs down the length of the Malay Peninsula to the Kra Isthmus, where neighboring Myanmar peters out and Thailand fronts the Andaman Sea (an arm of the Indian Ocean) as well as the Gulf of Thailand. In the entire country, no place lies farther from the capital than the southern end of this tenuous protrusion.

Consider this spatial situation in the context of Figure 10-4 (p. 530). Note that the Malay ethnic population group extends from Malaysia more than 300 kilometers (200 mi) into Thai territory. In Thailand's six southernmost provinces, 85 percent of the inhabitants are Muslims (the figure for Thailand as a whole is less than 4 percent). The political border between Thailand and Malaysia is porous; in fact, you can cross it at will almost anywhere along the Kolok River by canoe, and inland it is a matter of walking a forest trail. For more than a century, the southern provinces have had closer ties with Malaysia across the border than with Bangkok 1000 kilometers (600 mi) away, and the Thai government has not tried to restrict movement or impose onerous rules on the Muslim population here.

The Restive Peninsular South

But in the new era of the War on Terror and in view of rising violence in this southern frontier zone, Thailand's south is attracting national attention. The Pattani United Liberation Organization, named after the functional capital of the Muslim south, was active in the 1960s and 1970s but had been dormant since. Then a series of bombings at a Pattani hotel, several schools, a Chinese shrine, and a Buddhist temple in 2001 and 2002 raised fears that Muslim-inspired violence was on the rise again. In 2002, Bangkok newspapers reported that police had found bomb-making materi-

FIGURE 10-10   © H. J. de Blij, P. O. Muller, and John Wiley & Sons, Inc.

als in the house of a Muslim religious leader. In 2003, the notorious al-Qaeda terrorist Hambali, who worked to link al-Qaeda with its Southeast Asian counterpart, Jemaah Islamiyah, was captured in Thailand.

The middle years of this decade, however, brought the most ominous developments yet. A terrorist attack on an army base in 2004 killed four Buddhist soldiers and resulted in the theft of hundreds of rifles. And several dozen

people, including Buddhist monks and police officers, were killed in subsequent skirmishes when counterterrorism forces in the town of Takbai arrested six Muslims suspected of stealing weapons from the Thai armed forces. A riot outside the police station where the suspects were being held then led to a mass arrest and the deaths of nearly 80 Muslims during transfer to a military base. The government responded by declaring martial law in Thailand's

AMONG THE REALM'S GREAT CITIES . . . **Bangkok**

BANGKOK (6.8 MILLION), mainland Southeast Asia's second-largest city, is a sprawling metropolis on the banks of the Chao Phraya River, an urban agglomeration without a center, an aggregation of neighborhoods ranging from the immaculate to the impoverished. In this city of great distances and high tropical temperatures, getting around is often difficult because roadways are choked with traffic. A (diminishing) network of waterways affords the easiest way to travel, and life focuses on the busiest waterway of all, the Chao Phraya. Ferries and water taxis carry tens of thousands of commuters and shoppers across and along this bustling artery, flanked by a growing number of high-rise office buildings, luxury hotels, and ultramodern condominiums. Many of these modern structures reflect the Thais' fondness for domes, columns, and small-paned windows, creating a skyline unique in Asia.

Gold is Buddhist Thailand's symbol, adorning religious and nonreligious architecture alike. From a high vantage point in the city, you can see hundreds of golden spires, pagodas, and façades rising above the townscape. The Grand Palace, where royal, religious, and public buildings are crowded inside a white, crenellated wall 2 kilometers (more than one mile) long (see photo, p. 549), is a gold-laminated city within a city embellished by ornate gateways, dragons, and statuary. Across the mall in front of the Grand Palace are government buildings sometimes targeted by rioters in Bangkok's volatile politi-

© H. J. de Blij, P. O. Muller, and John Wiley & Sons, Inc.

cal atmosphere. Not far away are Chinatown and myriad markets; Bangkok has a throbbing commercial life.

Decentralized, dispersed Bangkok has major environmental problems, ranging from sinking ground (resulting in part from excessive water pumping) to some of the world's worst air pollution. Yet industries, port facilities, and residential areas continue to expand as the city's hinterland grows and prospers with Pacific Rim investment. Bangkok is the pulse of Southeast Asia's Mainland region.

three southernmost provinces (Fig. 10-10, inset map). By mid-2005, a series of renewed attacks by southern Muslims had provoked then-Prime Minister Thaksin to issue an emergency decree enabling him to more effectively combat the militants; but its effect was to worsen what was becoming a full-scale regional insurgency. When Thaksin, who had become the focal point of Muslim anger, was deposed by the military in 2006, the new regime immediately sought to negotiate a peace with the insurgents.

The border area nevertheless remained tense. The Thai government has claimed that some Muslim insurgents it arrested had received training in Malaysia's northern State of Kelantan. When several hundred Muslims crossed the border from Thailand into Malaysia to escape the fighting, the Malaysian government refused to return them despite Thailand's claim that some of the refugees were in fact militant separatists.

Religious tensions in this increasingly troubled Islamic outpost should be seen in the context of its geography.

Thailand's pronounced protrusion, its porous border with Muslim Malaysia (the central government now proposes the building of a "security fence" there), its dependence on foreign tourism, and the vulnerable location of its high-profile Andaman Sea coastal tourist facilities (which rebounded swiftly from the 2004 tsunami), combine to raise concern over the stability and security of this distant part of the national territory.

Coreland and Pacific Rim Prospects

As Figure 10-2 shows, Thailand occupies the heart of the Mainland region of Southeast Asia. While Thailand has no Red, Mekong, or Irrawaddy Delta, its central lowland is watered by a set of streams that flow off the northern highlands and the Khorat Plateau in the east. One of these streams, the Chao Phraya, is the Rhine of Thailand. From the head of the Gulf of Thailand to Nakhon Sawan, this river is a highway of traffic. Barge trains loaded with rice head for the coast, ferry boats head upstream, and freighters transport tin and tungsten (of which Thailand is among the world's largest producers). Bangkok sprawls on both sides of the lower Chao Phraya, here flanked by skyscrapers, pagodas, factories, boatsheds, ferry landings, luxury hotels, and modest dwellings in crowded confusion. On the right bank of the Chao Phraya, Bangkok's west side, lie the city's remaining *klong* (canal) neighborhoods, where waterways and boats form the transport system. Bangkok is still known as the Venice of Asia, although many klongs have been filled in and paved over to serve as roadways.

During the Pacific Rim boom of the 1990s, Thailand's economy grew rapidly. Relative location, natural environment, and social conditions all contributed to this growth. Situated at the head of the Gulf of Thailand, the country's core area opens to the South China Sea and hence the Pacific Ocean. The Gulf itself has long been a rich fishing ground and more recently a source of oil (the major refinery lies in the port area of Laem Chabang, an industrial zone just to the north of the tourist resort of Pattaya). But the key to Thailand's economic explosion was the Thai workforce, laboring under often dreadful conditions for low wages to produce goods sold on foreign markets at cheap prices. This labor force attracted major foreign investment that, in turn, stimulated service industries.

For a time, Thailand became a Pacific Rim economic tiger, producing goods ranging from plastic toys to "Japanese" cars. But while the workers toiled, government and financial mismanagement planted seeds of ruin. In 1997, the country's currency, the *baht*, began a slide in value that undermined the economy and led to numerous bank failures. Exports and worker productivity declined, real estate values plunged, and Thailand became the first casualty in a series of economic setbacks that would eventually affect the entire western Pacific Rim.

Although Thailand and the other Pacific Rim economies suffered severe downturns, we should remember that the boom period had yielded many infrastructural improvements not only in the core area but also in distant areas such as the north, centered on Chiang Mai. Thailand, like other Pacific Rim economies, has improved highways, ports, power systems, airports, and housing. Although a period of poor maintenance followed the events of 1997, Thailand's fundamental assets remained to serve the country well when the world economy revived. By 2005, Thailand's GNI per capita had recovered to surpass U.S. $7500 (it is almost $1000 higher today), its balance of trade had been favorable for seven consecutive years, and the tourist industry, buffeted by the post–2001 religious tensions and also by the 2004 tsunami (which took 5400 lives in Thailand), was recovering.

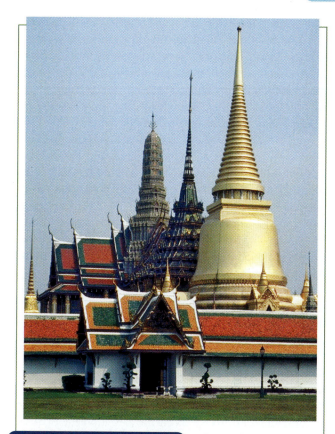

FROM THE FIELD NOTES

"Thailand exhibits one of the world's most distinctive cultural landscapes, both in its urban and its rural areas. Graceful pagodas, stupas, and spirit houses of Buddhism adorn city and countryside alike. Gold-layered spires rise above, and beautify cities and towns that would otherwise be drab and featureless. The architecture of Buddhism has diffused into public architecture, so that many secular buildings, from businesses to skyscrapers, are embellished by something approaching a national style. The grandest expression of the form undoubtedly is a magnificent assemblage of structures within the walls of the Grand Palace in the capital, Bangkok, of which a small sample is shown here. Climb onto the roof of any tall building in the city's center, however, and the urban scene will display hundreds of such graceful structures, interspersed with the modern and the traditional townscape." © H. J. de Blij.

www.conceptcaching.com

Concept Caching

Rebounding Tourism

Under normal circumstances, Thailand's natural environments and cultural landscapes continue to attract millions of visitors annually. The magnificent architecture of Buddhism (the country's old name, Siam, was a Chinese word meaning golden-yellow), swathed in gold and shining in the tropical sun, is without equal, as the photo on this page underscores. The warm coasts and beaches have long made this country a favorite destination for tourists from Germany to Japan. One of the world's most famous resorts is Phuket, on the distant Andaman coast near the southwestern end of Thailand's Malay Peninsula protrusion.

Tourism is Thailand's single leading source of foreign revenues, but there is a dark side to the industry. The Thai people's relaxed attitude toward sex has made Thai tourism in large part a sex industry, candidly advertised in such foreign markets as Japan where "sex tours" attract planeloads of male participants. When the AIDS pandemic reached Thailand in the 1980s, it changed no habits, and by the mid-1990s the country was suffering one of the world's worst outbreaks of the disease. After peaking at over 1 million early in this decade, the number of HIV cases today is about 600,000.

Thailand's territorial morphology creates both problems and opportunities. Problems include the effective integration of the national territory through surface communications, which must cover long distances to reach remote locales; the influx of refugees from neighboring countries dislocated by internal conflict; and control over contraband (opium) drug production and trade in the interior "Golden Triangle" where Myanmar, Thailand, and Laos meet. The government's efforts to speed the development of rural areas, notably the northeast, through megaprojects such as large dams have disrupted local life, damaged ecologies, and deepened the split between city and countryside. On the other hand, Thailand's lengthy Kra Isthmus creates

economic opportunities where southernmost Thailand lies close to Malaysia and Indonesia, raising hopes for a three-country development scheme in that area. Given political stability, Thailand's future as a regional leader appears secure.

MYANMAR'S MISERY

Thailand's also-protruded neighbor, Myanmar (still referred to as Burma in some anti-regime quarters), is one of the world's poorest countries where, it seems, time has stood still for centuries. Long languishing under one of the world's most corrupt and brutal military dictatorships, Myanmar shares with Thailand the neck of the Malay Peninsula—but look again at Figure 10-10 and note the contrast in surface communications.

Morphology and Structure

Myanmar's territorial morphology is complicated by a shift in the Burmese core area that took place during colonial times. Prior to the colonial period, the focus of embryonic Burma lay in the so-called dry zone between the Arakan Mountains and the Shan Plateau, which covers the country's triangular eastern extension toward the Laotian border (Fig. 10-3). The urban focus of the state was Man-

Myanmar (still called Burma by many of those opposed to the current regime that rules this poverty-stricken country) is the exception to Southeast Asia's generally positive picture: this could be a success story on a par with Thailand, Malaysia, and Vietnam. But a ruinous military dictatorship has blighted Myanmar for decades. This photograph of the main street of Yangon (Rangoon) was taken in December 2005. The apparently busy traffic does not reflect a thriving economy: cars here tend to belong to officials, buses are dilapidated, and not far from here bullock carts outnumber vehicles. What is significantly missing is the mass of bicycles (or, farther along on the economic ladder, scooters) filling the streets as in other cities of Southeast Asia. © EPA/Landov

dalay, which had a central situation and relative proximity to the non-Burmese highlands all around. Then the British developed the agricultural potential of the Irrawaddy Delta, and Rangoon (now called Yangon) became the capital and hub of the colony. The Irrawaddy waterway links the old and the new core areas, but the center of gravity has shifted to the south.

The political geography of Myanmar constitutes a particularly good example of the role and effect of territorial morphology on internal state structure (Fig. 10-11). Not only is Myanmar a protruded state: its Irrawaddy coreland is also surrounded on the west, north, and east by a horseshoe of great mountains—where many of the country's 11 minority peoples had their homelands before the British occupation. The colonial boundaries had the effect of incorporating these peoples into the domain of the Burman people, who today constitute about 60 percent of the population of 52.1 million. When the British departed and Burma became independent in 1948, the minorities traded one master for another. In 1976, nine of these indigenous peoples formed a union to demand the right of self-determination in their homelands.

As Figure 10-4 indicates, the peripheral peoples of Myanmar occupy a significant part of the state. A closer look at Figure 10-11 shows

FIGURE 10-11   © H. J. de Blij, P. O. Muller, and John Wiley & Sons, Inc.

that their domains have the status of internal State, of which there are seven; the Burman-dominated areas are designated as Divisions. The Shan of the northeast and far north, who are related to the neighboring Thai, account for almost 8 percent of the population, or 4 million. The Kayin (Karen), who constitute just over 8 percent

(4.3 million), live in the neck of Myanmar's protrusion and have proclaimed their desire to create an autonomous territory within a federal Myanmar. The Mon (2.3 percent, 1.2 million) were in what is today Myanmar long before the Burmans, and introduced Buddhism to the area; they want the return of ancestral lands from which they were ousted. Although the powerful military has dealt such aspirations a series of setbacks, centrifugal forces continue to bedevil the central regime. Its response has been to exert power by all available means rather than to accommodate these forces, even to the point of stifling political discourse—let alone opposition—among the Burman people themselves.

In 2005 media reports began describing the relocation of Myanmar's government offices from Yangon to newly constructed, largely underground facilities in the interior town of Pyinmana, which lies about halfway between Yangon and Mandalay. There, the regime proclaimed, the capital's central location would be more efficient administratively, and its below-ground offices would offer better protection against (unspecified) enemies. In early 2007, when journalists were invited to inspect the new capital, they photographed and reported on eight-lane avenues, large new ministry buildings, and pastel-colored new residential compounds—but were not shown any bunkers or tunnels. They also introduced the name "Nay Pyi Taw" to designate this complex located just west of Pyinmana, which as of mid-2007 is apparently what the regime has decided to call its new headquarters.

Myanmar today ranks among the world's poorest and least-developed countries not because of its intrinsic indigence but because a rapacious military regime, tolerated and even commercially engaged by the "international community," represses its peoples and destroys their aspirations.

● INSULAR SOUTHEAST ASIA

On the peninsulas and islands of Southeast Asia's southern and eastern periphery lie six of the realm's 11 states (Fig. 10-2). Few regions in the world contain so diverse a set of countries. Malaysia, the former British colony, consists of two major areas separated by hundreds of miles of South China Sea. The realm's southernmost state, Indonesia, sprawls across thousands of islands from Sumatera in the west to New Guinea in the east. North of the Indonesian archipelago lies the Philippines, a country that once was a U.S. colony. These are three of the most severely fragmented states on Earth, and each has faced the challenges that such politico-spatial division brings. This Insular region of Southeast Asia also contains two small but important sovereign entities: a city-state and a sultanate. The city-state is Singapore, once a part of Malaysia (and one instance in which internal centrifugal forces were too great to be overcome). The sultanate is Brunei, an oil-rich Muslim territory on the island of Borneo that seems transplanted from the Persian Gulf. In addition, a third small entity, East

Timor, achieved independence in 2002. Few parts of the world are more varied or more interesting geographically.

MAINLAND-ISLAND MALAYSIA

The state of Malaysia represents one of the three subtypes of fragmented states listed earlier: the mainland-island type, in which one part of the national territory lies on a continent and the other on an island. Malaysia is a colonial political artifice that combines two quite disparate components into a single state: the southern end of the Malay Peninsula and the northern part of the island of Borneo. These are known, respectively, as West Malaysia and East Malaysia (Fig. 10-2). The name *Malaysia* came into use in 1963, when the original Federation of Malaya, on the Malay Peninsula, was expanded to incorporate the areas of Sarawak and Sabah in Borneo. When the name *Malaya* is used, it refers to the peninsular part of the Federation, whereas Malaysia refers to the total entity.

Ethnic Components

The Malays of the peninsula, traditionally a rural people, displaced older aboriginal communities there and today make up about 58 percent of the country's population of 27.8 million. They possess a strong cultural identity expressed in adherence to the Muslim faith, a common language, and a sense of territoriality that arises from their perceived Malayan origins and their collective view of Chinese, Indian, European, and other foreign intruders.

The Chinese came to the Malay Peninsula and to northern Borneo in substantial numbers during the colonial period, and today they constitute about one-fourth of Malaysia's population (they are the largest single group in Sarawak).

Hindu South Asians were in this area long before the Europeans, and for that matter before the Arabs and Islam arrived on these shores. Today they still form a substantial minority of 8 percent of the population, clustered, like the Chinese, on the western side of the peninsula (Fig. 10-4).

The Dominant Peninsula

The populous peninsular part of Malaysia remains the country's dominant sector with 11 of its 13 States and over 80 percent of its population. Here the Malay-dominated government has strictly controlled economic and social policies while pushing the country's modernization. During the Asian economic boom of the 1990s, Malaysia's

STATES OF WEST MALAYSIA

0   50   100   150 Km.

0   50   100 Miles

FIGURE 10-12   © H. J. de Blij, P. O. Muller, and John Wiley & Sons, Inc.

Legend:

- "Multimedia Supercorridor"
- States with greatest Islamic strength
- States with largest Chinese and Hindu presence
- —— Highway
- —— Main road
- —— Railroad
- ✈ Airport

POPULATION
- • 50,000–250,000
- ● 250,000–1,000,000
- ⬤ 1,000,000–5,000,000

National capital is underlined

hamad, leader of the Malay-dominated party that forms the majority in government. Mahathir had the support not only of the great majority of the country's Malays, but also that of the important and influential ethnic Chinese minority, which saw him as the only acceptable alternative to the more fundamentalist Islamic party challenging his rule. But Malaysia's headlong rush to modernize caused a backlash among more conservative Muslims, which, in 2001, led to Islamist victories in two States, tin-producing Kelantan and energy-rich but socially poor Terengganu. As the fundamentalist governments in those two States imposed strict religious laws, Malaysians talked of two "corridors" marking their country: the Multimedia Supercorridor in the west and the Mecca Corridor in the east (Fig. 10-12). But after Mahathir's resignation and the appointment of Abdullah Badawi as his more moderate successor, Islamist fervor in the Mecca Corridor, which had featured calls for a *jihad* in Malaysia, began to wane. In the March 2004 elections, the Islamist party lost the two State legislatures it had gained in 2001, and its leader even lost his seat in the federal parliament. Democracy, flawed as it still may be in this Muslim country, was a victor.

The Malay Peninsula's primacy began long ago, during colonial times, when the British created a substantial economy based on rubber plantations, palm-oil extraction, and mining (tin, bauxite, copper, iron). The Strait of Malacca (Melaka) became one of the world's busiest and most strategic waterways, and Singapore, at the southern end of it, a prized possession (Singapore seceded from the Malaysian Federation in 1965).

Malaysia, despite the loss of Singapore and notwithstanding its recurrent ethnic troubles, became a major player on the burgeoning Pacific Rim. The strong skills and modest wages of the local workforce attracted many companies, and the government capitalized on its opportunities, for example, by encouraging the creation of a high-technology manufacturing complex on the island of Pinang (see box titled "Pinang: A Future Singapore?").

planners embraced the notion of symbols: the capital, Kuala Lumpur, was endowed with the world's tallest building; a space-age airport outpaced Malaysia's needs; a high-tech administrative capital was built at Putrajaya; and a nearby development was called Cyberjaya—all part of a so-called *Multimedia Supercorridor* to anchor Malaysia's core area (Fig. 10-12).

The chief architect of this program was Malaysia's long-term and autocratic head of state, Mahathir bin Mo-

FROM THE FIELD NOTES

"Walking along Tuanku Abdul Rahman Street in Kuala Lumpur, I had just passed the ultramodern Sultan Abdul Samad skyscraper when this remarkable view appeared: the old and the new in a country that seems to have few postcolonial hang-ups and in which Islam and democracy coexist. The British colonists designed and effected the construction of the Moorish-Victorian buildings in the foreground (now the City Hall and Supreme Court); behind them rises the Bank of Commerce, one of many banks in the capital. Look left, and you see the Bank of Islam, not a contradiction here in economically diversified Malaysia. And just a few hundred yards away stands St. Mary's Cathedral, across the street from still another bank. Several members of the congregation told me that the church was thriving and that there was no sense of insecurity here. 'This is Malaysia, sir,' I was told. 'We're Muslims, Buddhists, Christians. We're Malays, Chinese, Indians. We have to live together. By the way, don't miss the action at the Hard Rock Café on Sultan Ismail Street.' Now there, I thought, was a contradiction as remarkable as this scene." © H. J. de Blij.

Concept Caching

www.conceptcaching.com

Malaysian Borneo

The decision to combine the 11 Sultanates of Malaya with the States of Sabah and Sarawak on Borneo, creating the country now called Malaysia, had far-reaching consequences. These two States make up 60 percent of Malaysia's territory (although they represent only 19 percent of the population). They endowed Malaysia with major energy resources and huge stands of timber. They also complicated Malaysia's ethnic makeup because each State is home to more than two dozen indigenous groups (in fact, the immigrant Chinese form the largest single group in Sarawak). These locals complain that the federal government in Kuala Lumpur treats East Malaysia as a colony, and politics here

Pinang: A Future Singapore?

IN THE NORTHWESTERN corner of Malaysia's Malay Peninsula, opposite the northern end of Indonesian Sumatera, where the Strait of Malacca (Melaka) opens into the Indian Ocean, lies an island called Pinang (Fig. 10-12). A thriving outpost during the British colonial period, Pinang (capital: George Town), like Singapore, was one of the so-called Straits Settlements.

Today, Singapore no longer is a part of Malaysia, but Pinang's importance is growing rapidly. Now connected to the mainland by one of Asia's longest bridges (14 kilometers [9 mi]), the 250-square-kilometer (100-sq-mi) island has a population of over 1 million. Chinese outnumber Malays by 60 to 32 percent, and George Town, the old colonial port, has become the focus of an expanding high-technology manufacturing complex for the international computer industry. The 65-story Tun Abdul Razak Complex, housing government offices, businesses, department stores, and recreational facilities, is just one landmark in an ultramodern skyline. From the top you can see modern industrial parks with neon-lit signs proclaiming the presence of such companies as Intel, SONY, Philips, Motorola, and Hitachi.

The rapid development of Pinang is a calculated strategy by the Malaysian government. The emphasis is on building a highly skilled labor force that is locally nurtured in technical schools and training facilities financed by the multinational corporations themselves. Malaysia is making infrastructure investments ranging from the new toll bridge to a major airport. George Town's colonial and multicultural townscape is attracting a growing tourist trade, and new hotels and resorts are enlarging the service industry.

Malaysia hopes that Pinang will become the leading corner of a major *growth triangle* (a concept that is gaining popularity along the Pacific Rim), linking up with Indonesian Sumatera to the west and southern Thailand to the north. Meanwhile, Pinang is following in Singapore's economic footsteps. But Kuala Lumpur's politicians surely hope that Chinese-dominated Pinang will not follow Chinese-dominated Singapore's political course.

are contentious and fractious. It is likely that Malaysia will eventually confront devolutionary forces in East Malaysia.

SINGAPORE

In 1965, a fateful event occurred in Southeast Asia. Singapore, crown jewel of British colonialism in this realm, seceded from the recently independent (1963) Malaysian Federation and became a sovereign state, albeit a mini-state (Fig. 10-13). With its magnificent relative location, its human resources, and its firm government, Singapore then overcame the limitations of space and the absence of raw materials to become one of the economic tigers on the Pacific Rim.

With a mere 619 square kilometers (239 sq mi) of territory and a population of 4.6 million, space is at a premium in Singapore, and this is a constant worry for the government. The Jurong Reclamation Project has recently joined several islands and enlarged Singapore's territory (Fig. 10-13), but Singapore's leaders are always trying to find additional ways to add space. One plan is to ask Indonesia to, in effect, lease one of its many islands to Singapore, allowing Singapore to convert it into a giant industrial complex administered from the city-state. Already, Singapore has built factories on the nearby Indonesian islands of Batam and Bintan (Fig. 10-13), part of a string of industrial parks, ports, and resorts it has constructed from China to Thailand. The proximity of Batam and Bintan makes them especially attractive for large-scale development of this kind, and Indonesia may find Singapore's offer too

FIGURE 10-13   © H. J. de Blij, P. O. Muller, and John Wiley & Sons, Inc.

FROM THE FIELD NOTES

"My first visit to Singapore was by air, and I got to know the central city, where I was based, rather well. But the second was by freighter from Hong Kong, and this afforded quite a different perspective. Here was one reason for Singapore's success: I have simply never seen a neater, cleaner, better organized, or more modern port facility. Even piles of loose items were carefully stacked. Loading and offloading went quietly and orderly. Singapore is one of the world's leading *entrepôts*, where goods are brought in, stored, and transshipped. Nobody does it better, and Singapore carefully guards its reputation for dependability, on-schedule loading, and lack of corruption." © H. J. de Blij.

www.conceptcaching.com

lucrative to decline. In 2007, the first exploratory discussions (relating only to Bintan) were under way.

In any case, Singapore's expanding economy requires space-conserving high-technology and service industries in this new era. At first, benefiting from its relative location on the key corner of Asia, Singapore became one of the world's busiest ports (by number of ships served) even before independence. It thrived as an ***entrepôt*** between the Malay Peninsula, Southeast Asia, Japan, and other emerging economic powers on the Pacific Rim and beyond. Crude oil from Southeast Asia still is unloaded and refined at Singapore, then shipped to Asian destinations. Raw rubber from the adjacent peninsula and from Indonesia's island of Sumatera is shipped to Japan, the United States, China, and others. Timber from Malaysia, rice, spices, and other foodstuffs are processed and forwarded via Singapore. In return, automobiles, machinery, and equipment are imported into Southeast Asia through Singapore.

But that is the old pattern. Singapore's leaders are redirecting the city-state's economy and moving toward high-tech industries for the future. In Singapore the government tightly controls business as well as other aspects of life. (Some newspapers and magazines have been banned for criticizing the regime, and there are even fines for such things as eating on the subway and failing to flush a public toilet.) Its overall success after secession has tended to keep the critics quiet: while GNI per capita from 1965 to 2005 multiplied by a factor of more than 15—to U.S. $29,780—that of neighboring Malaysia reached just U.S. $10,320. Among other things, Singapore became (and for many years remained) the world's largest producer of disk drives for small computers.

To accomplish its revival, Singapore has moved in several directions. First, it is concentrating on three growth areas: information technology, automation, and biotechnology. Second, there are notions of a ***growth triangle*** involving Singapore's developing neighbors, Malaysia and Indonesia; those two countries would supply the raw materials and cheap labor, and Singapore the capital and technical know-how. Third, Singapore opened its doors to capitalists of Chinese ancestry who left Hong Kong when China took it over and who wanted to relocate their enterprises here. Singapore's population is 77 percent Chinese, 14 percent Malay, and 8 percent South Asian. The government is Chinese-dominated, and its policies have served to sustain Chinese control. Indeed, Singapore's combination of authoritarianism and economic success often is cited in China itself as proving that communism and market economies can coexist.

INDONESIA'S ARCHIPELAGO

The world's fourth-largest country in terms of human numbers is also the globe's most expansive **archipelago**. Spread across a chain of more than 17,000 islands, Indonesia's more than 230 million people live both separated and clustered—separated by water and clustered on islands large and small.

The complicated map of Indonesia (Fig. 10-14) requires close attention. Five large islands dominate the archipelago territorially, but one of these, *New Guinea* in the east, is not part of the Indonesian culture sphere, although its western

AMONG THE REALM'S GREAT CITIES . . . **Jakarta**

JAKARTA, CAPITAL OF Indonesia and the realm's largest city, sometimes is called the Kolkata (Calcutta) of Southeast Asia. Stand on the elevated highway linking the port to the city center and see the villages built on top of garbage dumps by scavengers using what they can find in the refuse, and the metaphor seems to fit. There is poverty here unlike that in any other Southeast Asian metropolis.

But there are other sides to Jakarta. Indonesia's economic progress has made its mark here, and the evidence is everywhere. Television antennas and satellite dishes rise like a forest from rusted, corrugated-iron rooftops. Cars (almost all, it seems, late-model), mopeds, and bicycles clog the streets, day and night. A meticulously manicured part of the city center contains a cluster of high-rise hotels, office buildings, and apartments. Billboards advertise planned communities on well-located, freshly cleared land.

Jakarta's population is a cross-section of Indonesia's, and the silver domes of Islam rise above the cityscape alongside Christian churches and Hindu temples. The city always was cosmopolitan, beginning as a conglomerate of villages at the mouth of the Ciliwung River under Islamic rule, becoming a Portuguese stronghold and later the capital of the Dutch East Indies (under the name Batavia). Advantageously situated on the northwestern coast of Jawa, Indonesia's most populous island, Jakarta bursts at the seams with growth. Sail into the port, and hundreds of vessels, carrying flags from Russia to Argentina, await berths. Travel to the outskirts, and huge shantytowns are being expanded by a constant stream

© H. J. de Blij, P. O. Muller, and John Wiley & Sons, Inc.

of new arrivals. So vast is the human agglomeration—nobody really knows how many millions have descended on this place (the official figure is 14.5 million)—that the majority live without adequate (or any) amenities.

A seemingly permanent dome of metallic-gray air, heavy with pollutants, rests over the city. The social price of economic progress is high, but Jakartans are willing to pay it.

half is under Indonesian control. The other four major islands are collectively known as the Greater Sunda islands: *Jawa* (Java), smallest but by far the most populous and important; *Sumatera* (Sumatra) in the west, directly across the Strait of Malacca from Malaysia; *Kalimantan*, the Indonesian sector of large, compact, minicontinent Borneo; and wishbone-shaped, distended *Sulawesi* (formerly Celebes) to the east. Extending eastward from Jawa are the Lesser Sunda Islands, including Bali and, near the eastern end, Timor. Another important island chain within Indonesia is the Maluku (Molucca) Islands, between Sulawesi and New Guinea. The central water body of Indonesia is the Java Sea.

As the map shows, Indonesia does not control all of the archipelago. In addition to East Malaysia's territory on Borneo, there also is Brunei (see box titled "Rich and Broken Brunei"). And on the eastern part of the island of Timor, a long struggle against Indonesian rule led, in 1999, to a referendum on independence. The overwhelming majority of

East Timorese voted in favor of sovereignty, and after guidance from the United Nations East Timor achieved sovereignty in 2002 (this country is discussed separately on pp. 564–565).

The Major Islands

Indonesia is a Dutch colonial creation, and the Hollanders chose Jawa as their colonial headquarters, making Batavia (now Jakarta) their capital. Today, Jawa remains the core of Indonesia. With more than 130 million inhabitants, Jawa is one of the world's most densely peopled places and one of the most agriculturally productive. It also is the most highly urbanized part of a country in which nearly 60 percent of the people still live on the land, and the Pacific Rim boom of the 1990s had strong impact here. The city of Jakarta (14.5 million), on the northwestern coast, became

FIGURE 10-14 © H. J. de Blij, P. O. Muller, and John Wiley & Sons, Inc.

the heart of a larger conurbation now known as **Jabotabek**, consisting of **Ja**karta as well as **Bo**gor, **Ta**ngerang, and **Bek**asi. During the 1990s the population of this megalopolis grew from 15 to 20 million, and it is predicted to exceed 30 million by 2010. Already, Jabotabek is home to over 10 percent of Indonesia's entire population and 25 percent of its urban population. Thousands of factories, their owners taking advantage of low prevailing wages, were built in this area, straining its infrastructure and overburdening the port of Jakarta. On an average day, hundreds of ships lie at anchor, awaiting docking space to offload raw materials and take on finished products.

Always in Indonesia, Jawa is where the power lies. As a cultural group (though itself heterogeneous), the Jawanese constitute about 45 percent of the country's population. In

the center of the island, a politically powerful sultanate centers on Yogyakarta; Jawa also is the main base for the two national Islamic movements, one comparatively moderate and the other increasingly revivalist.

Major active volcanoes in or near densely populated areas always raise concern, and on the island of Jawa such worries focus on Mount Merapi, located in the middle of the Semarang–Yogyakarta–Surakarta triangle (Fig. 10-14). In May 2006, Mount Merapi showed signs of an impending eruption (see photo, p. 527), and thousands of villagers and townspeople were moved from the slopes of the nearly 3000-meter (10,000-ft) high volcano that, in 1867, caused death and destruction over a wide area. Smaller, but also deadly, outbursts occurred in 1930 and 1994. The 2006 threat did not materialize, but the towering mountain may

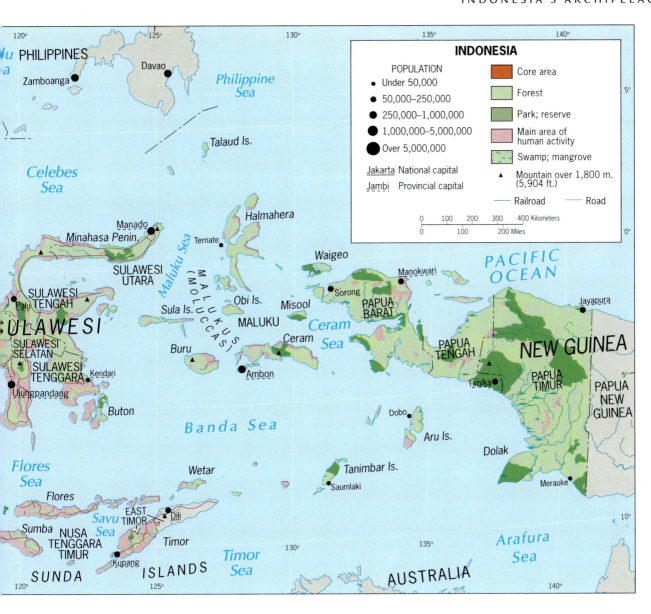

someday join Tambora and Krakatau in the annals of Indonesian catastrophes.

Sumatera, Indonesia's westernmost island, forms the western shore of the busy Strait of Malacca; Singapore lies across the Strait from approximately the middle of the island. Although much larger than Jawa, Sumatera has only about one-third as many people (49 million). In colonial times the island became a base for rubber and palm-oil plantations; its high relief makes possible the cultivation of a wide range of crops, and neighboring Bangka and Belitung yield petroleum and natural gas. Palembang is the key urban center in the south, but current attention focuses on the north. There, the Batak people accommodated colonialism and Westernization and made Medan one of Indonesia's Pacific Rim boom cities. Farther north the Aceh fought

the Dutch into the twentieth century and, after Indonesia became a sovereign country, demanded autonomy and even outright independence for their State. Rebels fought the Indonesian army to a costly stalemate, and thousands died in the conflict—which would probably continue today but for a dramatic turn of events. The seafloor epicenter of the December 26, 2004 Indian Ocean tsunami lay near the far northern coast of Sumatera, and Aceh was directly in the path of the most powerful ocean waves. Entire towns were swept away and tens of thousands died; Banda Aceh, the capital city, was devastated (Fig. 10-14). The international relief effort opened Aceh to foreigners in ways the Indonesians had long prevented, and the Indonesian army as well as the rebels were engaged in rescue missions rather than warfare. This combination of circumstances facilitated a

Rich and Broken Brunei

BRUNEI IS AN anomaly in Southeast Asia—an oil-exporting Islamic sultanate far from the Persian Gulf. Located on the north coast of Borneo, sandwiched between Malaysian Sarawak and Sabah (Fig. 10-7, upper-left map), the Brunei sultanate is a former British-protected remnant of a much larger Islamic kingdom that once controlled all of Borneo and areas beyond. Brunei achieved full independence in 1984. With a mere 5770 square kilometers (2228 sq mi)—slightly larger than Delaware—and only 430,000 people, Brunei is dwarfed by the other political entities of Southeast Asia (Fig. 10-1). But the discovery of oil here in 1929 (and natural gas in 1965) heralded a new age for this remote territory.

Today, Brunei is one of the largest oil producers in the British Commonwealth, and recent offshore discoveries suggest that production will increase. As a result, the population is growing rapidly through immigration (67 percent of Brunei's residents are Malay, 15 percent Chinese), and the sultanate enjoys one of the highest standards of living in Southeast Asia. In 1998, however, a financial scandal of gigantic proportions was revealed, involving the disappearance of more than U.S. $15 billion and the collapse of Brunei's largest private company, which had been under the direction of the sultan's youngest brother. This financial calamity continued to buffet the country's economy in this decade and was reflected by a steep drop in its per-capita GNI.

Most of Brunei's inhabitants live and work near the oil-fields in the western corner of the country and in the capital, Bandar Seri Begawan. Evidence of profligate spending is everywhere, ranging from sumptuous palaces to magnificent mosques to luxurious hotels. In this respect Brunei offers a stark contrast to neighboring East Malaysia, but even within Brunei there are spatial differences. Brunei's interior remains an area of subsistence agriculture (a small minority of indigenous groups survive here) and rural isolation, its villages a world apart from the modern splendor of the coast.

Brunei's territory is not only small: it is fragmented. A sliver of Malaysian land separates the larger west from the east (where the capital lies), and Brunei Bay provides a water link between the two parts. It has been proposed that Brunei purchase this separating corridor from the Malaysians, but no action has yet been taken.

---

truce and negotiations resulting in an agreement under which the rebels agreed to drop their demand for independence and the Indonesian army withdrew.

The settlement in Aceh Province had positive effects throughout Indonesia and was viewed by some cultural geographers as a template for the troubled provinces on the island of Papua (see below). But in Aceh itself the blessing was mixed. In 2003 the government of Aceh became the first of Indonesia's 33 provinces to put Sharia Islamic law onto the books, but it was Aceh's post-tsunami self-government that made implementation possible. So an area once known for its open society and bustling economy introduced a new force of "Sharia officers" to police social behavior, arresting unmarried couples, drunks, and gamblers and hauling them to public canings at mosques. In July 2006, the caning of a man arrested for drinking at the beach was televised nationally. Indonesia's moderate Muslims, used to the country's historic open-minded interpretation of Islam, are worried that Indonesia's secular 1945 Constitution will not be enough to protect the country from the kind of Muslim revivalism that is spreading elsewhere.

Kalimantan is the Indonesian part of the minicontinent of Borneo, a slab of the Earth's crystalline crust whose backbone of mountains is of erosional, not volcanic, origin. Compact and massive—at over 750,000 square kilometers (300,000 sq mi) it is larger than Texas—Borneo has a deep, densely forested interior that is a last refuge of some 35,000 orangutans. Elephants, rhinoceroses, and tigers still survive even as the loggers constrict their habitat (photo p. 563). Numerous other species of fauna—insects, reptiles, mammals, birds—inhabit the rich mountain and lowland forests. So much of Borneo's Pleistocene heritage has survived principally because of its comparatively small human population. The Indonesian part, Kalimantan, constitutes 28 percent of the national territory, but, with 14 million people, it contains barely 6 percent of Indonesia's population.

Figure 10-14 shows that the main areas of human activity lie in the west, southeast, and east. All towns of any size, including Pontianak in the west and Balikpapan in the east, are on or near the coast; the rivers still form important routes into the interior. Parks and reserves to protect flora and fauna cover but a small portion of the vast territory. Not only the flora and fauna, but also the human population has ancient roots. Indigenous peoples, including some of the Dayak clans, continue their slash-and-burn subsistence in the interior; others have become sedentary farmers. But Indonesia's transmigration policy has brought about half a million Jawanese and Madurese to Kalimantan, and they are laying out farms and digging drainage canals, bringing drastic change to rural areas. Of Kalimantan's four provinces, only the southernmost exports significant raw materials, including oil from offshore reserves, coal, iron ore, and some gold

FROM THE FIELD NOTES

"I drove from Manado on the Minahasa Peninsula in northeastern Sulawesi to see the ecological crisis at Lake Tondano, where a fast-growing water hyacinth is clogging the water and endangering the local fishing industry. On the way, in the town of Tomolon, I noticed this side street lined with prefabricated stilt houses in various stages of completion. These, I was told, were not primarily for local sale. They were assembled from wood taken from the forests of Sulawesi's northern peninsula, then taken apart again and shipped from Manado to Japan. 'It's a very profitable business for us,' the foreman told me. 'The wood is nearby, the labor is cheap, and the market in Japan is insatiable. We sell as many as we can build, and we haven't even begun to try marketing these houses in Taiwan or China.' At least, I thought, this wood was being converted into a finished product, unlike the mounds of logs and planks I had seen piled up in the ports of Borneo awaiting shipment to East Asia."   © H. J. de Blij.

Concept Caching

www.conceptcaching.com

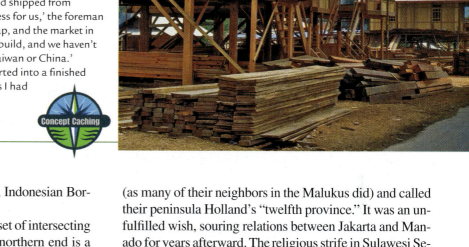

and diamonds; but in the national picture, Indonesian Borneo remains little developed.

Sulawesi is an island that looks like a set of intersecting mountain ranges rising from the sea. Its northern end is a 800-kilometer (500-mi) chain of extinct, dormant, and active volcanoes known as the Minahasa Peninsula, its future northern extension toward the Philippine Sea already marked by a ribbon of seafloor volcanoes just rising above the surface. So rugged is the relief on this island that you cannot take surface transportation from Palu to Ujungpandang. The ethnic mosaic is quite complex, with seven major groups inhabiting more or less isolated parts of the island as well as a sizeable cluster of Jawanese transmigrants who are concentrated in Sulawesi Tengah (Fig. 10-14). Sulawesi is unusual in Indonesia in that its Muslim and Christian numbers are just about equal. Since 1999, the level of strife between local Christians and immigrant Jawanese Muslims has risen and has destabilized this remote province of Sulawesi—forcing Christians who once lived peacefully among Muslim villagers to flee and culturally segregating settlements.

The two most populous and developed parts of Sulawesi are the southwestern peninsula centered on Ujungpandang and the eastern end of the Minahasa Peninsula, where Manado is the urban focus. Subsistence farming occupies most of the population of 18 million, but logging and wood products, some mining, and fishing augment the cash economy (see photo above). Cultural landscapes are varied; in Minahasa, Christian churches remain numerous and form reminders of the colonial period, when relations between this province and the Dutch were favorable. During the liberation period, Minahasans sided with the Dutch

(as many of their neighbors in the Malukus did) and called their peninsula Holland's "twelfth province." It was an unfulfilled wish, souring relations between Jakarta and Manado for years afterward. The religious strife in Sulawesi Selatan (Fig. 10-14), however, has been more enduring: by 2004 the Indonesian military had succeeded in pacifying the area, but animosities between Christians and Muslims, migrants and locals run deep, and only time and improving livelihoods are likely to soften them.

Papua, like East Timor, fell to Indonesia well after the Dutch colonial era had ended, but unlike East Timor, the United Nations approved Indonesia's takeover in 1969. Until 2002 this province was known by its Bahasa Indonesian name of *Irian Jaya* (Guinea West), but in that year the government of President Megawati Sukarnoputri bowed to the demands of indigenous leaders of the Papuan communities on the island and changed its name. As the map shows, the eastern half of the island of New Guinea comprises the independent state of Papua New Guinea. The Papuans are divided not only into numerous ethnolinguistic groups but also into two political entities.

Papua is an Indonesian province, but it lies in the Pacific geographic realm, not in Southeast Asia. As such, it has about 22 percent of the country's land area but just over 1 percent of its population, including more than 200,000 ethnic Indonesians, most from Jawa and many of them concentrated in the provincial capital with the non-Papuan name of Jayapura. Forested, high-relief, glacier-peaked Papua, lying as it does in a different geographic realm, is a world apart from teeming Jawa. It has what is reputedly the world's richest gold mine and second-largest open-pit copper mine;

its forests not only harbor the indigenous peoples but also sustain a major logging industry.

Sanctioned as Indonesia's rule in Papua may be, local opposition has been much in evidence for decades. The *Organisasi Papua Merdeka* (Free Papua Movement, or FPM) and occasional rebel attacks remind the Jakarta government of its role here. In 1999, during the turmoil in the other provinces, the FPM held rallies in Jayapura to promote its cause, and in 2000 a congress was held where independence was demanded and a Papuan flag, the Morning Star, displayed. Later that year, Indonesian troops killed demonstrators in Merauke on the province's south coast. In 2003, the Indonesian government divided Papua into three administrative provinces, and shortly thereafter several executives of a metals corporation accused of polluting Papuan waters and poisoning local residents were arrested, only to be exonerated months later, but not before the world was alerted to the allegations of corporate misdeeds in remote Papua. All this deepened Papuan opposition to Jakarta's rule, and the time may come when Indonesia will rue its acquisition of Papua, potential source of the kinds of centrifugal forces an archipelagic state must constantly confront.

Diversity in Unity

Indonesia's survival as a unified state is as remarkable as India's and Nigeria's. With more than 300 discrete ethnic clusters, over 250 languages, and just about every religion practiced on Earth (although Islam dominates), actual and potential centrifugal forces are powerful here. Wide waters and high mountains perpetuate cultural distinctions and differences. Indonesia's national motto is **bhinneka tunggal ika**: diversity in unity.

What Indonesia has achieved is etched against the country's continuing cultural complexity. There are dozens of distinct aboriginal cultures; virtually every coastal community has its own roots and traditions. And the majority, the rice-growing Indonesians, include not only the numerous Jawanese—who have their own cultural identity—but also

FROM THE FIELD NOTES

"Getting to Ambon was no easy matter, and my boat trip was enlivened by an undersea earthquake that churned up the waters and shook up all aboard. The next morning our first view of the town of Ambon, provincial capital of Maluku, showed a center dominated by a large mosque with a modern minaret and a large white dome to the right of the main street leading to the port. My host from the local university told me that, as on many of the islands of the Malukus, about half of the people were Christians, not Muslims, and that some large churches were situated in the outskirts. 'But the relationships between Muslims and Christians are worsening,' he said. It had to do with the arrival of Jawanese, some of whom were 'agitators' and Islamic fundamentalists, stirring up religious passions. 'Have you heard of the Taliban?' he asked. 'Well, we have similar so-called religious students here, educated in Islamic schools that teach extremism.' . . . We drove from the town into the countryside to the west, toward Wakisihu and past the old Dutch Fort Rotterdam. 'Let me show you something,' he said as we turned up a dirt road. At its end, in the middle of a field, sat a large single structure called University of Islam, distinguished by a pair of enormous stairways. 'Can you call a single building like this where all they teach is Islam a university?' he asked. 'What they teach here is Islamic fundamentalism and intolerance.' He was prophetic. Weeks later, religious conflict broke out, the mosque in town was burned along with Christian churches; his own university lay in ruins."   © H. J. de Blij.

Concept Caching

the Sundanese (who constitute 14 percent of Indonesia's population), the Madurese (8 percent), and others. Perhaps the best impression of the cultural mosaic comes from the string of islands that extends eastward from Jawa to Timor (Fig. 10-14). The rice-growers of Bali adhere to a modified version of Hinduism, giving the island a unique cultural atmosphere; the population of Lombok is mainly Muslim, with some Balinese Hinduism; Sumbawa is a Muslim community; Flores is mostly Roman Catholic. In western Timor, Protestant groups dominate; in the now-independent east, where the Portuguese ruled, Roman Catholicism prevails. Nevertheless, Indonesia nominally is the world's largest Muslim country: overall, 88 percent of the people adhere to Islam, and in the cities the silver domes of neighborhood mosques rise above the townscape. But Islam is not (perhaps not yet) the issue it is in Malaysia, where observance generally is stricter and where minorities fear Islamization as Malay power grows.

This does not mean that Indonesia is immune to extremism (photos p. 562) and even terrorism. Serious terrorist attacks have occurred in the tourist areas of Bali and in central Jakarta, mounted by a terrorist organization named **Jemaah Islamiyah (JI)**, apparently headquartered in eastern Jawa. In addition to Indonesians, many Australians, for whom Indonesia, and especially Bali, is a popular tourist destination, have lost their lives in these attacks. And in late 2005, the Australian government announced the capture of a JI cell active in Australia itself.

Indonesia still has vast areas of rainforest from Sumatera to Papua, but the forest is under assault in many parts of this populous country, with disastrous impact on biodiversity, soils, and rivers. High-quality wood fetches high prices in markets such as Japan and China, and for decades Indonesia's corrupt authoritarian regime actively participated in the plunder. More recently, democratically elected governments have sought to curb the rate of destruction, but corruption continues to impede these efforts. Here in Kalimantan, on the Indonesian side of Borneo, about 10,000 square kilometers (4,000 sq mi) of rainforest are lost every year. This depressing photograph shows the frontier of clear-cutting in Kalimantan Barat (West Kalimantan). © Wayne G. Lawler/Photo Researchers, Inc.

Transmigration and the Outer Islands

As noted earlier, Indonesia's population of 231.9 million makes it the world's fourth most populous country, but Jawa, we also noted, contains approximately 55 percent of it. With more than 130 million people on an island the size of Louisiana, population pressure is enormous here. Moreover, Indonesia's annual rate of population growth remains at the realm average of 1.4 percent. To deal with this problem, and at the same time to strengthen the core's power over outlying areas, the Indonesian government long pursued a policy known as **transmigration**, inducing Jawanese (and Madurese from the adjacent island of Madura) to relocate to other islands. Several million Jawanese have moved to locales as distant as the Malukus, Sulawesi, and Sumatera; many Madurese were resettled in Kalimantan.

In fact, the Dutch colonialists started the Transmigration Program, but then President Sukarno abandoned it and it was not revived until 1974, when President Suharto saw economic and political opportunities in it. Over the next 25 years, nearly 8 million people were relocated; the World Bank gave Indonesia nearly U.S. $1 billion to facili-

tate the program before it pulled out amid reports of forced migration.

The results were mixed. World Bank project reviews suggested that the vast majority of the transmigrants were happy with the land, schools, medical services, and other amenities with which they were provided, but independent surveys painted a different picture. About half of the migrants never managed to rise above a life of subsistence; many never received even the most basic needs for survival in their new environment; land was simply taken away from indigenous inhabitants, notably in Kalimantan; the newcomers did much damage to the traditional cultures they invaded; and the program led to massive forest and wetland destruction (see photo above). And thousands of migrants died in clashes with local inhabitants.

In 2001, the Transmigration Program was canceled as one of the final acts of the short-lived government of President Abdurrahman Wahid, but the damage it did will trouble Indonesia for generations. From northern Sumatera to western Kalimantan to central Sulawesi to the islands of the Malukus to Papua, cultural strife and its polarizing effect will challenge Indonesian governments of the future.

20

However, Indonesia's fluctuating economic fortunes, unstable and unproductive politics, and rampant graft and corruption tend to overwhelm such good intentions, and the country's future remains clouded.

EAST TIMOR

As Figure 10-14 shows, Timor is the easternmost of the major Lesser Sunda Islands, and during the Dutch colonial period the Portuguese maintained a colony on the eastern half of it. When Indonesia shook off Dutch rule in the late 1940s, the Portuguese hung on in Timor, but in 1975 Indonesian forces overran the colony and annexed it formally in 1976. Indonesian rule, however, was even less benign than the Portuguese had been, and soon a bitter struggle for independence was under way. In 1999, after much death and destruction, East Timor's 800,000 inhabitants were allowed, under UN supervision, to vote on independence. They voted overwhelmingly in favor, provoking Indonesia's armed forces to redouble their ruthless repression. East Timor's already modest infrastructure was devastated, the death toll mounted daily, and only foreign intervention, led by Australia (which, among all the world's states, had been Indonesia's sole supporter in the matter), established a degree of order. East Timor became an independent state in 2002, the 198th member of the United Nations.

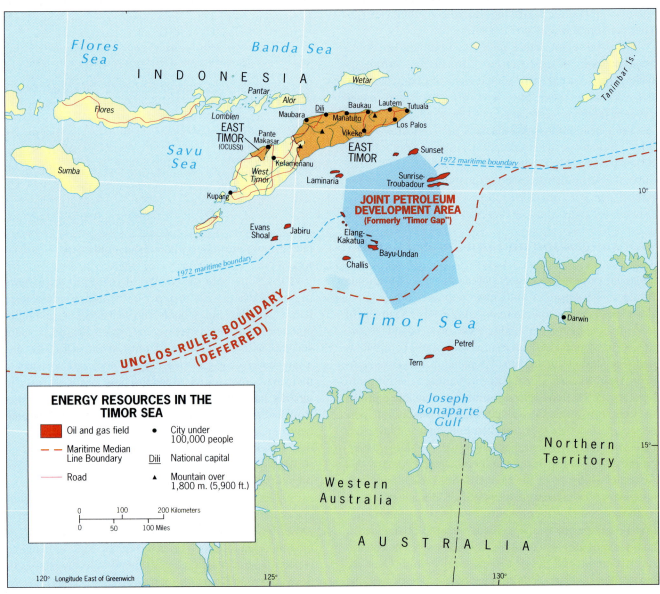

FIGURE 10-15 © H. J. de Blij, P. O. Muller, and John Wiley & Sons, Inc.

Subsequently East Timor's leaders proclaimed the official name of their country to be **Timor-Leste**, the Portuguese version of "Timor East." As in the case of Côte d'Ivoire in West Africa, the French name for Ivory Coast, we use the two options interchangeably.

Nation-Building Nightmare

Nation-building in this Connecticut-sized country continues to be a difficult proposition. Figure 10-15 helps explain why Australia supported Indonesia in its effort to absorb East Timor: by doing so, Australia would benefit from Indonesia's compliance in the delimitation of maritime boundaries in the Timor Sea, giving Australia a larger share of the known oil and gas reserves than an independent East Timor would be likely to relinquish. In the end, difficult negotiations between powerful and technologically capable Australia and weak and incapable East Timor produced a settlement that gave East Timor an acceptable share of future revenues—but not before Australia's international reputation for fairness was further damaged. On the map, the red maritime boundary is the one delimited under United Nations regulations; the blue boundary represents the bilateral agreement between Indonesia and Australia. At issue is the "gap" between them, most of it on East Timor's side but now a "joint petroleum development area."

But the billions of dollars in prospect cannot overcome the devastation and abject poverty that bog down the nation-building project here. As Figure 10-15 shows, East Timor is a fragmented country, with a dominant east (where the coastal capital, Dili, is located) and a small exclave on the north coast of Indonesian West Timor called *Ocussi* (sometimes mapped as "Ocussi-Ambeno Province"). Although there is a road from this exclave to the main territory, relations between East Timor and Indonesia make its use impractical. Worse, the three-year United Nations buildup to independence in 2002 launched a society totally unprepared to go it alone. Good government has been in short supply. The freedom fighters who were demobilized upon independence were organized into police and army units that soon fell apart; widely available weapons empowered gangs and criminals, ethnic hostilities erupted, returned exiles and locals were at odds, and East Timor's still-fragile institutions collapsed. East Timor is a predominantly agricultural country, but while United Nations experts were trying to erect political and social institutions (a judiciary, banking system, medical establishment, public education) farming was neglected and poverty worsened. Now East Timor, its population of 1.1 million growing explosively (in 2006, the country exhibited one of the highest fertility rates in the world with 6.3 children

per woman), has about 100,000 people concentrated in squalid refugee camps.

Given the limited dimensions of East Timor, its relatively small population, its prospective oil income, and foreign assistance, the failure of nation-building here is a warning to the world that independence and freedom do not guarantee well-being and progress.

FRAGMENTED PHILIPPINES

North of Indonesia, across the South China Sea from Vietnam, and south of Taiwan lies an archipelago of more than 7000 islands (only a few hundred of them larger than one square kilometer [0.4 sq mi] in area) inhabited by 90.1 million people. The inhabited islands of the Philippines can be viewed as three groups: (1) Luzon, largest of all, and Mindoro in the north, (2) the Visayan group in the center, and (3) Mindanao, second largest, in the south (Fig. 10-16). Southwest of Mindanao lies a small group of islands, the Sulu Archipelago, nearest to Indonesia, where a Muslim-based insurgency has kept the area in turmoil.

Few of the generalizations we have been able to make for Southeast Asia could apply in the Philippines without qualification. The country's location relative to the mainstream of change in this part of the world has had much to do with this situation. The islands, inhabited by peoples of Malay ancestry with Indonesian strains, shared with much of the rest of Southeast Asia an early period of Hindu cultural influence, which was strongest in the south and southwest and diminished northward. Next came a Chinese invasion, felt more strongly on the largest island of Luzon in the northern part of the Philippine archipelago. Islam's arrival was delayed somewhat by the position of the Philippines well to the east of the mainland and to the north of the Indonesian islands. The few southern Muslim beachheads were soon overwhelmed by the Spanish invasion during the sixteenth century. Today the Philippines, adjacent to the world's largest Muslim state (Indonesia), is 83 percent Roman Catholic, 9 percent Protestant, and only 5 percent Muslim.

Muslim Insurgency

The Muslim population, concentrated on the southeastern flank of the archipelago (Fig. 10-16), has long proclaimed its marginalization in this predominantly Christian country. Over the past 30 years, a half-dozen Muslim organizations have promoted the Muslim cause with tactics ranging from political pressure to violent insurgency. The campaigns of

**TERRITORIAL MORPHOLOGY:
THE PHILIPPINES**

Areas affected by Muslim insurgency

Core area

Railroad   Road

National capital is underlined

0   100   200   300 Kilometers

0   100   200 Miles

FIGURE 10-16   © H. J. de Blij, P. O. Muller, and John Wiley & Sons, Inc.

People and Culture

Out of the Philippines melting pot, where Malay, Arab, Chinese, Japanese, Spanish, and American elements have met and mixed, has emerged the distinctive Filipino culture. It is not a homogeneous or a unified culture, but in Southeast Asia it is in many ways unique. One example of its absorptive qualities is demonstrated by the way the Chinese infusion has been accommodated: although the "pure" Chinese minority numbers less than 2 percent of the population (far lower than in most Southeast Asian countries), a much larger portion of the Philippine population carries a decidedly Chinese ethnic imprint. What has happened is that the Chinese have intermarried, producing a sort of Chinese-mestizo element that constitutes more than 10 percent of the total population. In another cultural sphere, the country's ethnic mixture and variety are paralleled by its great linguistic diversity. Nearly 90 Malay languages, major and minor, are spoken by the 90 million people of the Philippines; only about 1 percent still use Spanish. At independence in 1946, the largest of the Malay languages, Tagalog, or Pilipino, became the country's official language, and the educational system promotes its general use. English is learned as a subsidiary language and remains the chief *lingua franca*; an English-Tagalog hybrid ("Taglish") is increasingly heard today, cutting across all levels of society.

violence have affected all of the Muslim areas, from southern Mindanao to Palawan, and the strategies of the rebels have ranged from bombings to kidnappings. Densely forested Basilan Island is the headquarters of a group known as Abu Sayyaf, which demanded a separate Muslim state and is believed to have received support through the al-Qaeda network. The War on Terror reached into this part of the Philippines when American troops joined Filipino forces to pursue Abu Sayyaf terrorists. Also active on behalf of Muslim causes in this area are the Moro National Liberation Front and the Moro Islamic Liberation Front; the Philippine government has entered into negotiations with both. Cease-fire agreements, however, have not held. The Muslim challenge is pirating a disproportionate share of Manila's operating budget.

The widespread use of English in the Philippines, of course, results from a half-century of American rule and influence, beginning in 1898 when the islands were ceded to the United States by Spain under the terms of the treaty that followed the Spanish-American War. The United States took over a country in open revolt against its former colonial master and proceeded to destroy the Filipino independence struggle, now directed against the new foreign rulers. It is a measure of the subsequent success of U.S. administration in the Philippines that this was the only dependency in Southeast Asia that sided against the Japanese during World War II in favor of the colonial power. The U.S. rule had its good and bad features, but the Americans did initiate reforms that were long overdue, and they were

AMONG THE REALM'S GREAT CITIES . . . Manila

MANILA, CAPITAL OF the Philippines, was founded by the Spanish invaders of Luzon more than four centuries ago. The colonists made a good choice in terms of site and situation. Manila sprawls at the mouth of the Pasig River where it enters one of Asia's finest natural harbors. To the north, east, and south a crescent of mountains encircles the city, which lies just 1000 kilometers (600 mi) southeast of China's Hong Kong.

Manila, named after a flowering shrub in the local marshlands, is bisected by the Pasig, which is bridged in numerous places. The old walled city, Intramuros, lies to the south. Despite heavy bombardment during World War II, some of the colonial heritage survives in the form of churches, monasteries, and convents. St. Augustine Church, completed in 1599, is one of the city's landmarks.

The CBD of Manila lies on the north side of the Pasig River (photo p. 568). Although Manila has a well-defined commercial center with several avenues of luxury shops and modern buildings, the skyline does not reflect the high level of energy and activity common in Pacific Rim cities on the opposite side of the South China Sea. Neither is Manila a city of notable architectural achievements. Wide, long, and straight avenues flanked by palm, banyan, and acacia trees give it a look similar to San Juan, Puerto Rico.

In 1948, a newly built city immediately to the northeast of Manila was inaugurated as the de jure capital of the Philippines and called Quezon City. The new facilities were eventually to house all government offices, but many functions of the national government never made the

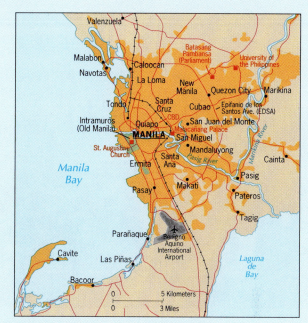

© H. J. de Blij, P. O. Muller, and John Wiley & Sons, Inc.

move. In the meantime, Manila's growth overtook Quezon City's, so that it became part of the Greater Manila metropolis (which today is home to 11.3 million). Although the proclamation of Quezon City as the Philippines' official capital was never rescinded, Manila remains the de facto capital of the country today.

---

already in the process of negotiating a future independence for the Philippines when the war intervened in 1941.

The Philippines' population, concentrated where the good farmlands lie in the plains, is densest in three general areas (Fig. G-9): (1) the northwestern and south-central part of Luzon, (2) the southeastern extension of Luzon, and (3) the islands of the Visayan Sea between Luzon and Mindanao. Luzon is the site of the capital, Manila-Quezon City (11.3 million, just over one-eighth of the entire national population), a major metropolis facing the South China Sea. Alluvial as well as volcanic soils, together with ample moisture in this tropical environment, produce self-sufficiency in rice and other staples and make the Philippines a net exporter of farm products despite a fairly high population growth rate of 2.1 percent.

Perhaps more than any other people, Filipinos take jobs in foreign countries in massive numbers, proving their capacity to succeed in jobs they cannot find at home. Not only

do hundreds of thousands of Filipinos work in the shipping industry around the world, but Filipino domestic workers find jobs from Dubai to Dubuque. They have a major impact on the Philippines' economy in the form of remittances, regularly ranking the country among the world's three or four leading recipients of such monetary inflows.

Prospects

The Philippines seems to get little mention in discussions of developments on the Pacific Rim, and yet it would seem to be well positioned to share in the Pacific Rim's economic growth. Governmental mismanagement and political instability have slowed the country's participation, but during the 1990s the situation improved. Despite a series of jarring events—the ouster of U.S. military bases, the damaging eruption of a volcano near the capital, the violence of

Manila, capital of the Philippines and directly across the South China Sea from Ho Chi Minh City (Saigon), is substantially more populous than its Vietnamese opposite. Manila, however, is far less dynamic today, although it has the better port and, unlike Saigon, has no domestic rival. Manila's metropolitan area (see map p. 567) includes the government headquarters at Quezon City, the parliament, and the University of the Philippines. This is the high-rise CBD, modest for a city of more than 11 million, but beyond these modern buildings and the shade trees of Greenbelt Square lie vast shantytowns housing a majority of this teeming city's inhabitants. © Jose Fuste Raga/Corbis.

Muslim insurgents, and a dispute over the nearby Spratly Islands in the South China Sea—the Philippines made substantial economic progress during the decade. Its electronics and textile industries (mostly in the Manila hinterland) expanded continuously, and more foreign investment arrived. The centrally positioned Visayan island of Cebu experienced particularly rapid growth, based on its central location, good port, expanded airport, large and literate labor force, and cadre of managers experienced in the service industries. A free-trade zone attracted foreign companies from the United States and Japan, and now products from toys to semiconductors flow from Cebu's manufacturers to world markets. But agriculture continues to dominate the Philippines' economy, unemployment remains high, further land reform is badly needed, and social restructuring (reducing the controlling influence over national affairs by a comparatively small group of families) must occur. However, progress is being made. The country now is a lower-middle-income economy, and given a longer period of stability and success in reducing the population growth rate, it will rise to the next level and finally take its place among Pacific Rim growth poles.

In recent years, the population issue has divided this dominantly Roman Catholic society, with the government promoting family planning and the clergy opposing it. But behind this debate lies another of the Philippines' assets: in a realm of mostly undemocratic regimes, the Philippines has come out of its period of authoritarian rule a rejuvenated, if not yet robust, democracy.

What You Can Do

RECOMMENDATION: Participate in National Geography Awareness Week! By an Act of Congress, a week in November is designated National Geography Awareness Week, and all over the country organizations ranging from colleges to corporations mark the occasion by offering programs highlighting geography's importance and usefulness. Activities include competitions, exhibits, lectures, GIS demonstrations, and field trips, and faculty and students alike make a special effort to reach out to the general public to show what professional geographers do. This always requires assistance, from posting announcements to meeting speakers and from supervising exhibits to planning field trips. How about asking your local Geography Department what is being planned for next November, and lending a hand?

GEOGRAPHIC CONNECTIONS

1 In this chapter we divide the Southeast Asian geographic realm into a Mainland region and an Insular region. A major factor in this regionalization is religion: Buddhism dominates in Mainland Southeast Asia, and Islam in Insular Southeast Asia. As Figure 7-2 reveals, however, the religious geography of Southeast Asia is more complicated than that, and important vestiges of one major religion cannot even be shown at this small scale. Look very carefully at the political boundaries of Southeast Asia and the religious regions in Figure 7-2, and identify those countries that experience significant internal religious division of a regional nature. How is such division finding expression? Is religious regionalism absent in countries dominated by a single faith?

2 When we studied colonialism and its legacies in Subsaharan Africa, it was noted that, although most African states have been independent for nearly half a century, the imprint of colonial power still remains—and continues to divide the realm. Southeast Asia, too, was occupied by colonial powers (with one significant exception), but the cultural impress of colonialism appears to be less durable here. To what geographic factors do you attribute this contrast between two formerly colonized realms?

Geographic Literature on Southeast Asia: The key introductory works on this realm were authored by Dixon, Fisher, Fryer, Leinbach & Ulack, Rigg, Spencer & Thomas, and Weightman. These works, together with a comprehensive listing of noteworthy books and recent articles on the geography of Mainland and Insular Southeast Asia, can be found in the *References and Further Readings* section of this book's website at **www.wiley.com/college/deblij**.

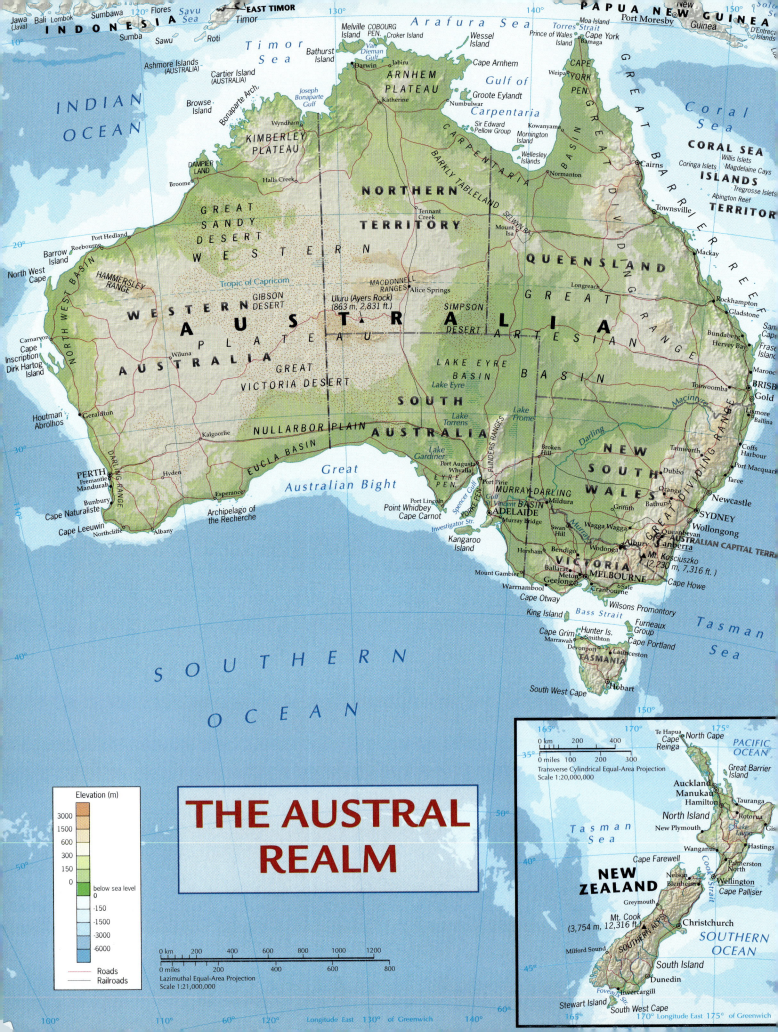

THE AUSTRAL REALM

11

REGIONS

- **AUSTRALIA**
  CORE AREA
  OUTBACK
- **NEW ZEALAND**

Where were these photos taken? Find out at **www.wiley.com/college/deblij**

In This Chapter

- Australia's amazing biogeography
- Australia's changing population
- Aboriginal claims to land and resources
- China covets Australian commodities
- Foreign policy dilemmas downunder
- New Zealand's matchless physiography

Photos: © H. J. de Blij

FIGURE 11-1 © H. J. de Blij, P. O. Muller, and John Wiley & Sons, Inc.

THE AUSTRAL REALM is geographically unique. It is the only geographic realm that lies entirely in the Southern Hemisphere. It is also the only realm that has no land link of any kind to a neighboring realm and is thus completely surrounded by ocean and sea. It is second only to the Pacific as the world's least populous realm. Appropriately, its name refers to its location (**Austral** means south)—a location far from the sources of its dominant cultural heritage but close to its newfound economic partners on the western Pacific Rim.

Defining the Realm

Two countries constitute the Austral Realm: Australia, in every way the dominant one, and New Zealand, physiographically more varied than its giant partner. Between them lies the Tasman Sea. To the west lies the Indian Ocean, to the east the Pacific, and to the south the frigid Southern Ocean.

This southern realm is at a crossroads. On the doorstep of populous Asia, its Anglo-European legacies are now infused by other cultural strains. Polynesian Maori in New Zealand and Aboriginal communities in Australia are demanding better terms of life. Pacific Rim markets are buying huge quantities of raw materials. Japanese and other Asian tourists fill hotels and resorts. Queensland's tropical Gold Coast resembles Honolulu's Waikiki. The streets of Sydney and Melbourne display a multicultural panorama unimagined just two generations ago. All these changes have stirred political debate. Issues ranging from immigration quotas to indigenous land rights dominate, exposing social fault lines (city versus Outback in Australia, North and South in New Zealand). Aborigines and Maori were here first, and the Europeans came next. Now Asia looms in Australia's doorway.

MAJOR GEOGRAPHIC QUALITIES OF

The Austral Realm

1. Australia and New Zealand constitute a geographic realm by virtue of territorial dimensions, relative location, and dominant cultural landscape.

2. Despite their inclusion in a single geographic realm, Australia and New Zealand differ physiographically. Australia has a vast, dry, low-relief interior; New Zealand is mountainous.

3. Australia and New Zealand are marked by peripheral development—Australia because of its aridity, New Zealand because of its topography.

4. The populations of Australia and New Zealand are not only peripherally distributed but also highly clustered in urban centers.

5. The realm's human geography is changing—in Australia because of Aboriginal activism and Asian immigration, and in New Zealand because of Maori activism and Pacific-islander immigration.

6. The export of livestock products dominates the economic geography of Australia and New Zealand (and in Australia, wheat production and mining also dominate).

7. Australia and New Zealand are being integrated into the economic framework of the western Pacific Rim, principally as suppliers of raw materials.

LAND AND ENVIRONMENT

Physiographic contrasts between massive, compact Australia and elongated, fragmented New Zealand are related to their locations with respect to the Earth's tectonic plates (consult Fig. G-4). Australia, with some of the geologically most ancient rocks on the planet, lies at the center of its own plate, the Australian Plate. New Zealand, younger and less stable, lies at the convulsive convergence of the Australian and Pacific plates. Earthquakes are rare in Australia and volcanic eruptions are unknown; New Zealand has plenty of both. This locational contrast is also reflected by differences in relief (Fig. 11-1). Australia's highest relief occurs in what Australians call the Great Dividing Range, the mountains that line the east coast from the Cape York Peninsula to southern Victoria, with an outlier in Tasmania. The highest point along these old, now eroding mountains is Mount Kosciusko, 2228 meters (7316 ft) tall. In New Zealand, entire ranges are higher than this, and Mount Cook reaches 3764 meters (12,349 ft).

West of Australia's Great Dividing Range, the physical landscape generally has low relief, with some local exceptions such as the Macdonnell Ranges near the center; plateaus and plains dominate (Fig. 11-2). The Great Artesian Basin is a key physiographic region, providing underground

AUSTRALIA: PHYSIOGRAPHY

0 200 400 600 800 Kilometers

0 100 200 300 400 Miles

INDONESIA

Timor
EAST TIMOR

Sumba

INDIAN
OCEAN

PAPUA
NEW GUINEA

Arafura Sea

Torres Strait

Cape
York
Penin.

Coral
Sea

Timor
Sea

Arnhem
Land

Gulf of
Carpentaria

Atherton
Plateau

Kimberley
Plateau

King Leopold
Ranges

Dampier
Land

Canning
Basin

Great Sandy
Desert

Gibson
Desert

NORTHERN
TERRITORY

Carpentaria
Barkly Tableland

Macdonnell Ranges

James Range

Selwyn Range

AUSTRALIA

Gregory Range

Great Barrier Reef

Great Dividing Range

QUEENSLAND

Simpson
Desert

Great
Artesian
Basin

Lake
Eyre

PACIFIC
OCEAN

Hamersley Range

North West Basin

WESTERN
AUSTRALIA

Yilgarn
Plateau

Musgrave Ranges

Stuart Range

Great
Victoria
Desert

SOUTH
AUSTRALIA

Grey Range

Flinders Range

Darling

NEW SOUTH
WALES

Lachlan

Murray

Murray-Darling
RiverBasin

Blue
Mts.

AUSTRALIAN
CAPITAL
TERRITORY

Darling Range

Nullarbor Plain

Eucla Basin

Eyre
Penin.

VICTORIA

Snowy
Mts.

Mt. Kosciusko
2,223 m.
(7,316 ft.)

Great Australian
Bight

Kangaroo
Island

Bass Strait

Tasman
Sea

SOUTHERN OCEAN

TASMANIA

Longitude East of Greenwich

INDIAN
OCEAN

Darwin

PACIFIC
OCEAN

Kimberley
Plateau

Carpentaria
Basin

Great
Sandy
Desert

Macdonnell
Ranges

Great
Artesian
Basin

Great Dividing Range

Brisbane

Great
Victoria Desert

Lake
Eyre

Darling R.

Perth

Nullarbor Plain

Eucla Basin

Murray-Darling
River Basin

Sydney

Adelaide

Murray R.

Canberra

Melbourne

Mt.Kosciusko
2,223 m.
(7,316 ft.)

SOUTHERN
OCEAN

Bass Strait

Tasman
Sea

TASMANIA

Hobart

Longitude East of Greenwich

PHYSICAL LANDSCAPES

Eastern Uplands   Central Lowlands   Western Plateau & Margins

0 400 800 1200 1600 Kilometers

0 500 1000 Miles

FIGURE 11-2 © H. J. de Blij, P. O. Muller, and John Wiley & Sons, Inc.

water sources in what would otherwise be desert country; to the south lies the continent's predominant river system, the Murray-Darling. The area mapped as *Western Plateau and Margins* in the inset map of Figure 11-2 contains much of Australia's mineral wealth.

Climates

Figure G-8 reveals the effects of latitudinal location and interior isolation on Australia's climatology. In this respect, Australia is far more varied than New Zealand, its climates ranging from tropical in the far north, where rainforests flourish, to Mediterranean in parts of the south. The interior is dominated by desert and steppe conditions, the steppes providing the grasslands that sustain tens of millions of livestock. Only in the east does Australia have an area of humid temperate climate, and here lies most of the country's economic core area. New Zealand, by contrast, is totally under the influence of the Southern and Pacific oceans, creating moderate, moist conditions, temperate in the north and colder in the south.

The Southern Ocean

2 Twice now we have referred to the **Southern Ocean**, but try to find this ocean on maps and globes published by famous cartographic organizations such as the National Geographic Society and Rand McNally. From their maps you would conclude that the Atlantic, Pacific, and Indian oceans reach all the way to the shores of Antarctica. Australians and New Zealanders know better. They experience the frigid waters and persistent winds of this great weather-maker on a daily basis.

For us geographers, it is a good exercise to turn the globe upside down now and then. After all, the usual orientation is quite arbitrary. Modern mapmaking started in the Northern Hemisphere, and the cartographers put their hemisphere on top and the other at the bottom. That is now the norm, and it can distort our view of the world. In bookstores in the Southern Hemisphere, you sometimes see tongue-in-cheek maps showing Australia and Argentina at the top, and Europe and Canada at the bottom. But this matter has a serious side. A reverse view of the globe shows us how vast the ocean encircling Antarctica is. The Southern Ocean may be remote, but its existence is real.

Where do the northward limits of the Southern Ocean lie? This ocean is bounded not by land but by a marine **3** transition called the **Subtropical Convergence**. Here the cold, extremely dense waters of the Southern Ocean meet the warmer waters of the Atlantic, Pacific, and Indian oceans. It is quite sharply defined by changes in temperature, chemistry, salinity, and marine fauna. Flying over it,

you can actually observe it in the changing colors of the water: the Antarctic side is a deep gray, the northern side a greenish blue.

Although the Subtropical Convergence moves seasonally, its position does not vary far from latitude 40° South, which also is the approximate northern limit of Antarctic icebergs. Defined this way, the great Southern Ocean is a huge body of water that moves clockwise (from west to east) around Antarctica, which is why we also call it the **West Wind Drift**. **4**

Biogeography

One of this realm's defining characteristics is its wildlife. Australia is the land of kangaroos and koalas, wallabies and wombats, possums and platypuses. These and numerous other *marsupials* (animals whose young are born very early in their development and then carried in an abdominal pouch) owe their survival to Australia's early isolation during the breakup of Gondwana (see Fig. 6-3). Before more advanced mammals could enter Australia and replace the marsupials, as happened in other parts of the world, the landmass was separated from Antarctica and India, and today it contains the world's largest assemblage of marsupial fauna.

Australia's vegetation also has distinctive qualities, notably the hundreds of species of eucalyptus trees native to this geographic realm. Many other plants form part of Australia's unique flora, some with unusual adaptation to the high temperatures and low humidity that characterize much of the continent.

The study of fauna and flora in spatial perspective combines the disciplines of biology and geography in a field known as **biogeography**, and Australia is a giant labora- **5** tory for biogeographers. In the Introduction (pp. 10–16), we noted that several of the world's climatic zones are named after the vegetation that marks them: tropical savanna, steppe, tundra. When climate, soil, vegetation, and animal life reach a long-term, stable adjustment, vegetation forms the most visible element of this ecosystem.

Biogeographers are especially interested in the distribution of plant and animal species, and in the relationships between plant and animal communities and their natural environments. (The study of plant life is called *phytogeography*; the study of animal life *zoogeography*.) These scholars seek to explain the distributions the map reveals. In 1876 one of the founders of biogeography, Alfred Russel Wallace, published a book entitled *The Geographical Distribution of Animals* in which he fired the first shot in a long debate: where does the zoogeographic boundary of Australia's fauna lie? Wallace's fieldwork in the area revealed that Australian forms exist not only in Australia itself but also in New Guinea and in some islands to the

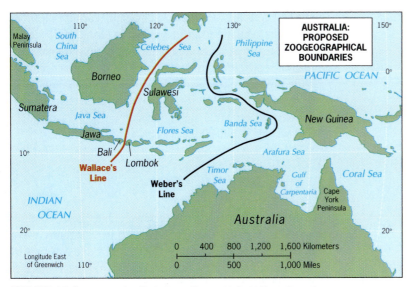

FIGURE 11-3 © H. J. de Blij, P. O. Muller, and John Wiley & Sons, Inc.

northwest. So Wallace proposed that the faunal boundary should lie between Borneo and Sulawesi, and just east of Bali (Fig. 11-3).

6   **Wallace's Line** soon was challenged by other researchers, who found species Wallace had missed and who visited islands Wallace had not. There was no question that Australia's zoogeographic realm ended somewhere in the Indonesian archipelago, but where? Western Indonesia was the habitat of nonmarsupial animals such as tigers, rhinoceroses, and elephants, as well as primates; New Guinea clearly was part of the realm of the marsupials. How far had the more advanced mammals progressed eastward along the island stepping stones toward New Guinea? The zoogeographer Max Weber found evidence that led him to postulate his own *Weber's Line*, which, as Figure 11-3 shows, lay very close to New Guinea.

The Human Impact

Not all research in zoogeography or phytogeography deals with such large questions. Much of it focuses on the rela-

tionships between particular species and their habitats, that is, the environment they normally occupy and of which they form a part. Such environments change, and the changes can spell disaster for the species. But it is not always clear whether environmental change or human intervention caused the problem. In Australia, the arrival of the **Aboriginal population** (about 50,000 years ago) appears to have caused an ecosystem collapse because there is no evidence of significant climate change at the time. In a recent article in *Science*, Gifford H. Miller and his colleagues reported that widespread burning of the existing forest, shrub, and grassland vegetation across Australia appears to have led to the spread of desert scrub and to the rapid extinction of most of the continent's large mammals soon after the human invasion occurred. The species that survived faced a second crisis when European colonizers introduced their livestock, leading to the further destruction of remaining wildlife habitats. Survivors include marsupials such as the koala "bear" and the wombat, but the list of extinctions is far longer.

7

POINTS TO PONDER

- Climate change seems to be worsening the droughts that transform ground fires into high-intensity firestorms, exacerbating an already-serious environmental hazard in Australia.

- Some geographers suggest that Australia by the middle of this century may have as many as 50 million inhabitants, and one of the most multicultural societies in the world—its water-supply problems notwithstanding.

- Australia's policy of "cooperative intervention" is helping the governments and peoples of Papua New Guinea as well as the Solomon Islands overcome persistent problems.

Regions of the Realm

Australia is the dominant component of the Austral Realm, a continent-scale country in a size category that also includes China, Canada, the United States, and Brazil. For two reasons, however, Australia has fewer regional divisions than do the aforementioned countries: Australia's relatively uncomplicated physiography and its diminutive human numbers. Our discussion, therefore, uses the core-periphery concept as a basis for investigating Australia and focuses on New Zealand as a region by itself.

● AUSTRALIA

On January 1, 2001, Australia celebrated its 100th birthday as a state, the Commonwealth of Australia, recognizing (still) the British monarch as the head of state and entering its second century with a strong economy, stable political framework, high standard of living for most of its people, and favorable prospects ahead. Positioned on the Pacific

MAJOR CITIES OF THE REALM

City	Population* (in millions)
Adelaide, Australia	1.2
Auckland, New Zealand	1.2
Brisbane, Australia	1.8
Canberra, Australia	0.4
Melbourne, Australia	3.7
Perth, Australia	1.5
Sydney, Australia	4.5
Wellington, New Zealand	0.4

*Based on 2008 estimates.

Rim, nine-tenths as large as the 48 contiguous U.S. States, well endowed with farmlands and vast pastures, major rivers, ample underground water, minerals, and energy resources, served by good natural harbors, and populated by 20.9 million mostly well-educated people, Australia is one of the most fortunate countries on Earth.

Sharing the Bounty

Not everyone in Australia shares adequately in all this good fortune, however, and the less advantaged made their voices heard during the celebrations. The country's indigenous (*Aboriginal*) population, though a small minority today of about 450,000, remains disproportionately disadvantaged in almost every way, from lower life expectancies to higher unemployment than average, from lower high school graduation rates to higher imprisonment ratios. But the nation is now embarked on a campaign to address these ills, with actions ranging from public demonstrations supporting reconciliation to official expressions of regret for past mistreatment and from enhanced social services to favorable court decisions involving Aboriginal land claims.

When Australia was born as a federal state, its per-capita GNI, as reckoned by economic geographers, was the highest in the world. Australia's bounty fueled a huge flow of exports to Europe, and the Australians prospered. That golden age could not last forever, and eventually the country's share of world trade declined. Still, Australia today ranks among the top 20 countries in the world in terms of GNI, and for the vast majority of Australians life is comfortable.

The statistics bear witness to Australia's current good fortune. In terms of the indicators of development discussed in Chapter 9, Australia is far ahead of all its western Pacific Rim competitors except Japan. As Australians celebrated

their first century, they were, on average, earning far more than Thais, Malaysians, or Koreans. In terms of consumption of energy per person, the number of automobiles and miles of roads, levels of health, and literacy, Australia had all the properties of a developed country. Australian cities and towns, where 91 percent of all Australians live, are not encircled by crowded shantytowns. Nor is the Australian countryside inhabited by a poverty-stricken peasantry.

Distance

Australians often talk about distance. One of their leading historians, Geoffrey Blainey, labeled it a "tyranny"—an imposed remoteness from without and a divisive part of life within. Even today, Australia is far from nearly everywhere on Earth. A jet flight from Los Angeles to Sydney takes 14 hours nonstop and is correspondingly expensive. Seaborne freighters carrying products to European markets take ten days to two weeks to get there. Inside Australia, distances also are of continental proportions, and Australians pay the price—literally. Until some upstart private airlines started a price war, Australians paid more per mile for their domestic flights than air passengers anywhere else in the world.

But distance also was an ally, permitting Australians to ignore the obvious. Australia was a British progeny, a European outpost. Once you had arrived as an immigrant from Britain or Ireland, there were a wide range of environments, magnificent scenery, vast open spaces, and seemingly limitless opportunities. When the Japanese Empire expanded, Australia's remoteness saved the day. When immigration became an issue, Australia in its comfortable isolation could adopt an all-white admission policy that was not officially terminated until 1976. When boat people by the hundreds of thousands fled Vietnam in the aftermath of the Indochina War, almost none reached Australian shores.

Immigrants

Today Australia is changing and rapidly so. Immigration policy now focuses on the would-be immigrants' qualifications, skills, financial status, age, and facility with the English language. With regard to skills, high-technology specialists, financial experts, and medical personnel are especially welcome. Relatives of earlier immigrants, as well as a quota of genuine asylum-seekers, also are readily admitted. In recent years, total immigration has been about 120,000 annually, which keeps the country's population growing. According to the latest data, Australia's declining natural rate of increase today stands at 0.6 percent.

Already, Australia's changed immigration policies have dramatically altered cultural landscapes, especially in the urban areas (see right photo, p. 584). The country is fast becoming a truly multicultural society. In Sydney, for instance, one in six residents is now of Asian ancestry, a ratio that will rise to one in five by 2010. Overall, a quarter of Australia's 2008 population is foreign-born, and another quarter consists of first-generation Australians.

Core and Periphery

Australia is a large landmass, but its population is heavily concentrated in a core area that lies in the east and southeast, most of which faces the Pacific Ocean (here named the Tasman Sea between Australia and New Zealand). As Figure 11-4 shows, this crescent-like Australian heartland extends from north of the city of Brisbane to the vicinity of

FIGURE 11-4   © H. J. de Blij, P. O. Muller, and John Wiley & Sons, Inc.

FROM THE FIELD NOTES

"My most vivid memory from my first visit to Alice Springs is spotting vineyards and a winery in this parched, desert environment as the plane approached the airport. I asked a taxi driver to take me there, and got a lesson in economic geography. Drip irrigation from an underground water supply made viticulture possible; the tourist industry made it profitable. None of this, however, is evident from the view seen here: a spur of the Macdonnell Ranges overlooks a town of bare essentials under the hot sun of the Australian desert. What Alice Springs has is centrality: it is the largest settlement in a vast area Australians often call 'the centre.' Not far from the midpoint on the nearly 3200-kilometer (2000-mi) Stuart Highway from Darwin on the Northern Territory's north coast to Adelaide on the Southern Ocean, Alice Springs also was the northern terminus of the Central Australian Railway (before the line was extended north to Darwin in 2003), seen in the middle distance. The shipping of cattle and minerals is a major industry here. You need a sense of humor to live here, and the locals have it: the town actually lies on a river, the intermittent Todd River. An annual boat race is held, and in the absence of water the racers carry their boats along the dry river bed. No exploration of Alice Springs would be complete without a visit to the base of the Royal Flying Doctor Service, which brings medical help to outlying villages and homesteads." © H. J. de Blij

Concept Caching

www.conceptcaching.com

Adelaide and includes the largest city, Sydney, the capital, Canberra, and the second-largest city, Melbourne. A secondary core area has developed in the far southwest, centered on Perth and its outport, Fremantle. Beyond lies the vast periphery, which the Australians call the **Outback**.

To better understand the evolution of this spatial arrangement, it helps to refer again to the map of world climates (Fig. G-8). Environmentally, Australia's most favored strips face the Pacific and Southern oceans, and they are not large. We can describe the country as a coastal rimland with cities,

towns, farms, and forested slopes giving way to the vast, arid, interior Outback. On the western flanks of the Great Dividing Range lie the extensive grassland pastures that catapulted Australia into its first commercial age—and on which still graze one of the largest sheep herds on Earth (over 160 million sheep, producing more than one-fifth of all the wool sold in the world). Where it is moister, to the north and east, cattle by the millions graze on ranchlands. This is frontier Australia, over which livestock have ranged for nearly two centuries.

Aborigines and the British

Aboriginal Australians reached this landmass as long as 50,000 years ago, crossed the Bass Strait into Tasmania, and had developed a patchwork of indigenous cultures when Captain Arthur Phillip sailed into what is today Sydney Harbor (1788) to establish the beginnings of modern Australia. The Europeanization of Australia doomed the continent's Aboriginal societies. The first to suffer were those situated in the path of British settlement on the coasts, where penal colonies and free towns were founded. Distance protected the Aboriginal communities of the northern interior longer than elsewhere; in Tasmania, the indigenous Australians were exterminated in just decades after having lived there for perhaps 45,000 years.

The Seven Colonies

Eventually, the major coastal settlements became the centers of seven different colonies, each with its own hinterland; by 1861, Australia was delimited by its now-familiar pattern of straight-line boundaries (Fig. 11-4). Sydney was the focus for New South Wales; Melbourne, Sydney's rival, anchored Victoria. Adelaide was the heart of South Australia, and Perth lay at the core of Western Australia. Brisbane was the nucleus of Queensland, and Hobart was the seat of government in Tasmania. The largest clusters of surviving Aboriginal people were in the so-called Northern Territory, with Darwin, on Australia's tropical north coast, its colonial city. Notwithstanding their shared cultural heritage, the Australian colonies were at odds not only with London over colonial policies but also with each other over economic and political issues. The building of an Australian nation during the late nineteenth century was a slow and difficult process.

A Federal State

On January 1, 1901, following years of difficult negotiations, the Australia we know today finally emerged: the Commonwealth of Australia, consisting of six States and two Federal Territories (Table 11-1; Fig. 11-4). The two

AMONG THE REALM'S GREAT CITIES . . . **Sydney**

MORE THAN TWO centuries ago, Sydney was founded by Captain Arthur Phillip as a British outpost on one of the world's most magnificent natural harbors (photo, p. 580). The free town and penal colony that struggled to survive evolved into Australia's largest city. Today metropolitan Sydney (4.5 million) is home to more than one-fifth of the country's entire population. An early start, the safe harbor, fertile nearby farmlands, and productive pastures in its hinterland combined to propel Sydney's growth. Later, as road and railroad links made Sydney the focus of Australia's growing core area, industrial development and political power augmented its primacy.

With its incomparable setting and mild, sunny climate, its many urban beaches, and its easy reach to the cool Blue Mountains of the Great Dividing Range, Sydney is one of the world's most liveable cities. Good public transportation (including an extensive cross-harbor ferry system from the doorstep of the waterfront CBD), fine cultural facilities headed by the multitheatre Opera House complex, and many public parks and other recreational facilities make Sydney attractive to visitors as well. A healthy tourist trade, much of it from Japan and other Asian countries, bolsters the city's economy. Sydney's hosting of the 2000 Olympic Games was further testimony to its rising visibility.

Increasingly, Sydney also is a multicultural city. Its small Aboriginal sector is being overwhelmed by the arrival of large numbers of Asian immigrants. The Sydney suburb of Cabramatta symbolizes the impact: more than half of its nearly 100,000 residents were born elsewhere, mostly in Vietnam. Unemployment is high, drug use is a problem, and crime and gang violence persist. Yet, de-

© H. J. de Blij, P. O. Muller, and John Wiley & Sons, Inc.

spite the deviant behavior of a small minority, tens of thousands of Asian immigrants have established themselves in some profession. As the photo and caption on p. 584 reveal, today things are definitely improving in this dynamic community.

These developments underscore Sydney's coming of age. The end of Australia's isolation has brought Asia across the country's threshold, and again the leading metropolis will show the way.

Federal Territories are the Northern Territory, assigned to protect the interests of the large Aboriginal population concentrated there, and the Australian Capital Territory, carved from southern New South Wales to accommodate the federal capital of Canberra, inaugurated in 1927.

Australia's six States are New South Wales (capital Sydney), at 7.1 million the most populous and politically powerful; Queensland (Brisbane), with the Great Barrier Reef offshore and tropical rainforests in its north; Victoria (Melbourne), small but populous by Australian standards with 5.3 million inhabitants; South Australia (Adelaide), where the Murray–Darling river system reaches the sea; Western Australia (Perth) with barely 2 million people in an area of more than 2.5 million square kilometers (nearly 1 million square miles); and Tasmania (Hobart), the island across the Bass Strait from the mainland and in the path of the storms of the Southern Ocean.

Successful Federation

In earlier chapters, we have referred to the concept of federalism, an idea with ancient Greek and Roman roots familiar to Americans, Canadians, South Asians, and, more recently, Russians. It is a notion of communal association whose name comes from the Latin *foederis*, implying alliance and coexistence, a union of consensus and common interest—a **federation**. It stands in contrast to the idea that states should be centralized, or *unitary*. For this, too, the ancient Romans had a term: *unitas*, meaning unity. Most European countries are **unitary states**, including the United Kingdom of Great Britain and Northern Ireland. Although the majority of Australians came from that tradition (a kingdom, no less), they managed to overcome their differences and establish a Commonwealth that was, in effect, a federation of States with different viewpoints, economies,

TABLE 11-1

States and Territories of Federal Australia, 2008

State	Area (1000 sq km [1000 sq mi])	Population (millions)	Capital	Population (millions)
New South Wales	801.6 (309.5)	7.1	Sydney	4.5
Queensland	1727.3 (666.9)	3.9	Brisbane	1.8
South Australia	983.9 (379.9)	1.5	Adelaide	1.2
Tasmania	67.9 (26.2)	0.5	Hobart	0.2
Victoria	227.7 (87.9)	5.3	Melbourne	3.7
Western Australia	2525.5 (975.1)	2.0	Perth	1.5
Territory				
Australian Capital Territory	2.3 (0.9)	0.4	Canberra	0.4
Northern Territory	1346.3 (519.8)	0.2	Darwin	0.1

and objectives, separated by vast distances along the rim of an island continent. And yet, the experiment succeeded.

An Urban Culture

During this century of federal association, the Australians developed an urban culture. Despite those vast open spaces and romantic notions of frontier and Outback, more than 90 percent of all Australians live in cities and towns. On the map, Australia's areal functional organization is similar to Japan's: large cities lie along the coast, the centers of manufacturing complexes as well as the foci of agricultural areas. Contributing to this situation in Japan was mountainous topography; in Australia, it was an arid interior. There, however, the similarity ends. Australia's territory is 20 times larger than Japan's, and Japan's population is more than six times that of Australia's. Japan's port cities are built to receive raw materials and to export finished prod-

ucts. Australia's cities forward minerals and farm products from the Outback to foreign markets and import manufactures from overseas. Distances in Australia are much greater, and spatial interaction (which tends to decrease with increasing distance) is less. In comparatively small, tightly organized Japan, you can travel from one end of the country to the other along highways, through tunnels, and over bridges with utmost speed and efficiency. In Australia, the overland trip from Sydney to Perth, or from Darwin to Adelaide using the newly completed railway link via Alice Springs, is time-consuming and slow. Nothing in Australia compares to Japan's high-speed bullet trains.

The Cities

For all its vastness and youth, Australia nonetheless developed a remarkable cultural identity, a sameness of urban and rural landscapes that persists from one end of the continent to the other. Sydney, often called the New York of Aus-

FROM THE FIELD NOTES

"See a scene like this, and you realize why Australians refer to their land as 'the lucky country.' Australia's periphery has much magnificent scenery ranging from spectacular cliffs to dune-lined beaches, but nothing matches Sydney Harbor on a sunny, breezy day when sailboats by the hundreds emerge from coves and inlets and, if you are fortunate enough to be on the water yourself, every turn around a headland presents still another memorable view. Ask the captains of ocean liners plying the world what port is their favorite, and most will point to this magnificent estuary with its narrow entrance and secluded bays as the grandest of all."
© H. J. de Blij

Concept Caching

www.conceptcaching.com

tralia, lies on a spectacular estuarine site, its compact, high-rise central business district overlooking a port bustling with ferry and freighter traffic. Sydney is a vast, sprawling metropolis of 4.5 million, with multiple outlying centers studding its far-flung suburbs; brash modernity and reserved British ways blend here. Melbourne (3.7 million), sometimes regarded as the Boston of Australia, prides itself on its more interesting architecture and more cultured ways. Brisbane, the capital of Queensland, which also anchors Australia's Gold Coast and adjoins the Great Barrier Reef, is the Miami of Australia; unlike Miami, however, its residents can find nearby relief from the summer heat in its immediate hinterland mountains (as well as at its beaches). Perth, Australia's San Diego, is one of the world's most isolated cities, separated from its nearest Australian neighbor by two-thirds of a continent and from Southeast Asia and Africa by thousands of kilometers of ocean.

And yet, each of these cities—as well as the capitals of South Australia (Adelaide), Tasmania (Hobart), and, to a lesser extent, the Northern Territory (Darwin)—exhibits an Australian character of unmistakable quality. Life is orderly and unhurried. Streets are clean, slums are few, graffiti rarely seen. By American and even European standards, violent crime (though rising) is uncommon. Standards of public transport, city schools, and health-care provision are high. Spacious parks, pleasing waterfronts, and plentiful sunshine make Australia's urban life more acceptable than that almost anywhere else in the world. Critics of Australia's way of life say that this very pleasant state of affairs has persuaded Australians that hard work is not really necessary. But a few days' experience in the commercial centers of the major cities contradicts that assertion: the pace of life is quickening. The country's cultural geography evolved as that of a European outpost, prosperous and secure in its isolation. Now Australia must reinvent itself as a major link in an Australo-Asian chain, a Pacific partner in a transformed regional economic geography.

Economic Geography

Because of its early history as a treasure trove of raw materials, Australia to this day is seen worldwide as a country whose economy depends primarily on its exportable natural resources. But in fact, Australia's economy depends overwhelmingly on services, not commodity exports. Tourism alone contributes around 5 percent—about the same as the value of mineral exports. The action is in those bustling coastal cities far more than in the Outback.

When Australia became established as a state, however, it needed goods from overseas, and here the tyranny of distance played a role. Imports from Britain (and later the United States) were expensive mainly because of transport costs. This encouraged local entrepreneurs to set up their own industries to produce such goods more cheaply. Economic geographers call such industries **import-substitution industries**, and this is how local industrialization got its start.

When prices of foreign goods became lower because transportation was more efficient and therefore cheaper, local businesses demanded protection from the colonial governments, and high tariffs were erected against imported goods. Local products now could continue to be made inefficiently because their market was guaranteed. If Japan could not afford this, how could Australia? We can see the answer on the map (Fig. 11-5). Even before federation in 1901, all the colonies could export valuable minerals whose earnings shored up those inefficient, uncompetitive local industries. By the time the colonies unified, pastoral industries were also contributing income. So the miners and the farmers paid for those imports Australians could not produce themselves, plus the products made in the cities. No wonder the cities grew: here were secure manufacturing jobs, jobs in state-run service enterprises, and jobs in the growing government bureaucracy. When we noted earlier that Australians once had the highest per-capita GNI in the world, this was initially achieved in the mines and on the farms, not in the cities.

But the good times had to come to an end. The prices of farm products fluctuated, and international market competition increased. The cost of mining, transporting, and shipping ores and minerals also rose. Australians like to drive (there are more road kilometers per person in Australia than in the United States), and expensive petroleum imports were needed. Meanwhile, the government-protected industries had been further fortified by strong labor unions. Not surprisingly, as the economy declined, the national debt rose, inflation grew, and unemployment crept upward. In 2002, only 21 percent by value of Australian exports were the kind of high-tech goods that have made East Asia's economic tigers so successful. That was more than double the 1985 figure but still far below what the country needed to produce.

Agricultural Abundance

And yet, Australia has material assets of which other countries on the Pacific Rim can only dream. In agriculture, sheep-raising was the earliest commercial venture, but it was the technology of refrigeration that brought world markets within reach of Australian beef producers. Wool, meat, and wheat have long been the country's big three income earners; Figure 11-5 displays the vast pastures in the east, north, and west that constitute the ranges of Australia's huge herds. The zone of commercial grain farming forms a broad crescent extending from northeastern New South Wales

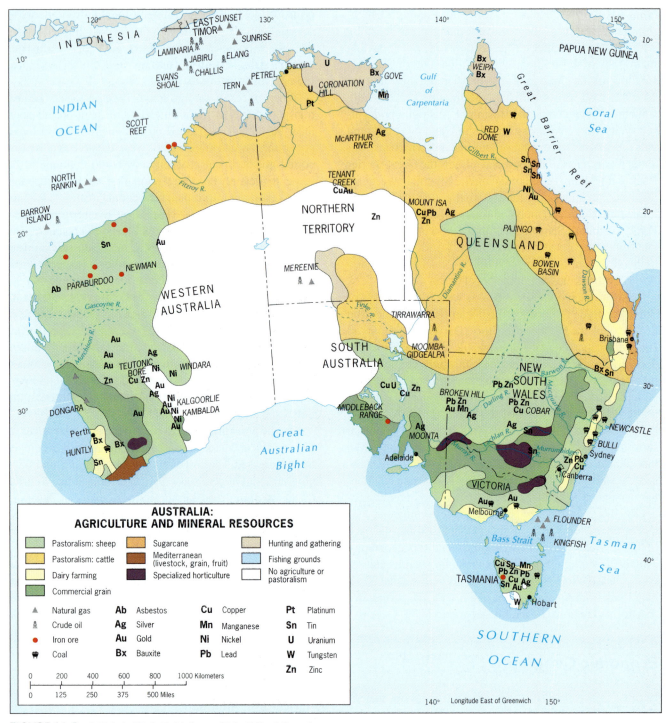

AUSTRALIA:
AGRICULTURE AND MINERAL RESOURCES

Pastoralism: sheep
Pastoralism: cattle
Dairy farming
Commercial grain
Sugarcane
Mediterranean (livestock, grain, fruit)
Specialized horticulture
Hunting and gathering
Fishing grounds
No agriculture or pastoralism

▲ Natural gas
⚒ Crude oil
● Iron ore
🛒 Coal

Ab Asbestos
Ag Silver
Au Gold
Bx Bauxite

Cu Copper
Mn Manganese
Ni Nickel
Pb Lead

Pt Platinum
Sn Tin
U Uranium
W Tungsten
Zn Zinc

0   200   400   600   800   1000 Kilometers
0   125   250   375   500 Miles

140°   Longitude East of Greenwich   150°

FIGURE 11-5 © H. J. de Blij, P. O. Muller, and John Wiley & Sons, Inc.

through Victoria into South Australia, and covers much of the hinterland of Perth. Keep in mind the scale of this map: Australia is only slightly smaller than the 48 contiguous States of the United States! Commercial grain farming in Australia is big business. As the climatic map would suggest, sugarcane grows along most of the warm, humid coastal strip of Queensland, and Mediterranean crops (in-

cluding grapes for Australia's highly successful wines) cluster in the hinterlands of Adelaide and Perth. Mixed horticulture concentrates in the basin of the Murray River system, including rice, grapes, and citrus fruits, all under irrigation. And, as elsewhere in the world, dairying has developed near the large urban areas. With its considerable range of environments, Australia yields a diversity of crops.

Mineral Wealth

Australia's mineral resources, as Figure 11-5 shows, also are diverse. Major gold discoveries in Victoria and New South Wales produced a ten-year gold rush starting in 1851 and ushered in a new economic era. By the middle of that decade, Australia was producing 40 percent of the world's gold. Subsequently, the search for more gold led to the discoveries of other minerals. New finds are still being made today, and even oil and natural gas have been found both inland and offshore (see the symbols in Fig. 11-5 in the Bass Strait between Tasmania and the mainland, and off the northwestern coast of Western Australia). Coal is mined at many locations, notably in the east near Sydney and Brisbane but also in Western Australia and even in Tasmania; before coal prices fell, this was a valuable export. Major deposits of metallic and nonmetallic minerals abound—from the complex at Broken Hill and the mix of minerals at Mount Isa to the huge nickel deposits at Kalgoorlie and Kambalda, the copper of Tasmania, the tungsten and bauxite of northern Queensland, and the asbestos of Western Australia. A closer look at the map reveals the wide distribution of iron ore (the red dots), and for this raw material as for many others, Japan was Australia's best customer for many years; today, however, China has become Australia's leading customer.

Manufacturing's Limits

Australian manufacturing, as we noted earlier, remains oriented to domestic markets. One cannot expect to find Australian automobiles, electronic equipment, or cameras challenging the Pacific Rim's economic tigers for a place on world markets—not yet, at any rate. Australian manufacturing is diversified, producing some machinery and equipment made of locally produced steel as well as textiles, chemicals, paper, and many other items. These industries cluster in and near the major urban areas where the markets are. The domestic market in Australia is not large, but it remains relatively affluent. This makes it attractive to foreign producers, and Australia's shops are full of high-priced goods from Japan, South Korea, Taiwan, and Hong Kong. Indeed, despite its long-term protectionist practices, Australia still does not produce many goods that could be manufactured at home. Overall, the economy continues to display symptoms of a still-developing rather than a fully developed country.

Australia's Future

The Commonwealth of Australia is changing, but its neighbors in Southeast and South Asia are changing even faster. Australia's European bonds are weakening and its Asian ties are strengthening, but Australia will be buffeted by the winds of change on the Pacific Rim. There is substantial risk for any economy that depends strongly on the export of raw materials and agricultural staples. But Australia faces challenges at home as well. These include: (1) Aboriginal issues, (2) immigration issues, (3) environmental issues, and (4) issues involving Australia's status and role.

Aboriginal Issues

The Aboriginal issues currently focus on two questions: the formal admission by government and majority of mistreatment of the Aboriginal minority with official apologies and reparations, and the ownership of land. Obviously, the second question has major geographic implications. The Aboriginal population of 450,000 (including many of mixed ancestry) has been gaining influence in national affairs, and in the 1980s Aboriginal leaders began a campaign to obstruct exploration on what they designated as ancestral and sacred lands. Until 1992, Australians had taken it for granted that Aborigines had no right to land ownership, but in that year the Australian High Court made the first of a series of rulings in favor of Aboriginal claimants. A subsequent court decision implied that vast areas (probably as much as 78 percent of the whole continent) could potentially be subject to Aboriginal claims (Fig. 11-6). Although

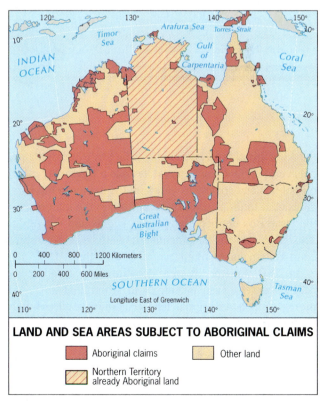

LAND AND SEA AREAS SUBJECT TO ABORIGINAL CLAIMS

Aboriginal claims — Other land — Northern Territory already Aboriginal land

FIGURE 11-6   © H. J. de Blij, P. O. Muller, and John Wiley & Sons, Inc.

FROM THE FIELD NOTES

"While in Darwin I had to visit a government office, and waiting in line with me was this Aboriginal girl, her brother, and her father. They lived about two hours from the city, she said (which I gathered was the length of their bus ride), and they had to come here from time to time for filing forms and visiting the doctor. When I asked whether they enjoyed coming to Darwin, all three shook their heads vigorously. 'It is not a friendly place,' she said. 'But we must come here.' In fact, I had been surprised at the comparatively small number of Aboriginal people I had seen in Darwin, far fewer than in Alice Springs. Certainly laid-back Alice Springs is a very different place from larger, busier, and more modern Darwin."
© H. J. de Blij

Concept Caching

www.conceptcaching.com

12 the **Aboriginal land issue** remains largely (though not exclusively) an Outback issue, it has had a polarizing effect on politics and society as views on the matter in the large cities tend to differ sharply from those in the interior (see the Issue Box entitled "Indigenous Rights and Wrongs"). This issue also has the potential to overwhelm Australia's court system and to inhibit economic growth.

In 2007, Australians were debating ways to bring so-called market-driven incentives to Aboriginal areas. At present, much Aboriginal land is administered communally by Aboriginal Land Councils, which makes the people land-rich but keeps them dirt-poor because the system in effect prevents private enterprise, including even the construction of family-owned housing. Some Aboriginal leaders are arguing that the system perpetuates an undesir-

able dependence on federal handouts, keeps jobs from becoming available, and restrains economic incentive. But not all Aboriginal leaders agree, worrying that privatization of any kind will undermine the Land Councils and create a small minority of better-off businesspeople, leaving the great majority of tradition-bound Aborigines even worse off. Still, the fact that the idea has some Aboriginal support is a sign of changing times in the relationships between Australian communities.

Immigration Issues

The immigration issue is older than Australia itself. Fifty years ago, when Australia had less than half the population it has today, 95 percent of the people were of European ancestry, and more than three-quarters of them came from the British Isles. Eugenic (race-specific) **immigration policies** **13**

The name Cabramatta conjures up varied reactions among Australians. During the 1950s and 1960s, many immigrants from southern Europe settled in this western suburb of Sydney, attracted by affordable housing. During the 1970s and 1980s, Southeast Asians arrived in large numbers, and during this period Cabramatta, in the eyes of many, became synonymous with gang violence and drug dealing. More recently, however, Cabramatta's ethnic diversity has come to be viewed in a more favorable light, and is now seen as the "multicultural capital of Australia," a tourist attraction and proof of Australia's capacity to accommodate non-Europeans. Meanwhile, Cabramatta has been spruced up with Oriental motifs of various kinds. This 'Freedom Gate' in the Vietnamese community is flanked by a Ming horse and a replica of a Forbidden City lion—all reflecting better times for an old transit point for immigrants and refugees. © Trip/Eric Smith

Indigenous Rights and Wrongs

THE LEAST WE SHOULD DO IS APOLOGIZE

"I can't believe this. Here we are more than a dozen years beyond our two-hundredth-year anniversary, the evidence of our mistreatment of the Aborigines is everywhere, and we can't, as a nation, say we're sorry? How could we get ourselves into a public debate over this? What does it take to offer sincere apologies to the original Australians and to tell them (and not only to tell them) that we'll do better?

"We should remind ourselves more often than we do of what happened here. When our ancestors arrived on these shores there were more than a million Aboriginal people in Australia, with numerous clans and cultures. They had many virtues, but one in particular: they didn't appropriate land. Land was assigned to them by the creator (what they call the Dreaming), and their relationship to it was spiritual, not commercial. They didn't build fences or walls. Neither had they adopted some bureaucratic religion. Just think, a world where land was open and free and religion was local and personal. Wouldn't that do something for the Middle East and Northern Ireland! Well, we showed up and started claiming and fencing off the land that, under British rules, was there for the taking. If the Aborigines got in the way, they were pushed out, and if they resisted, they got killed. You don't even want to think about what happened in Tasmania. A campaign of calculated extermination. Between 1800 and 1900 the Aboriginal population dropped from 1 million to around 50,000. And when we became a nation, they weren't even accorded citizenship.

"None of this stopped some white Australian men from getting Aboriginal women pregnant. But from 1910 on church and state managed to make things even worse. They took the young children away from their mothers and put them in institutions, where they would be 'Europeanized' and then married off to white partners, so that they would lose their Aboriginal inheritance. This, if you'll believe it, went on into the 1960s! Think of the scenes, these kids being kidnapped by armed officials from their mothers. And we have to argue about saying we're sorry?

"Well, at least we're discussing it in the open, and the misdeeds of the past are coming out. Aboriginal Australians finally got the vote in 1962, and they're able to get elected to parliament. I work in a State government office that assists Asian immigrants here in Sydney, and I only wish that we'd done for the Aborigines what we've done for the immigrants we admit."

Regional **ISSUE**

ENOUGH IS ENOUGH

"Prime Minister John Howard was right when he said that apologizing for history should be a private matter. I was born on this Outback sheep station 25 years ago, and we employ a dozen Aboriginal workers, most of whom were born in this area. I had nothing to do with what happened more than a century ago, or even the first 75 years of the past century. What I would have done is irrelevant, and what my great-great-grandparents may have done has nothing to do with me. All over the world people are born into situations not of their making, and they should all go around apologizing?

"And as a matter of fact I believe that this country has bent over backwards, in my time at least, to undo the alleged wrongs of the past. Look, the Aboriginal minority counts about 450,000 or 2 percent of the population. Compare this to the map: they've got the Northern Territory and other parts of the country too, and that's close to 15 percent of Australia. And now we're going to give them more? I could see disaster in the making when that guy Eddie Mabo got the High Court to agree to his so-called 'native title' claim to land with which he and his people hadn't done a thing. Since when do Aboriginal customs take precedence over the law of the land? But even worse was that decision supporting the Wiks up there in Cape York. That was unbelievable: native title on land already held in pastoral lease? That affects all of us. Pretty soon the whole of Australia will be targeted by these Aboriginal claimants and their lawyers. And by the way, in the old days those people moved around all the time. Who's to say what clans owned what land when it comes to claiming native title? These court cases are going to tie us up in legal knots for generations to come.

"And now I gather that some Aboriginal people are calling for a treaty between themselves and the government, for what purpose I'm not sure—I guess it'll be like the one the Kiwis made with the Maori there. Is this really what we need if we're going to put the past behind us? Look, I like the fellows working here, but you've got to realize that no laws or treaties are going to change all of the problems they have. They're getting preference for many kinds of government employment, remedial help in all kinds of areas, but still they wind up in jail, leaving school, giving up their jobs. Yes, I'm sure wrongs were done in the past. But apologizing isn't going to make any difference. They have their opportunity now. In a lot of countries they would have never gotten it: Aussies are pretty decent people. It's up to them to make the best of it."

Vote your opinion at www.wiley.com/college/deblij

maintained this situation until the 1970s. Today, the picture is dramatically different: of 20.9 million Australians, only about one-third have British-Irish origins, and Asian immigrants outnumber both European immigrants and the natural increase each year. During the early 1990s, nearly 150,000 legal immigrants arrived in Australia annually, most from Hong Kong, Vietnam, China, the Philippines, India, and Sri Lanka. Immigration quotas have since been reduced, most recently to 80,000, but Asian immigrants continue to outnumber those from Western sources. This situation roils Australian society, and restrictions on foreign ownership of Australian real estate reflect fears of what extremist groups call the "Asianization" of the country. But the success of many Asian settlers in Australia evinces the opportunities still available in this free and open society. Sydney, the main recipient of the Asian influx, has become a mosaic of ethnic districts, some of which have gone through periods of gang violence and drug dealing but have stabilized and even prospered over time (see photo, p. 584). Still, multiculturalism will remain a long-term challenge for Australia in the twenty-first century.

Environmental Degradation

14 **Environmental degradation** is almost synonymous with Australia. To both Aborigines and European settlers, its ecology paid a heavy price. Great stands of magnificent forest were destroyed. In Western Australia, centuries-old trees were simply "ringed" and left to die, so that the sun could penetrate through their leafless crowns to nurture the grass below. Then the sheep could be driven into these new pastures. In Tasmania, where Australia's native eucalyptus tree reaches its greatest dimensions (comparable to North American redwood stands), tens of thousands of acres of this irreplaceable treasure have been lost to chain saws and pulp mills. Many of Australia's unique marsupial species have been destroyed, and many more are endangered or threatened. "Never have so few people wreaked so much havoc on the ecology of so large an area in so short a time," observed a geographer in Australia recently. But awareness of this environmental degradation is growing. In Tasmania, the "Green" environmentalist political party has become a force in State affairs, and its activism has slowed deforestation, dam-building, and other "development" projects. Still, many Australians fear the environmentalist movement as an obstacle to economic growth at a time when the economy needs stimulation. This, too, is an issue for the future.

Another environmental problem involves Australia's wide and long-term climatic variability. In a dominantly arid continent, droughts in the moister fringes are the worst enemy, and Australia's history is replete with devastating dry spells. Australia is vulnerable to El Niño events (see p. 281), but recent global warming may also be playing a role in the process. One such serious drought, regarded as the worst in living memory, has been going on for the past seven years (as of mid–2007) and is imperiling the entire region watered by the Murray–Darling river system, Australia's breadbasket (Fig. 11-5). By some estimates, an area that accounts for 40 percent of Australia's agriculture and contains 85 percent of its irrigated land is on the verge of ruin. Part of this is caused by nature, but other factors have to do with the increasing demand for water in Aus-

Australia would seem to be one of the safest pieces of real estate on the planet, without active volcanoes, remote from tectonic-plate boundaries, occasionally but not frequently struck by high-intensity cyclones, sometimes touched by tornadoes but never by blizzards. But Australia is drought-prone, and when bush fires (from natural as well as human causes) break out, they can have devastating and even deadly consequences as high winds sweep the flames through tinder-dry forest and brush. These fires on the outskirts of Townsville, a coastal city facing the Great Barrier Reef in the State of Queensland, threaten homes on the hillside above the main road in the foreground. © Corbis.

tralia's burgeoning urban areas and by excessive damming, well-drilling, and water diversion in the Murray–Darling system's upstream tributaries. The Australian government wants to institute a coordinated drainage-basin-control program for the Murray–Darling region, but this requires cooperation from above (the State governments involved) and below (local farmers who survive by diverting water to raise their crops). So far, the plan has not been adopted, but if the drought continues to deepen, Australia's environmental crisis will leave no alternative—and even then it may be too little too late.

Status and Role

Several issues involving Australia's status at home, relations with neighbors, and position in the world are stirring up national debate as well. A persistent domestic question is whether Australia should become a republic, ending the role of the British monarch as the head of state, or continue its status quo in the British Commonwealth.

Relations with neighboring *Indonesia* and *East Timor* have been complicated. For many years, Australia had what may be called a special relationship with Indonesia, whose help it needs in curbing illegal seaborne immigration. It was also profitable for Australia to counter international (UN) opinion and recognize Indonesia's 1976 annexation of Portuguese East Timor, for in doing so Australia could deal directly with Jakarta for the oil reserves under the Timor Sea (see Fig. 10-15). Thus Australia gave neither recognition nor support to the rebel movement that fought for independence in East Timor. But this story had a relatively happy ending. When the East Timorese campaign for independence succeeded in 1999 and Indonesian troops began an orgy of murder and destruction, Australia sent an effective peacekeeping force and spearheaded the United Nations effort to stabilize the situation. Today, a new chapter has opened in Australia's relations with these northern neighbors.

Australia also has a long-term relationship with *Papua New Guinea (PNG)*, as we will note in Chapter 12. This association, too, has gone through difficult times. In recent years, the inhabitants of Papua New Guinea have strongly resisted privatization, World Bank involvement, and globalization generally. Australia assists PNG in several spheres, but its motives are sometimes questioned. In 2001, the construction of a projected gas pipeline from PNG to the Australian State of Queensland precipitated fighting among tribespeople over land rights, resulting in dozens of casualties, and Australian public opinion reflected doubts over the appropriateness of this venture.

When political violence and chaos overtook the *Solomon Islands* east of Papua New Guinea in 2003, Aus-

tralian forces intervened: a failing state in Australia's neighborhood could become a base for terrorist activity.

Immediately after the 9/11 attack on U.S. targets in New York and Washington, Australian leaders expressed strong support for the American campaign against terrorism, but they made it a point also to assure the Indonesian government that the War on Terror was not a war on Islam. Although the great majority of Australians supported this stance, their will was severely tested in 2002 when a terrorist attack on a nightclub on the Indonesian island of Bali killed 88 vacationing Australians—and again in 2005 (see p. 563) when another attack in Bali was followed by the apprehension of a terrorist cell preparing for an assault in Australia itself.

Territorial dimensions, raw-material wealth, and relative location have helped determine Australia's place in the world and, more specifically, on the Pacific Rim. Australia's population, still barely 20 million in the first decade of the twenty-first century, is smaller than Malaysia's and only slightly larger than that of the Caribbean island of Hispaniola. But Australia's importance in the international community far exceeds its human numbers.

● NEW ZEALAND

Two thousand four hundred kilometers (1500 miles) east-southeast of Australia, in the Pacific Ocean across the Tasman Sea, lies New Zealand. In an earlier age, New Zealand would have been part of the Pacific geographic realm because its population was Maori, a people with Polynesian roots. But New Zealand, like Australia, was invaded and occupied by Europeans. Today, its population of 4.2 million is almost 75 percent European, and the Maori form a minority of about 650,000, with many of mixed Euro-Polynesian ancestry.

New Zealand consists of two large mountainous islands and many scattered smaller islands (Fig. 11-7). The two large islands, with the South Island somewhat larger than the North Island, look diminutive in the great Pacific Ocean, but together they are larger than Britain. In contrast to Australia, the two main islands are mostly mountainous or hilly, with several peaks rising far higher than any on the Australian landmass. The South Island has a spectacular snowcapped range appropriately called the Southern Alps, with numerous peaks reaching beyond 3300 meters (10,000 ft). The smaller North Island has proportionately more land under low relief, but it also has an area of central highlands along whose lower slopes lie the pastures of New Zealand's chief dairying district. Hence, while Australia's land lies relatively low

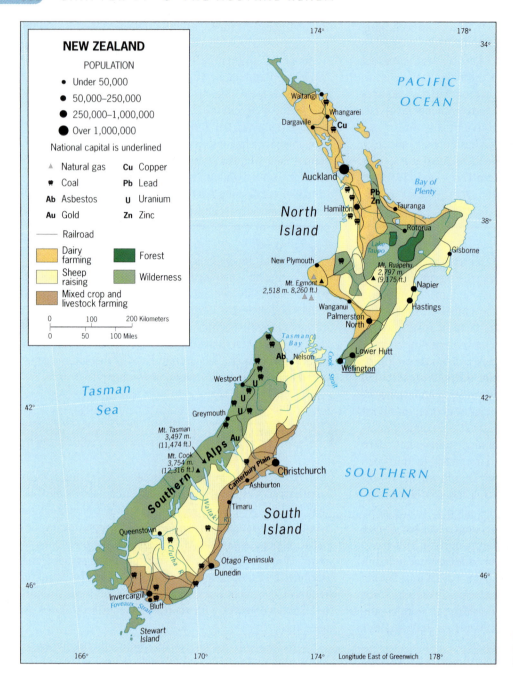

FIGURE 11-7 © H. J. de Blij, P. O. Muller, and John Wiley & Sons, Inc.

in elevation and exhibits much low relief, New Zealand's is on the average high and is dominated by rugged terrain.

Human Spatial Organization

Thus the most promising areas for habitation are the lower-lying slopes and lowland fringes on both islands. On the North Island, the largest urban area, Auckland, occupies a comparatively low-lying peninsula. On the South Island,

the largest lowland is the agricultural Canterbury Plain, centered on Christchurch. What makes these lower areas so attractive, apart from their availability as cropland, is their magnificent pastures. The range of soils and pasture plants allows both summer and winter grazing. Moreover, the Canterbury Plain, the chief farming region, also produces a wide variety of vegetables, cereals, and fruits. About half of all New Zealand is pasture land, and much of the farming provides fodder for the pastoral industry. Sixty million sheep and eight million cattle dominate these livestock-

FROM THE FIELD NOTES

"The drive from Christchurch to Arthur's Pass on the South Island of New Zealand was a lesson in physiography and biogeography. Here, on the east side of the Southern Alps, you leave the Canterbury Plain and its agriculture and climb into the rugged topography of the glacier-cut, snowcapped mountains. A last pasture lies on a patch of flatland in the foreground; in the background is the unmistakable wall of a U-shaped valley sculpted by ice. Natural vegetation ranges from pines to ferns, becoming even more luxuriant as you approach the moister western side of the island."
© H. J. de Blij

www.conceptcaching.com

raising activities, with wool, meat, and dairy products providing about two-thirds of the islands' export revenues.

Despite their contrasts in size, shape, physiography, and history, New Zealand and Australia have much in common. Apart from their joint British heritage, they share a sizeable pastoral economy, a small local market, the problem of great distances to world markets, and a desire to stimulate (through protection) domestic manufacturing. The high degree of urbanization in New Zealand (86 percent of the total population) again resembles Australia: substantial employment in city-based industries, mostly the processing and packing of livestock and farm products, and government jobs.

Spatially, New Zealand shares with Australia its pattern of **peripheral development** (Fig. G-9), imposed not by desert but by high rugged mountains. The country's major cities—Auckland and the capital of Wellington (together with its satellite of Hutt) on the North Island, and Christchurch and Dunedin on the South Island—are all located on the coast, and the entire rail and road system is peripheral in its configuration (Fig. 11-7). This is more pronounced on the South Island than in the north because the Southern Alps are New Zealand's most formidable barrier to surface communications.

The Maori Factor and New Zealand's Future

Like Australia, New Zealand has a history of difficult relations with its indigenous population. The Maori, who account for 15 percent of the country's population today, appear to have reached these islands during the tenth

FROM THE FIELD NOTES

"It was Sunday morning in Christchurch on New Zealand's South Island, and the city center was quiet. As I walked along Linwood Street I heard a familiar sound, but on an unfamiliar instrument: the Bach sonata for unaccompanied violin in G minor—played magnificently on a guitar. I followed the sound to the artist, a Maori musician of such technical and interpretive capacity that there was something new in every phrase, every line, every tempo. I was his only listener; there were a few coins in his open guitar case. Shouldn't he be playing before thousands, in schools, maybe abroad? No, he said, he was happy here, he did all right. A world-class talent, a street musician playing Bach on a Christchurch side street, where tourists from around the world were his main source of income. Talk about globalization."   © H. J. de Blij

www.conceptcaching.com

century AD. By the time the European colonists arrived, they had had a tremendous impact on the islands' ecosystems, especially on the North Island where most Maori lived. In 1840, the Maori and the British signed a treaty at Waitangi that granted the colonists sovereignty over New Zealand but guaranteed Maori rights over established tribal lands. Although the British abrogated parts of the treaty in 1862, the Maori had reason to believe that vast reaches of New Zealand, as well as offshore waters, were theirs.

As in Australia, judicial rulings during the 1990s supported the Maori position, which led to expanded claims and growing demands; one of the Maori's persistent complaints is the slow pace of integration of the minority into modern New Zealand society. Although Maori claims encompass much of the South Island, they also cover prominent sites in the major cities. Today, the Maori question is the leading national issue in the country.

Dominant cultural heritage and prevailing cultural landscape form two criteria on which the delimitation of the Austral Realm is based. But in both Australia and New Zealand, the cultural mosaic is changing, and the convergence with neighboring realms is proceeding.

What You Can Do

RECOMMENDATION: Become a member! Just by taking this course, you are joining a still-small minority of geographically literate Americans (Canadians are far ahead), and you may wish to consider joining some organizations that support this cause. The National Geographic Society (1145 Seventeenth Street N.W., Washington, DC 20036) invites members to join and sends them its famous Magazine with its remarkable foldout maps. Gamma Theta Upsilon is a students' geographical society (c/o Larry Handley, USGS/NWRC, 700 Cajundome Boulevard, Lafayette, LA 70506) that publishes *The Geographical Bulletin*, a journal that invites students to submit their work for publication. The American Geographical Society (120 Wall Street, Suite 100, New York, NY 10005-3904) publishes two journals that can be appreciated by the general public, *The Geographical Review* and *Focus*. And if you want to get really technical, there is the Association of American Geographers (1710 Sixteenth Street N.W., Washington, DC 20009-3198), which publishes the quarterly *Annals of the Association of American Geographers*, *The Professional Geographer*, and the monthly *AAG Newsletter* that tells you what's going on in the discipline and where the jobs are. Student members, at appropriately reduced annual dues, are especially welcome!

GEOGRAPHIC CONNECTIONS

1 Both Australia and New Zealand are plural societies, and there were indigenous peoples in both countries when the European settlers arrived. In virtually every respect, however, the historical geographies of contact and interaction in the two countries have progressed in very different ways. The differences relate to contrasts in indigenous as well as immigrant cultures, divergent population numbers, contrasting territorial dimensions and physiographies, and dissimilar governmental policies and practices. Comment on the differences between Australia and New Zealand in this context. Are there ways in which indigenous demands are similar? How are they expressed on the map?

2 Australia may have a modest population, but its territorial dimensions, resource base, economic wealth, military capacity, and relative location make it a regional power. Its relationships with its neighbors, however, have at times been problematic. To what extent does relative location affect Australia's interaction with Indonesia? How has Australia's relationship with East Timor, before and after the latter's independence in 2002, been affected by its links with Indonesia, and what was the key issue in this context? What interests does Australia maintain in Papua New Guinea? Why would Australia be involved in problems in the Solomon Islands?

Geographic Literature on the Austral Realm: The key introductory works on this realm were authored by Barrett & Ford, Heathcote, McKnight, Morton & Johnston, O'Connor et al., and Robinson et al. These works, together with a comprehensive listing of noteworthy books and recent articles on the geography of Australia and New Zealand, can be found in the *References and Further Readings* section of this book's website at **www.wiley.com/college/deblij**.

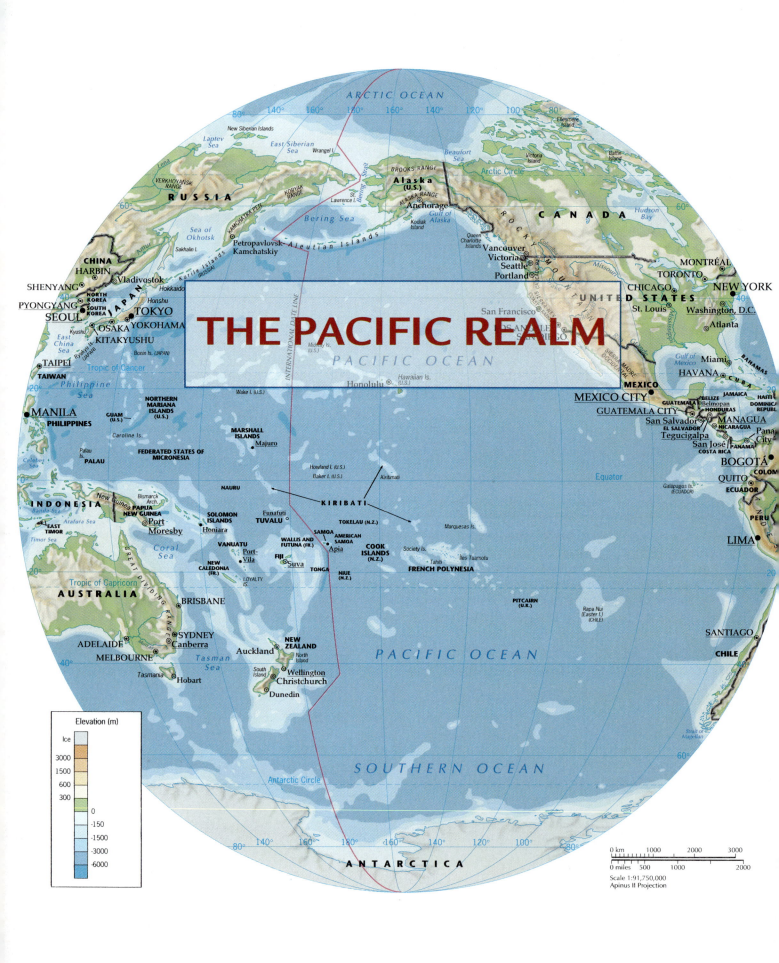

THE PACIFIC REALM

CONCEPTS, IDEAS, AND TERMS

1. Marine geography
2. Territorial sea
3. High seas
4. Continental shelf
5. Exclusive Economic Zone (EEZ)
6. Maritime boundary
7. Median-line boundary
8. High-island cultures
9. Low-island cultures
10. Antarctic Treaty

REGIONS

- MELANESIA
- MICRONESIA
- POLYNESIA

Where were these photos taken? Find out at www.wiley.com/college/deblij

In This Chapter

- Water, water everywhere—but who owns it?
- Islands high and low
- The conundrum of divided Papua
- Colonists and foreigners: Still there, still dominant
- The troubled legacy of Fiji
- Native Hawaiians hope for a sliver of sovereignty

Photos: © H. J. de Blij

FIGURE 12-1   © H. J. de Blij, P. O. Muller, and John Wiley & Sons, Inc.

BETWEEN THE AMERICAS to the east and the western Pacific Rim to the west lies the vast Pacific Ocean, larger than all the world's land areas combined. In this greatest of all oceans lie tens of thousands of islands, some large (New Guinea is by far the largest), most small (many are uninhabited). Together, the land area of these islands is a mere 975,000 square kilometers (377,000 sq mi), about the size of Texas plus New Mexico, and over 90 percent of this lies in New Guinea.*

Defining the Realm

The Pacific geographic realm—land and water—covers nearly an entire hemisphere of this world, the one commonly called the Sea Hemisphere (Fig. 12-1). This Sea Hemisphere meets the Russian and North American realms in the far north and merges into the Southern Ocean in the south. Despite the preponderance of water, this fragmented, culturally complex realm does possess regional identities. It includes the Hawaiian Islands, Tahiti, Tonga, and Samoa—fabled names in a world apart.

In terms of modern cultural and political geography, Indonesia and the Philippines are not part of the Pacific Realm, although Indonesia's political system reaches into it; nor are Australia and New Zealand part of it. Before the European invasion and colonization, Australia would have been included because of its Aboriginal population and New Zealand because of its Maori population's Polynesian affinities. But the Europeanization of their countries has engulfed black Australians and Maori New Zealanders, and the regional geography of Australia and New Zealand today is decidedly not Pacific. In New Guinea, on the other hand, Pacific peoples remain numerically and culturally the dominant element.

COLONIZATION AND INDEPENDENCE

The Pacific islands were colonized by the French, British, and Americans; an indigenous Polynesian kingdom in the Hawaiian Islands was annexed by the United States and is now the fiftieth State. Still, today, the map is an assemblage of independent and colonial territories (Fig. 12-1). Paris controls New Caledonia and French Polynesia. The United States administers Guam and American Samoa, the Line

Islands, Wake Island, Midway Islands, and several smaller islands; the United States also has special relationships with other territories, former dependencies that are now nominally independent. The British, through New Zealand, have responsibility for the Pitcairn group of islands, and

---

MAJOR GEOGRAPHIC QUALITIES OF

The Pacific Realm

1. The Pacific Realm's total area is the largest of all geographic realms. Its land area, however, is the smallest, as is its population.

2. The island of New Guinea, with 8.8 million people, alone contains over 80 percent of the Pacific Realm's population.

3. The Pacific Realm, with its wide expanses of water and numerous islands, has been strongly affected by United Nations Law of the Sea provisions regarding states' rights over economic assets in their adjacent waters.

4. The highly fragmented Pacific Realm consists of three regions: Melanesia (including New Guinea), Micronesia, and Polynesia.

5. Melanesia forms the link between Papuan and Melanesian cultures in the Pacific.

6. The Pacific Realm's islands and cultures may be divided into volcanic high-island cultures and coral-based low-island cultures.

7. In Micronesia, U.S. influence has been particularly strong and continues to affect local societies.

8. In Polynesia, local cultures nearly everywhere are severely strained by external influences. In Hawai'i, as in New Zealand, indigenous culture has been largely submerged by Westernization.

9. Indigenous Polynesian culture continues to exhibit a remarkable consistency and uniformity throughout the Polynesian region, its enormous dimensions and dispersal notwithstanding.

---

*The figures in Table G-1 do not match these totals because only the political entity of Papua New Guinea is listed, not the Indonesian Province of Papua that occupies the western part of the island. Here, as Figure 10-2 shows, the political and the realm boundaries do not coincide.

New Zealand administers and supports the Cook, Tokelau, and Niue Islands. Easter Island, that storied speck of land in the southeastern Pacific, is part of Chile. Indonesia rules Papua, the western half of the island of New Guinea.

Other island groups have become independent states. The largest are Fiji, once a British dependency, the Solomon Islands (also formerly British), and Vanuatu (until 1980 ruled jointly by France and Britain). Also on the current map, however, are such microstates as Tuvalu, Kiribati, Nauru, and Palau. Foreign aid is crucial to the survival of most of these countries. Tuvalu, for example, has a total area of around 25 square kilometers (10 sq mi), a population of some 10,000, and a per-capita GNI of about U.S. $1200, derived from fishing, copra sales, and some tourism. But what really keeps Tuvalu going is an international trust fund set up by Australia, New Zealand, the United Kingdom, Japan, and South Korea. Annual grants from that fund, as well as money sent back to families by workers who have left for New Zealand and elsewhere, allow Tuvalu to survive.

THE PACIFIC REALM AND ITS MARINE GEOGRAPHY

Certain land areas may not be part of the Pacific Realm (we cited the Philippines and New Zealand), but the Pacific Ocean extends from the shores of North and South America to mainland East and Southeast Asia and from the Bering Sea to the Subtropical Convergence. This means that several seas, including the Sea of Japan (East Sea), the East China Sea, and the South China Sea, are part of the Pacific Ocean. As we will see below, this relationship matters. Pacific coastal countries, large and small, mainland and island, compete for jurisdiction over the waters that bound them.

The Pacific Realm and its ocean, therefore, form an **[1]** ideal place to focus on **marine geography**. This field encompasses a variety of approaches to the study of oceans and seas; some marine geographers focus on the biogeography of coral reefs, others on the geomorphology of beaches, and still others on the movement of currents and drifts. A particularly interesting branch of marine geography has to do with the definition and delimitation of political boundaries at sea. Here geography meets political science and maritime law.

The State at Sea

Littoral (coastal) states do not end where atlas maps suggest they do. States have claimed various forms of jurisdiction over coastal waters for centuries, closing off bays and estuaries and ordering foreign fishing fleets to stay away

from nearby fishing grounds. Thus arose the notion of the **territorial sea**, where all the rights of a coastal state would **[2]** prevail. Beyond lay the **high seas**, free, open, and unfet- **[3]** tered by national interests.

It was in the interest of colonizing, mercantile states to keep territorial seas narrow and high seas wide, thus interfering as little as possible with their commercial fleets. In the seventeenth and eighteenth centuries the territorial sea was 3, 4, or at most 6 nautical miles wide, and the colonizing powers claimed the same widths for their colonies (1 nautical mile = 1.85 kilometers [1.15 statute miles]).*

In the twentieth century these constraints weakened. States without trading fleets saw no reason to limit their territorial seas. States with nearby fishing grounds traditionally exploited by their own fleets wanted to keep the increasing number of foreign trawlers away. States with shallow **continental shelves**, offshore continuations of **[4]** coastal plains, wished to control the resources on and below the seafloor, made more accessible by improving technology. States disagreed on the methods by which offshore boundaries, whatever their width, should be defined. Early efforts by the League of Nations in the 1920s to resolve these issues had only partial success, mainly in the technical area of boundary delimitation.

Scramble for the Oceans

In 1945, the United States helped precipitate what has become known as the "scramble for the oceans." President Harry S Truman issued a proclamation that claimed U.S. jurisdiction and control over all the resources "in and on" the continental shelf down to its margin, around 100 fathoms (600 ft or 183 m) deep. In some areas, the shallow continental shelf of the United States extends more than 200 miles (300 km) offshore, and Washington did not want foreign countries drilling for oil just beyond the 3-mile territorial sea.

Few observers foresaw the impact the Truman Proclamation would have, not only on U.S. waters but on the oceans everywhere, including the Pacific. It set off a rush of other claims. In 1952, a group of South American countries, some with little continental shelf to claim, issued the Declaration of Santiago, claiming exclusive fishing rights up to a distance of 200 nautical miles off their coasts. Meanwhile, as part of the Cold War competition, the Soviet Union urged its allies to claim a 12-mile territorial sea.

---

*Here, and in the rest of this section, we will give all distances involving maritime boundaries in nautical miles only. Throughout the modern history of maritime law, the latter unit has been the only one used for this type of boundary-making. Any distance can be converted to its kilometric equivalent by multiplying the number of nautical miles by 1.85.

UNCLOS Intervention

Now the United Nations intervened, and a series of meetings of UNCLOS (United Nations Conference on the Law of the Sea) began. These meetings addressed issues ranging from the closure of bays to the width and delimitation of the territorial sea, and after three decades of negotiations they achieved a convention that changed the political and economic geography of the oceans forever. Among its key provisions were the authorization of a 12-mile territorial sea for all countries and the establishment of a 200-mile **Exclusive Economic Zone (EEZ)** over which a coastal state would have total economic rights. Resources in and under this EEZ (fish, oil, minerals) belong to the coastal state, which could either exploit them or lease, sell, or share them as it saw fit.

These provisions had a far-reaching impact on the oceans and seas of the world (Fig. 12-2), and especially so on the Pacific. Unlike the Atlantic Ocean, the Pacific is studded with islands large and small, and a microstate consisting of one small island suddenly acquired an EEZ covering 160,000 square nautical miles. European colonial powers still holding minor Pacific possessions (notably France) saw their maritime jurisdictions vastly expanded. Small low-income archipelagos could now bargain with large, rich fishing nations over fishing rights in their EEZs. And

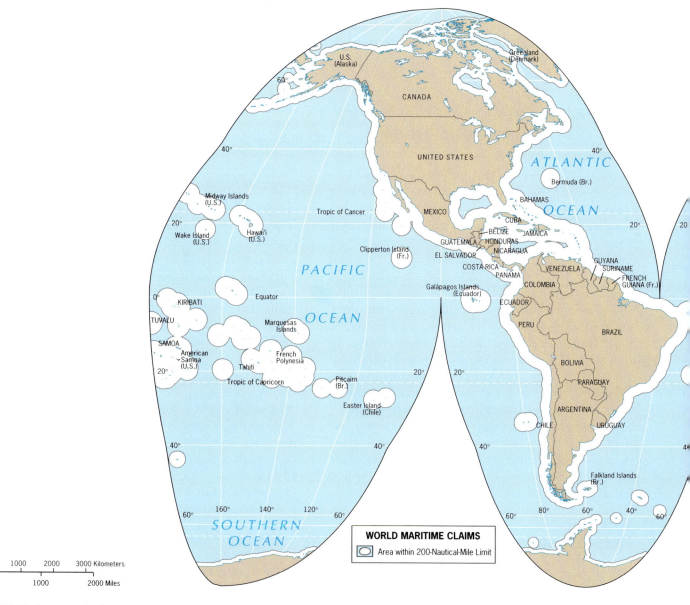

FIGURE 12-2 © H. J. de Blij, P. O. Muller, and John Wiley & Sons, Inc.

for all the UNCLOS Convention's provisions for the "right of innocent passage" of shipping through EEZs and via narrow straits, the world's high seas have obviously been diminished.

Maritime Boundaries

The extension of the territorial sea to 12 nautical miles and the EEZ to an additional 188 nautical miles created new

6 **maritime boundary** problems. Waters less than 24 nautical miles wide separate many countries all over the world,

7 so that **median lines**, equidistant from opposite shores, have been delimited to establish their territorial seas. And

even more countries lie closer than 400 nautical miles apart, requiring further maritime-boundary delimitation to determine their EEZs. In such maritime regions as the North Sea, the Caribbean Sea, and the Japan, East China, and South China seas, a maze of maritime boundaries emerged, some of them subject to dispute. Political changes on land can lead to significant modifications at sea.

A case in point involves newly independent East Timor and its neighbor across the Timor Sea, Australia. As we noted in Chapter 10, Australia had divided the waters and seafloor of the Timor Sea with Indonesia while recognizing Indonesia's 1976 annexation of East Timor. When East Timor achieved independence in 2002, the so-called

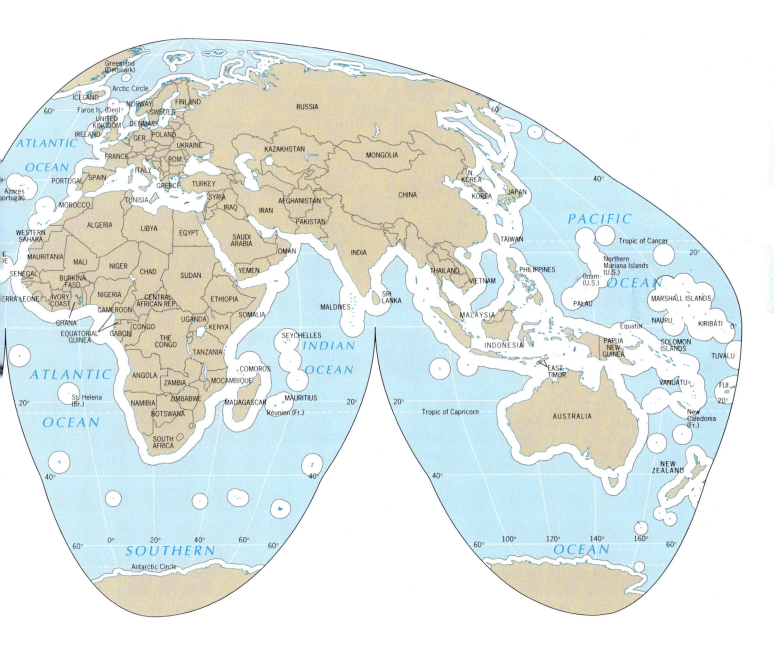

FROM THE FIELD NOTES

"Having arrived late at night, I did not have a perspective yet of the setting of Pago Pago—but even before I got up I knew what the leading industry here had to be. An intense, overpowering smell of fish filled the air. Sure enough, daylight revealed a huge tuna processing plant right across the water, with fishing boats arriving and departing continuously. American Samoa has a huge Pacific Ocean EEZ, one of the richest fishing grounds in the whole realm, and seafood exports rank ahead of tourism as the leading source of external income for the territory, an industry that is mostly American-owned. As the photograph shows, American Samoa is a group of 'high-islands' of volcanic origin, with considerable relief and environmental variety. Tuvalu, the country I had just left, was the complete opposite— a 'low-island' territory whose EEZ has been leased to Japanese fishing fleets, which do their processing on board and employ no locals." © H. J. de Blij

www.conceptcaching.com

Timor Gap (mapped in Fig. 10-15) became an issue: where was the median line that would divide Timorese and Australian claims to the oil and gas reserves in this zone? The Australians argued that since the line defined with Indonesia predated East Timor's independence, it should continue in effect, giving Australia the bulk of the energy resources. But UNCLOS regulations, to which Australia subscribed, required a redelimitation, giving East Timor a much larger share. This led Australia to withdraw from UNCLOS in 2002 so that it would not be bound by its rules. After difficult negotiations, the Australians in 2005 offered East Timor a half share of the energy revenues from the natural gas reserve that was in dispute, in addition to income already flowing from another major gasfield in the so-called Joint Petroleum Development Area. In return, East Timor's

government agreed to defer the maritime-boundary issue for 50 years, leaving a future generation to solve this question. It was estimated that, over the next 30 years, East Timor might receive U.S. $13 billion from these sources and, depending on the life and capacity of the reserves, perhaps even more. This will go a long way toward meeting the young state's most urgent needs.

EEZ Implications

The UNCLOS provisions created opportunities for some states to expand their spheres of influence. Wider territorial-sea and EEZ allocations raised the stakes: claiming an island now entailed potential control over a huge maritime area. In Chapters 9 and 10, we referred to several island disputes off mainland East Asia, which involve Japan and Russia, Japan and South Korea, Japan and China, China and Vietnam, and China and the Philippines. Ownership of many islands there is uncertain, and small specks of island territory have become large stakes in the scramble for the oceans. For example, in the case of the Spratly Islands (Fig. 10-2), six countries claim ownership, including both Taiwan and China. China's island claims in the South China Sea support Beijing's contention that this body of water is part and parcel of the Chinese state—a position that worries other states with coasts facing it.

Figure 12-2 reveals what EEZ regulations have meant to Pacific Realm countries such as Tuvalu, Kiribati, and Fiji, with nearly circular EEZs now surrounding the clusters of islands in this vast ocean space. Japan, Taiwan, and other fishing nations have purchased fishing rights in these EEZs from the island governments. Nonetheless, violations of EEZ rights do occur; recently, Vanuatu and the Philippines were at odds over unauthorized Filipino fishing in Vanuatu's EEZ.

The process of boundary delimitation continues. In an earlier edition of this book, we included a map of the South China Sea and nearby waters off East and Southeast Asia, showing the median-line boundaries delimited according to UNCLOS specifications. But that map changed when coastal states engaged in bilateral and multilateral negotiations—and argued over island ownership with major boundary implications. The Pacific—and world—map of maritime boundaries remains a work in progress (see the Issue Box entitled "Who Should Own the Oceans?").

This point is illustrated by a development that occurred late in 1999, when the implications of provisions in Article 76 of the 1982 UNCLOS Convention were reexamined. Where a continental shelf extends beyond the 200-mile limit of the EEZ, the Convention apparently allows a coastal state to claim that extension as a "natural prolongation" of its landmass. Although there is as yet no regulation that would also allow the state to extend its EEZ to the edge

Who Should Own the Oceans?

WE WHO LIVE HERE SHOULD OWN THE WATERS

"Without exclusive fishing rights in our waters up to 200 nautical miles offshore, our economy would be dead in the water, no pun intended, sir. As it is, it's bad enough. I'm an official in the Fisheries Department of the Ministry of Natural Resources Development here in Tuvalu, and this modest building you're visiting is the most modern in all of the capital, Funafuti. That's because we're the only ones making some money for the state other than our Internet domain (*tv*), which has been the big story around here since we commercialized it. We don't have much in the way of natural resources and we export some products from our coconut and breadfruit trees, but otherwise we need donors to help us and major fishing nations to pay us for the use of our 750,000 square kilometers of ocean.

"Which brings me to one of our big concerns. Three-quarters of a million square kilometers sound like a lot, but look at the map. It's a mere speck in the vast Pacific Ocean. Enormous stretches of our ocean seem to belong to nobody. That is, they're open to fishing by nations that have the technology to exploit them, but we don't have those fast boats and factory ships they use nowadays. Still, the fish they take there might have come to our exclusive grounds, but we don't get paid for them. It seems to us that the Pacific Ocean should be divided among the countries that exist here, not just on the basis of nearby Exclusive Economic Zones but on some other grounds. For example, islands that are surrounded by others, like we are, ought to get additional maritime territory someplace else. Islands that aren't surrounded should get much more than just 200 nautical miles. To us the waters of the Pacific mean much more than they do to other countries, not only economically but historically too. I don't like Japanese and American and Norwegian fishing fleets on these nearby 'high seas.' They don't obey international regulations on fishing, and they overfish wherever they go. When there's nothing left, they'll simply end their agreement with us, and we'll be left without income from a natural resource over which we never had control.

"I don't expect anything to change, but we feel that the era of the 'high seas' should end and all waters and subsoil should be assigned to the nearest states. The world doesn't pay much attention to our small countries, I realize. But we've put up with colonialism, world wars, nuclear weapons testing, pollution. It's time we got a break."

Regional ISSUE

THE OCEAN SHOULD BE OPEN AND UNRESTRICTED

"Ours is the Blue Planet, the planet of oceans and seas, and the world has fared best when its waters were free and open for the use of all. Encroachment from land in the form of 'territorial' seas (what a misnomer!) and other jurisdictions only interfered with international trade and commercial exploitation and caused far more problems than they solved. Already much of the open ocean is assigned to coastal countries that have no maritime tradition and don't know what to do with it other than to lease it. And when coastal countries lie closer than 400 nautical miles to each other, they divide the waters between them for EEZs and don't even leave a channel for international passage. Yes, I realize that there are guarantees of uninhibited passage through EEZs. But mark my words: the time will come when small coastal states will start charging vessels, for example cruise ships, for passage through 'their' waters. That's why, as the first officer of a passenger ship, I am against any further allocation of maritime territory to coastal states. We've gone too far already.

"In the 1960s, when those UNCLOS meetings were going on, a representative from Malta named Arvid Pardo came up with the notion that the whole ocean and seafloor beyond existing jurisdictions should be declared a 'common heritage of mankind' to be governed by an international agency that would control all activities there. The idea was that all economic activity on and beneath the high seas would be controlled by the UN, which would ensure that a portion of the proceeds would be given to 'Geographically Disadvantaged States' like landlocked countries, which can't claim any seas at all. Fortunately nothing came of this—can you imagine a UN agency administering half the world for the benefit of humanity?

"So you can expect us mariners to oppose any further extensions of land-based jurisdiction over the seas. The maritime world is complicated enough already, what with choke points, straits, narrows, and various kinds of prohibitions. To assign additional maritime territory to states on historic or economic grounds is to complicate the future unnecessarily, and to invite further claims. From time to time I see references to something called the 'World Lake Concept,' the preliminary assignment of all the waters of the world to the coastal states through the delimitation of median lines. Such an idea may be the natural progression from the EEZ concept, but it gives me nightmares."

Vote your opinion at www.wiley.com/college/deblij

of the continental shelf, it is not difficult to foresee such an amendment to the Convention. As it stands, states are now delimiting their proposed "natural prolongations" under a 2009 deadline, which puts poorer states at a disadvantage since the required marine surveys are very expensive. In any case, a recent article in *Science* reported that the big winners are likely to include the United States, Canada, Australia, New Zealand, Russia, and India—although a total of 60 coastal countries could ultimately benefit to some extent from the provision.

Thus the "scramble for the oceans" continues, and with it the constriction of the world's high seas and open waters.

POINTS TO PONDER

● Rising sea levels associated with global warming may threaten the very existence of entire Pacific nations.

● Kiribati transferred its political allegiance from China to Taiwan, expelling Beijing's embassy in 2006.

● Native Hawaiians and other Pacific Islanders today constitute only about 9 percent of the population of Hawai'i.

Regions of the Realm

Sail across the Pacific Ocean, and one spectacular vista follows another. Dormant and extinct volcanoes, sculpted by erosion into basalt spires draped by luxuriant tropical vegetation and encircled by reefs and lagoons, tower over azure waters. Low atolls with nearly snow-white beaches, crowned by stands of palm trees, seem to float in the water. Pacific islanders, where foreign influences have not overtaken them, appear to take life with enviable ease.

More intensive investigations reveal that such Pacific sameness is more apparent than real. Even the Pacific Realm, with its long sailing traditions, its still-diffusing populations, and its historic migrations, has a durable regional framework. Figure 12-3 outlines the three regions that constitute the Pacific Realm:

Melanesia: Papua (Indonesia), Papua New Guinea, Solomon Islands, Vanuatu, New Caledonia (France), Fiji

Micronesia: Palau, Federated States of Micronesia, Northern Mariana Islands, Republic of the Marshall Islands, Nauru, western Kiribati, Guam (U.S.)

Polynesia: Hawaiian Islands (U.S.), Samoa, American Samoa, Tuvalu, Tonga, eastern Kiribati, Cook and other New Zealand-administered islands, French Polynesia

MAJOR CITIES OF THE REALM

City	Population* (in millions)
Honolulu, Hawai'i (U.S.)	0.9
Nouméa, New Caledonia	0.2
Port Moresby, Papua New Guinea	0.3
Suva, Fiji	0.2

*Based on 2008 estimates.

Ethnic, linguistic, and physiographic criteria are among the bases for this regionalization of the Pacific Realm, but we should not lose sight of the dimensions. Not only is the land area small, but also in 2008 the population of this entire realm (including Indonesia's Papua Province) was only about 11.5 million—9 million without Papua—about the same as one very large city. Even fewer people live in this realm than in another vast area of far-flung settlements—the oases of North Africa's Sahara.

MELANESIA

The large island of New Guinea lies at the western end of a Pacific region that extends eastward to Fiji and includes the Solomon Islands, Vanuatu, and New Caledonia (Fig. 12-3). The human mosaic here is complex, both ethnically and culturally. Most of the 8.8 million people of New Guinea (including the Indonesian part, the Province of Papua, and the independent state of Papua New Guinea) are Papuans, and a large minority is Melanesian. Altogether there are as many as 700 communities speaking different languages; the Papuans are most numerous in the densely forested highland interior and in the south, while the Melanesians inhabit the north and east. The region as a whole has more than 7 million inhabitants, making this the most populous Pacific region by far.

With 6.3 million people today, Papua New Guinea (PNG) became a sovereign state in 1975 after nearly a century of British and Australian administration. Almost all of PNG's limited development is taking place along the coasts, while most of the interior remains hardly touched by the changes that transformed neighboring Australia. Perhaps four-fifths of the population lives in what we may describe as a self-sufficient subsistence economy, growing root crops and hunting wildlife, raising pigs, and gathering forest products. Old traditions of the kind lost in Australia persist here, protected by remoteness and the rugged terrain.

Welding this disparate population into a nation is a task hardly begun, and PNG faces numerous obstacles in addition to its cultural complexity. Not only are hundreds of languages in use, but close to half the population is illiterate. English, the official language, is used by the educated minority but is of little use beyond the coastal zone and its towns. The capital, Port Moresby, has less than 350,000 inhabitants, reflecting the very low level of urbanization (13 percent) in this developing economy.

Yet Papua New Guinea is not without economic opportunities. Oil was discovered in the 1980s, and by the late 1990s crude oil was PNG's largest export by value. Gold now ranks second, followed by copper, silver, timber, and several agricultural products including coffee and cocoa, reflecting the country's resource and environmental diversity. Pacific Rim developments have affected even PNG:

most exports go to the nearest neighbor, Australia, but Japan ranks close behind.

Turning eastward, it is a measure of Melanesia's cultural fragmentation that as many as 120 languages are spoken in the approximately 1000 islands that make up the Solomon Islands (about 80 of these islands support almost all the people, numbering about 500,000). Inter-island, historic animosities among the islanders were worsened by the events of World War II, when thousands of Malaitans were moved by U.S. forces to Guadalcanal. This started a postwar cycle of violence that led Australia to intervene in 2003.

New Caledonia, still under French rule, is in a very different situation. Only around 43 percent of the population of about 240,000 are Melanesian; 37 percent are of French ancestry, many of them descended from the inhabitants of the penal colony France established here in the nineteenth century. Nickel mines, based on reserves that rank among the world's largest, dominate New Caledonia's export economy (photo below). The

FROM THE FIELD NOTES

"Arriving in the capital of New Caledonia, Nouméa, in 1996, was an experience reminiscent of French Africa 40 years earlier. The French tricolor was much in evidence, as were uniformed French soldiers. European French residents occupied hillside villas overlooking palm-lined beaches, giving the place a Mediterranean cultural landscape. And New Caledonia, like Africa, is a source of valuable minerals. It is one of the world's largest nickel producers, and from this vantage point you could see the huge treatment plants, complete with concentrate ready to be shipped (left, under conveyor). What you cannot see here is how southern New Caledonia has been ravaged by the mining operations, which have denuded whole mountainsides. Working in the mines and in this facility are the local Kanaks, Melanesians who make up about 43 percent of the population of about 240,000. Violent clashes between Kanaks and French have obstructed government efforts to change New Caledonia's political status in such a way as to accommodate pressures for independence as well as continued French administration."   © H. J. de Blij

www.conceptcaching.com

Concept Caching

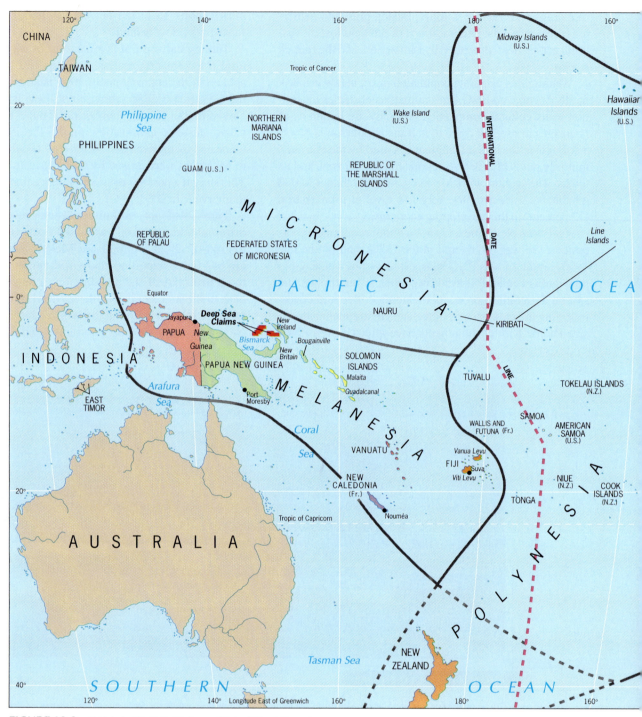

FIGURE 12-3 © H. J. de Blij, P. O. Muller, and John Wiley & Sons, Inc.

mining industry attracted additional French settlers, and social problems arose. Most of the French population lives in or near the capital city of Nouméa, steeped in French cultural landscapes, in the southeastern quadrant of the island. Melanesian demands for an end to colonial rule have led to violence, and the two communities are still in the process of coming to terms.

On its eastern margins, Melanesia includes one of the Pacific Realm's most interesting countries, Fiji. On two larger and over 100 smaller islands live nearly 800,000 Fijians, of whom 51 percent are Melanesians and 44 percent South Asians, the latter brought to Fiji from India during the British colonial occupation to work on the sugar plantations. When Fiji achieved independence in 1970, the native

140°   120°   100°   U.S.A.   80°

ATLANTIC OCEAN

Tropic of Cancer

MEXICO

CUBA

20°

HONDURAS
GUATEMALA
EL SALVADOR   NICARAGUA

COSTA RICA

PANAMA

VENEZUELA

COLOMBIA

Clipperton
(Fr. Poly.)

Equator

Galápagos
Islands
(Ecuador)

ECUADOR

0°

PACIFIC OCEAN

Marquesas
Islands

PERU

BRAZIL

FRENCH
POLYNESIA

Tahiti

Tuamotu
Archipelago

BOLIVIA

20°

PITCAIRN
(U.K.)

Easter
Island
(Chile)

Tropic of Capricorn

POLYNESIA

CHILE

ARGENTINA

URUGUAY

PACIFIC REGIONS

0   400   800   1200 Kilometers
0   400   800 Miles

SOUTHERN OCEAN

40°

Longitude West of Greenwich   100°   80°   60°

Fijians owned most of the land and held political control, while the Indians were concentrated in the towns (chiefly Suva, the capital; see photos, p. 604) and dominated commercial life. It was a recipe for trouble, and it was not long in coming when, in a later election, the politically active Indians outvoted the Fijians for seats in the parliament. A coup by the Fijian military was followed by a revision of the constitution, which awarded a majority of seats to ethnic Fijians. Before long, however, the Fijian majority splintered, but a coalition government for some time proved the constitution workable—until 2000. In 1999, Fiji's first prime minister of Indian ancestry had taken office, angering some ethnic Fijians to the point of staging another coup. The following year the prime minister and members of the government were taken hostage at the parliament building in Suva, and the perpetrators demanded that Fijians would henceforth govern the country.

This action, and Fiji's inability to counter it, had a devastating impact on the country. Foreign trading partners stopped buying Fijian products. The tourist industry suffered severely. And in the end, although the coup leaders were ousted and arrested, they secured a deal that gave ethnic Fijians the control they had sought. The coup continues to cast a shadow across this country. The sugar industry was damaged by the nonrenewal of Indian-held leases by Fijian landowners, which also resulted in a substantial movement of Indians from the countryside to the towns, where unemployment is already high. Fiji's future remains clouded.

Melanesia, the most populous region in the Pacific Realm, is bedeviled by centrifugal forces of many kinds. No two countries present the same form of multiculturalism; each has its own challenges to confront, and some of these challenges spill over into neighboring (or more distant foreign) islands.

● MICRONESIA

North of Melanesia and east of the Philippines lie the islands that constitute the region known as Micronesia (Fig. 12-3). The name (*micro* means small) refers to the size of the islands: the 2000-plus islands of Micronesia are not

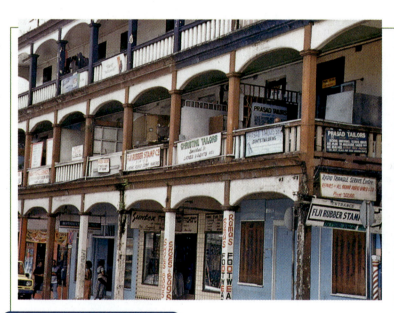

FROM THE FIELD NOTES

"Back in Suva, Fiji more than 20 years after I had done a study of its CBD, I noted that comparatively little had changed, although somehow the city seemed more orderly and prosperous than it was in 1978. At the Central Market I watched a Fijian woman bargain with the seller over a batch of taro, the staple starch-provider in local diets. I asked her how she would serve it. 'Well, it's like your potato,' she said. 'I can make a porridge, I can cut it up and put it in a stew, and I can even fry pieces of it and make them look like the French fries they give you at the McDonald's down the street.' Next I asked the seller where his taro came from. 'It grows over all the islands,' he said. 'Sometimes, when there's too much rain, it may rot—but this year the harvest is very good.' . . . Next I walked into the crowded Indian part of the CBD. On a side street I got a reminder of the geographic concept of agglomeration. Here was a colonial-period building, once a hotel, that had been converted into a business center. I counted 15 enterprises, ranging from a shoestore to a photographer and including a shop where rubber stamps were made and another selling diverse tobacco products. Of course (this being the Indian sector of downtown Suva) tailors outnumbered all other establishments." © H. J. de Blij

www.conceptcaching.com

Concept Caching

Tourists visiting Pacific islands see pristine beaches and meticulously maintained hotel facilities, but Pacific islanders live with other realities, including the pollution of their environment. From the distant deck of a cruise ship, the lagoons of Kiribati's Tarawa Atoll look unspoiled, but grab a bike and ride to some villages, and here is what you are likely to see. Already there is not enough room to dispose of local trash, and daily the tide contributes more. Look closely at piles like this, and you will find the spoils of "civilization" ranging from garbage-filled plastic bags thrown overboard to fish traps, netting, styrofoam, shards of glass, and pieces of wood. Pollution is one of this realm's most serious problems. © Caroline Penn/Corbis Images

only tiny (many of them no larger than 1 square kilometer [half a square mile]), but they are also much lower-lying, on an average, than those of Melanesia. Some are volcanic islands (**high islands**, as the people call them), but they are outnumbered by islands composed of coral, the **low islands** that barely lie above sea level. Guam, with 550 square kilometers (210 sq mi), is Micronesia's largest island, and no island elevation anywhere in Micronesia reaches 1000 meters (3300 ft).

The high-island/low-island dichotomy is useful not only in Micronesia, but also throughout the realm. Both the physiographies of these islands and the economies they support differ in crucial ways. High islands wrest substantial moisture from the ocean air; they tend to be well watered and have good volcanic soils. As a result, agricultural products show some diversity, and life is reasonably secure. Populations tend to be larger on these high islands than on the low islands, where drought is the rule and fishing and the coconut palm are the mainstays of life. Small communities cluster on the low islands, and over time many of these have died out. The major migrations, which sent fleets to populate islands from Hawai'i to New Zealand, tended to originate in the high islands.

Until the mid-1980s, Micronesia was largely a United States Trust Territory (the last of the post–World War II trusteeships supervised by the United Nations), but that status has now changed. As Figure 12-3 shows, today Micronesia is divided into countries bearing the names of independent states. The Marshall Islands, where the United States tested nuclear weapons (giving prominence to the name *Bikini*), now is a republic in "free association" with the United States, having the same status as the Federated States of Micronesia and (since 1994) Palau. The Northern Mariana Islands are a commonwealth "in po-

litical union" with the United States. In effect, the United States provides billions of dollars in assistance to these countries, in return for which they commit themselves to avoid foreign policy actions that are contrary to U.S. interests. There are other conditions: Palau, for example, granted the United States rights to existing military bases for 50 years following independence.

Also part of Micronesia is the U.S. territory of Guam, where independence is not in sight and where U.S. military installations and tourism provide the bulk of income, and the remarkable Republic of Nauru With a population of barely 12,000 and 20 square kilometers (8 sq mi) of land, this microstate got rich by selling its phosphate deposits to Australia and New Zealand, where they are used as fertilizer. Per-capita incomes rose to U.S. $11,500, making Nauru one of the Pacific's high-income societies. But the phosphate deposits ran out, and an island scraped bare faced an economic crisis.

In this region of tiny islands, most people subsist on farming or fishing, and virtually all the countries need infusions of foreign aid to survive. The natural economic complementarity between the high-island farming cultures and the low-island fishing communities all too often is negated by distance, spatial as well as cultural. Life here may seem idyllic to the casual visitor, but for the Micronesians it often is a daily challenge.

POLYNESIA

To the east of Micronesia and Melanesia lies the heart of the Pacific, enclosed by a great triangle stretching from the Hawaiian Islands to Chile's Easter Island to New Zealand.

This is Polynesia (Fig. 12-3), a region of numerous islands (*poly* means many), ranging from volcanic mountains rising above the Pacific's waters (Mauna Kea on Hawai'i reaches beyond 4200 meters [nearly 13,800 ft]), clothed by luxuriant tropical forests and drenched by well over 250 centimeters (100 in) of rainfall each year, to low coral atolls where a few palm trees form the only vegetation and where drought is a persistent problem. The Polynesians have somewhat lighter-colored skin and wavier hair than do the other peoples of the Pacific Realm; they are often also described as having an excellent physique. Anthropologists differentiate between these original Polynesians and a second group, the Neo-Hawaiians, who are a blend of Polynesian, European, and Asian ancestries. In the U.S. State of Hawai'i—actually an archipelago of more than 130 islands—Polynesian culture has been both Europeanized and Orientalized.

Its vastness and the diversity of its natural environments notwithstanding, Polynesia clearly constitutes a geographic region within the Pacific Realm. Polynesian culture, though spatially fragmented, exhibits a remarkable consistency and uniformity from one island to the next, from one end of this widely dispersed region to the other. This consistency is particularly expressed in vocabularies, technologies, housing, and art forms. The Polynesians are uniquely adapted to their maritime environment, and long before European sailing ships began to arrive in their waters, Polynesian seafarers had learned to navigate their wide expanses of ocean in huge double canoes as long as 45 meters (150 ft). They traveled hundreds of kilometers to favorite fishing zones and engaged in inter-island barter trade, using maps constructed from bamboo sticks and cowrie shells and navigating by the stars. However, modern descriptions of a Pacific Polynesian paradise of emerald seas, lush landscapes, and gentle people distort harsh realities. Polynesian society was forced to get used to much loss of life at sea when storms claimed their boats; families were ripped apart by accident as well as migration; hunger and starvation afflicted the inhabitants of smaller islands; and the island communities were often embroiled in violent conflicts and cruel retributions.

The political geography of Polynesia is complex. In 1959, the Hawaiian Islands became the fiftieth State to join the United States. The State's population is now 1.4 million, with over 80 percent living on the island of Oahu. There, the superimposition of cultures is symbolized by the panorama of Honolulu's skyscrapers against the famous extinct volcano at nearby Diamond Head. The Kingdom of Tonga became an independent country in 1970 after seven decades as a British protectorate; the British-administered Ellice Islands were renamed Tuvalu, and along with the Gilbert Islands to the north (now renamed Kiribati), they received independence from Britain in 1978. Other islands continued under French control (including the Marquesas Islands and Tahiti), under New Zealand's administration (Rarotonga), and under British, U.S., and Chilean flags.

In the process of politico-geographical fragmentation, Polynesian culture has suffered severe blows. Land developers, hotel builders, and tourist dollars have set Tahiti on a course along which Hawai'i has already traveled far. The Americanization of eastern Samoa has created a new society different from the old (see photo, p. 598). Polynesia has lost much of its ancient cultural consistency; today, the region is a patchwork of new and old—the new often bleak and barren, with the old under intensifying pressure.

The countries and cultures of the Pacific Realm lie in an ocean on whose rim a great drama of economic and political transformation will play itself out during the twenty-first century. Already, the realm's own former margins—in Hawai'i in the north and in New Zealand in the south—have been so recast by foreign intervention that little remains of the kingdoms and cultures that once prevailed. Now the Pacific world faces changes far greater even than those brought here by European colonizers. Once upon a time the shores and waters of the Mediterranean Sea formed an arena of regional transformation that changed the world. Then it was the Atlantic, avenue of the Industrial Revolution and stage of fateful war. Now it seems to be the turn of the Pacific as the world's largest country and next superpower (China) faces the richest and most powerful (the United States). Giants will jostle for advantage in the Pacific; how will the weak societies of the Pacific Realm fare?

A FINAL CAVEAT: PACIFIC AND ANTARCTIC

South of the Pacific Realm lies Antarctica and its encircling Southern Ocean. The combined area of these two geographic expanses constitutes 40 percent of the entire planet—two-fifths of the Earth's surface containing one one-thousandth of the world's population.

Is Antarctica a geographic realm? In physiographic terms, yes, but not on the basis of the criteria we use in this book. Antarctica is a continent, nearly twice as large as Australia, but virtually all of it is covered by a dome-shaped ice sheet 3 kilometers (nearly 2 mi) thick near its center. The continent often is referred to as the "white desert" because, despite all of its ice and snow, annual precipitation is low—less than 15 centimeters (6 in) per year. Temperatures are frigid, with winds so strong that Antarctica also is called the "home of the blizzard." For all its size, no functional regions have developed here, no towns, no transport networks except the supply lines of research stations. And Antarctica still is a frontier, even a scientific frontier still slowly giving up its secrets. Underneath all that ice lie some 70 lakes of which Lake Vostok is the largest at 14,000 square kilometers (more than 5400 sq mi). It may be as much as 600 meters (2000 ft) deep. No one has yet seen a sample of its water.

Like virtually all frontiers, Antarctica has always attracted pioneers and explorers. Whale and seal hunters destroyed huge populations of Southern Ocean fauna during the eighteenth and nineteenth centuries, and explorers planted the flags of their countries on Antarctic shores. Between 1895 and 1914, the quest for the South Pole became an international obsession; Roald Amundsen, the Norwegian, reached it first in 1911. All this led to national claims in Antarctica during the interwar period (1918–1939).

The geographic effect was the partitioning of Antarctica into pie-shaped sectors centered on the South Pole (Fig. 12-4). In the least frigid area of the continent, the Antarctic Peninsula, British, Argentinian, and Chilean claims overlapped—and still do. One sector, Marie Byrd Land (shown in neutral beige on the main map), was never formally claimed by any country.

Why should states be interested in territorial claims in so remote and difficult an area? Both land and sea contain raw materials that may some day become crucial: proteins in the waters, and fuels and minerals beneath the ice. Antarctica (14.2 million square kilometers [5.5 million sq mi]) is, as already noted, almost twice as large as Australia, and the Southern Ocean (see p. 574) is nearly as large as the North and South Atlantic. However distant actual exploitation may be, countries want to keep their stakes here.

But the claimant states (those with territorial claims) recognize the need for cooperation. During the late 1950s, they joined in the International Geophysical Year (IGY) that launched major research programs and established a number of permanent research stations throughout the continent. This spirit of cooperation led to the 1961 signing of the **Antarctic Treaty**, which ensures continued scientific collaboration, prohibits military activities, protects the environment, and holds national claims in abeyance.

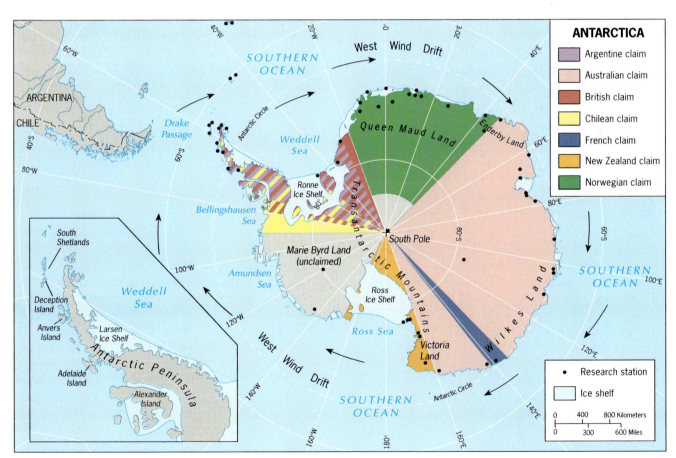

FIGURE 12-4 © H. J. de Blij, P. O. Muller, and John Wiley & Sons, Inc.

FROM THE FIELD NOTES

"The Antarctic Peninsula is geologically an extension of South America's Andes Mountains, and this vantage point in the Gerlach Strait leaves you in no doubt: the mountains rise straight out of the frigid waters of the Southern Ocean. We have been passing large, flat-topped icebergs, but these do not come from the peninsula's shores; rather, they form on the leading edges of ice shelves or on the margins of the mainland where continental glaciers slide into the sea. Here along the peninsula, the high-relief topography tends to produce jagged, irregular icebergs. The mountain range that forms the Antarctic Peninsula continues across Antarctica under thousands of feet of ice and is known as the Transantarctic Mountains. Looking at this place you are reminded that beneath all this ice lies an entire continent, still little-known, with fossils, minerals, even lakes yet to be discovered and studied."
© H. J. de Blij

Concept Caching

www.conceptcaching.com

In 1991, when the treaty was extended under the terms of the Wellington Agreement, concerns were raised that it does not do enough to control future resource exploitation.

In an age of growing national self-interest and increasing raw-material consumption, the possibility exists that Antarctica and its offshore waters may yet become an arena for international rivalry. Until now, its remoteness and its forbidding environments have saved it from that fate. The entire world benefits from this because evidence is mounting that Antarctica plays a critical role in the global environmental system, so that human modifications may have worldwide (and unpredictable) consequences.

What You Can Do

RECOMMENDATION: Attend an AAG Meeting! The Association of American Geographers is the largest organization for professional geographers and students in North America, and it holds an Annual Meeting every spring (its 103rd meeting took place in San Francisco in April 2007). The meeting sites rotate around North America: Boston in 2008 (April 15–20), Las Vegas in 2009 (March 22–26), and Washington, D.C. in 2010 (April 13–18), so one of them is likely to convene relatively close to your college or home in the not too distant future. Several thousand geographers and students participate, and while these meetings have their professional side, they are also a lot of fun, and there are jobs available as well, as the AAG can tell you. If you're within a reasonable distance, why not join the crowd and enjoy the exhibits, plenary lectures, and parties, and meet some of the people who keep geography in the limelight! Details are available on the AAG website at www.aag.org.

GEOGRAPHIC CONNECTIONS

1 One of the most important achievements of the United Nations has been a series of conferences out of which grew the UN Convention on the Law of the Sea (UNCLOS). This Convention redrew the maritime map of the world, stabilizing what had been a scramble for the oceans, limiting the width of the Territorial Sea, and introducing the concept of the Exclusive Economic Zone (EEZ). How has the EEZ changed the economic geography of the Pacific geographic realm? To what extent has the EEZ functioned to expand as well as limit states' offshore activities? Considering the map of the Pacific Realm, what countries are favored over others in terms of maritime holdings?

2 The Pacific Realm has the least land and the smallest population of all the geographic realms, but its cultural distinctiveness is second to none. Its largest territorial component and largest population cluster constitute the island of New Guinea, which is divided by a geometric, superimposed boundary between Indonesia and independent Papua New Guinea (PNG) (Fig. 10-7, lower-left map). Although Papuans on both sides of this border dream of a united homeland, the two entities are going in different directions. How did this division come about? Why was it allowed to continue during the period of decolonization? What demographic and cultural-geographic contrasts between the eastern (PNG) and western (Indonesian) sides of the boundary can you identify?

Geographic Literature on the Pacific Realm: The key introductory works on this realm were authored by Bier, Glassner, McKnight, Peake, and Woodcock. These works, together with a comprehensive listing of noteworthy books and recent articles on the geography of Melanesia, Micronesia, and Polynesia, can be found in the *References and Further Readings* section of this book's website at **www.wiley.com/college/deblij**.

METRIC (STANDARD INTERNATIONAL [SI]) AND CUSTOMARY UNITS AND THEIR CONVERSIONS

Appendix A provides a table of units and their conversions from older (British) units to metric (SI) units for the measures of length, area, and temperature that are used in this book.

NOTE: **Appendices B, C, and D may be found on the book's website at www.wiley.com/college/deblij**

Length

Metric Measure

1 kilometer (km)	= 1000 meters (m)
1 meter (m)	= 100 centimeters (cm)
1 centimeter (cm)	= 10 millimeters (mm)

Nonmetric Measure

1 mile (mi)	= 5280 feet (ft)
	= 1760 yards (yd)
1 yard (yd)	= 3 feet (ft)
1 foot (ft)	= 12 inches (in)
1 fathom (fath)	= 6 feet (ft)

Conversions

1 kilometer (km)	= 0.6214 mile (mi)
1 meter (m)	= 3.281 feet (ft)
	= 1.094 yards (yd)
1 centimeter (cm)	= 0.3937 inch (in)
1 millimeter (mm)	= 0.0394 inch (in)
1 mile (mi)	= 1.609 kilometers (km)
1 foot (ft)	= 0.3048 meter (m)
1 inch (in)	= 2.54 centimeters (cm)
	= 25.4 millimeters (mm)

Area

Metric Measure

1 square kilometer (km^2)	= 1,000,000 square meters (m^2)
	= 100 hectares (ha)
1 square meter (m^2)	= 10,000 square centimeters (cm^2)
1 hectare (ha)	= 10,000 square meters (m^2)

Nonmetric Measure

1 square mile (mi^2)	= 640 acres (ac)
1 acre (ac)	= 4840 square yards (yd^2)
1 square foot (ft^2)	= 144 square inches (in^2)

Conversions

1 square kilometer (km^2)	= 0.386 square mile (mi^2)
1 hectare (ha)	= 2.471 acres (ac)
1 square meter (m^2)	= 10.764 square feet (ft^2)
	= 1.196 square yards (yd^2)
1 square centimeter (cm^2)	= 0.155 square inch (in^2)
1 square mile (mi^2)	= 2.59 square kilometers (km^2)
1 acre (ac)	= 0.4047 hectare (ha)
1 square foot (ft^2)	= 0.0929 square meter (m^2)
1 square inch (in^2)	= 6.4516 square centimeters (cm^2)

Temperature

To change from Fahrenheit (F) to Celsius (C)

$$°C = \frac{°F - 32}{1.8}$$

To change from Celsius (C) to Fahrenheit (F)

$$°F = °C \times 1.8 + 32$$

GLOSSARY

Aboriginal land issue   The legal campaign in which Australia's **indigenous peoples*** have claimed title to traditional land in several parts of that country. The courts have upheld certain claims, fueling Aboriginal activism that has raised broader issues of indigenous rights.

Aboriginal population   See **indigenous peoples**.

Absolute location   The position or place of a certain item on the surface of the Earth as expressed in degrees, minutes, and seconds of **latitude**, 0° to 90° north or south of the equator, and **longitude**, 0° to 180° east or west of the *prime meridian* passing through Greenwich, England (a suburb of London).

Accessibility   The degree of ease with which it is possible to reach a certain location from other locations. *Inaccessibility* is the opposite of this concept.

Acculturation   Cultural modification resulting from intercultural borrowing. In **cultural geography**, the term refers to the change that occurs in the **culture** of **indigenous peoples** when contact is made with a society that is technologically superior.

Advantage   The most meaningful distinction that can now be made to classify a country's level of economic **development**. Takes into account geographic **location, natural resources**, government, political stability, productive skills, and much more.

Agglomeration   **Process** involving the clustering or concentrating of people or activities.

Agrarian   Relating to the use of land in rural communities, or to agricultural societies in general.

Agriculture   The purposeful tending of crops and livestock in order to produce food and fiber.

Alluvial   Refers to the mud, silt, and sand (collectively *alluvium*) deposited by rivers and streams. *Alluvial plains* adjoin many larger rivers; they consist of these renewable deposits that are laid down during floods, creating fertile and productive soils. Alluvial **deltas** mark the mouths of rivers such as the Nile and the Ganges.

Altiplano   High-elevation plateau, basin, or valley between even higher mountain ranges, especially in the Andes of South America.

Altitudinal zonation   Vertical regions defined by physical-environmental zones at various elevations (see Fig. 4-12), particularly in the highlands of South and Middle America. See *tierra caliente, tierra templada, tierra fría, tierra helada*, and *tierra nevada*.

American Manufacturing Belt   North America's near-rectangular Core Region, whose corners are Boston, Milwaukee, St. Louis, and Baltimore. Dominated the industrial geography of the U.S. and Canada during the industrial age; still a formidable economic powerhouse that remains the realm's geographic heart.

Animistic religion   The belief that inanimate objects, such as hills, trees, rocks, rivers, and other elements of the natural landscape, possess souls and can help or hinder human efforts on Earth.

Antarctic Treaty   International cooperative agreement on the use of Antarctic territory (see pp. 607–608).

Antecedent boundary   A political boundary that existed before the **cultural landscape** emerged and stayed in place while people moved in to occupy the surrounding area.

*Words in boldface type are defined elsewhere in this Glossary.

Anthracite coal   Hardest and highest carbon-content coal, and therefore of the highest quality.

Apartheid   Literally, *apartness*. The Afrikaans term for South Africa's pre-1994 policies of racial separation, a system that produced highly segregated socio-geographical patterns.

Aquaculture   The use of a river segment or an artificial pond for the raising and harvesting of food products, including fish, shellfish, and even seaweed (particularly in Japan).

Aquifer   An underground reservoir of water contained within a porous, water-bearing rock layer.

Arable   Land fit for cultivation by one farming method or another. See also **physiologic density**.

Archipelago   A set of islands grouped closely together, usually elongated into a *chain*.

Area   A term that refers to a part of the Earth's surface with less specificity than **region**. For example, *urban area* alludes generally to a place where urban development has occurred, whereas *urban region* requires certain specific criteria upon which such a designation is based (e.g., the spatial extent of commuting or the built townscape).

Areal functional organization   A geographic principle for understanding the evolution of regional organization, whose five interrelated tenets are applied to the spatial development of Japan on p. 506.

Areal interdependence   A term related to **functional specialization**. When one area produces certain goods or has certain raw materials and another area has a different set of raw materials and produces different goods, their needs may be *complementary*; by exchanging raw materials and products, they can satisfy each other's requirements.

Arithmetic density   A country's population, expressed as an average per unit area, without regard for its **distribution** or the limits of **arable** land. See also **physiologic density**.

Aryan   From the Sanskrit *Arya* ("noble"), a name applied to an ancient people who spoke an **Indo-European language** and who moved into northern India from the northwest.

Atmosphere   The Earth's envelope of gases that rests on the oceans and land surface and penetrates open spaces within soils. This layer of nitrogen (78 percent), oxygen (21 percent), and traces of other gases is densest at the Earth's surface and thins with altitude.

Austral   South.

Autocratic   A government that holds absolute power, often ruled by one person or a small group of persons who control the country by despotic means.

Balkanization   The fragmentation of a **region** into smaller, often hostile political units.

Barrio   Term meaning "neighborhood" in Spanish. Usually refers to an urban community in a Middle or South American city; also applied to low-income, inner-city concentrations of Hispanics in such western U.S. cities as Los Angeles.

Bauxite   Aluminum ore; usually deposited at shallow depths in the wet tropics.

Biogeography   The study of *flora* (plant life) and *fauna* (animal life) in spatial perspective.

Birth rate The *crude birth rate* is expressed as the annual number of births per 1000 individuals within a given population.

Bituminous coal Softer coal of lesser quality than **anthracite**, but of higher grade than **lignite**. When heated and converted to coking coal or *coke*, it is used to make steel.

Break-of-bulk point A location along a transport route where goods must be transferred from one carrier to another. In a port, the cargoes of oceangoing ships are unloaded and put on trains, trucks, or perhaps smaller river boats for inland distribution. An **entrepôt**.

Buffer state See **buffer zone**.

Buffer zone A country or set of countries separating ideological or political adversaries. In southern Asia, Afghanistan, Nepal, and Bhutan were parts of a buffer zone between British and Russian-Chinese imperial spheres. Thailand was a *buffer state* between British and French colonial domains in mainland Southeast Asia.

Caliente See *tierra caliente*.

Cartogram A specially transformed map not based on traditional representations of **scale** or area.

Cartography The art and science of making maps, including data compilation, layout, and design. Also concerned with the interpretation of mapped patterns.

Caste system The strict **social stratification** and segregation of people—specifically in India's Hindu society—on the basis of ancestry and occupation.

Cay A low-lying small island usually composed of coral and sand. Pronounced *kee* and often spelled "key."

Central business district (CBD) The downtown heart of a central city, the CBD is marked by high land values, a concentration of business and commerce, and the clustering of the tallest buildings.

Centrality The strength of an urban center in its capacity to attract producers and consumers to its facilities; a city's "reach" into the surrounding region.

Centrifugal forces A term employed to designate forces that tend to divide a country—such as internal religious, linguistic, ethnic, or ideological differences.

Centripetal forces Forces that unite and bind a country together—such as a strong national culture, shared ideological objectives, and a common faith.

Cerrado Regional term referring to the fertile savannas of Brazil's interior Central-West that make it one of the world's most promising agricultural frontiers (mapped in Fig. 5-4). Soybeans are the leading crop, and other grains and cotton are expanding. Inadequate transport links to the outside world remain a problem.

Chaebol Giant corporation controlling numerous companies and benefiting from government connections and favors, dominant in South Korea's **economic geography**; key to the country's **development** into an **economic tiger**, but more recently a barrier to free-market growth.

Charismatic Personal qualities of certain leaders that enable them to capture and hold the popular imagination, to secure the allegiance and even the devotion of the masses. Gandhi, Mao Zedong, and Franklin D. Roosevelt were good examples in the 20th century.

China Proper The eastern and northeastern portions of China that contain most of the country's huge population; mapped in Figure 9-4.

Choke point A narrowing of an international waterway causing marine traffic congestion, requiring reduced speeds and/or sharp turns, and increasing the risk of collision as well as vulnerability to attack. When the waterway narrows to a distance of less than 38 kilometers (24 mi), this necessitates the drawing of a **median line (maritime) boundary**. Examples are the Hormuz Strait between Oman and Iran at the entrance to the Persian Gulf, and the Strait of Malacca between Malaysia and Indonesia.

City-state An independent political entity consisting of a single city with (and sometimes without) an immediate **hinterland**. The ancient city-states of Greece have their modern equivalent in Singapore.

Climate The long-term conditions (over at least 30 years) of aggregate **weather** over a region, summarized by averages and measures of variability; a synthesis of the succession of weather events we have learned to expect at any given location.

Climate change theory An alternative to the **hydraulic civilization theory**; holds that changing **climate** (rather than a monopoly over **irrigation** methods) could have provided certain cities in the ancient Fertile Crescent with advantages over others.

Climate region A **formal region** characterized by the uniformity of the **climate** type within it. Figure G-8 maps the global distribution of such regions.

Climatology The geographic study of **climates**. Includes not only the classification of climates and the analysis of their regional distribution, but also broader environmental questions that concern climate change, interrelationships with soil and vegetation, and human-climate interaction.

Coal See **anthracite coal**, **bituminous coal**, **fossil fuels**, and **lignite**.

Collectivization The reorganization of a country's **agriculture** under communism that involves the expropriation of private holdings and their incorporation into relatively large-scale units, which are farmed and administered cooperatively by those who live there.

Colonialism Rule by an autonomous power over a subordinate and alien people and place. Although often established and maintained through political structures, colonialism also creates unequal cultural and economic relations. Because of the magnitude and impact of the European colonial thrust of the last few centuries, the term is generally understood to refer to that particular colonial endeavor. Also see **imperialism**.

Command economy The tightly controlled economic system of the former Soviet Union, whereby central planners in Moscow assigned the production of particular goods to particular places, often guided more by socialist ideology than the principles of **economic geography**.

Commercial agriculture For-profit **agriculture**.

Common market A **free-trade area** that not only has created a **customs union** (a set of common tariffs on all imports from outside the area) but also has eliminated restrictions on the movement of capital, labor, and enterprise among its member countries.

Compact state A politico-geographical term to describe a state that possesses a roughly circular, oval, or rectangular territory in which the distance from the geometric center to any point on the boundary exhibits little variance. Poland and Cambodia are examples of this shape category.

Complementarity Exists when two regions, through an exchange of raw materials and/or finished products, can specifically satisfy each other's demands.

Confucianism A philosophy of ethics, education, and public service based on the writings of Confucius (*Kongfuzi*); traditionally regarded as one of the cornerstones of Chinese **culture**.

Congo See the footnote on p. 297, which makes the distinction between the two Equatorial African countries, Congo and The Congo.

Coniferous forest A forest of cone-bearing, needleleaf evergreen trees with straight trunks and short branches, including spruce, fir, and pine. Also see **taiga**.

Contagious diffusion The distance-controlled spreading of an idea, innovation, or some other item through a local population by contact from person to person—analogous to the communication of a contagious illness.

Conterminous United States The 48 **contiguous** or adjacent States that occupy the southern half of the North American realm. Alaska is not contiguous to these States because western Canada lies in between; neither is Hawai'i, separated from the mainland by over 3000 kilometers (2000 mi) of ocean.

Contiguous Adjoining; adjacent.

Continental drift The slow movement of continents controlled by the processes associated with **plate tectonics**.

Continental shelf   Beyond the coastlines of many landmasses, the ocean floor declines very gently until the depth of about 660 feet (200 m). Beyond the 660-foot line the sea bottom usually drops off sharply, along the *continental slope*, toward the much deeper mid-oceanic basin. The submerged continental margin is called the continental shelf, and it extends from the shoreline to the upper edge of the continental slope.

Continentality   The variation of the continental effect on air temperatures in the interior portions of the world's landmasses. The greater the distance from the moderating influence of an ocean, the greater the extreme in summer and winter temperatures. Continental interiors also tend to be dry when the distance from oceanic moisture sources becomes considerable.

Conurbation   General term used to identify a large multi-metropolitan complex formed by the coalescence of two or more major **urban areas**. The Atlantic Seaboard **Megalopolis**, extending along the northeastern U.S. coast from southern Maine to Virginia, is a classic example.

Copra   The dried-out, fleshy interior of a coconut that is used to produce coconut oil.

Cordillera   Mountain chain consisting of sets of parallel ranges, especially the Andes in northwestern South America.

Core   See **core area**; **core-periphery relationships**.

Core area   In geography, a term with several connotations. *Core* refers to the center, heart, or focus. The core area of a **nation-state** is constituted by the national heartland, the largest population cluster, the most productive region, and the part of the country with the greatest **centrality** and **accessibility**—probably containing the capital city as well.

Core-periphery relationships   The contrasting spatial characteristics of, and linkages between, the *have* (core) and *have-not* (periphery) components of a national or regional **system**.

Corridor   In general, refers to a spatial entity in which human activity is organized in a linear manner, as along a major transport route or in a valley confined by highlands. More specifically, the politico-geographical term for a land extension that connects an otherwise **landlocked state** to the sea.

Cultural diffusion   The **process** of spreading and adoption of a cultural element, from its place of origin across a wider area.

Cultural ecology   The multiple interactions and relationships between a **culture** and its **natural environment**.

Cultural environment   See **cultural ecology**.

Cultural geography   The wide-ranging and comprehensive field of geography that studies spatial aspects of human **cultures**.

Cultural landscape   The forms and artifacts sequentially placed on the **natural landscape** by the activities of various human occupants. By this progressive imprinting of the human presence, the physical (natural) landscape is modified into the cultural landscape, forming an interacting unity between the two.

Cultural pluralism   See **plural (istic) society**.

Cultural revival   The regeneration of a long-dormant **culture** through internal renewal and external infusion.

Culture   The sum total of the knowledge, attitudes, and habitual behavior patterns shared and transmitted by the members of a society. This is anthropologist Ralph Linton's definition; hundreds of others exist.

Culture area   See **culture region**.

Culture hearth   Heartland, source area, innovation center; place of origin of a major **culture**.

Culture region   A distinct, culturally discrete spatial unit; a **region** within which certain cultural norms prevail.

Customs union   A **free-trade area** in which member countries set common tariff rates on imports from outside the area.

Death rate   The *crude death rate* is expressed as the annual number of deaths per 1000 individuals within a given population.

Deciduous   A deciduous tree loses its leaves at the beginning of winter or the onset of the dry season.

Definition   In **political geography**, the written legal description (in a treaty-like document) of a boundary between two countries or territories. See **delimitation**.

Deforestation   The clearing and destruction of forests (especially tropical rainforests) to make way for expanding settlement frontiers and the exploitation of new economic opportunities.

Deglomeration   Deconcentration.

Delimitation   In **political geography**, the translation of the written terms of a boundary treaty (the **definition**) into an official cartographic representation (map).

Delta   Alluvial lowland at the mouth of a river, formed when the river deposits its alluvial load on reaching the sea. Often triangular in shape, hence the use of the Greek letter whose symbol is Δ.

Demarcation   In **political geography**, the actual placing of a political boundary on the **cultural landscape** by means of barriers, fences, walls, or other markers.

Demographic transition model   Multi-stage **model**, based on Western Europe's experience, of changes in population growth exhibited by countries undergoing industrialization. High **birth rates** and **death rates** are followed by plunging death rates, producing a huge net population gain; this is followed by the convergence of birth and death rates at a low overall level. See Figure 8-7.

Demography   The interdisciplinary study of population—especially **birth rates** and **death rates**, growth patterns, longevity, **migration**, and related characteristics.

Desert   An arid area supporting sparse vegetation, receiving less than 25 centimeters (10 in) of precipitation per year. Usually exhibits extremes of heat and cold because the moderating influence of moisture is absent.

Desertification   **Process** of **desert** expansion into neighboring **steppelands** as a result of human degradation of fragile semiarid environments.

Development   The economic, social, and institutional growth of national **states**.

Devolution   The **process** whereby regions within a **state** demand and gain political strength and growing autonomy at the expense of the central government.

Dhows   Wooden boats with characteristic triangular sails, plying the seas between Arabian and East African coasts.

Dialect   Regional or local variation in the use of a major language, such as the distinctive accents of many residents of the U.S. South or New England.

Diffusion   The spatial spreading or dissemination of a **culture** element (such as a technological innovation) or some other phenomenon (e.g., a disease outbreak). For the various channels of outward geographic spread from a source area, see **contagious, expansion, hierarchical**, and **relocation diffusion**.

Distance decay   The various degenerative effects of distance on human spatial structures and interactions.

Diurnal   Daily.

Divided capital   In **political geography**, a country whose central administrative functions are carried on in more than one city is said to have divided capitals. The Netherlands and South Africa are examples.

Domestication   The transformation of a wild animal or wild plant into a domesticated animal or a cultivated crop to gain control over food production. A necessary evolutionary step in the development of humankind: the invention of **agriculture**.

Domino theory   The belief that political destabilization in one **state** can result in the collapse of order in a neighboring state, triggering a chain of events that, in turn, can affect a series of **contiguous** states.

Double cropping   The planting, cultivation, and harvesting of two crops successively within a single year on the same plot of farmland.

Dry canal   An overland rail and/or road **corridor** across an **isthmus** dedicated to performing the transit functions of a canalized waterway. Best adapted to the movement of containerized cargo, there must be a port at each end to handle the necessary **break-of-bulk** unloading and reloading.

Ecology   The study of the many interrelationships between all forms of life and the natural environments in which they have evolved and continue to develop. The study of *ecosystems* focuses on the interactions between specific organisms and their environments. See also **cultural ecology**.

Economic geography   The field of geography that focuses on the diverse ways in which people earn a living, and how the goods and services they produce are expressed and organized spatially.

Economic restructuring   The transformation of China into a market-driven economy in the post-Mao era, beginning in the late 1970s.

Economic tiger   One of the burgeoning beehive countries of the western **Pacific Rim**. Following Japan's route since 1945, these countries have experienced significant modernization, industrialization, and Western-style economic growth since 1980. Three leading economic tigers today are South Korea, Taiwan, and Singapore. The term is increasingly used more generally to describe any fast-developing economy.

Economies of scale   The savings that accrue from large-scale production wherein the unit cost of manufacturing decreases as the level of operation enlarges. Supermarkets operate on this principle and are able to charge lower prices than small grocery stores.

Ecosystem   See **ecology**.

Ecumene   The habitable portions of the Earth's surface where permanent human settlements have arisen.

Elite   A small but influential upper-echelon social class whose power and privilege give it control over a country's political, economic, and cultural life.

El Niño-Southern Oscillation (ENSO)   A periodic, large-scale, abnormal warming of the sea surface in the low latitudes of the eastern Pacific Ocean that has global implications, disturbing normal **weather** patterns in many parts of the world, especially South America.

Elongated state   A **state** whose territory is decidedly long and narrow in that its length is at least six times greater than its average width. Chile and Vietnam are two classic examples.

Emigrant   A person **migrating** away from a country or area; an out-migrant.

Empirical   Relating to the real world, as opposed to theoretical abstraction.

Enclave   A piece of territory that is surrounded by another political unit of which it is not a part.

Endemism   Refers to a disease in a host population that affects many people in a kind of equilibrium without causing rapid and widespread deaths.

Entrepôt   A place, usually a port city, where goods are imported, stored, and transshipped; a **break-of-bulk point**.

Environmental degradation   The accumulated human abuse of a region's **natural landscape** that, among other things, can involve air and water pollution, threats to plant and animal ecosystems, misuse of **natural resources**, and generally upsetting the balance between people and their habitat.

Epidemic   A local or regional outbreak of a disease.

Epochs of metropolitan evolution   The five-stage Borchert model that conceptualizes the growth **process** of the U.S. national urban system. Consists of the Sail-Wagon Epoch, Iron Horse Epoch, Steel-Rail Epoch, Auto-Air-Amenity Epoch, and Satellite-Electronic-Jet Propulsion Epoch.

Eras of intraurban structural evolution   The four-stage Adams model that conceptualizes the growth **process** of the U.S. city. Consists of the Walking-Horsecar Era, Electric Streetcar Era, Recreational Automobile Era, and Freeway Era.

Escarpment   A cliff or very steep slope; frequently marks the edge of a plateau.

Estuary   The widening mouth of a river as it reaches the sea; land subsidence or a rise in sea level has overcome the tendency to form a **delta**.

Ethanol   The leading U.S. biofuel that is essentially alcohol distilled from corn mash. Most of it is produced in the historic Corn Belt centered on Iowa. Many risks and problems accompany this energy source, however, and they are discussed on p. 193.

Ethnic cleansing   The slaughter and/or forced removal of one **ethnic** group from its homes and lands by another, more powerful ethnic group bent on taking that territory.

Ethnicity   The combination of a people's **culture** (traditions, customs, language, and religion) and racial ancestry.

European state model   A **state** consisting of a legally defined territory inhabited by a population governed from a capital city by a representative government.

European Union (EU)   **Supranational** organization constituted by 27 European countries to further their common economic interests. In alphabetical order, these countries are: Austria, Belgium, Bulgaria, Cyprus, the Czech Republic, Denmark, Estonia, Finland, France, Germany, Greece, Hungary, Ireland, Italy, Latvia, Lithuania, Luxembourg, Malta, the Netherlands, Poland, Portugal, Romania, Slovakia, Slovenia, Spain, Sweden, and the United Kingdom.

Exclave   A bounded (non-island) piece of territory that is part of a particular **state** but lies separated from it by the territory of another state. Alaska is an exclave of the United States.

Exclusive Economic Zone (EEZ)   An oceanic zone extending up to 200 **nautical miles** from a shoreline, within which the coastal **state** can control fishing, mineral exploitation, and additional activities by all other countries.

Expansion diffusion   The spreading of an innovation or an idea through a fixed population in such a way that the number of those adopting grows continuously larger, resulting in an expanding area of dissemination.

Extraterritoriality   The politico-geographical concept suggesting that the property of one **state** lying within the boundaries of another actually forms an extension of the first state.

Failed state   A country whose institutions have collapsed and in which anarchy prevails. Afghanistan during the late-1990s rule of the Taliban is a recent example.

Fatwa   Literally, a legal opinion or proclamation issued by an Islamic cleric, based on the holy texts of Islam, long applicable only in the *Umma*, the realm ruled by the laws of Islam. In 1989 the Iranian ayatollah Khomeini extended the reach of the *fatwa* by condemning to death a British citizen and author living in the United Kingdom.

Favela   Shantytown on the outskirts or even well within an urban area in Brazil.

Fazenda   Coffee plantation in Brazil.

Federal state   A political framework wherein a central government represents the various subnational entities within a **nation-state** where they have common interests—defense, foreign affairs, and the like—yet allows these various entities to retain their own identities and to have their own laws, policies, and customs in certain spheres.

Federation   See **federal state**.

Fertile Crescent   Crescent-shaped zone of productive lands extending from near the southeastern Mediterranean coast through Lebanon and Syria to the **alluvial** lowlands of Mesopotamia (in Iraq). Once more fertile than today, this is one of the world's great source areas of **agricultural** and other innovations.

First Nations   Name given Canada's **indigenous peoples** of American descent, whose U.S. counterparts are called Native Americans.

Fjord   Narrow, steep-sided, elongated, and inundated coastal valley deepened by glacier ice that has since melted away, leaving the sea to penetrate.

Floodplain   Low-lying area adjacent to a mature river, often covered by **alluvial** deposits and subject to the river's floods.

Forced migration   Human **migration** flows in which the movers have no choice but to relocate.

Formal region   A type of **region** marked by a certain degree of homogeneity in one or more phenomena; also called *uniform region* or *homogeneous region*.

Forward capital   Capital city positioned in actually or potentially contested territory, usually near an international border; it confirms the **state's** determination to maintain its presence in the region in contention.

Fossil fuels   The energy resources of **coal**, natural gas, and petroleum (oil), so named collectively because they were formed by the geologic compression and transformation of tiny plant and animal organisms.

Four Motors of Europe   Rhône-Alpes (France), Baden-Württemberg (Germany), Catalonia (Spain), and Lombardy (Italy). Each is a high-technology-driven region marked by exceptional industrial vitality and economic success not only within Europe but on the global scene as well.

Fragmented state   A **state** whose territory consists of several separated parts, not a **contiguous** whole. The individual parts may be isolated from each other by the land area of other states or by international waters. The United States and Indonesia are examples.

Francophone   French-speaking. Quebec constitutes the heart of Francophone Canada.

Free-trade area   A form of economic integration, usually consisting of two or more **states**, in which members agree to remove tariffs on trade among themselves. Usually accompanied by a **customs union** that establishes common tariffs on imports from outside the trade area, and sometimes by a **common market** that also removes internal restrictions on the movement of capital, labor, and enterprise.

Free Trade Area of the Americas (FTAA)   The ultimate goal of **supranational** economic integration in North, Middle, and South America: the creation of a single-market trading bloc that would involve every country in the Western Hemisphere between the Arctic shore of Canada and Cape Horn at the southern tip of Chile.

Fría   See *tierra fría*.

Frontier   Zone of advance penetration, usually of contention; an area not yet fully integrated into a national **state**.

FTAA   See **Free Trade Area of the Americas**.

Functional region   A **region** marked less by its sameness than its dynamic internal structure; because it usually focuses on a central node, also called *nodal region* or *focal region*.

Functional specialization   The production of particular goods or services as a dominant activity in a particular location. See also **local functional specialization**.

Fundamentalism   See **revivalism** (**religious**).

Gentrification   The upgrading of an older residential area through private reinvestment, usually in the downtown area of a central city. Frequently this involves the displacement of established lower-income residents, who cannot afford the heightened costs of living, and conflicts are not uncommon as such neighborhood change takes place.

Geographic change   Evolution of **spatial** patterns over time.

Geographic realm   The basic spatial unit in our world regionalization scheme. Each realm is defined in terms of a synthesis of its total human geography—a composite of its leading cultural, economic, historical, political, and appropriate environmental features.

Geography of development   The subfield of economic geography concerned with spatial aspects and regional expressions of **development**.

Geometric boundaries   Political boundaries **defined** and **delimited** (and occasionally **demarcated**) as straight lines or arcs.

Geomorphology   The geographic study of the configuration of the Earth's solid surface—the world's landscapes and their constituent landforms.

Ghetto   An intraurban region marked by a particular **ethnic** character. Often an inner-city poverty zone, such as the black ghetto in U.S. central cities. Ghetto residents are involuntarily segregated from other income and racial groups.

Glaciation   See **Pleistocene Epoch**.

Globalization   The gradual reduction of regional contrasts at the world **scale**, resulting from increasing international cultural, economic, and political exchanges.

Green Revolution   The successful recent development of higher-yield, fast-growing varieties of rice and other cereals in certain developing countries.

Gross domestic product (GDP)   The total value of all goods and services produced in a country by that state's economy during a given year.

Gross national product (GNP)   The total value of all goods and services produced in a country by that state's economy during a given year, plus all citizens' income from foreign investment and other external sources.

Growth pole   An urban center with certain attributes that, if augmented by a measure of investment support, will stimulate regional economic **development** in its **hinterland**.

Hacienda   Literally, a large estate in a Spanish-speaking country. Sometimes equated with the **plantation**, but there are important differences between these two types of agricultural enterprise (see pp. 214–216).

Heartland theory   The hypothesis, proposed by British geographer Halford Mackinder during the early twentieth century, that any political power based in the heart of Eurasia could gain sufficient strength to eventually dominate the world. Further, since Eastern Europe controlled access to the Eurasian interior, its ruler would command the vast "heartland" to the east.

Hegemony   The political dominance of a country (or even a region) by another country. The former Soviet Union's postwar grip on Eastern Europe, which lasted from 1945 to 1990, was a classic example.

Helada   See *tierra helada*.

Hierarchical diffusion   A form of **diffusion** in which an idea or innovation spreads by trickling down from larger to smaller adoption units. An urban **hierarchy** is usually involved, encouraging the leapfrogging of innovations over wide areas, with geographic distance a less important influence.

Hierarchy   An order or gradation of phenomena, with each level or rank subordinate to the one above it and superior to the one below. The levels in a national urban hierarchy are constituted by hamlets, villages, towns, cities, and (frequently) the **primate city**.

High island   Volcanic islands of the Pacific Realm that are high enough in elevation to wrest substantial moisture from the tropical ocean air (see **orographic precipitation**). They tend to be well watered, their volcanic soils enable productive agriculture, and they support larger populations than **low islands**—which possess none of these advantages and must rely on fishing and the coconut palm for survival.

High seas   Areas of the oceans away from land, beyond national jurisdiction, open and free for all to use.

Highveld   A term used in Southern Africa to identify the high, grass-covered plateau that dominates much of the region. The lowest-lying areas in South Africa are called *lowveld*; areas that lie at intermediate elevations are the *middleveld*.

Hinterland   Literally, "country behind," a term that applies to a surrounding area served by an urban center. That center is the focus of goods and services produced for its hinterland and is its dominant urban influence as well. In the case of a port city, the hinterland also includes the inland area whose trade flows through that port.

Historical inertia   A term from manufacturing geography that refers to the need to continue using the factories, machinery, and equipment of heavy industries for their full, multiple-decade lifetimes to cover major initial investments—even though these facilities may be increasingly obsolete.

Holocene   The current *interglaciation* epoch (the warm period of glacial contraction between the glacial expansions of an **ice age**); extends from 10,000 years ago to the present. Also known as the *Recent Epoch*.

Human evolution   Long-term biological maturation of the human species. Geographically, all evidence points toward East Africa as the source of humankind. Our species, *Homo sapiens*, emigrated from this hearth to eventually populate the rest of the **ecumene**.

Humus   Dark-colored upper layer of a soil that consists of decomposed and decaying organic matter such as leaves and branches, nutrient-rich and giving the soil a high fertility.

Hydraulic civilization theory   The theory that cities able to control **irrigated** farming over large **hinterlands** held political power over other cities. Particularly applies to early Asian civilizations based in such river valleys as the Chang (Yangzi), the Indus, and those of Mesopotamia.

Hydrologic cycle   The **system** of exchange involving water in its various forms as it continually circulates between the **atmosphere**, the oceans, and above and below the land surface.

Ice age   A stretch of geologic time during which the Earth's average atmospheric temperature is lowered; causes the equatorward expansion of continental ice sheets in the higher latitudes and the growth of mountain glaciers in and around the highlands of the lower latitudes.

Iconography   The identity of a region as expressed through its cherished symbols; its particular **cultural landscape** and personality.

Immigrant   A person **migrating** into a particular country or area; an in-migrant.

Immigration policies   Australia's policies to regulate immigration. The issue continues to roil Australian society and is discussed on pp. 584–586.

Imperialism   The drive toward the creation and expansion of a colonial empire and, once established, its perpetuation.

Import-substitution industries   The industries local entrepreneurs establish to serve populations of remote areas when transport costs from distant sources make these goods too expensive to import.

Inaccessibility   See **accessibility**.

Indentured workers   Contract laborers who sell their services for a stipulated period of time.

Indigenous peoples   Native or *Aboriginal* peoples; often used to designate the inhabitants of areas that were conquered and subsequently colonized by the **imperial** powers of Europe.

Indo-European languages   The major world language family that dominates the European **geographic realm** (Fig. 1-8). This language family is also the most widely dispersed globally (Fig. G-11), and about half of humankind speaks one of its languages.

Industrial Revolution   The term applied to the social and economic changes in agriculture, commerce, and especially manufacturing and urbanization that resulted from technological innovations and specialization in late-eighteenth-century Europe.

Informal sector   Dominated by unlicensed sellers of homemade goods and services, the primitive form of capitalism found in many developing countries that takes place beyond the control of government.

Infrastructure   The foundations of a society: urban centers, transport networks, communications, energy distribution systems, farms, factories, mines, and such facilities as schools, hospitals, postal services, and police and armed forces.

Insular   Having the qualities and properties of an island. Real islands are not alone in possessing such properties of **isolation**: an **oasis** in the middle of a **desert** also has qualities of insularity.

Insurgent state   Territorial embodiment of a successful guerrilla movement. The establishment by antigovernment insurgents of a territorial base in which they exercise full control; thus a **state** within a state.

Intercropping   The planting of several types of crops in the same field; commonly used by **shifting cultivators**.

Interglaciation   See **Pleistocene Epoch**.

Intermontane   Literally, between mountains. The location can bestow certain qualities of natural protection or **isolation** to a community.

Internal migration   **Migration** flow within a country, such as ongoing westward and southward movements toward the **Sunbelt** in the United States.

International migration   **Migration** flow involving movement across an international boundary.

Intervening opportunity   In trade or **migration** flows, the presence of a nearer opportunity that greatly diminishes the attractiveness of sites farther away.

Inuit   **Indigenous** peoples of North America's Arctic zone, formerly known as Eskimos.

Irredentism   A policy of cultural extension and potential political expansion by a **state** aimed at a community of its nationals living in a neighboring state.

Irrigation   The artificial watering of croplands.

Islamic Front   The southern border of the African Transition Zone that marks the religious **frontier** of the **Muslim** faith in its southward penetration of Subsaharan Africa (see Fig. 6-18).

Islamization   Introduction and establishment of the **Muslim** religion (see Fig. 7-5). A **process** still under way, most notably along the **Islamic Front**, that marks the southern border of the African Transition Zone.

Isohyet   A line connecting points of equal rainfall total.

Isolated state   See **von Thünen's Isolated State model**.

Isolation   The condition of being geographically cut off or far removed from mainstreams of thought and action. It also denotes a lack of receptivity to outside influences, caused at least partially by poor **accessibility**.

Isotherm   A line connecting points of equal temperature.

Isthmus   A **land bridge**; a comparatively narrow link between larger bodies of land. Central America forms such a link between Mexico and South America.

Jakota Triangle   The easternmost region of the East Asian realm, consisting of *Ja*pan, (South) *Ko*rea, and *Ta*iwan.

Juxtaposition   Contrasting places in close proximity to one another.

Karst   The distinctive natural landscape associated with the chemical erosion of soluble limestone rock.

Land alienation   One society or culture group taking land from another. In Subsaharan Africa, for example, European **colonialists** took land from **indigenous** Africans and put it to new uses.

Land bridge   A narrow **isthmian** link between two large landmasses. They are temporary features—at least in terms of geologic time—subject to appearance and disappearance as the land or sea level rises and falls.

Land Hemisphere   The half of the globe containing the greatest amount of land surface, centered on Western Europe (Fig. 1-3). In **geomorphology**, this can also refer to the position of the African continent, which lies central to the world's landmasses (Fig. 6-3).

Land reform   The spatial reorganization of **agriculture** through the allocation of farmland (often expropriated from landlords) to **peasants** and tenants who never owned land.

Land tenure   The way people own, occupy, and use land.

Landlocked   An interior **state** surrounded by land. Without coasts, such a country is disadvantaged in terms of **accessibility** to international trade routes, and in the scramble for possession of areas of the **continental shelf** and control of the **exclusive economic zone** beyond.

Language family   Group of languages with a shared but usually distant origin.

"Latin" American city model   The Griffin-Ford model of intraurban spatial structure in the Middle American and South American realms, diagrammed in Figure 5-7.

Latitude   Lines of latitude are **parallels** that are aligned east-west across the globe, from 0° latitude at the equator to 90° North and South latitude at the poles.

Leached soil   Infertile, reddish-appearing, tropical soil whose surface consists of oxides of iron and aluminum; all other soil nutrients have been dissolved and transported downward into the subsoil by percolating water associated with heavy rainfall.

Leeward   The protected or downwind side of a **topographic** barrier with respect to the winds that flow across it.

Lignite   Low-grade, brown-colored variety of **coal**.

Lingua franca   A "common language" prevalent in a given area; a second language that can be spoken and understood by many peoples, although they speak other languages at home.

Littoral   Coastal or coastland.

Llanos   The interspersed **savanna** grasslands and scrub woodlands of the Orinoco River's wide basin that covers much of interior Colombia and especially Venezuela.

Local functional specialization   A hallmark of Europe's **economic geography** that later spread to many other parts of the world, whereby particular people in particular places concentrate on the production of particular goods and services.

Location   Position on the Earth's surface; see **absolute location** and **relative location**.

Location theory   A logical attempt to explain the locational pattern of an economic activity and the manner in which its producing areas are interrelated. The agricultural location theory that underlies the **von Thünen model** is a leading example.

Loess   Deposit of very fine silt or dust that is laid down after having been windborne for a considerable distance. Notable for its fertility under **irrigation** and its ability to stand in steep vertical walls.

Longitude   Angular distance (0° to 180°) east or west as measured from the *prime meridian* (0°) that passes through the Greenwich Observatory in suburban London, England. For much of its length across the mid-Pacific Ocean, the 180th meridian functions as the *international date line*.

Low island   Low-lying coral islands of the Pacific Realm that—unlike **high islands**—cannot wrest sufficient moisture from the tropical ocean air to avoid chronic drought. Thus productive agriculture is impossible, and their modest populations must rely on fishing and the coconut palm for survival.

Lusitanian   The Portuguese sphere, which by extension includes Brazil.

Madrassa   **Revivalist (fundamentalist)** religious school where the curriculum focuses on Islamic religion and law and requires rote memorization of the Qu'ran (Koran), Islam's holy book. Founded in former British India, these schools were most numerous in present-day Pakistan but have **diffused** as far as Turkey in the west and Indonesia in the east.

Maghreb   The region occupying the northwestern corner of Africa, consisting of Morocco, Algeria, and Tunisia.

Main Street   Canada's dominant **conurbation** that is home to nearly two-thirds of the country's inhabitants; extends southwestward from Quebec City in the mid-St. Lawrence Valley to Windsor on the Detroit River.

Mainland-Rimland framework   The twofold regionalization of the Middle American realm based on its modern cultural history. The Euro-Amerindian *Mainland*, stretching from Mexico to Panama (minus the Caribbean coastal strip), was a self-sufficient zone dominated by **hacienda land tenure**. The Euro-African *Rimland*, consisting of that Caribbean coastal zone plus all of the Caribbean islands to the east, was the zone of the **plantation** that heavily relied on trade with Europe. See Figure 4-6.

Maquiladora   The term given to modern industrial plants in Mexico's U.S. border zone. These foreign-owned factories assemble imported components and/or raw materials, and then export finished manufactures, mainly to the United States. Import duties are disappearing under **NAFTA**, bringing jobs to Mexico and the advantages of low wage rates to the foreign entrepreneurs.

Marchland   An area or **frontier** of uncertain boundaries that is subject to various national claims and an unstable political history. Refers specifically to the movement of various armies across such zones.

Marine geography   The geographic study of oceans and seas. Its practitioners investigate both the physical (e.g., coral-reef **biogeography**, ocean-**atmosphere** interactions, coastal **geomorphology**) as well as human (e.g., **maritime boundary-making**, fisheries, beachside development) aspects of oceanic environments.

Maritime boundary   An international boundary that lies in the ocean. Like all boundaries, it is a vertical plane, extending from the seafloor to the upper limit of the air space in the atmosphere above the water.

Median-line boundary   An international **maritime boundary** drawn where the width of a sea is less than 400 **nautical miles**. Because the **states** on either side of that sea claim **exclusive economic zones** of 200 nautical miles, it is necessary to reduce those claims to a (median) distance equidistant from each shoreline. **Delimitation** on the map almost always appears as a set of straight-line segments that reflect the configurations of the coastlines involved.

Medical geography   The study of health and disease within a geographic context and from a spatial perspective. Among other things, this geographic field examines the sources, **diffusion** routes, and distributions of diseases.

Megacity   Informal term referring to the world's most heavily populated cities; in this book, the term refers to a **metropolis** containing a population of greater than 10 million.

Megalopolis   When spelled with a lower-case *m*, a synonym for **conurbation**, one of the large coalescing supercities forming in diverse parts of the world. When capitalized, refers specifically to the multimetropolitan corridor that extends along the northeastern U.S. seaboard from north of Boston to south of Washington, D.C. (Fig. 3-9).

Mercantilism   Protectionist policy of European **states** during the sixteenth to the eighteenth centuries that promoted a **state's** economic position in the contest with rival powers. Acquiring gold and silver and maintaining a favorable trade balance (more exports than imports) were central to the policy.

Meridian   Line of **longitude**, aligned north-south across the globe, that together with **parallels** of **latitude** forms the global grid system. All meridians converge at both poles and are at their maximum distances from each other at the equator.

Mestizo   Derived from the Latin word for *mixed*, refers to a person of mixed European (white) and Amerindian ancestry.

Metropolis   Urban **agglomeration** consisting of a (central) city and its suburban ring. See **urban (metropolitan) area**.

Metropolitan area   See **urban (metropolitan) area**.

Migration   A change in residence intended to be permanent. See also **forced, internal, international**, and **voluntary migration**.

Migratory movement   Human relocation movement from a source to a destination without a return journey, as opposed to cyclical movement (see **nomadism**).

Model   An idealized representation of reality built to demonstrate its most important properties. A **spatial** model focuses on a geographical dimension of the real world, such as the **von Thünen model** that explains agricultural location patterns in a commercial economy.

Modernization   In the eyes of the Western world, the Westernization **process** that involves the establishment of **urbanization**, a market (money) economy, improved circulation, formal schooling, adoption of foreign innovations, and the breakdown of traditional society. Non-Westerners mostly see "modernization" as an outgrowth of **colonialism** and often argue that traditional societies can be modernized without being Westernized.

Monsoon   Refers to the seasonal reversal of wind and moisture flows in certain parts of the subtropics and lower-middle latitudes. The *dry monsoon* occurs during the cool season when dry offshore winds prevail. The *wet monsoon* occurs in the hot summer months, which produce onshore winds that bring large amounts of rainfall. The air-pressure differential over land and sea is the triggering mechanism, with windflows always moving from areas of relatively higher pressure toward areas of relatively lower pressure. Monsoons make their greatest regional impact in the coastal and near-coastal zones of South Asia, Southeast Asia, and East Asia.

Mosaic culture   Emerging cultural-geographic framework of the United States, dominated by the fragmentation of specialized social groups into homogeneous communities of interest marked not only by income, race, and **ethnicity** but also by age, occupational status, and lifestyle. The result is an increasingly heterogeneous socio-spatial complex, which resembles an intricate mosaic composed of myriad uniform—but separate—tiles.

Mulatto   A person of mixed African (black) and European (white) ancestry.

Multilingualism   A society marked by a mosaic of local languages. Constitutes a **centrifugal force** because it impedes communication within the larger population. Often a *lingua franca* is used as a "common language," as in many countries of Subsaharan Africa.

Multinationals   Internationally active corporations capable of strongly influencing the economic and political affairs of many countries they operate in.

Muslim   An adherent of the Islamic faith.

Muslim Front   See **Islamic Front**.

NAFTA (North American Free Trade Agreement)   The **free-trade area** launched in 1994 involving the United States, Canada, and Mexico.

Nation   Legally a term encompassing all the citizens of a **state**, it also has other connotations. Most definitions now tend to refer to a group of tightly-knit people possessing bonds of language, **ethnicity**, religion, and other shared **cultural** attributes. Such homogeneity actually prevails within very few states.

Nation-state   A country whose population possesses a substantial degree of **cultural** homogeneity and unity. The ideal form to which most **nations** and **states** aspire—a political unit wherein the territorial state coincides with the area settled by a certain national group or people.

NATO (North Atlantic Treaty Organization)   Established in 1950 at the height of the Cold War as a U.S.-led **supranational** defense pact to shield postwar Europe against the Soviet military threat. NATO is now in transition, expanding its membership while modifying its objectives in the post-Soviet era.

Natural hazard   A natural event that endangers human life and/or the contents of a **cultural landscape**.

Natural increase rate   Population growth measured as the excess of live births over deaths per 1000 individuals per year. Natural increase of a population does not reflect either **emigrant** or **immigrant** movements.

Natural landscape   The array of landforms that constitutes the Earth's surface (mountains, hills, plains, and plateaus) and the physical features that mark them (such as water bodies, soils, and vegetation). Each **geographic realm** has its distinctive combination of natural landscapes.

Natural resource   Any valued element of (or means to an end using) the environment; includes minerals, water, vegetation, and soil.

Nautical mile   By international agreement, the nautical mile—the standard measure at sea—is 6076.12 feet in length, equivalent to approximately 1.15 statute miles (1.85 km).

Near Abroad   The 14 former Soviet republics that, together with the dominant Russian Republic, constituted the U.S.S.R. Since the 1991 breakup of the Soviet Union, Russia has asserted a sphere of influence in these now-independent countries, based on its proclaimed right to protect the interests of ethnic Russians who were settled there in substantial numbers during Soviet times.

Neocolonialism   The term used by developing countries to underscore that the entrenched **colonial** system of international exchange and capital flow has not changed in the postcolonial era—thereby perpetuating the huge economic advantages of the developed world.

Network (transport)   The entire regional **system** of transportation connections and nodes through which movement can occur.

Nevada   See *tierra nevada*.

New World Order   A description of the international system resulting from the collapse of the Soviet Union in which the balance of nuclear power theoretically no longer determines the destinies of **states**.

Nomadism   Cyclical movement among a definite set of places. Nomadic peoples mostly are **pastoralists**.

North American Free Trade Agreement   See **NAFTA**.

Nucleation   Cluster; **agglomeration**.

Oasis   An area, small or large, where the supply of water (from an **aquifer** or a major river such as the Nile) permits the transformation of the immediately surrounding **desert** into productive cropland.

Occidental   Western. Also see *Oriental*.

Offshore banking   Term referring to financial havens for foreign companies and individuals, who channel their earnings to accounts in such a country (usually an "offshore" island-state) to avoid paying taxes in their home countries.

Oligarchs   Opportunists in post-Soviet Russia who used their ties to government to enrich themselves.

OPEC (Organization of Petroleum Exporting Countries)   The international oil *cartel* or syndicate formed by a number of producing countries to promote their common economic interests through the formulation of joint pricing policies and the limitation of market options for consumers. The 11 member-states (as of late 2007) are: Algeria, Indonesia, Iran, Iraq, Kuwait, Libya, Nigeria, Qatar, Saudi Arabia, United Arab Emirates (UAE), and Venezuela.

Organic theory   Friedrich Ratzel's theory of **state** development that conceptualized the state as a biological organism whose life—from birth through maturation to eventual senility and collapse—mirrors that of any living thing.

Oriental   The root of the word *oriental* is from the Latin for *rise*. Thus it has to do with the direction in which one sees the sun "rise"—the east; *oriental* therefore means Eastern. *Occidental* originates from the Latin for fall, or the "setting" of the sun in the west; occidental therefore means Western.

Orographic precipitation   Mountain-induced precipitation, especially when air masses are forced to cross **topographic** barriers. Downwind areas beyond such a mountain range experience the relative dryness known as the **rain shadow effect**.

Outback   The name given by Australians to the vast, peripheral, sparsely-settled interior of their country.

Outer city   The non-central-city portion of the American **metropolis**; no longer "sub" to the "urb," this outer ring was transformed into a full-fledged city during the late twentieth century.

Overseas Chinese   The more than 50 million ethnic Chinese who live outside China. Over half live in Southeast Asia (see Fig. 10-5), and many have become quite successful. A large number maintain links to China, and as investors played a major economic role in stimulating the growth of **SEZs** and Open Cities in China's **Pacific Rim**.

Pacific Rim   A far-flung group of countries and parts of countries (extending clockwise on the map from New Zealand to Chile) sharing the following criteria: they face the Pacific Ocean; they evince relatively high levels of economic development, industrialization, and urbanization; their imports and exports mainly move across Pacific waters.

Pacific Ring of Fire   Zone of crustal instability along **tectonic plate** boundaries, marked by earthquakes and volcanic activity, that ring the Pacific Ocean basin.

Paddies (paddyfields)   Ricefields.

Pandemic   An outbreak of a disease that spreads worldwide.

Pangaea   A vast, singular landmass consisting of most of the areas of the present-day continents. This supercontinent began to break up more than 200 million years ago when still-ongoing **plate** divergence and **continental drift** became dominant **processes** (see Fig. 6-3).

Parallel   An east-west line of **latitude** that is intersected at right angles by **meridians** of **longitude**.

Pastoralism   A form of **agricultural** activity that involves the raising of livestock.

Peasants   In a **stratified** society, peasants are the lowest class of people who depend on **agriculture** for a living. But they often own no land at all and must survive as tenants or day workers.

Peninsula   A comparatively narrow, finger-like stretch of land extending from the main landmass into the sea. Florida and Korea are examples.

Peon (*peone*)   Term used in Middle and South America to identify people who often live in serfdom to a wealthy landowner; landless **peasants** in continuous indebtedness.

Per capita   Capita means *individual*. Income, production, or some other measure is often given per individual.

Perforated state   A **state** whose territory completely surrounds that of another state. South Africa, which encloses Lesotho and is perforated by it, is a classic example.

Periodic market Village market that opens every third day or at some other regular interval. Part of a regional network of similar markets in a preindustrial, rural setting where goods are brought to market on foot and barter remains a major mode of exchange.

Peripheral development Spatial pattern in which a country's or region's development (and population) is most heavily concentrated along its outer edges rather than in its interior. Australia, with its peripheral population distribution, is a classic example (note its **core area** in Fig. 11-4), and nearby New Zealand exhibits a similar configuration of people and activities.

Periphery See **core-periphery relationships**.

Permafrost Permanently frozen water in the near-surface soil and bedrock of cold environments, producing the effect of completely frozen ground. Surface can thaw during brief warm season.

Physical geography The study of the geography of the physical (natural) world. Its subfields include **climatology, geomorphology, biogeography, soil geography, marine geography**, and water **resources**.

Physical landscape Synonym for **natural landscape**.

Physiographic political boundaries Political boundaries that coincide with prominent physical features in the **natural landscape**—such as rivers or the crest ridges of mountain ranges.

Physiographic region (province) A **region** within which there prevails substantial **natural-landscape** homogeneity, expressed by a certain degree of uniformity in surface **relief, climate**, vegetation, and soils.

Physiography Literally means *landscape description*, but commonly refers to the total **physical geography** of a place; includes all of the natural features on the Earth's surface, including landforms, **climate**, soils, vegetation, and water bodies.

Physiologic density The number of people per unit area of **arable** land.

Pilgrimage A journey to a place of great religious significance by an individual or by a group of people (such as a pilgrimage to Mecca for **Muslims**).

Plantation A large estate owned by an individual, family, or corporation and organized to produce a cash crop. Almost all plantations were established within the tropics; in recent decades, many have been divided into smaller holdings or reorganized as cooperatives.

Plate tectonics Plates are bonded portions of the Earth's mantle and crust, averaging 100 kilometers (60 mi) in thickness. More than a dozen such plates exist (see Fig. G-4), most of continental proportions, and they are in motion. Where they meet one slides under the other, crumpling the surface crust and producing significant volcanic and earthquake activity; a major mountain-building force.

Pleistocene Epoch Recent period of geologic time that spans the rise of humankind, beginning about 2 million years ago. Marked by *glaciations* (repeated advances of continental ice sheets) and milder *interglaciations* (ice sheet contractions). Although the last 10,000 years are known as the **Holocene** Epoch, Pleistocene-like conditions seem to be continuing and we are most probably now living through another Pleistocene interglaciation; thus the glaciers likely will return.

Plural(istic) society A society in which two or more population groups, each practicing its own **culture**, live adjacent to one another without mixing inside a single **state**.

Polder Land reclaimed from the sea adjacent to the shore of the Netherlands by constructing dikes and then pumping out the water trapped behind them.

Political geography The study of the interaction of geographic space and political **process**; the spatial analysis of political phenomena and processes.

Pollution The release of a substance, through human activity, which chemically, physically, or biologically alters the air or water it is discharged into. Such a discharge negatively impacts the environment, with possible harmful effects on living organisms—including humans.

Population decline A decreasing national population. Russia, which now loses nearly 1 million people per year, is the best example. Also see **population implosion**.

Population density The number of people per unit area. Also see **arithmetic density** and **physiologic density** measures.

Population distribution The way people have arranged themselves in geographic space. One of human geography's most essential expressions because it represents the sum total of the adjustments that a population has made to its natural, cultural, and economic environments.

Population explosion The rapid growth of the world's human population during the past century, attended by accelerating *rates* of increase.

Population geography The field of geography that focuses on the spatial aspects of **demography** and the influences of demographic change on particular places.

Population implosion The opposite of the **population explosion**; refers to the declining populations of many European countries and Russia in which the **death rate** exceeds the **birth rate** and **immigration** rate.

Population movement See **migration**; **migratory movement**.

Population projection The future population total that demographers forecast for a particular country. For example, in Table G-1 such projections are given for all the world's countries for 2025.

Population (age-sex) structure Graphic representation (*profile*) of a population according to age and gender.

Postindustrial economy Emerging economy, in the United States and a handful of other highly advanced countries, as traditional industry is increasingly eclipsed by a higher-technology productive complex dominated by services, information-related, and managerial activities.

Primary economic activity Activities engaged in the direct extraction of **natural resources** from the environment such as mining, fishing, lumbering, and especially **agriculture**.

Primate city A country's largest city—ranking atop the urban **hierarchy**—most expressive of the national culture and usually (but not always) the capital city as well.

Process Causal force that shapes a spatial pattern as it unfolds over time.

Productive activities The major components of the spatial economy. For individual components see: **primary economic activity, secondary economic activity, tertiary economic activity, quaternary economic activity**, and **quinary economic activity**.

Protruded state Territorial shape of a **state** that exhibits a narrow, elongated land extension (or *protrusion*) leading away from the main body of territory. Thailand is a leading example.

Push-pull concept The idea that **migration** flows are simultaneously stimulated by conditions in the source area, which tend to drive people away, and by the perceived attractiveness of the destination.

Qanat In **desert** zones, particularly in Iran and western China, an underground tunnel built to carry **irrigation** water by gravity flow from nearby mountains (where **orographic precipitation** occurs) to the arid flatlands below.

Quaternary economic activity Activities engaged in the collection, processing, and manipulation of *information*.

Quinary economic activity Managerial or control-function activity associated with decision making in large organizations.

Rain shadow effect The relative dryness in areas downwind of mountain ranges caused by **orographic precipitation**, wherein moist air masses are forced to deposit most of their water content as they cross the highlands.

Rate of natural population increase See **natural increase rate**.

Realm See **geographic realm**.

Refugees People who have been dislocated involuntarily from their original place of settlement.

Region A commonly used term and a geographic concept of central importance. An **area** on the Earth's surface marked by specific criteria, which are discussed on pp. 5–6.

Regional boundary In theory, the line that circumscribes a **region**. But razor-sharp lines are seldom encountered, even in nature (e.g., a coastline constantly changes depending upon the tide). In the **cultural landscape**, not only are regional boundaries rarely self-evident, but when they are ascertained by geographers they most often turn out to be **transitional** borderlands.

Regional character The personality or "atmosphere" of a **region** that makes it distinct from all other regions.

Regional complementarity See **complementarity**.

Regional concept The geographic study of **regions** and regional distinctions, as discussed on pp. 5–6.

Regional disparity The spatial unevenness in standard of living that occurs within a country, whose "average," overall income statistics invariably mask the differences that exist between the extremes of the wealthy **core** and the poorer **periphery**.

Regional geography Approach to geographic study based on the spatial unit of the **region**. Allows for an all-encompassing view of the world, because it utilizes and integrates information from geography's topical (**systematic**) fields; diagrammed in Figure G-14.

Regional state A "natural economic zone" that defies political boundaries, and is shaped by the global economy of which it is a part; its leaders deal directly with foreign partners and negotiate the best terms they can with the national governments under which they operate.

Regionalism The consciousness of and loyalty to a **region** considered distinct and different from the **state** as a whole by those who occupy it.

Relative location The regional position or **situation** of a place relative to the position of other places. Distance, **accessibility**, and connectivity affect relative location.

Relict boundary A political boundary that has ceased to function, but the imprint of which can still be detected on the **cultural landscape**.

Relief Vertical difference between the highest and lowest elevations within a particular area.

Religious revivalism See **revivalism (religious)**.

Relocation diffusion Sequential **diffusion process** in which the items being diffused are transmitted by their carrier agents as they relocate to new areas. The most common form of relocation diffusion involves the spreading of innovations by a **migrating** population.

Restrictive population policies Government policy designed to reduce the **rate of natural population increase**. China's one-child policy, instituted in 1979 after Mao's death, is a classic example.

Revivalism (religious) Religious movement whose objectives are to return to the foundations of that faith and to influence state policy. Often called *religious fundamentalism*; but in the case of Islam, **Muslims** prefer the term revivalism.

Rift valley The trough or trench that forms when a thinning strip of the Earth's crust sinks between two parallel faults (surface fractures).

Rural-to-urban migration The dominant **migration** flow from countryside to city that continues to transform the world's population, most notably in the less advantaged geographic realms.

Russification Demographic resettlement policies pursued by the central planners of the Soviet Empire, whereby ethnic Russians were encouraged to emigrate from the Russian Republic to the 14 non-Russian republics of the U.S.S.R.

Sahel Semiarid **steppeland** zone extending across most of Africa between the southern margins of the arid Sahara and the moister tropical **savanna** and forest zone to the south. Chronic drought, **desertification**, and overgrazing have contributed to severe famines in this area since 1970.

Savanna Tropical grassland containing widely spaced trees; also the name given to the tropical wet-and-dry climate type (*Aw*).

Scale Representation of a real-world phenomenon at a certain level of reduction or generalization. In **cartography**, the ratio of map distance to ground distance; indicated on a map as a bar graph, representative fraction, and/or verbal statement. *Macroscale* refers to a large area of national proportions; *microscale* refers to a local area no bigger than a county.

Scale economies See **economies of scale**.

Secondary economic activity Activities that process raw materials and transform them into finished industrial products. The *manufacturing* sector.

Sedentary Permanently attached to a particular area; a population fixed in its location. The opposite of **nomadic**.

Separate development The spatial expression of South Africa's "grand" **apartheid** scheme, whereby nonwhite groups were required to settle in segregated "homelands." The policy was dismantled when white-minority rule collapsed in the early 1990s.

Sequent occupance The notion that successive societies leave their cultural imprints on a place, each contributing to the cumulative **cultural landscape**.

Shantytown Unplanned slum development on the margins of cities in disadvantaged countries, dominated by crude dwellings and shelters mostly made of scrap wood and iron, and even pieces of cardboard.

Sharecropping Relationship between a large landowner and farmers on the land whereby the farmers pay rent for the land they farm by giving the landlord a share of the annual harvest.

Sharia The criminal code based in Islamic law that prescribes corporal punishment, amputations, stonings, and lashing for both major and minor offenses. Its occurrence today is associated with the spread of **religious revivalism** in **Muslim** societies.

Shatter belt Region caught between stronger, colliding external cultural-political forces, under persistent stress, and often fragmented by aggressive rivals. Eastern Europe and Southeast Asia are classic examples.

Shifting agriculture Cultivation of crops in recently cut and burned tropical-forest clearings, soon to be abandoned in favor of newly cleared nearby forest land. Also known as *slash-and-burn agriculture*.

Sinicization Giving a Chinese cultural imprint; Chinese **acculturation**.

Site The internal locational attributes of an urban center, including its local spatial organization and physical setting.

Situation The external locational attributes of an urban center; its **relative location** or regional position with reference to other non-local places.

Social stratification See **stratification (social)**.

South American culture spheres The fivefold Augelli scheme of cultural regionalization in the South American geographic realm. Consists of the tropical-plantation sphere, European-commercial sphere, Amerind-subsistence sphere, Mestizo-transitional sphere, and undifferentiated sphere.

Southern Ocean The ocean that surrounds Antarctica (see discussion on p. 574).

Spatial Pertaining to space on the Earth's surface. Synonym for *geographic(al)*.

Spatial diffusion See **diffusion**.

Spatial interaction See **complementarity, intervening opportunity**, and **transferability**.

Spatial model See **model**.

Spatial process See **process**.

Spatial system The components and interactions of a **functional region**, which is defined by the areal extent of those interactions. Also see **system**.

Special Administrative Region (SAR) Status accorded the former dependencies of Hong Kong and Macau that were taken over by China, respectively, from the United Kingdom in 1997 and Portugal in 1999. Both SARs received guarantees that their existing social and economic systems could continue unchanged for 50 years following their return to China.

Special Economic Zone (SEZ) Manufacturing and export center within China, created since 1980 to attract foreign investment and technology transfers. Seven SEZs—all located on China's Pacific coast—currently operate: Shenzhen, adjacent to Hong Kong; Zhuhai; Shantou; Xiamen; Hainan Island, in the far south; Pudong, across the river from Shanghai; and Binhai New Area, next to the port of Tianjin.

Squatter settlement See **shantytown**.

State   A politically organized territory that is administered by a sovereign government and is recognized by a significant portion of the international community. A state must also contain a permanent resident population, an organized economy, and a functioning internal circulation system.

State boundaries   The borders that surround **states** which, in effect, are derived through contracts with neighboring states negotiated by treaty. See **definition, delimitation,** and **demarcation.**

State capitalism   Government-controlled corporations competing under free-market conditions, usually in a tightly regimented society. South Korea is a leading example. Also see **chaebol.**

State formation   The creation of a **state** based on traditions of human **territoriality** that go back thousands of years.

State planning   Involves highly centralized control of the national planning process, a hallmark of communist economic systems. Soviet central planners mainly pursued a grand political design in assigning production to particular places; their frequent disregard of the principles of economic geography contributed to the eventual collapse of the U.S.S.R.

State territorial morphology   A **state's** geographical shape, which can have a decisive impact on its spatial cohesion and political viability. A **compact** shape is most desirable; among the less efficient shapes are those exhibited by **elongated, fragmented, perforated,** and **protruded** states.

Stateless nation   A national group that aspires to become a **nation-state** but lacks the territorial means to do so; the Palestinians and Kurds of Southwest Asia are classic examples.

Steppe   Semiarid grassland; short-grass prairie. Also the name given to the semiarid climate type (*BS*).

Stratification (social)   In a layered or stratified society, the population is divided into a **hierarchy** of social classes. In an industrialized society, the working class is at the lower end; **elites** that possess capital and control the means of production are at the upper level. In the traditional **caste system** of Hindu India, the "untouchables" form the lowest class or caste, whereas the still-wealthy remnants of the princely class are at the top.

Subduction   In **plate tectonics,** the **process** that occurs when an oceanic plate converges head-on with a plate carrying a continental landmass at its leading edge. The lighter continental plate overrides the denser oceanic plate and pushes it downward.

Subsequent boundary   A political boundary that developed contemporaneously with the evolution of the major elements of the **cultural landscape** through which it passes.

Subsistence   Existing on the minimum necessities to sustain life; spending most of one's time in pursuit of survival.

Subtropical Convergence   A narrow marine **transition zone,** girdling the globe at approximately latitude 40°S, that marks the equatorward limit of the frigid **Southern Ocean** and the poleward limits of the warmer Atlantic, Pacific, and Indian oceans to the north.

Suburban downtown   In the United States (and increasingly in other advantaged countries), a significant concentration of major urban activities around a highly accessible suburban location, including retailing, light industry, and a variety of leading corporate and commercial operations. The largest are now coequal to the American central city's **central business district (CBD).**

Sunbelt   The popular name given to the southern tier of the United States, which is anchored by the mega-States of California, Texas, and Florida. Its warmer climate, superior recreational opportunities, and other amenities have been attracting large numbers of relocating people and activities since the 1960s; broader definitions of the Sunbelt also include much of the western U.S., particularly Colorado and the coastal Pacific Northwest.

Superimposed boundary   A political boundary emplaced by powerful outsiders on a developed human landscape. Usually ignores preexisting cultural-spatial patterns, such as the border that still divides North and South Korea.

Supranational   A venture involving three or more **states**—political, economic, and/or cultural cooperation to promote shared objectives. The **European Union** is one such organization.

System   Any group of objects or institutions and their mutual interactions. Geography treats systems that are expressed spatially, such as in **functional regions.**

Systematic geography   Topical geography: **cultural, political, economic geography,** and the like.

Taiga   The subarctic, mostly **coniferous** snowforest that blankets northern Russia and Canada south of the **tundra** that lines the Arctic shore.

Takeoff   Economic concept to identify a stage in a country's **development** when conditions are set for a domestic Industrial Revolution.

Taxonomy   A **system** of scientific classification.

Technopole   A planned techno-industrial complex (such as California's Silicon Valley) that innovates, promotes, and manufactures the products of the **postindustrial** informational economy.

Tectonics   See **plate tectonics.**

Templada   See *tierra templada.*

Terracing   The transformation of a hillside or mountain slope into a step-like sequence of horizontal fields for intensive cultivation.

Territoriality   A country's or more local community's sense of property and attachment toward its territory, as expressed by its determination to keep it inviolable and strongly defended.

Territorial sea   Zone of seawater adjacent to a country's coast, held to be part of the national territory and treated as a segment of the sovereign **state.**

Tertiary economic activity   Activities that engage in *services*—such as transportation, banking, retailing, education, and routine office-based jobs.

Tierra caliente   The lowest of the **altitudinal zones** into which the human settlement of Middle and South America is classified according to elevation. The *caliente* is the hot humid coastal plain and adjacent slopes up to 750 meters (2500 ft) above sea level. The natural vegetation is the dense and luxuriant tropical rainforest; the crops include sugar and bananas in the lower areas, and coffee, tobacco, and corn along the higher slopes.

Tierra fría   Cold, high-lying **altitudinal zone** of settlement in Andean South America, extending from about 1800 meters (6000 ft) in elevation up to nearly 3600 meters (12,000 ft). **Coniferous** trees stand here; upward they change into scrub and grassland. There are also important pastures within the *fría,* and wheat can be cultivated.

Tierra helada   In Andean South America, the highest-lying habitable **altitudinal zone**—ca. 3600 to 4500 meters (12,000 to 15,000 ft)—between the tree line (upper limit of the *tierra fría*) and the snow line (lower limit of the *tierra nevada*). Too cold and barren to support anything but the grazing of sheep and other hardy livestock.

Tierra nevada   The highest and coldest **altitudinal zone** in Andean South America (lying above 4500 meters [15,000 ft]), an uninhabitable environment of permanent snow and ice that extends upward to the Andes' highest peaks of more than 6000 meters (20,000 ft).

Tierra templada   The intermediate **altitudinal zone** of settlement in Middle and South America, lying between 750 meters (2500 ft) and 1800 meters (6000 ft) in elevation. This is the "temperate" zone, with moderate temperatures compared to the *tierra caliente* below. Crops include coffee, tobacco, corn, and some wheat.

Topography   The surface configuration of any segment of **natural landscape.**

Toponym   Place name.

Transculturation   Cultural borrowing and two-way exchanges that occur when different **cultures** of approximately equal complexity and technological level come into close contact.

Transferability   The capacity to move a good from one place to another at a bearable cost; the ease with which a commodity may be transported.

Transhumance   Seasonal movement of people and their livestock in search of pastures. Movement may be vertical (into highlands during the summer and back to lower elevations in winter) or horizontal, in pursuit of seasonal rainfall.

Transition zone   An area of spatial change where the **peripheries** of two adjacent realms or regions join; marked by a gradual shift (rather than a sharp break) in the characteristics that distinguish these neighboring geographic entities from one another.

Transmigration   The now-ended policy of the Indonesian government to induce residents of the overcrowded, **core-area** island of Jawa to move to the country's other islands.

Treaty ports   **Extraterritorial enclaves** in China's coastal cities, established by European colonial invaders under unequal treaties enforced by gunboat diplomacy.

Triple Frontier   The turbulent and chaotic area in southern South America that surrounds the convergence of Brazil, Argentina, and Paraguay. Lawlessness pervades this haven for criminal elements, which is notorious for money laundering, arms and other smuggling, drug trafficking, and links to terrorist organizations, including money flows to the Middle East.

Tropical deforestation   See **deforestation**.

Tropical savanna   See **savanna**.

Tsunami   A seismic (earthquake-generated) sea wave that can attain gigantic proportions and cause coastal devastation. The tsunami of December 26, 2004, centered in the Indian Ocean near the Indonesian island of Sumatera, produced the first great natural disaster of the twenty-first century.

Tundra   The treeless plain that lies along the Arctic shore in northernmost Russia and Canada, whose vegetation consists of mosses, lichens, and certain hardy grasses.

Turkestan   Northeasternmost region of the North Africa/Southwest Asia realm. Known as Soviet Central Asia before 1992, its five (dominantly Islamic) former S.S.R.'s have become the independent countries of Kazakhstan, Uzbekistan, Turkmenistan, Kyrgyzstan, and Tajikistan. Today Turkestan has expanded to include a sixth state, Afghanistan.

Unitary state   A **nation-state** that has a centralized government and administration that exercises power equally over all parts of the state.

Urbanization   A term with several connotations. The proportion of a country's population living in urban places is its level of urbanization. The **process** of urbanization involves the movement to, and the clustering of, people in towns and cities—a major force in every geographic realm today. Another kind of urbanization occurs when an expanding city absorbs rural countryside and transforms it into suburbs; in the case of cities in disadvantaged countries, this also generates peripheral **shantytowns**.

Urban (metropolitan) area   The entire built-up, nonrural area and its population, including the most recently constructed suburban appendages. Provides a better picture of the dimensions and population of such an area than the delimited municipality (central city) that forms its heart.

Urban realms model   A spatial generalization of the contemporary large American city. It is shown to be a widely dispersed, multicentered metropolis consisting of increasingly independent zones or *urban realms*, each focused on its own **suburban downtown**; the only exception is the shrunken central realm, which is focused on the central city's **central business district** (see Figs. 3-11 and 3-12).

Veld   See **highveld**.

Voluntary migration   Population movement in which people relocate in response to perceived opportunity, not because they are forced to **migrate**.

Von Thünen's Isolated State model   Explains the location of **agricultural** activities in a commercial economy. A **process** of spatial competition allocates various farming activities into concentric rings around a central market city, with profit-earning capability the determining force in how far a crop locates from the market. The original (1826) Isolated State model (see Fig. 1-5) now applies to the continental scale (see Fig. 1-6).

Wahhabism   A particularly virulent form of (Sunni) **Muslim revivalism** that was made the official faith when the modern **state** of Saudi Arabia was founded in 1932. Adherents call themselves "Unitarians" to signify the strict fundamentalist nature of their beliefs.

Wallace's Line   As shown in Figure 11-3, the zoogeographical boundary proposed by Alfred Russel Wallace that separates the marsupial fauna of Australia and New Guinea from the nonmarsupial fauna of Indonesia.

Weather   The immediate and short-term conditions of the **atmosphere** that impinge on daily human activities.

West Wind Drift   The clockwise movement of water around Antarctica in the **Southern Ocean**.

Wet monsoon   See **monsoon**.

Windward   The exposed, upwind side of a **topographic** barrier that faces the winds that flow across it.

World geographic realm   See **geographic realm**.

INDEX

A Reference with an italicized number refers to figures, tables, boxes and captions.

List of Maps and Figures

List of Maps and Figures (continued)